ENVIRONMENTAL AND METABOLIC ANIMAL PHYSIOLOGY

Comparative Animal Physiology, Fourth Edition

D0904789

ENVIRONMENTAL AND METABOLIC ANIMAL PHYSIOLOGY

Comparative Animal Physiology, Fourth Edition

Edited by
C. LADD PROSSER
Department of Physiology and Biophysics
University of Illinois
Urbana, Illinois

WILEY-LISS

A JOHN WILEY & SONS, INC., PUBLICATION
New York • Chichester • Brisbane • Toronto • Singapore

Address all Inquiries to the Publisher
Wiley-Liss, Inc., 41 East 11th Street, New York, NY 10003

Copyright © 1991 Wiley-Liss, Inc.

Printed in United States of America

Under the conditions stated below the owner of copyright for this book hereby grants permission to users to make photocopy reproductions of any part or all of its contents for personal or internal organizational use, or for personal or internal use of specific clients. This consent is given on the condition that the copier pay the stated per-copy fee through the Copyright Clearance Center, Incorporated, 27 Congress Street, Salem, MA 01970, as listed in the most current issue of "Permissions to Photocopy" (Publisher's Fee List, distributed by CCC, Inc.), for copying beyond that permitted by sections 107 or 108 of the US Copyright Law. This consent does not extend to other kinds of copying, such as copying for general distribution, for advertising or promotional purposes, for creating new collective works, or for resale.

Library of Congress Cataloging-in-Publication Data

Environmental and metabolic animal physiology / edited by C. Ladd
  Prosser.
    p. cm.
   "Comparative animal physiology, fourth edition."
   "This book, Environmental and metabolic animal physiology, and its
  companion volume, Neural and integrative animal physiology, together
  comprise the fourth edition of Comparative animal physiology"–
  –Pref.
   Includes bibliographical references and index.
   ISBN 0-471-85767-X : $44.95
   1. Metabolism. 2. Homeostasis. 3. Adaptation (Physiology)
4. Physiology, Comparative. I. Prosser, C. Ladd (Clifford Ladd),
1907– . II. Comparative animal physiology.
QP171.E67 1991
591.1'33 – dc20   90-27165
                                                 CIP

CONTRIBUTORS

Warren Burggren Department of Zoology, University of Massachusetts, Amherst, Massachusetts 01003

James Wayne Campbell Department of Biochemistry and Cell Biology, Rice University, P. O. Box 1892, Houston, Texas 77251

Edward J. DeVillez Zoology Department, Miami University, Oxford, Ohio 45056

Anthony P. Farrell Department of Biological Science, Simon Fraser University, Barnaby, British Columbia, Canada V5A 1S6

James Heath Department of Physiology and Biophysics, University of Illinois at Urbana-Champaign, 524 Burrill Hall, 407 South Goodwin Avenue, Urbana, Illinois 61801

Peter W. Hochachka Department of Zoology, University of British Columbia, 6270 University Blvd., Vancouver, British Columbia, Canada V6T 2A9

Leonard B. Kirschner Department of Zoology, Washington State University, Pullman, Washington 99164

Brian A. McMahon Department of Biology, University of Calgary, 2500 University Drive, Calgary, Alberta, Canada T2N 1N4

James G. Morris Department of Physiological Sciences, School of Veterinary Medicine, University of California, Davis, California 95616

Dennis Powers Hopkins Marine Laboratory, Stanford University, Pacific Grove, California 93950

C. Ladd Prosser Department of Physiology and Biophysics, University of Illinois at Urbana-Champaign, 524 Burrill Hall, 407 South Goodwin Avenue, Urbana, Illinois 61801

John L. Roberts Department of Zoology, University of Massachusetts, Amherst, Massachusetts 01003

George N. Somero Environmental, Population, and Organismic Biology, University of Colorado at Boulder, 122 Ramaley, Boulder, Colorado 80309-0334

PREFACE

This book, **Environmental and Metabolic Animal Physiology,** and its companion volume, **Neural and Integrative Animal Physiology,** together comprise the fourth edition of **Comparative Animal Physiology.** Previous editions of **Comparative Animal Physiology** were published in 1951, 1961, and 1973. Each book is designed to serve as a study text for upper division and graduate courses in comparative physiology; together they constitute a reference work that provides a comprehensive introduction to the literature in specific areas of comparative physiology, biochemistry, and biophysics.

The chapters in **Environmental and Metabolic Animal Physiology** are arranged in three groups. The first four chapters deal with environmental physiology and are oriented toward ecological physiology: "Theory of Adaptation," "Water and Ions," "Temperature," and "Hydrostatic Pressure." The next four chapters deal mainly with biochemical aspects of physiology: "Nutrition," "Digestion," "Nitrogen Metabolism" and "Energy Transfer." Three chapters on holistic physiology follow: "Respiration," "O_2 and CO_2 Transport," and "Circulation."

Traditionally, animal physiology has been concerned with function in organs and cells. Whole-animal physiology has led to many medical applications; other kinds of animal physiology are veterinary physiology, and physiology of fishes, insects, parasites, and other animal groups. Cellular and molecular physiology, like biochemistry and biophysics, are reductionist approaches to function.

Comparative physiology has been defined in several ways:

(1) *The functional analysis of non-laboratory or unfamiliar animals.* Routine physiological measurements, although made on different species, contribute little to biological principles.

(2) As experimental variables, comparative physiology uses, in addition to physical and biotic parameters, different kinds of animal. *Comparative physiology is thus the study of physiological diversity.*

(3) *Comparative physiology deals with highly conserved functions and substances.* The essential biochemical and biophysical properties of organisms evolved before there were organisms we would recognize as such. Thus, comparative physiology analyzes the functioning of genetically replicating molecules; passive membrane properties and active transport reactions; and energy transfer. To these it adds the study of variation on common themes and adaptive modulation.

(4) *Another characteristic of comparative physiology is the recognition of the uniqueness of each kind of animal.* We do not speak of "higher" and "lower" animals but of degrees of complexity. Each kind of animal is adapted to its physical and biotic environment if it survives and reproduces.

Comparative physiology has made important contributions to human and veterinary medicine and to agriculture. However, the main objective of comparative physiology is to contribute to basic biological theory.

Evolution is the most unifying concept of biology. Comparative physiology, by comparing the biochemistry and physiology of related animals, can provide clues to earlier life forms. Extrapolation to ancient forms requires some knowledge of previous physical environments. Comparative physiology contributes to assessing the relative importance of gradualism and saltation, the relative importance of selection and neutral change. For example, what is the meaning of small differences in amino acid sequences in families of proteins? Physiological adaptation is a key to phylogeny.

Speciation as identification of kinds of animals is aided by comparative physiology. Several bases for the definition of species are: (a) classification by systematists based on best judgments of specialists; (b) cladistic, quantitative differences between populations; (c) biological species based on observed gene exchange; (d) new species formed by hybridization or by parthenogenesis and separated spatially; and (e) physiological species based on adaptations to ecological niche or geographic range. A modern systematist makes use of all of these concepts in describing species.

Animal distribution, ecological location, is determined by physiological adaptation to the environment—both biotic and physical. Stress tests and other physiological measurements are useful in monitoring animal distribution, especially in diverse and disturbed environments.

Medical applications have been derived from knowledge of comparative physiology. A few of many examples of contributions of comparative physiology to medicine are: (a) establishment of filtration-reabsorption and secretion in kidney function; (b) skin and gill models of active ion transport; (c) mechanisms of O_2 transport in reduced O_2, as at high altitude; and (d) induction of digestive enzymes by diet.

An organism constantly interacts with its microenvironment, physiological and biological; hence the physiology of an organism cannot be described without considering its environmental interactions. A whole organism is not equal to the sum of its parts. Out of the whole organism there emerge unique characteristics not present in any of the isolated parts. It is important, therefore, to analyze the relation between components of the environment and the whole organism and to analyze these interactions in terms of organ and cell physiology.

This book is a comprehensive survey of function systems in diverse animals. Each chapter has an extensive reference list. Most text statements are supported by citations to specific references. Selection in the reference lists has been mainly for recent research papers, reviews, and classical papers.

A certain factual background of the reader is assumed. The reader needs to have acquaintance with the principal phyla and classes of animals; a guide to animal groups is furnished in Chapter 1. A knowledge of elementary cellular physiology is assumed, and cellular phenomena are used to arrive at comparative generalizations. Background of elementary biochemistry is essential for understanding the chapters on nutrition, digestion, nitrogen metabolism, and energy transfer. Some knowledge of organ function as presented in mammalian and human physiology textbooks is assumed.

The chapters have been written by specialists. Many consultants have critically read sections and entire chapters. It is the hope of the authors and editor that this volume will continue a tradition in comparative animal physiology, will present the relevance of molecular biology to comparative problems, and will strengthen the position of holistic biology.

C. LADD PROSSER

Urbana, Illinois
September 1990

CONTENTS

Chapter 1

Introduction: Definition of Comparative Physiology: Theory of Adaptation

C. Ladd Prosser

The science of physiology is the analysis of function in living organisms. Physiology is a synthesizing science that applies physical and chemical methods to biology. Physiology developed as the theoretical basis for medical practice. Comparative physiology goes beyond medical physiology to contribute to the understanding of basic biology. Comparative physiology requires some background in zoology (classification and structure of animals), biochemistry and molecular biology, cellular physiology, and human physiology.

Comparative physiology (including comparative biochemistry, pharmacology, and biophysics) is defined in several ways according to usage. One definition of comparative physiology is that it studies the functions of unfamiliar, nonlaboratory animals. However, routine measurements on several related species contribute little toward establishing general biological principles. It has been useful to examine the physiology of some groups of animals in detail. Traditional

physiology is that of mammals, particularly of humans. Physiology of higher plants gives a basis for agriculture. Other kinds of special animal physiology are fish physiology, insect physiology, and the physiology of parasites. Several-volume treatises on physiology of crustaceans, molluscs, and fishes are available.

A second definition of comparative physiology is that, in addition to familiar physical and biotic parameters of the environment, this discipline uses *kind* of animal as an experimental variable. Comparative physiology considers diversity in modes of solving life problems.

A third definition is that comparative physiology recognizes the diversity and modulation of highly conserved properties of organisms. Comparative physiology provides an intellectual bridge between molecular and whole-animal biology. Comparative physiology emphasizes the integration of the organism at all levels of organization.

Comparative physiology contributes to ecology in describing mechanisms of adaptation to diverse environments. Ecological correlates are presented in several chapters of this book.

Comparative physiology contributes to evolutionary theory in providing correlations of function with structure and chemical composition from which relatedness and origins can be deduced. Evolution over long times considers properties of phyla that have been dated by paleontological methods. Experiments on fossils are not possible but measurements on modern relatives of extinct forms make possible extrapolations to ancient animals. Analyses of adaptations elucidate the relative importance of gradualism and saltation in evolution. Evolution can be deduced from phyletic trees constructed from physiological and biochemical comparisons of closely related animals. For determining phylogenetic relatedness, sequences of amino acids in proteins and of nucleic acids in genes are examined by comparative physiologists and these measurements are correlated with function. The relative importance of neutral changes and adaptive alterations are also considered by comparative physiologists. When closely examined, small changes in amino acid sequences can be seen to have adaptive effects on enzyme kinetics.

A related subject to which comparative physiology contributes is speciation. Classification may, but need not, reflect phylogeny. The concept of species can be looked at in five ways according to context:

1. The most general meaning of species is that of the systematist and is based on morphological and biochemical identity of individuals in a group.
2. Cladistic species are determined by computer-based quantitative morphological and biochemical similarities.

3. The most general definition is the biological species according to which species are populations of similar organisms which can interbreed; reproductive isolation of species may be by several mechanisms, some of which are physiological.
4. The biological definition cannot be used for parthenogenetic animals or for those that show much hybridization or introgression; plants hybridize extensively and show population and subspecies differences that make determination of reproductive species difficult.
5. A functional definition of species is adaptational or physiological; each species must be adapted to its ecological niche and geographic range throughout its life cycle.

Modern systemists avail themselves of all five ways of assessing species. In estimates of decline of endangered species, stress studies on populations in critical environments supplement conclusions of systematists.

Animal Phylogeny

The physiology of an animal group reflects the evolutionary history of that group. A phylogenist uses data from paleontology supplemented by data from taxonomy, comparative morphology and, in increasing amount, from comparative physiology and biochemistry. The presence of homologous structures indicates a common ancestral gene pool. Physiological *homology* refers to a similar function, for example, use of a sodium pump in dissimilar organs or use of rhodopsin in eyes of diverse animals. Physiological *analogy* refers to evolutionary convergence or similar solution of a given life problem by different means, such as the

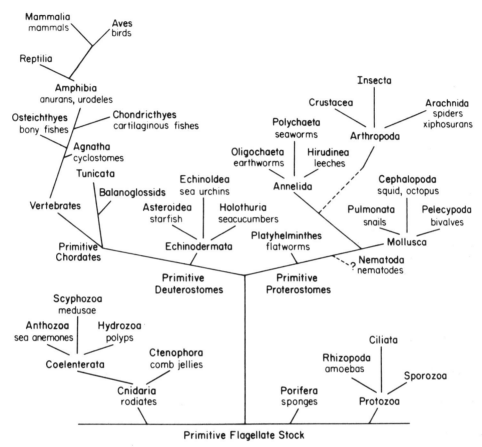

Figure 1. Phylogenetic tree of animals based on adult morphologies, fossil evidence and developmental patterns. Only those phyla and classes that are frequently used in comparative physiology are included. Common names are given in lower case. (Modified from M. J. Greenberg, Fig. 1, in *Comparative Animal Physiology*, 3rd ed.)

use of different metalloproteins for O_2 transport.

Since comparative physiologists use taxonomic types—from subspecies to phyla—as experimental variables, it is important to know something of the relationships among organisms. Physiological analyses are useful in elucidating evolutionary relationships. As an aid to students whose background in zoology may be limited, an abbreviated phyletic chart is given in Figure 1; this is based mainly on morphological, fossil, and de-velopmental characters (5). The common names of phyla and some classes are added to the systematic names. Only the groups of animals that are frequently used in comparative physiology are included.

Classical phylogeny relies mainly on gross anatomy and embryology, and bio-chemical data provide general substanti-ation for classical phylogenetic trees. The principal phyla evolved more or less si-multaneously during the Cambrian era, hence phyletic trees are polyphyletic. The most primitive of protozoans are plant–animals, the flagellates. Other modern protozoans and sponges are represented as offshoots from the direct phylogenetic line. It is postulated that cnidarians (coelenterates) and ctenophores evolved from primitive flagellate stock. At a primitive level acoelomate animals such as Nemathelminthes and Platyhelminthes emerged, although their points of origin on the phyletic tree are uncertain. Near this level, a branching resulted in two parallel lines. On one side are the proterostomes—annelids, arthropods, and molluscs. On the other side are deuterostomes—echinoderms and chordates. The cephalopods show greatest specialization among molluscs, insects among arthropods, and birds and mammals among chordates. The annelids–arthropods–molluscs show determinate or spiral cleavage, that is, the blastomeres arrange themselves in a stereotyped pattern and there is little cellular equipotentiality, each cell in cleavage—blastula stages having a fixed prospective role. In echinoderms–chordates cleavage is indeterminate. In the proterostomes the mesoderm begins with a particular cell in the blastula that starts two mesodermal bands; in these animals the blastopore gives rise to the mouth and the anus is opened secondarily. In the deuterostomes, cleavage is indeterminate, mesoderm arises as an outpouching from the archenteron, that is, from endoderm, the blastopore becomes the anus and the mouth opens secondarily.

Many essential biochemical and biophysical properties evolved during prebiotic evolution before there were organisms we would recognize as such. Structure of replicating molecules, coding of protein synthesis, mechanisms of osmotic and ionic balance, of electron transport reactions are common to all living animals. Most of the properties of nerve conduction and synaptic transmission found in complex animals occur also in the nerve nets of cnidarians. Mechanisms of osmotic, ionic, and volume regulation evolved before modern phyla, and modulation of these functions occurred at a later date; these general physiological properties are not useful in establishing origins of phyla.

There were undoubtedly many kinds of Precambrian organisms. One group of Precambrian animals, the ediacarans (700–600 My b.p.) were soft-bodied creatures that left impressions in several regions of the earth. Little is known of their physiology although they were undoubtedly anaerobic. During a long Precambrian period the oxygen concentration in the environment gradually increased by action of photosynthetic blue-green algae, modern remnants of which are stromatolites. Once oxygen levels rose to ~1%, animal diversification was extensive. Comparative physiology considers the transition from anaerobic to aerobic metabolism.

Phylogenetic relations have been deduced from amino acid sequences in similar proteins of the same family and from nucleotide sequences in ribosomal RNAs and DNAs. Proteins are aligned for similar amino acids at corresponding sites. The most extensive analyses have been with globins—mostly O_2-carrying molecules; other proteins that have been sequenced are calmodulins, Ca-binding proteins, fibrinopeptides, and cytochromes c. Comparisons have been mostly between classes and orders of vertebrates. Evolutionary relations based on protein homologies are well correlated with paleontological evidence (2–4). An evolutionary cladogram of globins including several phyla of invertebrates as well as vertebrates shows close relations for vertebrate α and β hemoglobins (but not myoglobin), between hemoglobins of several molluscs and insects. Insects are distantly related to crustaceans and annelids are at a distance from arthropods (4). Proteins are related to life habits and are adaptive to the environment, hence are specific for subphyletic groups and are not much used for relationships of phyla.

Ribosomes are of very ancient origin and are essential to protein synthesis in all organisms, hence are relatively unchanged in their characteristics. The ribosome comprises two subunits. The smaller of these contains a single (16S) RNA species, of ~1500 nucleotides length; the larger contains several RNA species, the small 5S RNA (~120 nucleotides in length) and the much larger 23S rRNA (~2900 nucleotides in length). Each ribosomal RNA (rRNA) folds back upon itself to produce a large number of loop structures—base paired stalks; the 16S rRNA, for example, contains ~50 loops (9). Two studies of 5S RNAs from a few species of many phyla have given very different cladograms. It was concluded that 5S RNA is not useful for tracing relations between phyla (6). Archaebacteria, eubacteria, and eukaryotes examined for homologies in 16S RNAs show clear differences (9). A cladogram based on 18S RNAs from 22 classes in 10 phyla indicates that rapid early radiation resulted in divergence of four major groups: chordates, echinoderms, arthropods, and eucoelomate proterostomes consisting of annelids, molluscs, brachiopods, and sipunculids. This cladogram, is very different from the ones based on 5S RNA but differs in only a few details from the classical phyletic trees based on embryology (1). Mitochondrial DNAs are useful for elucidating relations of closely related species but they mutate so rapidly as to be useless for phyla.

Physiological Variation

Three determinants of biological variation are genetic, environmental, and developmental. Environmentally induced variation can take place only within the limits set for an animal by its genotype. An adult character can develop only if critical genes are expressed at appropriate times and in appropriate tissues during development. Genetically and environmentally based variation can be distinguished by acclimatization and ultimately by cross-breeding. Acclimation refers to compensatory changes in an organism under maintained deviation of a single environmental factor (usually in the laboratory). Acclimatization refers to compensatory changes in an organism under multiple natural deviations of the environment—seasonal or geographical.

Adaptive Variations

Adaptive variations, both genetically determined and environmentally induced, are measured as functional differences in animals in altered, often stressful, environments. Two general classes of adaptive variations are (1) resistance adaptations, which permit function at or near tolerable limits of environmental stress; and (2) capacity adaptations, which are alterations that permit relatively normal function over a mid-range of environmental conditions (7, 8).

Examples of resistance adaptations are:

1. *Survival at Environmental Limits.* Survival tests include median lethal values for stress such as heat, cold, salinity, and oxygen supply. A prelethal effect may be a state of coma from which recovery is possible. Time-stress measurements of death or coma give quantitative criteria. The criteria of survival are different for intact animals, for tissues, for isolated enzymes (inactivation), and for lipids (phase change). Survival limits are wide for molecules (e.g., enzymes), narrower for tissues, still narrower for intact organisms, and are narrowest for populations. Integration reduces the range of

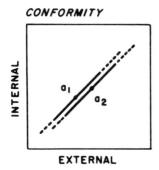 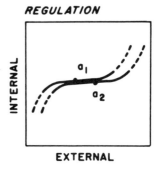

Figure 2. Diagram representing internal state as a function of external state for a given parameter, for example, temperature or salinity. Patterns of conformity and of regulation (homeostasis). Two levels of acclimation are indicated as a_1 and a_2. Solid lines indicate the range of normal tolerance and broken lines indicate the range of tolerance for brief periods. Conformers tolerate a wider range of internal variation than do regulators but regulators tolerate wider external ranges. [From Fig. 2 in Chapter 1 of (7)].

tolerance. Not all parts of an animal are equally subject to functional failure; frequently the nervous system is the most sensitive. Survival limits of organisms or parts of organisms can be modified (acclimated) by prior experience of the environmental factor.

2. *Reproduction.* Environmental limits for reproduction or for embryos are often narrower than for adults. Limits for completion of a full life cycle may differ from limits for short-term survival.

Adaptive variations of internal state as a function of the environment that are classified as capacitative are conformity and regulation.

3. *Adaptive Variations.* Some animals change internally to conform to the environment, for example, poikilothermic or poikilosmotic animals (Fig. 2); these are conformers. Other animals are regulators and maintain relative internal constancy in a changing environment (Fig. 2); these are regulators, for example, homeotherms. Regulation fails at extreme limits of an environmental stress. In general, conformers tolerate wide internal varia-

tion but narrow environmental limits, whereas regulators tolerate only narrow internal variation but wide environmental range (Fig. 2). Acclimation can shift the tolerated internal limits for a conformer; in a regulator, acclimation can change the critical limits for activation or failure of homeostatic controls. Both patterns, conformity and regulation, are homeostatic in the sense of permitting survival in a changing environment and most animals show elements of both patterns. One parameter may show regulation (ionic composition) while another shows conformity (osmotic concentration), for example, in marine and brackish water invertebrates.

4. *Recovery.* Recovery from a stressed state varies with the kind of stress and is best shown by regulators. Some osmoregulating crustaceans maintain relative constancy of hemolymph osmotic concentrations in both hyper- and hypo-osmotic media; beyond critical concentrations in the milieu at either high or low levels they lose the ability to regulate themselves and concentrations of hemolymph rise or fall in parallel with the medium (Chapter 2). Oxygen regulators maintain constant levels of oxygen consumption as

P_{O_2} is decreased to critical concentrations below which metabolism declines steeply. Homeothermic animals (birds and mammals) maintain their body temperature by a variety of behavioral, circulatory, metabolic, and insulative means over a range of ambient temperature. Below some critical temperature, regulation fails, O_2 consumption increases, and body temperature is maintained, until at a low ambient temperature metabolism is insufficient and body temperature falls (Fig. 3). At high temperatures, circulatory and insulative means maintain relative constancy of T_b until at a critical temperature metabolism increases as it does in poikilotherms.

5. *Rate Functions.* Rates can be measured for movement, for metabolism, and for enzymatic reactions measured *in vivo* or *in vitro*. For enzymatic rates, two methods are useful: (a) measurement of maximum velocity where substrate is saturating and enzyme activity is rate limiting and (b) measurement of Michaelis constants (K_ms), which give rates in the physiological range of substrate concen-

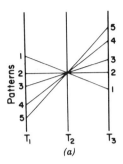

Figure 4a. Patterns of rate functions as altered by temperature acclimation in poikilothermic animals. T_2 is intermediate temperature. T_1 is lower and T_3 is higher temperature. Pattern 4 shows direct response and no compensation, merely Q_{10} effect. Pattern 3 shows partial acclimation, pattern 2 shows perfect compensation, pattern 1 shows overcompensation, pattern 5 shows inverse or undercompensation. [From Fig. 1-4 in (8)].

tration. Several patterns of rate function have been described for temperature conformers; they are applicable for other parameters as well. Figures 4 and 5 diagram patterns of acclimation after transfer from an intermediate temperature (T_2) to a lower temperature (T_1) or to a higher temperature (T_3). Immediately after transfer the rate declines or rises in direct response (pattern 4). After a period of ac-

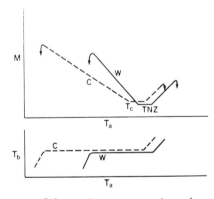

Figure 3. Schematic representation of metabolism M and body temperature T_b in homeothermic animals at different ambient temperatures T_a. TNZ is thermaneutral zone. T_c is critical temperature below which M increases in T_b maintenance. C and W refer to cold and warm acclimated animals. [From Fig. 1-7 in (8)].

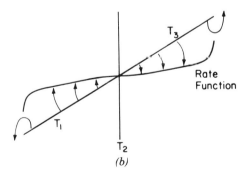

Figure 4b. Diagrammatic representation of reaction rates at different temperatures for animals acclimated at the three temperatures T_1, T_2, and T_3. [From Fig. 1-6 in (8)].

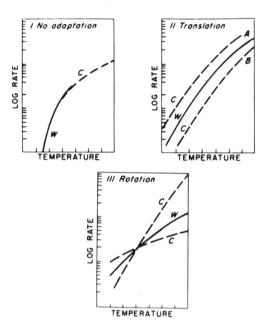

Figure 5. Three patterns of rate function in cold (C) and warm (W) acclimated poikilothermic animals. Rate functions measured at different temperatures. [From Fig. 1-5 in (8)].

climation, the rate at T_1 rises and at T_3 declines toward the original rate (T_2). If compensation is complete at steady state the rate is the same at all three temperatures (pattern 2); overcompensation occurs in some systems (pattern 1) but partial compensation is the more common pattern (pattern 3). No compensation occurs when the rate remains the same as it was immediately after the transfer (pattern 4); some rate functions show undercompensation or inverse acclimation (pattern 5). Another way of treating the data is to plot the entire rate–temperature curve. On acclimation there may be translation of the curve to right or left (constant Q_{10}) or rotation (change in Q_{10}) or a combination of the two responses (Fig. 5). Comparable changes in metabolic rates have been described during acclimation of marine and estuarine animals to var-

ious salinites and during acclimation to various levels of O_2.

Changes in the environment bring about not only metabolic acclimation but also behavioral acclimation. These are shown by taxic (direct) responses, selection of "preferred" environments in gradients, and feeding and reproductive changes. Each of these can be altered by acclimation to an environmental change.

Three time courses can be distinguished in response to environmental alteration. First, there are direct reactions to the environmental change. These may be direct metabolic responses (Q_{10}'s) or may be reflex behaviors. Changes in ions, temperature, oxygen, or foodstuffs may lead to immediate alterations in reaction rates. Capacity adaptations are often hereditary; for example, homeotherms are more sensitive in reaction rates than poikilotherms to changes in body temperature. The time for the direct response to a new rate is usually minutes or hours.

The second time period in response may require days or weeks. This is the period of acclimation or compensation and the physiological adjustment is related to the amount of environmental change and the genotype. In seasonal changes, compensatory alterations may develop gradually. In the laboratory, compensatory acclimations take place by a number of biochemical alterations, the net effect of which is homeokinesis or constancy of energy output. Homeokinesis is to be distinguished from homeostasis which is constancy of internal state.

A third time course of biological response occurs over many generations, the time for selection of genetic variants. Evolutionary changes correlated with environment alterations are influenced by (1) behavior—competition, predation, and social interaction, and (2) biochemical changes—alterations in protein structure, enzyme kinetics, and membrane lipids.

Mechanisms of acclimatization are multiple. Local populations are adapted in north–south gradients, (temperature), or estuarine gradients (salinity) and altitude (oxygen). Populations in a cline differ in kinetic properties of their enzymes, in tolerance of stresses. Isolation of populations at the extremes of clines leads to speciation.

Extension of range exploits new foods, new niches. Sense organs are the first detectors of environmental changes. Much adaptive behavior is programmed in inherited patterns of neuronal networks that are modifiable within fixed limits. Behavioral comparative physiology examines variations on common themes— visual transduction in photoreceptors, movement by fast and slow muscles, chemical signaling by neurotransmitters and pheromones, and nonmuscle movements. Cellular neurobiology is concerned with membrane properties of excitable cells; one aspect of comparative cognitive science examines complex behaviors that are mediated by neural networks.

Molecular biology has the potential to provide explanations of many of the adaptive phenomena of comparative physiology. Analysis of diversity in molecular terms starts with sequencing of amino acids in proteins. Differences in structure of proteins are correlated with functions of the cells in which they are contained. Proteins are made by translation of messenger RNAs, which are coded genetically. Description of base sequences in DNA reveals much greater genetic diversity than has been inferred from physiological measurements.

An essential feature of living cells is integration, a poorly understood and complex property. An enzyme in isolation has properties very different from its properties in an organelle and cell. Important questions are how sequences in proteins as coded genetically act to produce biological variations. Specific genes are expressed at different times in development and in different amounts in various tissues. Little is known of the nature of feedback control, how specific environmental factors bring about quantitative and qualitative differences in protein synthesis.

An organism is adapted as a unified whole and it would be disadvantageous for component parts to be adapted by different amounts or at different rates. Natural selection acts on whole phenotypes rather than on genotypes.

Hierarchical Organization of Biological Systems

Biological systems can be arranged by levels of organization as follows:

Communities
Independent organisms
Dependent (colonial and parasitic) organisms
Organs
Tissues
Cells
Organelles
Molecules
Atoms

Critical properties and kinds of measurement differ according to the level of organization. One aim of comparative and cellular physiology is to describe a more complex level of function solely in terms of one nearer the molecular level. Complexity is determined by spatial interactions and temporal sequences of actions. Complexity is a statistical concept; the probability is low that a given state will occur as a result of random events. A few examples of differences between levels are social behaviors that cannot be described solely in terms of synaptic interactions in the brain. Regulation of blood pressure is described by measures that are different from those for describing contractions of vascular smooth muscle. Isolated enzymes of electron transport differ from those enzymes in mitochondria. A hemoglobin molecule in solution has affinity for oxygen different from its affinity when it is in a red blood cell.

The principle of emergent properties states that at each more complex level some properties are different from those at simpler levels. The properties of the more complex organization could not have been inferred from the less organized components. Knowledge of all the nucleotides and proteins in an organism does not make it possible to synthesize the organism. The photochemical reactions in retinal cones are essential to color vision but the total sensations of color are based on higher levels of integration. Some theoreticians have argued that it is not possible to extrapolate from one level of biological organization to another. Analysis or breakdown from complex to simple is easier than is synthesis of complexity. One goal of comparative physiology is to show that extrapolation between levels of biological organization is possible, to provide conceptual bridges between the whole organism and molecular levels, and to derive causations from correlations.

Practical applications of comparative physiology are found in ecology and medicine. Physiological adaptations to physical and biotic factors in the environment have given insights into animal distribution and animal–plant interactions. With increasing numbers of endangered species and drastic environmental changes as a consequence of human ac-

tivity there is need for more information regarding physiological adaptations of stressed species. Also, changes in climate and the atmosphere will have great effects on flora and fauna, and predictions based on capacity for physiological adaptation are needed.

Comparative physiology makes important contributions to medical science. Comparative approaches have shown the diverse ways in which different kinds of animals solve physiological problems. For example, renal physiologists have reached important conclusions regarding the relative importance of filtration, reabsorption, and secretion by a study of these functions in fishes. Comparative physiology has provided tools for diagnosis and therapy, for example, the toxins in treating neuromuscular disease.

Comparative physiology provides a viewpoint that adds to our understanding of human nature. The comparative approach puts mankind in its proper biological perspective. This approach emphasizes our evolutionary heritage. It combines the viewpoints of genetic and cultural inheritance. Every kind of animal that has been successful during its period in the earth's history has been adapted to its physical and biotic environment. Man is currently successful because he excels in cognition, in communication, and social organization. Other animals, for example, insects, are successful in other ways. Recognition of our evolutionary heritage and its modulation by experience and environment provides a satisfying personal philosophy.

References

1. Field, K. G., G. Olsen, D. Lane, S. Giovannoni, M. Ghiselin, E. Raff, N. Pace, and R. Raff. *Science* 239:748–752, 1988. Molecular phylogeny of the animal kingdom.

2. Goodman, M. *Prog. Biophys. Mol. Biol.* 38:105–164, 1981. Animal phylogenies shown by amino acid sequences in proteins.

3. Goodman, M. *Syst. Zool.* 31:376–379, 1982. Molecular evolution above the species level.

4. Goodman, M., J. Pedwaydon, J. Czelnsniak, T. Suzuki, T. Gotoh, L. Molns, J. Shishikura, D. Walz, and S. Vinogradov. *J. Mol. Evol.* 27:236–249, 1988. Evolutionary tree for globin sequences.

5. Greenberg, M. J. Phylogenetic tree of animal kingdom. Personal communication.

6. Hendricks, L. et al. *J. Mol. Evol.* 24:103–109, 1986. Primary structure of 5S rRNAs and their use in metazoan phylogeny.

7. Prosser, C. L., ed. Introduction, In *Comparative Animal Physiology*, 3rd ed., pp. xv–xxii. Saunders, Philadelphia, PA, 1973.

8. Prosser, C. L. *Adaptational Biology: Molecules to Organisms.* Wiley, New York, 1986.

9. Woese, C. *Microbiol. Rev.* 51:221–271, 1987. Bacterial evolution.

Chapter 2 | *Water and Ions*

Leonard B. Kirschner

It is commonly believed that living cells developed in the ocean, although it is impossible to obtain unequivocal evidence for this. Whatever the ancestral home, it seems likely that the basic chemical composition of cells has been stable through a billion or more years of evolution. The major osmotically active intracellular cation in most animal, plant, and microbial cells is K^+, with Na^+ and Cl^- ions present at lower, usually much lower, concentration. The main anionic component is polyvalent protein with some charge contributed by smaller organic anions like ATP. There are some variations on this theme. We will see that the total amount of these intracellular solutes varies in different animals, and also that amino acids may or may not be a dominant molecular species. But these modifications appear to satisfy a particular set of conditions. They are added to the basic pattern without changing it. One can safely suppose that maintaining this pattern within some narrow range is essential for the normal function of cells.

If the composition of cells has been stable over the course of evolution, the same cannot be said of their habitat. Living organisms now inhabit every body of water and every land mass on earth. The physical and chemical diversity of these environments is enormous. A few inland bodies of water are nearly saturated brines, others have an unusual ionic composition and sometimes a very high pH. The oceans are about one-half molar NaCl solutions, while most freshwater lakes and streams contain only a few millimolar total solutes, and glacial melt may be nearly ion-free. Access to ions and water has a different meaning on land where an organism is bathed by neither. A general theme of this chapter is how animals can maintain recognizably similar cellular composition (and hence normal function) in such diverse environments. As we examine the question in more detail, it will become apparent that in none of these environments are an animal's cells in chemical equilibrium with the medium bathing them; work must be done to maintain cells in a steady state. Moreover, the biochemical, morphological, and behavioral adaptations that work in one environment may be inappropriate in another. For instance, the mechanisms used by a fish to maintain its cells in a chemical steady state in fresh water would be completely maladaptive in the ocean.

Thus, a common problem facing animals in any environment is to maintain cellular volume and composition in a steady state or to return them to some "preferred" condition after a perturbation. This would be simple, whatever the preferred condition, if cells were closed systems unable to exchange materials with their extracellular environments. However, all cells must be open for purposes of obtaining nutrients and eliminating metabolic end products, and the activities of many also depend on movements of molecules or ions for specialized activities. In general, then, the system cell-environment is open for purposes of

nutrient and energy exchange, but the exchange surfaces are usually permeable to at least some nonnutrient molecules.

The medium bathing most animal cells is either an extracellular fluid (ECF) derived from a hemolymph (blood) or the external environment in some aquatic forms. If we compare concentrations of the major intracellular solutes with their concentrations in the bathing medium, as is done schematically in Figure 1, it is immediately apparent that the two are rarely the same and often are very different (i.e., by one to several orders of magnitude). Such concentration differences represent a gradient of potential capable of causing the solute to diffuse from the region of higher to the one at lower concentration, providing only that the membrane is permeable to the solute. As noted above, membranes are permeable to many of the solutes on both sides. Diffusion is a "dissipative" flow in that it reduces the concentration gradient and if allowed to proceed to equilibrium may abolish (dis-

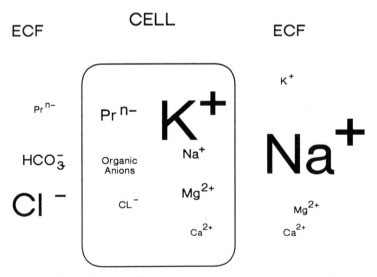

Figure 1. Intracellular and extracellular solutes in animal cells. The relative size of the symbols within a compartment suggests relative concentrations *within that compartment* (e.g., within the cell). However, in osmoconforming animals, organic compounds like amino acids or urea may exceed intracellular K^+ as shown in Figure 4.

sipate) it. One of the problems addressed in this chapter is how cells maintain a relatively constant internal composition in the face of spontaneous dissipative flows that act to change it.

The other problem addressed here is how cells maintain constant volumes. Since water is incompressible, net water entry will cause cells to swell while water loss will cause them to shrink. Diffusive water movement (osmosis is a familiar example) is also caused by a difference in the chemical potential (energy) or activity of water between two regions. For reasons described below, the chemical potential of water in an aqueous system is not expressed in terms of its concentration. The two factors that are important in determining water activity in living organisms are (a) the hydrostatic pressure applied to the water and (b) the total concentration of solute molecules in solution (the *osmotic pressure*, which is described in more detail later). Gradients of water due to hydrostatic pressure differences are important in only a few cases in animals (e.g., filtration of fluid from blood into a kidney tubule in some excretory organs). But the marked asymmetry in solute patterns between the inside and outside of cells leads to the question of whether there may exist osmotic pressure differences across membranes and, if so, how cells cope with the resulting water flow. Since the osmotic pressure of a solution is due to the *total* dissolved solute, it is obviously determined by those species present in relatively high concentrations [at least millimolar (mM)], and these comprise only the handful shown in Figure 1. For instance, although total intracellular Ca^{2+} is in the millimolar range, most of it is sequestered in organelles or bound to protein and contributes almost nothing to the chemical potential of intracellular water. This is true for any solute present in very low concentration and

includes many inorganic and organic species that may, in other ways, be crucial to cellular function, for example, trace nutrients and metabolic intermediates.

We will see that animals employ two strategies to cope with the problems of dissipative flows of materials, both solute and water. One of these is to minimize such flows by manipulating either the driving force or the membrane permeability or both. We can call such maneuvers "evasive" since they act to minimize or avoid the problem. However, since cells must remain open for exchange there always is some dissipative movement. The second strategy is to balance each dissipative flow by an equivalent movement in the opposite direction. Such a flow we can call "compensatory" to denote its effect on the problem. It is worth noting that a compensatory flow, since it occurs from a region of lower to one of higher potential, must require an input of metabolic energy. This is one of the reasons such mechanisms are interesting to physiologists.

Physical Principles

Properties of Water

The uniqueness of water as the solvent within which biological reactions occur has been frequently emphasized, classically by L. J. Henderson in his *Fitness of the Environment* (106). It may be argued that the unique properties of water permitted the chemical evolution that led to life as we know it. *Water is remarkable in being a liquid over the range from 0 to 100°C,* much higher than other hydrides (e.g., NH_3, CH_4, and SiH_4). The range spans the temperatures of most parts of the earth where life can occur. *Water has a very high specific heat;* that is, much heat must be added or withdrawn to alter its temperature. *It has a high latent heat of vaporization;*

hence it evaporates slowly from ponds and lakes, and when it evaporates from a body surface, heat is removed. *In its solid state water is less dense than in its liquid state* and, in fact, its density is maximum at 3.98°C above its freezing point. This property has permitted life to develop in temperate and subpolar regions where ice, lower in density than water, forms in winter only on the surfaces of bodies of water.

Water is also notable as a solvent for electrolytes and for most organic nonelectrolytes, as well as for oxygen and carbon dioxide. Nonpolar compounds such as some lipids do not dissolve in water. Although water is a neutral compound the oxygen is relatively negative while the hydrogen atoms are relatively positive. Such a dipolar structure favors its orientation toward charged or polar groups on other molecules with the formation of weak bonds between oppositely charged regions. The dissolution of inorganic salts depends on such bonding of water to the ions, a process liberating energy that is used to dissociate the salt into ions that are thus hydrated. In addition, water molecules in ice are hydrogen bonded together. The energy in a hydrogen bond is low, only 5 kcal/mol as compared to 50–100 kcal/mol for many covalent bonds. Although hydrogen bonds are broken when ice melts, there remains considerable hydrogen bonding in the liquid phase, and the ordering this causes probably accounts for many of the remarkable physical properties of this substance. In steam all such bonds are broken. The unpredictability of the degree of H bonding in solutions leads to uncertainty concerning the state of water in the intracellular compartment. Nuclear magnetic resonance (NMR) measurements show that more water is in a "crystalline" state in tissue than in the pure liquid (27% in muscle) and may not be available as solvent water (123).

Colligative Properties of Solutions

The number of moles of water in a solution, either within an animal or outside it, gives little indication of its actual thermodynamic activity (usually approximated by concentration in biological systems). Yet this is an important quantity that determines, among other things, the driving force for water movement. The chemical potential, or activity, of the water in a solution is related to the *total* concentration of dissolved solute molecules and is independent of the particular species involved. Determining this quantity by chemical analysis in a complex solution would be a tedious procedure even if minor components were ignored. Fortunately, several physical characteristics of solutions, collectively termed the *colligative properties*, are simple functions of the total dissolved solute concentration. Colligative properties include the freezing point (i.e., temperature), the boiling point, the vapor pressure, and the osmotic pressure. Each of these is modified by the presence of dissolved solute. For example, the freezing temperature of pure water under standard conditions is 0°C. When solute is dissolved in the water the freezing temperature is depressed, and the amount of depression (ΔT_f) bears a simple relation to the total solute concentration, $[c]$:

$$\Delta T_f = K_f[c]$$

The proportionality constant K_f varies for different solvents; for H_2O the value is 1.86°C/mol. If the freezing point depression, ΔT_f, is measured, the solute concentration can easily be calculated.

Solute also depresses the vapor pressure and elevates the boiling point and osmotic pressure. These changes are simple linear functions of the solute concentration. In principle, any of these changes can be used to estimate concentration. In practice, vapor pressure and freezing point depressions are used, and commercial instruments are available employing one or the other of these measurements. There are also simple techniques for measuring ΔT_f that require practically no equipment apart from a good, low-temperature thermometer.

It should be emphasized that the significant variable affecting the chemical potential of water is the *total* solute concentration. Each dissolved particle counts, and chemical species is of no moment. A liter of solution containing 1 mol of glucose and 1 mol of the undissociated amino acid glycine is 2 *M* in total solute (often denoted 2 osmolar or os*M* because of the effect on the solution osmotic pressure). Moreover, individual ions each count for one particle. Thus, 1 mol of NaCl gives 2 mol of solute when dissolved; 1 mol of $AlCl_3$ gives 4 mol of total solute, and so on.

Solute Movement: Diffusion and Active Transport

The biophysics of ion and water transfer is treated in more or less detail in nu-

merous reviews (8, 31, 121, 298), monographs (123, 134, 156, 256, 285), and at least one textbook of general physiology (53). Some of these aspects will be considered here, both because they will be encountered when the primary literature is consulted and because they provide a useful general description of solute and water movements and the parameters controlling them.

A solute may move in solution under the influence of any potential energy gradient. Two of the most important are gradients of chemical potential (approximated by concentration) and electrical potential. A representative situation is depicted schematically in Figure 2, for two cations. An interface (a cell membrane or an epithelium) separates two compartments. The solution in the outer compartment contains a higher concentration of cation A, but a lower concentration of cation B than the inner compartment. In addition, there is a voltage across the interface; the inner compartment is at the lower (negative) potential. Both gradients favor inward diffusion of ion A. The situation is more complex for cation B. The voltage gra-

> Concentration Gradient
⋮⋰⋅ Electrical Gradient

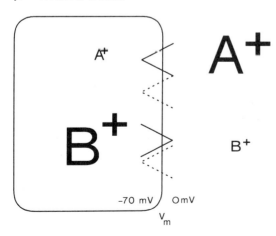

Figure 2. The membrane potential is a diffusion gradient for ions. It may be oriented in the same direction as the concentration gradient; both promote inward diffusion of A[+] in this figure. Outward diffusion of B[+] is favored by its concentration gradient but opposed by the membrane potential.

dient favors diffusion into compartment 1, but the concentration gradient favors diffusion in the opposite direction. Whether flow will occur and in which direction depends on which force is greater and can only be determined by quantitative analysis (e.g., using Eqs. 3 or 4 below). Figure 1 shows that such a distribution portrays the status of the two most abundant cations in animal body fluids. Sodium concentration is high in the extracellular fluid, low in the cell, while the reverse is true for potassium (see Table 2 for many examples). There is also a potential difference, like the one shown in Figure 2, across the cell membrane. The principle is actually more general than suggested by these figures. It applies to the asymmetric distribution of a solute between any two compartments separated by a barrier, not merely to cells and cell membranes. The gill epithelium of a freshwater fish separates blood, with a high sodium concentration from the environment with a lower concentration, and there is often a potential difference across the gill. The rate and direction of sodium diffusion depends on the magnitude and polarity of these forces, as well as on other factors to be described.

For diffusion in one dimension, as through a membrane or across an epithelium, the rate of solute movement (called a flux and denoted J) is expressed by the general diffusion equation:

$$J_{net} = -AD \times \left[\frac{dc}{dx} + \frac{z\mathcal{F}}{RT} c \frac{d\psi}{dx} \right]$$

$$\text{rate of solute flow} = \text{conductance for solute} \times \text{driving force}$$

$$(1)$$

where A = area (cm^2)

D = diffusion coefficient (cm^2/s)

c = solute concentration (mol/cm^3)

z = charge on the solute

$\mathcal{F}$ = the Faraday (96,500 coulombs/mol)

R = gas constant (8.314 V · coulomb/K · mol)

T = absolute temperature (K)

ψ = electric potential (V)

x = distance (cm)

There is no general analytical solution for this equation, so it is never used in biological research. Nevertheless, it is worth inspecting because it displays in compact form the variables that determine the rate of movement of materials across interfaces like membranes and epithelia.

The two terms within the brackets represent the forces driving a solute. The concentration gradient dc/dx is determined by the solute concentrations on the two sides of the interface and the second term, a voltage gradient (and some constants), is generated by any difference in electrical potential across the interface. Although the bracketed term appears complex, it is worth noting that the equation as a whole has the simple form: Flow is proportional to force, the same relationship as is expressed by Ohm's law for electric current flow. The proportionality constant is the product of the area available for diffusion and a constant called the diffusion coefficient. The former is easy to apprehend; for a given "push" the rate of solute movement is proportional to the area available for diffusion. The diffusion coefficient for a given solute is determined by the physical and chemical makeup of the interface. These are quantities that an animal can manipulate in controlling the passive flow of solute between two compartments, and many of the adaptations that we will encounter below involve such manipulations.

Although there is no general solution for the diffusion equation there are a number of useful special solutions, each appropriate for a particular set of "boundary conditions." For example, a nonelectrolyte has no charge ($z = 0$), and the entire second term in the bracket becomes zero, giving a form of the diffusion equation known as Fick's law:

$$J = -AD(dc/dx) \qquad (2a)$$

where J is flux.

If, in addition, the concentration gradient is linear through the membrane (i.e., $dc/dx = \Delta c/\Delta x = (c_2 - c_1)/x$, where x is the membrane thickness), then

$$J = -AD[(c_2 - c_1)/x] \qquad (2b)$$

We rarely know the value of D for a particular solute and membrane and may also be uncertain about membrane thickness. However, they are fixed quantities, and hence $D/x = P$ should have a constant value for a set of defined conditions. P is called the permeability coefficient. Equation 2b is useful for comparing quantitatively the permeability of different interfaces to a given solute, or of a single interface to several solutes. But to use this solution it is essential that the boundary conditions be observed. It can only be used for uncharged solutes, and the experimental conditions must ensure that the concentration gradient is linear and constant in time. It is often used by respiratory physiologists to compare the movements of a gas like oxygen across gills in different aquatic animals. Oxygen fulfills the first boundary condition; it is uncharged. Moreover, with water flowing past the gill exterior and blood past the gill interior, the O_2 concentration re-

mains constant in both compartments, even though the gas is moving out of one and into the other. This satisfies the other boundary condition, that $c_2 - c_1$ be constant in time.

Many solutes of biological interest are ions (i.e., are charged; $z \neq 0$), and the electrical term in the diffusion equation cannot be neglected. Furthermore, we seldom know the form of the concentration gradient, and the linear case must be rare, especially for ions moving through membranes containing charges. However, other restricted solutions of the diffusion equation have been developed for these cases. (a) The "constant field" equation has been used to relate the potential difference across a membrane to the concentrations and permeabilities of permeant ions in the bathing solutions (cf. 327): (b) Another is the flux ratio equation. In this case, it was shown (295, 302) that if, instead of considering the net movement of an ion, one considers two ion streams (unidirectional fluxes), one passing from compartment 1 to compartment 2 (J_{12}) and the other from 2 to 1 (J_{21}):

$$\frac{J_{12}}{J_{21}} = \frac{c_1}{c_2} e^{(-z's/RT)V_m} \qquad (3)$$

where V_m is the potential difference across the interface. If an ion conforms to the diffusion equation (is moving only by diffusion), the ratio of unidirectional fluxes can be calculated from the concentrations and potential differences across the membrane. Two additional boundary conditions lead to useful simplifications. (c) For nonelectrolytes $z = 0$ and the electrical term $= 1$; the flux ratio for a diffusing nonelectrolyte should equal the ratio of concentrations on the two sides of the interface. (d) In addition, if the system is in a steady state ($J_{12} = J_{21}$; $J_{net} = 0$), the flux ratio is unity and Eq. 3, solved for V_m,

TABLE 1. Permeability Coefficients for Water and Diffusible Solutes

Animals	Tissue	Permeabilities ($\times 10^6$ cm/s)				References
		Pos	P_K	P_{Cl}	$P_{aa}{}^a$	
Chaos chaos	Protozoan	36				215
Amoeba proteus	Protozoan	57				180
Squid	Axon	1,081				304
Crayfish	Axon	3,203				308
Crayfish	Muscle	6,540	3	3–7		228,325
Crab	Muscle	9,609			0.02	79,283
Frog	Muscle	22,156	1–2	2–4		112,325
Mammalian	Ascites cell	14,281	0.1	0.04	0.1	105,118,119

^{a}aa = amino acids.

reduces to

$$V_m = \frac{RT}{z\mathcal{F}} \ln \frac{c_1}{c_2} \qquad (4)$$

This is the Nernst equation, which shows how a diffusing ion is distributed *at equilibrium* across an interface with a potential difference V_m. But again, the boundary condition must be observed; the equation can be applied only when there is no net movement of the ion from one compartment to the other.

Such equations have two important uses. The first is to *estimate the permeability of an interface to a given molecule,* which might be water, a nonelectrolyte, or an ion. The ability to express this parameter quantitatively makes it possible to compare the permeabilities of an interface to different solutes or of different interfaces to one solute; for example, to compare the permeabilities of different cells and epithelia to water (cf. Tables 12 and 14–16). The procedure is simple, at least in principle. The system is set up and experimental conditions arranged so that they conform to the boundary conditions of one of the solutions of the diffusion equation. Then one measures the force applied (two concentrations and a voltage if an

ion is involved) and the flow, J. Then:

$$P = \frac{\text{flow}}{A \cdot \text{force}}$$

If the flow is in moles per second, area in square centimeters, and force (e.g., concentration difference) is in moles per centimeter cubed, P will have the units centimeters per seconds. There are the units shown in Tables 1, 8, 12, and 14–16. As an example, you can imagine using Eq. 2b to measure the permeability of a gill to oxygen. A few examples of solute and water permeabilities are shown in Table 1. The effect of cell membranes in restraining solute movement is apparent on comparing membrane permeability coefficients for water with the coefficient for ions.

The other application of the diffusion equation(s) is in *defining the metabolically energized transfer of solute that we call "active transport."* Active transport systems have several characteristic kinetic behaviors, and two of them are useful in distinguishing them from diffusion. One is dependence on metabolic energy. Metabolic inhibitors should block active transport without much affecting diffusion. The other is nonconformity with the behavior predicted by the diffusion equa-

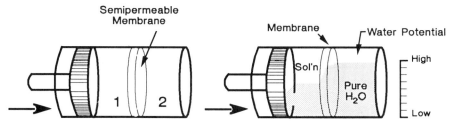

Figure 3. An operational definition of the chemical potential of water in aqueous systems. The description in the text supposes that there is a route for entrance and exit of water in each filled compartment, although these are not shown. This will permit water to leave one chamber and enter the other when the piston moves. In the figure on the right, compartment 2 contains pure water and compartment 1 contains a solution (water plus solute). The chemical potential of water is suggested by the height of the shaded area in each compartment.

tion. A solute subject only to diffusion must obey the equation; one being actively transported will not. Suppose we are interested in whether a muscle cell membrane has an active transport system for Cl^- or whether the ion is simply diffusing. We note that both intracellular and extracellular Cl^- concentrations are constant. For a diffusing ion this can only be because the ion is at equilibrium. If it is not, diffusion will occur from higher to lower electrochemical potential, and such a net movement would change the concentrations. Therefore, Eq. 4 provides an appropriate test. First we measure the concentrations on each side of the membrane, and from these we *calculate* what membrane potential (V_m) is required for equilibrium. Then we *measure* V_m. If calculated and measured values agree, diffusion is all that is occurring. If they disagree, then some nondiffusive transfer must exist; we call this active transport. The flux ratio test (Eq. 3) is used in the same manner for nonequilibrium situations. Many instances of active transport will be mentioned below. In nearly every case, they were first demonstrated because they did not conform to predictions of the diffusion equation.

Water Movement: Osmosis

Characterization of water flow across biological interfaces follows the same general rule as that for solutes: Flow is proportional to force. However, the concentration of water in aqueous solutions is extremely high and water molecules do not act independently of each other. As a result, the chemical potential is not even approximated by its concentration. In only one type of experiment, called the self-diffusion of water, is the treatment described above adequate. Here a tracer quantity of labeled water (e.g., ^2H_2O or ^3H_2O) is added to the aqueous solution. The tracer concentration is very low, and the isotopic water can be treated like any uncharged solute, so its movement is described by Eq. 2a. Just as for solutes, D and x are unknown, and the diffusional permeability to water is expressed in terms of a permeability coefficient, $P_d = D/x$.

For other types of water flow experiments the potential gradient must be expressed differently. Figure 3 (on the left) shows a membrane separating two compartments filled with water. If the membrane is water permeable but impermeable to solute, and pressure is applied by the piston on the left, the difference in hydrostatic pressure (ΔP) will cause water to flow from compartment 1 into 2. The volume that flows is proportional to the pressure difference; $J_v \propto \Delta P$. If solute is added to the compartment 1 (on the left) water will flow from right to left even in the absence of a hydrostatic pressure difference. The reason is that the solute lowers the chemical activity, or potential, of water in compartment 1, so that an activity difference exists between the two compartments. As a result flow will occur from pure water in compartment 2 into the solution in compartment 1. This is osmotic flow or osmosis, and the gradient causing it is suggested in Fig. 3 (on the right). The question is, how can we express this gradient? Flow can be abolished by applying a hydrostatic pressure to the compartment on the left. *The chemical activity of water in the presence of a solute can be expressed as the hydrostatic pressure necessary to abolish water flow under the conditions specified above.* This is the osmotic pressure (π) of the solution. Notice that the osmotic pressure of pure water is 0, and that water flows from lower to higher osmotic pressure. The volume flow is proportional to the osmotic pressure difference: $J_v \propto \Delta\pi$. Thus, the forces that can cause water to flow in this system are ΔP and $\Delta\pi$, and a number of important points have emerged from the study of the flow–force relationships. First, *equal osmotic and hydrostatic pressures generate equal water flows.* Therefore the appropriate proportionality constant is the same for both and is usually designated L_p, and we can write

$$J_v = -AL_p(\Delta P - \Delta\pi) \qquad (5a)$$

where J_v is the volume flow of water (cm^3/s), and L_p is the hydraulic or osmotic conductivity (cm/s · atm). Second, *osmotic pressure increases with solute concentration,* and the relationship is described by the gas law, $\pi = c_sRT$, where c_s is the total solute concentration. As noted earlier, the chemical species and characteristics of the solute play no special role. As long as the membrane is impermeable, 1 mol of any solute, even a single ion, in 1 L of solution under standard conditions exerts an osmotic pressure of 22.4 atm. It follows that a mixture of solutes exerts the same osmotic pressure as a single solute of equal concentration providing only that the membrane is impermeable to all of them. Therefore Eq. 5a can be written (remembering the gas law relationship):

$$J_v = -AL_p(\Delta P - \Delta c_sRT) \qquad (5b)$$

And finally, *the definition of osmotic pressure used above requires that the interface be completely impermeable to the solute(s).* Most biological membranes and epithelia are not completely solute impermeable, and it is important to be able to characterize the osmotic behavior of a "leaky" system. This is done by introducing a third membrane coefficient (the first two characterize water and solute permeabilities) to describe the interaction of the two diffusive flows. The "reflection coefficient," σ, was introduced (284) for this purpose and can be defined by simple experiment. If the membrane in Figure 3 is solute permeable we find that the pressure needed to abolish J_v is reduced; the effective osmotic pressure difference is less. In general, the leakier the membrane is to solute, the lower is the effective osmotic pressure difference. The reflection coefficient is simply the ratio of the $\Delta\pi$ we measure to the value we calculate from the gas law

$$\sigma = \frac{\Delta\pi \text{ (measured)}}{\Delta\pi \text{ (calculated)}}$$

and for a solute-permeable membrane

$$J_v = -AL_p(\Delta P - \sigma\Delta c_sRT) \qquad (5c)$$

If the membrane is impermeable to solute, $\sigma = 1$, and Eq. 5c reduces to Eq. 5b; if it is freely permeable to the solute, $\sigma = 0$, there is no effective osmotic pressure difference, hence no osmosis. It is clear tht if L_p is to be calculated from osmotic flow data the values of σ for any solute present must be known. Otherwise the correct value would be overestimated. This a a real problem for biological interfaces, few of which are completely impermeable to the solutes bathing both sides.

Note that either of two parameters can be used to characterize water flow across an interface; the diffusional permeability, P_d (cm/s) measured with tracers, and the hydraulic conductivity, L_p (cm/s · atm) obtained in an osmotic experiment. Since they have different units they cannot be compared directly, but L_p can be easily converted to a parameter, called the osmotic permeability (P_{os}). The parameter P_{os} has the same dimensions as P_d, which makes it possible to compare water permeability estimated by the two procedures. Unfortunately, these usually differ even for the same membrane or epithelium. The reason for this difference is still unsettled (123) but will not concern us.

Coupling of Flows

The movement of one molecular species, by volume flow, diffusion, or by active transport, sometimes creates a force that acts on a second species causing the latter to move. A particularly simple case is involved in filtration where the application of a hydraulic pressure difference causes bulk flow of solution. If the interface is permeable to any of the solutes in the solution, it will be carried in the direction of the flow even when the usual forces promoting solute movement (concentration and electrical gradients) are absent. Conversely, two compartments separated by an interface may be at osmotic equilibrium, but diffusion or active

transport of a solute across the interface will dilute one solution and make the other more concentrated, and the resulting osmotic gradient will cause water to flow in the same direction. In addition, an active transport system may couple the movements of more than one solute so they move together either in the same direction (cotransport) or in opposite directions (countertransport). Examples of both kinds of coupling occur frequently in the regulation of volume and composition of animal body fluids.

Regulation of Intracellular Volume and Composition

Solute Composition in the Steady State

All animal cells are bathed by an aqueous medium. In a few cases, for example, protozoa and porifera, the cells are exposed directly to the external medium, but in nearly all metazoans the extracellular fluid is blood (hemolymph) or a protein-free ultrafiltrate with solute pattern virtually identical with blood (plasma). Table 2 shows the concentration of the major solutes in muscle cells from animals living in different environments, together with corresponding values in the extracellular medium. Several generalizations are immediately obvious:

1. The most abundant intracellular solute is K^+, and the intracellular concentration is much higher than the extracellular concentration; that is, $K_i/K_o > 1$. High cytoplasmic K^+ is a biochemical characteristic common to all nucleated cells, animal, plant, and microbial. Steinbach called this "the prevalence of K" (286). The corollary, that extracellular K^+ is low, is equally general.

2. In contrast, intracellular solutes Na^+ and Cl^- are low. Nearly all measurements show that Na_i/K_i and $Cl_i/K_i <$

TABLE 2. Solute Composition of Muscle Cells[a] and Extracellular Fluids

Species	Na$^+$	K$^+$	Ca^{2+}	Mg^{2+}	Cl$^-$	NPN[b]	Other Solutes	Total Measured Solute[c]	References
Typical seawater	480	10	10	55	560			1033	11
Marine Animals									
Enchinoderms									
Parastichopsis tremulus	219	169	11	36	267	489	55	1246	239
Coelomic fluid	473	10	10	53	548	1	32	1117	
Echinus esculentus	211	131	16	35	216	549	22	1190	239
Coelomic fluid	444	10	10	51	519	2	30	1065	
Annelids									
Neanthes succinea	125	195	14	22	124	412	53	948	72
Coelomic fluid	483	14	13	44	545	9		1108	
Molluscs									
Sepia officinalis	31	189	2	19	45	678	132	1098	244
Hemolymph	465	22	12	58	591	4	13	1160	
Eledone cirrhosa	33	167	3	17	55	484	143	901	244
Hemolymph	432	14	11	54	516	3	30	1061	
Loligo forbesi	78	152	3	15	91	578	143	1060	244
Hemolymph	419	21	11	52	522			1032	
Ostrea edulis	180	165	9		109	314	299		30
Mytilus edulis	79	152	7	34	94	289	39	694	210
Hemolymph	490	13	13	56	573	3		1148	
Arthropods									
Callianassa californiensis	19	144	37		26	397			71
Hemolymph	465	12	59		506				
Emerita analoga	11	160	69		26	482			71
Hemolymph			15		499				
Cancer antennarius	79	154	33		30	464			71
Hemolymph	470	13	36		503				
Callinectes sapidus	40	186			46	780		1052	78
Hemolymph	470	13			420			1100	
Carcinus maenus	54	146	5	17	53	617	91	983	258
Hemolymph	468	12	35	47	524			1000	
Nephrops norvegicus	25	188	4	20	53	602	142	1037	243
Hemolymph	512	9	16	10	527	4	25	1108	
Limulus polyphemus	126	100	4	21	159	110	535	1060	237
Serum	445	12	10	46	514		15	1042	
Vertebrates									
Myxine glutinosa	32	142	2	13	41	706	250	1186	242,245
Plasma	487	8	4	9	510	35	1	1059	
Gadus morhua	21	119	4	25	15		70	224	293
Plasma	174	6	6.6	3	150		77	308	
Muraena helena	25	165	9	7	24		77	308	242
Plasma	212	2	4	2	188		11	419	
Chimera monstrosa	70	105	3	11	79	767	79	1106	245
Plasma	338	12	4	6	353	478	8	1114	
Squalus acanthias	18	130	2	14	13	905	107	1189	238
Plasma	296	7	3	4	276	504	6	1096	

TABLE 2. (*Continued*)

Species	Na$^+$	K$^+$	Ca^{2+}	Mg^{2+}	Cl$^-$	NPN[b]	Other Solutes	Total Measured Solute[c]	References
			Fresh Water and Terrestrial Animals						
Mollusks									
Anodonta cygnaea	5	21	12	5	2	11	19	75	210
Hemolymph	15	0.5	8	0.2	11	0.5	31	66	
Arthropods									
Potomon niloticus	44	111			32	212	78	477	261
Hemolymph	259	8	13		242			522	
Astacus fluviatilis	11	122	9		14	153		309	325
Hemolymph	208	5	14		250			477	
Vertebrates									
Malaclemys centrata	48	102			32	27			80,81
Plasma	152	6			95	8			
Rana esculenta	10	124	5	14	2	68	31	248	44
Plasma	109	2	2	1	78	9	38	245	
Rattus norvegieus	16	152	2	16	5	70	80	340	44
Plasma	150	6	3	2	119	10	34	324	

[a]The upper line of each pair shows values for muscle.
[b]NPN = nonprotein nitrogen.
[c]Total solute concentration, usually from freezing point depression.

1. A few values giving ratios around 1 have been reported, but the estimates of Na_i and Cl_i may be too high. It is very easy to overestimate the intracellular concentration of these ions because of their high concentration in the ECF present in the tissue of interest.

3. In most animals Na_i/Na_o and $Cl_i/Cl_o < 1$. In fact, the major extracellular solutes in most animal bloods are Na$^+$ and Cl$^-$. Two notable exceptions to this generalization are, on the one hand, protozoans, poriferans, and coelenterates living in fresh water (FW) where the cells are exposed directly to a nearly ion-free external medium, and, in addition, a number of herbivorous larval and adult insects that have high K$^+$ and low Na$^+$ bloods. But even in these groups $K_i/K_o > 1$.

4. Table 2 shows that the intracellular diffusible cation concentration [Na$^+$ + K$^+$] far exceeds that of the main diffusible anion Cl$^-$. Intracellular electroneutrality must be maintained, and the apparent anion deficit is accounted for by organic anions to which the cell membrane is impermeable. Most of these are proteins that are negatively charged around pH 7, but a fraction of the total is accounted for by compounds such as ATP, phosphorylated metabolic intermediates, and organic (carboxy) acids.

5. In marine invertebrates and some vertebrate groups the electrolyte pattern is as described above. However, they also have high intracellular concentrations of small organic compounds such as amino acids and urea, which might comprise more than one half of the osmotically active solute. This is a specialization involved in cellular volume regulation.

These chemical features are illustrated in Figures 1 and 4. Figure 4 also shows that one vertebrate group, the elasmobranchs, has a high concentration of organic solute in the blood.

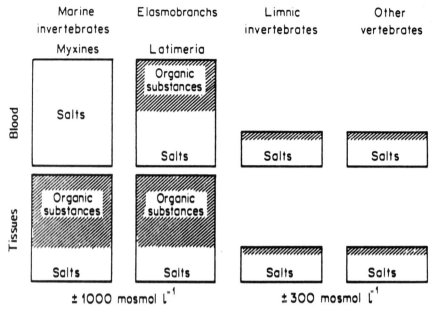

Figure 4. Organic and inorganic solutes in ECF and cells. Both total solute concentration and the relative concentrations of organic and inorganic solutes vary among animals and environments. However, intracellular and extracellular concentrations are the same in each case (same size rectangle). Exceptions to this rule (isosmotic compartments) are the few FW groups in which the cells are exposed directly to an FW medium (e.g., protozoa, poriferans, and coelenterates). [Reproduced from Giles (79) with permission of John Wiley and Sons Ltd.]

Steady-State Control of Intracellular Solute Concentrations

The first question we might ask is whether maintaining the ion configuration shown in Figure 1 requires the expenditure of metabolic energy. The answer will depend on whether it can be sustained by diffusion alone, or whether cells must actively transport one or more of the diffusible ions to maintain such steady-state distributions. An unchanging distribution between intra- and extracellular compartments could be maintained by diffusion only if the ion is at equilibrium; that is, if a concentration gradient promoting diffusion in one direction were balanced precisely by an electrical gradient promoting movement in the opposite direction. As noted earlier, the test is that a solution of the diffusion equation appropriate for diffusion equilibrium, namely, Eq. 4, the Nernst equation, describes the distributions. Using values for blood and muscle ions from Table 2 we calculate first the equilibrium potential for Cl^- in resting muscle from the crayfish *Astacus*. The calculated equilibrium potential, -73 mV, is close to the measured membrane potential, -77 mV (the negative sign indicates that the cell interior is negative). The difference is considered to be within the limits of experimental error, and the conclusion is that chloride distribution between intra- and extracellular compartments is purely diffusive.

The same procedure applied to K^+ gives an equilibrium potential of -86 mV for the crayfish muscle. This too, is not very different from the measured resting potential, and it was originally believed that the distribution of K^+ is also passive. However, the discrepancy has been noted in many cell types and suggests that intracellular K^+ is a little higher than can be accounted for by diffusion alone. We will see that there is independent evidence for inward active transport of K^+, in addition to diffusion, in most cells.

The analysis repeated for Na^+ gives an equilibrium potential of $+74$ mV; even the polarity is wrong, indicating that there is no correspondence between the Na^+ equilibrium potential and the measured membrane potential. Low intracellular Na^+ requires a mechanism capable of actively extruding the ion from the cell.

Such a "flux-force" analysis has been applied to a variety of cells from many animals, and the conclusions are, with some exceptions, the same. *Low intracellular Na^+ requires the presence of an active extrusion mechanism in the membrane, while low intracellular Cl^- can be accounted for by passive distribution across a permeable membrane.* High intracellular K^+ would be expected on the basis of diffusion alone, but *the concentration of K^+ usually exceeds that expected for diffusion equilibrium and argues for the presence of an active accumulation mechanism as well.*

The active transport of Na^+ and K^+ across cell membranes has been intensively investigated for more than 30 years. A detailed description of this work is more appropriate for a book on general cellular processes, but the system is so important in regulating the composition of cells that a brief description is warranted here (also see Ref. 82). The following points, illustrated in Figure 5, are central:

1. Cell membranes are permeable to the three major diffusible ions. In general, membrane permeabilities to K^+ and Cl^- are relatively high and of the same order of magnitude; permeability to Na^+ is usu-

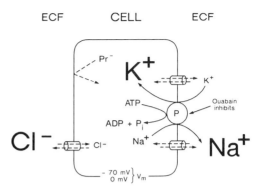

Figure 5. The pump and leak model for maintaining solute asymmetry. The salient features are described in the text. Dashed line and cylinder is a diffusion pathway. Solid line represents active transport. Although not shown, several other membrane solute pumps have been described and are important; for example, in nutrient transfer, intracellular pH regulation, and maintaining intracellular Ca^{2+} very low.

ally lower, much lower in excitable cells like nerve and muscle (113).

2. Active transport of Na^+ and K^+ requires an energy input by the cell. Consistent with this, it has been found that metabolic inhibitors reduce the efflux of Na^+ and influx of K^+ (114, 206). Significantly, one of these inhibitors is 2,4-dinitrophenol, which blocks the aerobic synthesis of ATP.

3. The active transports of Na^+ (out) and K^+ (in) are obligatorily coupled together (206, 207). In the absence of K^+ at the outer face of the membrane, Na^+ extrusion stops and cell Na^+ begins to rise. Conversely, low Na^+ at the inner face of the membrane will reduce K^+ uptake. This coupled biochemical membrane mechanism is called the Na^+-K^+ pump. Thus, maintaining intracellular ions in a steady state requires an expenditure of metabolic energy in the form of ATP

breakdown. It is probably one of the larger components contributing to basal energy metabolism in animals.

4. The operation of the Na^+-K^+ pump in inhibited by low concentrations of a class of compounds called cardiac glycosides; the most frequently used is named *ouabain*. The action of this class of compounds is restricted to the Na^+-K^+ pump, and is so specific that they are used as probes to decide whether the Na^+-K^+ pump plays a role in an uncharacterized ion transport problem. As shown in Figure 5 they act at the outer face of the cell membrane.

5. If cells are broken up, the membrane fragments will hydrolyze added ATP (273, 274). The rate of hydrolysis is increased when Na^+ and K^+ are added to the medium, and the ion-stimulated increment is blocked by added ouabain. This is the test tube equivalent of the Na^+-K^+ pump in intact cells. It is usually called the Na^+-K^+ ATPase.

6. Some herbivorous insect larvae and adults have relatively high K^+ concentrations and very little Na^+ in the blood. For example, the lepidopteran *Manduca sexta* was reported (59) to have blood Na^+, K^+, and Cl^- (in mM) of 17, 25, and 29; concentrations in muscle cells were 4.7, 100, and 12. The intracellular pattern is not atypical for muscle, but the blood ions are different from most. Using the Nernst equation as outlined earlier it appears that both Na^+ and K^+ must be actively extruded in the steady state and probably Cl^- as well. Obviously the Na^+-K^+ pump cannot account for this pattern; in fact, ouabain has no effect on metabolically driven ion movement in these muscles. The nature of the pump(s) is not yet known.

Cellular Volume Regulation in the Steady State

While the solute patterns of intra- and extracellular fluids are very different, the *total* solute concentration (the osmotic concentration) in nearly all animal cells is the same in both compartments. The reason for this is not hard to appreciate in light of our earlier discussion of the flow–force relationship for water. The osmotic pressure of a solution is determined by its total solute concentration, and a difference in osmotic pressures will cause water to move in the appropriate direction. If the solute concentration of the extracellular fluid is lower than that within the cells ($\pi_{out} < \pi_{in}$) water will diffuse into the cell, and there are three possible outcomes. If the cell surface is both rigid and strong, the entry of water will cause the hydrostatic pressure in the cell to rise. This "turgor" pressure (ΔP in Eqs. 5) acts to oppose water entry; when it is equal to the osmotic pressure difference, net water movement stops, and the system remains in a steady state in spite of the osmotic asymmetry. This solution is found in plant and microbial cells where a rigid extracellular wall (cellulose in plants) confers the necessary strength. Alternatively, an active water transport system might extrude water against an osmotic gradient as rapidly as it diffused in. This would require an expenditure of metabolic energy such as we find when solutes are transported against gradients, but is in principle feasible. However, in spite of intensive research no evidence for active water transport has been found in animal cells, and the general consensus is that such a mechanism probably does not exist. The third possibility for avoiding water flow is to maintain the same solute concentrations in the cellular and extracellular compartments. Table 2 shows the results of reasonably complete analyses of blood and intracellular solutes in muscle from a number of animals inhabiting terrestrial, fresh water (FW) and seawater (SW) environments. Blood concentrations vary over a wide range; they are highest in marine invertebrates and one vertebrate group and are substantially lower in terrestrial and FW forms. But the pattern is nearly always the same. The major hemolymph solute is NaCl with much lower concentrations of other, mostly ionic, constituents. There are two notable exceptions to this generalization. Elasmobranch fish and one amphibian that can live in seawater have ionic concentrations similar to other vertebrates (total 400–500 mos$M$). But blood urea and trimethylamine oxide concentrations are so high that total blood osmotic concentration approximates that of the seawater outside. This is shown in Figure 4 and by the values for *Chimera* and *Squalus* in Table 2. The second exception comprises a broad group of herbivorous insects in which blood K^+ is high and Na^+ low. But what is significant in the context of water movement is that, whatever the composition of the ECF, total ECF and cell solute concentrations are the same. Cells and their ECFs appear always to be at osmotic equilibrium.

Within the limits of analytical accuracy, the total solute concentration in the cells is the same as outside, but two distinct patterns are notable. In FW and terrestrial animals most of the intracellular solute comprises inorganic ions with lower concentrations of negatively charged proteins and phosphorylated organic compounds. But as suggested in Figure 4, the major osmotically active solute in many marine animals comprises a group of small, organic nitrogen compounds among which amino acids and urea are notable. Figure 6 provides a basis for understanding this difference. In the first group, extracellular Na^+, which accounts for nearly one half of the osmotic pressure, varies between 20–200 mM. Over this range intracellular K^+ rises and, together with smaller quantities of Na^+, Cl^- and organic anions, is able to match the extracellular osmotic pressure. But there appears to be an upper limit to intracellular K^+ around 200 mM for reasons that are not understood. In marine and some

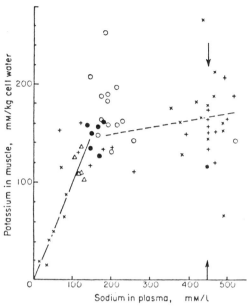

Figure 6. Intracellular $[K^+]$ and extracellular $[Na^+]$ in aquatic animals. Each symbol shows the mean intracellular (muscle) $[K^+]$ and extracellular $[Na^+]$ for a single species of arthropod (+), other invertebrates (x), fish (○), amphibia (△), and other vertebrates (●). In nearly all cases, the major blood solute is NaCl (elasmobranchs and a few other marine vertebrates are exceptions). In FW animals, blood $[Na^+]$ is <200 mM and intracellular $[K^+]$ matches it as also suggested in Figure 1. However, in marine invertebrates and in hagfish, blood $[Na^+]$ is much higher, usually approximating SW $[Na^+]$. Intracellular $[K^+]$ apparently cannot rise much above 200 mM in any animal. Therefore, when blood $[Na^+]$ exceeds this value the two compartments can be maintained isosmotic only by increasing another intracellular solute (amino acids and other small nitrogenous compounds). [Reproduced from Burton (35b) with permission of Pergamon Press.]

estaurine invertebrates and in the vertebrate hagfish, blood Na^+ is much higher, often matching that in the external environment (~450 mM). Since intracellular K^+ can rise only to ~200 mM, the concentration of organic nitrogen compounds becomes elevated, so that total intra- and extracellular solute concentrations remain the same. While this strategy for maintaining an osmotic steady state is clear enough, the mechanisms involved are not completely understood. A central role is certainly played by the Na^+-K^+ pump in most cells, since if this system did not operate Na^+ would diffuse into the cell and K^+ out. This would depolarize the membrane and allow Cl^- to diffuse in. The resulting rise in intracellular total solute would promote water entry and cell swelling (297). Such a system, employing a single ion pump and fixed permeabilities to the three major ions, is called the "pump–leak" model for maintaining the cell volume in a steady state. It probably applies in this simple form to most cells in FW and terrestrial animals (156, 182). It is also thought to be important in marine animals, but here an additional pump (or pumps) and leak(s) must be incorporated into the model in order to account for maintenance of high intracellular organic solutes such as amino acids (68, 79). Inhibition of such a pump would cause the cell to lose organic solute and shrink, but no difference in principle is involved.

Volume Recovery after a Perturbation: Volume Regulation

The model presented above, with fixed permeabilities and more or less fixed pump rates, can account for maintenance of volume in a steady state, but would not be very flexible should the cell, for some reason, swell or shrink. Recent work has shown that animal cells are able to regulate their volume even when the latter is perturbed by a change in osmotic pressure in the bathing medium. Moreover, several systems for accomplishing this have been described, and it is impossible to present a unified picture applicable to all animal cells; indeed different mechanisms may be involved in

specialized cell types. Two well-studied examples, one involving primarily the movement of amino acids and the other mainly of inorganic ions, will be described in order to illustrate how such systems may operate.

Some invertebrates are capable of living as adults in either SW or FW. Notable among these are several decapod crustaceans like the blue crab, *Callinectes sapidus*, and the "Chinese crab," *Eriocheir sinensis*. Figure 7 shows an experiment in which *Eriocheir* was first acclimated to SW. In this medium blood and muscle cells were at osmotic equilibrium with solute concentrations ~1000 mos*M*. At the beginning of the experiment the animals were transferred to FW. Although not shown in the figure, salt was rapidly lost to, and water gained from, the dilute external medium with the result that blood osmotic pressure dropped. The resulting osmotic gradient between blood and muscle caused water to enter the cells. Cell swelling (open circles in the figure) was, on this time scale, too fast to follow, but recovery was also rapid, and cell volume was restored to the original level within the first day after transfer. In the absence of active water extrusion, volume recovery can only be explained by loss of solute from the muscle cells. It is clear that the amino acid content of the cells (open squares) dropped rapidly and substantially during the volume-recovery phase. At equilibrium blood and muscle are again in osmotic balance but at a much lower concentration than in SW (~590 mos*M*). More detailed data have been obtained from marine invertebrate muscle tissue examined *in vitro*. Table 3 shows intracellular patterns of the major solutes from two marine crabs and a polychaete worm. The analyses were performed on isolated muscle cells incubated in SW and in dilute SW. Only the main inorganic solutes and small nitrogenous compounds are shown, and it is apparent that both

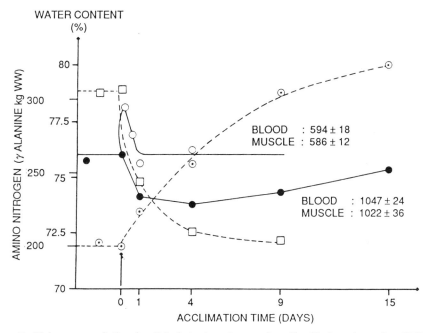

Figure 7. Volume regulation by *Eriocheir sinensis* muscle cells. Crabs adapted to SW were abruptly transferred to FW, and tissue water content (O——O) and amino acid concentration (□——□) followed. Total blood and cell solute was ~1000 mos*M* in SW and dropped to ~590 mos*M*. Another group was adapted to FW and abruptly transferred to SW. Tissue water content (●——●) decreased rapidly and recovered very slowly as intracellular amino acid concentration rose (⊙——⊙). After recovery blood and tissue solute concentrations were ~1000 mos*M* and isosmotic with the medium. [Reproduced from Gilles (79) with permission of John Wiley and Sons Ltd].

TABLE 3. The Effect of Extracellular Osmolarity on Intracellular Solute Concentrations[a]

Solute	*Callinectes sapidus*[b]		*Neanthes succinea*[c]		*Carcinus maenus*[d]	
	SW	0.5 SW	SW	0.5 SW	SW	0.4 SW
Na^+	39.5	28.4	125	58.4	54	35
K^+	186.0	162.2	195	153	146	90
Cl^-	45.6	25.9	124	56.7	53	32
Amino acids[e]	780	514	412	155	617	377
Total solute	1052	731	856	423	870	534
Blood solute	1100	850	1106	562	1000	~650

[a]Concentration in m*M*.
[b]See Gérard and Gilles (78).
[c]See Freel et al. (72).
[d]See Shaw (259, 260).
[e]Includes other nitrogenous organic compounds (e.g., trimethylamine oxide).

are lower when the tissue finally comes into equilibrium with the dilute medium, but much more organic solute than inorganic leaves the cells in this situation.

Obviously, a crab adapted to FW, and returning to SW (e.g., to breed), will face precisely the opposite problem. Figure 7 shows that such a transfer is followed by rapid shrinking of the cells (filled circles). This is followed by a sustained rise in amino acid concentration and recovery of the original volume, although both changes are much slower than in the SW to FW transition.

Fresh water and most terrestrial animals have lower blood and cell osmotic concentrations, and organic compounds are a smaller proportion of the intracellular solute concentration (compare the values for frog muscle in Table 2). One mammalian cell that has provided information on volume regulation is the Ehrlich ascites tumor cell. When these cells are transferred from isosmotic physiological saline to a more dilute saline they swell very rapidly (Fig. 8, upper left panel), but the volume begins to decrease almost immediately back toward the control level. The middle and lower left panels of Figure 8 show that K^+ and Cl^- are lost during this regulatory volume decrease (RVD) phase. Not shown in this figure is a substantial loss of the amino acid derivative taurine and lesser amounts of neutral amino acids (117, 118). The loss of cell solute with accompanying water accounts for the recovery toward the initial volume. In contrast to marine invertebrate cells where most of the solute lost compromises amino acids and their derivatives, these compounds account for only about one third of the total in the ascites cells. The rest is KCl. At the end of this first experimental period the cells, only slightly swollen, are at osmotic equilibrium with the diluted medium. At this time they are transferred back into normal physiological saline,

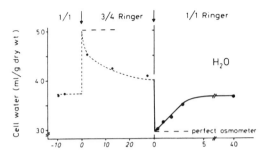
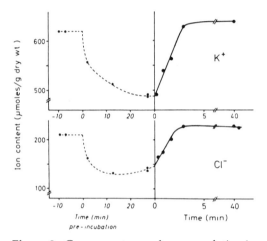

Figure 8. Compensatory volume regulation in Ehrlich ascites cells. Regulatory volume decrease in Ehrlich ascites cells after reduction of external osmolarity, and regulatory volume increase after restoration of external osmolarity. The cells were preincubated in 300 mosmol Cl^- medium for 40 to 50 min and at zero time were resuspended in 225 mosmol Cl^- medium. Cell water, K^+, and Cl^- content were followed with time after the reduction of external osmolarity. The cell water content calculated for a perfect osmometer is indicated by the horizontal broken line. At second zero time (solid vertical line) a tonicity of 300 mosmol was restored by addition of one-fourth volume of a double-strength saline solution, and cell water, K^+, and Cl^- content were followed with time. The cell water content calculated for a perfect osmometer is indicated as before. [Reproduced from Hoffman (115) with permission of *Federation Proceedings*].

whereupon they immediately shrink (Fig. 8, upper right panel). Shrinking is followed by a gradual increase in volume back to the control value (relative volume increase, RVI). The middle and lower panels show that the RVI is caused by an increase in intracellular KCl. It may be surmised that intracellular taurine and amino acid concentrations also increase.

A detailed description of the cellular mechanisms involved in such compensatory solute shifts is outside the scope of this book. However, the following brief summary is based on work with the ascites tumor cell. As noted earlier, because of the operation of the Na^+-K^+ pump in a cell at osmotic equilibrium with blood or a physiological saline, the intracellular $[K^+]$ is above the equilibrium value (i.e., higher than predicted by the Nernst equation for the existing membrane potential). Thus, there is a diffusion gradient favoring loss of K^+. Normally, outward leakage of K^+ is just balanced by active inward transport through the Na^+-K^+ pump, and the ion is maintained in a steady state. At the same time, Cl^- is at diffusion equilibrium; no energy need be expended to maintain its concentration steady. Intracellular taurine and amino acid concentrations are much higher than extracellular and these compounds diffuse out of the cell. This leak is just balanced by active inward transport to maintain intracellular concentrations constant. When the cell swells, P_K and P_{Cl} increase (116, 119), the diffusive efflux of K^+ increases and exceeds inward transfer through the pump. The net loss of K^+ will increase the membrane potential and create a diffusion gradient favoring outward Cl^- movement (an example of one kind of flux-coupling mentioned on pp. 23). This, together with the increase in P_{Cl}, will allow Cl^- to leak from the cell in company with K^+. The initial swelling also increases the amino acid permeability, and the augmented efflux is no longer bal-

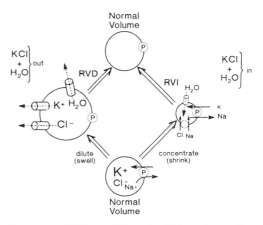

Figure 9. RVD and RVI in Ehrlich ascites cells. Swelling after transfer to dilute medium is followed by KCl loss and return to normal volume (RVD). Shrinking is followed by net gain of KCl and swelling to normal volume. Amino acids play a lesser role and are not shown in the figure. [Modified after Hoffman (115).]

anced by active uptake, so there is a net loss of organic solute at the same time. *Note that no extra energy input is required to account for solute loss during RVD.* The solute gradients are poised to favor efflux, and only changes in membrane permeability are needed to generate these flows. Somehow swelling acts as a trigger. Less is known about the movement of solute during the RVI following shrinking, although Figure 8 shows clearly that the net effect is uptake of KCl from the medium. Present work suggests (115, 120) that shrinking activates a mechanism for cotransport inward of NaCl followed by extrusion of the Na^+, with concomitant accumulation of K^+, through the Na^+-K^+ pump. Such a two-step process would account for the KCl accumulation noted. The sequence of changes attendant on swelling or shrinking is illustrated in Figure 9. It must be emphasized that the model is based on the behavior of a single cell type, and an aberrant (tumor) cell at that. More research will be required to determine how broadly applicable it is.

Several reviews summarize what we know about this system as well as variations found in other cells (150, 303).

Solute and Volume Regulation in Protozoans

The cells of most metazoans are bathed by an ECF derived from blood, and the latter may (many animals in SW) or may not (all animals in FW) resemble that of the external medium. Even those tissues exposed on one side to the dilute external environment in FW (e.g., a skin or gill) are bathed on the other side by hemolymph. But this is not true for several taxa, animals whose cells are exposed directly to the external medium. Of the three large groups, *Protozoa, Porifera,* and *Coelenterata,* to which this applies only the first has received more than passing attention. The data on protozoans are far from exhaustive, but it is possible to outline the beginnings of a picture on regulation of chemical composition and volume. Table 4 shows the intracellular concentrations of the three main diffusible inorganic ions for three FW and one marine organism. Total intracellular osmotic concentration, measured by physical methods (freezing point and vapor pressure) varies from 70 to 120 mM in FW protozoa. Thus, these animals are more dilute than most metazoans, but they are considerably more concentrated than the medium. There is obviously a volume regulatory problem to which we will return. *Miamiensis,* a marine ciliate, is presumably in osmotic equilibrium with SW, although the inorganic ion and amino acid concentrations measured comprise only ~60% of the necessary intracellular solute (no attempt was made to identify other solutes).

Intracellular K^+ is always considerably higher than extracellular K^+ and appears to be regulated fairly closely in the range 25–35 mM in FW species. The same pattern is seen in *Miamiensis,* but the intracellular concentration is higher. Intracellular Na^+ and Cl^- are more variable. In dilute media concentrations of both are always lower than intracellular K^+, but the examples in Table 4 show that they may be lower than extracellular NaCl in some cases, higher in others. In the marine ciliate the $[Na^+]$ and $[Cl^-]$ are not very different from $[K^+]$, but they are nearly an order of magnitude below those in the medium, so at least one of them is not at equilibrium (we need to know the membrane potential to decide which one). The data for *Tetrahymena pyriformis* look most like the metazoan cells discussed earlier, and if we assume that chloride is distributed passively by diffusion

TABLE 4. Intracellular and Extracellular Ion Concentrations[a] in Protozoans

Species	K^+	Na^+	Cl^-	Amino Acids	Total Solute	References
Tetrahymena pyriformis	28.8	6.5	9.2	56.5	111	288
Medium	5.0	30.0	36.0	0		
Chaos chaos	28.3	0.35	16.5		117	32
Medium	0.43	0.1	1.0			
Amoeba proteus	24.8	1.1	9.7		101	220
Medium	0.14	0.02	0.16			
Miamiensis avidus	73.7	87.9	60.8	316.6	539	132
Medium	10	450	550	0	1010	

[a]Intracellular concentrations in mmol/kg of wet weight. Extracellular concentrations in mM.

the situation would be as follows. From the Nernst equation we can calculate what the membrane potential must be to account for the chloride distribution. This is -35 mV. Such a membrane potential could account for an intracellular K^+ of just under 20 mM if K^+ were moving solely by diffusion. The actual concentration in the cell is substantially greater than this, therefore K^+ must be actively transported inward. To account for the Na^+ distribution by diffusion alone the membrane potential would have to be $+39$ mV. The polarity is wrong for excluding a cation, therefore Na^+ is actively extruded from the cell. This, in fact, is exactly the picture obtained with many metazoan cells. Unfortunately, the same analysis applied to the other organisms in Table 4 leads to different conclusions, some of them untenable. This is especially so for *C. chaos* and *A. proteus* where the accumulation of Cl^- by diffusion would require a positive membrane potential, and this has never been observed in nucleated cells. In spite of a number of studies showing a relationship between metabolism and ionic pattern, and dependence of the latter on the composition of the medium, mechanisms are still largely unknown. It can be concluded that intracellular solute patterns, especially in FW forms, are similar to those in metazoan cells bathed by an internal circulating medium, but questions relating to the nature of solute pumps and leaks remain to be resolved.

Unlike most cells, FW protozoans are never at osmotic equilibrium with their environment. Intracellular solute concentration is relatively low (e.g., only 25–30% those in FW vertebrate or crustacean cells). This tends to minimize the osmotic gradient promoting water entry. In addition, a few measurements indicate (180) that P_{os} ($\sim5 \times 10^{-5}$ cm/s) is about two orders of magnitude lower than in typical metazoan cells, as shown in Table 1. Remembering the diffusion equation, this too will minimize osmotic flow. We called such mechanisms evasive, and they will be discussed again later in connection with regulating the composition of extracellular fluids in larger animals. Such evasive strategies reduce the magnitude of dissipative flows, and hence the cost of managing them. However, with an inward osmotic gradient and a finite water permeability these animals face water loading, and a compensatory mechanism is needed to extrude water from the cell as fast as it enters. Water extrusion is carried out by an organelle called the contractile vacuole, which is found in FW protozoans and poriferans and in some marine species as well. Details of vacuolar structure and function have been described elsewhere (199). It is a more or less spherical organelle that fills over a period of time (diastole), then fuses with the cell membrane and discharges its aqueous contents into the medium (systole). The rate of fluid extrusion varies inversely with the osmotic pressure of the medium (143), which was an early indication of its role in maintaining a steady state cell volume in dilute medium. As shown in Figure 10 this has been observed in some marine forms where vacuolar output increases when an organism is transferred from SW to dilute SW (132). In many FW species the output also decreases when external solute concentration is increased. Some observations indicate that protozoa are capable of regulating cell volume after a perturbation in a manner similar to metazoan cells. When the marine ciliate *M. avidus* is transferred from SW to 50% SW there is an immediate increase in cell volume as shown in Figure 11. However, within ~5 min cell volume decreases; 20 min later volume is back to the control value, which then appears to be maintained. Conversely, when the animal is transferred to 175% SW it shrinks rapidly to a minimum at ~30 min, after

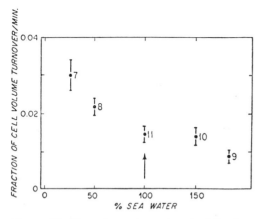

Figure 10. Vacuolar water output in *Miamiensis avidus*. Vacuolar water excretion is not negligible (>1% cell volume/min) even in SW. It increases in more dilute media and decreases in more concentrated. [Repduced from Kaneshiro et al. (132) with permission of *Biological Bulletin*.]

which there may be a slow recovery toward the original volume. This behavior is reminiscent of the RVD and RVI that occurred after similar transfers in ascites cells. In *M. avidus* transfer to the more dilute medium induced a marked increase in vacuolar output together with a rapid loss of amino acids (133). Na$^+$ and Cl$^-$ concentrations also drop substantially. The solute loss reduces the osmotic gradient and slows water entry from dilute medium, while the augmented vacuolar output increases water efflux from the cell. The two changes act together to restore cell volume to normal in the dilute medium. In the transfer to a more concentrated medium vacuolar output is reduced and intracellular amino acid concentration rises, but both changes are small, which probably accounts for the slow and incomplete recovery of volume following this change.

The osmotic pressure of vacuolar fluid is lower than that of the cytoplasm in FW protozoans; that is, the vacuole excretes a dilute fluid (230, 252). This anticipates

the fact that many FW metazoans excrete a urine that is more dilute than the body fluids. In addition, [Na$^+$] in the vacuole is higher than in cytoplasm and the reverse is true for [K$^+$]. This suggests the operation of ion transport mechanisms in forming the fluid. Metabolic energy is required for operating the vacuoles, which stop functioning in the presence of inhibitors like cyanide (142,143). In one experiment the contractile vacuole was isolated from *A. proteus* and was shown to contract and expel its contents when ATP was added (219). Thus, an energy input might be required both for forming the fluid (i.e., in actively transporting ions) and to support the mechanics of emptying. There is no comparable information for the organelles in marine species.

Coelenterates are metazoans with two tissue layers, the outer bathed by the external medium and the inner lining a central gastrovascular cavity or enteron.

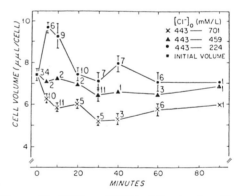

Figure 11. RVD and RVI in *Miamiensis avidus*. The cells were in an osmotic steady state in SW. At time zero one group was transferred to a more concentrated medium (701 mM Cl$_o^-$, x——x), another to a dilute medium (229 mM Cl$_o^-$, ●——●), and a third to a nearly isosmotic solution (459 mM Cl$_o^-$, ▲——▲). Volume recovery after swelling was complete within 20 to 40 min after transfer. Recovery after shrinking was much slower. [Reproduced from Kaneshiro et al. (132) with permission of *Biological Bulletin*.]

They occur in both SW and FW, but hydromineral regulation has been studied in only a few FW species (*Hydra* and *Chlorohydra* sp. and the medusa *Craspedacusta sowerbyi*). Tissue composition resembles that in protozoans: Total solute concentration is low (~50–120 mos$M$) and with cell K$^+$ > Na$^+$ (19, 102, 159). However, cell ion concentrations are much higher than ambient, and since the animals are water permeable (19, 159), they face both ion loss by diffusion and water loading by osmosis. Uptake of the major ions (Na$^+$, K$^+$, and Cl$^-$) is almost certainly active, although the mechanism is still unknown. The water problem is solved by forming a hypoosmotic fluid in the enteron and ejecting this by contractions of the body wall, which are more rapid in dilute media (19). The enteron appears to play the same role as the contractile vacuole in protozoans, and in both solute-linked water "secretion" seems to be involved. But in neither is the mechanism known.

Regulation of the Extracellular Fluids

The foregoing should make it clear that both composition and volume of animal cells are dependent on the chemical makeup and osmotic pressure of the ECF. Although cells have the ability to compensate for changes in a steady state, that is, for concentration or dilution of the ECF, such changes are likely to engender costs, both energetic and in terms of functional efficiency, which are better avoided. For this reason, maintaining a constant ECF composition with some optimal pattern would appear to be a reasonable and economic strategy.

The range of habitats occupied by animals varies enormously in physiochemical characteristics. Aquatic environments vary from near solute-free water to crystallizing brine as in the Great Salt Lake or the Dead Sea, and all of them differ again

from a terrestrial situation. The body surface must be open for nutrient and gas exchange, and since it is permeable to water and at least some nonnutrient solutes, it is clearly impossible to maintain the ECF at diffusion equilibrium with the environment in all of these situations, although it is feasible in some. An ECF at solute and osmotic equilibrium with FW, for example, would be far too dilute for the cells of metazoans with an extracellular fluid compartment. Evaporative water loss in a terrestrial animal would also lead to unacceptable changes in ECF osmotic pressure in the opposite direction. Remember that the relative concentrations of osmotically active solutes in the ECF is qualitatively similar in nearly all metazoans; bloods are high in Na$^+$ and Cl$^-$, with lower concentrations of other ions and organic nutrients (cf. Table 2). But Table 5 shows that the total solute concentration, expressed by the freezing point depression (cf. p. 6), differs in animals from different environments and frequently in different animals in the same environment. In many animals ECF concentration changes when environmental concentration changes, in others it is substantially constant. Thus, study of the composition of the ECF is an important element in understanding osmotic and solute regulation in animals. Indeed, this aspect has received much more attention than the cellular behavior described earlier. The patterns in different environments are varied and were fairly completely characterized by the 1940s. Since then attention has been directed in one of two directions. *One class of questions deals with the environment as a variable.* Why are some groups confined to a narrow range of salinities (stenohaline animals) while others can cope with and exploit a wide range (euryhaline animals)? What changes in osmoregulatory parameters made it possible for animals to move out of the ocean into FW or onto land? *Another*

TABLE 5. Ranges of Variation in Respect to Osmoconcentration in Body Fluids, in Aquatic Environments, and in Respect to Water Concentration of Air

Medium	ΔT_F (°C) (Medium)	Animal	ΔT_F (°C) (Body Fluids)
Freshwater	-0.01	FW mussel	-0.08
<5‰		*Pelomyxa*	-0.14
		FW fish	-0.50 to -0.55
		Amphibians	-0.45
		Crayfish	-0.82
		Crabs (a few)	-1.1
Brackish water	-0.2 to -0.6	Euryhaline invertebrates	-0.5 to -1.8
5–15‰			
Seawater	-1.85	Marine fish	-0.65 to -0.7
35‰		Fiddler crab	-1.3
		Marine invertebrates	-1.8 to -1.9
		Marine Elasmobranchs	-1.85 to -1.92
Salt lakes	-13.5 to -15	Brine shrimp (*Artemia*)	-1.2 to -1.6
50–250‰			
Humid niches		Earthworms	-0.3 to -0.4
(80–95% Relative		Snails, slugs	-1.2 to -1.6
humidity)		Amphibians, mammals	-0.5 to -0.58
		Insects	-0.8 to -1.2
Dry air		Reptiles	-0.6 to -0.7
(<10% Relative humidity)		Insects	-0.8 to -1.2
		Mammals	-0.5 to -0.6

class inquires into the mechanisms involved in regulating ECF composition. This entails studying the physiology of specialized organs both as integrated structures and at the cellular and molecular level.

Effector Organs in Osmoregulatory Control: General principles

1. The Body Surface

Material exchanges between ECF and environment may occur across the general body surface or be restricted to specialized organs such as gills. For aquatic animals passive or dissipative flows will be governed by the diffusion equations for solutes and water. Such flows can only change the composition of the ECF in the direction of equilibrium, but rates of change will depend on the driving forces, as well as the permeability of the ex-

change surface and its area. Active transport does not conform to these rules and can therefore compensate for dissipative losses or gains, but only with the expenditure of metabolic energy.

2. The Excretory System

As the name suggests, excretory organs are specialized to facilitate the efflux of material from animals. Their primary role may or may not be in osmotic regulation, but they have an impact on it in all cases, because excreta are never devoid of the major osmotically active materials, water and NaCl; in most animal groups the final urine is fluid and contains substantial amounts of salt.

Filtration kidneys are widely distributed; examples are found in the major phyla (Chordata, Arthropoda, Mollusca, and Annelida) as well as in less well studied

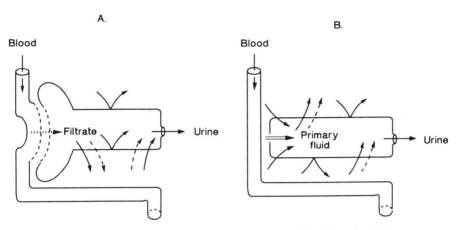

Figure 12. Filtration and secretion excretion organs. (*A*) In a filtration kidney presumptive urine is formed by convective flow (----) from plasma into tubule. Tubular fluid is then modified by active transport (→) and diffusion (----). The tubular epithelium is impermeable (↛) to some solutes. (*B*) In a secretory organ solutes are actively transported from hemolymph into tubule (→). This creates an osmotic gradient favoring water flow (⇒) into the tubule. Tubular fluid is subsequently modified as in a filtration kidney.

taxa. They have been intensively studied in vertebrates where their physiology is well worked out. The principles are illustrated schematically in Figure 12*A*. The basic structural unit is a long tube with the proximal end separated from the blood and ECF by an epithelium, and the distal end open to the environment. A bladder may be present distally. A primary or presumptive urine is formed at the proximal end by filtration. This may then be modified by reabsorption of some tubular solutes and/or water back into the blood, and by transport of solutes into the tubule, a process called secretion. As a result of these processes the final urine may bear almost no resemblance to the initial filtrate or to blood.

The basic ideas underlying the formation of a filtrate are expressed in Eq. 5c. If there exists across the filtration epithelium a hydrostatic pressure difference in excess of any effective osmotic pressure difference, filtration (i.e., bulk movement of fluid) will occur from blood into the tubule. Such a hydrostatic pressure difference will exist as long as the blood pressure at this site exceeds the back pressure of the fluid in the tubule. In mammals the pressure difference is ~40 mmHg: It is probably considerably lower in most animals but is rarely measured. The total osmotic pressure of animal bloods is very large by comparison with blood pressure (~1000 mmHg even in the most dilute hemolymphs), but filtration epithelia are supposed to be freely permeable ($\sigma = 0$) to all solutes smaller than polymers. They will exert no osmotic pressure, and hence the *effective* osmotic pressure (π_{eff}) is due largely to blood protein to which the epithelium is impermeable, and this will oppose filtration. To continue with the mammalian example, π_{eff}, resulting from plasma protein, opposes water movement into the tubule, and is ~25 mmHg; the net force is therefore 40-mm hydrostatic–25-mm osmotic or ~15 mmHg favoring filtration. The rate at which filtrate is formed will be determined by this value together with the area of epithelium and its hydraulic conductivity L_p. Since the epithelium is freely permeable to most of the blood solutes, convective water movement will carry them into the tubule. In this case, passive solute movement is not described by the diffusion equation. Instead the flux of any solute past the filtration locus is simply

$$J_{sol} = c_{bl} \times J_v^f \qquad (6)$$

where c_{bl} is the concentration of that solute in blood and J_v^f is the rate of fluid entry into the tubule from the blood, that is, the filtration rate. Note that this flux is simply the *amount* of a solute entering the tubule in a unit of time.

It is often important to know the filtration rate, and this is rarely accessible to direct measurement. The quantities that are accessible are (1) the concentrations of solute in blood and urine and (2) the urine flow rate, J_v^u. These can be used to calculate the filtration rate from the *clearance* of a test solute. The test compound must have several special characteristics: it must be freely filterable ($\sigma = 0$) in the filtration epithelium, and it must not be reabsorbed, secreted, or metabolized by the tubular epithelium (i.e., once within the tubule it must be completely inert). When such a compound is introduced into the blood and allowed to equilibrate, its rate of entry into the excretory tubule is given by Eq. 6. The *amount* excreted in a unit of time is $c_u \times J_v^u$. Conservation of mass requires that the rate of excretion equal the rate of entry. Writing out the equality and solving it for J_v^f:

$$J_v^f = J_v^u c_u/c_b \qquad (7)$$

where all quantities on the right can be determined experimentally. The calculated flow is the volume of blood "cleared" of the test compound, hence the name for the procedure. But for a compound with the specific qualities described above it is also equal to the rate of filtration. The test compound most fre-

quently used is *inulin,* a polymer of fructose with a molecular weight of ~5000.

Clearance studies can give additional information about the physiology of these organs. For example, since filtration markers like inulin are inert within the tubule their concentration in the urine will differ from that in blood only if water moves across the tubular wall. Thus, $c_u/c_b = 2$ indicates that one half of the filtered water was reabsorbed from the tubule, while $c_u/c_b = 1$ indicates that no water was reabsorbed and that urine flow equals the filtration rate. In addition, if the clearance of some compound of interest differs from that of inulin it shows that that compound was either reabsorbed from or secreted into the tubule. Variations of this procedure permit us to assess the rates of filtration, of excretion, and of reabsorption or secretion of any blood-borne substance. In fact, most of the information we have about the excretory physiology of lower vertebrates and invertebrates is based on clearance-type studies, although much current work on mammals (occasionally on other animals) uses much more sophisticated procedures.

Secretory organs (shown schematically in Fig. 12b) are found in all insects and in some fish that do not have a filtration apparatus. The basic difference from filtration kidneys lies in the mode of forming a primary urine. The process is initiated by the transport of a solute into the tubule. This creates an osmotic pressure difference that causes water to move into the tubule as well. If tubular permeability to other solutes is high they will be transferred into the tubule with the water. In general, however, molecules as large as the filtration markers (e.g., inulin) cannot move freely through the tubular epithelium, and they will be absent, or present at very low concentration, in the presumptive urine. The rest of the operation is as in filtration organs; some solutes may be reabsorbed as the fluid moves down the tubule, others may be secreted into the tubule. Water may also be transferred between the forming urine and hemolymph. Again, the final urine may bear no resemblance to the primary fluid or to hemolymph.

3. The Gastrointestinal (GI) Tract

Many aquatic animals drink appreciable quantities of the external medium. In some of them the ingested water is essential in maintaining an osmotic steady state, as it also is for most terrestrial animals. Ingested water is rarely distilled, so drinking also entails an influx of solutes. Since drinking is a convective movement like filtration, solute influx is described by a modified Eq. 6; $J_{sol} = c_m \times J_v^d$, where c_m is the concentration of that solute in the medium and J_v^d is the drinking rate. In the stomach or intestine solutes may be absorbed into the blood, while others might be secreted from blood into the lumen and later excreted. In overview, the functional behavior of this system is not very different from that of a filtration kidney: fluid influx is convective and unselective for solutes; the luminal fluid composition is then modified by selective absorption and secretion and the product excreted into the environment. The main difference concerns the source of the initial fluid: blood in the case of the kidney and the environment in the case of the GI tract. But this difference is of utmost importance for the roles played by these organs in osmotic regulation. It means that the GI tract is involved in material influx from the environment. Some material may be excreted from the ECF by this route, but this is not normally an important component of osmotic regulation. The kidney is involved in efflux of material from the ECF. The consequences, in adapting animals to particular environmental constraints, will be seen later.

Sometimes it is useful to know an animal's drinking rate in a laboratory situation. This is often straightforward for a terrestrial animal where the volume removed from the source can be measured directly. For an aquatic animal the procedure is indirect but not difficult. A test compound that cannot be absorbed from the gut is incorporated into the medium (inulin is often used). The rate of its entry into the gut can be determined by measuring how much is in the organ after a specified time (J_{sol}). Its concentration in the medium is also determined, and from these two quantities J_v^d can be calculated. The influx of any other solute can then be determined, since this is the product of its concentration in the medium and J_v^d. Much of this general information can be summarized graphically as in Figure 13. The key points are as follows:

1. The cells in most metazoan animals exchange solute and water with a circulating internal medium rather than directly with the environment.
2. The internal medium, comprising blood and extracellular fluid, has a composition compatible with cell viability. This may or may not resemble the composition of the environment and is regulated by passive and active flows across the body surface (gills or skin), GI tract, and kidneys.
3. As regards osmotically active material, water and permeant solutes, the body surface is a site of uptake from and/or loss to the environment, hence an exchange organ.
4. The GI tract is a route for uptake of osmolytes.
5. The nephridial organs (kidneys) are a route for excretion of osmolytes.

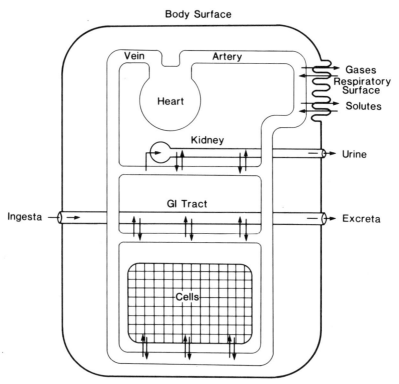

Figure 13. Compartments and exchanges in animals with a circulating hemolymph. Solute and water exchanges occur between cells and hemolymph. Most cells are never exposed directly to the external medium. The hemolymph *is* in contact and exchanges material with the environment across the body surface (skin or gills) and to some extent in the GI tract. Hemolymph and environment also communicate via the excretory organ, but the molecular traffic is one way (a few exceptional insects are described later).

Marine Osmoconformers

In most marine invertebrates, the osmotic concentration of the body fluids is at equilibrium, or nearly so, with the ocean. Recent measurements show that the ECF is actually slightly hyperosmotic in most marine animals (196, 226). Table 2 shows the composition of "typical" seawater, and the equivalence can be appreciated by comparing the values with those of the hemolymphs of the marine animals in the table. The table also shows that concentrations of the major osmolytes are also nearly the same in the ECF and in SW. Less abundant ions, particularly Mg^{2+} and SO_4^{2-}, are often regulated well below environmental concentrations. Sipuncu-

lid worms, cephalopod molluscs, ascidians, and most coelenterates are limited in habitat to near full-strength SW. Animals restricted to such a narrow salinity range are called *stenohaline*. The osmotic concentration of marine jellyfish is the same as that of the ocean where they occur, and they are relatively intolerant of dilution (28, 167). Corals and ctenophores are said to tolerate a 20% reduction in salinity. This is also around the maximum dilution tolerated in a number of crabs [e.g., *Maja* (257), *Palinurus*, *Cancer antennarius*, *Portunus*, *Hyas*, *Pagurus* (151), and *Libinia* (47)]. A typical response to dilution is shown in Figure 14 for *Maja squinado*. Immediately after transfer from SW to 80% SW the crab gained weight as water en-

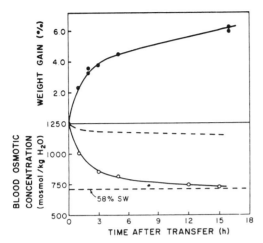

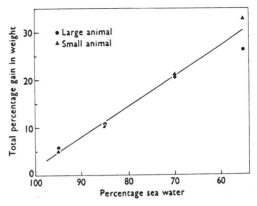

Figure 14. Water and solute permeability in a marine osmoconformer. At 0 h the crab (*Maja*) was transferred from SW to 58% SW. Weight gain reflects water entry. The upper dashed line shows how blood concentration would have changed due solely to dilution by the incoming water. The observed change was both larger and faster, showing that rapid diffusive salt loss occurred in the dilute medium; that is, the crab was permeable to salt as well as to water. Hemolymph was in equilibrium with the dilute SW within 10 to 15 h. [Drawn from data in Schwabe (257).]

Figure 15. Lack of volume regulation in the starfish *Asterias*. Weight-gain reflects entry of water. At each dilution of the SW entry contributes toward establishing a new osmotic equilibrium between body fluids and medium. [From Binyon (22). Reprinted by permission of Cambridge University Press.]

tered and also rapidly lost solute. After a relatively short time it came into equilibrium with the diluted medium (hence osmoconformer), but within 16 h most of the animals were dead.

Many marine osmoconformers can tolerate substantial dilution, and their distribution often reflects this. An animal capable of surviving over a broad range of salinities is called *euryhaline,* and the span available to some euryhaline osmoconformers is remarkable. When echinoderms are transferred to 80% SW they gain weight because of osmotic water entry and at the same time lose salt. Within 2–4 h the osmotic concentration of the body fluids is essentially the same as that

of the diluted medium, but the animal's volume has increased and remains stable; that is, there is no volume recovery. Figure 15 shows the equilibrium volumes of the starfish *Asterias rubens* (22) transferred to different dilutions of SW (the weight change is due to water entry). The ability to survive dilution has permitted the North Sea population of *Asterias* (salinity 35‰) to spill over into the Baltic (salinity 15‰). The biological cost of conformity is considerable in this animal (25, 148). In the dilute environment it has a soft integument, increased water content, lower heat tolerance, and reduced metabolism. Most significantly, it does not breed in the Baltic, and populations there represent a surplus from the North Sea population [G. Thorson in ref. 23].

A number of intertidal and estuarine species are osmoconformers, in spite of the low salinities encountered. Many annelid worms can survive transfer to ~25% SW in the laboratory and are found to conform at each dilution (196). Although found in estuaries, they do not have extensive distributions in the lower salinity regions. Marine molluscs are also osmoconformers, but many brackish water species have been described. When bivalves such as *Crassostrea*, *Modiolus*, and *Mytilus* are transferred to diluted SW, the valves remain closed for many days, and the hemolymph is diluted little if at all (76, 175, 248). However, if the valves are forced open, there is weight gain and rapid dilution as shown for *Modiolus* in Figure 16. The data also show that the whole animal exhibits volume regulation;

animals exposed to 75% SW regained their original volume within 24 h, while those in 10% took 48 h. It also appears that volume recovery does not occur in more concentrated media. In fact, animals that lost solute in acclimating to a dilute medium shrank when returned to the original solution (36‰) and died within 12 days without recovering the original volume. An extreme case of volume regulation is shown (204) by the opisthobranch mollusk, *Elysia chlorotica*, an intertidal animal that survives in 2.5‰. Transfer to dilute SW is followed by rapid

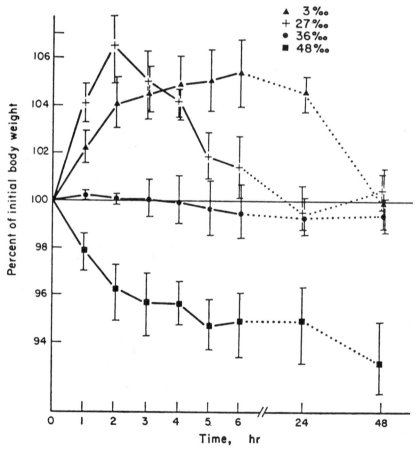

Figure 16. Volume regulation in *Modiolus*. Animals were transferred at time 0 from SW (36‰) to the concentrations shown. Volume recovery was rapid in 27‰ but much slower in 3‰. Shrinking occurred in 48‰, but was not followed by a RVI. [Reproduced from Pierce (203a).]

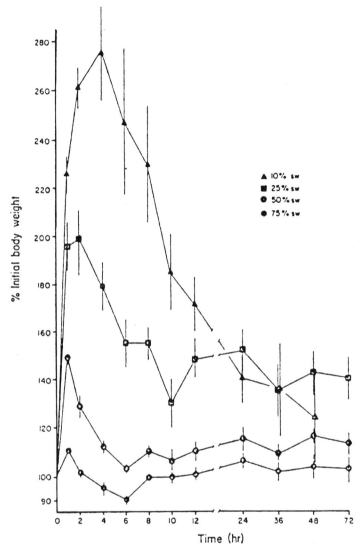

Figure 17. Volume regulation in *Elysia*. Volume changes were larger and faster than for *Modiolus*. [Reproduced from Pierce et al. (204) with permission of The University of Chicago Press.]

swelling, the extent of which is remarkable in this soft bodied animal (Fig. 17). But recovery is equally striking, and the diluted animals are only a little swollen after 48 h.

Most euryhaline crustaceans and vertebrates are osmoregulators. An interesting exception is the barnacle *Balanus improvisus*, which conforms but is slightly hyperosmotic down to 3‰. This animal,

like *Elysia*, is intertidal. It remains active even in the most dilute medium and has been described as the most estuarine barnacle in the San Francisco Bay system (191). Among vertebrates, only the myxinoids, coelocanths, and elasmobranches are osmoconformers, and only the last has some euryhaline representatives. They will be treated in more detail later.

It should be noted that the ability to

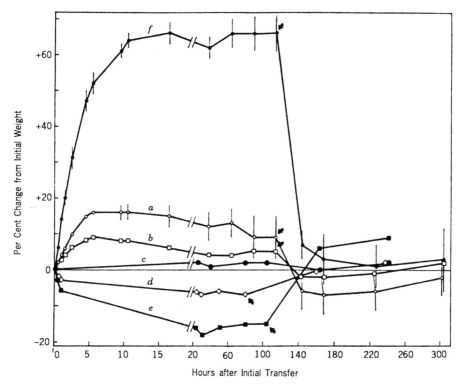

Figure 18. Osmotic equilibration in *Themiste dyscritum*. Time course of weight changes in *T. dyscritum* after transfers from 658 m*M* of Cl⁻ (98% SW) to 139% SW (■———■), 111% SW (◇———◇), 98% SW (●———●) 89% SW (□———□), 80% SW (○———○), 49% SW (●———●). At the points indicated by the black arrows, the worms were returned to 98% SW. Note change in scale of time axis after 20 h. [Reproduced from Oglesby (195).]

acclimate to diluted SW in the laboratory does not ensure than an animal will, in fact, be euryhaline in nature. Figure 18 shows weight changes in the sipunculid worm *Themiste dyscritum* on transfer to dilute media (195). There is apparently only little volume regulation for small (10–18%) dilutions, and none for 50% dilution or any change to a more concentrated medium. Nevertheless, the worms live for long periods in >50% SW and recover their original volume when returned to SW. But the capability appears not to be exploited; the sipunculids are considered to be a stenohaline marine group, and *Themiste* seems to occur only in full SW.

Since these animals are not actively regulating the osmotic composition of the ECF there is only a little to note concerning mechanisms involved in volume regulation. The process of cellular regulation, described earlier, undoubtedly contributes substantially, since all marine invertebrate cells examined exhibit the behavior. However, the situation is more complicated in the intact animal. A number of worms, like *Themiste*, show little or no regulation of whole body volume. If the cellular compartment *is* regulating (i.e., reducing its volume after first swelling), then another compartment must be expanding. This is probably the coelomic cavity in most cases, although sequestering of considerable fluid in vacuoles of special cells has been described in the flatworm *Procerodes ulvae* (14). In good volume regulators the nephridial organs may play an important role in first minimizing, then reversing the swelling that occurs on dilution. *Libinia emarginata* is a stenohaline crab which is, however, an excellent volume regulator (47). On transfer to 80% SW it swells <1% and is close to its original weight within a couple of hours. If the nephropores are blocked, the initial weight gain is more rapid and continues for 6 h by which time the crab has gained 5% over its control weight. The same phenomenon, exaggerated swelling on dilution, has also been noted in several worms with nephridia ligated or removed (94).

At this point it is worth looking more generally at the function(s) of nephridial organs (kidneys) in osmoconformers. For many years filtration kidneys were believed to be found exclusively in vertebrates where they purportedly developed to cope with the osmotic water load imposed by a very dilute milieu (276, 277). In fact, the presence of a glomerular or filtration kidney was presented as evidence that the vertebrates must have evolved in FW, although all other chordates are exclusively marine. The story is worth pursuing (138), because it shows how a compelling argument based on the evidence available to about 1945 was shown by later research to be untenable. By 1965 application of clearance studies (Eq. 7 and accompanying discussion) made it clear that nephridial organs in many invertebrates are filtration-type organs (178, 212). Furthermore, some of these animals are stenohaline marine osmoconformers with no history or close relatives in FW. Table 6 shows filtration rates in a number of marine and FW invertebrates with four FW vertebrates for comparison. There are no systematic differences between vertebrates and invertebrates or between marine and FW

animals. In one of the more thorough studies of invertebrate excretion it was shown that renal function in the octopus is remarkably similar to that in vertebrates (101). Filtration from the blood occurs in an accessory (branchial) heart where high pressure can develop. The solution entering the renal organ has the composition of a protein-free ultrafiltrate of blood which is then modified by reabsorption of useful compounds (e.g., glucose) and secretion of others (*p*-amino hippuric acid) into the tubular urine. Ammonia, the major end product of protein metabolism in aquatic animals, is also excreted, in part, by the octopus kidney where its behavior resembles that in the vertebrate nephron (208). Similar observations, although rarely as thorough, have been made on other groups of osmoconformers, notably crustaceans (232).

If the fluid formed at the filtration site is essentially a protein-free filtrate of blood, what can one say about the composition of the final urine excreted? Table 7 shows the ionic composition of urine from many animals, the concentration of each ion expressed as a fraction of its blood concentration, and a few general points are apparent:

1. Remembering that blood osmotic concentration is determined primarily by its NaCl concentration, the final urine of marine osmoconformers is isosmotic with blood.
2. Filtered nutrients (e.g., glucose) are partially or completely recovered by active transport from the tubular fluid (not shown in the table).
3. There is evidence that magnesium becomes concentrated in the urine. This correlates with a tendency of some of these animals to regulate blood Mg^{2+} at concentrations below those in SW. However, this ability is impressive only in brackish water crustaceans and marine fish.

TABLE 6. Filtration Rate and Urine Flow in Osmoconformers and Hyperregulators

Animal	Phylum[a]	Medium	$J_v^{f,b}$	$J_v^{u,b}$	Reference
		Osmoconformers			
Homarus americanus	C	SW	2	2	34
Pugettia producta	C	SW	4.0	2.7	45
Carcinus maenus	C	SW	5.7	4.4	21
Gammarus duebeni	C	SW	1.5	1.5	163
Octopus vulgarus	M	SW	2.6	2.6	101
Strombus gigas	M	SW	3.0	3.0	161
		Hyperregulators			
Carcinus maenus	C	40% SW	21.1	21.1	21
Gammarus duebeni	C	2% SW	24	21	163
Orconnectes virilis	C	FW	2.7	1.3	233
Pseudoteophasa jouyi	C	FW	0.13	0.13	296
Viviparous viviparus	M	FW	15	5.5	160
Anodonta cygnaea	M	FW	12	19	211
Carassius auratus	V	FW	20.4	13.7	174
Salmo gairdneri	V	FW	7.6	4.7	122
Lampetra fluviatilis	V	FW	11.2	6.5	109
Ambystoma gracile (larval)	V	FW	10.4	7.8	287

[a]C, Arthropoda (crustaceans); M, Mollusca; V, Chordata (vertebrates).
[b]J_v^f (filtration rate) and J_v^u (urine flow) in ml/kg · h. Filtration rates were measured by marker (usually inulin) clearance. It is possible that measured urine/blood (U/B) values in the range 1.1–1.5 are artifacts and should be 1.0 (235). If so, some of the filtration rates above are overestimated and should equal urine flow.

4. In osmoconformers urine flow rates are similar to or only a little less than filtration rates determined by inulin clearances. This indicates that little of the filtered water is reabsorbed before being excreted. The corollary, since NaCl concentrations are unchanged, is that these ions are neither reabsorbed from nor secreted into the kidney lumen.

Hyperosmotic Regulation: The Invasion of Dilute Environments

1. Overview

Although some osmoconformers do well in estuarine environments, and a few, like *Elysia* and *Balanus*, can tolerate nearly fresh water, most animals living in dilute environments regulate their ECF concentrations well above those of the external media. Regulating the ECF out of equilibrium with the environment is energy requiring and should be selected against unless it confers a substantial advantage of some sort. However, the loss and resynthesis of amino acids in cell volume regulation is also likely to require metabolic energy, and it may be more economical to regulate the ECF than all of the cells it bathes. In addition, a high ECF concentration probably leaves cells better able to function, rather than merely survive. *Asterias* survives in the Baltic Sea at 15‰ but with obviously impaired function. This is also the case for *Mytilus*, which tolerates even greater dilution (247). Altered physiological behaviors

TABLE 7. Urine/Blood (U/B) Ratios for Ions and Total Solutes

Animal	Medium	Na^+	K^+	Ca^{2+}	Mg^{2+}	Cl^-	SO_4^{2-}	Total Solute	Reference
				Osmoconformers					
Sepia officinalis	SW	0.79	0.50	0.70	0.68	1.00	2.15	1.02	240
Nephrops norvegicus	SW	0.98	0.83	0.81	1.30	1.01	1.06	~1	240
Homarus vulgaris	SW	0.96	~1	0.81	1.7	1.01		0.98	240
				Euryhaline Hyperregulators					
Carcinus maenus	SW	0.95	0.78	0.94	3.90	0.98	2.24	1.01	234
Pachygrapsus crassipes	SW	0.70	0.94	1.01	6.1			0.93	95
Eriocheir sinensis	SW		1.15	0.93	2.17	1.14		1.02	157
E. sinensis	2% SW		1.39	0.47	0.22	0.95		1.04	157
				Stenohaline FW					
Pheretima posthuma ^earthworm^	FW	0.13	0.40	0.54	0.14	0.05	0.71	0.19	9
Pacifastacus leniusculus ^crayfish^	FW	0.07				0.06		0.095	218
Esox lucius ^fish^	FW	0.03	1.35	0.05		0.05		0.09	109
				Hyporegulators					
Uca sp.	SW	0.84	1.45	1.01	2.35	1.15	1.12	1.17	87
Palaemon macrodactylus	SW	0.82	0.86	0.94	6.70	1.06	3.8	1.00	26
Lophius americanus	SW	0.94	0.4	~2	62.0	1.1		0.90	109
Paralichthys lethostigma	SW	0.10	0.50	7.07	125	0.83	360	0.96	109
				Terrestrial					
Dixippus morosus	T	0.45	8.1	0.29	0.17	0.75		1.00	223
Homo sapiens	T	1.16	7.5	1.87	2.41	1.63	~50	~5 (varies)	96

have been described in several cells such as nerve and muscle when they are diluted, in spite of the fact that they can regulate their volumes (203, 205, 321).

Brackish water crustaceans provide much of the evidence for the events permitting marine animals to regulate ECF osmotic pressure above ambient as they invade more dilute environments. Figure 19 shows how ECF concentration varies in the shore crab *Carcinus maenas* as it moves into more dilute media (263). The animal is isosmotic with SW, but when the medium is diluted by >15%, ECF concentration drops less than ambient and remains above it through a range of dilutions. Regulation breaks down when the medium is more dilute than ~20% SW, and the crab does not survive further dilution. *Carcinus* is a weak regulator in the sense that ECF concentration falls with environmental, although remaining above it, and especially because the regulatory system fails at a relatively high environmental concentration. Figure 20 shows the behavior of ECF from four species of the amphipod genera *Gammarus* and *Marinogammarus* over a series of environmental concentrations up to SW (313). *Gammarus oceanicus* and *M. Finmarchicus* are, like *Carcinus*, basically marine crustaceans capable of living in brackish

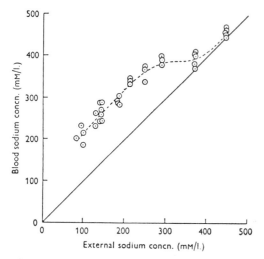

Figure 19. Hyperosmotic regulation in *Carcinus maenus*. Steady-state hemolymph [Na$^+$] in media of different salinities. The total osmotic concentration is a little more than double that for Na$^+$ and is ~1 os$M$ (1000 mosmol/L) in SW. In any medium between ~20 and 80% SW the animal maintains ECF constant at the value shown. In this situation there is a gradient favoring Na$^+$ loss and one favoring inward water movement. [Reproduced from Shaw (263) with permission of the Company of Biologists Ltd.]

water. Their body fluids begin to be regulated above ambient below ~80% SW, and regulation is somewhat better than in the crab down to 10% SW. But body fluid concentration falls markedly at lower concentrations and neither animal survives in FW.

A similar pattern is shown by *G. tigrinus* (Fig. 20) and crabs like *Eriocheir sinensis* and *Callinectes sapidus* (Fig. 21). These crustaceans are isosmotic with SW and begin hyperregulating when the latter is diluted ~25%. Regulation is very good; ECF concentration drops only a little between 70% SW and FW, and the adults are often found naturally in FW, although they must return to the ocean to breed. These are still marine animals but are much stronger regulators than the brackish water group. With ECF varying little, the cells see little change in *their* environment as the external milieu becomes more dilute. Even more significant, the range of regulation is extended into FW (e.g., ~5 mos$M$).

The picture in annelid worms resembles that in crustaceans. Most polychaetes are osmoconformers, but some, notably among the nereidae, are hyperregulators in dilute media (196). Figure 22 shows ECF concentrations for *Nereis limnicola* in external media between FW and SW (194). This worm is isosmotic in SW and remains so until the medium reaches 30–40% SW. Below this point the body fluids are regulated hyperosmotically. Regulation is good to very dilute media but breaks down at ~4 mM Cl$^-$, and internal concentration falls on further dilution. Nevertheless, the animal is normally found in FW. Another nereid, *N. diversicolor*, does almost as well. In contrast, *N. succinea* regulates less tightly down to ~10% SW, below which regulation breaks down. *Nereis limnicola* falls into the strong regulator group, *N. succinea* into the weak, and their distribution in the San Francisco Bay system reflects this distinction (194). The main difference between hyperregulating crustaceans and annelids is the level at which the body fluids are maintained in dilute environments; 300–400 mM NaCl in the former but 100–200 mM in the latter. This may be due to a difference in permeability of the body surface to ions and water as will be discussed later, but in both cases, the animals are not merely viable but active in dilute environments.

We have less information about molluscs in brackish water. A number of them are osmoconformers. A few brackish water gastropods have been described (162), but these may be of FW origin, in which case they would give little insight into how the group invaded dilute media in the first place. Extant aquatic verte-

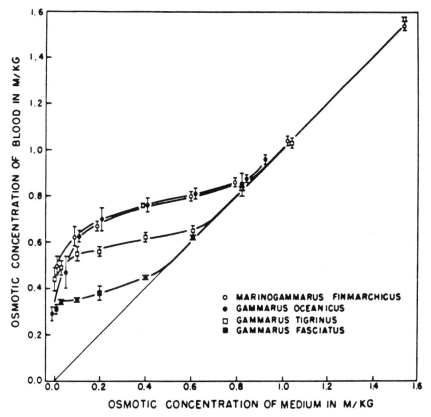

Figure 20. Hyperregulation in gammarids. Osmotic concentration of hemolymph as the medium osmotic concentration varies. [Reproduced from Werntz (313).]

brates are basically a FW group and their presence in brackish water and in the ocean represents a secondary radiation which will be discussed later. Other groups have not been sufficiently characterized to warrant discussion, but there is no reason to think that generalizations drawn from studies of the crustaceans and annelids would not hold for them as well.

The weak and strong hyperregulators are essentially marine animals and the differences between them are only quantitative. A third group of hyperregulators is more distinct in that they are stenohaline FW animals breeding in this medium and unable to live at concentrations above ~30–50% SW. Body fluid concen-

trations in these animals range from ~50–500 mosM when they are in "normal" FW (e.g., NaCl ~1 mM, $CaCl_2$ ~0.5 mM). The examples in Table 2 indicate that FW molluscs have body fluids at or below 100 mosM, while vertebrates are usually between 200 to 300 mosM as are FW annelids. The range for crustaceans is wider and related to size. Very small animals (branchiopods or amphipods) are between 100–300 mosM, while the decapods are 400–500 mosM. The change in ECF concentration with salinity is illustrated by the behavior of *G. fasciatus* in Figure 20. These animals can live in media as dilute as 0.1 mosM or lower, but regulation of the ECF breaks down at high, rather than at low, concentrations.

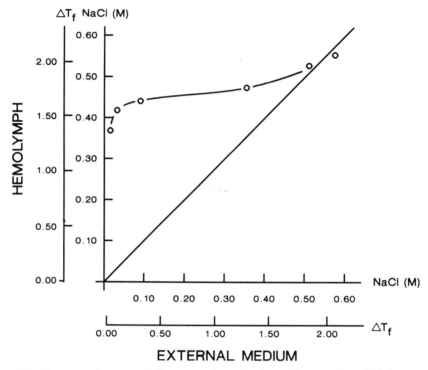

Figure 21. Hyperregulation in *Callinectes sapidus*. Similar to Figures 19 and 20, but osmotic concentrations are expressed in terms of equivalent [NaCl]. Freezing point depressions are also shown to demonstrate the relationship between these two ways of expressing osmotic concentrations. [Drawn from data in Ballard and Abbott (10a).]

Two general points that emerge from this survey are worth emphasizing. As shown in Table 2, the main osmotic components of the extracellular compartment in all aquatic animals is NaCl, and the same thing is true for most of their environments (some saline lakes are high in Mg^{2+} instead of Na^+, and/or carbonate or sulfate rather than chloride). For regulators this means that a salt gradient in one direction is accompanied by a water gradient in the other. What is less obvious is that in compensating for one of the resulting dissipative flows (water) the other (ion loss) is usually aggravated. The other general point is illustrated by Figure 20: When we compare a series of similar animals blood concentrations are usually

higher in weak regulators than in strong regulators and tend to be lowest in the stenohaline FW representatives of such a series.

2. Mechanisms Allowing Hyperosmotic Regulation

The range of strategies available for maintaining body fluids out of equilibrium with the environment is limited, and it is not surprising that the manner of coping is similar in all animals facing a common problem. It was pointed out earlier that there are really only two useful modifications to the original (marine osmoconformer) state. One, called evasive, is to reduce dissipative flows of ions and water. This would include behavioral

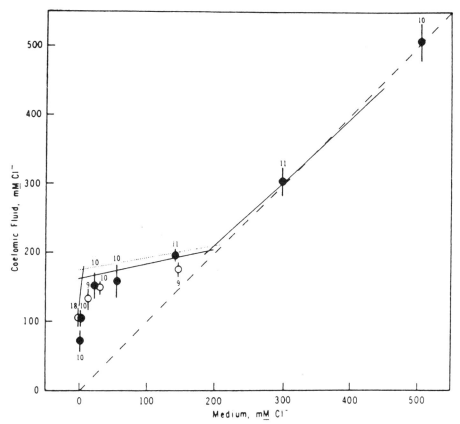

Figure 22. Hyperregulation in *Nereis limnicola*. Hemolymph concentrations at different medium concentrations (both expressed as mM Cl⁻). SW [Cl⁻] is ~500 mM. The data at the extreme left show the breakdown in regulation in very dilute solutions. Animals from two locations (● and ○) show the same behavior. [Reproduced from Oglesby (195a) with permission of *Biological Bulletin*.]

adaptations as well as physiological changes. The second strategy, which was called compensatory, requires flows equivalent to the dissipative movements but in the opposite direction. Osmotic water uptake must be balanced by an equal water excretion, and diffusive salt loss requires equal uptake from the medium. But these movements are against potential energy gradients and hence are energy requiring. This is precisely why the evasive strategy is so important. The slower the leaks the less energy is expended in recovery.

3. Reducing Dissipative Flows of Ions and Water

The diffusion equations show that hypersomotic regulation can be accomplished by reducing the effective permeability or by reducing the diffusion gradient between blood and medium. *Carcinus maenas* was called a weak regulator, in part because body fluid concentration dropped constantly as the medium was diluted. However, the change can also be viewed as adaptive, since it reduced the driving forces favoring water loading and salt loss at each external concentration.

As long as the cells can function efficiently, the lower the osmotic concentration of the ECF the better. Another expression of the tendency to reduce gradients is seen in the series of gammarids described above; more effective hyperregulators tend to have lower ECF osmotic concentrations. The phenomenon is clearly adaptive. However, such comparisons must be made cautiously. An annelid regulating weakly has a lower osmotic concentration than a strong crustacean regulator like *Eriocheir*, and small crustaceans clearly have lower osmotic concentrations than large ones. The annelids and very small amphipods may have relatively higher permeabilities, the former because it lacks an exoskeleton, the latter because the surface/volume ratio is large. If so, lower ECF concentrations would reduce dissipative fluxes, hence would be adaptive.

Many studies show that animals regulating the body fluids out of equilibrium with the environment are less permeable to ions and water than comparable animals in equilibrium. Table 8 shows water and Na^+ permeabilities (P_{osm} and P_{Na}) for a number of crustaceans and vertebrates in the size range 10–1000 g, and the tendency toward lower permeabilities in more dilute media is clear. When *Carcinus* is in 40% SW, the ECF osmotic concentration is reduced to about one half, and the P_{osm} about three- to fourfold. As a result, osmotic water loading in the dilute medium is nearly an order of magnitude smaller than it would be if the crab remained isosmotic with SW and P_{osm} were as large as in conformers. In *Eriocheir* P_{osm} is reduced by two orders of magnitude and osmotic concentration nearly to one half. Both parameters are still lower in *P. jouyi* and *P. niloticus*. The $[Na^+]$ and P_{Na} show the same general trend. Table 9 shows the rate of Na^+ loss in a series of gammarids and of nereid worms, and the tendency

toward lower effluxes in dilute habitats is apparent. The relative effects of lowered ECF concentration and P_{Na} were not separated, but both probably play a role in reducing loss by about an order of magnitude. In addition, there is evidence that ion and water permeabilities in a single animal decrease when it is transferred from a more to a less concentrated medium (164), although this is not apparent in the data for *Carcinus* (P_{Na} is about the same in 100 and 40% SW).

4. Compensatory Active Fluxes of Ions

In spite of the fact that they lose NaCl by diffusion across the body surface and in the urine, successful hyperregulators can maintain blood concentrations in a steady state even in the absence of food. Net absorption of salt across the body surface of FW annelids, molluscs, crustaceans, and vertebrates was clearly demonstrated in the 1930s. Influx against such a steep gradient of concentration was called "active transport" (125), and some of its characteristics were described during that period. The study of transport mechanisms continues today, but at a molecular level to which we will give only a little attention.

In some of the earliest work on hyperregulators, Krogh showed (152–154) that FW animals can absorb Na^+ and Cl^- independently of each other (i.e., Na^+ from Na_2SO_4, and Cl^- from, e.g., $MgCl_2$; neither sulfate nor magnesium can be absorbed across epithelia like skins or gills). This means that in order to establish which ion is actively transported each one must be tested separately by one of the solutions of the diffusion equation. It is not sufficient to note that it is moving against a concentration gradient, because a voltage across the body surface, if it has the appropriate polarity and is large enough, can cause diffusive movement from lower to higher concentrations. The

TABLE 8. Osmotic and Ionic Permeabilities in Some Crustaceans and Vertebrates

Animal	Body Weight (g)	Medium	Test Medium (ΔosM)	J_v (ml/g · h)	P_{osm} (10⁴ cm/s)	ΔNa⁺ (μmol/ml)	J_{Na} (μeq/g · h)	P_{Na} (10⁶ cm/s)	Ref.
Crustaceans									
Libinia emarginata	152	SW	0.19	0.030	12.8				45,46
Pugettia producta	121	SW	0.20	0.070	25.7	82	8.0	13.2	45,46
Porcellana platychales	5	SW	0.38	0.068	4.7	190	4.2	1.0	52
Eupagurus bernhardus	5	SW	0.38	0.18	12.4	190	6.6	1.6	52
Maja verrucosa	60	SW	0.42	0.024	3.4	250	30	12.9	257
Carcinus maenus	50	50% SW	0.20	0.010	2.8	340	16.9	5.0	263
Cancer magister	85	50% SW	0.25	0.005	1.3	350	7.8	2.7	127
Eriocheir sinensis	150	FW	0.60	0.0017	0.22	280	1.7	0.87	264
Potomon niloticus	15	FW	0.54	0.0001	0.007	260	0.8	0.21	262
Pseudotelphusa jouyi	12	FW	0.46	0.0002	0.015	240	1.5	0.39	296
Astacus fluviatilis	40	FW	0.45	0.0034	0.88	200	0.2	0.10	33
Vertebrates									
Eptatretus stoutii	50	SW	0.57	0.013	1.3				181
Carassius auratus	100	FW	0.25	0.014	3.9	110	0.15	0.36	74
Salmo gairdneri	300	FW	0.31	0.005	1.6	120	0.30	0.32	74,136
Anguilla anguilla	130	FW	0.25	0.004	1.2	140	0.42	0.59	188

^{a}In order that P_{osm} have conventional units (cm/s) it is necessary to express osmotic water movement in ml · s. The exchange area is known only for a few fish, amphibians and crustaceans, and water fluxes are usually expressed per unit body weight. In this and following tables the regression: Area = 10 (body weight)$^{2/3}$ has been used to estimate the area. The exponent obviously relates to the relationship between surface (cm²) and volume (cm³) for a regular geometric object (e.g., a sphere or cylinder). The geometry of animals is anything but regular, and the relationship also ignores other factors that may cause water movement to vary among animals of different size. However, the equation gives reasonably good agreement with measured surface–weight relationships. For example, the regression among 13 species in four families of amphibians was 13.52 (body weight)$^{0.579}$ (127a). The areas calculated from the two regressions differed by only 3% for a 20-g animal and 11% for one of 100 g.

TABLE 9. Salt Loss in Gammarids and Annelids

Animal	Habitat[a]	Blood Na+	Medium Na+	Sodium Loss Rate[b]
		Gammarids[c]		
M. finmarchicus	SW	~300	60–120	57.0
M. obtusus	SW		60–120	80.0
G. tigrinus	BW	220	10	20.0
G. duebeni	BW	273	10	10.9
G. zaddachi	BW	252	10	13.7
G. pulex	FW	135	0.3–0.5	4.9
G. lacustris	FW	135	0.3–0.5	4.0
		Nereids		
N. succinea	BW		0	54
N. diversicolor	BW		0	13
N. limnicola	FW		0	7.8

[a]SW (seawater), BW (brackish water), FW (fresh water).
[b]Na+ loss rate for gammarids in μmol/ml of blood $\cdot$ h; for nereids in μmol/g $\cdot$ h.
[c]Data on gammarids (164); on nereids (281).

form used most often is the flux ratio equation (Eq. 3), and the results for some FW vertebrates and invertebrates are shown in Table 10. It can be seen that both Na+ and Cl− are actively transported in these animals and this is probably generally true.

Another corollary of the independence of Na+ and Cl− transports is that each ion must be exchanged for an ion of like sign coming out of the animal in order to preserve electrical neutrality (151). It is generally found that Na+ absorption is accompanied by an equivalent efflux of H+ (75a) or NH_4^+ (173), while Cl− uptake is balanced by HCO_3^- excretion. Although we will not pursue it here, it is worth noting that regulating ECF ion concentration is coupled in this way to acid–base regulation of the body fluids, a subject that is receiving current attention, especially in fish (64, 67, 103).

Transports of both ions increase with environmental concentration (of that ion) but reach an upper limiting value, J_{in}(max); that is, they show "saturation kinetics" described by

$$J_{in}^{ion} = \frac{J_{in}(max)\,[ion]_{ext}}{K_m + [ion]_{ext}} \qquad (8)$$

where $[ion]_{ext}$ is the environmental concentration, and the parameters J_{in}(max) and K_m describe the system's behavior. The term K_m is the external concentration at which transport is half-maximal $[J_{in} = 0.5J_{in}(max)]$, and the lower the value of K_m, the lower the range of concentrations over which transport develops its maximum rate. There is a striking correlation between the value of K_m and habitat, at least among vertebrates and crustaceans (90, 164, 290). Animals adapted to more dilute media tend to have an uptake system (for sodium) with a low K_m as shown in Table 11. However, it is not merely the value of K_m, but the interactions among K_m, J_{in}(max) and dissipative losses that determine whether an animal will be able to adapt to a given external salinity (264). To illustrate this, consider three animals

TABLE 10. Active Transport of Na$^+$ and Cl$^-$: The Flux Ratio Test

Animal	c_{out}/c_{in}	TEP (V)	J_{in} (nmol/g $\cdot$ h)[a]	J_{out}	J_{in}/J_{out} Calc.	J_{in}/J_{out} Exp.	Reference
			Sodium				
Salmo gairdneri	0.0067	0.008	193	143	0.005	1.35	136
Ambystoma tigrinum	0.0130	0.014	130	160	0.007	0.81	5
Rana pipiens	0.0045	0.016	144	142	0.002	1.01	3
Astacus fluviatilis	0.0098	0.005	410	410	0.008	1.00	33
Lumbricus terrestris	0.0067	−0.020	32	37	0.016	0.86	57
Carunculina texasensis	0.067	−0.006	1114	990	0.086	1.13	56
			Chloride				
S. gairdneri	0.007	−0.012	110	129	0.004	0.85	135
A. gracile	0.013	0.014	90	90	0.023	1.00	4
R. pipiens	0.0053	0.016	73	95	0.014	0.77	3
L. terrestris	0.011	−0.020	88	90	0.005	0.98	57
C. texasensis	0.089	−0.006	1100	1380	0.069	0.80	58

[a]A nanomole (nmol) = 10^{-9} mol.

whose uptake curves are shown in Figure 23. Animal 1 has a maximum transport rate of 10/h (arbitrary units) but a high affinity for Na$^+$; K_m = 1 mM. Animals 2 and 3 have much higher maximum rates [J_{in}(max) = 100] but lower affinities K_m = 5 and 25 mM. To maintain a steady state the transport system must be able to generate an influx exactly equal to the rate of ion loss (LR), and the latter will be governed by the passive factors described earlier (permeability, gradient, and urine loss). Assume a series of different loss rates (LR$_1$ to LR$_5$ in Fig. 23). If the animals

TABLE 11. The Sodium Influx K_m and Habitat in Hyperregulators

Vertebrates Species	Habitat[a]	K_m	Reference	Invertebrates Species	Habitat[a]	K_m	Reference
Rana cancrivora	FW/SW	0.40[b]	89	*Carcinus maenus*	BW	20	263
Bufo boreus	ST	0.25	3	*Mesidotea entomon*	BW	9	50
Rana pipiens	ST	0.20	89	*Cyathura carinata*	BW	5.6	164
Hyla regilla	ST	0.14	3	*Gammarus duebeni*	BW	1.5	289
Salmo gairdneri	FW	0.46	136	*G. locustris*	FW	0.15	292
Carassius auratus	FW	0.26	173	*G. pulex*	FW	0.15	291
Ascaphus truei	FW	0.06	3	*Limnaea stagnalis*	FW	0.25	88
Xenopus laevis	FW	0.05	89	*Hirudo medicinalis*	FW	0.14	41

[a]FW (fresh water); ST (semiterrestrial); BW (brackish water).
[b]Units of K_m are mM.

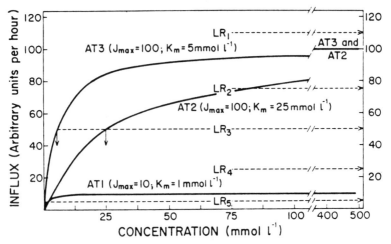

Figure 23. Sodium influx as a function of external concentration. Hypothetical patterns are shown for three animals (AT1, AT2, and AT3). All show "saturation kinetics," but with different values for the parameters K_m and J_{max}. The relationship shown by AT1 resembles that for animals found in FW, although most have a K_m lower than 1 mM. AT2 and AT3 have the same J_{max} but different K_m values. A series of five loss rates is shown (LR$_1$ to LR$_5$ on the ordinate to the right). The vertical arrows on AT2 and AT3 show the minimal external concentration for Na$^+$ balance if both animals are losing Na$^+$ at the rate LR$_3$. [Reproduced from (140)].

are very leaky (LR$_1$ = 110 units/h) none of them could maintain a steady state because net loss exceeds the highest possible rate of absorption. If the animals were less permeable (LR$_2$ = 75 units/h) animals 2 and 3 could regulate, but animal 1 could not, even though it has the highest affinity for Na$^+$. Further reduction in efflux (LR$_3$ and LR$_4$) still find animal 1 unable to regulate, because losses still exceed the maximum uptake it can generate, and it is only at the lowest rate of loss that it can generate an adequate influx. This emphasizes the importance of the evasive mechanisms in reducing losses so that active transport can cope with them. However, the K_m is important in determining how dilute a medium an animal can invade and still maintain a steady state. Consider animals 2 and 3 in Figure 23 operating at LR$_3$ (50 units/h). Animal 2 can develop the necessary compensatory influx down to a concentration of 25 mM but not below this. Animal 3, with a lower K_m, can move into an envi-

ronment as dilute as 5 mM and still remain in a steady state. This is the explanation of the correlation shown in Table 11. As long as J_{in}(max) is >LR, the lower the K_m the more dilute the medium in which the necessary influx can develop.

The mechanisms underlying ion transport are best known in the stenohaline FW group, but studies on a few BW forms suggest that these are similar. We have already noted that transports of Na$^+$ and Cl$^-$ are both active, saturable, independent of each other, and coupled in some fashion to the efflux of an ion of like sign. One other characteristic noted in the stenohaline group (141), and in at least one BW crab (*Callinectes sapidus*) (216), is that J_{in}^{Na} is blocked by a synthetic compound called amiloride when it is added to the external medium. This is essentially all that can be learned from studies on intact animals, but the degree of similarity in gross function, especially for Na$^+$ transport, among annelids, molluscs, crustaceans and vertebrates, provides hope that analytical studies on any of them will be relevant to all.

The preparation from which most has been learned is the frog skin (and a "relative," the toad urinary bladder, which performs the same role in transporting ions from its lumen back into the blood). The isolated skin can be mounted as a sheet separating two chambers. The inner (serosal) chamber is filled with a Ringer's solution having the composition of ECF. The outer can be filled with an artificial pond water (NaCl ~1 mM). When both solutions are stirred and aerated transports of Na$^+$ and Cl$^-$ occur just as *in vivo*. Moreover, the former is blocked by adding amiloride to the outer (mucosal) medium. We can now add an observation that could not be made in the intact animal; if ouabain is added to the serosal Ringer's solution it blocks Na$^+$ transport (145). We noted early that ouabain is a potent inhibitor of the membrane Na$^+$-K$^+$ pump; the experiment suggests that this enzyme is present in the serosal membrane and plays a role in Na$^+$ absorption by an animal. Other relevant studies have shown (1) that there is a potential difference across both mucosal and serosal membranes: The interior of each membrane is negative to the solution bathing it (104, 190), and (2) the [Na$^+$] in the epithelial cell is low as in all other cells, probably 1–5 mM (229). This information, coupled with biophysical and biochemical studies of the two membranes, has led to the transport model shown in Figure 24. The main features are

1. The Na$^+$ diffuses into the cell through an apical membrane channel that is chemically specific. Only Li$^+$ can also enter, and other ions, even other alkali metal cations, are excluded. The diffusion gradient is provided mainly by the large membrane potential across the apical membrane. The Na$^+$ channel is blocked

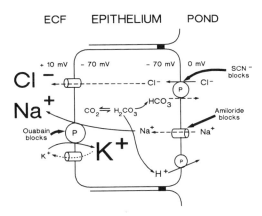

Figure 24. Epithelial NaCl transport model in FW animals. Symbols are as in Figure 5. However, in this case the ions move across a sheet of cells rather than simply between bathing medium and the intracellular compartment. This requires transfer across two membranes. Details of the transfer across each membrane are described in the text.

by amiloride in the external solution, which is how this compound inhibits transport.

2. Inward Na$^+$ movement is accompanied by outward H$^+$ efflux. The latter requires a pump to move it against the large membrane potential (63). The nature of the coupling between these ions need not concern us. It suffices that the system remains in an electrical steady state as long as the two fluxes are equal as they are in both intact frog and isolated skin.

3. At the serosal membrane electrochemical gradient is large and oriented against passive outward diffusion of a cation. Extrusion of the Na$^+$ ion is mediated by the Na$^+$-K$^+$ pump in the serosal membrane, which accounts for the action of ouabain. Energy for this transport system is provided by the hydrolysis of ATP by the pump.

Details of the Cl$^-$ transport system are far less well known. As shown in Figure 24, movement from the dilute external medium into the cell is an active step because the large membrane potential opposes passive influx of an anion. Outward movement of HCO_3^- and inward Cl^- movement are probably mediated by the same "carrier," rather than through independent pathways as in the Na–H exchange. The Cl–HCO_3 exchange in a number of FW animals (7, 38, 58, 135), and in the isolated frog skin (6, 75), is inhibited by thiocyanate ion (SCN^-) added to the external medium. The mechanism for extrusion from the cell across the serosal membrane is not known but is probably diffusive, with the large membrane potential favoring passive exit of an anion from the cell.

5. Compensatory Water Excretion

When BW hyperregulators, like C. maenas and G. duebeni, are in SW their filtration rates are about the same as those in conformers (Table 6). There is no evidence for reabsorption of water in the kidney (insulin U/B = 1). However, when the crabs move into a more dilute medium the rate of urine production increases markedly (Table 6), and it is this response that maintains a constant volume in the face of osmotic water influx during hyperregulation. The P_{osm} for G. duebeni is reduced in dilute medium (165); otherwise urine flow would be 2–3 times faster. An increase in J_v^u is suggested for polychaete worms when they move into a more dilute milieu, although the evidence is less direct (282).

Table 7 shows that urine is isosmotic with ECF in most BW hyperregulators, and the major osmotic solute remains NaCl. This is even true for Eriocheir and other crabs regulating in very dilute media. However, there is a clear tendency for Mg^{2+} to become more concentrated in BW and in SW. The Mg^{2+} ion is a neuromuscular depressant, and an important role for the kidney of active animals is to excrete this ion when its environmental concentration is high (241). Since inulin U/B = 1, the higher Mg^{2+} U/B is evidence for secretion of this ion by the renal epithelium; it could not have become concentrated by water reabsorption. There is similar evidence for sulfate secretion.

Excretion of a near isoionic urine will maintain volume, but it aggravates the problem of maintaining blood solute levels above ambient. A hyperregulator already faces NaCl loss by diffusion across the body surface. In these animals there is additional loss through the kidney ($= c_{sol} \times J_v^u$ as already discussed for convective movements like filtration and drinking). In a conformer this poses no problem, since there is no gradient to maintain between blood and medium. But in a regulating animal, ions lost in the urine must be transported across the body surface against a gradient, with obvious implications for energy expenditure. Carcinus maenas provides a useful illustration. When this crab is in 40% SW diffusive salt loss is ~5.6 mmol/kg · h. Osmotic water entry amounts to 8–10 ml/kg · h. If P_{osm} in Carcinus were as high as in some of the osmoconformers, urine would flow 3 to 4 times faster and renal salt loss would exceed that by diffusion; the crab would be "leaky" even if salt permeability were reduced to a very low value. For this reason lowering P_{osm} is as important for ionic regulation as for volume regulation. In fact, the two cannot be separated so long as the major osmotic solute in blood and medium is the same, and we will meet this interrelationship again in hypoosmotic regulators.

The problem of renal salt loss is minimized in a number of aquatic animals that produce a dilute, rather than an isoionic, urine. This is an adaptation characteristic of stenohaline FW animals (250) as suggested by examples such as Pheretima, Pacifastacus, and Esox in Table 7. It is found in nearly all representatives of this environmental group and is virtually restricted to them. Like many generalizations, this one admits of a few exceptions. The ability to produce dilute urine has also been reported in two BW animals, *G. duebeni* (163) and *N. diversicolor* (280). Their urines are isoosmotic with blood when they are in SW, but in near FW urine osmotic concentration is only about one half that of the blood. In addition, several species of stenohaline FW crabs (e.g., *Potomon niloticus* and *Pseudotelphusa jouyi*) are unable to reduce urine concentration (262, 296). In these crabs P_{osm} is so reduced (Table 8) that osmotic water influx, and hence urine flow, is very slow (Table 6). The resulting renal ion loss is small even though the urine is isoosmotic with blood. The ability to produce a hypoosmotic urine is not confined to animals with filtration-type kidneys. The malpighian tubule–gut complex of insects is a secretory system as are the

nephridia of the leech *Hirudo medicinalis*, and both have the capability of producing very dilute urines in the appropriate environment (27, 221, 326).

Excretory organs in FW animals are flexible, adjusting both urine composition and the rate of output in an adaptive manner. On transfer to a less dilute medium urine volume decreases and solute concentration increases. Figure 25A shows osmotic regulation by the leech is very good in media up to 200 mosM (~20% SW), and the urine is very dilute. But urine volume (Fig. 25B) is reduced to 20% of the original rate at the higher concentration (a response to reduction of the osmotic gradient between worm and medium). Further increasing the external concentration results in a rise in both blood and urine concentrations, and the

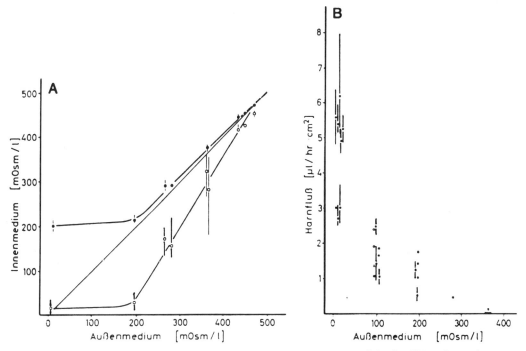

Figure 25. The role of urine in hyperregulation in *Hirudo medicinalis*. Urine is copious and dilute in FW, thus playing a role in excreting water and conserving ions. As the osmotic gradient decreases, urine volume is reduced (*B*), but the fluid is still more dilute than blood (*A*). At external concentrations >200 mosM, ECF concentration rises and the nephridia respond by excreting solutes, probably NaCl. [Reproduced from Boroffka (27) with permission of Springer-Verlag.]

two become isosmotic with each other and with the medium at ~450 mosM. Urine flow was too low to measure at this point (326). Such a reduction of urine flow and increased concentration has been described in many such animals as the ionic and osmotic gradients are reduced. The ability to reduce renal flow and solute reabsorption is probably only minimally useful as long as the animal stays in FW but has great adaptive significance if it moves into a potentially desiccating situation (e.g., on to land). Movement into a more dilute medium (e.g., from 1 to 0.1 mosM) does not much modify the osmotic driving force and urine volume changes little. Ionic balance is upset, because the influx system at the body surface is working at a much lower concentration, and J_{in} is reduced while

diffusive efflux is unchanged (Fig. 23). The ensuing salt loss initiates compensatory changes, one of which is to reduce the ionic concentration of the urine to a still lower value.

After the introduction of methods for obtaining and analyzing minuscule volumes of fluid in the 1920s it became possible to work out the way in which presumptive urine is processed after its formation by filtration or secretion. Some salient features for the vertebrate nephron (*Necturus maculosa* and *Rana pipiens*) are shown in Figure 26, which is one of the classics in renal physiology. Concentrations of glucose (not shown here), chloride, and the osmotic concentration of the forming fluid are the same at the filtration site as in the blood (U/B = 1), characteristic of a filtrate. The glucose is

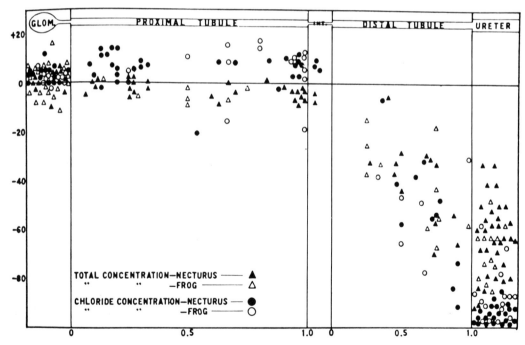

Figure 26. Tubular solute concentrations in the amphibian nephron. Fluid was obtained by micropuncture from different sites along the nephron and from the ureter of *Necturus* (▲, ●) and the frog (△, ○). The position of each puncture is indicated by the schematic drawing of nephron at the top. Chloride (●, ○) and total osmotic concentrations (▲, △) are expressed as the percent above or below plasma concentration. [Reproduced from Walker, A. M., C. L. Hudson, T. Findley, and A. N. Richards (305a).]

reabsorbed in the first part of the proximal segment of the nephron, and reabsorption is blocked by the drug phlorizin. In contrast, little change in Cl^- and osmotic concentrations occurs in this region, but both drop markedly in the distal tubule as salt is reabsorbed in excess of water. Salt reabsorption is an active process since it occurs against a gradient, and this segment must have a very low P_{osm} in order to prevent water from moving back as the osmotic gradient favoring reabsorption increases. The product is the urine characteristic of stenohaline FW animals, low in ion concentration and more dilute than the blood. Similar *micropuncture* studies involving other solutes have provided a fairly complete picture of function in the vertebrate nephron, and current research is focused on the biophysics and biochemistry of transport mechanisms.

Fewer systematic studies of invertebrate excretory organs have been carried out, but the information available (232, 314) suggests a surprising degree of analogy with the vertebrate kidney. For example:

1. A filtration locus, often with distinct ultrastructure has been clearly identified in molluscs, crustaceans, and vertebrates, and the fluid formed has the characteristics of an ultrafiltrate of blood. The last is also true in annelids (coelomic fluid has the appropriate composition) although the locus is still unknown. Even the protonephridia in trematodes, cestodes, and rotifers appear to have a filtration apparatus, and the final urine in the FW rotifer *Asplanchia* is much more dilute than the body fluids.

2. The proximal region of the post filtration "tubule" in many invertebrates, as well as vertebrates, has a brush border with high alkaline phosphatase activity in the cell membranes bordering the lumen (138). Glucose absorption (and phlorizin inhibition) has been shown to occur here in a number of crustaceans and molluscs, including the lobster and octopus, which do not have a distal segment. Apparently nutrient recovery from the filtrate takes place in the proximal region in the invertebrates as well as vertebrates. Little water or salt is reabsorbed here in aquatic animals, although this changes in terrestrial species.

3. Fresh water vertebrates, decapod crustaceans, and oligochaete worms have a distinct distal tubular segment. Micropuncture shows that reabsorption of ions without accompanying water takes place in this region and in the urinary bladder in both the crayfish and earthworm, and the same distal dilution is seen in FW molluscs where the structural analogy with the vertebrate nephron is less clear. Even in secretory kidneys (FW leech and insects) dilution of the initial urine takes place distally. A distal diluting segment is a good diagnostic character for the stenohaline FW condition, even though it is absent from the few crabs mentioned above.

One could ask what is gained by reabsorbing ions in the kidney to produce a dilute urine, since the net flux must be active. It might appear to be as effective to excrete the salt and recover the ions by the transport systems in the body surface. However, it is energetically less costly to transport the ions against a smaller gradient between tubular fluid and blood than against the very steep gradient between FW and blood (209).

Hypoosmotic Hypoionic Regulation

1. Overview

In addition to osmoconformers and euryhaline hyperregulators, the oceans, as well as some inland saline bodies of water, are inhabited by crustaceans and vertebrates that regulate ECF solute concentration well *below* environmental. Such "hyporegulation" is found in a number of estuarine and shore crabs, some prawns and isopods, a few branchiopods, and many species of fish. Other marine vertebrates, reptiles and mammals, are also hypoosmotic. When hemolymph concentrations are examined in media of different salinities hyporegulators appear to fall into two groups (Fig. 27). One group, all crustaceans, regulates the ECF at relatively high concentrations, ~300–400 mM NaCl. All are also fair hyperregulators in dilute media. Some are isoosmotic in SW and regulate only a little below ambient at higher concentrations [*Pachygrapsus crassipes* (246) and *Crangon crangon* (69)]. Others appear to be isosmotic in 65–75% SW, and body fluids are even better regulated at higher concentrations. The shore crab *Uca* sp. maintains ECF NaCl at 400 mM in 10% SW, and this rises only to 450 mM in 175% SW (10, 65). Thus, this group shows moderately good regulation over a wide range. The other pattern is shown by marine teleost fish and by some branchiopods. The ECF NaCl concentration in SW is lower than

in the first group (150–250 mM) and some of them regulate over an extraordinary range of concentrations [e.g., Ref. 49]. In the branchiopod *Parartemia zeitziana* the hemolymph was 130 mM in 16% SW (hyperregulating) and rose only to 157 mM in SW and 330 mM in a medium of 3900 mM (77). These are very strong regulators! Many fish are also capable of hyper- as well as hypoosmotic regulation; that is, they are extremely euryhaline. In the eel, *anguilla anguilla*, total plasma solute was 330 mM (Na$^+$ 121 mM) in FW and rose only to 381 mM (Na$^+$ 174 mM) in 150% SW (188). These species are generally better hyperregulators than the branchiopods and do well in FW with NaCl < 1 mM. Most are not as good at hyporegulating in extremely concentrated water. However, it should be noted that such euryhalinity, while not rare, is not invariable. The majority of ocean fishes are stenohaline marine and hypoionic to the environment.

Hyperosmotic regulators present a reasonably graded series of adaptive stages

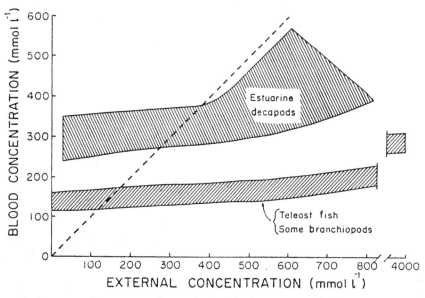

Figure 27. Patterns of hypoosmotic regulation. Both groups include euryhaline species capable of regulating both hyperosmotically and hypoosmotically. However, some are incapable of hyperregulation (e.g., many marine fish).

(progressive reduction of permeabilities, development of ion transport capabilities, and finally the ability to reabsorb ions from the urine), and it might be supposed that the invasion of FW involved such a serial development of the mechanisms. That kind of progression does not emerge for hyporegulators, and for some of them the adaptive significance of hyporegulation is not clear. The ability probably has a different basis in the two groups. Strong hyporegulators are descendants of the strongest of the FW regulators, the stenohaline FW group, with comparable ECF concentrations and a kidney that derives from such progenitors [the kidney picture is clear in the vertebrates (109, 232), only suggested in branchiopods (299, 300)]. It is arguable (and argued!) whether the first recognizable vertebrate developed in SW or in FW, but it is virtually certain that all extant vertebrates, except the hagfishes, arose from a FW stock. The cellular osmotic concentration was probably genetically fixed during a long sojourn in FW, and the cells fail to function at higher concentrations. Thus, if this group is to return to the ocean it must develop mechanisms for maintaining the ECF ions near FW levels. The weaker regulators are basically marine animals with none of the diagnostic features of stenohaline FW forms. However, their estuarine or littoral habitats are extremely variable, differing from place to place with proximity to FW and from time to time with the tides. Some may have developed the mechanisms necessary for maintaining body fluids at a relatively constant value (65–70% SW is common) where average osmotic work is minimized.

Like the hyperregulators, these animals have to cope with dissipative flows generated by the gradients between blood and environment. The difference is that the gradients are reversed, so flows are in the opposite direction. They face osmotic desiccation rather than water loading and salt loading rather than loss. But the same strategies are available; reduce dissipative flows as much as possible, and develop equivalent active flows in the opposite direction.

2. Reducing Dissipative Flows

The high rates of urine flow that are so important for regulation in dilute media are clearly maladaptive where water loss is the primary problem, unless an animal can produce urine more concentrated than the blood. Such an ability is restricted to mammals, birds, and some insects and will be discussed later. For hyporegulators, urine flow is an additional route of water loss. The response in vertebrates is well known and is shown in Figure 28 for several species of euryhaline fish. Urine flow is reduced in SW by an order of magnitude and becomes a minor component (10% or less) of the total water efflux in most species. Comparable data are not available for

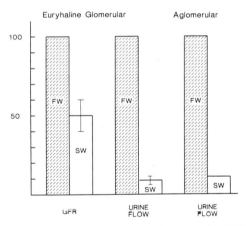

Figure 28. Vertebrate renal function in FW and SW. Urine flow is much lower in SW than in FW in both glomerular and aglomerular fish reflecting the need to conserve water when hyporegulating. In glomerular fish lower filtration rate (GFR) as well as increased tubular fluid reabsorption contribute to the reduction in urine flow. Less is known about the components in aglomerular fish. [The figure is drawn from data in Hickman and Trump (109).]

branchiopods. A few measurements show that urine flow in the weaker crustacean regulators (e.g., *Palaemonetes varians*) is much lower when these animals are hyporegulating than when they are hyperregulating (198).

More detail about renal function in these animals will follow; in the present context the low excretion rate is an adaptive response to the dissipative loss of water.

Table 12 shows water flux and gradient data, and P_{osm} calculated from them for several strong and a few weak hyporegulators. Drinking (J_v^d) is the sole route for net entry of water. If it is assumed that urine flow can be neglected, then drinking just balances osmotic water loss in the steady state. The permeability is obtained from this net flux and the osmotic gradient. It can be seen that P_{osm} is much lower than in osmoconformers in every case except *Penaeus duorarum*; in *Penaeus* the osmotic driving force is relatively small. Even more impressive is the fact that data for *S. gairdneri* and *A. anguilla* were obtained when the fish were hyperregulating (Table 8) as well as when they were hyporegulating, and it is clear that the low permeability in FW is still

further reduced in SW. This is adaptive, since the outward osmotic gradient is much larger in SW. However, the mechanism by which permeability is reduced in this situation is unknown.

Estimates of P_{Na} for hyporegulators (not shown) are much higher than for strong hyperregulators and in the range for osmoconformers (Table 8), a surprising result. The evidence indicates that gills in these animals are very cation permeable. How then is rapid salt loading prevented? The answer appears to be that the gills in most fish (139, 214) and in *Artemia* (279) are much less permeable to Cl^-, and the low anion permeability restricts the influx of NaCl, since electroneutrality requires that the ion pair move together whatever the mechanism.

3. Compensatory Water Influx

This has been studied thoroughly only in the strong regulators. By incorporating a marker compound in the medium (inulin and phenol red have been used because they are not absorbed from the gut or across the body wall), it is possible to show that these animals drink the medium. From the amount of marker in the intestine and its concentration in the me-

TABLE 12. Osmotic Permeability in Hyporegulators

Species	Body Weight (g)	J_v^d (mL/kg · h)	$\Delta\pi$ (osM)	P_{osm} (10^4 cm/s)	Reference
Salmo gairdneri	200	4.5	0.63	0.63	265
Anguilla anguilla	116	3.3	0.82	0.30	189
Platichthys flesus	220	1.5	0.83	0.17	189
Pholis gunnellus	6	0.5	0.6	0.02	66
Serranus sp.	92	2.1	0.82	0.18	189
Paralichthys lethostigma	1000	4.6	0.65	1.07	108
Artemia salina[a]	0.0063	21.6	0.85	0.10	280
Parartemia zeitziana	0.035	13.9	0.63	0.11	77
Uca sp.	2	5.0	0.41	0.24	10
Penaeus duorarum	5.8	17.3	0.16	0.48	98
Metapenaeus bennetlae	5	6.0	0.35	0.45	51

[a]The exchange area calculated from body weight (Table 8) is 0.33 cm². Smith (280) estimated gill area from microscopic observation and found 0.03 cm². His value for P_{osm} is 0.96×10^{-4} cm · s.

dium the drinking rate can be calculated. The rates, shown in Table 12, vary among fish from ~0.5–5 ml/kg · h and for the two brachiopods 3 to 4 times faster reflecting a proportionately larger surface in these much smaller animals. In the steady state, drinking must balance osmotic loss plus renal fluid loss. It is usually assumed that the urine volume is small enough to be neglected and that the drinking rate is a measure of osmotic water loss; these values, together with the osmotic gradients, were used to calculate the osmotic permeabilities in Table 12. Thus, convective flow of water into the gut is the compensatory mechanism used for volume regulation in these animals. A few measurements on weaker hyporegulators show that they also drink the medium at comparable rates [5.0 ml/kg · h for *Uca* (10), 17.3 ml/kg · h for *Penaeus duorarum* (51, 98)].

Balance studies showing the fate of the ingested medium have been carried out in fish. The volume drunk is measured as described above, and from this the rate of entry of any solute can be calculated $(c_{sol} \times J_v^d)$. The volume of fluid leaving the gut is measured as is its concentration of each solute. The difference between rates of entry and exit give the rates of absorption from the digestive (GI) tract. More than 75% of the ingested water and >95% of ingested Na^+, K^+, and Cl^- are absorbed (265). Divalent ions are poorly absorbed; most are excreted, although enough Mg^{2+} enters the animal to create a problem for the kidney.

Since the SW ingested is markedly hyperosmotic to blood, water is absorbed against an osmotic gradient, essentially an active transport process. In addition, absorption of the monovalent ions adds to the salt-loading problem created by the large gradient across the permeable gills. The last is another example of how solving the problem of volume regulation adds to the difficulties of maintaining an ionic steady state.

4. Mechanisms of Intestinal Ion and Water Transfer

Information on mechanisms underlying absorption of the ingested SW is available for fish but for none of the invertebrates. The first step occurs in the esophagus. This structure, which is impermeable in FW forms, becomes very permeable to NaCl in marine teleosts but remains impermeable to water (111, 137). Since there is a steep salt gradient favoring diffusion into the blood, much of the ingested salt is absorbed across the esophageal wall. Whether absorption is really passive or may involve an active transport step is still undecided (197), but the result is not. The osmotic gradient against which water must be absorbed lower in the gut is much reduced.

In the anterior small intestine NaCl is actively absorbed from lumen to blood (124, 269). Both ions are needed for transport of either to occur; and if one is replaced by a nonabsorbable ion, transport of the other stops as does water absorption. The cellular mechanism suggested (61, 73, 270) is shown in Figure 29. The Na^+ and Cl^- ions move from intestinal lumen into the cell through a coupled membrane transporter, which will only move the pair together. Chloride transfer is active (uphill) at this step. The energy is provided by the electrochemical gradient for Na^+, which favors passive inward movement of the ion. The Na^+ is then extruded across the *lateral* membrane by the Na^+-K^+ pump and Cl^- follows, because the potential difference across this membrane favors passive exit of an anion.

Entry of the ions into the narrow intracellular space creates a local osmotic gradient favoring water movement from the lumen into this channel. Since the channel is open to the ECF at the basal surface, convective flow carries the fluid the rest of the way. It is supposed that extrusion of the ions into the lateral intercellular space permits a local osmotic

ECF   EPITHELIUM   LUMEN

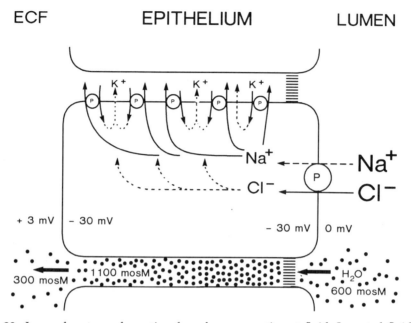

Figure 29. Ion and water reabsorption from hyperosmotic gut fluid. Ingested fluid has a concentration of ~1000 mos$M$. Even after some salt is absorbed in the esophagus, fluid arriving in the anterior intestine may be hyperosmotic to ECF, and water absorption takes place against this osmotic gradient. A key role is played by intercellular channels, two of which are shown here. Movement of NaCl from gut lumen to intercellular channels requires two "pumps," the dispositions of which are shown. Chloride actually moves from the cell into the intercellular channels, but the dashed lines are not *shown* extending that far. The heavy stippling in the lower intercellular channel represents the high solute concentration as ions are transported from the cell into this compartment. The resulting osmotic gradient across the apical junction favors water movement from lumen (600 mos$M$) into channel (1100 mos$M$). There is no resistance to flow at the basal end of the channel; intercellular fluid flows into the ECF as fast as it forms.

gradient to generate water flow (269), even though the osmotic gradient across the entire epithelium is unfavorable. Both net salt and net water movement have been measured in some fish, and the ratio J_{NaCl}/J_v gives the concentration of the reabsorbate. This is very hyperosmotic (~1100 mos$M$) and supports the idea of a local compartment of very high concentration. In this model, the entire cost of fluid absorption is paid by ATP expended in the Na^+-K^+ pump on the basolateral membrane. Thus, this step is the same as in salt absorbing epithelia in hyperregulators (Fig. 24). However, the apical membrane step is totally different as is the water permeability of the intercellular junction. The latter is much lower where only ion absorption occurs as in hyperregulators.

5. Mechanism of Salt Extrusion

The machinery for extruding the salt load is still not entirely clear, and the most reasonable model comes from studies on fish and *Artemia* gills (67, 140, 213) and on "salt glands" in elasmobranchs (267, 268) and marine birds. In most marine hyporegulators extrusion occurs across the gills, and the cells involved, called chloride cells, share a novel structural feature. The basolateral membrane is exten-

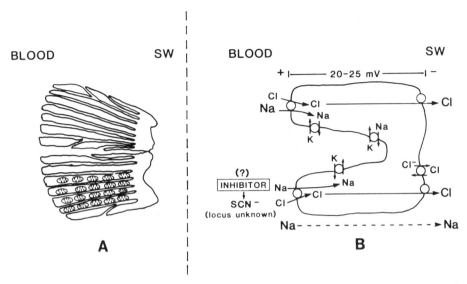

Figure 30. The "chloride cell" in gills and salt glands. (*A*) Depicts, schematically, the ultrastructure of a chloride cell in fish gills and avian salt glands. In reptilian salt glands the membrane infoldings are lateral instead of basal. (*B*) Disposition of the two transport systems on the basal membrane and infolded membrane. Chloride movement across the apical membrane is probably active, but this is still uncertain. Transgill movement of Na^+ is passive, perhaps through intercellular channels as depicted. [Reproduced from (140).]

sively infolded, and the infoldings form a system of minute channels that interdigitate with the cytoplasm and run nearly to the apical pole of the cell. The presence of such channels means that the ECF penetrates far into the cell but is prevented by the infolded membrane from mixing with the cytoplasm. The ultrastructure is suggested in Figure 30*A*. The cells have a high concentration of the Na^+-K^+ pump located on the infolded membranes and also appear to have a coupled NaCl carrier, like the one on the apical membrane of intestinal cells, but here also located on the basolateral membrane. The model is depicted in Figure 30*B*. Sodium chloride is transferred into the cytoplasm by the coupled carrier; the necessary energy is provided by the large gradient favoring diffusion of Na^+ in this direction. The Na^+ is extruded back into the ECF by the Na^+-K^+ pump leaving Cl^- to move, perhaps passively, across the apical membrane into SW. Since the gill is very cation permeable, a Na^+ will diffuse passively from blood to SW for each Cl^- extruded. Once more, the energy for the process is provided by splitting ATP in the Na^+-K^+ pump located, as usual, on the basolateral membrane. The location of the NaCl carrier, and the selective cation permeability of the gill provide the other details needed to make such a model work. It should be noted that transbranchial Na^+ transfer is passive (diffusive) in this model, even though work is done on Na^+ at the basolateral membrane. For a few fish a flux–force analysis shows that extrusion of both Na^+ and Cl^- is active (67, 213). The basis for the difference is unknown.

6. The Role of the Kidney

From an osmoregulatory point of view the kidney in hyporegulators can only be called a loser. It aggravates the main problem, water loss, without excreting enough salt to help with the ionic problem these animals face. Yet it has an important job to do. There was already an indication that the excretory organ in marine animals played a role in keeping ECF Mg^{2+} lower than ambient by excretion the ion in the urine. This function becomes more pronounced in hyporegulators, probably because they must drink SW with its high Mg^{2+} concentration. Since Mg^{2+} becomes concentrated in the urine, there must be a mechanism for actively secreting it. This is certainly located in the proximal tubular segment, since secretion occurs in both crustaceans and fish that have only this segment. Little more is known about the mechanism of secretion.

While their kidneys are an osmotic liability, hyporegulators have developed a number of adaptations that minimize renal water loss. First, the distal tubular segment, which we have seen is concerned with water excretion and salt conservation, is missing in many marine fishes (70, 109). In addition, the glomerular filtering apparatus is also missing in two entire families of marine fish, and the nephron is reduced to a segment of the proximal tubule. It operates as a secretory organ, producing very high U/B Mg^{2+} ratios, but low urine flows. In several fish the urinary bladder absorbs NaCl and H_2O, but the osmolarity of the reabsorbate is only 400 mM (227). A milliliter of fluid absorbed here will not have to be recovered by drinking, with absorption in the gut generating three times the solute load. Thus, there are, in the vertebrates, both structural and physiological modifications in the kidney acting to reduce fluid output in a desiccating environment.

There is practically no comparable information on adaptive changes in excretory function among invertebrate hyporegulators, but it is worth noting that most of them have no distal, diluting segment. A comparative study of renal structure and function in *Artemia* and other, stenohaline FW, branchiopods would be especially interesting.

Isosmotic, Hypoionic Regulation

1. Overview

We have seen that volume and ionic regulation are practically inseparable in aquatic animals, and that coping with one of the problems adds to the other. A few marine vertebrates, the elasmobranches (276, 278), holocephalans and the sole surviving crossopterygean *Laterimeria chalumnae* (92) among fishes, and the "crab eating" frog *Rana cancrivora* (84, 85), have succeeded in separating the problems by eliminating one of them. Table 13 shows that ECF (plasma) osmolarity in the dogfish is nearly the same as ambient, resembling in this regard the hagfish. It is, in fact, slightly hyperosmotic, so these animals do not face the problem of osmotic water loss; rather, there is flow into the animal. But when we examine the components of the ECF it turns out that, although NaCl is still an important osmotically active solute, it is no longer the only one of consequence. Ionic concentrations are not much different from marine teleosts, and nearly one half of the ECF osmotic concentration is contributed by the nitrogenous nonelectrolytes urea and tetramethylamine oxide (TMAO). These animals still have to extrude ions entering by diffusion from SW, but they do not have to drink. The consequences are significant. Energy is not expended in the gut to absorb a hyperosmotic fluid, and the salt load is thereby much reduced. This means that energy is also spared in extruding ions. It is curious, in this regard, that *R. cancrivora* tadpoles appear to hyporegulate in SW like a teleost fish (83).

2. Reducing Dissipative Flows

The use of urea and trimethylamine oxide (TMAO) eliminates the osmotic efflux that

TABLE 13. Renal Function in the Dogfish, *Squalus acanthias*

Solute	Solute Concentration (mM)		Renal Turnover (μmol/kg · h)		
	Plasma	Urine	Filtered	Excreted	Reabsorbed
Na⁺	250	240	875	271	604
K⁺	4	2	14	2.3	11.7
Ca²⁺	3.5	3	12.3	3.5	8.8
Mg²⁺	1.2	40	4.2	46	
Cl⁻	240	240	840	276	564
SO₄²⁻	0.5	70	3.4	80.5	
Glucose	14	3	49	3.5	45.5
Urea	350	100	1225	115	1110
TMAO	70	10	245	11.5	233.5
Osmolality	1000	800	(local SW = 930 osM)		
H₂O[a]			3.50	1.15	2.35

[a]Units are in ml/kg · h. From data compiled in Hickman and Trump (109).

occurs in most marine fishes. Synthesis of the compounds takes place in the liver and will be considered in detail in a later chapter. It suffices to note here that the syntheses are energy requiring, and minimizing losses is obviously important. To this end, the gills are nearly impermeable, and little is lost by this route. Table 13 shows that a large quantity of urea is filtered in the kidney. This is measured as described earlier; filtration rate × plasma concentration. But of the 1225 μmol filtered per hour only 115 μmol are excreted. Therefore, >90% are reabsorbed, and since the urine concentration is lower than ECF, reabsorption is an active process. The mechanism is unknown, but we might expect that it will prove to cost less ATP/urea than resynthesis in the liver.

3. Compensatory Mechanisms

The inward water flow across the body surface is, as in FW fish, excreted by the kidney. The nephron in cartilaginous fish is also similar to the FW unit, with a fully developed glomerulus, proximal, and distal segments and a collecting duct (109). Filtration rates are higher than for other marine fish (Table 13). Urine flows are variable, but the value in Table 13 is representative and represents a large part of the water absorbed across the body surface.

There remains the need to regulate blood ionic levels. Table 13 shows that plasma Mg²⁺ and SO₄²⁻ ions are much lower than in SW. Both ions become concentrated in the urine. The transport systems are likely the same as those found in other marine animals. Although plasma NaCl is higher in these fish than in teleosts, it is still much lower than ambient, and salt loading is to be expected. A substantial amount of this load is excreted in the urine, although nearly 70% of the NaCl filtered is reabsorbed. Reabsorption of NaCl appears to be counterproductive but may be related to the urea/TMAO reabsorption. The transfer of many organic compounds across epithelia is linked to transport of sodium in the same direction.

These fish have a large "rectal" gland, an appendage of the posterior intestine emptying into the cloaca, which provides an extrarenal route for salt excretion (35).

The main cell type in this gland is similar to the chloride cell in the gill, and much of our information about the molecular machinery and its control comes from experiments on rectal glands (267, 268). It secretes intermittently, but flow rates appear to be in the same range as for urine flow. The fluid produced is approximately isosmotic with blood but comprises NaCl at ~500 mM with much smaller concentrations of K^+ (7 mM) and urea (18 mM). Thus, it appears to be a vehicle for excreting an important part of the salt load, and the fluid volume also accounts for some of the water excreted. It has also been suggested that the gills may actively extrude NaCl, since chloride cells have been found there as well [reviewed in Evans (67)].

Terrestrial Regulation

Most of the major phyla are represented on land (echinoderms and cnidarians are absent), but studies of osmotic regulation are virtually confined to the arthropods, chordates, molluscs, and annelids. There is no simple, unequivocal way to characterize success in adapting to life on land, but in terms of some combination among number of species represented, independence of the need for immersion at any stage, mobility and metabolic activity, insects are the most successful and annelids the least.

Terrestrial life permits animals easier access to oxygen, which is present in much higher concentration in air than in water. The gas is more mobile when respiration is based on diffusion and is energetically less costly when convective movement of the medium is required. However, if there is profit in exploiting this new milieu, there is also a price that must be paid in coping with a different set of problems. One of the most serious is the highly variable, but generally severe, dehydrating conditions on land.

Physical Principles

In regulating the body fluids out of equilibrium with an aqueous environment, dissipative flows of both ions and water are problems that must be solved. The latter is somewhat more serious, and where it is eliminated (as by the elasmobranchs), regulation is much less costly. On invading land, animals face a situation in which ionic regulation is reduced to matching the outflow, usually through the excretory organs (but sometimes through glands as well), with adequate ingestion. In most cases, movement across the body surface is not a factor. But if the magnitude of the ionic problem is reduced, that concerning water is much more severe, for the environment is very dehydrating, and the body surface presents a large area for water loss in addition to losses through kidneys and glands.

Movements of water (and ions, as well) *within* a terrestrial animal occur from one aqueous phase to another, for example, from gut lumen to blood, and they obey the laws of osmosis and diffusion described earlier. But water moves across the body surface of a terrestrial animal between an aqueous phase (body fluid) and a gaseous phase (air). The physical law governing this movement is based on the same fundamental principle as before; dissipative flow occurs from a region of high chemical potential to one of lower chemical potential. But the expression of chemical potential differs in the two kinds of system. For water transfer between two aqueous media, chemical potentials were represented by osmotic pressures (but expressed by the solute concentrations in the two media). For transfer across an aqueous–air interface we find new expressions; the vapor pressure of water in air and the relative humidity, and the *mole fraction* of water in a solution. We will introduce them here without undertaking a detailed treatment.

The chemical potential of water in dry air is obviously zero. If a container of pure water is in contact with dry air in a closed system, water molecules will escape across the interface and enter the gas phase; the chemical potential of water in the latter increases (Fig. 31B). Some molecules from the gas phase will now begin to reenter the water, but as long as the chemical potential is higher in the latter, evaporation will exceed condensation, and the water concentration (and chemical potential) in the air will continue to rise. When the potentials in the two phases are equal the system is at equilibrium (Fig. 31A), and the air is said to be saturated with water.

The chemical potential of water vapor in a gas phase is related to its concentration (sometimes called the *vapor density*) there, but also, from the gas law, to its partial pressure, called the *vapor pressure*. If the vapor pressure of air in equilibrium with a water surface is P_0, unsaturated air will have a vapor pressure less than this, P. The chemical potential of water in air is a function of P/P_0. The ratio, called the *relative humidity* (RH), expresses the water concentration in air as a fraction of the saturation concentration.

If we are to examine the driving force for water movement between an animal (an aqueous system) and air, it is necessary to express water potential in the body fluids in the same terms described for the gas phase. The presence of a solute depresses the vapor pressure in a solution; if we repeat the experiment in Figure 31, but with a solution instead of pure water, we will find that the vapor pressure at equilibrium is P_0', which is $<P_0$. The higher the solute concentration, the lower the equilibrium vapor pressure P_0'. The relationship between concentrations in solution and equilibrium vapor pres-

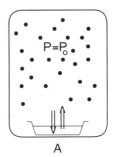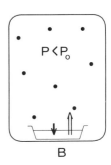

Figure 31. Evaporation at a water–air interface. (A) The chemical potential of water is the same in the gas and aqueous phases. There is no net water movement. The two-phase system is at equilibrium. The chemical potential is represented by the equilibrium vapor pressure (P_0). (B) The chemical potential of water (reflected by its vapor pressure, P) in the gas phase is lower than P_0. The difference in chemical potentials between the two phases can be expressed by the "saturation deficit," $P_0 - P$. This is the force causing net movement of water into the gas phase.

sure in the gas phase is

$$P_0'/P_0 = N_w/(N_w + N_s) \qquad (9)$$

where N_w and N_s are the number of moles of water and solute in the solution. The ratio on the right is called the *mole fraction* of water in the solution, but it also equals the RH of air in equilibrium with the solution. Since the chemical potentials of the two phases are equal at equilibrium, we have the needed expression. If the RH of air around an animal is lower than P_0'/P_0 calculated from Eq. 9 water will evaporate from the animal; if higher, then the animal will gain moisture from the air. Two interesting corollaries follow. First, most terrestrial animals have body fluid solute concentrations <0.6 os$M$, while the water concentration is $\sim$55 M. From Eq. 9 $P/P_0 = 55/(55 + 0.6) = 0.989$. At any RH $<98.9\%$ the animal faces water loss by evaporation. Furthermore, no achievable increase in body fluid solute concentration will permit animals to come into water equilibrium with air at relative humidities usually encountered. For example, to be at equilibrium with air at 70% RH the body fluid solute concentration would have to be 23.8 M! The elasmobranch's solution of the osmotic problem is simply not available to terrestrial animals.

There remains only the question of how to express the gradient driving water in order, for example, to calculate and compare permeabilities. This is usually expressed by the difference between the vapor pressure equivalent of the body fluid and that of the ambient air. The relationship between water flux by evaporation and the vapor pressure difference (often called the saturation deficit) has the form of a flow $-\Delta P$ function like that in Eq. 5, and a practical permeability is found in the same way as for aquatic animals

$$\text{permeability} = \text{flux/force}$$

The units often used are mg of H_2O per gm of body weight $\cdot$ h $\cdot$ mmHg vapor pressure difference, but if the animal's surface area is known, the rational units, cm/s, can be calculated.

Evaporative water loss in unsaturated air is increased by convective air movement and also by a rise in temperature. The reason is that both increase the vapor pressure gradient between animal and ambient air. One adaptive strategy available to animals in a dehydrating environment is to seek a microhabitat that minimizes both variables.

The Transition to Land: Semiterrestrial Animals

Land as been invaded from FW and from SW, but the main osmotic problem, dissipative water loss, is independent of origin. We have seen that aquatic hyporegulators manipulate the osmotic permeability of the body surface and respiratory epithelium to minimize passive outflow. Reduced urine volume also serves this end. The same evasive maneuvers are available to terrestrial animals, which also employ a remarkable range of behaviors to minimize water loss.

Many species resident on land show only little to moderate ability to cope with a dehydrating environment. We call them semiterrestrial to indicate this. Table 14 shows rates of evaporative water loss in a few representative examples, and some general features emerge:

1. Water is lost by these animals at rates between 1–5% body weight/h in unsaturated air. The earthworm (*Lumbricus*), slug (*Limax*), and frog (*Rana*) have a moist body surface with no structure external to the limiting epithelium (skin), but a thin chitinous layer in onychophorans and a thick exoskeleton in the crustaceans obviously does not substantially impede evaporation. Removal of the skin in frogs had no effect on evaporative water loss (2), showing that the limiting resistance to water movement is in the vapor phase. Values of P_{osm} like those shown in Table 14 for the frog, slug, and earthworm really indicate that there is no structural barrier to water loss.

2. The evaporation rate depends on ambient RH as expected (compare values for the slug, earthworm, and frog at different humidities).

3. Values for the two crabs, *Uca* and *Ocypode*, suggest that transpiration per gram of body weight (the weight-specific loss) might be lower in larger animals.

The last point was studied in crabs ranging from very small (0.2 g) to large (>500 g), and Figure 32 shows that the weight-specific water loss (in air at 25%

TABLE 14. Permeability and Transpiration in Semiterrestrial Animals

Species	Wt (g)	Temper-ature (°C)	RH	ΔVP (mm)	Transpiration (mg/g · h)	P_{osm} 10^4 cm/s	Reference
Peripatopsis							
sedgwicki	0.4–0.8	24	0	22	38	667	177
Lumbricus							
terrestris	2.7	13	70–80	3	24	3100	39
Lumbricus							
terrestris[a]	1.8	13	100	0	4		39
Limax maximus	1–4	20	70–80	12	40	1260	217
Limax maximus	1–4	20	25–30	15	17	1280	217
Uca pugilator	1–2	25	25	18	30	530	107
Ocypode quadrata	10	25	25	18	15	500	107
Rana pipiens	28	20	0	17	32	1575	2
Rana pipiens	28	20	50	8.5	9	890	2
Rana pipiens	28	20	96	1.7	2		2

[a]Transpiration should be zero. The loss is probably in urine and mucus. It is apparent that P_{osm} values are huge compared with even the largest in Tables 8 and 12 for aquatic animals. This is because the driving forces (osmotic and vapor pressures) are very different, even though both have the same dimensions (change in atmospheres across an interface). A frog sitting on a lotus leaf has a P_{osm} ~1000 × 10^4 cm/s as shown above. But when it jumps into the pond P_{osm}, now calculated as in Table 8, is ~1 × 10^4 cm/s. The lesson to learn is that we can compare osmotic permeabilities among aquatic animals and among terrestrial animals, but not between the two groups. Not, in any event, when the coefficients are calculated as they are here.

RH, 25°C) decreased logarithmically with body weight. The equation describing this relationship is shown, and the exponent, -0.31, is almost exactly what is expected if water loss is surface limited. It follows that, if other factors are equal, larger animals can sustain a longer period of dehydration than smaller ones; the adaptive significance of this is obvious.

Such rates of water loss cannot be sustained for many hours, and physiological mechanisms to reduce them have been described in many animals. In Figure 32, most of the values for *Gecarcinus* fall well below the general regression line, and values for a group of aquatic and "mud" crabs (not shown) were well above the line. Since *Gecarcinus* is the most terrestrial of these crabs, and the last group is the least terrestrial, restriction of evaporative loss correlates well with habitat. In another arthropod group, the isopods, a similar correlation between transpiration and habitat has been described (62). Sim-

ilar examples have been noted in other taxa. Thus, most amphibians have high rates of water loss, which restrict them to moist or humid environments. The rate of cutaneous water loss in terrestrial pulmonates with shells is not notably different from that in slugs when the animals are active, but is markedly reduced when they withdraw into the shell. Water loss in *Helix aspera* was 40 mg/g · h during activity. It was reduced to 0.17 mg/cm · h when the snails withdrew into their shells and to 0.10 mg/cm · h when the shell aperture was sealed by a membrane, the epiphragm (168). This means of reducing water loss has been exploited by some snails that survive in even the most xeric (arid) environments. *Sphincterochila boissieri* is found in Middle Eastern deserts, such as the Negev, where daytime temperatures exceed 50°C and the scant rainfall is confined to a short period during the winter. It lost water at <0.5 mg/day, and internal water stores would permit it

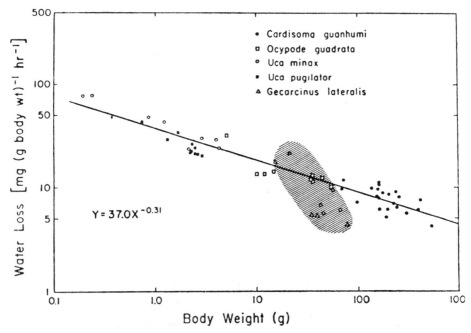

Figure 32. Weight-specific water loss for semiterrestrial crabs in air. Evaporative water loss per gram of body weight (bw) is shown for five species of crab. The relationship for four of them is shown by the regression equation in the lower left (y = water loss; X = bw). Most of the values for *G. lateralis* (shaded region) are well below the line. [Reproduced from Herreid (107) with permission of Pergamon Press.]

to survive even a drought lasting more than a year (255). However, while the snail has adapted to *survive* in a xeric environment, it is no better adapted to *function* under such conditions than other animals in this category. Activities like feeding and reproducing must await the hygic winter.

The contribution of the respiratory surface to total water loss is infrequently measured. In animals like annelids, where gas exchange takes place across the body surface, the distinction has no meaning. But many of them have a well differentiated respiratory organ that is a potential site of evaporation. There is no evidence that respiratory water loss is significant in most terrestrial amphibians, and it comprises <1% of the total water loss in the land crab *Cardisoma guanhumi* (107). Structural modification of respira-

tory surfaces in the more terrestrial crabs and isopods may serve to reduce evaporation. However, closure of the pneumostome (opening to the lung) in the slug *Limax* restricts water loss under dehydrating conditions; animals with pneumostomes kept open dehydrated nearly twice as fast as controls (217). And it was suggested that water is lost primarily from the trachae, rather than from the chitin-covered body surface, in the onychophoran *Peripatus* (177). To summarize: In most semiterrestrial species, water loss from the body surface is so rapid that respiratory evaporation comprises a fraction that ranges from negligible to modest.

The other potential route of water loss is through the excretory organ. This has not been extensively investigated, but some observations suggest that volume flow is reduced under dehydrating conditions, at least in some species. In the snails *Achatina ventricosa* and *Helix pomatia* less urine is produced in the absence of food and water than in their presence (179). Urine in hydrated snails is only slightly (*Achatina*) or moderately (*Helix*) diluted (305). The crabs *Carcinus maenas* (234) and *Cardisoma* (100, 107) become anuric in air. Both molluscs and crustaceans have filtration kidneys, and there is evidence that water is reabsorbed from the tubular fluid. Thus, in hydrated *Achatina fulica* the ratio of inulin concentration in the urine to that in blood (U/B_{in}) was often >1 when urine flow rates were low (179). Hydrated *Achatina achatina* reabsorbed 35% of the filtered water; when dehydrated they reabsorbed 70% (272). The U/B_{in} was ~2 when *Carcinus* was in SW, and it increased when the animal was transferred to air, even though the RH was 100%. In frogs and toads both filtration rate *and* tubular water reabsorption are involved in renal water conservation. When *Rana clamitans* was in water GFR was 34 ml/kg · h and urine flow was 13 ml/kg · h; ~60% of the filtrate volume was reabsorbed. After 5 h in air GFR was reduced to 5.1 ml/kg · h and urine flow to 0.8 ml/kg · h; reabsorption had increased to 84% (249).

Nitrogen metabolism and excretion are described in Chapter 8, but the relationship with volume regulation should be noted briefly here. Most aquatic animals produce and excrete NH_3 as the major end product of N metabolism. The excretory mechanism was alluded to earlier, because it is coupled to Na^+ ion absorption in FW. In aquatic animals, an extensive exchange surface, for example, a gill, and an infinite environmental sink ensures that rates of removal are adequate to keep blood concentration low, usually <1 mM. This is important because NH_3 is toxic. On land the problem of NH_3 excretion is formidable, even though it is volatile and could be excreted as a gas. Only the is-

opods, among terrestrial animals, are known to use this mechanism to excrete a large proportion of their total nitrogen content (317), although some terrestrial snails also excrete gaseous ammonia. It may be that in larger, more active animals the blood concentration needed to sustain diffusive excretion at an adequate rate would be lethal. Many of the transitional species switch to urea as the end product. This compound is even more soluble than ammonia and is nontoxic (remember the high plasma concentration in elasmobranches). However, it is not volatile, and hence must be excreted in solution. Its elimination thereby increases water loss. Nevertheless, it is one form in which nitrogen atoms are excreted in annelids (NH_3 and amino acids are others), and there is a suggestion that the proportion of urea increases when the worms are in air. Most terrestrial amphibians excrete urea in the urine. But when animals are in air the demands of nitrogen excretion and water conservation are clearly at odds. A distinct step toward resolving these competing demands is incorporation of waste nitrogen into end products like uric acid and the purines xanthine and guanine. These are very insoluble in neutral or acidic solutions and can either be excreted as a crystalline precipitate or stored as "concretions" until water becomes available. In either case, a minimum of water is lost when it is in short supply. This mode of excretion is best developed in the insects and in reptiles and birds. But it is also characteristic of the more terrestrial crabs and snails discussed here, as well as in two genera of frogs.

An astonishing range of behaviors are also employed to minimize water loss in these animals. Thus, individuals may adopt postures that decrease evaporative surface area as in "curling up" seen in some isopods (309). Groups of individuals may aggregate (slugs and isopods) with the same result. Transpiration in the hermit crab *Clibanarius vittatus* was 2–3 mg/min when it was in its snail shell but >10 mg/min without the shell (107). Snails and slugs are active at night and retreat, during the day, to a moist microhabitat. Isopods in mesic environments live under logs or leaf mold where the humidity is high. In such a niche convective air movements that would tend to reduce humidity are absent. In addition, many of these animals aestivate when they begin to dehydrate; this is a characteristic response of both gastropods and annelids. Inactivity is particularly important in both, because substantial amounts of water are lost in the form of shed mucus during locomotion. Many groups create an appropriately moist habitat by burrowing into the sand or soil.

By employing some combination of these strategies such animals have been able to occupy a range of habitats ranging from hygric in the less well adapted, through mesic in many, to xeric in a few. Finally, however, water lost must be replaced by drinking or osmotic transfer across the body surface, or by water in the food.

Fully Terrestrial Animals: The Arthropods

Most crustaceans are only marginally successful on land, but a few isopods, and many species of insects, myriapods, and arachnids make their livings in habitats ranging from moderately dry to severely arid. Here, as before we will emphasize

TABLE 15. Evaporative Water Loss in Terrestrial Arthropods[a]

Species	Habitat	Transpiration Rate (μg/cm^2 · h · mmHg)	P_{osm} (10^4 cm/s)
Isopod Crustaceans			
Porcellio scaber	Hygric	110	323
Hemelepistus reaumuri	Xeric	23	68
Insects			
Blatta orientalis	Mesic	48	141
Periplanata americana	Mesic	55	162
Arenivaga investigata	Xeric	12	36
Locusta migratoria	Xeric	22	65
Thermobia domestica	Xeric	15	44
Tenebrio molitor (larva)	Xeric	5	15
Manduca sexta (larva)	Xeric	40	116
Myriapods			
Glomeris marginata	Hygric	200	588
Orthopterus ornatus	Xeric	8	23
Arachnids			
Ixodes ricinus	Mesic	60	176
Pandinus imperator	Mesic	76	223
Buthotus minax	Xeric	1	3
Ornithodorus moubata	Xeric	4	12

[a]Values selected from Edney (62). These *can* be compared with those in Tables 14 and 16.

the problems of maintaining water balance under dehydrating conditions.

1. Transpiration across the Body Surface
Table 15 shows weight-specific transpiration rates, and the water permeabilities deduced from them, in selected arthropods that occupy a range of habitats from mesic through xeric. Since transpiration rate depends on conditions (e.g., temperature, RH) that vary in different experiments, as well as on the animal's size, it is more useful to compare permeabilities. First, it is apparent that these are generally lower than those shown in Table 14, in some cases by two orders of magnitude. The first big step in becoming independent of hygric conditions is clearly to restrict transpiration. Secondly, there is good correlation between habitat and resistance to water loss. A step in this direction was noted among terrestrial crabs, but the differences here are larger. The desert cockroach, *Arenivaga investigata* is considerably more resistant than is *Blatta orientalis,* and the desert scorpion, *Pandinus imperator,* is still less permeable by an order of magnitude. The sole exception in this group is the larval tobacco hornworm, *Manduca sexta,* which can function under very dry conditions, even though it is moderately water permeable. This larva feeds constantly, and the water content of the ingested leaves apparently is adequate to compensate for substantial transpiration. There is also, among these small animals, the negative correlation between size and weight-specific transpiration rate that we saw in crabs (Fig. 32), and its significance is the same; the fraction of the water reserves depleted by evaporation is smaller in larger animals. If other factors are equal, larger animals can sustain longer periods of desiccation than smaller ones.

The exoskeleton is an important factor in restricting water loss in these animals. Transpiration is low in arthropods that inhabit dry environments, and if the cu-

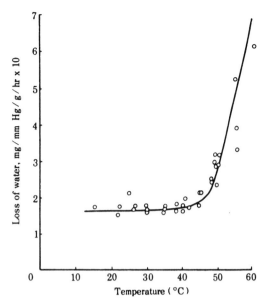

Figure 33. Water loss through the cuticle of *Locusta migratoria.* The sharp transition around 45°C is due to a change in water permeability. [Reproduced from Loveridge (165a) with permission of the Company of Biologists Ltd.]

ticle is abraded, water is lost at a higher rate (15, 318). Moreover, exposing the insect cuticle to a lipid solvent like chloroform also increases water loss (16, 318). It appears that "waterproofing" involves a layer of cuticular lipid. If ambient temperature is increased, transpiration rates increase only slightly over a substantial range, but above some "transition temperature" the rate increases greatly, even in dead insects. This is shown for *Locusta* in Figure 33. The change may depend on the melting temperature of the lipid (15). Transition temperatures vary between ~30–60°C in different animals, perhaps due to a difference in lipids. This is a significant variable, since many arid regions experience high daytime temperatures.

Respiratory water loss can also be substantial in animals with a tracheal gas exchange system. The large superficial trunks open to the outside through spiracles, and these can be closed by mus-

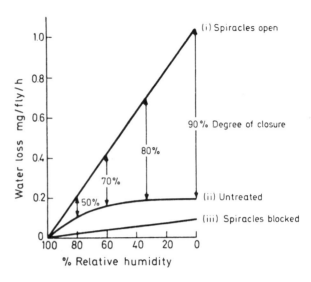

Figure 34. The role of spiracles in evaporative water loss. Water loss from the tsetse fly (*Glossina morsitans*) is shown at different RH. (i) Flies are exposed to 15% CO_2 in air and the spiracles are open. (ii) Flies are exposed to air (no CO_2). The differences in water loss at each RH is supposedly due to partial closure of the spiracles (vertical arrows). (iii) In air with the spiracles blocked by paraffin wax. [Reproduced from Bursell (35a) with permission of Academic Press.]

cular valves in many species. The importance of controlling ventilation of the trachea can be seen in Figure 34. The rate of water loss in untreated tsetse flies increases when RH drops from saturation to ~60% but is little changed at still lower RH. When the spiracles are blocked, water loss increases as RH drops, but rates are very low even in dry air. In the presence of high ambient CO_2 the spiracles are always open. Water loss is rapid at all humidities, and there is no evidence of control in drier air. Flies normally control spiracle opening to minimize water loss. Spiracles may also be kept partly closed when the flies are dehydrated. In *Locusta*, ventilatory movements also augment water loss (166), and this is probably the general case in active animals. Similar observations have been made on many insects and arachnids, and two conclusions follow. The respiratory tract is potentially a site of major water loss. In addition, the competing demands of water conservation and adequate gas exchange cannot always be optimized. In dry air the first is critical, especially at high ambient temperature (high vapor pressure deficit). But at high temperature or during activity, metabolism is rapid

and gas exchange becomes dominant. Many studies show that water loss is greatly increased under such conditions.

2. The Excretory System

Most of our information on excretory physiology in arthropods comes from studies on the malpighian tubule–hindgut complex in insects [reviewed in Bradley (29) and Maddrell (170)]; relatively little is known about the contribution of maxillary glands (isopods) and coxal glands (arachnids). Most of these animals excrete several nitrogenous compounds. Uric acid predominates in insects, guanine in arachnids, and the water-sparing feature of these insoluble compounds was mentioned earlier. A few insects excrete a copious, dilute urine, at least when they are feeding. However, in most, the excreta are more concentrated than hemolymph even when they are hydrated. Faced with desiccation, many insects dry the excreta to a remarkable extent by reabsorbing nearly all the water from a presumptive urine.

Tracheolar respiration in insects eliminates the need for a high-pressure circulation of hemolymph between gas exchange surface and the cells. This, in turn, precludes use of pressure-driven filtration in the first stage of urine formation. Instead, secretion (active transport) of solutes into the tubule provides the basis for tubular fluid formation. Subsequent steps are analogous to those in filtration organs: The tubular fluid becomes modified by selective reabsorption of some materials and by addition of others through diffusion or transport. In secretion, as in filtration organs, the final urine usually bears little resemblance to the initial tubular fluid.

A generalized anatomy of the insect excretory system is shown in Figure 35. The Malpighian tubules are closed at one end and open into the anterior hindgut. The number varies in different animals. In most, the blind end of the tubule lies free in the hemocoele as shown here, but an exception will be described below.

The presumptive urine forms in the distal (closed) end of the Malpighian tubule. In most cases, tubular fluid has a higher [K^+] than hemolymph; for example, 150 versus 18 mM in the stick insect, *Dixipus morosus* (224). It is usually electrically positive to hemolymph by 20–40 mV, so K^+ must be actively transported into the tubule against a large gradient. Fluid formation (i.e., water movement) is strongly dependent on hemolymph [K^+] and virtually stops in its absence, even though [Na^+] in the bathing fluid is normal. The rate of fluid formation varies inversely with hemolymph osmotic pressure, indicating that osmosis plays a role. The picture that has emerged is that the (distal) tubule actively transports K^+ from hemolymph into lumen. Chloride diffuses into the lumen (the gradient is favorable). Increasing luminal solute concentration induces osmotic water flow in the same direction. The resulting fluid is about isosmotic with hemolymph and has KCl as the major solute. The Na^+ ion is also transported into the tubule, but more slowly than K^+ and to a much lower final

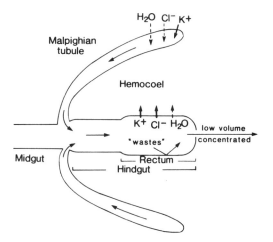

Figure 35. Generalized insect excretory system. Active transport is shown by solid arrows, passive movement by dashed arrows.

concentration. The tubular epithelium is permeable to organic molecules (171), and these can diffuse into the tubule, although their concentrations in the presumptive urine are lower than in hemolymph and vary inversely with flow rates (faster flow allows them less time to diffuse to equilibrium). Urates enter as the Na and K salts and are converted to insoluble uric acid by acidification of the fluid, either lower in the tubule or later in the hindgut. Transport systems for transferring organic acids and organic bases into the tubule exist in insects as in the proximal tubule of filtration organs in other animals. In addition, some Malpighian tubules have transport mechanisms specially adapted to a particular life style and not seen in other insects.

The tubular fluid flows into the anterior hindgut, and some modification may occur by selective secretion and/or reabsorption in this region, but adaptive fine tuning is the major role of the rectum. In normally hydrated cockroaches (*Periplanta*) (307) and locusts (*Schistocerca*) (202) the rectal fluid is moderately hyperosmotic to hemolymph, indicating that water is reabsorbed (remember that tubular urine is isosmotic). The ionic concentration is also reduced (e.g., rectal fluid $[K^+]$ ~20 mM; $[Na^+]$ and $[Cl^-]$ even lower) and the volume is much less than that entering from the anterior chamber (202). The rectum absorbs ions and water, reducing the excreta to a semisolid paste. In dehydrated animals water can be absorbed against even steeper osmotic gradients; fecal pellets become still drier as a greater fraction of the incoming water is absorbed.

The mechanism of fluid absorption from the rectum is energy requiring, hence inhibited by metabolic inhibitors and involves the active reabsorption of Na^+, K^+, and Cl^- as well as H_2O (201). The accepted model has the ions being pumped from lumen through cell into thin, intercellular spaces. As a result of this transport, the solute concentration in the space rises above that in the lumen. This creates a local osmotic gradient favoring passive water movement from lumen into the intercellular space, and the fluid formed flows into a larger channel that communicates freely with the hemocoele (306). This is another example of solute-coupled water transfer against an osmotic gradient. In principle, it is the same as is thought to occur in the absorption of SW from the marine teleost intestine, although details differ in the two systems.

The mealworm (larval *Tenebrio molitor*) is one of several insects that live in stored food products and without drinking. It can excrete a dry pellet in equilibrium with RH 88% (occasionally even drier), so is even more effective at absorbing water than the insects described above (93, 225). The tubules, instead of being in the general hemocoele, lie close to the hindgut with the distal (blind) end of the tubule apposed to the posterior hindgut. The distal part of the tubules and the hindgut are enclosed by a thick "perinephric"

membrane, and the "perirectal" chamber so formed is fluid filled. This "cryptonephridial system" is depicted in Figure 36 with the presumed pathways for solute and water movements indicated. The K^+ ion is transported from hemocoele, through perinephric sheath, into the tubule, its concentration rising to 2–3 os$M$! The sheath appears to be water impermeable, so that water cannot follow from the hemocoele. Instead, water enters the tubule from the perirectal fluid, which becomes concentrated in the posterior region. It is supposed that a radial osmotic gradient is formed in this way, one that favors the passive flow of water from the vapor phase in the rectum, through the perirectal fluid, into the Malpighian tubule. There remain details to work out, especially whether the *osmotic* gradient between rectum and perirectal fluid can account for passive water movement at this step, but the broad features of the model are probably correct. In consequence, this cryptonephridial complex can recover nearly all the water entering the hindgut; little is lost with the dry excretory pellets. In fact, it can do more than this, as we will see later.

The role of the excretory system in water conservation has been emphasized, but the system is at least as flexible as filtration-based kidneys in modifying its output according to the demands of different environments. Two very different examples will illustrate this flexibility. (a) *Rhodnius prolixus* (172, 222, 319) feeds intermittently on mammalian blood with an ionic composition and osmotic pressure similar to its own hemolymph. It may ingest more than 10 times its body weight at a single feeding and rapid volume adjustment is essential. The Malpighian tubules respond by producing an isosmotic fluid at a high rate. Tubular fluid is unusual only in containing nearly equal concentrations of Na^+ and K^+. Most of the latter, together with some Cl^- and a little

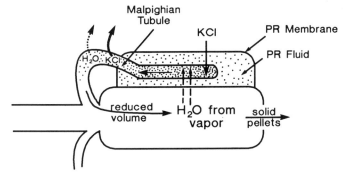

Figure 36. The crytonephridial complex in *Tenebrio molitor*. The distal end of the malpighian tubule lies within a fluid-filled perirectal (PR) chamber and is closely apposed to the hindgut. The entire chamber is separated from the hemolymph by a PR membrane. Much of the initial tubular fluid is recovered in the mid-section of the tubule after it emerges from the PR chamber and traverses the hemocoele. Other details are given in the text.

water are reabsorbed in the lower region of the tubule. The resulting fluid is large in volume, somewhat hypoosmotic to hemolymph, and has a composition similar to the ingesta. It is excreted with little further modification in the hindgut. (b) Larvae of the mosquito *Aedes campestris* (200) can live in very hypoosmotic media (tap water), but can also regulate their hemolymph below ambient in very saline water. In the former, their problem is that of any hyperregulator, and the excretory system plays a similar role in excreting a copious, dilute urine; water is excreted, ions conserved. The Malpighian tubules function as described earlier. The important change is that ions, but not water, are reabsorbed in the gut. When these larvae are hyporegulating they face osmotic desiccation, and they respond like a teleost fish; they drink the medium to replace water lost across the body surface, and absorb the fluid from the gut. But in this case the excess ions ingested are excreted in a concentrated, fluid urine. Water conservation is not a problem, since it is abundant as long as the ions can be excreted. The basic operation of the Malpighian tubules is unchanged.

What is new in this situation is the presence of ion transport systems capable of secreting excess ions *into the rectum*. It is notable that when the ambient water contains high $MgSO_4$, both of these ions are also secreted into the urine, so their hemolymph concentrations remain low. This is also reminiscent of marine fish.

In overview, the Malpighian tubule is roughly equivalent to the filtration locus in filtration kidneys (but the process is *not* filtration, and the tubular urine in virtually all cases does not have the composition of a filtrate). It provides a primary fluid that is suitably modified by more posterior regions of the system. To pursue the analogy, the rectum plays the same role as the tubular regions of a filtration kidney, modifying the primary fluid in a manner that is adaptive for a given environment and pattern of behavior.

3. Water Uptake

As before, when account is taken of all adaptations minimizing water loss, there remains the problem of how residual loss is balanced. Liquid water is ingested by many terrestrial arthropods. Standing water is an obvious source, but one that may not be available, for example, to animals living in the desert or in stored products. Some desert insects collect droplets of moisture deposited as dew or by advective fog. Food is another water source. Animals that ingest blood, plant sap, or succulent leaves occupy a moist microhabitat, even when the surroundings are arid. Even much drier food, the meal or bran in which *Tenebrio* develops, for instance, may contain substantial preformed water.

Another source is water formed in the metabolism of foods. Oxidation of carbohydrates produces ~0.6 mg of H_2O/mg of sugar or 0.7 mg of H_2O/ml of O_2 used. For fats the values are 1.1 mg of H_2O/mg of fat and 0.5 mg of H_2O/ml of O_2. At 25°C a fed locust consumes ~0.35 ml of O_2/g h^{-1} using about half-carbohydrate, half-fat. This produces ~0.2 mg of H_2O/ g · h. The transpiration rate at 70% RH is ~10 mg/g · h, so the contribution of metabolic water is small. Similar results have been obtained in other insects. However, animals like desert beetles and stored-product insects have permeabilities an order of magnitude lower than *Locusta* but similar metabolic rates, and the contribution of metabolic water may be crucial for them. It is also important in flying insects. An aphid, *Aphis fabae* (42) weighing 770 µg, lost 65 µg during a 6-h flight at 25°C. About 26 µg represented metabolized fat and 39 µg was H_2O. Since oxidation of the fat produced 28 µg of H_2O, it reduced total water loss by >40%. Locusts (*Locusta* and *Schistocerca*) (312) can do even better, remaining in water balance while flying at moderate temperatures and humidities.

These modes of water gain might be seen in any animal group, but a fourth is, as far as we know, found only among arachnids (mites and ticks) and insects. It was first noted that larval *Tenebrio* could gain weight if the RH were high (>88%),

even in the absence of food (36). This could only occur if the animals were absorbing water from the unsaturated air. Since then net water absorption from unsaturated air has been noted in several dozen arthropod species; see Table 33 in Edney (62). Most of the individuals with this ability are wingless; thus, wingless *Arenivaga* nymphs and adult females absorb from air at RH >82%, winged adult males cannot. They fall into two behavioral categories. "Gorgers and fasters" feed on mammalian blood but fall off the host after feeding and may spend a long time waiting for another host to appear (e.g., ticks). Others simply spend long periods in a dry environment. For example, stored product insects may spend their entire developmental period without access to adequate water. For all of these animals there is a lower limit to the RH from which they can absorb. This critical equilibrium humidity (CEH) varies among species. For many it is >85%, but is much lower in a few; the firebrat *Thermobia* can absorb water down to RH 45%.

The ability to absorb water from vapor is obviously adaptive. It is also impressive. The chemical potential of water at 88% RH is equal to that in a 7.6 osM solution. Net absorption of water is equivalent to moving it from such a solution into one at 0.6 osM (the concentration of many arthropod hemolymphs), that is, against a 12-fold osmotic gradient. A solution equivalent to vapor at RH 45% would have more solute than H_2O molecules and *concentration* is probably a meaningless term (but we can calculate that it would be 67.9 osM!). In *Tenebrio* the rectum is the site of absorption. We have already seen that the cryptonephridial complex can absorb water from the rectum until the feces are in equilibrium with RH 88%. It remains only to ventilate the rectal chamber through the anus to exploit this for absorbing water from the environment. *Thermobia* also absorbs vapor in the rectum (192, 193), although the structures involved (rectal sacs) bear no resemblance to the cryptonephridial arrangement. In fleas, ticks, and the cockroach *Arenivaga* blocking the anus does not stop water absorption, but blocking the mouth does (144). Glands in the oral region produce a material (still uncharacterized) that is very hygroscopic (315). When dried it is solid, but if exposed to water vapor it becomes moist and finally dissolves even though liquid H_2O is absent. An animal produces this material, exposes it to air (at the appropriate RH) and then swallows (drinks?) the resulting solution. How the water is then freed for absorption into the hemolymph is unknown. It is reasonable to suppose that such a material is involved in anal absorption in *Thermobia* and perhaps in *Tenebrio*, but this has not been demonstrated.

Fully Terrestrial Animals: The Vertebrates

1. Transpiration across the Body Surface
Reptiles, birds, and mammals have representatives in terrestrial habitats ranging from hygric to extremely arid, and a few amphibian species have also adapted to cope with water loss. The range of adaptations is similar to those seen in arthropods. Table 16 shows rates of pulmocutaneous water loss in a number of vertebrates, most of them adapted to xeric conditions. The tree frogs are arboreal. A hylid has a lower water permeability than most nonarboreal anurans (by about a factor of two), but still loses water rapidly when exposed to dehydrating conditions. In the other two, *Phyllomedusa pailoma* and *Chiromantis petersi*, permeability is reduced by an order of magnitude (60, 266). The low water permeability in *Phyllomedusa* is due to a layer of wax secreted onto the skin, but this is not the case in *Chiromantis* where the mechanism

is unknown. Nearly all amphibians can be included in the semiterrestrial group, at least in terms of their physiological machinery for coping with dehydration. But these two genera clearly have undergone adaptive changes (another will be described later) that characterize fully terrestrial animals.

Reptiles were the first vertebrate group entirely adapted to life on land, and while some species have secondarily become aquatic (or amphibious), most are still tied to land, at least to lay eggs. Skin water permeability is generally low. Resistance to cutaneous water transfer is greater in species normally exposed to arid environments, but even an aquatic reptile, like the snake *Natrix sipedon*, loses water much less rapidly in air than a typical amphibian (Table 16). The desert animals shown (a turtle, two lizards, and a snake) have the lowest permeabilities; they are in the same range as those of arid living arthropods. Total transpiration has been partitioned in a few reptiles [Table XII in Minnich (186)], and cutaneous is larger than pulmonary in most (50–80%). But the latter is not a negligible fraction as it generally is in amphibians, and it may increase with ventilation rate during activity.

In birds and mammals evaporative water loss is smaller than in semiterrestrial species but considerably larger than in many terrestrial reptiles (Table 16). A new variable, high-temperature hemeothermy, complicates the equation for water loss. Body temperature in most resting birds and mammals is maintained constant somewhere in the range 37–42°C, and this impacts transpiration in at least three ways. First, even at low to moderate environmental temperatures (e.g., <35°C) the vapor pressure deficit between a homeotherm and the environment is larger than it would be for a poikilotherm at thermal equilibrium with the same environment. The corollary is

TABLE 16. Evaporative Water Loss in Terrestrial Vertebrates

Species	Habitat	T	RH%	Wt (g)	J_{evap}[a]	P_{osm}[b]	Reference
			Amphibians				
Hyla cinerea	Xeric	25	38	2.3	49.8	1340	310
Phyllomedusa pailoma	Xeric	26	0	12.7	0.568	14	266
Chiromantis petersi	Xeric	26	0	30	1.10	37	60
			Reptiles				
Natrix sipedon	Hygric	27	0–10	100	2.62	138	18
Terrepene carolina	Mesic	23	0–10	305	0.44	41	20
Sauromalus obesus	Xeric	23	0–10	134	0.14	10	20
Dipsosaurus dorsalis	Xeric	39	56	50	0.23	8	184
Crotalus atrox	Xeric	25	0–10	123	0.18	11	40
			Birds				
Gallus gallus	Mesic	20		3100	0.92	368	55
Poephila guttata	Xeric	25	0	11.5	8.58	229	37
Stellula caliope	Mesic	23–25	0	3	14.0	261	155
Melopsittacus undulatus	Xeric	25	0	35	2.08	80	91
Phalaenoptilus nutallii	Xeric	25	0	40	2.50	101	12
Lophortyx gambelii	Xeric	24	11–23	149	1.33	104	183
			Mammals (Rodents)				
Spermophilus beecheyi	Mesic	28	40	400–600	0.8	106	13
S. lateralis	Xeric	29	0	164	1.6	83	324
Dipodomys merriami	Xeric	26–27	0	30–37	1.5	54	40
Rattus rattus	Mesic	27	0	132	3.7	198	43

[a]Units are in mg/g · h.
[b]Units are in 10^4 cm/s.

that pulmocutaneous water loss will be larger in the former, other factors being equal. Second, at moderate temperatures the higher metabolic rate in homeotherms requires more gas flow over the respiratory epithelium than in heterotherms. The potential for rapid pulmonary water loss is apparent. In addition, when ambient temperature is high, both birds and mammals use evaporative cooling to maintain body temperature within tolerable limits. Sweating (cutaneous evaporation) is the dominant mode in large mammals, panting (pulmonary evaporation) in smaller mammals and in birds. In either case, evaporative water loss in both groups increases many times when the environment is hot (>30°C) and dry.

Reptiles, birds, and mammals employ a range of strategies to modify the parameters controlling evaporative water loss. The catalog is too extensive to be described fully, but some examples follow:

1. Cutaneous water loss usually accounts for 50–80% of the total in resting animals, but it is variable. In all three groups it is lower in species adapted to arid habitats than in those from more mesic conditions. In some birds [Table III

in Willoughby and Peaker (320)] and mammals (97) the same animal acclimated to an arid environment (or previously dehydrated) loses water less rapidly than when there is ready access to water. The mechanism underlying such a change is unknown. It may involve a change in skin structure or in cutaneous blood flow.

2. Pulmonary evaporation accounts for less than one half of the total transpiration in resting animals under moderate conditions. Oxygen extraction is greater in the lungs of at least some birds than in mammals of comparable size. This permits lower ventilation rates and less evaporative loss during respiration. Increased oxygen extraction, allowing reduced ventilation volumes, has also been reported for snakes and lizards.

In desert rodents and birds incoming dry air is warmed and humidified in the nasal passages, so that its RH is high on arrival at the gas exchange surface. At the same time, the nasal epithelium is cooled. During exhalation the air leaves the exchange surface at body temperature and saturated with water, but on passing the cooler nasal surface loses water (and heat) by condensation to the cooler epithelium. As much as 75% of the potential water loss in respiration is averted in this manner [discussed in Schmidt-Nielsen et al. (253, 254)]. This may also take place in reptiles.

3. In both birds and mammals evaporative cooling is a common feature at high environmental temperatures where it becomes the dominant mode of water loss. If body temperature must be maintained nearly constant, then evaporative cooling will become important when ambient temperature approaches that of the body. In contrast, if body temperature is allowed to rise with ambient, evaporative cooling can be *postponed* or minimized. Such an adaptation has been described in the hydrated dromedary camel. In a resting state, hydrated camel body temperature fluctuates daily between ~37–39°C. But when the animal is dehydrated the fluctuation is from 35–41°C. As the air heats up during the day the animal's body temperature is allowed to rise rather than being maintained within a narrow range. Heat is stored instead of being dissipated at substantial cost to the animal's water economy. A similar phenomenon is seen in other large desert mammals and in some birds that can tolerate hyperthermia to 46°C. Such heat storage is obviously adaptive in hot, arid regions.

4. Behavioral adaptations are also aimed at reducing transpiration; these are not different in principle from those seen in semiterrestrial species. Many reptiles and smaller mammals live in burrows where temperature is lower and RH higher than ambient. Especially in deserts, they may forage at night when temperature is lower and the chance of finding water (e.g., from condensation) higher. Shape changes (coiling in snakes) reduce the surface from which water evaporates. While birds do not burrow, some may, in hot environments, fly higher where the air is cooler. They can also manipulate their plumage in order to increase conductive and convective cooling.

2. Urinary Water Loss

We have seen that insects facing desiccation conserve water by excreting a urine hyperosmotic to the hemolymph. This enables them to extract water from food (or a hyperosmotic aquatic medium) without becoming solute loaded. Among vertebrates, only birds and mammals have the ability to produce hyperosmotic urine, and the mechanism, based on a filtration kidney, is totally different than in insects. However, even in amphibia and reptiles renal function is modulated to minimize water loss when facing dehydration.

The reptilian nephron is similar in structure to the one found in fish and amphibia; fluid is filtered out of the glomerular capillaries and enters, first a proximal tubular segment, then a distal tubule, collecting duct and ureter. The last empties into the cloaca (hindgut). Terrestrial reptiles have low filtration rates (J_v^f) in comparison to their aquatic relatives or to terrestrial birds and mammals. A few values in Table 17 illustrate how J_v^f can be adapted to deal with desiccation. First, filtration rates are lower in species inhabiting dry environments. In addition, when an individual becomes dehydrated, filtration is markedly reduced. The ureteral urine is usually somewhat hypoosmotic in normally hydrated animals, but tubular water reabsorption results in an isoosmotic urine (osmolar U/P = 1) when they are dehydrated. Reptiles, like amphibians, cannot produce urine more concentrated than the blood. Both low filtration rate and tubular water reabsorption contribute to reduced urinary water loss.

Most reptiles, like insects, excrete a major fraction of their urinary nitrogen as uric acid or urate salts. Turtles are an exception; they excrete a mixture of urea and urate (and sometimes NH_3, as also seen in the crocodilia). But even among turtles, individuals adapted to land usually produce substantial urate. This has

TABLE 17. Renal Function in Terrestrial Vertebrates

Species	Habitat	Hydration State	$J_v^{f\ a}$	$J_v^{u\ a}$	$(U/P)_{osm}$	References
		Reptiles				
Hemidactylus sp.	Mesic	N	10.4	2.7	0.64	236
		D	3.3	1.3	0.74	
Phrynosoma cornutum	Xeric	N	3.5	2.0	0.93	236
		D	2.1	0.8	0.97	
Tropidurus sp.	Xeric	N	3.6	1.9	0.96	236
		D	1.2	0.6	0.97	
		Birds				
Gallus gallus	Mesic	N	127	17.9	0.4	149,270
		D	104	1.1	1.6	
Melopsiticus undulatus	Xeric	N	266	6.3	0.7	149
		D	194	1.7	2.3	
		Mammals				
Camelus dromedarius	Xeric	N	54	0.17	5	176
		D	37	0.07	8	
Homo sapiens	Mesic	N	106	0.58	1.9	24
		D	93	0.40	3.4	
Notomys alexis	Xeric	N		2.14	1.2	169
		D		0.02	18.2	
Notomys cervinus	Xeric	N		3.20	1.1	169
		D		0.12	9.3	
Chinchilla laniger	Xeric	D			10.7	311

$^a J$ in ml/k · h.

two consequences. First, there is a powerful active transport mechanism in the proximal tubule, which secretes urate into the tubular fluid; urate clearance is much higher than the filtration rate (inulin clearance). This means that even a carnivore with a high rate of nitrogen metabolism has no trouble excreting waste nitrogen at low filtration rates. In addition, urates exist as a colloidal suspension or moist precipitate, hence they contribute nothing to the osmotic pressure of the urine. As a result, little water is obligated in excreting nitrogen. This is most important in animals that cannot produce a concentrated urine, for the nitrogen can be excreted in a small volume. In carnivorous animals facing water deprivation the saving can be significant.

Table 17 shows that urine flow also tends to be low in birds and mammals, and especially that it is reduced, in some cases substantially, in dehydrated animals. But there are some notable differences from values in reptiles. Glomerular filtration rates are much higher, and a much larger proportion of the filtrate is reabsorbed. Since reabsorption is energetically costly, these are expensive kidneys to operate. In addition, these filtration kidneys *can* produce a hyperosmotic urine. Birds can produce urines with total solute concentrations about twice that of the plasma (the dehydrated chicken and a budgerigar in Table 17 are typical; see also 271), and the salt-marsh sparrow, *Passerculus sandwichensis*, has an osmolar U/P = 4.5. Among mammals, U/P < 1 is seen only in water-loaded animals; concentrated urine is the rule, and the degree of concentration substantially exceeds that seen in birds. Even the few examples in the table show that osmolar U/P increases in dehydration. Moreover, there is an excellent correlation between available water and the ability to concentrate the urine. The maximal osmolar U/P in the mountain beaver, *Aplodontia*

rufa, is ~2. It is ~5 in man and other mesic-living mammals, but may exceed 20 in some desert rodents. This means that a given solute load can be excreted in minimum water even if all urinary solutes are osmotically active. Urine flow is so low in dehydrated desert rodents that collecting enough to analyze is a challenging technical problem. The mechanism is unique to birds and mammals and will be described later.

Urates are the major form in which nitrogen is excreted by most birds. Since they cannot produce a very concentrated urine, the advantage in using an insoluble product is the same as in reptiles (and insects); little water is osmotically obligated. Mammals produce relatively little urate and package most of their nitrogen in urea (some ammonia) for excretion. Urea is nontoxic, but it is soluble, and hence osmotically active. Moreover, the urea molecule contains only two N atoms (four in uric acid), so twice as many molecules are required to eliminate a given nitrogen load. The ability to concentrate the urine allows mammals to excrete osmotically active solute loads in a minimum volume of urine. Conditions that produce dehydration (in the field) are often those which also increase evaporative water loss in maintaining body temperature constant. It is obviously advantageous to reduce urinary water loss as much as possible.

3. The Renal Countercurrent System
The mammalian nephron has proximal and distal segments homologous with those in other vertebrates and simlar in plan to those in FW invertebrates. These lie in the superficial region of the kidney, the cortex. But intercalated between them is a tubular loop that dips into the medulla, the Loop of Henle. In some nephrons the loop reaches the tip of the papilla (long loop); in others it may be confined to the outer zone of the medulla (short

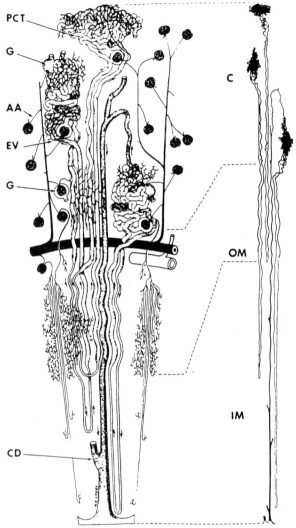

Figure 37. The mammalian nephron. Diagram of renal vascular and tubular organization. Only three nephrons are shown (a human kidney contains ~1 million nephrons, a rat kidney ~50,000), and the vascular structures have been extensively simplified and the vertical scale compressed. (To the right, the same nephrons are shown in their true proportions.) Major zones include cortex (C), outer medulla (OM), and inner medulla (IM). Afferent arterioles (AA), glomeruli (G), and efferent vessels (EV) are shown together with parts of the peritubular capillary network. Outer medullary vascular bundles arise from the efferent vessels of inner cortical glomeruli. From the glomeruli, fluid flows first into proximal convoluted tubules (PCT) and enters the descending parts of Henle loops. In the outer medulla the thin descending parts of Henle loops begin. Thin ascending limbs begin at the loop turning points in the inner medulla. Thick ascending limbs begin at the border between outer and inner medulla and continue upward into the cortex. In the outer medulla, descending thin limbs of short loops are close to the vascular bundles, whereas thin limbs of long loops are generally located together with collecting ducts and thick ascending limbs in the region between vascular bundles. Distal convoluted tubules (dark hatching) lie among proximal convolutions. Connecting segments join distal convolutions to collecting ducts (CD), which descend through the cortex and medullary regions to open at the papillary tip. [Reproduced from Beeukes (17).]

loop). The descending limb of the loop segment is fairly uniform throughout its length, thin walled, and with a narrow diameter. The initial segment of the ascending limb is similar in appearance (thin ascending limb), but the segment traversing the outer medullary zone is larger in caliber, with cells that appear to be metabolically active (thick ascending limb). The macroscopic anatomy is shown in Figure 37.

The roles of proximal and distal tubules are the same as in other terrestrial vertebrates. But it was clear many years ago that the ability to produce a concentrated urine, unique to birds and mammals among animals with filtration kidneys, was somehow related to the presence of the Loop of Henle. The picture to about 1950 has been summarized (279). Little was known about the Loop's role, but comparative studies before and since that time have reinforced the correlation. For example:

1. Insects excepted, only birds and mammals produce a concentrated urine and the loop segment is found only in those two groups.

2. Most of the nephrons in bird kidneys are of the lower vertebrate (reptilian) type without a loop segment. A small fraction has loops, but these are short compared to mammals. Correspondingly, the ability to concentrate urine is only modest. The U/P osmolarity in birds is on average ~2 and the maximum ~4.

3. Among mammals a correlation was noted earlier between availability of habitat water and renal concentrating ability. In the same animals there is good correspondence between loop length and concentrat-

ing ability. The beaver (maximum U/P osmolarity ~2) has only short loops. Desert rodents (maximum U/P ~25) have only long loops that extend into the renal papilla (251). This was a signal observation from a laboratory that has contributed much to comparative renal physiology (251a). Most mammals are intermediate in concentrating ability and have mixtures of short- and long-looped nephrons.

Two other general features have to be incorporated into any detailed mechanism. First, the production of a concentrated urine requires either that water be transported out of the urine against an osmotic gradient, in this way concentrating solutes, or that solutes (largely urea) be transported into the forming urine against a concentration gradient. Either process is energy requiring, and a model will require at least one process linked to metabolism. In addition, there are indications that urea plays a role in the process of concentrating urine in mammals as well as being the major urinary solute (158).

Detailed information about the concentrating mechanism can be found in several excellent reviews (1, 17, 130). Microsamples from both proximal and distal tubules show that cortical tubular fluids are isosmotic with the general body fluids (~300 mosM). This result shows that the urine finally becomes concentrated in the collecting duct as it courses through outer and inner medullary zones to the renal papilla, an inference that has been amply confirmed experimentally (110). A volume marker like inulin also becomes concentrated showing that the process involves water removal rather than solute secretion. It would appear that water is reabsorbed against an osmotic gradient that becomes steeper through the medulla. We have seen that water is absorbed against an osmotic gradient in the marine teleost gut, but that this involves a clever arrangement (Fig. 29) for creating a local region within the organ where the osmotic gradient favors passive water flow. If we reason by analogy, the fluid at any point in the collecting duct should face a medullary interstitial fluid that is at least slightly hyperosmotic. At this point water should flow out of the tubule into the interstitium as long as the collecting duct wall is permeable to water and not to solutes. But since the tubular fluid becomes more and more concentrated through the medulla, such a model requires that the same be true of the in-

terstitial fluid. That is, there must be a vertical concentration gradient within the medulla, from isosmotic at the cortical–medullary border to a value approximating that of the final urine at the tip of the papilla. In fact, such a gradient was postulated together with a generalized mechanism for producing it with a tubular loop (99). Experimental confirmation was not long in coming (322, 323) as shown in Figure 38. Panel *B* of the figure shows that NaCl is the main solute through the outer medullary zone while urea comprises one half of the total solute in the inner zone. Thus, the mammalian kidney is another organ in which apparent active water

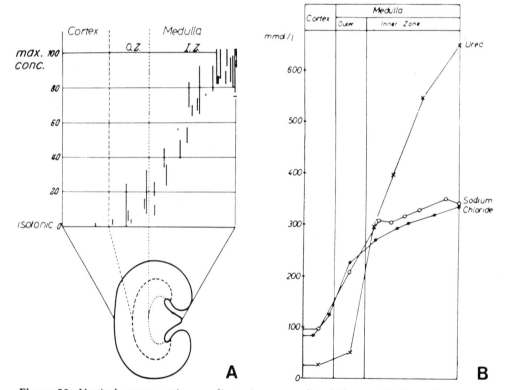

Figure 38. Vertical concentration gradients in mammalian kidney. (*A*) Osmotic concentration in tissue slices was estimated from measurement of ΔT_f. The total solute concentration begins to rise in the outer medulla, but most of the increase occurs in the inner zone (323). (*B*) Sodium chloride and urea concentrations account for the vertical gradient in total solute. Most of the change in the outer zone is due to an increase in NaCl concentration. Urea concentration rises steeply through the medullary inner zone. [Reproduced from (301) with permission of W. B. Saunders.]

TABLE 18. Permeability and Transporting Properties of the Nephron

Tubular Segment	Active NaCl Transport	Permeability to		
		H_2O	NaCl	Urea
Descending thin limb	0	+ +	0	+
Ascending thin limb	0	0	+ +	+
Ascending thick limb	+ +	0	0	0
Distal tubule and outer medullary collecting duct	+	+ +	0	0
Inner medullary collecting duct	+	+ +	0	+ +

transport finds water moving passively with an osmotic gradient formed by an unusual distribution of solute.

How, then, is the medullary osmotic gradient set up. The permeability and transport properties of the system, loop plus collecting duct, differ remarkably from one region to another. In order to measure these properties it was necessary to isolate tiny segments of tubule and to perfuse them with an artificial tubular fluid while they were bathed in an artificial ECF. Developing the necessary techniques was a signal accomplishment (86, 146). Table 18 summarizes the characteristics in different regions of the system in the rabbit. The behavior (i.e., the model) inferred from such information follows and is illustrated in Figure 39.

1. A powerful active transport system transfers NaCl from the lumen of the thick ascending limb of Henle's Loop into the outer medullary interstitium. This, of course, is energy requiring and is the main *motor* for the whole operation (some energy is also expended in transporting NaCl out of the collecting duct).

2. The resulting rise in interstitial solute concentration causes water to move out of the descending thin limb, so the tubular fluid becomes more concentrated.

3. These events, repeated along the length of the thick ascending limb, cause tubular fluid and interstitium to become more concentrated vertically through the outer medullary zone. Note (Figure 38*B* and 39) that the main solute in both the descending tubule and interstitium is NaCl; urea concentrations in this region are relatively low by comparison.

4. In the inner medulla total solute continues to rise but for a different reason. There is no NaCl transport out of the thin ascending limb and NaCl increases only a little in this region. But there is a marked rise in interstitial urea (301). The basis for the modest increase in NaCl is explained in point (6) and for urea in point (10).

5. Osmotic equilibration across the water permeable wall of the descending thin limb ensures that tubular fluid in this lowest segment of the loop continues to become more concentrated, to reach a maximum at the hairpin tip. The main *tubular* solute is still NaCl.

6. The tubular fluid now enters the thin ascending limb. It is at osmotic equilibrium with the interstitium but has a much higher NaCl concentration, because one half of the interstitial solute in this region is urea. Since this tubular segment is ion permeable, NaCl diffuses out accounting for the small rise in interstitial NaCl through the inner medullary interstitium. At the same time, the tubular fluid becomes more dilute as it rises.

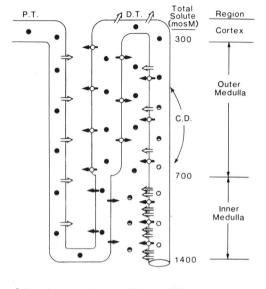

Solute Concentrations
- ● Predominantly NaCl
- ○ Predominantly Urea
- ◑ Mixed NaCl + Urea

Solute and Water Movements
Passive H₂O ⟹
Passive NaCl ➡
Passive Urea ⇒
Active NaCl ⟶⬤

Figure 39. Fluxes and permeabilities in the mammalian countercurrent system. Operation of the countercurrent loop involves three fluxes (NaCl, urea, and H₂O), as well as the corresponding permeabilities and concentrations. All of these change from one region to another and each is crucial in generating a concentrated urine. An attempt to represent them in a single diagram results in a complicated, cluttered figure like the one shown. The mechanism is best worked out by using the figure together with information in Table 18 and in the text, preferably one step at a time.

7. As the ascending fluid enters the thick limb more NaCl is lost to the interstitium, this time through the transport system mentioned in point (1). Since this segment is impermeable to H₂O and urea it becomes slightly, but distinctly, hypoosmotic to the surrounding interstitium.

8. The hypoosmotic fluid then enters the distal tubule. There is now an osmotic gradient between tubular fluid and cortical interstitial fluid, and water is reab-

sorbed across the tubular wall. By the time the fluid leaves the distal tubule it is at osmotic equilibrium with the cortical interstitium (~300 mosM), and its volume is substantially reduced.

9. On entering the cortical collecting duct the major tubular solute is NaCl although urea concentration is elevated. As it passes down through the outer medullary zone, water is continually reabsorbed by osmosis into the increasingly concentrated interstitium. In addition, NaCl is reabsorbed by active transport. This tubular segment is impermeable to urea with the result that tubular urea concentration rises to become the dominant solute.

10. These processes continue to concentrate the tubular urine (and urea) through the inner medullary zone. But this part of the collecting duct is urea permeable, and some urea diffuses out into the interstitium where it comprises a major fraction of the total solute.

11. The final urine exits the collecting duct. In antidiuresis it has a concentration 5–6 times that of the body fluids (in man), and the main solute is urea.

One remaining detail is that blood flow through the renal medulla requires a special countercurrent arrangement of capillaries to avoid washing out the vertical osmotic gradient. K. Schmidt-Nielsen gives a clear description of the role of countercurrent systems in minimizing dissipative losses (253).

It must also be emphasized that the model described is an inference based on properties of the rabbit nephron. Some differences in detail have been described in other species. Of particular interest to us is that the descending thin limb is reported to be quite ion permeable in *Psammomys*, one of the superstars in concentrating urine (54, 131). This will require a different mechanism to account for the final stage of concentration in the inner medulla.

Finally, the mammalian kidney is as flexible as those described earlier, adjusting both volume and concentration over a wide range to cope with varying intake of water and solutes. The description above refers to urine production when water must be conserved. But when water intake is excessive the final urine is copious and dilute rather than scanty and concentrated. A key variable is the H_2O permeability (P_{osm}) of the distal tubule and collecting duct. The P_{osm} of these epithelia is variable and controlled by a neurohypophyseal hormone, the antidiuretic hormone (ADH). When the hormone is present the walls are water permeable. In its absence they are impermeable; water absorption from the distal tubule and collecting duct, shown in Figure 39, does not occur, and a large volume of dilute urine is excreted.

4. *Water Ingestion and Salt Glands*

Nearly all terrestrial vertebrates drink when standing water is available and in this way maintain themselves in water balance. By employing some of the strategies for reducing water loss, animals can withstand dehydrating conditions, at least for an extended period of time. This situation in camels and donkeys, so ably explored and described by K. Schmidt-Nielsen and colleagues, is in some ways the simplest. During water deprivation urine flow is reduced as are evaporative losses, but equally important is the fact that both animals can tolerate water loss that would be lethal to most animals (nearly 30% of the body weight; almost one half of the total body water). The ability to withstand dehydration has been described in many desert animals. It does not eliminate the need for standing water but extends the time available for finding it when it is scarce. A number of desert

reptiles, birds and mammals have reduced dissipative losses so much that they can subsist on the preformed water in their food plus that produced in metabolism. The budgerigar (30 g), for example, consumes 4.9 g of air-dried seeds/day. This produces 2.8 ml of H_2O, which is enough to balance an equivalent daily loss.

In some habitats water is present, but in a form that requires the intake of very large quantities of salt. Food, described above as a water source, is not an unalloyed blessing, because it introduces a solute load that must be disposed of. The ability of desert rodents to excrete urine with osmolal U/P = 10–20 enables them to excrete salt loads even though the ions are in solution, and some of them can subsist without drinking or by drinking saline. The kidney also may play an important role in birds. The seeds eaten by a budgerigar contain little Na^+ and Cl^-, but a substantial amount of K^+. Ingestion of such a diet is equivalent to drinking a 200 mM K^+ solution. Free water becomes available because the ingested K^+ is excreted in the urine. Trapping of K^+ in the insoluble urates permits the bird's kidney to handle this load, even though it produces only a moderately concentrated urine.

But the ability of bird and reptile kidneys to excrete solute loads is limited, and in both groups the limitation would not allow the use of water or food sources with large solute content. Instead some members of both groups use a *salt gland* to excrete ions at very high concentration, and hence with minimum water loss. Some desert reptiles can maintain themselves in water balance by ingesting plants with little or no standing water to drink. The desert iguana, *Dipsosaurus dorsalis*, eats plant food containing ~50% water, but the ionic content of the food is high; expressed in terms of concentration in the ingested water K^+ = 364 mM,

Na^+ = 34 mM, and Cl^- = 105 mM. The total ionic concentration exceeds that in the body fluids and is very unbalanced (K^+/Na^+ ~ 11 versus 0.03 in plasma). A balance sheet for the iguana is shown in Table 19. The animal is in water balance but at the expense of a large intake of ions. About one half of the ingested Na^+ and K^+, but little Cl^- is excreted through the cloaca (the apparent anion deficit comprises urate), but the other one half of the ingested cations is excreted through a nasal salt gland together with most of the ingested Cl^- (there is also an apparent anion deficit in this secretion). It is worth noting that the concentration in this fluid may be well over 3 os$M$ or nearly 10 times the body fluid concentration. Salt glands have been described in many terrestrial reptiles, and in a number of marine species. The organ appears to be able to adjust the composition of the secretion to the nature of the ingested solutes; K^+/Na^+ may be either > or <1.0.

Salt glands are found in many marine and estuarine birds, in some falconiforms, and in other species living in arid regions. The paired glands open into the nasal cavity and secrete, in all reported cases, a concentrated NaCl solution with a much lower $[K^+]$. The size of the glands varies with the size of the bird. More important, the concentrating ability of the gland reflects the salt load imposed by habitat and food; the highest $[Na^+]$ has been measured in the petrel, *Oceanodroma leucorrhoa* which is pelagic and eats only marine invertebrates. Secretion is intermittent; the stimulus appears to be a rise in plasma osmotic pressure, and initiation of flow occurs within minutes after ingestion of a salt load. Measurement of the excretion of a salt load in gulls and ducks show that the gland eliminates much (50–90%) of the NaCl. The rest leaves through the cloaca together with most of the ingested K^+.

Summary of Terrestrial Osmotic Regulation

The terrestrial environment is much more complex than either FW or SW. The number of variables impacting on osmotic regulation and their ranges are beyond anything experienced by animals in an aquatic milieu. Accordingly, the number and complexity of physiological and behavioral adaptations is very large; the foregoing description is by no means a complete catalog. But the adaptive responses are variations on a few basic themes, and a brief summary of the themes might be useful.

1. The driving force for water loss in terrestrial animals is usually much greater than for any aquatic animal. Transpiration may occur from either the general body surface or from the respiratory surfaces.

2. Cutaneous water loss is much lower in fully adapted terrestrial animals than in semiterrestrial species. Among the fully adapted animals water permeability

TABLE 19. Water and Ion Balance in *Dipsosaurus dorsalis*[a]

	H_2O[b]	Na^+[c]	K^+[c]	Cl^-[c]
Gain				
Food	2.69	92	974	280
Metabolic	0.36			
Total	3.05	92	974	280
Loss				
Transpiration	0.86			
Feces	1.86	28	169	17
Urine	0.08	19	384	4
Salt gland	0.25	45	421	259
Total	3.05	92	974	280

[a]Units are in ml/100 g · day.
[b]Units are in μeq/100 g · day.
[c]Data from Minnich and Shoemaker (185, 187).

is lower in those adapted to more arid environments.

3. Respiratory water loss is only a small proportion of the total in semiter-restrials, but it may be as large or larger than cutaneous in the better adapted groups.

4. In birds and mammals, but not in-sects or reptiles, evaporative water loss is involved in dissipating heat loads. The demands of thermal regulation and water balance compete when ambient temper-ature is high, and they cannot be com-pletely reconciled.

5. Behaviors (e.g., burrowing and noc-turnal activity) are at least as important as physiological adaptations in reducing pulmocutaneous water loss.

6. The excretory system is another route of water loss in nearly all animals under any set of conditions. Several strat-egies are employed by terrestrial animals to reduce this loss. Insects, reptiles, and birds excrete nitrogen as uric acid, which is insoluble, hence osmotically inactive. Many insects can also concentrate soluble ions to levels well above those in hemo-lymph. Among vertebrates, birds concen-trate the urine a little, but the ability to produce concentrated, fluid urine is best developed in mammals. Generally, arid-living species produce the most concen-trated excreta (i.e., lose the least water per unit solute excreted).

7. Finally, water lost by any of these routes must be replaced. Most animals will drink when FW is available. Some have reduced losses sufficiently to subsist on preformed water in their food plus water produced when the food is metab-olized. Some reptiles and birds have de-veloped special salt-excreting glands that enable them to excrete salt loads associ-ated with water intake (food or drinking saline water). A few insects can absorb water from unsaturated air, but no ver-tebrates have this ability.

Reinvasion of Water: Aquatic Mammals

Three orders of mammals moved from land back into water. Pinnipeds (seals, walruses, and sea lions) spend much of their time in the ocean but bear and nurse their young on land. Cetaceans (whales and dolphins) and Sirens (manatees and sea cows) are entirely aquatic. With the exception of a few cetacean species the last two groups are marine, and our ac-count will deal with marine cetaceans, be-cause they have received most of the little attention paid these animals.

Their blood chemistry is typically mammalian (231) with a total plasma os-motic concentration ~330 mM and with Na$^+$ and Cl$^-$ the major solutes. They ob-viously face the same desiccating envi-ronment as marine teleosts. Until a few years ago it was assumed that cetaceans do not drink seawater, and that the body surface is impermeable to both ions and water. There is no convincing evidence for a salt gland, which sets this group apart from others, like pelagic birds, that share the same environment. For dol-phins and whales osmotic balance seemed to require simply that water from their food (preformed and oxidative) match the sum of that lost in respiration, through the urine and with the feces.

Several adaptations function to mini-mize dissipative flows. Respiratory rate is lower than in most mammals, oxygen ex-traction greater (129), and the expired air may not be saturated with water (48). A number of early measurements [sum-marized by Krogh (151)] showed that urine might be either hypo- or hyperos-motic. Freezing points depression up to 4°C (>2000 mos$M$) were reported in seals, and [Cl$^-$] in whales varied between 75 and 820 mM. The kidney appeared to be able to produce urine more than twice as concentrated as SW. No water is lost in thermoregulation, and an impermeable skin would be a major advantage (but see

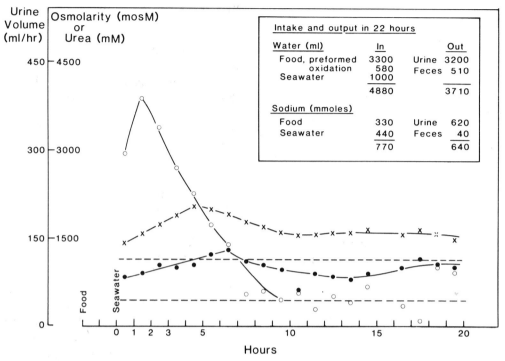

Figure 40. Excretion of solutes by the dolphin after feeding. A female *T. truncatus* (101 kg) was fed 4.7 kg of mackerel and 2 h later given 1 L SW by stomach tube (shown as hour 0). The first urine sample was taken 1 h after the SW, but the urine was forming throughout this period and analytical values are placed midway between 0 and 1 h. The urine was collected periodically and assayed for volume (O——O), osmolality (x——x) and urea (●——●) concentration. Values were also obtained for Na$^+$ and Cl$^-$ (not shown). The lower dashed line represents mean urine volume of a fasted animal, and the upper dashed line is the average osmolarity during fasting. [Drawn from data in Ridgway (231).]

below). Based on such a combination of assumptions and measurements Irving and collaborators calculated that the harbor seal, *Phoca vitulina*, could maintain water balance on a diet of herring by producing urine with $\Delta T_f \sim 2.7°C$ (128). Krogh extended the calculation (151) to the walrus and whale, which eat marine invertebrates and incur a large salt load per unit water ingested.

More recent research has challenged some of the older assumptions, and while the general conclusions of the earlier investigators must be correct (after all, seals do survive on a diet of fish, and whales

on invertebrates!), some details may require modification. When radioactive markers (e.g., ^{24}Na) were added to SW the isotopes appeared in the blood of dolphins (126, 294), although the skin was shown to be impermeable to Na$^+$ (126). These cetaceans appear to drink SW. In addition, the skin was shown to be permeable to water using ^{3}H$_2$O as a marker (126). If confirmed and extended, this means that one route of water loss was ignored in the older calculation. At the same time, if these animals can produce free water from SW, an additional input also exists.

A single study (231) provides interesting detail on the renal response to ingestion of food plus SW in the dolphin, *Tursiops truncatus*. A fasting animal excreted urine with an osmotic concentration of 1100–1200 mos$M$ at ~45 ml/h. About 80% of the solute was urea, and NaCl was nearly isoionic with plasma. A meal of fish was followed within an hour by a marked increase in flow rate that lasted for ~12 h with elevated clearances of urea, Na$^+$, K$^+$, and Cl$^-$. Total urine solute reached 2000 mos$M$. In another experiment, shown in Figure 40, a dolphin was fed 4.7 kg of fish and 2 h later was given 1 L of SW by stomach tube. Urine samples were taken hourly beginning 1 h after the SW infusion (3 h after the fish). Urine flow reached nearly 400 ml/h but dropped gradually and was near normal after the tenth sample. Urine osmotic concentration in this animal also reached 2000 mos$M$ and remained higher than in the fasting animal for the duration of the experiment. Urea contributed about one half of the urine solute, and [Na$^+$] and [Cl$^-$] were nearly double those in an animal given only fish. Rough input–output data for the 20-h experiment are shown as an inset in the figure. They show that the kidney can excrete the solutes in a mixture of food and SW and suggest that some free water might be left to balance water lost across the body surface.

In females, postpartum nursing engenders additional water loss. One adaptive response is to produce milk with less than one half of the water content of milk from terrestrial mammals and more than 10 times the fat and protein concentration (147, 316).

References

1. Abramow, M. In *Mechanisms of Osmoregulation in Animals* (R. Gilles, ed). pp. 349–412. Wiley, New York, 1979.
2. Adolph, E. F. *Biol. Bull.* (Woods Hole, Mass.) 62:112–125, 1932.
3. Alvarado, R. H. In *Comparative Physiology of Osmoregulation in Animals* (G. M. O. Maloiy, ed.). pp. 261–303. Academic Press, London, 1979.
4. Alvarado, R. H. and T. H. Dietz. *Comp. Biochem. Physiol.* 33:93–110, 1970.
5. Alvarado, R. H. and L. B. Kirschner. *Comp. Biochem. Physiol.* 10:55–67, 1963.
6. Alvarado, R. H., T. H. Dietz, and T. L. Mullen. *Am. J. Physiol.* 229:869–876, 1975.
7. Alvarado, R. H., A. M. Poole, and T. L. Mullen. *Am. J. Physiol.* 229:861–869, 1975.
8. Andreoli, T. E., J. F. Hoffman, and D. D. Fanestil. *Membrane Physiology*. Plenum Medical Book Co., New York, 1980.
9. Bahl, K. N. *Biol. Rev. Cambridge Philos. Soc.* 22:109–147, 1947.
10. Baldwin, G. and L. B. Kirschner. *Physiol. Zool.* 49:158–171, 1976.
10a. Ballard, B. S. and W. Abbott. *Comp. Biochem. Physiol.* 29:671–687, 1969
11. Barnes, H. *J. Exp. Biol.* 31:582–588, 1954.
12. Bartholomew, G. A., J. W. Hudson, and T. R. Howell. *Condor* 64:117–125, 1962.
13. Baudinette, R. V. *J. Comp. Physiol.* 81:57–72, 1972.
14. Beadle, L. C. *J. Exp. Biol.* 11:382–396, 1934.
15. Beament, J. W. L. *J. Exp. Biol.* 21:115–131, 1945
16. Beament, J. W. L. *J. Exp. Biol.* 32:514–538, 1955.
17. Beeukes, R., III. In *A Companion to Animal Physiology* (E. R. Taylor, K. Johansen, and L. Bolis, eds.). pp. 266–288. Cambridge Univ. Press, London and New York, 1982.
18. Bennett, A. L. and P. Licht. *Comp. Biochem. Physiol. A* 52A:213–215, 1975.
19. Benos, D. J. and R. D. Prusch. *Comp. Biochem. Physiol. A* 43A:165–171, 1972.
20. Bentley, P. J. and K. Schmidt-Nielsen. *Science* 151:1547–1549, 1966.
21. Binns, R. *J. Exp. Biol.* 51:11–16, 1969.
22. Binyon, J. *J. Mar. Biol. Assoc. U.K.* 41:161–174, 1961.

23. Binyon, J. *Physiology of the Echinodermata*. Wiley (Interscience), New York, 1966.

24. Black, D. A. K., R. A. McCance, and W. F. Young. *Nature (London)* 150:461, 1942.

25. Bock, K. J. and C. Schlieper. *Kiel. Meeresforsch.* 9:201–221, 1953.

26. Born, J. W. *Biol. Bull. (Woods Hole, Mass.)* 134:235–244, 1968.

27. Boroffka, I. *Z. Vergl. Physiol.* 57:348–375, 1968.

28. Bottazzi, F. *Ergeb. Physiol.* 7:161–402, 1908.

29. Bradley, T. J. In *Comparative Insect Physiology, Biochemistry and Pharmacology* (G. A. Kerkut and L. I. Gilbert, eds.). Vol. 4, pp. 421–465. Pergamon, Oxford, 1985.

30. Bricteux-Gregoire, S., G. K. Duchateau-Bosson, C. Jeuniaux, and M. Florkin. *Arch. Int. Physiol. Biochim.* 72:267–275, 1964.

31. Bronner, F. and A. Kleinzeller. *Current Topics in Membranes and Transport*, Vols. 1–32. Academic Press, New York, 1970–1988.

32. Bruce, D. L. and J. M. Marshall. *J. Gen. Physiol.* 49:151–178, 1965.

33. Bryan, G. W. *J. Exp. Biol.* 37:83–99, 1960.

34. Burger, J. W. *Biol. Bull. (Woods Hole, Mass.).* 113:207–223, 1957.

35. Burger, J. W. and W. N. Hess. *Science* 131:670–671, 1960.

35a. Bursell, E. *An Introduction to Insect Physiology*. Academic Press, New York, 1970.

35b. Burton, R. F. *Comp. Biochem. Physiol.* 27:763–777, 1968.

36. Buxton, P. A. *Proc. R. Soc. London, Ser. B* 106:560–577, 1930.

37. Cade, T. J., C. A. Tobin, and A. Gold. *Physiol. Zool.* 38:9–33, 1965.

38. Cameron, J. J. *Comp. Physiol.* 133:219–225, 1979.

39. Carley, W. W. *J. Exp. Zool.* 205:71–78, 1978.

40. Chew, R. M. and A. E. Dammann. *Science* 133:384–385, 1961.

41. Cobbold, P. H. *J. Physiol. (London)* 219:10P–12P, 1971.

42. Cockbain, A. J. *J. Exp. Biol.* 38:175–180, 1961.

43. Collins, B. G. and S. D. Bradshaw. *Physiol. Zool.* 46:1–21, 1973.

44. Conway, E. J. *Physiol. Rev.* 37:84–132, 1957.

45. Cornell, J. C. *Biol. Bull. (Woods Hole, Mass.)* 157:221–233, 1979.

46. Cornell, J. C. *Biol. Bull. (Woods Hole, Mass.)* 157:422–433, 1979.

47. Cornell, J. C. *Biol. Bull. (Woods Hole, Mass.)* 158:16–25, 1980.

48. Coulombe, H. N., S. H. Ridgway, and W. E. Evans. *Science* 149:86, 1965.

49. Croghan, P. *J. Exp. Biol.* 35:234–242, 1958.

50. Croghan, P. and A. P. M. Lockwood. *J. Exp. Biol.* 48:141–158, 1968.

51. Dall, W. *Comp. Biochem. Physiol.* 21:653–678, 1967.

52. Davenport, J. *J. Mar. Biol. Assoc. U. K.* 52:863–877, 1972.

53. Davson, H. *A Textbook of General Physiology*. Williams & Wilkins, Baltimore, MD, 1970.

54. de Rouffignat, C. and F. Morel. *J. Clin. Invest.* 48:474–486, 1969.

55. Dicker, S. E. and J. Haslam. *J. Physiol. (London)* 224:515–520, 1972.

56. Dietz, T. H. *Am. J. Physiol.* 235:R35–R40, 1978.

57. Dietz, T. H. and R. H. Alvarado. *Biol. Bull. (Woods Hole, Mass.)* 138:247–261, 1970.

58. Dietz, T. H. and W. D. Branton. *Physiol. Zool.* 52:520–527, 1979.

59. Dow, J. A. T., B. L. Gupta, T. A. Hall, and W. R. Harvey. *J. Membr. Biol.* 77:223–241, 1984.

60. Drewes, R. C., S. S. Hillman, R. W. Putnam, and D. M. Sokol. *J. Comp. Physiol.* 116:257–267, 1977.

61. Duffey, M. E., S. M. Thompson, R. A. Frizzell, and S. G. Schultz. *J. Membr. Biol.* 50:331–341, 1979.

62. Edney, E. B. *Water Balance in Land Arthropods*. Springer-Verlag, Berlin, 1977.

63. Ehrenfeld, J., F. Garcia Romeu, and

B. J. Harvey. *J. Physiol. (London)* 359:331–355, 1985.

64. Evans, B. H. *Comp. Biochem. Physiol. A* 51A:491–495, 1975.

65. Evans, B. H., K. Cooper, and M. B. Bogan. *J. Exp. Biol.* 49:158–171, 1976.

66. Evans, D. H. *J. Exp. Biol.* 50:179–190, 1969.

67. Evans, D. H. In *Comparative Physiology of Osmoregulation in Animals* (G. M. O. Maloiy, ed.). pp. 305–390. Academic Press, New York, 1979.

68. Florkin, M. *Bull. Cl. Sci. Acad. R. Belg.* 48:687–694, 1962.

69. Flügel, H. *Kiel. Meeresforsch.* 16:186–200, 1960.

70. Forster, R. P. *Fortschr. Zool.* 23(2/3):232–247, 1975.

71. Freel, R. W. *J. Exp. Biol.* 72:107–126, 1978.

72. Freel, R. W., S. G. Medler, and M. E. Clark. *Biol. Bull. (Woods Hole, Mass.)* 144:289–303, 1973.

73. Frizzell, R. A., P. L. Smith, E. Vosburgh, and M. Field. *J. Membr. Biol.* 46:27–39, 1979.

74. Fromm, P. O. *Comp. Biochem. Physiol.* 10:121–128, 1963.

75. Garcia Romeu, F. and J. Ehrenfeld. *Am. J. Physiol.* 228:845–849; 1975.

75a. Garcia Romeu, F., A. Salibian, and S. Pezzani-Hernandez. *J. Gen. Physiol.* 53:816–835, 1969.

76. Garry, W. E. *Biol. Bull. (Woods Hole, Mass.)* 8:257–270, 1905.

77. Geddes, M. C. *Comp. Biochem. Physiol. A* 51A:561–571, 1975.

78. Gérard, J. F. and R. Gilles. *J. Exp. Mar. Biol. Ecol.* 10:125–136, 1972.

79. Gilles, R., ed. In *Mechanisms of Osmoregulation in Animals.* pp. 111–154. Wiley, New York, 1979.

80. Gilles-Baillen, M. *J. Exp. Biol.* 59:39–43, 1973.

81. Gilles-Baillen, M. *J. Exp. Biol.* 59:45–51, 1973.

82. Glynn, I. and C. Ellory. *The Sodium Pump.* The Company of Biologists Limited, Cambridge, England, 1985.

83. Gordon, M. S. and V. A. Tucker. *J. Exp. Biol.* 42:437–445, 1965.

84. Gordon, M. S. and V. A. Tucker. *J. Exp. Biol.* 49:185–193, 1968.

85. Gordon, M. S., K. Schmidt-Nielsen, and H. M. Kelly. *J. Exp. Biol.* 38:659–678, 1961.

86. Grantham, J. J. and M. B. Burg. *Am. J. Physiol.* 211:255–259, 1966.

87. Green, J. W., M. Harsch, L. Barr, and C. L. Prosser. *Biol. Bull. (Woods Hole, Mass.)* 116:76–87, 1959.

88. Greenaway, P. *J. Exp. Biol.* 53:147–163, 1970.

89. Greenwald, L. *Physiol. Zool.* 44:149–161, 1972.

90. Greenwald, L. *Physiol. Zool.* 45:229–237, 1972.

91. Greenwald, L., W. B. Stone, and T. J. Cade. *Comp. Biochem. Physiol.* 22:91–100, 1967.

92. Griffith, R. W., B. L. Umminger, B. F. Grant, P. K. T. Pang, and G. E. Pickford. *J. Exp. Zool.* 187:87–102, 1974.

93. Grimstone, A. V., A. M. Mullinger, and J. A. Ramsay. *Philos. Trans. R. Soc. London, Ser. B* 253:343–382, 1968.

94. Gross, W. J. *J. Exp. Biol.* 31:402–423, 1954

95. Gross, W. J. and R. L. Capen. *Biol. Bull. (Woods Hole, Mass.)* 131:272–291, 1966.

96. Guyton, A. C. *Textbook of Medical Physiology.* Saunders, Philadelphia, PA, 1956.

97. Haines, H., C. F. Shield, and C. Twitchell. *Life Sci.* 8:1063–1068, 1969.

98. Hannan, J. V. and D. H. Evans. *Comp. Biochem. Physiol. A* 44A:1199–1213, 1973.

99. Hargitay, B. and W. Kuhn. *Z. Elektrochem.* 55:539, 1951.

100. Harris, R. R. and G. A. Kormanik. *J. Exp. Biol.* 218:107–116, 1981.

101. Harrison, F. M. and A. W. Martin. *J. Exp. Biol.* 42:71–98, 1965.

102. Hazelwood, D. H., W. T. W. Potts, and W. R. Fleming. *Z. Vergl. Physiol.* 67:186–191, 1970.

103. Heisler, N. *Fish Physiol.* 10:315–401, 1984.

104. Helman, S. I. and R. S. Fisher. *J. Gen. Physiol.* 69:571–604, 1977.

105. Hempling, H. G. *J. Gen. Physiol.* 44:365–379, 1960.

106. Henderson, L. J. *Fitness of the Environment.* Macmillan, New York, 1913.

107. Herreid, C. F., II. *Comp. Biochem. Physiol.* 28:829–839, 1969.

108. Hickman, C. P., Jr. *Can. J. Zool.* 46:439–455, 1968.

109. Hickman, C. P., Jr. and B. F. Trump. *Fish Physiol.* 1:91–239, 1969.

110. Hilger, H. H., J. D. Klümper, and K. J. Ullrich. *Pflüegers Arch. Gesamte Physiol. Menschen Tiere* 267:218–237, 1958.

111. Hirano, T. and N. Meyer-Gostan. *Proc. Natl. Acad. Sci. U.S.A.* 73:1348–1350, 1976.

112. Hodgkin, A. L. and P. Horowicz. *J. Physiol.* (*London*) 148:127–160, 1959.

113. Hodgkin, A. L. and B. Katz. *J. Physiol.* (*London*) 108:37–77, 1949.

114. Hodgkin, A. L. and R. Keynes. *J. Physiol.* (*London*) 128:28–60, 1955.

115. Hoffman, E. K. *Fed. Proc., Fed. Am. Soc. Exp. Biol.* 44:2513–2519, 1985.

116. Hoffman, E. K. *Mol. Physiol.* 8:167–185, 1985.

117. Hoffman, E. K. and K. B. Hendil. *J. Comp. Physiol.* 108:279–286, 1976.

118. Hoffman, E. K. and I. H. Lambert. *J. Physiol.* (*London*) 338:613–625, 1983.

119. Hoffman, E. K., L. O. Simonsen, and C. Sjøholm. *J. Physiol.* (*London*) 296:61–84, 1979.

120. Hoffman, E. K., C. Sjøholm, and L. O. Simonsen. *J. Membr. Biol.* 76:269–280, 1983.

121. Hoffman, J. F., ed. *Membrane Transport Processes*, Vol. 1. Raven Press, New York, 1978.

122. Holmes, W. N. and I. M. Stainer. *J. Exp. Biol.* 44:33–46, 1966.

123. House, C. R. *Water Transport in Cells and Tissues.* Edward Arnold, London, 1974.

124. House, C. R. and K. Green. *J. Exp. Biol.* 42:177–189, 1965.

125. Huf, E. *Pflügers Arch Gesamte Physiol. Menschen Tiere* 237:143–166, 1936.

126. Hui, C. A. *Physiol. Zool.* 54:430–440, 1981.

127. Hunter, K. C. Ph.D. Thesis, Univ. of Oregon, Eugene, 1973.

127a. Hutchison, V. H., W. G. Whitford and M. Kohl. *Physiol. Zool.* 41:65–85, 1968.

128. Irving, L., K. C. Fisher, and F. C. McIntosh. *J. Cell. Comp. Physiol.* 6:387–391, 1935.

129. Irving, L., P. F. Scholander, and S. W. Grinnell. *J. Cell. Comp. Physiol.* 17:145–168, 1941.

130. Jamison, R. L. In *The Kidney*, Vol. 1, pp. 391–476. (B. Brenner and F. C. Rector, eds.), Saunders, Philadelphia, PA, 1981.

131. Jamison, R. L., N. Roinel, and de Rouffignat, C. *Am. J. Physiol.* 236:F448–F453, 1979.

132. Kaneshiro, E. S., P. B. Dunham, and G. G. Holz, Jr. *Biol. Bull.* (*Woods Hole, Mass.*) 136:63–75, 1969.

133. Kaneshiro, E. S., G. G. Holz, Jr., and P. B. Dunham. *Biol. Bull.* (*Woods Hole, Mass.*) 137:161–169, 1969.

134. Katchalsky, A. and P. F. Curran. *Nonequilibrium Thermodynamics in Biophysics.* Harvard Univ. Press, Cambridge, MA, 1965.

135. Kerstetter, T. H. and L. B. Kirschner. *J. Exp. Biol.* 56:263–272, 1972.

136. Kerstetter, T. H., L. B. Kirschner, and D. Rafuse. *J. Gen. Physiol.* 56:342–359, 1970.

137. Kirsch, R., D. Guinier, and R. Meens. *J. Physiol.* (*Paris*) 70:605–626, 1975.

138. Kirschner, L. B. *Annu. Rev. Physiol.* 29:169–196, 1967.

139. Kirschner, L. B. and D. Howe. *Am. J. Physiol.* 240:R364–R369, 1981.

140. Kirschner, L. B. In *Mechanisms of Osmoregulation in Animals* (R. Gilles, ed.). Wiley, pp. 157–222. New York, 1979.

141. Kirschner, L. B., L. Greenwald, and T. H. Kerstetter. *Am. J. Physiol.* 224:832–837, 1973.

142. Kitching, J. A. *J. Exp. Biol.* 13:11–27, 1936.

143. Kitching, J. A. *J. Exp. Biol.* 15:143–151, 1938.

144. Knülle, W. *Zool. Beitr.* 30:393–408, 1987.

145. Koefoed-Johnson, V. *Acta Physiol. Scand.* 42(Suppl. 145):87–88, 1957.

146. Kokko, J. P. and C. Tisher. *Kidney Int.* 10:64–81, 1976.

147. Kooyman, G. L. and C. M. Drabek. *Physiol. Zool.* 41:187–194, 1968.

148. Kowalski, R. *Kiel. Meeresforsch.* 11:201–213, 1955.

149. Krag, B. and E. Skadhauge. *Comp. Biochem. Physiol. A* 41A:667–683, 1972.

150. Kregenow, F. H. *Annu. Rev. Physiol.* 43:493–505, 1981.

151. Krogh, A. *Osmotic Regulation in Aquatic Animals.* Cambridge Univ. Press, London and New York, 1939.

152. Krogh, A. *Skand. Arch. Physiol.* 76:60–73, 1937.

153. Krogh, A. *Z. Vergl. Physiol.* 24:656–666, 1937.

154. Krogh, A. *Z. Vergl. Physiol.* 25:335–350, 1938.

155. Lasiewski, R. C. *Physiol. Zool.* 37:212–223, 1964.

156. Leaf, A. *Ann. N.Y. Acad. Sci.* 72:396–404, 1959.

157. Leersnyder, M. de *Cah. Biol. Mar.* 8:295–321, 1967.

158. Levinsky, N. G. and R. W. Berliner. *J. Clin. Invest.* 38:741–748, 1959.

159. Lilly, S. *J. Exp. Biol.* 32:423–439, 1955.

160. Little, C. *J. Exp. Biol.* 43:39–54, 1965.

161. Little, C. *J. Exp. Biol.* 46:459–474, 1967.

162. Little, C. *J. Molluscan Stud.* 47:221–247, 1981.

163. Lockwood, A. P. M. *J. Exp. Biol.* 38:647–658, 1961.

164. Lockwood, A. P. M. In *Transport of Ions and Water in Animals* (B. L. Gupta, R. B. Moreton, J. L. Oschman, and B. J. Wall, eds.). pp. 673–707. Academic Press, London, 1977.

165. Lockwood, A. P. M., C. B. E. Inman, and T. H. Courtenay. *J. Exp. Biol.* 58:137–148, 1973.

165a. Loveridge, J. P. *J. Exp. Biol.* 49:1–13, 1968.

166. Loveridge, J. P. *J. Exp. Biol.* 49:15–29, 1968.

167. Macallum, A. B. *J. Physiol. (London)* 29:213–241, 1903.

168. Machin, J. In *Pulmonates*, Vol. 1, pp. 105–163. (V. Fretter and J. Peak, eds.), Academic Press, London, 1975.

169. MacMillan, R. E. and A. K. Lee. *Comp. Biochem. Physiol.* 28:493–514, 1969.

170. Maddrell, S. H. P. In *Transport of Ions and Water in Animals* (B. L. Gupta, R. B. Moreton, J. L. Oschman, and B. J. Wall, eds.). pp. 541–569. Academic Press, London, 1977.

171. Maddrell, S. H. P. and B. O. C. Gardiner. *J. Exp. Biol.* 60:641–652, 1974.

172. Maddrell, S. H. P. and J. E. Phillips. *J. Exp. Biol.* 62:671–683, 1975.

173. Maetz, J. *J. Exp. Biol.* 58:255–275, 1973.

174. Maetz, J. *Symp. Zool. Soc. London* 9:107–140, 1963.

175. Maloeuf, N. S. R. *Z. Vergl. Physiol.* 25:1–28, 1938.

176. Maloiy, G. M. O. In *Comparative Physiology of Desert Animals* (G. M. O. Maloiy, ed.). pp. 243–259. Academic Press, London, 1972.

177. Manton, S. M. and J. A. Ramsay. *J. Exp. Biol.* 14:470–472, 1937.

178. Martin, A. W. *Annu. Rev. Physiol.* 20:225–242, 1958.

179. Martin, A. W., D. M. Stewart, and F. M. Harrison. *J. Exp. Biol.* 42:99–123, 1965.

180. Mast, S. O. and C. Fowler. *J. Cell. Comp. Physiol.* 6:151–167, 1935.

181. McFarland, W. N. and F. W. Munz. *Comp. Biochem. Physiol.* 14:383–398, 1965.

182. McKnight, A. D. C. and A. Leaf. In *Membrane Physiology* (T. E. Andreoli, J. F. Hoffman, and D. D. Fanestil, eds.). pp. 315–334. Plenum Medical Book Co., New York, 1980.

183. McNabb, F. M. A. *Comp. Biochem. Physiol.* 28:1045–1058, 1969.

184. Minnich, J. E. *Copeia* 1970:575–578, 1970.

185. Minnich, J. E. *Comp. Biochem. Physiol.* 35:921–933, 1970.

186. Minnich, J. E. In *Comparative Physiology*

of Osmoregulation in Animals (G. M. O. Maloiy, ed.). pp. 391–641. Academic Press, New York, 1979.

187. Minnich, J. E. and V. H. Shoemaker. *Am. Midl. Nat.* 84:496–509, 1970.

188. Motais, R. *Ann. Inst. Oceanogr. (Paris)* 45:1–84, 1967.

189. Motais, R., J. Isaia, C. Rankin, and J. Maetz. *J. Exp. Biol.* 51:529–546, 1969.

190. Nagel, W. *Pflüegers Arch.* 365:135–143, 1976.

191. Newman, W. A. *Proc. Symp. Crustacean Biol.* Part 3:1038–1066, 1967.

192. Noble-Nesbitt, J. *J. Exp. Biol.* 50:745–769, 1970.

193. Noble-Nesbitt, J. In *Transport of Ions and Water in Animals* (B. L. Gupta, R. B. Moreton, J. L. Oschman, and B. J. Wall, eds.). pp. 571–597. Academic Press, London, 1977.

194. Oglesby, L. C. *Comp. Biochem. Physiol.* 14:621–640, 1965.

195. Oglesby, L. C. *Comp. Biochem. Physiol.* 26:155–177, 1968.

195a. Oglesby, L. C. *Biol. Bull. (Woods Hole, Mass.)* 134:118–138, 1968.

196. Oglesby, L. C. In *Physiology of Annelids* (P. J. Mills, ed.). pp. 555–658. Academic Press, New York, 1978.

197. Parmelee, J. T. and J. L. Renfro. *Am. J. Physiol.* 245:R888–R893, 1983.

198. Parry, G. *J. Exp. Biol.* 32:408–422, 1955.

199. Patterson, D. J. *Biol. Rev. Cambridge Philos. Soc.* 55:1–46, 1980.

200. Phillips, J. E. and T. J. Bradley. In *Transport of Ions and Water in Animals* (B. L. Gupta, R. B. Moreton, J. L. Oschman, and B. J. Wall, eds.). pp. 709–734. Academic Press, New York, 1977.

201. Phillips, J. E. *J. Exp. Biol.* 41:39–67, 1964.

202. Phillips, J. E. *J. Exp. Biol.* 41:69–80, 1964.

203. Pichon, Y. and J. E. Treherne. *J. Exp. Biol.* 65:553–563, 1976.

203a. Pierce, S. K. *Comp. Biochem. Physiol. A* 39A:103–117, 1971.

204. Pierce, S. K., M. K. Warren, and H. H. West. *Physiol. Zool.* 56:445–454, 1983.

205. Pilgrim, R. L. C. *J. Exp. Biol.* 30:297–316, 1953.

206. Post, R. L. In *Biophysics of Physiological and Pharmacological Actions,* Publ. No. 69; 19–30. Am. Assoc. Adv. Sci., Washington, D.C., 1961.

207. Post, R. L. and P. C. Jolly. *Biochim. Biophys. Acta* 25:119–128, 1957.

208. Potts, W. T. W. *Comp. Biochem. Physiol.* 14:339–355, 1965.

209. Potts, W. T. W. *J. Exp. Biol.* 31:618–630, 1954.

210. Potts, W. T. W. *J. Exp. Biol.* 35:749–764, 1958.

211. Potts, W. T. W. *J. Exp. Biol.* 31:614–617, 1954.

212. Potts, W. T. W. and G. Parry. *Osmotic and Ionic Regulation in Animals.* Oxford Univ. Press, London and New York, 1964.

213. Potts, W. T. W. In *Transport of Ions and Water in Animals* (B. L. Gupta, R. B. Moreton, J. L. Oschman, and B. J. Wall, eds.). pp. 453–480. Academic Press, London, 1977.

214. Potts, W. T. W., C. R. Fletcher, and B. Eddy. *J. Comp. Physiol.* 87:21–28, 1973.

215. Prescott, D. M. and E. Zeuthen. *Acta Physiol. Scand.* 28:77–94, 1953.

216. Pressley, T. A. and J. E. Graves. *J. Exp. Biol.* 226:45–52, 1983.

217. Prior, D. J., M. Hume, D. Varga, and S. D. Hess. *J. Exp. Biol.* 104:111–127, 1983.

218. Pritchard, A. W. and D. E. Kerley. *Comp. Biochem. Physiol.* 35:427–437, 1970.

219. Prusch, R. D. and P. B. Dunham. *J. Cell Biol.* 46:431–434, 1970.

220. Prusch, R. D. and P. B. Dunham. *J. Exp. Biol.* 56:551–563, 1972.

221. Ramsay, J. A. *J. Exp. Biol.* 27:145–157, 1950.

222. Ramsay, J. A. *J. Exp. Biol.* 29:110–126, 1952.

223. Ramsay, J. A. *J. Exp. Biol.* 32:183–199, 1955.

224. Ramsay, J. A. *J. Exp. Biol.* 32:200–216, 1955.

225. Ramsay, J. A. *Philos. Trans. R. Soc. London, Ser. B* 248:279–314, 1964.

226. Remmert, H. *Z. Vergl. Physiol.* 65:424–427, 1969.

227. Renfro, J. L. *Am. J. Physiol.* 228:52–61, 1975.

228. Reuben, J. P., L. Girardier, and H. Grundfest. *J. Gen. Physiol.* 47:1141–1174, 1964.

229. Rick, R., A. Dörge, E. von Arnim, and K. Thurau. *J. Membr. Biol.* 39:313–331, 1978.

230. Riddick, D. H. *Am. J. Physiol.* 215:736–740, 1968.

231. Ridgway, S. H. In *Mammals of the Sea* (S. H. Ridgway, ed.). Thomas, Springfield, IL, 1972.

232. Riegel, J. A. *Comparative Physiology of Renal Excretion.* Oliver & Boyd, Edinburgh, 1972.

233. Riegel, J. A. *J. Exp. Biol.* 38:291–299, 1961.

234. Riegel, J. A. and A. P. W. Lockwood. *J. Exp. Biol.* 38:491–499, 1961.

235. Riegel, J. A., A. P. M. Lockwood, J. R. W. Norfolk, N. C. Bulleid, and D. A. Taylor. *J. Exp. Biol.* 60:167–181, 1974.

236. Roberts, J. S. and B. Schmidt-Nielsen. *Am. J. Physiol.* 211:476–486, 1966.

237. Robertson, J. D. *Biol. Bull. (Woods Hole, Mass.)* 138:157–183, 1970.

238. Robertson, J. D. *Biol. Bull. (Woods Hole, Mass.)* 148:303–319, 1975.

239. Robertson, J. D. *Comp. Biochem. Physiol. A* 67A:535–543, 1980.

240. Robertson, J. D. *J. Exp. Biol.* 26:182–200, 1949.

241. Robertson, J. D. *J. Exp. Biol.* 30:277–296, 1953.

242. Robertson, J. D. *J. Exp. Biol.* 37:879–888, 1960.

243. Robertson, J. D. *J. Exp. Biol.* 38:707–728, 1961.

244. Robertson, J. D. *J. Exp. Biol.* 42:153–175, 1965.

245. Robertson, J. D. *J. Zool.* 178:261–277, 1976.

246. Rudy, P. P. *Comp. Biochem. Physiol.* 18:881–907, 1966.

247. Schlieper, C. *Biol. Rev. Cambridge Philos. Soc.* 10:334–359, 1935.

248. Schlieper, C. *Z. Vergl. Physiol.* 9:478–514, 1929.

249. Schmidt-Nielsen, B. and R. P. Forster. *J. Cell. Comp. Physiol.* 44:233–246, 1954.

250. Schmidt-Nielsen, B. and D. Laws. *Annu. Rev. Physiol.* 25:631–658, 1963.

251a. Schmidt-Nielsen, B. *Physiol. Zool.* 61:312–321, 1988.

252. Schmidt-Nielsen, B. and C. R. Schrauger. *Science* 139:606–607, 1963.

253. Schmidt-Nielsen, K. *Animal Physiology.* Cambridge Univ. Press, London, 1975.

253. Schmidt-Nielsen, K. *Animal Physiology.* Cambridge Univ. Press, London, 1975.

254. Schmidt-Nielsen, K., F. R. Hainsworth, and D. E. Murrish. *Respir. Physiol.* 9:263–276, 1970.

255. Schmidt-Nielsen, K., C. R. Taylor, and A. Shkolnik. *Symp. Zool. Soc. London* 31:1–13, 1972.

256. Schultz, S. G. *Basic Principles of Membrane Transport.* Cambridge Univ. Press, London, 1980.

257. Schwabe, E. *Z. Vergl. Physiol.* 19:183–236, 1933.

258. Shaw, J. *J. Exp. Biol.* 32:383–396, 1955.

259. Shaw, J. *J. Exp. Biol.* 32:664–680, 1955.

260. Shaw, J. *J. Exp. Biol.* 35:920–929, 1958.

261. Shaw, J. *J. Exp. Biol.* 36:145–156, 1959.

262. Shaw, J. *J. Exp. Biol.* 36:157–176, 1959.

263. Shaw, J. *J. Exp. Biol.* 38:135–152, 1961.

264. Shaw, J. *J. Exp. Biol.* 38:153–162, 1961.

265. Shehadeh, Z. H. and M. S. Gordon. *Comp. Biochem. Physiol.* 30:397–418, 1969.

266. Shoemaker, V. H. and L. L. McClanahan. *J. Comp. Physiol.* 100:331–345, 1975.

267. Siegel, N. S., D. A. Schon, and J. P. Hayslett. *Am. J. Physiol.* 203:1250–1254, 1976.

268. Silva, P., J. Stoff, M. Field, L. Fine, J. N. Forrest, and F. H. Epstein. *Am. J. Physiol.* 233:F298–F306, 1979.

269. Skadhauge, E. *J. Exp. Biol.* 60:535–546, 1974.

270. Skadhauge, E. *J. Physiol. (London)* 204:135–158, 1969.

271. Skadhauge, E. and B. Schmidt-Nielsen. *Am. J. Physiol.* 212:793–798, 1967.

272. Skelding, J. M. *Malacologia* 14:93–96, 1973.

273. Skou, J. C. *Biochim. Biophys. Acta* 23:394–401, 1957.

274. Skou, J. C. *Physiol. Rev.* 45:596–617, 1965.

275. Smith, H. W. *Am. J. Physiol.* 98:296–310, 1931.

276. Smith, H. W. *Q. Rev. Biol.* 7:1–26, 1932.

277. Smith, H. W. *From Fish to Philosopher.* Little, Brown, Boston, MA, 1953.

278. Smith, H. W. *Biol. Rev. Cambridge Philos. Soc.* 11:49–82, 1936.

279. Smith, H. W. *The Kidney.* Oxford Univ. Press, New York, 1951.

280. Smith, P. G. *J. Exp. Biol.* 51:739–757, 1969.

281. Smith, R. I. *Biol. Bull. (Woods Hole, Mass.)* 125:332–343, 1963.

282. Smith, R. I. *J. Exp. Biol.* 53:101–108, 1970.

283. Sorenson, A. L. *J. Gen. Physiol.* 58:287–303, 1971.

284. Staverman, A. J. *Recl. Trav. Chim. Pays Bas* 70:344–352, 1951.

285. Stein, W. D. *Transport and Diffusion Across Cell Membranes.* Academic Press, London, 1986.

286. Steinbach, H. B. *Perspect. Biol. Med.* 5:338–355, 1962.

287. Stiffler, D. F. and R. H. Alvarado. *Am. J. Physiol.* 226:1243–1249, 1974.

288. Stoner, L. C. and P. B. Dunham. *J. Exp. Biol.* 53:391–399, 1970.

289. Sutcliffe, D. W. *J. Exp. Biol.* 46:529–550, 1967.

290. Sutcliffe, D. W. *J. Exp. Biol.* 48:359–380, 1968.

291. Sutcliffe, D. W. *J. Exp. Biol.* 55:345–355, 1971.

292. Sutcliffe, D. W. and J. Shaw. *J. Exp. Biol.* 46:519–528, 1967.

293. Sutton, A. H. *Comp. Biochem. Physiol.* 24:149–161, 1968.

294. Telfer, N., L. H. Cornell, and J. H. Prescott. *J. Am. Vet. Med. Assoc.* 157:555–558, 1970.

295. Teorell, T. *Arch. Sci. Physiol.* 3:205–219, 1949.

296. Thompson, L. *Osmoregulation of the Freshwater Crabs* Metapaulius depressus *and* Pseudotelphusa jouyi. University Microfilms, Ann Arbor, MI, 1970.

297. Tosteson, D. C. and J. F. Hoffman. *J. Gen. Physiol.* 44:169–194, 1960.

298. Tosteson, D. C., Y. V. Ovchinnikov, and R. Latorre, eds. *Membrane Transport Processes,* Vol. 2. Raven Press, New York, 1978.

299. Tyson, G. Z. *Zellforsch. Mikrosk. Anat.* 86:129–138, 1968.

300. Tyson, G. *Zeit. Zellforsch. Mikrosk. Anat.* 93:151–163, 1969.

301. Ullrich, K. J. and K. H. Jarausch. *Pflügers Arch. Gesamte Physiol. Menschen Tiere* 262:537–550, 1956.

302. Ussing, H. H. *Acta Physiol. Scand.* 19:43–56, 1949.

303. Ussing, H. H. *Renal Physiol.* 9:38–46, 1986.

304. Villegas, R. and G. M. Villegas. *J. Gen. Physiol.* 43(5):73–104, 1960.

305. Vorwohl, G. Z. *Vergl. Physiol.* 45:12–49, 1961.

305a. Walker, A. M., C. L. Hudson, T. Findley, and A. N. Richards. *Am. J. Physiol.* 118:121–129, 1937.

306. Wall, B. J. and J. L. Oschman. *Fortschr. Zool.* 23(2/3):193–222, 1975.

307. Wall, B. J. and J. L. Oschman. *Am. J. Physiol.* 218:1208–1215, 1970.

308. Wallin, B. G. *J. Gen. Physiol.* 54:462–478, 1969.

309. Warburg, M. R. *Am. Zool.* 8:545–559, 1968.

310. Wygoda, M. L. *Physiol. Zool.* 57:329–337, 1984.

311. Weisser, F., F. B. Lacy, H. Weber, and R. L. Jamison. *Am. J. Physiol.* 219:1706–1730, 1970.

312. Weis-Fogh, T. *J. Exp. Biol.* 47:561–587, 1967.

313. Werntz, H. O. *Biol. Bull. (Woods Hole, Mass.)* 124:225–239, 1963.

314. Wessel, A., ed. *Fortschr. Zool.* 23(2/3):1–362, 1975.

315. Wharton, G. In *Comparative Physiology: Water, Ions and Fluid Mechanics* (K. Schmidt-Nielsen, L. Bolis, and S. H. P. Madrell, eds.). pp. 79–95. Cambridge, Univ. Press, London, 1978.

316. White, J. C. D. *Nature (London)* 171:612, 1953.

317. Wieser, W. and G. Schweizer. *J. Exp. Biol.* 52:267–274, 1970.

318. Wigglesworth, V. B. *J. Exp. Biol.* 21:97–114, 1945.

319. Wigglesworth, V. B. *J. Exp. Biol.* 8:415–427, 1931.

320. Willoughby, E. J. and M. Peaker. In *Comparative Physiology of Osmoregulation in Animals* (G. M. O. Maloiy, ed.). pp. 1–55. Academic Press, New York, 1979.

321. Wilmer, P. G. *J. Exp. Biol.* 77:181–205, 1978.

322. Wirz, H. *Helv. Physiol. Pharmacol. Acta* 14:353–362, 1956.

323. Wirz, H., B. Hargitay, and W. Kuhn. *Helv. Physiol. Pharmacol. Acta* 9:196–207, 1951.

324. Yousef, M. K. and W. G. Bradley. *Comp. Biochem. Physiol. A* 39A:671–682, 1971.

325. Zachar, J. *Electrogenesis and Contractility in Skeletal Muscle Cells.* University Park Press, Baltimore, MD, 1971.

326. Zerbst-Boroffka, I. and J. Haupt. *Fortschr. Zool.* 23(2/3):33–47, 1975.

327. Prosser, C. L. In *Neural and Integrative Animal Physiology* (C. Ladd Prosser, ed.). pp. 1–66. Wiley-Liss, Inc., New York, 1991.

Chapter 3 | *Temperature*

C. Ladd Prosser; contributions by J. E. Heath

Heat results from molecular agitation and temperature is a measure of heat content. Environmental temperature limits the distribution of living organisms and is an important determinant of their activity. The range of temperatures on earth is greater than the range within which life is possible. Air temperatures range from −70°C in polar regions to +85°C in hot deserts. Surface waters range from −1.82°C in Arctic and Antarctic waters to +30°C in tropical landlocked bodies of water. In general, life activities occur within a range of ~0 to +40°C; most animals live within much narrower limits. Some animals can, at certain stages in their life history, become dormant and survive in an inactive state well below 0°C or above 40°C. A few animals tolerate freezing; some animals live in thermal springs as hot as 70°C. Living bacteria have been reported from deep oceanic vents at +250°C under pressure but growth appears to cease above 105°C (Chapter 4). The temperature for development is usually more limited than for survival of adults.

The thermal properties of water determine the thermal relations of animals. The heat conductivity of water at 20°C is 0.00143 in cal/s · cm³ · °C, low compared with metal—1.0 for copper—but higher than that of many liquids (ethanol, 0.000399). The specific heat of water at 20°C is high—1.0 cal/g · °C compared with Cu, 0.09; ethanol, 0.535. Animal tissues, except for compact bone, require 0.7 to 0.9 cal to raise the temperature of 1 g of tissue 1°C. Thermal diffusivity is given by the heat conductivity divided by the product of density and specific heat. Low thermal diffusivity results in slow warming and cooling and in limited conduction of heat within an animal; fat is a good insulator but is not as effective as air. Large animals are slow to warm or cool; most of their transfer of heat is regulated by circulating fluid; sluggish circulation makes for slow heat transfer. A 1-g fish cools at the rate of 1.8°C/min · °C gradient; the rate for a 100-g fish is only 0.4°C/min · °C. Rates of heating and cooling differ according to circulatory responses; a lizard heats faster than it cools

(49). Water breathing and air breathing animals differ in their thermal relationships. Because of more rapid thermal than gaseous diffusion, heat exchange across respiratory surface in water is rapid and water breathers cannot attain whole-body homeothermy–endothermy as can air-breathers.

Natural waters (hot springs and thermal vents excepted) because of their high specific heat and low heat conductivity seldom have a temperature above the limits for most aquatic animals. Ice has a lower specific gravity than water; in fresh water, aquatic animals do not freeze as long as they remain in water below ice. Terrestrial poikilotherms are subject to much greater fluctuations in temperature than are aquatic animals and in the former, body temperature is largely controlled by factors related to their water balance such as evaporation from body surface.

Water has a high heat of fusion (79.7 cal/g). Aqueous solutions supercool by several degrees before freezing; bound water is resistant to freezing. Lowering of freezing point is a colligative property and the lowering of the freezing point of a solution is proportional to the concentration of solute particles. Some animals supercool; dehydrated animals or those with some of their water replaced by polyhydric organic solvents can withstand temperatures well below the freezing point. Water loss by evaporation has a cooling effect on any moist surface (585 cal lost/g of water evaporated at 20°C) and in a hot environment, animal temperature control is limited by heat of vaporization of water.

Many kinds of animals—poikilotherms—conform in body temperature (T_b) to the environment; T_b equals ambient temperature, T_a, and T_b rises or falls with T_a. Homeotherms regulate or maintain relative constancy of body temperature in a varying ambient temperature. Heterotherms regulate either spatially in certain body regions or temporally at certain times. Some poikilotherms, for example, basking insects or reptiles, raise body temperature by external heat (ectothermy); some poikilotherms (flying insects and large swimming fishes) use internally produced heat (endothermy). Homeotherms—birds and mammals—use both endothermy and heat conservation to maintain body temperature. Poikilothermy and homeothermy refer to variability or constancy of internal temperature; ectothermy and endothermy refer to the major heat source—external or internal. Some ectotherms (deep sea fishes) are homeothermic because they live in constant T_a.

There are two general ways by which the temperature of both poikilotherms and homeotherms is controlled: behavioral and metabolic–circulatory. Many animals regulate behaviorally by aggregating at "preferred" temperatures. Behavioral regulation of temperature in ciliates, various invertebrates, and fishes consists of a coordinated sequence of sensory, locomotor, and integrative actions. Metabolic control in both poikilotherms and homeotherms may be by the same receptors and integrators (neural and hormonal) as behavioral control or control may be at the cellular level. The metabolic effector system modulates heat production and controls heat loss by circulatory, insulative, and evaporative means.

The T_b of many mammals is 37–39°C, T_b of birds is slightly higher (38–40°C), and the T_b of primitive mammals (monotremes and marsupials) is somewhat lower (32–36°C) (Table 1). Body temperature of most homeotherms cycles diurnally. Regulation of T_b develops slowly in most newly hatched birds and newborn

TABLE 1. Body Temperature (T_b) During Activity and Thermoneutral Zones for Selected Mammals and Birds[a]

Animals	T_b(°C)	Thermoneutral Temperatures	
		Low	High
Large Mammals			
Man	37	27	32
Baboon	38.1		
Mountain sheep	37.9 night		
	39.8		
Goat	37–40		
Fur seal	38	20 water	
		0–3 air	
Humpback whale	36		
Rodents and Related Mammals			
Baiomys	35	29	36
Perognathus	36–38	23.5	
Peromyscus eremicus	36.6	30	35
Peromyscus californicus	36.4	27	34.5
Tamias	36–40.3	28.5	32
Citellus	35.5–39.5		
Hyrax *Procavia*	37–38.5	20	30
Bats			
Macroderma	35–39	30	35
Pteropus polycephalus	35–39	18	35
Dobsonia	37		34
Myotis		32.5	34.5
Eptesicus	35.6		
Miniopterus	37–39.1		
Primitive Mammals			
Armadillo *Dasypus*	34–356	30	
Echidna	30.7	20	30
Opossum *Marmosa*	33.2 day	28	
	35.7 night		
Phalanger Cercaertus	32–38	31	35
Large Birds			
Quail *Lophortyx*	40.6	27.3	37.5
Inca dove	38.8–42.7	35	
Pigeon		25	30
Red-footed booby	40.3 day		
	38.0 night		
Night hawk	36–40	35	
Chicken	39.8		

TABLE 1. (*Continued*)

Animals	T_b(°C)	Thermoneutral Temperatures	
		Low	High
Ostrich	39.3		
Owl	38	25	37
Cormorant	39–40		40
Shearwater	39.5 day		
	27.7 night		
Ptarmigan	39.6	4	36
Evening grosbeak	41 day	16	34
	38.5 night		
Crossbill Loria	38.5–40	15	28.5
Gray jay *Prisoreus*	42.3 day	36 summer	
	41 night	7 winter	
Passeriformes			
Zebra finch	41.5–42.2	36	42
Zebra finch	39.8–42.4	30	40
House sparrow	41.1 day		
	38.3 night		
Hummingbirds			
Giant Patagona		27	
Eugenes		31	
Blue-throated hummingbird			
Eulampia	38.5–40	31–33	
Stellula	40	30	
	35–40	27–30	

[a]References given in Table 9-3 in *Comp. Anim. Physiol.*, 3rd ed (119b).

mammals. Active insects and large fishes tend to regulate the temperature of critical organs in the mid-30°C range. Why this temperature range has been selected so generally is not clear. One hypothesis is that it represents an economical balance between heat loss and heat production. Another hypothesis relates this temperature to properties of water. A more reasonable hypothesis is that in this temperature range $\Delta G^{\pm}$ (activation free energy) is minimal and contributions of entropy and enthalpy are equal. Above 45°C many proteins show changes in conformation resulting in denaturation. Male gametogenesis is limited to a narrower range than is tolerated by other tissues.

Since temperature is a measure of molecular agitation it controls the rate of chemical reactions and is one factor limiting energy liberation and organismic growth. Kinetic activity, frequency of molecular collisions, is proportional to absolute temperature and increases ~3%/10°C rise in temperature, whereas enzyme activity increases many times more. One method of quantifying the effect of temperature on reaction rates is the Q_{10} approximation. The Q_{10} term is the factor by which reaction velocity is increased by a rise of 10°C.

$$Q_{10} = \left(\frac{k_2}{k_1}\right) \frac{10}{(t_2 t_1)}$$

where k_1 and k_2 are velocity constants corresponding to temperatures t_1 and t_2.

Velocity (V_1 and V_2) is normally used in place of the rate constant. Some functions are linear but most are logarithmic with respect to temperature. The Q_{10} of many biological processes is not constant over the normal temperature range and is larger in low rather than in high ranges of temperature. Hence, the range for which a Q_{10} is calculated must be specified. Physical properties of solutions are less affected by temperature changes than are catalyzed reactions. A Q_{10} of 2.5 means an increase of 9.6% per degree (119).

The critical incremental energy of activity of a reaction is given by the Arrhenius constant μ or E_a:

$$\mu = RT \ln \frac{k_2}{k_1} (T_2 - T_1)$$

where k_1 and k are velocity constants (proportional to measured velocities) at absolute temperatures T_1 and T_2 and R is the universal gas constant (1.98 cal/mol). When the logarithm of a velocity (or rate constant) is plotted against the reciprocal of the absolute temperature, most biological processes give a straight line, the slope of which is $-\mu/2.3R$ or $-\mu/4.6m$ (Fig. 1). This relation permits the conclusion that the number of molecules exceeding a critical energy may double for a 10°C rise in temperature even though the kinetic change, proportional to the absolute temperature, is less. Some reactions show breaks in the Arrhenius curve, usually with steeper slope in the low temperature range.

Temperature as a Factor in Ecological and Geographic Distribution

Temperature is an important environmental parameter determining animal distribution. Modifying factors are night–day cycles, seasonal sequences, availability of water, hydrostatic pressure, and mode of reproduction. Correlations of animal occurrence have been made with maximum and minimum temperature, mean temperature, and constancy of temperature.

Regions of relative constancy are the

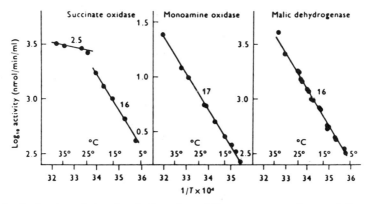

Figure 1. Arrhenius plots of activity rates of three enzymes from rat liver. Slopes are kcal/mol of substrate. (Raison, J. K. et al., *Symp. Soc. Exp. Biol.* 27:488–491. Copyright © 1973, Cambridge University Press, Cambridge, UK.)

deep oceans and polar seas where temperature is 2 to 4°C at all times, and tropical forests where there are diurnal cycles but little or no seasonal change in temperature. It has been suggested that heterogeneous and unstable environments lead to greater variety of animals than do environments of uniform and stable temperature. This has been corroborated for some kinds of animals but not for others. A diversified environment, especially one with high productivity, provides many ecological niches and the number of species may be great despite thermal stability (e.g., in coral reefs). Polar environments, both terrestrial and aquatic, are less varied and species diversity is less than in the tropic or temperate zones. Average heterozygosity of euphausid crustaceans in the tropics is three times greater than in the temperate or Antarctic waters. *Drosophila* speciation, however, is less in the tropics than in temperate climates. The deep sea is a uniform environment where nutrients are obtained as they fall from upper layers of ocean, temperature is low and hydrostatic pressure high. Aquatic animals of polar seas (Arctic and Antarctic fishes, crustaceans) have special adaptations of metabolism and behavior. Air-breathing animals such as penguins and seals in polar regions are adapted by insulation and behavior for extremes of cold and for short summer seasons. Some subarctic insects tolerate freezing, some are torpid during long winter periods. Arctic and Antarctic birds and mammals are well insulated and their reproduction is timed for a short summer.

Desert animals have behavioral and morphological adaptations to heat and desiccation (58). Cacti and desert insects have (1) cellular tolerance of high temperature, (2) decrease of thermal load by orientation and heat reflection, (3) periodic opening and closing of apertures for gas exchange, (4) lipid or waxy coats that reduce water loss, (5) thermolability (het-

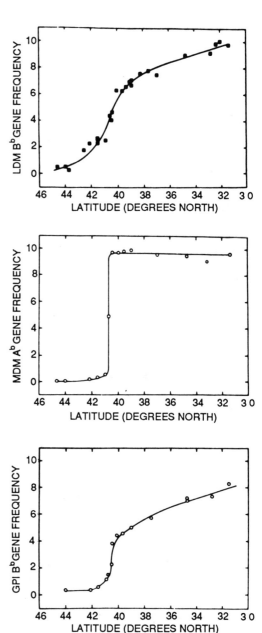

Figure 2. Distribution for one of two allotypes of three enzymes: LDH-B[a], MDH-A[b], and GPI-B[b] in killifish from northern to southern ends of its range, Halifax to Florida. [Powers, and Place (118). Copyright © 1978, Plenum Publishing, New York.]

erothermy), (6) absorption and utilization of water from air and substrate, (7) utilization of metabolic water, and (8) tolerance of cell and tissue desiccation (57).

Temperature can trigger reproduction and development. Sea lampreys (*Petromyzon maximus*) migrate upstream for spawning; spawning occurs over the range 11–25°C, embryonic development at 18–23°C. The temperature preferendum of adult lampreys is ~13.6°C (61, 109). The spawning range is similar for the anadromous parasitic *Lampetra fluviatilis* and the nonparasitic *L. planeri* (60, 61). Toads from Arizona have wide temperature limits of development, 19.1–32.0°C, toads from tropical Guatemala a narrower range 32.1–34.8°C (41a). *Rana pipiens* from the northern United States have a lower temperature range for development than *Rana sphenocephala* populations (107). The milkweed bug, *Oncopeltus*, has a lower limit of 5°C for egg survival, 14°C for embryonic development, and 18°C for emergence as adults (122).

Many animal groups occur in clines that may correlate with mean annual temperature or minimal winter temperature (62). In the killifish *Fundulus heteroclitus* two alleles are known for one isozyme of lactate dehydrogenase (LDH-B). At the northern end of the range (Halifax) the fish are homozygous for one allele (B^b); at the southern end (Florida) populations are homozygous for another allele (B^a); the two genotypes occur in different proportions related to latitude in the mid-Atlantic states. Clines for the enzymes MDH-A^a, LDH-B^b and GPI-B^b in Fundulus are shown in Figure 2 (118). The enzyme differences are genetically determined. Clines for fishes with several allozymes and isozymes of some 13 enzymes have been established for large-mouth bass; northern (Minnesota) and southern (Florida) populations differ and intergrades are found in intermediate regions (Fig. 3) (117b).

Figure 3. Distribution of allotypes of cytosolic malic dehydrogenase (MDH) in large mouth bass. Solid symbols represent frequency of B^1 allele, open symbols B^2. [Phillip, et al. (117b). Copyright © 1981, Department of Fisheries and Oceans—Scientific Information and Publications Branch, Ottawa, Ontario, Canada.]

There are biological compensations for seasonal changes in temperature. Oxygen consumption by wolf spiders shows a steeper temperature dependence (Q_{10} = 4.5) in summer than in winter (Q_{10} = 1.6–2.9). Temperature for heat coma in a snail *Physa* is 8.4°C higher in summer than in winter (104).

Resistance Adaptations; Lethal Limits of Heat and Cold

Whole animals survive and are active over a narrower range of temperature than isolated tissues and cells; temperature limits for proteins (denaturation) and lipids (melting or solidification) are wider than for cells. A mitochondrion is more limited in its temperature of function than the enzymes contained in it. Integrated systems are more temperature limited than their component parts. Lethal limits of intact animals are only partly determined by temperature inactivation of enzyme proteins. A frequent cause of cell death is increase in membrane permeability; potassium leaks out and sodium enters. Another cause of cell death is reduction of energy liberation below basal needs. An effect on cell nuclei is the depolymerization or melting of nucleic acids. An important cause of death in terrestrial poikilotherms (insects and reptiles) is dehydration. At the organismic level some organ systems are more sensitive to heat or cold than others and some tissues are more readily inactivated. Neurons fail to function with less heating or cooling than cells of skin or liver. Consequently, criteria of heat or cold death are imprecise and a legal definition of the lethal state does not correspond to a biological definition. Integration by central nervous systems can fail at temperatures well below lethal temperatures for heat and above lethality for cold. Animals may enter a state of chill or heat coma. Mortality curves plotted as percent of deaths as a function of time at low or high temperatures show that there are multiple causes of thermal death, each operating within a certain time–temperature combination. Temperature–mortality curves

are species specific (Fig. 4). The temperatures of heat death after 15-min exposure of *Rana pipiens* were intact tadpole 37.5°C, intact frog 38.6°C, gastrocnemius muscle 40.2°C, heart *in situ* 42°C, and sciatic nerve 43°C (104a, 117).

Individuals of a given species can be acclimated to several temperatures; then groups from each acclimation temperature can be transferred to a series of high temperatures for heat death determination and to a series of low temperatures for cold death determination. The percentage mortality or coma is plotted as a function of time at each temperature (Fig. 5), and time to death or coma of 50% of each group is plotted against exposure temperature. The median tolerance temperature (LD_{50}) falls at longer times until after an appreciable time (e.g., 48 h) no further deaths occur; this is the incipient lethal temperature (high or low). Two lines are plotted, incipient lethal temperatures against acclimation temperatures; these enclose a tolerance polygon (Fig. 5).

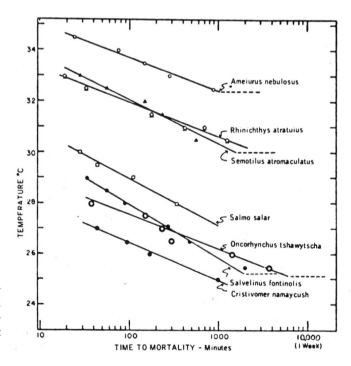

Figure 4. Median high-temperature resistance times for seven species of freshwater fishes acclimated to 20°C and tested at various temperatures (16b).

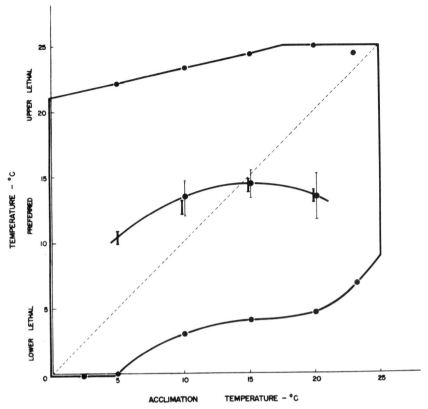

Figure 5. Thermal tolerance and preferred temperatures of young sockeye salmon in relation to acclimation temperature (16c).

Acclimation occurs at different rates according to species and acclimation to cold is usually slower than to warmth. In sockeye salmon the upper lethal temperature changed ~1°C for every 4°C change in acclimation temperature. For goldfish the upper lethal temperature increased ~10°C for every 30°C rise in acclimation temperature up to 36.5°C; the high lethal temperature was then 41°C and could not be increased further. The low lethal temperature decreased 2°C for every 3°C fall in acclimation temperature down to 17°C where the lethal temperature was 0°C (44).

The tolerance polygon is characteristic of each genotype. Within a tolerance polygon are smaller polygons for devel-opment and locomotor activity, functions that are more limited in range than survival. Fourteen species of North American freshwater fishes were examined for racial differences in tolerance polygons. Of these, three gave different polygons for populations from Ontario and Tennessee (62). For the barnacle *Balanus balanoides* from North Carolina the upper tolerance limit is 5°C higher than for the same species from New England. Genetic strains of thermophilic and psychrophilic bacteria, flagellates, and animals such as nematodes and insects have been selected and reared for heat and cold tolerance. Lethal temperature is influenced not only by acclimation temperature and genetic makeup, but also by age, hormonal state, diet, and environmental factors such as oxygen, salinity, and photoperiod. In lobsters the upper lethal temperature is reached sooner in low salinity and in low O_2 levels than in high salinity and O_2 (103). *Mytilus* in low salinity showed decreased tolerance of high temperature. Blowfly larvae on a diet of saturated fat showed increased heat tolerance (78a). Lethal temperature of an intertidal crab *Hemigrapsus* is ~20°C higher in 75% seawater than in 25% seawater (137). In the guppy *Lebistes*, the temperature for heat coma increased 1.4°C per 3°C rise in temperature of rearing and the cold coma temperature decreased 1°C for 3°C fall in rearing temperature. A series of intertidal snails showed temperatures of heat coma that correlated with their ecology and genotype; they also showed acclimation effects. The temperatures for coma in *Littorina* are essentially the same as for heat block of spontaneous electrical activity in the central nervous system (58a). Snails from a hot Israeli desert showed a median lethal temperature of 55°C.

As temperatures for heat coma are approached for crayfish, the potassium concentration in the hemolymph rises and central nervous function becomes impaired (16a). However, synaptic transmission in crayfish fails with less warming than is required to block postsynaptic depolarization by directly applied transmitter (glutamate). Synaptic block appears to result from decrease in transmitter liberation at temperatures above a critical level (145). After sunfish are cooled or warmed by 5–10°C beyond their acclimation temperature the maximum velocity of swimming is decreased. In goldfish, conditioned responses to flashes of light are blocked by less cooling then will block direct responses to light; still more cooling blocks spinal reflexes. Temperature changes have more effect on synaptic transmission than on axon con-

duction. Fish from intermediate temperatures (20–22°C), when cooled or heated by 8–10°C, become hyperactive and hypersensitive to touch. With several more degrees of cooling or heating the fish show motor disturbances and with still more temperature change, control of equilibrium is lost. Breathing becomes shallow and the fish may enter a state of coma (43). The temperature for each of these effects is raised or lowered by acclimation. Death appears to be due not to temperature per se but to asphyxia. This sequence of initial hyperactivity followed by incoordination, loss of equilibrium, and respiratory failure has been noted in many kinds of animals. Recordings from fish cerebellum during brain cooling or heating showed a sequence of failure of neural function in the following order: inhibitory synaptic transmission, patterned spontaneous activity, rhythmic endogenous activity, disynaptic excitatory transmission, monosynaptic transmission, and axon conduction (43). In other preparations such as crayfish neuromuscular junctions, it has been shown that inhibition is more thermolabile than is synaptic excitation (145). In sunfish skeletal muscle, the resting conductance at 20°C for chloride ions is six to seven times greater than for potassium; after cold acclimation the conductance for chloride is lowered more than for potassium and the membrane potential is determined by K^+. Membrane resistance increases on cooling and this may increase the shunting of synaptic potentials, resulting in increased likelihood of spike electrogenesis. These membrane changes may make possible swimming at low temperature (88). Conductance of ion channels may be related to molecular fluidity, which depends on both lipids and proteins in the membrane. Membranes of synaptosomes have greater fluidity in the series: Arctic cod > temperate zone sunfish > desert pupfish (29). Lipid composition and membrane fluidity can be altered by temperature acclimation.

Thermal Effects on Enzymes

Heat inactivation of enzymes, especially those of energy liberation, has been correlated with habitat and life style. Enzymes of subtidal invertebrates are more heat sensitive than corresponding enzymes from high intertidal species. For several species of frogs, temperature for enzyme inactivation, reflex failure, and loss of motor coordination are closely correlated with the temperature of habitat. There may be selection of open and flexible tertiary structures in cold environments and more rigid molecular structures in warm environments. An outcome of this is more thermolabile proteins in cold species and more heat stable proteins in species from warm environments. Thermal denaturation is a dramatic measure of rigidity of higher protein structure.

Myosin ATPases (adenosine triphosphatase) from many species of fishes from various habitats differ in half-time for inactivation at 37°C by more than an order of magnitude. Inactivation time at 37°C is shortest for enzymes from Antarctic species, intermediate for Mediterranean species, and longest for ATPases of fishes from African hot springs (82) (Fig. 6). Conformation of a protein is established by weak bonds; secondary, tertiary, and quaternary structures give flexibility to protein molecules. Inactivation of enzymes can result from changes in conformation such as are caused by mild heating. The breaking of weak bonds and consequent changes in conformation occur with little heating and a subtle relation between primary and higher-order conformation may account for inactivation by less heating than is normally experienced by the organisms containing the enzymes.

The primary structure of collagen is correlated with thermal transition of the molecules. When strips of collagen are

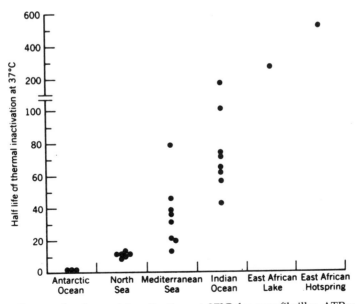

Figure 6. Half-times for thermal inactivation at 37°C for myofibrillar ATPase from fish living in different environments (83a).

warmed under slight tension they show a critical temperature for shrinkage that reflects intermolecular and intramolecular changes. Comparisons of shrinkage temperature for collagens from many animals correlate with the sum of proline plus hydroxyproline content. Transition temperatures are higher than lethal temperatures but fall in the same sequence for a number of species.

A number of polymeric enzymes depolymerize into subunits at low temperatures. Depolymerization of enzymes from cold-blooded vertebrates occurs at temperatures lower than for mammals. In the aquarium fish *Idus* the enzyme G6PDH, a tetrameric enzyme, dissociates into doublets and becomes inactive below 4°C; a tetramer phosphofructokinase (PFK) also dissociates when cooled.

Thermostability of rRNA is such that the rRNA of thermophilic bacteria melts at higher temperature (69°C) than the rRNA of mesophiles (64°C). Prostomians (ciliates, molluscs, flatworms, and insects) have a 28s rRNA, which is converted to an 18s form by heating to 45°C for 10 min; deuterostomian (mostly vertebrate) rRNA is converted to units of unequal size (79).

Heat-Shock Proteins*

Organisms of many kinds, exposed briefly to sublethal heat, produce heat-shock proteins (hsps).

As body temperature increases, the central nervous system of most animals begins to dysfunction, resulting in irreversible coma and ensuing death. Before critical thermal maximum is reached, however, a threshold body temperature is exceeded, which causes the selective activation of a highly specific gene family. Because expression of this gene family was first noted after *Drosophila* larvae

*Section contributed by Michael Koban.

were acutely treated with heat, the genes and their encoded proteins have been termed heat-shock genes and heat-shock proteins (hsps) (95). Exposure to high temperatures is not the only stress that induces heat-shock gene activity. A variety of physical and chemical stresses such as temperature, ionizing radiation, amino acid analogs, lysergic acid (LSD), sodium arsenate, and viral infections, to name a few, can result in hsp synthesis. Heat-shock proteins may therefore be induced as a generalized nonspecific cellular response to stress. The most prominent hsp in *Escherichia coli*, yeast, slime molds, and several tissues of mammals has a molecular weight (MW) approximating 70,000 (Table 2). In addition, there are some large (80 kD) and small (20–30 kD) hsps; a 3°C fever in a rabbit induces synthesis of a 74-kD hsp.

Heat-shock genes are universally found in all organisms, animal or plant, prokaryote or eukaryote. The genes from a variety of organisms representing diverse taxa have been cloned, sequenced, and characterized for their regulatory aspects. There is a remarkable degree of evolutionary conservation of gene (and protein) structure from *E. coli* to humans. Hence, heat-shock genes are an ancient set of genes whose biological function(s) must be very general and fundamental to cellular homeostasis. Nevertheless, the function(s) of hsps is not known. With the exception of a 28-kD hsp of yeast, hsps do not have any enzymatic activity. They may play a structural role, however, because hsps accumulate around specific organelles such as the cytoskeletal network, necleolus, and others during periods of intense stress. If heat stress is applied to induce hsp synthesis and accumulation, cells and organisms can tolerate a temperature that would otherwise be lethal in the absence of hsps. Thus, hsp can confer a transient thermotolerance.

TABLE 2. The Most Prominent Heat-Shock Protein in *E. coli*, Yeast, Slime Molds, Salmon Eggs, *Drosophila* Tissues, Chick Embryos, and Several Tissues of Mammals Has an M_5 Approximating 70,000. In Addition There Are Some Large (80 + kD) and Small (20–30 kD) hsp's

Organism or Cell	Growth or Adaptation Temperature	hsp Synthesis Temperature Range	hsp Classes (kD)	Reference
E. coli	30	34–42	76,73,64	153a
Yeast		40	70	146
Drosophila	25	36	7 hsp 83–22	95
Fundulus tissues	20	32.5–40	27–15 76,74	M. Koban (unpublished)
Catfish tissues	15	32.5–37.5	76,74	M. Koban (unpublished)
Rainbow trout cells	22	27–29	70	92
Tilapia ovary cells	31	43	70	25b
Rana cell explants	22	32–36.5	65	86a
African locust fat body	27	39–45	73,68	146
Quail red blood cells	37	43	90,70,26	32
Rat tissues	37	42 (15 min)		32

During periods of intense thermal or other stress, heat-shock genes are rapidly transcribed and heat-shock mRNA is translocated to the cytoplasm. Heat-shock mRNA is usually preferentially translated compared to normal preexisting mRNAs. Non heat-shock mRNA is rapidly degraded in yeast, but in most other organisms it is somehow rendered nontranslatable (perhaps by sequestration) during the period of stress. As a result, normal protein synthesis is significantly reduced during heat-shock while hsp synthesis occurs at very high rates. Regulation of the heat-shock response is complex. Purely transcriptional control is found only in *E. coli* and yeast, whereas translational control is found in *Xenopus* oocytes (10). In *Drosophila* salivary glands transcription increases within 3 min of heating (95). In most organisms, regulation resides at both transcription and translation levels.

Although the function of hsps is not known, an interesting correlation can be made between the temperature span over which hsps are synthesized and the natural temperature range of a given organism (see Table 2). Stenothermal organisms synthesize hsps at much lower temperature than eurythermal organisms that tolerate high-temperature stress. Thus there is a positive correlation between hsp synthesis and the thermal ecology of an organism. Heat-shock proteins may play a role in evolutionary adaptation of organisms to different thermal environments, but hsp synthesis is not altered by short-term acclimation temperature (91).

Freeze Tolerance and Freeze Avoidance

Damage by freezing of plants or poikilothermic animals may be caused in several ways and may be averted by protective mechanisms. Freezing damages by disruption by ice of cytoplasm or cell membranes. Rapid freezing results in a vitreous state without formation of ice crystals and rapid thawing permits recovery. When tissues freeze, ice crystals form first in extracellular fluid (ECF); this leaves the remaining ECF hypertonic and cells may lose water osmotically with resulting changes in ionic strength and enzyme activity. Cell membranes provide some protection against freezing of cell contents. In plant and animal cells, chill damage may cause leakage of potassium from cytoplasm. Prior dehydration protects against freezing.

Freeze Tolerance

A few invertebrate animals—intertidal molluscs and barnacles, subarctic insects—tolerate freezing of extracellular fluid. The plasma membrane resists penetration of ice crystals into cells. The amount of free water in cytoplasm that can freeze may be small relative to bound water that does not freeze. When intertidal mussels and barnacles survive freezing only 55 to 65% of body water is frozen; the remainder is bound to proteins and other solutes.

Some insects, for example, the hornet *Vespula* and beetle *Dendroides,* have protein nucleators that induce formation of extracellular ice at upper subzero temperatures (7, 8, 36, 37). These insects tolerate extracellular freezing and the amount of intracellular water that is bound increases at low temperatures (133). Nucleators are proteins that inhibit supercooling and promote extracellular freezing. Removal of nucleators from the gut in the fall enhances supercooling and cryoprotection in the winter (155).

A few species of frogs and turtle hatchlings tolerate extracellular freezing when hibernation behavior is ineffective. Freeze tolerant frogs show some regulation of cell volume by colligative agents. Also the frozen frogs are capable of some anaerobic production of ATP when in the frozen state. Formation of ice crystals initiates synthesis of cryoprotectants glucose and glycerol. Blood of supercooled unfrozen frogs had glucose concentrations of 1.6 µM/mL, blood of a frozen frog had glucose at 1881 µM/mL. In the first 50 min at subfreezing temperature, glucose increased in concentration by 100%. The increased glucose lowered the freezing point from -1.5 to $-3.0°C$; cryoprotected frogs at -2 to $-4°C$ survive freezing (134). Turtle hatchlings survive at $-4°C$ with 53% of body water as ice, but $-11°C$ with 69% as ice is lethal (135).

Freeze Avoidance

Avoidance of freezing may be by colligative or noncolligative means. Some fishes have increased NaCl concentration in plasma in winter; other fishes have increased plasma solutes such as glucose (140).

Avoidance of freezing may be behavioral. Frogs and toads burrow below the freezing level in soil.

Many subarctic insects synthesize large amounts of polyhydric alcohols, which lower the freezing point of body fluids. For example, larvae of a gall-fly on goldenrod synthesize glycogen in quantity during late summer and when the air cools activity of phosphorylase *a* of fat bodies increases. As glycogen decreases, glycerol is formed and sorbitol is unchanged (Fig. 7) (135).

Frogs, particularly those that do not tolerate freezing, synthesize glucose that acts colligatively to protect against freezing. Liver phosphorylase activity is increased when ice crystals begin to form.

Noncolligative Antifreeze
An important means of protection against freezing is by synthesis of glycoproteins and polypeptides that bring about depression of the freezing point by noncolligative means. There is hysteresis in that the temperature of freezing is lower than the temperature of melting (Fig. 8). Fish antifreezes act as protectors against

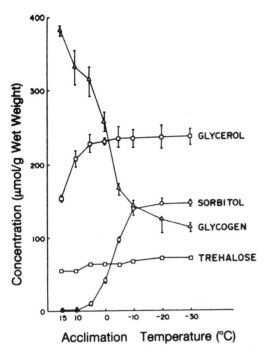

Figure 7. Acclimation of glycogen, glycerol, sorbitol, and trehalose in larvae of gallfly *Eurosta solidaginis* during acclimation to temperatures indicated. Temperature was lowered 1°C/day. [Storey (133a).]

resemble those of Antarctic fishes, indicating either evolutionary convergence or common ancestry.

In some north temperate and subarctic marine fishes, polypeptides rather than glycoproteins serve as antifreeze compounds. Most of these polypeptides are rich in alanine. Urine and endolymph of Antarctic fishes do not contain antifreeze glycoproteins; the kidneys are aglomerular. Some northern glomerular fishes lack antifreeze peptides in urine because of a charge repulsion mechanism that prevents the acidic peptides from filtration barrier. The antifreezes are synthesized in large amounts as winter approaches; a winter flounder *Pseudopleuronectes* has an antifreeze content <1 ng/mL of blood in summer and 25 mg/mL in winter. All antifreezes are expanded in secondary structure. One hypothesis for the freezing depression is that by adsorption inhibition, they retard ice formation. The strongest bonding occurs in antifreezes with the most hydrogen, carboxyl, and amino groups that can form hydrogen

freezing rather than as supercooling agents and fishes that contain antifreezes live in the presence of ice and at temperatures lower than those at which their colligative properties would permit freezing. In the Antarctic fish *Trematomus* the freezing point of serum is −2.75°C and the melting point −0.98°C; this contrasts with the temperate water black perch *Embiuu* in which both freezing and melting points are −0.7°C (34). The serum of an Antarctic notothenid fish contains eight glycoproteins with MW = 2600 to 33,700 D; each consists of a skeleton of two alanines and one threonine as repeating units with a disaccharide attached to each threonine (34a,b) (Fig. 9). In some antifreeze glycoproteins, proline replaces alanine. Some fish from Bering Sea or Labrador have glycoprotein antifreezes that

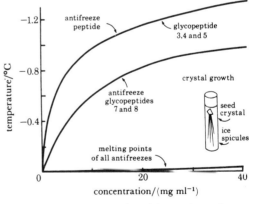

Figure 8. Freezing and melting points of peptide and glycopeptides at several concentrations. The large glycopeptides and the polypeptides have similar antifreeze activity, low molecular weight glycopeptides (numbers 7 and 8) have less action in lowering freezing points. Melting points are the same for all sizes of antifreezes. [DeVries (34b).]

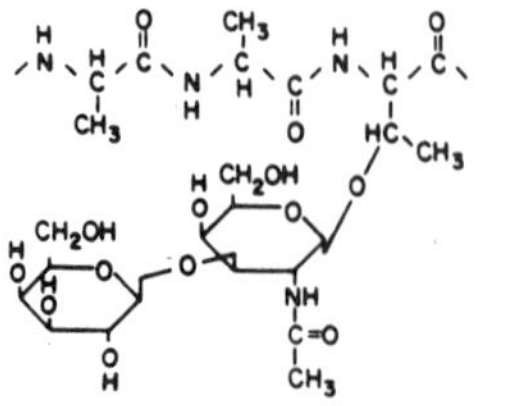

(a)

Ala-Ala-Thr-Ala-Ala-Thr-Ala-Ala-Thr-Ala-Ala-Thr-Ala-Ala-Thr-Ala-Ala
 Naga   Naga   Naga   Naga   Naga
  Gal   Gal   Gal   Gal   Gal

Asp-Thr-Ala-Ser-Asp-Ala-Ala-Ala-Ala-Ala-Leu-Thr-Ala-Ala-Asp-
  Ala-Ala-Ala-Ala-Ala-Ala-Leu-Thr-Ala-Ala-Asp-
  Ala-Ala-Ala-Ala-Ala-Ala-Ala-Thr-Ala-Ala-X

|←——— 4.5 Å ———→|

Figure 9. (*a*) Polymer unit of glycopeptide with antifreeze properties. Eight glycopeptides contain repeating units ranging in molecular weight from 2600 to 33700 Da. The polypeptide consists of a backbone of alanine and threonine (ala-Ala-Thr) with a disaccharide joined to every threonine (34). (*b*) The primary structure of a glycoprotein from the Antarctic mot.hetheniid, *Dissostichus mawsoni*. Naga—N-acetyl galactose amine. (34b) (*c*) Shows the structure of one of the antifreeze peptides isolated from the blood of the winter flounder, *Pseudopleuronectes americanus*. In the conformation of a helical rod, the aspartic and threonine residues are separated by 4.5 Å, a distance that also separates the oxygen atoms on the prism faces parallel to the *a* axes in hexagonal ice (34b).

bonds with oxygen or hydrogen atoms in the ice lattice (34b,41). The cDNA for flounder antifreeze polypeptides has been cloned and sequenced. The genes for antifreeze peptides are linked in tandem in the genome of winter flounder.

Capacity Adaptations; Homeokinesis

Capacity adaptations are properties that permit constancy of animal activity (homeokinesis) when T_b changes. Capacity adaptations may be genetically deter-

mined or environmentally induced within genetic limits.

Genetically Determined Capacity Adaptations

Several kinetic properties of proteins are adaptive for temperature experienced by tissue of animals. These are properties that are adaptive for direct responses to temperature, properties that have been accumulated by selection over long periods. The change in kinetic activity of many chemical reactions is ~0.3%/10°C; yet, most enzyme-catalyzed reactions change E_a by >100 to 300%/10°C. Apparent energy of activation E_a is usually based on reaction velocity at saturating concentrations of substrate (V_{max}). Enzyme reactions in poikilotherms are relatively temperature independent, that is, they have less steep slopes on Arrhenius plots than enzymes of homeothermic animals. Enzyme reactions of homeotherms may show breaks, slope changes at temperatures well below T_b; curves for the same enzymes of poikilotherms may not show breaks. An apparent K_m (Michaelis constant) is the substrate concentration corresponding to half-maximal velocity, which is generally near or equivalent to the physiological concentration of substrate. In general, K_m rises with an increase in temperature, more steeply at high than at low temperatures. The position of K_m–temperature curves for poikilothermic animals that live at high temperature lie to the right of those for low-temperature animals (Fig. 10). The energy of catalyzed reactions can come from internal kinetic activity—enthalpy ($\Delta H^\ddagger$) or from environmental heat entropy ($\Delta S^\ddagger$). The change in free energy is given by $\Delta G^\ddagger = T\Delta H^\ddagger - T\Delta S^\ddagger$. For similar changes in free energy the entropic contribution is greater in ectotherms and the enthalpic contribution is greater in endotherms. The higher activation energies for homeotherms than for poikilotherms

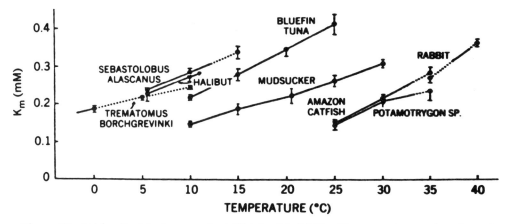

Figure 10. Michaelis–Menton constant (K_m) for M_4-LDH with pyruvate as a function of temperature for species living in Antarctic (*Trematomus*, *Sebastolobus*), temperate (tuna, mudsucker), and tropical waters (Amazon catfish, *Potamotrygon*). All measurements made at pH values corrected for temperatures. [Yancey and Somero (154).]

indicate that enzymes from homeotherms are more temperature sensitive than those from poikilotherms (77). Activation free energy, entropy, and enthalpy of enzymes from ectotherms are lower than of those from birds and mammals (98, 99):

	$\Delta G^{\ddagger}$ (cal/mol)	$\Delta H^{\ddagger}$ (cal/mol)	Animal
LDH-A[4]	13,200	12,550	Rabbit
	12,500	8,800	Tuna
G6PDH	15,300	18,450	Rabbit
	14,700	13,900	Tuna

The changes in free energy $\Delta G^{\ddagger}$ for animals living in cold environments are lower than for the same enzymes of those living in the tropics. Entropy and enthalpy change in the same direction so that there is compensation for low heat content of cold organisms (93a, 129c,d).

Kinetic properties of enzymes may be better than gross metabolism as measures of differences correlated with phylogeny and ecology. Higher activation energies for homeotherms than for poikilotherms suggest that enzymes from homeotherms are more temperature sensitive than those from poikilotherms. The $\Delta G^{\ddagger}$ values for enzymes of species living in cold environments are lower than for the same enzymes of tropical species—that is, enthalpy differences are less in cold adapted forms—and there is compensation for the low heat content (129c). An example is given by myofibrillar ATPases from fishes (82) (Fig. 11).

Source of Fish	V_m (mM) (mmP_i/mg · min) at Habitat Temperature (°C)	$\Delta H^{\ddagger}$ (cal/mol)	$\Delta S^{\ddagger}$ (Entropy Units)	$\Delta G^{\ddagger}$ (cal/mol)
Antarctic	7.7 at −1–2	7,400	−31	15,870
North Sea	3.5 at 3–12	13,530	−9.8	16,290
Indo Pacific	0.31 at 18–26	26,500	+32.5	17,620

The position of the K_m–temperature curve shifts adaptively with the ecotype.

An example is fructose diphosphatase which in a fish *Genypterus* is modulated

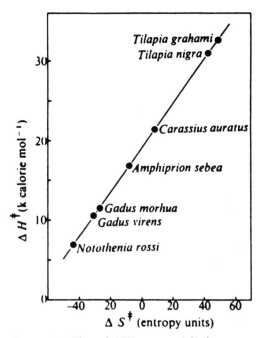

Figure 11. Plot of ΔH‡ versus ΔS‡ for myofibrillar ATPase for fishes from cold water (*Notothenia, Gadus*), temperate water (*Amphiprion, Carassius*), and warm water (*Tilapia*). [Johnston, I. A., and G. Goldspink. *Nature (London)* 257:622, 1975.) (82)

23°C; this enzyme from several fish shows no such break. Enzymes from rat liver—succinic dehydrogenase, cytochrome c reductase—show breaks but toad liver shows no breaks for any of these enzymes (121). In an alpine mammal (chamois) leg muscle is heterothermic and has one LDH, which is relatively independent of temperature; internal tissues are homeothermic and have two LDHs for which the K_m is very temperature dependent (9).

Other adaptive properties of enzymes include rigidity of tertiary and quaternary structure. Folding at weak bonds is sensitive to environmental parameters such as temperature. Thermal stability of a protein is correlated with the cell temperature at which the protein normally functions (Fig. 6). Temperatures for denaturation or for 50% inactivation of pyruvate kinase from animals that function within designated temperature ranges are (129a,b):

Animal	Normal Temperature (°C)	Inactivation Temperature (°C)
Rabbit	30–40	62
Mullet	18–30	58
Bufo	25–32	57
Shark	8–17	52
Abalone	8–15	45
Ice fish	1–2	42

Conformational changes often occur when ligands bind; an enzyme must be sufficiently flexible to permit changes at biological temperatures. Since protein structure may become more rigid at low temperatures it may be adaptive for the inherent rigidity to be low for enzymes that function in the cold. This is reflected by lower heat inactivation temperatures; there is a compromise between rigidity and flexibility.

by AMP such that its activity at 7–13°C is similar to that of the same enzyme in rabbit at 37°C (101). LDH from the benthic fish *Antimora* is less sensitive to temperature than LDH from beef. Adaptive differences in the abundance of two allozymes of liver LDH in *Fundulus* living in a north–south cline were described previously. It appears that one allotype has an advantage in warm water and is disadvantageous at temperatures below 12.4°C (118). Rearing *Fundulus* at different temperatures showed that the population differences are genetic. The K_m–temperature curves of four species of Pacific barracuda show correlation of K_m for LDH-A^4 with habitat temperature; K_m values from cold water species are higher than from warm water species (54).

Malic dehydrogenase from rabbit shows a break in its Arrhenius plot at

Breaks in Arrhenius plots are most frequent for enzymes of mammals and chill-sensitive plants, especially for membrane-bound enzymes. After phospholipids had been detergent extracted from rabbit tissue mitochondria or from chill-sensitive plants such breaks were no longer found (121). Na^+-K^+ ATPase from rabbit kidney showed a break for maximum activity at 16.7°C; the same enzyme showed a break for ouabain binding at 25°C. The breaks in Arrhenius plots were abolished by treatment with phospholipase A (24). Membrane-bound enzymes are influenced by the phospholipid of the membrane to which they are bound. Phospholipid composition is different in homeotherms and chill-sensitive plants than that found in poikilotherms and chill-insensitive plants (33).

Transitions may also be due to the occurrence of two forms of an enzyme, forms that function at either of two temperature ranges. An example is rabbit muscle aldolase, the two forms of which differ in kinetic properties above and below a transition temperature (93a).

Lipids are important in both direct and acclimatory responses to temperature. Storage fats become fluid at high temperatures, solid at low temperatures. Increased unsaturation makes for greater fluidity at a given temperature; fluidity of a membrane can be measured by rotation of fluorescent dyes (polarization fluorescence) and unsaturation/saturation ratios as measured by gas chromatography.

Phospholipids (PL) make up 90% of the composition of plasma membranes and organelle membranes. Phospholipids differ in proportion of PL classes—phospha-tidylcholine (PC), P-ethanolamine (PE), P-inositol, P-serine, and of cardiolipins and plasmalogens. Membranes also differ in relative proportions of cholesterol. Phospholipids differ not only in classes (head groups) but also in fatty acid moieties—chain lengths and unsaturated bonds. Organisms that live at low temperatures have more unsaturated fatty acids in their PLs than organisms that live at high temperatures. Lipid composition differs in thermophilic and psychrophilic bacteria and protozoans. Freshwater crustaceans that overwinter as adults have high proportions of long-chain polyunsaturated fatty acids; crustaceans that overwinter as eggs have smaller proportions of polyunsaturated acids (39). Planktonic copepods from deep cold water have a higher proportion of unsaturated fatty acids than temperate-zone species. The lipid content in marine copepods decreased from 2.1 mg/individual in September to 0.4 mg/individual in June. Pelagic copepods have 25–40% wax esters in their lipids; a high content of waxes provides buoyancy (94). Lipid composition of an animal is in part due to differences in *de novo* synthesis and in part to diet. Temperate shallow-water shrimp contain 32.5% of 16:0 fatty acid and 22% of 18:1 fatty acid while in deep-water crustaceans the corresponding percentages are 9.8 and 52 (93). Similarly, deep water oceanic fish have higher u/s ratios in body lipids than shallow water fishes. Composition of synaptosomal membranes from Arctic, temperate, and hot-springs fishes are (28):

| | Phosphatidylcholine | | | Phosphatidylethanolamine | | | Phosphatidyl-inositol | |
	Arctic Sculpin	Desert Pupfish	Goldfish	Arctic Sculpin	Desert Pupfish	Goldfish	Arctic Sculpin	Pupfish
Monounsaturated	26	34	32	22	22	41	22	31
Polyunsaturated	35	14	23	57	38	55	46	29
Saturated	37	49	45	21	34	34	31	37

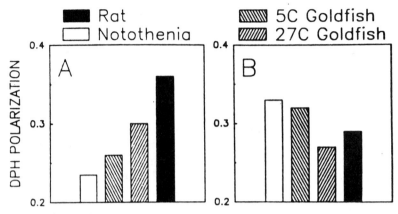

Figure 12. Membrane fluidity as indicated by DHB fluorescence polarization in synaptosomal membranes of rat, Antarctic *Notothenia,* temperature acclimated goldfish. (*A*) Polarization values are compared at assay temperature of 20°C. (*B*) Polarization compared at body or acclimation temperature. [Hazel (63a).]

Membrane fluidity as indicated by fluorescence polarization is correlated with lipid composition (Fig. 12). At equivalent temperature of measurement arctic and temperate-zone fishes acclimated to low temperatures have membranes that are more fluid than warm-water fishes or warm-acclimated fishes (Fig. 13). The net effect is homeovisious adaptation and membrane-bound metabolic enzymes (28, 29a).

Properties of proteins in ion channels may be altered according to the fluidity of their molecular environment. Measurements of lipids in total or bulk membranes only approximate the microenvironment. Methods are now available for examining protein–lipid interactions in artificial membranes.

Mechanisms that correlate with lipid differences are digestion and absorption of dietary fats, enzymes that catalyze incorporation of fatty acids, enzymes of fatty acid desaturation and elongation, and enzymes of acylation–deacylation reactions.

Environmentally Induced Homeokinetic Adaptations

Acclimation under controlled laboratory conditions and acclimatization under complex natural conditions induce phenotypic changes that permit activity over a wide range of temperatures. Biochemical acclimatory properties develop during days or weeks at altered temperatures in contrast to direct responses. Ability for capacity acclimation is genetically limited. Changes in rate functions such as enzyme activity, heart rate, or metabolic rate may be compensatory, noncompensatory, or there may be no change after transfer from one temperature to another or between summer and winter. For patterns of acclimation see Chapter 1. Thermal acclimation may be modified by many environmental factors. The time to death of menhaden exposed to 33.7°C lengthened from 52 min on a short day to 96.6 min on a long day. Acclimatization of mammals outdoors in winter is not equivalent to acclimation to cold in the laboratory. Vasomotor changes are more marked in

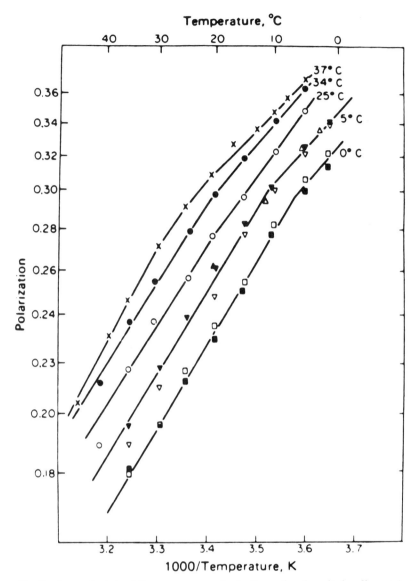

Figure 13. Reciprocal plot of fluorescence polarization of a dye dipheylhexatrine incorporated into synaptic membranes of fishes. Low values of polarization indicate high fluidity. Arctic sculpin from 0°C, goldfish acclimated at 5 and 25°C, desert pupfish from 34°C, rat 37°C. [Cossins and Prosser (29).]

wild animals, tolerance of extreme cold is greater, there is less shivering and more nonshivering thermogenesis. Eskimo fishermen have higher finger temperature when handling cold nets than non-Eskimos. Catfish show compensatory changes in several energy-yielding enzymes during laboratory acclimation to different temperatures; similar fish collected at different times of the year show more complex responses. Photoperiod, nutrition, reproductive phase, endocrine state, and social factors combine to make acclimatization different from acclimation when only temperature is changed.

Some fishes, amphibians, and reptiles enter a state of lethargy in winter and these animals are capable of little temperature capacity acclimation although they may show resistance changes. Turtles show little seasonal acclimatization (124). Of two lizards, *Anolis* shows noncompensatory acclimation of both aerobic and anaerobic metabolism while *Sceloporus* shows positive (compensatory) aerobic but noncompensatory anaerobic acclimation (46). An intertidal amphipod *Talorchestia* shows a higher Q_{10} of O_2 consumption in winter than in summer. Catfish acclimated at low temperatures consume more O_2 at that temperature and fish acclimated at high temperatures consume less O_2 at high temperature than the converse (Fig. 14) (86).

A lungfish *Neoceradotus* shows higher V_{O_2} and lower Q_{10} of O_2 consumption when cold acclimated (56). Oxygen consumption by the heart of goldfish is greater when taken from cold- than from warm-acclimated animals (139).

Animals that live in a relatively constant environment have less capacity for acclimation than animals that live in environments with seasonal cycles. Two species of California rockfish *Sebastes* differ in that one lives in a relatively constant temperature throughout the year. They live near the surface in winter and in deeper water in summer; another species is subject to seasonal cycles near the surface. The species of fish from the more constant environment are less capable of acclimation than the fish from the cycling environment (149). Of 29 species of amphibians, 22 that live in the temperate zone showed acclimation of metabolism; of 7 tropical species only one showed some thermal acclimation (40).

Isolated cells can show some temperature acclimation but the acclimation pat-

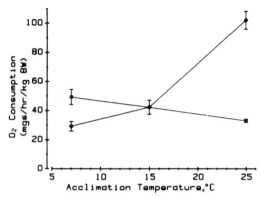

Figure 14. The rate of O_2 consumption (mg of O_2/h · kg of body weight) in 25, 15, and 7°C acclimated channel catfish measured both at their temperatures of acclimation (■) and at an intermediate temperature 15°C (◆). All animals were 24-h post-absorptive at the time of measurement and were acclimated for a minimum of 4 weeks at each temperature prior to use. Each point represents the mean ± SE (*n* = 4) of rate measurements at each of the three acclimation temperatures. [Kent et al. (86).]

tern differs from that *in vivo*. Primary cultures of catfish hepatocytes maintained at three temperatures showed positive acclimation 7 > 15 > 25°C in their activity of citrate synthase, cytochrome *c* oxidase, and G6PDH, whereas *in vivo* cytochrome *c* oxidase showed no acclimation (Fig. 15) (86,90); the pattern of protein synthesis at the three temperatures *in vitro* was in the following sequence 15 > 25 > 7°C, whereas that *in vivo* was 7 > 15 > 25°C (90). Factors (hormones) in serum modify protein synthesis when added to hepatocytes in culture. Hamster ovary cells were cultured at several temperatures; those cultured at 39°C tolerated 43°C heat better than those cultured at 12°C. There is considerable specificity of enzymes and of organs for thermal acclimation. In aerobic organisms enzymes that function in energy liberation, particularly enzymes of the TCA cycle and electron transport systems

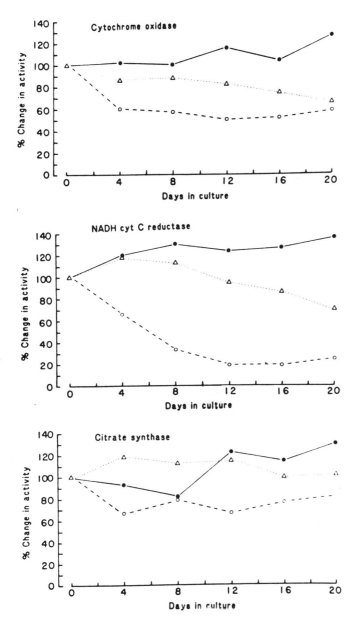

Figure 15. Changes in activity of three enzymes in cultured hepatocytes of catfish. All curves normalized to initial values. Solid lines are cells cultured at 7°C, dotted lines at 15°C, and dashed lines cultures at 25°C. [Koban (90) by permission of *Amer. Phys. Soc.*]

show the most temperature compensation; glycolytic enzymes show less compensation, sometimes none; degradative enzymes, hydrolytic enzymes, phosphatases, and enzymes of nitrogen metabolism show no or inverse acclimation (63a,101). Enzymes of ion transport show much variation of response according to the tissue and kind of animal. Table 3 shows both organ and enzyme differences in acclimation of green sunfish (126). Comparisons of respiratory quotients, of the effects of enzyme inhibitors, and of metabolic intermediates in striped bass acclimated to low or high temperatures indicate the following: increased he-

TABLE 3. Enzyme Activities in Tissues of Green Sunfish Acclimated to 5° and to 25°C at Ambient Dissolved Oxygen Concentrations Measured at 15°C[a]

Enzyme	Muscle		Liver		Brain	
	25°C	5°C	25°C	5°C	25°C	5°C
Aldolase	1.55	0.59	0.0827	0.100		
Pyruvate kinase	2.04	1.50	0.189	1.138	1.07	1.57
Lactate dehydrogenase	2.06	1.86	0.035	0.0287	1.18	1.44
Succinic dehydrogenase	3.16	4.26	0.176	0.536	2.79	4.02
Cytochrome oxidase	0.0115	0.0223	0.0061	0.0237		
Cytochrome *c*	0.99	1.46				

[a]See Shaklee et al. (126).

patic activity of pentose shunt in cold, more deposition of fat in liver in cold, and an increase in the number of red oxidative fibers in muscle in cold (82). White muscle of brook trout shows compensatory acclimation of myofibrillar ATPase but red muscle does not show this change (128a). Enzymes of aerobic metabolism of pickerel muscle, *Esox,* are more active from winter fish than from summer fish, from 5 rather than from 25°C acclimated ones; LDH shows the reverse effects (87). LDH of glycolytic white muscle of the fish *Gillichthys,* shows noncompensatory change with temperature. The rate of protein synthesis in muscle of carp, a eurythermal fish, showed positive acclimation; in rainbow trout, a stenothermal fish, no acclimation occurred (87). Cytochrome *c* oxidase of goldfish red muscle showed no change in K_m but positive acclimation of specific activity on acclimation from 30 to 10°C (150). Since the cytochrome *c* oxidase is membrane bound and since the protein appears unchanged,

positive acclimation of specific activity may be due to membrane lipids. Hepatocytes of European eel, unlike hepatocytes of goldfish, show no acclimation of oxygen consumption; however, eel hepatocytes show a positive acclimation of glycogenolysis, an increase in glycogen content, and an increase in ketogenesis in cold acclimation (80).

When enzyme activities are expressed on a protein basis, the acclimatory differences are much less than when expressed per total tissue weight. This indicates that one mechanism of change in metabolism is increase in the amount of catalytic protein in cold and a decrease in warm acclimation. In green sunfish muscle the concentration of cytochrome *c* was 50% more in fish acclimated to 5°C compared to that in fish from 25°C (127). In eels acclimated to 7 and 25°C the mitochondrial protein of liver was 1.5 times greater and in muscle 2 times more in the 7°C acclimated fish (152). Total protein in these tissues did not change much. The

Figure 16. Quantitation of G6PDH in liver from 25 and 15°C acclimated catfish. Rocket immunoelectrophoresis with specific antibody. Higher peaks for 15°C samples indicate greater quantities than from 25°C acclimated fish (J. Kent, unpublished).

TABLE 4. 15/25°C Ratios of Steady-State Physical and Biochemical Parameters of Liver from 15 and 25°C Acclimated Channel Catfish

Body Weight	hsi[b]	mg Protein/ g Wet Weight	mg Protein/ Total Liver	mg DNA/ g Wet Weight	mg DNA/ Total Liver	DNA/ Protein	% Dry Weight	Total Lipid
0.90± 0.07	2.19±[c] 0.09	0.93± 0.06	2.18±[b] 0.16	0.50±[c] 0.06	1.04± 0.11	0.43±[c] 0.05	0.97	0.73[b]

[a]From Kent et al. (86).
[b]hsi = hepatosomatic index.
[c]$p = 0.05$.
[d]15 to 25°C ratios are expressed as $\overline{X} \pm SEM$.

specific activity of cytochrome oxidase of liver decreased at 7°C but total organ activity increased (152). Increase in absolute concentration of G6PDH and decreases in LDH are shown by rocket immunophoresis (Fig. 16).

Changes in concentrations of substrates and cofactors also occur in temperature acclimation. Data for muscle LDH activity with pyruvate and lactate as substrates in the fish *Gillichthys* at 15 and 25°C are (142)

Acclimated (°C)	Pyruvate (mM/kg) (wet weight)	Lactate (mM/kg) (wet weight)
15	0.021	3.61
25	0.043	4.79

In channel catfish the hepatosomatic index (ratio of liver to body weight) of 15°C acclimated fish is approximately double that of 25°C acclimated fish (Table 4, Fig. 17). The DNA is unchanged and the average cell size is increased at 15°C. Total protein increases in proportion to the increase in cell size, that is, protein hypertrophy occurs. Green sunfish showed little cell hypertrophy but liver glycogen decreased while liver protein increased. In cold acclimation there appears to be increase in synthesis of specific enzyme proteins and there is also some increase in total protein synthesis. Ventricular hypertrophy occurs in cold-

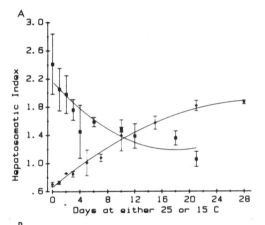

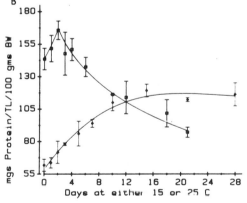

Figure 17. The time course of change in hepatosomatic index (*A*) and liver protein (*B*) following the transfer of channel catfish from either 15 to 25°C (♦) or reciprocally from 25 or 15°C (■). Each time point represents the mean ± SE of either liver mass or protein content determined for 3 to 4 fish. A 2-day transient increase in liver protein content followed the transfer of fish to 25°C from 15°C. [Kent et al. (86).]

acclimated carp (50) and kidney hypertrophy in catfish. Cytochrome c oxidase mitochondria in carp compensates much more in muscle than in liver (151).

Ratios of Activities 7°/25°

	Liver	Muscle
Activity/mg of protein	0.64	1.85
Activity/g (wet weight)	0.98	3.48
Activity/organ	1.82	5.14

One mechanism of compensatory acclimation is selective synthesis of certain allozymes and isozymes at different temperatures. In trout liver the proportion of two isozymes of pyruvate kinase changes with the season. In trout brain, acetylcholine esterase occurs in two isozymal forms, one predominant in winter, the other in summer. The K_m of the winter form is lower at all temperatures than the K_m of the summer form (4). Warm acclimation induces an increase in myofibrillar ATPase in goldfish muscle (81, 81a, 141). The most extensive changes in enzyme patterns have been described for polyploid species of salmonid and cyprinid fishes (148). In diploid species (Gillichthys, sunfish) polymorphism is virtually lacking and isozymes of such polymeric enzymes as LDH show no correlation with the temperature of acclimation. In the mullet, *Mugil*, G6PDH occurs in two forms, each a tetramer. Form I is inhibited by co-factor NADPH, more at high than at low temperature; the relative affinity of form I for substrate decreases in the cold and its affinity for co-factor increases. Form II has a K_m that is virtually independent of temperature; form I increases relative to form II in warm water (76).

Lipids

Since proteins are genetically coded, their amino acids do not change during temperature acclimation; proteins do change

adaptively by genetic selection. The changes in lipids during acclimation are similar to those described previously for genetic adaptations. Lipids change in fatty acid composition and in proportion to different classes of phospholipids (PL). There may be changes in other related compounds such as cholesterol and prostaglandins. Storage fats are more unsaturated when laid down at low than at high temperatures. Superficial fats such as subcutaneous fat and fat in exposed legs of mammals have lower melting points in winter than in summer and lower melting points than visceral storage fats. The most important adaptive function of lipid changes is maintenance of relatively constant fluidity of membranes at all temperatures (63).

The changes in fatty acid components of phospholipids are mainly an increase of unsaturated as compared to saturated fatty acids in cold acclimation; also there is some tendency to have longer fatty acids in the cold. Data for trout liver mitochondrial membranes follow (63):

	5°C Acclimation	25°C Acclimation
16:1/18:1	39	23
16:1/18:1	5	83

The temperature of breaks in the Arrhenius plots of Na^+-K^+ ATPase from gill membranes of acclimated eels correlates with fatty acid composition (136a):

Acclimation Temperature	12°C	5°C
Ratio of saturated and monounsaturated to polyunsaturated fatty acids	1.6	1.1
Percentage saturated fatty acids	33.6	23.7
Temperature of break in Arrhenius plot	2°C	11°C

Microsomes from goldfish intestine show an increase in percentage of fatty acids 22:6 and 20:4 in phosphatidylethanolamine in cold acclimation (104b). When *Tetrahymena* are shifted from 39.5 to 15°C for culture there are no cell divisions during 10 h but the ratio of unsaturated to saturated fatty acids increases with a peak at 6–8 h (110). An Arrhenius plot of the fluidity of *Tetrahymena* cell membranes shows a break at 16°C for 38°C cultured cells and at 9°C for cells from 20°C cultures (26, 35). Microsomal membranes show breaks in fluidity curves, ciliary membranes show no breaks. In trout kidney, desaturation products of the $(n - 3)$ series (20.5) and of the $(n - 6)$ series (20.4) are preferentially incorporated into PLs at low temperatures, and U/S ratios for incorporation are (65):

Assay Temperature (°C)	Cold Acclimated	Warm Acclimated
5	2.55	3.2
20	1.48	3.8

Mechanisms that may account for fatty acid changes with temperature are enzymes of fatty acid desaturation, elongation, enzymes of acylation, and deacylation such as acyl transferases and phospholyceride transferases.

Classes of phospholipids change during temperature acclimation. When trout are transferred from 20 to 5°C, the plasma membranes of kidney cells show an initial increase in the proportion of PE with a peak at 24 h and a decrease in PC. Changes in percentage of unsaturated fatty acids occur later. The head groups of PC are large and cylindrical, those of PE are smaller and conical in shape. In gill membranes of trout PE increases in cold acclimation and PC decreases; the change in PE is faster than in PC (63, 64). In *Tetrahymena*, the changes in acyl chains

is faster than in PL classes. Replacement of PC by PE at cold temperatures may exert a disordering effect on membrane organization since the conical shape of PE does not pack efficiently in the lamellar phase. In crustaceans, biosynthesis of PC is more temperature dependent than that of PE. Conversion of PE to PC was twofold higher in hepatopancrease at 23°C compared with 13°C acclimated *Carcinus*. Changes in membrane fluidity may alter the activity of bound enzymes and the permeability of plasma membranes. Succinic dehydrogenase (SDH) of goldfish mitochondria is activated more by PL extracted from 5°C acclimated goldfish than from 25°C acclimated fish. The amount of activation depends on the ratio of unsaturated to saturated fatty acids in the membrane PL; the threshold concentration of PL depends on the proportion of classes of PL (84). Mitochondrial membranes of carp liver have more PE and less PC when cold acclimated and the membranes are more negatively charged in the cold (150, 151). In general, phosphatidylserine (PS), phosphatidylinosine (PI), and sphingomyelin changes less than phosphatidylethanolamine (PE) and phosphatidylcholine (PC).

An additional temperature-induced restructuring of fish membranes in temperature acclimation is an increase in the proportion of plasmalogens (ether containing PLs) as compared with diacyl forms of PLs at warm temperatures. Changes in proportion of plasmalogens are particularly evident in neural tissue. Cholesterol synthesis is increased 10-fold in the microsomes of carp that are acclimated to low temperature (151). Cholesterol–PL interactions are related to membrane fluidity. Ciliary membranes of *Tetrahymena* have more of a triterpenoid when cultured at low temperatures.

The precise manner in which membrane fluidity as determined by PLs affects protein function is poorly understood. Evidence from X-ray diffraction measurements on *Tetrahymena* membranes indicate that microsomal membranes have both fluid and ordered states of lipids and that the proportion changes with temperature, the percentage in the fluid state being greater at high temperatures (26).

In addition to effects of lipids on membrane structure and on activities of membrane-bound enzymes, changes occur in lipid utilization as compared with carbohydrate metabolism. In striped bass muscle, palmitate oxidation is twofold higher and glucose oxidation 38% lower in 5°C acclimated than in 25°C acclimated fish (85).

Acid–Base Balance

A property of aqueous solutions that modifies the effects of temperature is acid–base balance. Water declines in pH by 0.016 pH unit per degree C rise in temperature; water is neutral at 22°C, at a lower pH when warmed, and higher pH when cooled (Fig. 18). The most common buffer in biological fluids is the imidazole group of histidine in which, for different compounds $\Delta pH/\Delta T$ is 0.016–0.20 (120a). The pH–temperature curve for blood parallels that of water but the pH is higher than that of water at each temperature. Curves of intracellular pH are similar in slope to those for blood although the curves vary along the pH axis. In hibernating mammals a relatively constant blood pH is maintained at different temperatures; hibernators are acidotic at low body temperatures. The slopes of pH–temperature relations vary according to a combination of buffers and according to the method of acid–base regulation in a particular animal.

Changes in temperature tend to result in a constant $[OH^-]/[H^+]$ ratio rather than a constant pH. When enzyme activities are measured at different temperatures at a constant $[OH^-]/[H^+]$ rather than at a constant pH, the results are

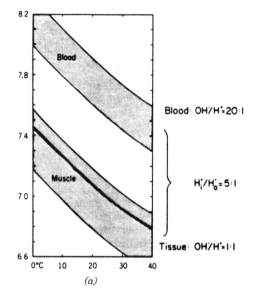

(a)

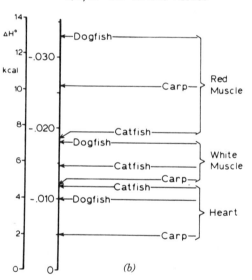

$\Delta pH/\Delta T$ for Various Tissues

(b)

Figure 18. (*a*) Range of pH for blood and muscle of vertebrates as a function of body temperature. Solid line is muscle in relation to water. [Rahn et al. (120a).] (*b*) Observed $\Delta pH/\Delta T$ for red muscle, white muscle, and heart from dogfish, carp, and catfish. Calculated $\Delta H°$ in kcal at far left. [Cameron (20) by permission of *Amer. Phys. Soc.*]

more physiological and K_m values are less sensitive to temperature (154). At constant pH, K_m for LDH_{pyr} varied according to temperature from 0.07 to 0.8 mM, whereas at constant $[OH^-]/[H^+]$ the range of K_m was only 0.15 to 0.35 mM (154). Respiratory and excretory control of acid–base balance become adjusted toward the passive pH–temperature shifts. Air-breathing animals maintain a constant $[OH^-]/[H^+]$ ratio by adjusting blood CO_2 while maintaining a constant $[HCO_3^-]$. Water breathing animals regulate acid–base balance mainly by varying blood HCO_3^-. A water breathing shark *Scyliorhinus* adjusts acid–base balance by 90% in <1 h and completely in 10 h; the slope of the pH–temperature curve is 0.012 for plasma and 0.018 for white muscle (Fig. 18*b*) (72).

A turtle *Chrysemys picta* was acclimated at temperatures from 5 to 30°C; its V_{O_2} increased by 6.7 times, ventilation by 4.4 times, arterial P_{CO_2} increased from 13 to 32 mmHg, and plasma pH decreased with a slope of 0.01 pH unit/°C (48). Iguanas held at different temperatures showed serum $\Delta pH/\Delta°C$ of -0.004 between 16 and 24°C, -0.015 between 24 and 35°C; change in pH after temperature change occurred in 18–24 h (1). In the lizard *Disosaurus*, pH of muscle is lower at night than at day and hypercapnia occurs at night (9a).

Muscle and Central Nervous Adaptation

Acclimatory changes in the properties of muscle membranes and contractile proteins and of central nervous integration control locomotor activity over a range of temperatures. Muscles from Antarctic fishes contract two times faster at low temperatures (2–5°C) than muscles from temperate fishes but at higher temperatures (14–16°C) the muscles from temperate zone fish are faster. Velocity of contraction but not force correlates with normal body temperature (129, 129b). As winter approaches, some freshwater fishes migrate to warmer water, others enter a state of lethargy but many freshwater fish continue to swim and feed beneath the ice. Many marine fish migrate seasonally and experience relative constancy of temperature, others remain in shallow water and experience an annual thermal cycle.

Acclimatory responses in the muscles of intact fish muscle include changes in contractile properties as seen in isolated muscle, changes in volume occupied by mitochondrial changes in proportions of red slow and white fast fibers, and changes in neural recruitment during swimming (Tables 5 & 6). Muscles of cold acclimated cyprinids such as carp and goldfish are capable of faster contraction and of more force than of warm-acclimated carp when measured at an intermediate temperature. Volume percentage occupied by mitochondria in goldfish muscle is greater at 5°C acclimation than in 25°C acclimated fish. Proliferation of mitochondria is greater in red than in white muscle fibers (Table 6); carp muscles show compensatory changes in activities of citrate synthase, cytochrome *c* oxidase and in myofibrillar Mg^{2+}, Ca^{2+} ATPase. In carp, the maximum force developed is nearly independent of temperature ($Q_{10} = 1$), but energetic cost is sensitive to temperature ($Q_{10} = 3$) (97, 123). Carp acclimated at 10°C can swim at >30 cm/s before recruiting white fibers (123); white fiber recruitment compensates for lower power output in the cold. Speed for white muscle fiber recruitment when swimming at 10°C is (123):

Speed (cm/s)	Acclimation Temperature (°C)
31	5
26	15
22	26

A different strategy is found in striped bass, yellow perch, and pickerel (noncyprinids). Myofibrillar ATPase from red

TABLE 5.

Parameter	Acclimation Temperature	
	7°C	23°C
Oxidative (Red) Fibers		
	(9)	(10)
Isometric force (kN/m²)	76 ± 9	47 ± 6
Unloaded contraction velocity, V_{max} (L_0/s)[b]	0.95 ± 0.08	0.42 ± 0.05
Maximum paver output (W/kg)	8.5 ± 0.8	2.4 ± 0.5
Glycolytic (White) Fibers		
	(16)	(10)
Isometric force (kN/m²)	178 ± 15	76 ± 10
Unloaded contraction velocity, V_{max} (L_0/s)	1.8 ± 0.5	0.62 ± 0.04
Maximum power output (W/kg)	19.3 ± 2.2	6.9 ± 0.6

[a]Data are from Johnston, Sidell, and Driedzic (84). All values shown are mean ± SEM, for the number of fibers indicated in parentheses.
[b]L_0 = initial muscle length.

and white muscle differs in kinetics (Fig. 19). Velocity of a contraction, force developed, and activity of myofibrillar Mg^{2+}, Ca^{2+} ATPase are not influenced by acclimation but swimming is sustained in cold acclimation by proliferation of red muscle fibers. The percentage of muscle volume occupied by mitochondria in cold acclimation increases more in red than in white fibers in both goldfish and striped bass. Diffusion path lengths between sarcoplasmic and mitochondrial compartments decrease for the lower diffusivity of O_2 at low temperature (38a). In both

TABLE 6. Mitochondrial Volume as Estimated from Stereologic Measurements[a]

	Mitochondria	Acclimation Temperature	
		25°C	5°C
Goldfish			
Red fibers	Volume percentage	6.3 ± 1.3	18.1 ± 1.3
White fibers	Volume percentage	0.8 ± 0.1	1.5 ± 0.2
Striped Bass			
Red fibers	Volume percentage	0.286	0.45
White fibers	Volume percentage	0.027	0.040

[a]See Egginton and Sidell (38a) and Sisson and Sidell (129b).

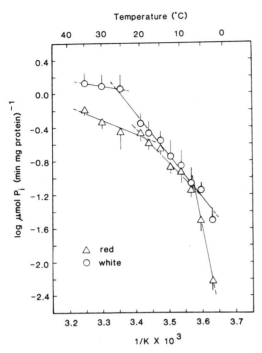

Temperature (°C)

Figure 19. Arrhenius plot of maximally activated ATPase of myofibrils prepared from the red (△) and white (○) skeletal muscle of striped bass. Data are pooled from animals acclimated to 5 and 25°C. ATPase from red muscle shows 19 two breaks, that from white muscle one break. E_a values specific for red or white muscle and independent of acclimation. [Moreland and Sidell (105).]

types of fish in cold acclimation, mitochondria increase in number, not in size. In striped bass, red fibers are added in cold acclimation resulting in 1.7 times more power output in sustained swimming (129b). Recruitment of motor neurons is compressed into a narrower range of speeds in cold than in warm water. Individual muscle fibers do not acclimate in force and velocity but addition of red fibers provides increased power (129a). Some proliferation of red fibers occurs in cold acclimation in cyprinids as well as in noncyprinids (83, 85, 128).

In many fish muscles, contractions are triggered at moderate temperatures by excitatory junction potentials rather than by spikes. When green sunfish were acclimated to 7°C, muscle membrane resistance was higher than when acclimated to 15 or 25°C and the threshold for spike generation was reduced, hence swimming in cold-acclimated fish may be triggered by spikes, while swimming in warm-acclimated ones may be triggered by junction potentials (88).

When green sunfish were swum at different speeds over a range of temperatures, the temperatures at which swimming was induced on warming and at which it failed on cooling, and the maximum speed of swimming were higher for 25°C acclimated than for 15°C acclimated fish; 5°C acclimated fish could not be induced to swim (123a). The velocity of a lizard, if running, is maximal at the preferred temperature of 35°C, and is limited by the twitch contraction time of leg muscles (101a). Increase in the proportion of red fibers occurs in lizards in the cold as in fishes (123).

Thermoreception and Thermoregulation

All thermoregulation, ectothermic or endothermic, is initiated by temperature receptors, integrated by the central nervous systems, and effected by motor actions or metabolic-circulatory responses. Some thermoreceptors are peripheral, some central; peripheral receptors may be multimodal—thermal and mechanoreceptors; central receptors may be sensitive to both temperature and chemical stimulation. In a thermal gradient, most motile animals go to a "preferred" temperature. In a gradient, parasitic worms that invade warm hosts select higher temperatures than free-living worms. Ciliate protozoans and worms detect a temperature gradient by probing the environment with alternating movements, left and right. In *Tetrahymena* the amplitude of helical swimming decreases and velocity increases as temperature is raised (26). In some insects, the antennae perceive air temperature, the tarsi perceive ground temperature. In a beetle, antennal receptors discharge spontaneously and on cooling, impulse frequency increased 24/s · °C; on warming, frequency increased by 14 impulses/s · °C (96). Receptors on the antennae of a cockroach respond to both the rate of temperature change and to ambient temperature; steady-state responses were 0.6 impulses/s · °C, rate of change responses were 200 impulses/s · rate of cooling in degrees per second (96). Temperature-sensitive pads on the tarsi of cockroaches showed increased activity on cooling below 10°C or on warming above 30°C. Many insects have thermosensitive neurons in thoracic ganglia. A cockroach has thermosensitive neurons in the prothoracic ganglion whose firing rate increases at temperatures above the preferred 26–27°C and other neurons whose firing rate increases at temperatures below the preferred temperature (111). Removal of antennae and tarsi has little effect on temperature selection; the central nervous system is the essential thermoreceptor (111). Moths with connectives cut between head and thorax respond behaviorally at the same temperature as unoperated controls (102). Local heating or cooling of the thoracic ganglia causes a cecropia moth to shift to and from flight and warm-up motor sequences in the thoracic muscles (59).

Fish have thermoreceptors in the skin and the preoptic hypothalamus. Trout have been trained to respond to rapid changes of water temperature as small as 0.1°C. After 15°C acclimation, *Leuciscus* mechanoreceptors were maximally sensitive to touch at 22°C and after 5°C acclimation at 18°C (129e). Elasmobranch fishes have large sensory bulbs, ampullae of Lorenzini, on the surface of the head from which nerve fibers discharge autonomously with maximum frequency at the usual habitat temperature. Rapid cooling causes a transient increase in frequency, warming a decrease in frequency. Ampullae of Lorenzini are highly sensitive to mechanical stimulation and to electric fields as well as to temperature.

Pit vipers (*Crotalida*) have facial temperature-sensitive pits; boas have labial pits. The function of the pit organ is to find warm-blooded prey in the dark. The membrane at the base of a pit has many free nerve endings (17). An air space behind the membrane may reduce heat loss. The organ detects radiant heat and responds to long infrared wavelengths; heat detection is directional and shadows passing over the pit are readily perceived. From modulation of the steady discharge in a sensory nerve from the pit it was calculated that as small a change as 0.003°C in the pit membrane can be detected (111a). A temperature rise of 0.4°C at the membrane increased the frequency of nerve impulses from 18 to 68/s (17). Central responses show convergence, center-surround sensory fields (131a). A snake can strike within 5° of the center of a sensory field (116a).

Selected temperatures of bullfrog tadpoles increase from 24°C in early stages to 32°C in late stages; temperature for maximum steady firing by cutaneous thermoreceptors also increases but is not the same as selected temperature (120).

In mammals, heat and cold receptors are distributed in definite patterns in the skin; warmth receptors usually lie deeper than cold receptors; the latter may be more abundant. Thermoreceptors cooled or heated show either tonic or phasic changes in impulse frequency. Cold and warmth receptors show endogenous activity; fibers from warmth receptors are active over the range from 20 to 47°C with maximum frequency between 38 and 43°C; fibers from cold receptors are active from 20 to 40°C, maximum activity is between 20 and 34°C (130, 131). When the skin temperature is abruptly raised there is a brief increase of discharge from heat receptors and when skin temperature is lowered a cold fiber gives a brief discharge. Some cold fibers remain silent at temperatures between 35 and 45°C, but discharge at between 45 and 50°C, a paradoxical response to heat. The beaks and tongues of birds contain thermosensitive nerve endings.

One hypothesis for stimulation of cold receptors in the skin of frogs is that cold blocks the Na^+-K^+ pump and thus depolarizes; treatment with ouabain abolishes the response to cold (130, 131). In cat nasal skin the temperature for maximum activity of cold receptors is 27°C, of warmth receptors 46°C (75). Some temperature receptors in mammals are located on veins.

Cooling the spinal cord induces shivering and vasoconstriction in the dog; warming the spinal cord stops shivering and brings about panting and vasodilation. The spinal cord may be the principal thermoreceptor in birds (pigeon) (5).

The preoptic anterior hypothalamus (POAH) is a temperature sensing and regulating center in fishes, birds, and mammals and likely in other vertebrates as well. Some neurons of this region of the brain receive input from temperature-sensitive cells—peripheral receptors, and receptor neurons in other regions of the brain and spinal cord. Some neurons of the POAH are highly temperature sensitive. In rabbit, brainstem stimulation affects 64% of cold sensitive neurons in the POAH, 60% of warm sensitive, and 17% of temperature-insensitive neurons. Hippocampal stimulation similarly affected thermosensitive neurons of the hypothalamus (14, 16). In rat, some 30% of POAH neurons are warm sensitive with a $Q_{10} = 2.0$, 10% are cold sensitive with $Q_{10} = 0.5$, and 60% are temperature insensitive with $Q_{10} = 0.5$–2 (3). In rabbit hypothalamus, a distinction between receptors and pas-

sive temperature effects is that warm-sensing units increase the firing rate by 0.7 impulse/s · °C with a $Q_{10} = 2$, cold sensing units decrease firing as temperature rises with a slope of -0.6 impulse/s · °C (78). Responses have been examined both *in vivo* and from brain slices. Approximately one half of the temperature-sensitive neurons are affected also by other stimuli such as osmotic concentration or sex hormones. Passive effects of temperature and responses leading to thermoregulation are not readily separated. In rabbit hypothalamus a distinction between receptors and passive temperature effects is that warm-sensing units increase the firing rate by 0.7 impulse/s · °C to a $Q_{10} = 2$, cold-sensing units decrease firing as temperature rises with a slope of -0.6 impulses/s · °C (15). A receptor for heat or cold activates regulatory responses; this is in contrast with other neurons that show simple kinetic effects of temperature (14). Hypothalamic neurons that receive input from outside the hypothalamus show synaptic potentials; temperature-receptor cells are not synaptically activated (15).

Cold receptors in brain are stimulated either by inhibition of warm receptors or by the temperature optimum for firing being below ambient temperature. Receptor responses are (1) phasic transient, *on* and *off* or (2) tonic, prolonged change in firing rate.

Figure 20 diagrams the firing rate as a function of hypothalamic temperature and presents a circuit model for temperature regulation. Responses of mammals to heating include panting, behavioral seeking of cold, vasodilation, and sweating. Responses to cold are vasoconstriction, increased insulation (by erection of hair or feathers), behavior, shivering, or nonshivering thermogenesis (15). Lesions to the preoptic hypothalamus disrupt thermoregulation (89). Local warming of the anterior hypothalamus of

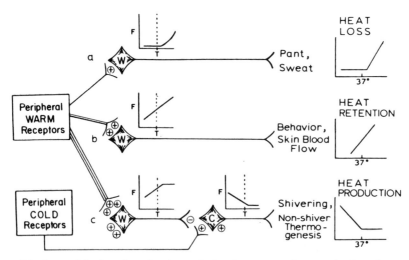

Figure 20. A model showing the hypothalamic neuronal control of various thermo-regulatory responses. W = warm-sensitive neuron; C = cold-sensitive neuron; (+) = excitatory input; (−) = inhibitory input. Hypothalamic thermoresponse curves of different neurons and thermoregulatory responses are shown. F = neuronal firing rate; T = hypothalamic temperature; dashed line = thermoneutral temperature. [Boulant and Dean (15).]

a cat causes peripheral vasodilation and increase in breathing rate. At a T_a of 25°C, local cooling of the POAH leads to hyperthermy; at a T_a of 35°C, local cooling is ineffective. Rats cooled and trained to press a bar for heat, pressed when the hypothalamus was cooled. Hypothalamic neurons are 10 times more sensitive to temperature than skin receptors. Hypothalamic units respond to temperature change of face and scrotum. Cold sensitive neurons may be dopaminergic (125).

Ectothermic thermoregulation is behavioral. Insects and reptiles at low T_b bask in the sun and orient the body to absorb heat. These animals can move only after the body temperature rises to a critical range. When the ambient temperature becomes too high, reptiles remain in the shade or in a burrow. In a thermal gradient, reptiles and insects go to the temperature that corresponds to optimal activity. Reptiles can be trained to regulate their body temperature by turning on or off a heat lamp according to T_a. The body temperature response of lizards to warming is faster than to cooling; turtles cool faster than they warm. This hysteresis in heat loss or gain is primarily by dermal vasomotor means. Large reptiles are very slow to change temperature, hence they show some constancy of T_b (57). Box turtles in high T_a keep the brain as much as 10.5°C below ambient temperature by evaporative cooling (108). Behavioral thermoregulation is a general property of vertebrates and in fishes the same preoptic hypothalamic region of the brain integrates thermal behavior as functions in metabolic-circulatory regulation in mammals (113). Cyclostomes, cartilagenous and bony fishes, regulate body temperature behaviorally when given a choice of water temperature. Early vertebrates probably evolved in heterothermic fresh water or brackish environments where behavioral thermoregulation was essential for feeding and escape. In a temperature gradient, the initially preferred temperature is strongly influenced by previous temperature acclimation (115). However, if fish are left in a gradient for some hours, they go to a temperature (final preferendum) that is species specific and different from the initial selected temperature (30, 31, 44). After small lesions are placed in the medial and lateral preoptic regions (POAH) of green sunfish, the choice of temperature in a gradient becomes disrupted. Recordings from neurons in the POAH of green sunfish shows warm-sensitive, cold-sensitive, and temperature-insensitive units (55, 115). Peripheral input from cutaneous thermoreceptors reaches neurons in this region. Three types of responses to central warming are observed (114): (1) linear increase in firing frequency with temperature, (2) rapid increase in firing rate on initial warming, (3) slow initial increase followed by rapid increase in frequency (Fig. 21*A*); some neurons respond linearly, others exponentially to skin warming. Firing of cold-sensitive neurons increases nonlinearly as temperature is lowered. Some warmth-responsive interneurons are spontaneously active and show excitatory synaptic potentials that are responses to input from either central or peripheral receptors. Cold-responsive interneurons show both excitatory and inhibitory synaptic potentials; these neurons are evidently excited by temperature insensitive central neurons and inhibited by warmth responding neurons; the inhibition on warming increases as temperature rises. A diagrammatic circuit for sunfish thermoregulation is given in Figure 21*B*.

Lizards regulate body temperature by posture, orientation, and microclimate selection. Elevated posture avoids hot substrate. Lizards pant if heated, presumably cooling by evaporation. Dermal vasomotion and thermal shunting in the head regulate heat transfer. Some lizards regulate absorption of radiant heat by skin chromatophore constriction or expan-

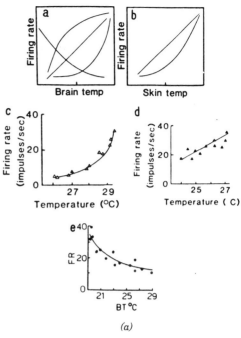

(a)

Figure 21*A*. Patterns of responses in green sunfish of single units in preoptic hypothalamus. (*a*) Four types of response to brain warming or cooling. (*b*) Two types of responses to warming of skin of mouth or operculum. (*c*) Nonlinear response of one unit to brain warming, (*d*) linear response, and (*e*) response of a unit to cooling. [Prosser et al. (119a).]

sion. A few large reptiles (pythons) raise body temperature for incubation of eggs by muscle activity (78b). Behavioral regulation during cooling or heating is important also in birds and mammals; altricial birds and mammals are essentially poikilothermic when first hatched or born.

Endothermic insects—butterflies, moths, bees, beetles, some dragonflies produce heat oxidatively in thoracic (th) muscles, with or without visible movement (23). Heat production raises the temperature of the nervous system and muscles to the temperature necessary for flight (23, 85a). A moth *Hyalophora* warms endothermically at 4°C/min and main-

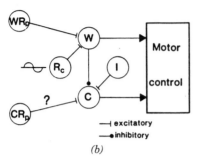

(b)

— excitatory
— inhibitory

Figure 21B. Neuronal model summarzing intracellular recordings form preoptic neurons. Warm-sensitive, endogenously active units (R$_c$) drive other neurons (W) by excitatory synapses; W cells show excitatory miniature synaptic potentials and receive inputs from peripheral warm receptors (WR$_p$) and from central warm receptors (R$_c$). W cells presumably send axons to motor centers. Cold-sensitive interneurons (C) show both excitatory and inhibitory synaptic potentials and are not endogenously active; activity of (C) cells results form excitatory synaptic input from temperature-sensitive cells (I) and inhibitory synaptic input from W cells; excitatory synaptic input from peripheral cold receptors (CR$_p$) is postulated. Output from C cells also presumably goes to motor centers [Nelson and Prosser. *Science* 213:787–789, 1981; also Nelson et al. (113).]

tains its T_{th} at 32 to 36°C over an ambient range of 7 to 29°C during which the moth's $\dot{V}_{O_2}$ increases by 2.3 times; body fat is the primary fuel. Flight is possible only if the nerve cord is at a critical temperature; if the thorax is warmed by a thermode, flight can be initiated without warm-up contractions (59, 67, 102).

Bumblebees warm the thorax by muscle contractions; more heat is produced at low thoracic temperature than at high temperature (Fig. 22). During incubation, heat is transferred from thorax to abdomen in workers but no heat transfer occurs during flight and perching (68). Heat transfer to the abdomen is via hemolymph; temperature of the abdomen changes with ventilatory movements; no heat transfer occurs if the heart is lesioned. Queen bumblebees regulate T_{th} at 36 to 45°C during flight over an ambient temperature range of 2 to 35°C. During foraging on inflorescences of *Spirea* or *Solidago* the T_{th} of workers is allowed to fall below the minimum for flight and on flying to another flower T_{th} rewarms (70). In flight 80–90% of the energy produced by muscles is converted to heat. *Bombus* is insulated by an abdominal air-sac; epidermal scales insulate moths. At high T_a

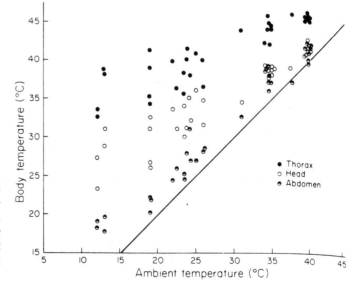

Figure 22. Thorax, head, and abdominal temperatures during continuous free flight of a carpenter bee, during free flight as a function of ambient temperature. [Heinrich and Buchmann (69).]

(46°C) evaporative cooling and water secretion keep the T_{th} well below ambient temperature. In a cluster brood, cooling occurs by fanning and increased evaporation. Warming occurs by clustering; winter bees at the center of a cluster may be degrees warmer than peripheral bees (68, 71).

Bees in the desert regulate T_{head} and T_{th} by regurgitating droplets of nectar from which water evaporates (27). Honeybees and termites maintain a relatively constant nest temperature throughout the year. In winter, bees generate heat by muscle activity and form clusters, a bee periodically changes position between interior and peripheral of the cluster. In summer, bees cool the hive by bringing in water and fanning. In cicadas, for simple behavior, the permissive range is wider than for complex behavior: 31–35°C for chorusing, 20–37°C for flight, 13–40°C for walking, and 10–43°C for feeding (66, 67a, 67b) (Fig. 23).

Large marine fishes such as tuna and lamnid sharks are fast, long-distance swimmers that maintain brain, muscle, and visceral temperatures several degrees higher than ambient water temperatures (12, 22). They have countercurrent heat exchangers consisting of retia between lateral arteries and veins so that heat is transferred directly from muscles to viscera without going to skin (Fig. 24). In tuna, the temperature of the red swimming muscle is maintained at ~30°C in water temperature varying from 10 to 22°C. The tail of the fish is more nearly at ambient water temperature. In lamnid sharks, the temperature of the brain can be maintained 14°C above T_a (12, 13).

In billfish (marlin, swordfish) ocular muscles are transformed into specialized heater tissue. The muscles associated with thermogenesis lose their myofilaments and have densely packed mitochondria, extensive sarcoplasmic reticulum (SR), and some transverse (T) tubules (12, 13) (Figs. 24, 25). The SR

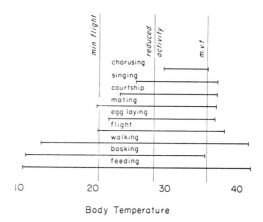

Figure 23. Temperature ranges of different behaviors of cicadas. Complex behaviors more restricted in temperature range than simple behaviors (J. E. Heath, unpublished).

membranes are rich in Ca^{2+} ATPase and the inner membranes of the mitochondria area are a source of ATP (13). Heat results from cycling of Ca^{2+} and high activity of the Ca^{2+} ATPase with excess heat coming from ATP hydrolysis (12).

There are several methods used by animals to produce heat. Each depends on the utilization of ATP for other functions. The most familiar source of heat is from ATP used to power the myosin crossbridges of contracting muscle. This is well known from exercise and shivering. In nonshivering thermogenesis, heat is derived from ATP in three ways: tonic levels of activity for myosin ATPase without muscle contraction, activation of Ca^{2+} ATPase in sarcoplasmic reticulum, and use of ATP by Na^+-K^+ ATPase. Thyroid hormone results in an increased amount and activity of $Na^+ = K^+$ ATPase, an important part of basal metabolism. Brown adipose tissue (BAT) of young mammals and hibernators produces heat by uncoupling phosphorylation from oxidation; BAT has a specialized uncoupling protein. The P/O ratios are low and production of ATP is reduced; oxidation of fatty acids accounts for most of the heat. The BAT response is stimulated by norepinephrine.

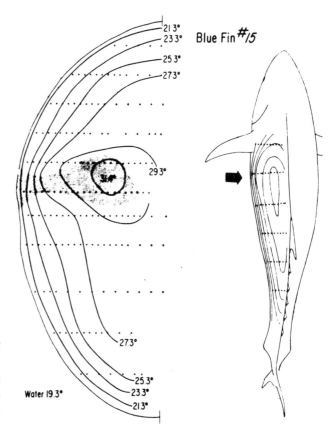

Figure 24. Temperature profiles of bluefin tuna in water of 19.3°C. Isotherms show that the warmest parts of the fish are in the dark muscle on each side of the body. [Carey et al. (21a).]

Thermal Regulation; Reptiles, Birds, and Mammals

In terrestrial vertebrates regulation of body temperature is accomplished by heat production and by control of heat transfer from the body surface (19, 45). Heat flow (Q) to or from the environment is via four channels:

Radiation	$Q = \delta(T_{skin}^4 - T_{ambient}^4)$	δ = Bolzmann constant
Conduction	$Q = k(T_{skin} - T_{ambient\ surface})$	k = conductance
Convection (natural or forced)	$Q = h(T_{skin} - T_{air})$	h = convection coefficient
Evaporation	$Q = e(P_{skin} - P_{air})$	e = latent heat of evaporation
		P = partial pressure of water at temperature of skin and air

In general, terrestrial animals control these fluxes by maintaining or changing the temperature of the surface or skin. Animals further regulate the total heat flow from the surface by controlling the amount of surface area exposed by changes of posture. Birds and mammals can control the depth of the feathers or

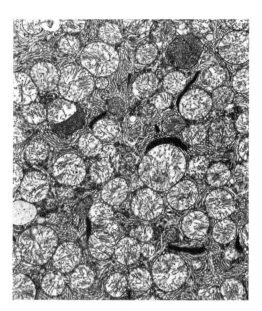

Figure 25. Electron micrograph of a heater cell showing numerous mitochondria and extensive smooth endoplasmic reticulum. (Courtesy of Barbara Block.)

fur and thereby can alter the magnitude of the conductive and convective constants. The effective temperature (radiative and convective gain or loss) may differ substantially from the surrounding air temperature. Forced movement or wind greatly enhance the heat loss due to convection. Animals can enhance the effective partial pressure of water at the evaporating surface by keeping the surface as warm as possible. Movement of the surface or air flow increases the rate of removal of saturated air from the evaporating surface, which increases the rate of heat loss. Protrusion of the tongue and exposure of the inner lips (dogs) increases the evaporative surface in panting.

Net thermal conductance (C) has been expressed in terms of metabolism (M) in mL of $O_2/g \cdot h \cdot °C$ gradient

$$C = \frac{M}{T_b - T_a}$$

The quantity C is large for small mammals that have a high surface-to-volume ratio and small for mammals of small surface-to-volume ratio. The value C is converted to $kcal/g \cdot h \cdot °C$ by multiplying by 4.74, the caloric equivalent of 1 L of O_2. However, C is not conductance as used by heat engineers, but is a measure of net heat transfer by the preceding methods. True thermal conductance is the flux of heat flow in a material in a specified gradient.

The use of metabolic-based models of *insulation* and *conductance* continues without regard for its limitations, probably because of the simplicity of their application and measurements as well as the large amount and diversity of data that have accumulated. However, insulation and conductance as measured metabolically correspond neither to Newton's law of cooling nor to accurate heat transfer models. Conductance derived from the metabolic model is a size or volume dependent measurement of aerobic heat production while Newton's law described a heat flux unrelated to the size of the organism or source of heat production. The metabolic-based models assume uniform insulation and simplistic geometric shape for animals and ignore radiative heat loss by adopting a linear relationship between gradients. It is only between species of similar size and shape that metabolism-based models are useful, but these limits are rarely observed.

Further constraints on metabolic-based models arise with large animals whose heat content varies substantially with temperature and with animals that have organs or areas specialized to facilitate heat transfer. Some exposed skin surfaces like the corneal surface are kept as close to T_b as possible at all temperature gradients, while other bare surfaces of animals must be defended against freezing at low temperatures by increasing blood flow, thus, decreasing overall insulation (Fig. 20).

Figure 26. Summaries of the thermal responses and heat exchanges of medium and small mammals. Each diagram shows the regulated deep and superficial temperatures and channels of heat production and heat loss in relation to ambient temperature. Critical environmental temperatures are indicated by capital letters: A, zone of cold failure; B, critical or failure point; B′, freezing point; C, lower critical temperature; C–D, zone of vasomotor regulation of heat loss; D–E, zone of evaporative control of body temperature; E, upper failure point. (*a*) Medium-sized (10 kg) mammal shows good body insulation, the ability to regulate heat flow from uninsulated and poorly insulated surfaces, and only small dependence on heat production over its normal active temperature range. (*b*) Small (100 g) mammals show limited insulation, poor control of heat dissipation, dependence on heat production over its activity range. Very large mammals reduce insulation and are nearly independent of heat production. Man, a tropical medium-large mammal, has reduced insulation, but shows some dependence on heat production at the lower end of his native temperature range and enormously enhanced system of evaporative cooling. (Courtesy of J. E. Heath.)

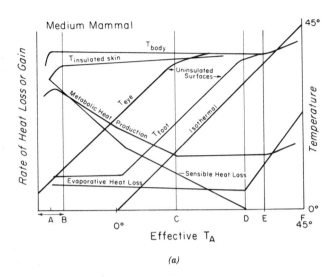

(a)

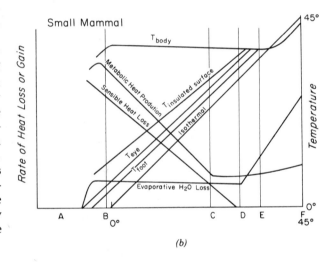

(b)

Finally, the measured aerobic metabolism of larger animals shows marked discrepancies with heat transfer from whole body calorimetry when measured over short intervals of a few minutes to a few hours or over 24 h. These differences relate to changes in heat content and metabolic shifts in the course of a day.

More quantitative, complex, and comprehensive measures should be employed to estimate and compare control of body temperature and of heat transfer.

Differences between large mammals that have heat storage and small–medium sized mammals are described in Figure 26.

The first defense against cooling or warming is behavioral, this is followed by vasomotion and changing insulation. At low ambient temperatures redistribution of blood flow away from the skin lowers skin temperature, which decreases heat loss. In the pinnae of mammals, such as rabbits or elephants, the difference in flow between full heat-dissipating dilatation and the minimal flow for metabolic needs of the tissues may be as great as 100-fold. In the limbs and in regions between blood source and snout and brain, countercurrent heat exchange vessels further conserve heat, even when the minimal blood flow is substantial (2). Deep arteries may pass through a network of veins in a rete. The exchange of heat between artery and vein (ΔE) is calculated as the difference in artery $-$ vein temperatures ($T_A - T_V$) times the specific heat of blood (S) times the flow rate (F):

$$\Delta E = (T_A - T_V) \times S \times F$$

Countercurrent heat exchange may reduce heat loss by 10-fold in a large limb. During heat stress a massive increase in flow and opening of A–V shunts to superficial veins allows distal segments (toes and fingers) to warm nearly to core temperature (6). Many large mammals—camels, giraffes, elephants, and hippopotami have thick skin that stores heat in mid-day and radiates it at night. Body temperature is thus buffered against overheating. These mammals are heterotherms (92a, 124a,b).

Countercurrent exchanges permit independent regulation of the temperature of the head and brain in reptiles, birds, and mammals (2). During exercise the increased evaporation in the head (especially in nasal membranes) coupled with countercurrent heat exchange between the carotid arteries and jugular veins may hold head temperature nearly unchanged while the core body temperature is rising (2). Thus, the brain is buffered from major temperature changes brought about by exercise (51, 52, 138). In high T_a, reindeer inhale through the mouth, exhale through the nose. In heat a segment of facial artery is occluded, in cold these vessels dilate, thus the brain is maintained at a relatively constant temperature (80b).

The temperature of the testes is regulated to a lower temperature than the core by both countercurrent exchange in heat and by muscles that pull the testes against the body wall in cold. An elevated testicular temperature causes a ram to pant and results in an anomolously lowered core temperature (140b). In severe cold, both birds and mammals increase flow to the limbs and pinnae and keep these regions above freezing; this shift in flow increases heat loss.

The enormous increases in dermal blood flow in a hot environment coupled with enhanced blood flow to muscles in exercise can result in potentially dangerous reduction of blood flow to the head and brain. Men, after extensive exercise training, show up to a 10% increase in serum volume, possibly due to increase in the amount of fluid available for dermal blood flow (112).

Large changes in skin blood flow produce substantial changes in skin temperature, which skews central perception and control of temperature in that those regions most likely to serve as estimators of the effective environmental temperature show the greatest temperature changes in response to altered blood flow (51, 52). Since the blood flow to the skin pulses at a frequency of once to twice a minute, the preoptic or spinal centers may distinguish between the rising or motor phase and falling or sensory phase in deciphering effective ambient temperature (105). Pulsation continues in both the fully dilated and fully constricted condition.

Reptiles have countercurrent heat exchangers between carotid and jugular vessels (65a, 66, 67). Panting or open-mouth breathing in heat occurs in reptiles, birds, and mammals as a method of cooling by increased evaporative loss of heat. Turtles, like many mammals, spread saliva or urine over appendages when overheated.

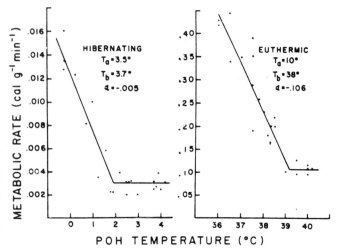

Figure 27. Metabolic responses of euthermic and hibernating golden-mantled ground squirrels as a function of temperature of the preoptic hypothalamus. [Heller, H. C. et al. (74), by permission of *Amer. Physiol. Soc.*]

As in insects, many animals regulate body temperature by movement. A tropical lizard alternates between basking in the sun and foraging in the shade; the T_b thresholds for shuttling by a desert lizard are 39.4 and 34.5°C (140a).

Body heat is retained by feathers, hair, and subcutaneous fat. It has been suggested that the transition from low to high metabolism, from reptiles to birds was made possible by insulative feathers. Erection of feathers or of hair in ambient cold is under control of spinal reflexes together with hypothalamic modulation. Heat loss by mammals with or without hair can be twice as great in water as in air.

When ambient temperature rises, the cutaneous heat receptors are stimulated and reflex responses favoring heat dissipation result. Cutaneous vessels dilate, there is increased blood flow in the skin and increased heat transfer by as much as five- to sixfold. Insensible water loss also increases. With further heating, reflex sweating occurs in those mammals that have sweat glands; increased respiratory loss of heat occurs in other mammals and birds. Panting may result in alkylosis. In humans the output by sweat increases by 20 g/h for each degree rise in T_a. Not only is water lost in sweat, but also salt, and with increased sweating, urine production decreases. Strains of cattle indigenous to hot climates lose much less heat by skin than temperate zone cattle. Some mammals salivate profusely and lick their fur when heated. The horns of goats function in the control of heat loss or retention.

When cooling mechanisms fail during heating and T_b rises, O_2 consumption increases because of the direct cellular effects of heat. Heat shock results if body temperature rises critically. Heat exhaustion can result in part from dehydration. Heat stroke results from brain damage. Humans tolerate heat less well if heated externally than internally (as by a fever or exercise).

Desert animals tolerate a limited hyperthermia (Fig. 27). The T_b of a camel can range from 34 to 40°C and T_b rises in sunlight faster after shearing of fur than before. A diurnal ground squirrel can tolerate an environmental effective temperature 30°C above its upper critical temperature for a few minutes before seeking shade. It can drop T_b from 41.7 to 37.8°C in 3 min on transfer to 25°C T_a.

Rate of heat gained from the environment may be 3–4 times greater than metabolic heat production. A ground squirrel shades its back with its tail and thus lowers T_b by 3°C (23a).

When physical cooling continues and insulative mechanisms are inadequate to maintain body temperature, muscle activity and oxygen consumption increase. Heat production is first by shivering, then by nonshivering (thermogenesis). Nonshivering thermogenesis in chilled mammals is normally activated by norepinephrine. A general increase in metabolism ultimately involves the thyroid. A normal rat regulates its body temperature in air as cold as $-10°C$; a rat lacking either adrenals or thyroid regulates down to $T_a = -2°C$; rats lacking both glands regulate body temperature only to $\pm10°C$. Continued exposure to cold activates the adrenal cortex. Cooled rats may show a twofold increase in blood epinephrine and a corresponding increase in O_2 consumption by muscle. When metabolism (M) is measured at different ambient temperatures, a neutral range of minimum O_2 consumption is found, usually at air temperature slightly below T_b (Chapter 1). The thermoneutral zone may be broad in large mammals or may be a minimum point on the M/T curve in small mammals and birds.

Acclimatization to low temperatures, or to winter conditions, shifts the lower part of the thermoneutral zone downward. When insulative and heat-producing reactions fail to maintain constant T_b, birds and mammals go into coma and may die.

Models of temperature-regulating circuits have been made to resemble physical systems for controlling temperature (14). It is assumed that in the control circuit there is a set point (T_{set}), which is the threshold temperature for regulating responses. The set point may vary with physiological state, time of day, lunar cycle, or season. In hibernation or torpor the set-point is reduced much below normal. Regulatory responses are the net sum of all mechanisms of heat loss or retention and of heat production.

Many thermal effectors, especially behavioral responses, are under all-or-none, or on–off control (66). An effector response occurs when T_b is higher or lower than T_{set}. Responses are activated like a simple bimetallic thermostat in physical systems. On–off control is a switching of responses in a stepwise fashion. Efferent patterns are unlike afferent input; this indicates central integration.

In proportional control the effector responses are determined by an error signal, the deviation from T_{set}:

$$R - R_0 = \alpha(T_{hypo} - T_{set})$$

where R = regulatory response
R_0 = basal level of response
α = proportionality constant
T_{hypo} = hypothalamic temperature

In rate or differential control the effector response is triggered by a changing temperature and is proportional to the rate at which T_b changes

$$R - R_0 = \gamma\frac{dT_b}{dt}$$

where γ is a proportionality constant and

$$\frac{dT_b}{dt} = \text{change in } T_b \text{ with time}$$

Differential control can be feed-forward, and hence anticipatory.

The evolution of homeothermy in vertebrates can be inferred from properties of existing classes (Table 7). Peripheral receptors, mostly free nerve endings, and central neurons sensitive to warmth or cold are present in all vertebrates. Behavioral thermoregulation is evidenced by temperature selection in a gradient and by conditioned reflex methods. Behavioral regulation is important in niche selection, in daily vertical migrations (fishes), and in keeping nocturnal species separated from diurnal species. Another behavioral mechanism is change in heat absorption or reflection by pigment changes that may also alter visibility, for example, in lizards. Many adaptations favor heat dissipation or retention. Variation in peripheral blood flow controls heat transfer, vasomotor responses are general for regulation. Fishes that have local muscle warmers restrict skin flow by shunting. Large reptiles show hysteresis of cooling and heating. In reptiles, dermal circulation is enhanced during warming and restricted during cooling. Part of the net redistribution in flow to the skin is under nervous and to a lesser extent hormonal control. Forebrain control of behavioral thermoregulation is general. A series of semiautonomous subsystems, each with thermosensitivity and each probably appropriating and elaborating preexisting motor patterns provided diversity in modes of temperature regulation. Evaporation for control in high temperatures may have been first by skin, later by respiratory organs.

Most important in the transitions from reptiles to birds and mammals was the increase in metabolism and consequent heat production. There is difference of opinion as to whether the early reptile-birds as well as dinosaurs were "warm-blooded." Fossil evidence from Jurassic and Cretaceous periods indicates that the first mammals may have been nocturnal insectivores with variable T_b (32). Some present-day monotremes have lower and more variable T_b than eutherian mammals. Metabolic heat production by lizards is one fifth to one third that of mammals even when measured at comparable body temperatures. A balance must have been struck between the greater energy production required for thermal regulation and its resulting behavior and the necessity for expending more effort in food procurement.

The mechanisms of homeothermic regulation have not evolved de novo; rather, temperature regulation makes use of modifications of preexisting mechanisms. This is well illustrated in reptiles [Table 7 (113)].

The primary function of behaviors used in temperature regulation may have been aggressive encounters and sexual displays. Panting is clearly a modification of respiratory exchange. Vasomotor control is derived from skin respiration, especially in amphibians. Changes in *albedo* are derived from concealment behavior. Heat production by muscles is a derivative of locomotion. Vasomotion for heat exchange derives from nervous and hormonal control of blood flow to organs of heat exchange.

Temperature regulation is by a hierarchical arrangement of controllers, each operating a set of regulating mechanisms in cascade (113). The importance of cutaneous receptors, of spinal, medullary, midbrain, and forebrain controllers varies among vertebrate classes. First, mammals probably with low T_b and low M, may have been nocturnal insectivores (Jurassic and Cretaceous). By Paleocene, diurnal mammals had higher T_b and M (32). The preoptic hypothalamus appears to have retained its primary function as an integrator in the thermoregulation of all classes of vertebrates. In birds and mammals, the large response to spinal thermal stimulation and the segmental responses to local dermal heating and cooling may argue for semiautonomous segmental controllers that are modulated by oscillating preoptic inputs (51). Cutaneous receptors are minimal or lacking in sauropsid reptiles and in birds. Spinal circuits are especially important in birds. In the reptilian line leading to mammals the

TABLE 7. Potential Origins of Thermoregulator Effectors in Reptiles[a]

System	Effector	Primary Function	Thermoregulatory Function
Behavior	Elevated posture (spinal)	Aggressive display Sexual display	→ Substrate avoidance
	Burrowing (telecephalic)	Concealment	→ Heat or cold avoidance
Evaporative mechanism	Panting (medullary-spinal)	Respiratory exchange	→ Evaporative heat loss
Vasomotor	Dermal vasomotion (spinal)	Respiratory exchange in amphibians	→ Control at heat transfer
	Eye bulging (jugular constrictions) (medullary)	Shedding	→ Bypass carotid-jugular Heat exchange
Albedo	Color change (hypothalamic-hormonal)	Concealment Aggressive display	→ Control of radiant energy absorption
Heat production	Muscle contraction (telencephalic)	Locomotion	→ Heat for incubation (python)

[a]Modified from Nelson et al. (113).

preoptic–hormonal axis was elaborated; thyroid and adrenal hormones opened the possibility of nonshivering thermogenesis as a supplement to exercise and shivering for heat production. Catecholamines, hormonal and neural in source, have multiple effects on the vasomotor state, cardiac activity, piloerection, and heat production. Sweating appears to have been incorporated into thermoregulation in mammals where it released the respiratory system from the alkalosis resulting from hyperventilation.

Hibernation and Torpidity

Many animals enter a state of dormancy as adverse climatic conditions—heat, cold, and drought—approach. Ciliates and many aquatic invertebrates encyst; lungfish spend a dry season encased in mud cocoons. Some frogs overwinter in mud, some frogs can survive several years of drought. Daily torpor conserves energy in hummingbirds and bats (144). A dasyurid marsupial enters torpor at T_a

below 27°C (47). In most mammals, body temperature is slightly reduced during sleep. A few mammals hibernate with T_b declining as T_a falls; usually their T_b remains slightly higher than T_a. Body temperature of hibernants can go as low as 2°C, whereas nonhibernators die at 28 to 30°C body temperature. Brain cooling is rapid (53). Only a few kinds of mammals—some monotremes, insectivors, rodents, and bats—are capable of true hibernation. Hibernation can be entered by a series of test drops in body temperature (100). Some large mammals such as bears go for long periods without food, but their body temperature does not drop below about 31°C; true hibernators periodically awaken to urinate and defecate; bears are able to reabsorb and use urea from the bladder (116). Bears are protected against uremia by biochemical changes in nitrogen metabolism that precede entry into winter sleep; also bears do not become acidotic during winter sleep. There is evidence for an annual cycle underlying hibernation in some species. Ground squirrels, *Spermophilus*, maintained at 12°C for 2 to 4 years on a 12-h photoperiod showed annual cycles of hibernation, more animals entering hibernation if food was limited.

Some hibernators such as *Spermophilus* deposit increased body fat before entering hibernation; others such as the hamster, *Mesocricetus*, do not deposit body fat, but build a nest and store food. In some hibernators (hedgehogs), blood levels of insulin become elevated. In others, the thyroid becomes involuted. Adrenalectomized animals do not hibernate and those in hibernation do not respond by thermogenesis to injections of norepinephrine.

As body temperature falls in hibernation, the heart slows and may become irregular at ~3°C. The interval between T and P waves of the electrocardiogram lengthens. The heart of a hibernator continues to beat at low temperatures (1°C for *Citellus*); the heart of a nonhibernator stops at 16–18°C (99a). In a hedgehog at T_b below 6.4°C the vagus has little effect on heart rate or blood pressure; on rewarming the heart is unaffected by the vagus until a temperature of 20°C is reached.

Hibernating animals assume characteristic postures, often rolled into a ball. The metabolic and heat-producing responses of hamsters and ground squirrels to low T_a are very different from those in the same animals when euthermic [Fig. 27; (74)]. Oxygen consumption levels are lower in the hibernating animal and the minimum temperature at which metabolism increases (T_c) is lower. The basal metabolism in a hibernating marmot is 0.1–0.12 W/kg, in a euthermic marmot it is 12 W/kg; T_c in hibernation is ~5–6°C, in euthermia it is 37°C (42, 74). Activity in the brain does not entirely stop when the hibernating body cools. In a hibernating ground squirrel some electrical waves were recorded from forebrain at 5°C and at 6.1°C the animal could localize

sound, erect pinnae, vocalize, and move (136). The threshold temperature for the hypothalamus to stimulate metabolism declined by 0.2–0.4°C/day. During entry into hibernation, threshold as measured by a thermode in the hypothalamus changes and indicates a decline in T_{set}. In hamsters some neurons of the preoptic anterior hypothalamus (POAH) are active over a 30°C range, whereas in a guinea pig, all temperature-sensitive neurons are blocked at 28–30°C. Slow-wave sleep increases and rapid eye movement sleep (REM) decreases during entry into hibernation (74a). Brain serotonin (5HT) in hibernating ground squirrels and hamsters is more concentrated than in the active state. Stimulation of 5HT receptors decreases heat production (117c).

The proportional model for regulation of T_b applies to hibernators as it does to euthermic animals. An essential step in the evolution of hibernation was change in temperature sensitivity of thermoregulating neurons of the hypothalamus. The role of hormones in hibernation varies with species. Administration of melatonin prolongs the duration of seasonal hibernation in dormice (132). Melatonin content of pineal of hamsters rises during daily arousal from sleep in warm animals but shows no diurnal rhythm in hibernating animals (80a).

Changes in energy yielding enzymes occur in parallel with changes in membranes of hypothalamic neurons. Oxidative metabolism may be reduced by as much as 20 to 200 times in hibernation. Fat is an important fuel and body fat may decrease by 50 to 70% during a winter of hibernation. Kidney slices from *Citellus* show enhanced capacity for gluconeogenesis and glycolysis at a low temperature, properties that may be related to tolerance of hypoxia (18). Respiration by mitochondria from hamster heart shows cold acclimatory responses. Inner mitochondrial membranes of liver from hibernating ground squirrels show no compensatory changes in fluidity. Hamster brain maintains a normal level of ATP but ATP content declines slowly in the heart and in red blood cells. Rate–temperature curves for succinate oxidation in ground squirrels show breaks in the Arrhenius plot in summer, not in hibernation (143, 144); thyroidectomy reduces the temperature of the break in summer animals. Nonshivering thermogenesis by the brown adipose tissue (BAT) yields 27% of the heat produced by chilled hamsters in the summer, 59% in winter (73).

Hibernants maintain ionic gradients at low temperatures (147). Slices of ground squirrel kidney retain potassium concentration during several days of storage at 3°C, but slices from rat kidney lose potassium under comparable conditions (147). The Na^+-K^+ ATPase from the brain of a hibernating hedgehog and from hamster kidney is cold resistant and the activity of this enzyme from a hibernating hamster is twice that from awake hamsters, measured at the same temperature (25). Hibernators regulate acid–base balance so that blood and tissue pH values do not rise with cooling as in nonhibernators, that is, hibernators are acidotic at low temperatures.

Arousal from hibernation is a rapid awakening; warming starts in the thorax. A ground squirrel's temperature may rise from 4 to 35°C in 4 h; the T_b of a chipmunk rose at a rate of 0.72°C/min. Arousal is endothermic and most of the heat is produced by thoracic brown fat (adipose tissue or BAT). Oxidation of triglycerides in BAT is activated mainly by adrenergic sympathetic nerves. In the first hour of arousal, production of CO_2 increases by 15-fold. Both shivering and nonshivering thermogenesis participate in heat production. Heart rate accelerates rapidly; in a spermophile asleep heart rate at 1°C was 3 beats per min, at 8°C it was 20 beats per min, and at 20°C was 300 beats per min.

A chipmunk arousing from hibernation increases its heart rate by 10- to 23-fold and its $\dot{V}_{O_2}$ by 7- to 128-fold (144). Brain waves are much reduced (136).

Stimulation for arousal arises from numerous factors—neural and hormonal. Many kinds of hibernators arouse periodically and rid themselves of urine and feces and feed; cooling the hypothalamus can awaken a hibernating animal. It is concluded that no single or specific biochemical differences make for hibernation; rather there are quantitative differences in many cellular functions. The adaptive advantage of hibernation is considerable. It is estimated that hibernation provides an energy saving of 88% during the 8.5 months of hibernation by a ground squirrel (144).

Temperature Summary

Environmental temperature is critical in limiting the distribution of animals and in controlling metabolic reactions. The effects of temperature both environmental and in organisms are limited by the physical properties of water. Aquatic environments range in temperatures from thermal vents and hot springs to polar oceans; terrestrial environments range from hot deserts to ice fields. The regions of the world occupied by polar, temperate, and tropical areas have changed during geological time. Climatic change in mid-latitudes have been responsible for changes in flora and fauna.

In poikilotherms, the body temperature, T_b, is the same as the environment; homeotherms maintain constancy of internal temperature within limits; heterotherms show differences in temperature with time or between body regions. Ectotherms (usually poikilothermic) gain warmth from the environment; endotherms produce warmth metabolically. Regulation of temperature may be by behavior or by metabolic–circulatory and insulative means. Resistance adaptations take place within the tolerated range of temperature—high or low; capacity adaptations permit activity over a range of body temperatures. Constancy of body temperature is homeostasis; constancy of energy production at different body temperatures is homeokinesis.

Lethal limits at either low or high temperatures are wider for proteins (denaturation) and lipids (melting) than for cells and are wider for tissues than for intact integrated organisms. Death of an intact animal may result from synaptic failure in the central nervous system. Cellular death may result from breakdown of the selective permeability of plasma membranes. Enzyme inactivation in extreme heat or cold may be due to conformational changes in protein molecules. Exposure of cells to sublethal high temperatures induces synthesis of hsps, which confer tolerance of more severe heating. Synthesis stimulation of the highly conserved hsps results in cessation of synthesis of other proteins. At extremes of cold some terrestrial insects and amphibians synthesize cryoprotectant compounds (e.g., glycerol) that act colligatively, or synthesize of ice nucleators that act extracellularly. Fish from polar waters synthesize glycoproteins or polypeptides that lower the freezing point of tissues beyond colligative values and below melting points. Other adaptations for enzyme activity are modifications of structure such that kinetic properties—activation energy, Michaelis constant, and inactivation tempeature—permit functioning at temperature extremes. Direct responses are to be distinguished from acclimatory responses that take many days.

Homeokinetic adaptations, both the genetic and the environmentally induced ones, include the following: selective synthesis of adaptive isoforms (isozymes and allozymes), synthesis of increased amounts of energy-yielding enzymes, general protein hypertrophy, changes in metabolic pathways, changes in membrane lipids that result in alterations of membrane fluidity, percent unsaturated fatty acids or phospholipid classes, and changes in ion and pH balance. Poikilotherms have smaller activation coefficients, E_a values, than do homeotherms. Acclimation of poikilotherms may be positive compensations of metabolism; acclimation may be neutral or negative in reducing energy expenditure at low temperatures. In muscles, the proportion of red and white fibers can vary with temperature. Acclimation to low or high temperature occurs in periods that vary with the acclimatory process. Acclimatory changes differ with the tissue of an animal and with the measured process. Acclimatization in natural environments is influenced not only by temperature but by photoperiod, nutrition, hormonal and reproductive state, and social interaction. Kinetic alterations in acclimation are made evident by occurrence or absence of breaks in Arrhenius plots and by shifts in K_m–temperature curves. Endothermic animals have higher relative dependence on enthalpy and ectotherms on entropy for constant free energy.

Temperature sensing may be by peripheral receptors for cold or warmth and by central thermosensitive neurons. Temperature regulation requires complex neural circuits. In a temperature gradient, most animals go to a "selected" temperature, which is influenced by previous acclimation; after a period in a gradient, the final preferendum may differ from the initial selected temperature. Some insects can fly only after thoracic temperature has reached a critical value. Some large fishes have special muscles or heater tissues that produce the heat required by the nervous system for swimming. All vertebrates have thermoregulating regions in the hypothalamus and central receptors in other regions of the nervous system as well as as peripheral receptors. Neural, vascular, and metabolic mechanisms may have been adapted for thermoregulation from other functions in birds and mammals. Heat gain or loss is balanced in endotherms against metabolic costs of heat production.

Large tropical mammals are heterotherms with heat storage in their thick skin. A few endothermic vertebrates become torpid periodically—diurnally or seasonally. Hibernation of mammals is a lowering of the "set point" of the hypothalamic temperature regulator. Several metabolic, hormonal, and insulative adaptations prepare for hibernation. Arousal may be accompanied by an abrupt rise in body temperature and heat production by brown adipose tissue (BAT).

Temperature is the most pervasive of environmental factors. Adaptations for coping with variations in temperature and with extremes of heat and cold make use of all physiological systems.

References

1. Ackerman, R. A. and F. N. White. *Respir. Physiol.* 39:133–147, 1980. Effect of temperature on acid–base balance and regulation in iguana.
1a. Anderson, R. L. and Mutchmor. *J. Insect Physiol.* 14:243–251, 1968. Temperature acclimation and the nervous system, cockroach.
2. Baker, M. A. *Annu. Rev. Physiol.* 44:85–96, 1982. Brain cooling in endotherms in heat and exercise.
3. Baldino, F. and H. Geller, *J. Physiol. (London)* 327:173–184, 1982. Sensitivity of rat preoptic hypothalamus cells in culture.
4. Baldwin, J. *Comp. Biochem. Physiol.* 40:181–187, 1971; *Biochem J.* 116:883–887, 1970. Isozyme of AchEs in trout brain.
5. Barnas, G. M., M. Geeson, and W. Rau-tenberg. *J. Exp. Biol.* 114:415–426, 1985. Temperature regulating center in chicken spinal cord.
6. Barton, A. C. *Am. J. Physiol.* 127:437–453, 1939. The range and variability of the blood flow in human fingers and the vasomotor regulation of body temperature.
7. Baust, J. In *Living in the Cold* (H. C. Heller, X. J. Musacchia, and L. C. H. Wang, eds.), pp. 125–130. New York, 1986. Insect cold hardiness.
8. Baust, J. D. and R. T. Brown. *Comp. Biochem. Physiol. A* 67A:447–452, 1980. Heterothermy and cold acclimatization in Arctic ground squirrels.
9. Behrisch, H., I. Ostas, and W. Wieser. *J. Therm. Biol.* 2:185–189, 1977. Temperature regulation of LDH in an alpine mammal (chamois).
9a. Bickler, P. E. *J. Comp. Physiol. B* 156:853–857, 1986. Day–night pH variations in Dipsosaurus.
10. Bienz, M. and J. B. Gurdon. *Cell (Cambridge, Mass.)* 19:811–819, 1982. Heat shock in Xenopus oocytes controlled at translation.
11. Bligh, J. *Front. Biol.* 30:1–39, 1973. Temperature regulation in mammals and other vertebrates.
12. Block, B. A. and F. G. Carey. *J. Comp. Physiol. B* 156:229–236, 1985. Warm brain and eye temperatures in sharks.
13. Block, B. A. *J. Cell Biol.* 107:1099–1112; 2587–2600, 1988. Structure of heater organ of oceanic fishes.
13a. Block, B. A. *Physiol. Rev.* 64:1–65, 1988. Heat release by modified muscle in large fishes.
14. Boulant, J. *Brain Res.* 120:367–372, 1977; *J. Physiol. (London)* 240:639–660, 1974; *Yale J. Biol. Med.* 59:179–185, 1986. Activity of temperature regulating neurons in hypothalamus.
15. Boulant, J. and J. Dean. *Annu. Rev. Physiol.* 48:639–644, 1986. Temperature receptors in CNS.
16. Boulant, J. and H. Demieville. *J. Neurophysiol.* 40:1356–1368, 1977. Re-

sponses of rabbit preoptic anterior hypothalamus to stimulation of hippocampus and brain stem.

16a. Bowler, K. et al. *Comp. Biochem. Physiol. A* 45A:441–450, 1973. Cellular heat death in crayfish.

16b. Brett, J. R. *Q. Rev. Biol.* 31:75–87, 1956. Some principles in the thermal requirements of fishes.

16c. Brett, J. R. *Am. Zool.* 11:99–113, 1971. Energetic responses of salmon to temperature.

17. Bullock, T. H. *Fed. Proc., Fed Am. Soc. Exp. Biol.* 12:666–672, 1953. Bullock, T. H. and R. Diecke. *J. Physiol.* 134:47–89, 1956. Pit viper perception of heat.

18. Burlington, R. *Comp. Biochem. Physiol.* 17:1049–1052, 1966. Biochemistry of kidney from cold-exposed rats and hamsters.

19. Cabanac, M. *Annu. Rev. Physiol.* 38:415–439, 1975. Temperature regulation.

20. Cameron, J. *Am. J. Physiol.* 246:R452–R459, 1984. Effects of temperature on acid-base balance in fishes.

21. Cannon, B. et al. In *Living in the Cold* (H. C. Keller, X. J. Musacchia, and L. C. H. Wang, eds.), pp. 71–81. Elsevier, New York, 1986.

21a. Carey, F. G. *Am. Zool.* 11:137–145, 1973. Warm bodied fish.

22. Carey, F. G. In *Companion to Animal Physiology* (C. R. Taylor et al., eds.), pp. 216–234. Cambridge Univ. Press, London, 1982; *Comp. Biochem. Physiol.* 28:199–213, 1969. Warm fish.

23. Casey, T. et al. *J. Exp. Biol.* 94:119–135, 1981. Pre-flight warm-up in tent caterpillar moth Molacosoma.

23a. Chappell, M. and G. Bartholomew. *Physiol. Zool.* 54:81–83, 1981. Standard operative temperature and thermal energetics of desert ground squirrel.

24. Charnock, J. S. and L. P. Simonson. *Comp. Biochem. Physiol. B* 59B:223–229, 1978; 60B:433–439, 1978. Lipid effects on Na^+-K^+ ATPase in hibernating ground squirrels.

25. Chen, J. and G. Li. *J. Cell Physiol.*

134:189–199, 1988. Heat shock proteins in ovary cells of Tilapia.

26. Connolly, J. G. et al. *Comp. Biochem. Physiol. A* 81A:287–292, 303–310, 1985. Temperature dependent changes in swimming behavior of Tetrahymena (NTI) and electrophysiology.

27. Cooper, P. D. et al. *J. Exp. Biol.* 114:1–15, 1985. Temperature regulation of honey bees in Sonoran desert.

28. Cossins, A. R., M. Friedlander, and C. L. Prosser. *J. Comp. Physiol.* 120:109–121, 1977. Temperature acclimation of synaptic membrane.

29. Cossins, A. R. and C. L. Prosser. *Proc. Natl. Acad. Sci. U.S.A.* 75:2040–2043, 1978. Evolutionary adaptation of membranes to temperature.

29a. Cossins, A. R. and K. Bowler. *Temperature Biology of Animals.* Chapman & Hall, London, 1987.

30. Crawshaw, L. I. et al. *Comp. Biochem. Physiol. A* 47A:51–60, 1974; 51A:11–14, 1975; 52A:171–173, 1975. Thermal preferendum of fish.

31. Crawshaw, L. I. et al. *Am. Sci.* 69:543–561. Evolutionary development of vertebrate thermoregulation.

32. Crompton, A. W., T. W. Taylor, and J. A. Jagger. *Nature (London)* 272:333–336, 1978. Evolution of homeothermy in mammals.

32a. Currie, W. and L. White. *Can. J. Bioch.* 61:438–446, 1983. HSPs in rat tissues.

32b. Dean, P. L. and B. Atkinson. *Comp. Bioch. Physiol.* 81B:185–191, 1985. HSPs in quail red blood cells.

33. de la Roche, A. I. In *Comparative Mechanisms of Cold Adaptation* (L. S. Underwood, ed.), pp. 235–253. Academic Press, New York, 1979. Temperature resistance and cold hardening of plants.

34. De Vries, A. L. *Oceanus* 23–31, 1976; in *Animals and Environmental Fitness* (R. Gilles, ed.), pp. 583–607. Pergamon, Oxford, 1980; *Comp. Biochem. Physiol. A* 73A:627–640, 1982. Antifreeze in Anarctic fish.

34a. DeVries, A. L. *Science* 172:1152, 1971.

Glycoproteins as biological antifreeze agents in antarctic fishes.

34b. DeVries, A. L. *Philos. Trans. R. Soc. London, Ser.* B304:507–588, 1984; *Comp. Biochem. Physiol. A* 73A:627–640, 1982. Biological antifreeze agents in fish.

35. Dickens, B. F. and G. A. Thompson. *Biochemistry* 21:3604–3611, 1982. Phospholipids in microsomal membranes in cold acclimation of Tetrahymena.

36. Duman, J. and K. L. Horwath. *Annu. Rev. Physiol.* 45:261–270, 1983. Role of hemolymph proteins in cold tolerance of insects.

37. Duman, J. et al. *J. Comp. Physiol. B* B151:233–240, 1983; B154:79–83, 1984; *J. Exp. Zool.* 230:355–361, 1984. Antifreeze protein and nucleation protein in beetle Dendroides and hornet Vespula.

38. Dupre, R. K. et al. *Physiol. Zool.* 59:254–262, 1986. Temperature preference and responses of cutaneous temperature-sensitive neurons during bullfrog development.

38a. Egginton, S. and B. D. Sidell. *Am. J. Physiol.* 256:R1–R10, 1989. Thermal acclimation of subcellular structure of fish muscle.

39. Farkas, T. *Comp. Biochem. Physiol.* 64B:71–76, 1979. Adaptation of fatty acids to temperature in crustaceans.

40. Feder, M. E. et al. *Ecology* 63:1657–1664, 1982; *J. Therm. Biol.* 7:23–28, 1982; *Physiol. Zool.* 56:513–552, 1983; *J. Therm. Biol.* 9:255–260, 1984. Field body temperatures and thermal acclimation of metabolism in salamanders.

41. Feeney, R. E. and T. S. Burcham. *Annu. Rev. Biophys. Chem.* 15:59–78, 1986. Antifreeze glycoproteins from polar fish blood.

41a. Fitzpatrick, L. C. and A. V. Brown. *Comp. Biochem. Physiol. A* 50A:733–737, 1975. Geographic variation in salamander.

42. Florant, G. L. and H. C. Heller. *J. Physiol. (London)* 232:R203–R208, 1977. CNS regulation of TB in hibernating mammals.

43. Friedlander, M. J. et al. *J. Comp. Physiol.* 112:19–45, 1976. Temperature effects on fish brain.

44. Fry, F. E. *Publ. Ont. Fish. Res. Lab.* 66:1–35, 1946; 68:1–52, 1947. Environmental effects on activity of fish.

45. Gates, P. M. *Biophysical Ecology.* Springer-Verlag, Berlin, 1980.

46. Gatten, R. E. *J. Therm. Biol.* 10:209–215, 1985. Activity metabolism of lizards after thermal acclimation.

47. Geiser, F. *J. Comp. Physiol. B* 156:751–757, 1986. Torpor metabolism strategies.

48. Glass, M. L., N. Heisler, Et al. *J. Exp. Biol.* 114:37–51, 1985. Effects of TB on respiration, blood gases, and acid-base status in Chysemys picta.

49. Gonzalez, J. G. and L. V. Porcell. *Comp. Biochem. Physiol. A* 83A:709–713, 1986. Thermoregulation in lizard Galliotin.

50. Goolish, E. M. *J. Therm. Biol.* 12:203–206, 1987. Cold acclimation increases size of ventricle of carp.

51. Gorden, C. J. and J. E. Heath. *Annu. Rev. Physiol.* 48:595–612, 1986. Integration and central processing in temperature regulation.

52. Gorden, C. J. and J. E. Heath. *Comp. Biochem. Physiol. A* 74A:479–489, 1983. Reassessment of the neural control of body temperature: importance of oscillating neural and motor components.

53. Gorden, C. J., A. Rezvani, A. A. Fruin, S. N. Trautwein, and J. E. Heath. *J. Appl. Physiol.* 51:1349–1354, 1981. Rapid brain cooling in the free running hamster (*Mecocricetus auratus*).

54. Graves, J. and G. N. Somero. *Evolution (Lawrence, Kans.)* 36:97–100, 1982. Enzyme evolution in four species of eastern Pacific barracuda from different thermal environments.

55. Greer, G. L. and D. R. Gardner. *Comp. Biochem. Physiol. A* 48A:189–203, 1974. Responses from temperature–sensitive units in trout brain.

56. Grigg, G. C. *Aust. J. Zool.* 13:407–411, 1965. Thermal acclimation of lungfish Neoceratodus.

57. Hadley, N. F. *Am. Sci.* 60:338–347, 1972. Desert species and adaptation.

58. Hadley, N. F. *The Adaptive Role of Lipids in Biological Systems.* Wiley, New York, 1985.

58a. Hamby, R. J. *Biol. Bull. (Woods Hole, Mass.)* 149:331–347, 1975. Heat effects in a marine snail.

59. Hannegan, J. L. *J. Exp. Biol.* 53:629–639, 1970. Temperature control of the moth flight system.

60. Hardisty, M. W. and J. C. Potter. In *Biology of Lampreys* (M. W. Hardisty and J. C. Potter, eds.), pp. 85–125. Academic Press, New York, 1971. Behaviour, ecology, and growth of larval lampreys.

61. Hardisty, M. W. and J. C. Potter. In *Biology of Lampreys* (M. W. Hardisty and J. C. Potter, eds.), Academic Press, New York, 1971. General biology of adult lampreys.

62. Hart, J. S. *Publ. Ont. Fish Res. Lab.* 72:1–79, 1952. Lethal temperature of fish from different latitudes.

63. Hazel, J. R. *Am. J. Physiol.* 246:R460–R470, 1984. Effects of temperature on structure and metabolism of cell membranes in fish.

63a. Hazel, J. R. In *Comp. Physiol. Physiological Regulation of Membrane Fluidity*, Ch. 6, pp. 149–188. Liss, New York, 1988. Homeoviscons adaptation in cell membranes.

63b. Hazel, J. R. and C. L. Prosser. *Z. Vergl. Physiol.* 67:217–238, 1970; *Physiol. Rev.* 54:620–677, 1974. Temperature adaptation of enzymes.

64. Hazel, J. R. and P. A. Sellner. *J. Exp. Zool.* 209:105–114, 1979. Fatty acid synthesis by hepatocytes of thermally acclimated trout.

65. Hazel, J. R. et al. *J. Comp. Physiol., B* 155:597–602, 1985; 156:665–674, 1986. Changes in phospholipid composition of membranes from gills and liver of thermally acclimated trout.

65a. Heath, J. E. *Physiol. Zool.* 36:30–35, 1966. Venous shunts in cephalic sinuses of horned lizard.

66. Heath, J. E. *Physiologist* 13:399–410, 1970. Behavioral regulation of body temperature in poikilotherms.

67. Heath, J. E. and P. A. Adams. *J. Res. Lepid.* 3:69–72, 1964. An evaporative cooling mechanism in the Sphinx moth *Pholus achemon*.

67a. Heath, J. E. et al. *Am. Zool.* 11:147–158, 1970. Adaptation of thermal responses of insects.

67b. Heath, J. E. and R. K. Josephson. *Biol. Bull. (Woods Hole, Mass.)* 138:272–285, 1970. Body temperature and singing in katydids.

68. Heinrich, B. In *Insect Thermoregulation* (B. Heinrich, ed.), pp. 235–302. Wiley, New York, 1981. Ecological and evolutionary perspective.

69. Heinrich, B. and S. Buchmann. *J. Comp. Physiol., B* 156:557–562, 1986. Thermoregulatory physiology of carpenter bee Xylocopa.

70. Heinrich, B. and E. McClain. *Physiol. Zool.* 59:273–282, 1986. Laziness and hypothermia as foraging strategy in flower scarabs.

71. Heinrich, B. et al. *J. Exp. Biol.* 64:561–585, 1976; *J. Comp. Physiol.* 96:155–166, 1975; *Physiol. Zool.* 56:563–567, 1983. Thermoregulation in bumblebees.

72. Heisler, N. et al. *Respir. Physiol.* 33:145–160, 1978; *J. Exp. Biol.* 85:89–110, 1980; *Am. J. Physiol.* 246:R441–R451, 1984. Acid-base adjustments to temperature in dogfish.

73. Heldmaier, G. and S. Buchberger. *J. Comp. Physiol., B* 156:237–245, 1985. Nonshivering thermogenesis in hamsters.

74. Heller, H. C. et al. *Comp. Biochem. Physiol. A* 41A:349–359, 1972; *Pfluegers Arch.* 369:55–59, 1977; *Am. J. Physiol.* 227:576–589, 1974. Central nervous system in the control of hibernation, ground squirrels.

74a. Heller, H. C. In *Strategies in Cold.* (L. Wang and J. Hudson, Eds.), pp. 225–265, Academic, New York, 1978. Sleep and hibernation.

74b. Hellon, R. F. *J. Therm. Biol.* 8:7–9, 1983. Central projections and processing of skin-temperature signals.

75. Hensel, H. *Annu. Rev. Physiol.* 36:233–249, 1974. Thermoreceptors.

76. Hochachka, P. W. et al. *Mar. Biol. (Berlin)* 18:251–259, 1973; *Nature (London)* 260:648–650, 1976. Thermal acclimation of enzymes.

77. Hochachka, P. W. and G. N. Somero. *Strategies of Biochemical Adaptation.* Saunders, Philadelphia, PA, 1973; *Biochemical Adaptation.* Princeton Univ. Press, Princeton, NJ, 1984.

78. Hori, T. Pfluegers Arch. 389:297–299, 1981; *Am. J. Physiol.* 231:1573–578, 1986; *Brain Res.* 186:203–207, 1980. Electrical activity of neurons in temperature sensitive units in midbrain of rabbit.

78a. House, H. L. et al. *Can. J. Zool.* 36:629–632, 1958. Effect of diet on temperature resistance, insects.

78b. Huey, R. B. In *Biology of the Reptilia* (C. Gans and F. H. Pough, eds.), Vol. 12, pp. 25–91. Academic Press, New York, 1982. Temperature, physiology and ecology of reptiles.

79. Ishikawa, H. *Comp. Biochem. Physiol. B* 46B:217–227, 1973; 50B:1–4, 1975. Thermal stability of ribosomal RNAs from sea anemones.

80. Jankowsky, D. et al. *Am. J. Physiol.* 246:R471–R478, 1984. Thermal acclimation of adult eels measured by hepatocytes.

80a. Jansky, L. In *Living in the Cold* (H. C. Heller, X. J. Musacchia, and L. C. H. Wang, eds.), pp. 331–340. Elsevier, New York, 1986. Hormones and hibernation.

80b. Johnsen, H. *Amer. J. Physiol.* 253:R848–R853, 1987. Countercurrent from nasal passages to brain in reindeer.

81. Johnston, I. A. *J. Comp. Physiol.* 129:163–177, 1979. Temperature changes in calcium regulatory proteins and in thin filaments of goldfish muscle.

81a. Johnston, I. A. In *Cellular Acclimation to Environmental Changes* (A. Cossins, ed.), pp. 121–144. Cambridge Univ. Press, London and New York, 1983. Altered expression of protein isoforms.

82. Johnston, I. A. et al. *Nature (London)* 254:74–75, 1975; 257:620–622, 1975; *J.*

Comp. Physiol. 119:195–206, 1977. Temperature adaptation of muscle ATPase.

83. Johnston, I. A. and M. Lucking. *J. Comp. Physiol.* 124:111–116, 1978. Temperature induced variation in fiber types in muscles of goldfish.

83a. Johnston, I. A. and W. A. Ublesby. *J. Comp. Physiol.* 119:195–206, 1977. Adaptive properties of myofibrillar ATPase.

84. Johnston, I. A. et al. *J. Exp. Biol.* 111: 179–189, 1984; 119:239–249, 1985. Temperature dependence of contractile properties of muscle and of myosin ATPase in acclimated carp and *Myoxocephalus*.

85. Jones, P. L. and B. D. Sidell. *J. Exp. Zool.* 219:163–171, 1982. Metabolic response of fish muscle to temperature.

85a. Kammer, A. E. *Z. Vergl. Physiol.* 70:45–56, 1970. Pre-flight warm-up in moths.

86. Kent, J., M. Koban, and C. L. Prosser. *J. Comp. Physiol., B* 158:185–198, 1988. Protein hypertrophy as a strategy for adaptation to cold in fishes.

86a. Ketola-Pirie, C. A. and B. G. Atkinson, *Can. J. Biochem. Cell Biol.* 61:462–471, 1983. Heat shock in cultured amphibian cells.

87. Kleckner, N. W. and B. D. Sidell. *Physiol. Zool.* 58:18–28, 1985. Maximum activities of the enzymes of thermally acclimated and acclimatized chain pickerel (Esox niger).

88. Klein, M. J. *J. Exp. Biol.* 114:563–579, 581–598, 1985. Electrical properties of muscle of sunfish from different temperatures.

89. Kluger M. J. and J. E. Heath. *Am. J. Physiol.* 221:144–149, 1971. Effect of anterior hypothalamus-preoptic lesions on temperature in the bat *Eptesicus juscus*.

90. Koban, M. *Am. J. Physiol.* 250:R211–R220, 1986. *In vitro* thermal acclimation of teleost hepatocytes.

91. Koban, M., G. Graham, and C. L. Prosser. *Physiol. Zool.* 60(2):290–296, 1987. Induction of heat-shock protein synthesis in teleost hepatocytes: effects of acclimation temperature.

92. Kothary, R. K. et al. *Biochim. Biophys. Acta* 783:137–143, 1984. Four heat shock phenomenon in cultured cells of rainbow trout.

92a. Langman, V. A. et al. *Respir. Physiol.* 50:141–152, 1982. Respiration and metabolism in a giraffe.

93. Lee, R. F. *Mar. Biol. (Berlin)* 26:313–318, 1974. Lipid composition of Calanus from Arctic Ocean.

93a. Lehrer, G. and R. Baker. *Biochemistry* 9:1533–1541, 1970. Energetics of aldolases.

94. Lewis R. W. *Comp. Biochem. Physiol.* 6:75–89, 1962. Temperature effects on fatty acid of marine organisms.

95. Lindquist, S. *Dev. Biol.* 77:463–479, 1980; *Nature (London)* 293:311–314, 1981. Synthesis of heat shock proteins in Drosophila and yeast.

96. Loftus, R. *J. Comp. Physiol.* 143:443–452, 1981. Cave beetle Speophyes antennae have thermoreceptors.

97. Lorylina, P. T. and G. Goldspink. *J. Exp. Biol.* 118:267–276, 1985. Muscle protein synthesis during temperature acclimation in eurythermal *Cyprinus carpio* and Stenothermal *Salmo*.

98. Low, P. S. and G. N. Somero. *J. Exp. Zool.* 198:1–12, 1976. Temperature effects and kinematics of enzymes.

99. Low, P. S. and G. N. Somero. *Comp. Biochem. Physiol. B* 49B:307–312, 1974. Proposed molecular basis for different catalytic efficiencies of ectotherm and endotherm enzymes.

99a. Lyman, C. P. and D. C. Blinks. *J. Cell. Comp. Physiol.* 54:53–63, 1959. Effects of cooling on hearts of hibernators and non-hibernators.

100. Lyman, C. P., J. S. Willis, A. Morlan, and L. C. Wang. *Hibernation and Torpor in Mammals and Birds.* Academic Press, New York, 1982.

101. Marcus, F. and F. Villanueva. *J. Biol. Chem.* 249:745–749, 1974. Temperature adaptation of fish enzymes.

101a. Marsh, R. L. and A. F. Bennett. *J. Exp. Biol.* 126:79–87, 1986. Thermal dependence of sprinting of Sceloporus.

102. McCrae, M. E. and J. E. Heath. *J. Exp. Biol.* 54:415–435, 1971. Dependence of flight on temperature regulation in the moth, Manduca sexta.

103. McLeese, D. W. *J. Fish. Res. Board Can.* 13:247–272, 1956. Effects of temperature, salinity, and oxygen on survival of lobster.

104. McMahon, R. F. *Comp. Biochem. Physiol. A* 55A: 23–28, 1976. Thermal tolerance of population of snail Physa.

104a. Miller, L. C. and S. Mizell. *Comp. Biochem. Physiol. A* 42A:773–779, 1972. Seasonal effects on frog heart.

104b. Miller, N. G. A. et al. *Biochim. Biophys. Acta* 455:644–654, 1976. Phospholipid adaptation of goldfish membrane.

105. Moerland, T. S. and B. D. Sidell. *J. Exp. Zool.* 238:287–296, 1986. Biochemical responses to temperature in contractile protein complex of striped bass Morone.

106. Mohler, F. S. and J. E. Heath. *Anim. J. Physiol.* 254:R309–R395, 1988. Oscillating heat flow from the rabbit pinna.

107. Moore, J. A. *Evolution (Lawrence, Kans.)* 3:1–21, 1949. Geographic variation of temperature adaptation in Rana pipiens.

108. Morgareidge, K. and H. T. Hammel. *Science* 187:366–368, 1975. Evaporative water loss in turtles.

109. Morman, R. H., S. Cuddy, and P. C. Rugen. *Can. J. Fish. Aquat. Sci.* 37:1811–1826, 1980. Factors influencing distribution of sea lamprey (*Petromyzon mrainus*) in Great Lakes.

110. Morris, G. J. and A. Clark. In *The Biochemistry and Physiology of Tetrahymena* (D. L. Hill, ed.), pp. 55–82. Academic Press, New York, 1981. Effects of low temperatures on biological membranes.

111. Murphy, B. F. and J. E. Heath. *J. Exp. Biol.* 105:305–315, 1983. Temperature sensitivity in the prethoracic ganglion of cockroach.

112. Nagel, E. R. *Am. Sci.* 73:334–343, 1985. Physiological adaptations to aerobic training.

113. Nelson, D. O., J. E. Heath, and C. L.

Prosser. *Am. Zool.* 24:791–804, 1984. Evolution of temperature regulation in vertebrate nervous systems.

114. Nelson, D. O. and C. L. Prosser. *Annu. Rev. Physiol.* 43:281–300, 1981. The role of nervous system in temperature adaptation of poikilotherms.

115. Nelson, D. O. and C. L. Prosser. *J. Comp. Physiol.* 129:193–197, 1979; *Science* 213:787–789, 1981; *Am. J. Physiol.* 241:R259–R263, 1981. Temperature-sensitive neurons in the preoptic region of sunfish.

116. Nelson, R. A. et al. *Science* 226:841–842, 1984; *Mayo Clin. Proc.* 50:141–146, 1975; *J. Comp. Physiol. B* B155:75–79, 1984. Nitrogen metabolism in winter sleep of black bears.

116a. Newman, E. A. and P. Hartline. *Sci. Amer.* 246:116–127, 1982. Infrared "vision" of snakes.

117. Orr, P. R. *Physiol Zool.* 28:290–302, 1955. Heat death, whole animal and tissues.

117a. Pengelley, E. T. and K. C. Fisher. *Can. J. Zool.* 41:1103–1120, 1965; Pengelley, E. T., and S. M. Asmundson. *Comp. Biochem. Physiol.* 30:177–183, 1969. Circadian and annual rhythms in ground squirrels.

117b. Phillip, D. P. et al. *Can. J. Fish. Aquat. Sci.* 38:1715–1723, 1981. Genetic variations in races of largemouth bass.

117c. Popova, W. In *Living in the Cold* (H. C. Heller, X. J., Musacchia, and L. C. H. Wang, eds.), pp. 193–205. Elsevier, New York, 1986. Thermoregulating effect of serotonin in hibernation.

118. Powers, D. and A. Place. *Biochem. Genet.* 16:593–607, 1978. Geographic variations in temperature effects on enzymes of Fundulus.

119. Prosser, C. L. In *Adaptational Biology*, pp. 260–321. Wiley, New York, 1986. Temperature.

119a. Prosser, C. L. et al. In *Evolutionary Biology of Primitive Fishes* (R. E. Forman, ed.), pp. 203–215. Plenum, New York, 1985.

119b. Prosser, C. L. ed. *Comparative Animal Physiology*. 3rd ed. Saunders, 1973.

120. Putnam, R. W. and A. F. Bennett. *Anim. Behav.* 29:502–509, 1981. Thermal dependence of behavior of anurans.

120a. Rahn, H. et al. *Am. Rev. Respir. Dis.* 112:167, 1975. Acid–base balance and pH regulation.

121. Raison, J. K. et al. *Arch. Biochem. Biophys.* 142:83–90, 1971; *J. Biochem. (Tokyo)* 246:4036–4040, 1971; *Symp. Soc. Exp. Biol.* 27:485–512, 1973. Temperature-induced phase change in membrane lipids of plants.

122. Richards, A. G. *Physiol. Zool.* 37:199–211, 1964. Temperature effects on development and oxygen consumption in insect eggs.

122a. Roberts, J. L. *Helgol. Wiss. Meeresunters.* 9:459–573, 1964. Responses of sunfish to photoperiod and temperature.

123. Rome, L. C. *Am. J. Physiol.* 247:R272–R279, 1984; *Science* 228:194–196, 1985; In *Living in the Cold* (H. C. Heller, X. J. Musacchia, and L. C. H. Wang, eds.), pp. 485–495. Elsevier, New York, 1986; *J. Exp. Biol.* 97:411, 1982; *Science* 228:194, 1985. Temperature effects on muscle and locomotor performance in fishes and reptiles.

123a. Roots, B. I. and C. L. Prosser. *J. Exp. Biol.* 39:617–629, 1962. Behavioral temperature acclimation in fish.

124. Santos-Pinto, F. W. et al. *Comp. Biochem. Physiol. A* 82A:859–861, 1985. Temperature acclimation of oxygen consumption in turtle Geocheline.

124a. Schmidt-Nielsen, K. *Desert Animals; Physiological Problems of Heat and Water*. Oxford Univ. Press. 1964.

124b. Schmidt-Nielsen, K. *Am. J. Physiol.* 212:341–346, 1967. Body temperature and water relations in camels.

125. Scott, S. M. and J. Boulant. *Brain Res.* 306:157–163, 1984. Dopaminergic neurons in hypothalamus brain slices.

126. Shaklee, J. B. et al. *J. Exp. Zool.* 201:1–20, 1977. Molecular temperature acclimation in sunfish.

127. Sidell, B. D. *J. Exp. Zool.* 199:233–250, 1977. Turnover of cytochrome c in muscle of green sunfish.

128. Sidell, B. D. *Physiol. Zool.* 53:98–107, 1980. Goldfish muscle enzymes.

129. Sidell, B. D. and I. A. Johnston. *Can. J. Zool.* 63:811–816, 1984. Thermal sensitivity of contractile function in muscle of pickeral Esox.

129a. Sidell, B. D. and T. S. Moreland. In *Advances in Environmental and Comparative Physiology* (C. P. Mangum, ed.), pp. 116–158. Springer-Verlag, Berlin and New York, 1988. Effects of temperature on muscular function and locomotor performance in teleost fishes.

129b. Sisson, J. and B. D. Sidell. *Physiol. Zool.* 60:310–320, 1987. Effect of thermal acclimation on recruitment of muscle fibers in striped bass.

129c. Somero, G. N. In *The Enzymes* (P. D. Boyer, ed.), 3rd ed., Vol. 11, pp. 221–235. Academic Press, New York, 1975; Somero, G. N. with P. S. Law. *Nature (London)* 266:276–278, 1977. Energetics of LDHA.

129d. Somero G. N. and J. Low. *J. Exp. Zool.* 194:175–188, 1975; 198:1–12, 1976. *Annu. Rev. Ecol. Syst.* 9:1–29, 1978. Adaptations of kinetic properties of enzymes from poikilotherms.

129e. Späth, M. *Zeit. Vergl. Physiol.* 56:431–462, 1967. Action of temperature on mechanoreceptors of fish *Leuciscus*.

130. Spray, D. C. *J. Physiol. (London)* 237:15–38, 1974; *Brain Res.* 72:354–359, 1974; *Comp. Biochem. Physiol.* 50:391–395, 1975. Properties of cutaneous cold receptors in frog Rana.

131. Spray, D. C. *Annu. Rev. Physiol.* 48:625–638, 1986. Cutaneous temperature receptors.

132. Stanton, T. et al. In *Living in the Cold* (H. C. Heller, X. J. Musacchia, and L. C. H. Wang, eds.), Elsevier, New York, 1986. Pineal melatonin in ground squirrel.

133. Storey, K. B. *Cytobiology* 20:365–379, 1983. Metabolism and bound water in overwintering insects.

133a. Storey, K. B. *Verh. Dtsch. Zool. Ges.* 80:77–91, 1987. Biochemical mechanisms of adaptation: strategies of winter survival.

134. Storey, K. B. and J. M. Storey. *Can. J. Zool.* 64:49–56, 1986; *Comp. Biochem. Physiol. A* 83A:613–617, 1986; *J. Comp. Physiol., B* 156:191–195, 1985. Freeze tolerance in hibernating amphibians; cryoprotectant compounds in frog.

135. Storey, K. B. and J. M. Storey. *Physiol. Rev.* 68:27–84, 1988. Freeze tolerance in animals.

136. Strumwasser, R. *Am. J. Physiol.* 196:8–30, 1959. Brain activity in hibernating Citellus.

136a. Thompson, A., J. Sargent, and J. Owen. *Comp. Bioch. Physiol.* 56B:223–228, 1977. Acclimation of all ATPase in eels.

137. Todd, M. E. and P. A. Dehnel. *Biol. Bull. (Woods Hole, Mass.)* 118:150–172, 1960. Effects of temperature and salinity on heat tolerance in grapsoid crabs.

138. Trautwein, S. N., C. J. Gorden, and J. E. Heath. *Comp. Biochem. Physiol. A* 80A:199–204, 1985. Changes in brain and body temperature of the lizard Sceloporus during rest and exercise.

139. Tsukuda, H. et al. *Comp. Biochem. Physiol. A* 82A:281–283, 1985. Heart rate and oxygen consumption of isolated hearts of goldfish acclimated to different temperatures.

140. Umminger, B. L. In *Physiological Ecology of Estuarine Organisms* (F. J. Vernberg, ed.), Univ. of South Carolina Press, Columbia, 1975. Cold resistance in *Fundulus*.

140a. van Berkum, F., R. Huey, and B. Adams. *Physiol. Zool.* 59:464–472, 1986. Thermoregulation in tropical lizard Ameira festiva.

140b. Waites, G. M. H. *Q. J. Exp. Physiol.* 47:314–323, 1962. Effect of heating scrotum of ram on respiration and temperature regulation.

141. Walesby, N. J. and I. Johnston. *Experientia* 37:716–718, 1981. Temperature acclimation of myofibrillar ATPase in trout and goldfish.

142. Walsh, P. J. and G. N. Somero. *Can. J. Zool.* 60:1293–1299, 1982. Purified LDH in cold and warm acclimated *Gillichthys*.

143. Wang, L. C. H. In *Strategies in the Cold* (L. C. H. Wang and J. W. Hudson,

eds.), pp. 109–146. Academic Press, New York, 1978. Energetic and field aspects of mammalian torpor: Richardson's ground squirrel.

144. Wang, L. C. H. and J. W. Hudson. *Comp. Biochem. Physiol.* 38:59–90, 1971; *Cryo-Lett.* 6:257–274, 1985. Hibernation in ground squirrels.

145. White, R. *Physiol. Zool.* 56:174–194, 1983. Temperature block of neuromuscular transmission in crayfish.

146. Whyard, S., G. R. Wyatt, and V. K. Walker. *J. Comp. Physiol., B* 156:813–817, 1986; *Cell* 48:507–515, 1987. Heat shock proteins in Locusta and yeast.

147. Willis, J. S. In *Living in the Cold* (H. C. Heller, X. J. Musacchia, and L. C. H. Wang, eds.), pp. 27–34. Elsevier, New York, 1986. Membrane transport at low temperatures.

148. Wilson, F. R., G. Whitt, and C. L. Prosser. *Comp. Biochem. Physiol. B* 46B:105–116, 1973. LDH and MDH patterns in temperature acclimation in goldfish.

149. Wilson, F. R., C. L. Prosser, and G. Somero. *Comp. Biochem. Physiol. B* 47B:485–491, 1974. Temperature metabolism of two species of Sebastes.

150. Wodtke, E. *Biochim. Biophys. Acta* 640:698–709, 1981. Phospholipids of mitochondrial membranes of carp acclimated to 20 and 32 degrees C.

151. Wodtke, E. *Biochim. Biophys. Acta* 640:710–720, 1981. Temperature adaptation of biological membranes of carp.

152. Wodtke, E. *J. Comp. Physiol.* 91:277–307, 309–332, 1974. Properties of mitochondria of eel acclimatized to 7 and 25 degrees C.

153. Wodtke, E. et al. In *Living in the Cold* (H. C. Heller, X. J. Musacchia, and L. C. H. Wang, eds.), Elsevier, New York, 1986; *J. Comp. Physiol.* 110:145–157, 1976. Homeoviscous adaptation in Arrhenius plots of enzymes from temperature-adapted carp.

153a. Yamamon, T. and T. Yura. *Proc. Nat. Acad. Sci.* 79:860–864, 1982. Heat shock proteins in *E. coli*.

154. Yancey, P. H. and G. N. Somero. *J. Comp. Physiol.* 125:129–134, 1978. Temperature dependence of intracellular pH.

155. Zachariassen, K. E. *Physiol. Rev.* 65:799–822, 1985. Cold tolerance in insects.

4 | Hydrostatic Pressure and Adaptations to the Deep Sea

George N. Somero

Introduction: The Deep-Sea Environment

The deep sea often is regarded as an exotic environment whose conditions are very different from those of most of the biosphere. However, when viewed in terms of volume, by far the greatest portion of the biosphere is comprised by the deep sea (Fig. 1). Approximately 70% of earth is covered by oceans, 86% of which are deeper than 2000 m. The average depth of the oceans is ~3800 m, and a depth of 10,790 m has been recorded in the Challenger Deep in the Marianas Trench. Although there is no hard and fast definition of where the deep sea begins—definitions of "deep" will be seen to depend on the type of depth-related stress to biological systems we are considering—if we take 1000 m as the start of the deep sea, then ~78.5% of the biosphere's volume is within the deep sea. Most of the remainder of the biosphere is in the shallow oceans; the terrestrial fraction of the biosphere is <1%. Thus, the deep sea is by far the largest of all earthly habitats.

Animals and bacteria are found throughout the water column, even at depths near 11,000 m. The adaptations that enable deep-sea organisms to exploit an environment that is inimical to the survival of most organisms can be grouped into two general classes. One class comprises *resistance adaptations*, the traits of deep-sea organisms that allow survival under the environmental conditions, especially the high pressures, found in the deep sea. The other broad class of adaptations are *capacity adaptations*. These adaptations establish appropriate rates of activity, for example, of respiration, locomotion, and growth, for life in the deep sea.

Physical Characteristics of the Deep Sea

The physical characteristics of the deep sea—high pressures, low temperatures,

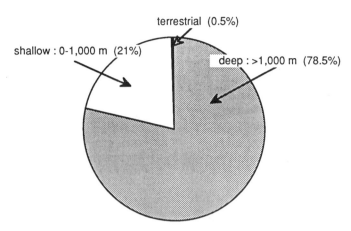

Figure 1. The biosphere subdivided in terms of volume. Note that almost eight tenths of the biosphere is comprised by oceanic waters having depths equal to or greater than 1000 m. The terrestrial volume is based on the assumption of a 50-m height of living material [Modified after Childress (16).]

and an absence of light (other than bioluminescence)—have favored selection for numerous resistance and capacity adaptations. To appreciate these adaptations one must comprehend the nature of the physical stresses that have elicited them. We begin with hydrostatic pressure, a stress unique to deep environments.

Hydrostatic Pressure

The deepest regions of the oceans are trenches associated with subduction zones, whose depths may reach ~11,000 m below sea level. The highest elevations on land are ~8500 m. Because of the vast difference in density between air and water, the entire atmospheric envelope of the earth impresses upon the surface of the planet and its oceans a pressure of only 1 atm (760 mmHg), whereas the mass of seawater lying above the deepest trenches establishes a pressure of nearly 1100 atm at these great depths. That is, for each 10-m increase in depth in the marine water column, pressure increases by 1 atm. The vertical displacements undergone by terrestial animals and most freshwater organisms create negligible problems due to pressure changes, except for effects on gas-filled spaces, for ex-

ample, swimbladders. In the marine realm, however, pressure gradients play major roles in determining organisms' distribution limits.

Units of Pressure[a]

(psi)	(atm)	(kg/cm²)	(N/m² × 10⁵)
14.696	1	1.033	1.01325
1,000	68	70	68.0
5,000	340	352	344.5
8,000	544	562	551.2
10,000	680	703	689
15,000	1021	1055	1034.5

[a]psi: pounds per square inch; atm: atmosphere; N/m²: newtons per squared meter [1 N/m² = 1 pascal (Pa)]. Most studies of the biology of high pressure have expressed pressure in atmospheres, and this chapter will follow this convention. Pressures are sometimes expressed as "atmospheres absolute" (ATA) to emphasize the pressure increment imposed on the system. Thus, in a system on which an additional 100-atm pressure has been imposed, the true pressure is 101 ATA. The slight deviation from linearity in pressure with depth reflects the increase in density of seawater as a consequence of compression with increasing depth.

All effects of pressure, whether in gas- or water-filled spaces, are a result of volume changes. Gas-filled spaces offer the intuitively clearest examples of pressure–volume interactions. Boyle's law (pressure × volume = constant) states that the volume occupied by a gas is inversely proportional to the pressure exerted on the gas. In the absence of any regulatory effort by the animal, the volume of a fish's swimbladder would decrease by 50% as the fish migrated from the surface to a depth of 10 m (2 ATA). Migration to 30 m (4 ATA) would lead to a further halving of the volume, and so forth. Regulating the volumes of gas-filled spaces like swimbladders thus is a problem for an organism such as a vertical migrator that moves through the water column. Also, the high absolute pressures of the deep sea establish severe problems at pressures of several hundred atm for fishes that use gas-filled buoyancy devices [buoyancy mechanisms used by marine animals are reviewed by Macdonald (64)].

Aqueous solutions are much less compressible than gases. The volume of seawater decreases by only ~4% for each 1000 atm increase in pressure. The low compressibility of water might seem to imply that volume changes in aqueous phases play only minor roles in establishing the pressure sensitivities of organisms. This is incorrect, however. The sensitivities to pressure of metabolic processes often result from volume changes in the aqueous phase of the cell. These volume changes arise from alterations in water organization, that is, in water density, near the surfaces of biomolecules. These changes in system (water + bio-

molecules) volume accompany a wide spectrum of biochemical processes (Table 1), including the dissociation of weak acids, and the unfolding and disassembly of proteins.

The relationship between the volume change accompanying a process and the effect of pressure on the equilibrium constant of the process is given by the equation:

$$K_p = K_0 \exp(-P\Delta V/RT)$$

K_0 is the equilibrium constant at 1 atm; K_p is the value at pressure P. The pressure effects tabulated in part A of Table 1 were calculated with this equation.

The effects of hydrostatic pressure on the velocities of physiological and biochemical processes obey a similar exponential relationship. If we substitute rate constants (k_p, k_0) for equilibrium constants, and activation volumes ($\Delta V\ddagger$, the change in system volume that accompanies the transition from the ground state to the activated complex; designated by the double dagger symbol) for reaction volumes, we can express pressure effects on rates as:

$$k_p = k_0 \exp(-P\Delta V\ddagger/RT)$$

Table 1, part B lists activation volumes for a variety of enzymatic reactions. Cau-

TABLE 1. Part A. Relationships of Volume Changes to Changes in Equilibrium Constants and Rate Constants Under Pressure[a,b]

ΔV (cm³/mol)	100 atm	500 atm	1000 atm
+25	−10	−42	−67
+50	−20	−67	−89
+100	−35	−89	−99
+150	−48	−96	− >99
−25	+12	+73	+199
−50	+25	+199	+795
−100	+55	+795	+7,900
−150	+93	+2577	+71,567

TABLE 1. Part B. Volume Changes Associated with Representative Biochemical Reactions[a,c]

Type of Reaction	ΔV (cm³/mol)	References
Proton Dissociation		
$H_2O = H^+ + OH^-$	21.3	7
Imidazole-H^+ = imidazole + H^+	+1.1	55, 86
$H_2PO_4^- = HPO_4^{2-} + H^+$	−24	72
$H_2CO_3 = HCO_3^- + H^+$	−25.5	50, 51
Protein—COOH = Protein—COO⁻ + H^+	−10	55, 86
Protein—NH_2 + H_2O = Protein—NH_3^+ + OH^-	−20	55, 86
Hydrogen-bond Formation		
Poly-lysine helix formation	−1.0	74
Poly(A + U) helix formation	+1.0	110
Hydrophobic Hydration		
$C_6H_6 = (C_6H_6)$ in H_2O	−6.2	54
(CH_4) in hexane = (CH_4) in H_2O	−22.7	54
Polar Group Hydration		
n-Propanol = (*n*-propanol) in H_2O	−4.5	36
Protein Subunit Dissociation		
Microtubules (Tubulin)	−90	91
Actin	−9 to −139	113
Enolase	−65	76
Lactate dehydrogenase	−170 to −220	60, 61
Mellitin	−150	115
Ribosomes	−242	77
Protein Denaturation		
Myoglobin	−98	125
Ribonuclease	−43	8
Activation Volumes		
Acetylcholinesterase	28	47
Alkaline phosphatase	14–24	38
ATPase-mitochondrial	30	71
Catalase	5	71
Chymotrypsin	−6	71
Fumarase	28	71
Invertase	−4	71
Luciferase	22	71
Pepsin	22	71
Pyruvate carboxylase	47	79
Trypsin	−10	71
Xanthine oxidase	−40	71

[a]The percent change listed for each combination of pressure and volume change applies to both equilibrium and rate constants. The temperature used for the calculations is 5°C.
[b]Percent change in K_{eq} or rate constant at pressure (atm) relative to 1-atm equilibrium (or rate) constant.
[c]Reaction volumes and activation volumes may depend on solution conditions such as measurement temperature, salt composition, pH, ionic strength, and range of measurement pressure.

tion must be taken in using published $\Delta V\ddagger$ values to predict pressure effects *in vivo*, however. The sensitivity of an enzymatic reaction to pressure may depend on the temperature, pH, ionic strength, and salt composition of the solution (38, 63, 71). Many studies of pressure effects have not attempted to simulate physiological conditions. The term $\Delta V\ddagger$ may vary with measurement pressure, so a realistic estimate of $\Delta V\ddagger$ must be made at physiological pressures. Thus, $\Delta V\ddagger$ values often are only approximate indexes of how pressure affects reaction rates in the cell.

Even with these limitations, the data in Table 1 allow the following generalizations: (1) Enzymatic activities frequently are strongly affected by pressure. The $\Delta V\ddagger$ values often range in absolute value between 10 and 40 cm^3/mol. The $\Delta V\ddagger$ values of this magnitude cause large changes in rate constants under biologically realistic pressure regimes (Table 1, part A). (2) Pressure effects differ among enzymes both qualitatively (pressure stimulates some reactions while inhibiting others) and quantitatively ($\Delta V\ddagger$ values differ in absolute values among reactions). Thus, the imposition of a change in pressure on an organism may affect the total rate of metabolic activity and the relative activities of different metabolic pathways.

Table 1 lists the changes in volume associated with a number of types of biological reactions. An increase in exposure to water of a charged side chain of a protein may lead to a constriction of water around this group and, therefore, to a decrease in system volume. Conversely, withdrawal of a water-constricting group into the interior of a protein may necessitate the group's dehydration and an expansion of the system's volume. These types of volume changes accompany many alterations in protein conformation, for example, those that occur during substrate binding and catalysis.

Another example of a water-structure-dependent volume change is the dissociation of weak acids and weak bases at elevated pressure. Because an increase in the total number of charged particles in an aqueous solution usually leads to a decrease in system volume, increases in pressure typically favor the dissociation of weak acids. The effects of pressure on acid–base equilibria vary considerably among weak acids and bases. These differences reflect the net change in number of charged particles in solution. Phosphate buffers are much more pressure sensitive than imidazole buffers (Table 1); dissociation of $H_2PO_4^-$ to $HPO_4^{2-} + H^+$ increases the number of charged, water-constricting species in solution, while dissociation of the imidazole group results in no net change in the number of charged species in solution. Thus, while pressure will have only a very small effect on the pH of solutions buffered by imidazole buffers, the pH of a solution buffered by phosphate buffers will decrease by ~0.4 pH unit per 1000-atm increase in pressure. Because the major intracellular buffering compounds are imidazole compounds (105), changes in pressure may have only small effects on intracellular pH. However, there has been no experimental study of pressure effects on *in vivo* pH values.

Noncharged polar molecues and nonpolar molecules affect the organization and volume of water. Nonpolar (hydrophobic) regions on a protein's surface may be surrounded by relatively dense water. When two hydrophobic surfaces are brought together and dense water around these surfaces is displaced into the bulk water of the solution, a volume increase may result. This type of phenomenon may explain in part why subunit assembly equilibria are so pressure sensitive, and why volume decreases accompany protein denaturation (Table 1).

Hydrogen-bond formation is relatively insensitive to pressure, as indicated by the small volume changes accompanying the formation of alpha helices (e.g., polylysine helices) and nucleic acid double helices (e.g., poly(adenine–uridine) helices) (Table 1).

The nucleic acid–based systems of organisms appear to be much less perturbed by pressure than systems based primarily on proteins or lipids. Lipids are fairly compressible, and pressure effects on membrane-associated processes often show extremely strong pressure dependencies, as discussed later.

Temperature

The effects of temperature on biological processes are treated in detail in Chapter 3. Deep-sea temperatures typically are very stable and extremely low, near 1–4°C. The cold water found in most of the deeper regions of the oceans originates at polar latitudes, where cold, dense surface waters sink to the sea floor and flow towards the equator. However, in some deep-sea areas temperatures may be appreciably warmer, notably at deep-sea spreading centers where geothermally heated waters with temperatures up to ~400°C mix with cold bottom water (52). When mixing occurs beneath the seafloor at spreading centers, water with variable temperatures bathes the resident animals and bacteria. In the very warm (45–56°C) and highly saline brine pools of the Red Sea, at depths near 2000 m, only bacterial life exists.

WATER TEMPERATURES IN THE DEEP SEA

Water Mass	Temperature (°C)
Typical deep sea	1–4
Mediterranean sea	13
Red Sea brine pools	45–56
Hydrothermal vents at spreading centers	
Black Smoker chimneys	up to ~400
Warm water vents	2–25

The interactions of temperature and pressure may create unique problems for deep-living organisms because some of the stresses of high pressure are exacerbated by low temperatures, particularly in the case of the membrane-based functions examined next.

Resistance Adaptations: Pressure Effects on Membrane-Based Functions

Behavioral and Physiological Effects of Pressure

Cellular functions dependent on membrane-localized processes often are sensitive to pressure. Pressure-sensitive membrane functions include ionic and osmotic regulation, maintenance of nerve resting potentials, conduction of nerve impulses, synaptic transmission, and the activities of membrane-localized enzymes such as ion-dependent ATPases. Undoubtedly other membrane-localized processes, for example, neurotransmitter and hormone binding, signal transduction, and transmembrane protein transport, will also be found to be significantly affected by pressure.

Adaptations of membrane systems to pressure are important even at modest depths because pressures of only 50 atm or less are sufficient to perturb many membrane-localized functions (reviewed by Macdonald (65)). Recent studies have begun to link the long-known effects of pressure on the behavioral and neurophysiological characteristics of organisms—effects that were first discovered by pioneering work of Regnard (87) a century ago—to changes at the molecular level.

The High-Pressure Neurological Syndrome
The high sensitivities of membrane-localized functions to pressure are reflected in striking behavioral and physiological changes in response to pressure changes. One example is the High-Pressure Neu-

rological (or Nervous) Syndrome (HPNS) observed in terrestrial and shallow-water animals subjected to pressures of ~20–100 atm (9). The HPNS is marked by tremors and convulsions and, for animals subjected to anesthesia, by an increase in the amount of anesthetic required to exert a given level of anesthesic effect. For many shallow-living animals, uncoordinated movements are noted at pressures of 50 atm or less; convulsions occur at 50–100 atm, and reversible paralysis is found at higher pressures (66).

Deep-living animals have much higher pressure thresholds for the HPNS symptoms. Brauer et al. (11) observed that the average threshold pressure for the appearance of convulsions was 88 atm for shallow-living cottid fishes and 120–270 atm for deep-living cottids of Lake Baikal (greatest depth 1600 m). Decompression of deep-living cottids to 1-atm induced aberrant behavioral and neurological symptoms that were reversed by recompression to original *in situ* pressures. Other studies of deep-living animals have reported similar effects of decompression and recompression on freshly captured specimens (10, 12, 66, 69, 123).

The cellular basis of the HPNS is not well understood. Neither is it known whether the different pressure thresholds for the onset of HPNS found between deep- and shallow-living species are the result of acclimations to pressure or, instead, to fixed, genetically based differences between the species. The possible role of acclimatory change in membrane lipids is considered later.

Pressure Effects on Transmembrane Ion Flux
Increased pressure leads to rapid and large-scale alterations in the exchanges of ions between the organism and its environment and between the cells and the extracellular fluids (90). Threshold pressures at which significant effects are first observed are low relative to the pressures encountered by deep-living animals. For

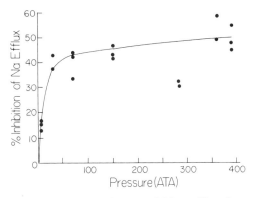

Figure 2. The inhibition of Na^+ efflux from human erythrocytes as a function of pressure. The percent inhibition values are based on the ratios of the rate constants determined at elevated pressures and at ambient pressure (1 ATA). [Modified after Goldfinger et al. (37).]

example, in the freshwater crayfish *Procambarus clarkii*, only 15 to 25 atm were adequate to initiate inhibition of Na^+ uptake, and 50 or 100 atm led immediately to an ~80% decrease in uptake rate (88). In human red blood cells (RBCs) pressures in the range of 30–150 atms led to 40% inhibition of active sodium extrusion (Fig. 2). Isolated gills from seawater-acclimated eels (*Anguilla anguilla*) showed a complex response to pressure in their passive Na^+ permeability (80). As pressure was increased from 1 to 250 atm at 36% decrease in passive Na^+ influx was found; at 500 atm a 150% increase in Na^+ occurred.

Ion fluxes in deep-living animals have received little study. Studies of the effects of pressure on Na^+ influx in shallow- and deep-living gammarid amphipods from Lake Baikal showed that pressures of 100 atm led to an ~35% decrease in Na^+ influx in the shallow-living gammarids (89). Deeper-living species, which were acclimatized to depths of 600–1400 m, exhibited a 40% inhibition of active Na^+ uptake when brought to 1 atm. These responses of ion flux to pressure suggest that membrane transport systems of deep-living organisms are perturbed by reductions in pressure, much as the systems of shallow-living species are upset by increased pressure.

The effects of pressure on ion fluxes across membranes derive from two primary sources: effects on permeability and effects on the activities of enzyme systems involved in active transport. Both sources of pressure effects lead to complex interactions between pressure and transmembrane movement of ions, and both passive and active ion movements across membranes typically show marked interactions between pressure and temperature (65, 119). The passive efflux of K^+ and Na^+ from liposomes decreased linearly with pressure, suggesting a reduction in permeability with increasing pressure (65). However, the passive permeability of human erythrocytes to K^+ increased with rising pressure, and these pressure effects were more pronounced at high than at low temperatures (40). In voltage-sensitive ion channels, the rates of movement of Na^+ and K^+ within the channels are little affected by pressure, which exerts its major influence on the opening and closing (gating) of the channels (65).

Active ion transport by ion-dependent adenosine triphosphatase (ATPase) enzyme systems, like the Na^+-K^+ ATPase, is strongly affected by pressure. As shown for passive permeability of ions, the effects of pressure on ion transporting enzyme systems often are complex, and strong interactions between temperature and pressure may occur (65). High pressures often inhibit the Na^+-K^+ ATPase, but examples of pressure activation of the enzyme have also been reported (65, 81). These effects are discussed later in the context of adaptations in membrane fluidity.

Pressure Effects on Neural and
Contractile Functions
Several different aspects of electrophysiological function in neural and contractile

tissues are affected by increased pressure. In mammalian heart elevated pressure caused membrane depolarization and a reduction in the velocity of the conducted impulse (31). At 150 atm, the conduction time, the time taken for an impulse to propagate between two fixed sites in the heart, increased by 40%. The slowing of conduction velocity was a result in part of the reduced sodium current. Pressure also slowed the rate of repolarization and increased the duration of the action potential in mammalian heart (32). Under high pressure the beating frequency of the mammalian heart is reduced (hyperbaric bradycardia) because of retardation by pressure of the rate at which the cardiac pacemaker cells attain the correct membrane potential required for initiating a burst of activity (75).

In diving mammals like seals, the heart rate may also decrease when the animal dives, the so-called "diving bradycardia" response. However, this response is not a result of the effects of pressure, since it can be elicited by merely submerging the head of a diver slightly beneath the surface of the water.

The apparent activation volumes for several cardiac electromechanical events in mammals have been estimated (49). Only one event, twitch tension in cat papillary muscle, was stimulated by pressure. All other events were characterized by positive activation volumes of 24 to 105 cm^3/mol. Recall that activation volumes of +50 and +100 cm^3/mol yield 20 and 35% rate inhibition, respectively, as pressure is increased by 100 atm (Table 1). Thus, maintenance of adequate cardiac function in deep-sea animals and in certain diving mammals (some whales and seals may dive to depths of at least 1000 m) is apt to require pervasive adaptations to pressure. The nature of these cardiac adaptations has not been explored.

Pressure effects on several steps in impulse propagation in a variety of species have been studied (56, 65). Similarities in the effects of pressure exist among species for a given type of neurophysiological event. The kinetics of action potential propagation is slowed by pressure in all vertebrates and invertebrates so far studied (65). In squid giant axon the activation of the Na$^+$ current was slowed by increased pressure; the apparent activation volume for the sodium transport process was 32 cm^3/mol (26, 27). A second general effect of pressure is seen on excitatory synaptic transmission, which is decreased by pressure. This inhibition has been seen in squid giant axon (42), in vertebrate (4) and crustacean (13) neuromuscular junctions, in molluscan ganglion cells (78), and in vertebrate sympathetic ganglia (57). This inhibitory effect apparently is a result of a decrease in transmitter release (56).

Synaptic transmission not only is inhibited by pressure, but the sizes of pressure effects may be great, and the threshold pressures at which effects appear may be low. In rat diaphragm synaptic transmission fails at 100 atm (58). The pressures at which synaptic transmission is impeded are among the lowest pressure thresholds known. The spontaneous release of acetylcholine from the frog neuromuscular junction is reduced when pressure is raised by only a few atm (4), and in lobster muscle the release of transmitter at the inhibitory neuromuscular synapse is reduced by only 4-atm pressure (25). Post-synaptic effects of pressure are also evident. The binding of acetylcholine to its receptors is inhibited by pressure (92). Acetylcholinesterase is also very sensitive to pressure (46, 47).

There has been little study of the neurophysiology of deep-living animals. Campenot (13) compared the effects of pressure on neuromuscular excitatory junctional potentials (ejps) of lobster *Homarus americanus*, which occurs to depths of ~520 m and of red crab, *Geryon quinquedens*, found at depths between 300 and 1100 m. Pressures of 50–200 atm re-

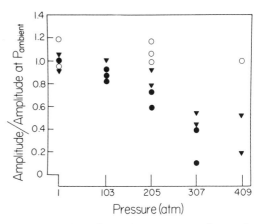

Figure 3. The effect of pressure on the peak amplitude of the compound action potential of the vagus nerve of marine fishes adapted to different depths: *Gadus morhua*, cod-shallow water, 10°C [●]; *Mora moro*, 900 m, 6°C [▼]; *Coryphaenoides armatus* and *Bathysaurus mollis*, 4000–4200 m, 2.5°C [○; both species]. [Modified after Forbes et al. (35).]

versibly depressed the amplitudes of the ejps of the lobster neuromuscular junction, but did not depress the ejps of the red crab in the normal physiological stimulatory frequency range. A decrease in the amount of transmitter released at the lobster neuromuscular junction under pressure was suggested.

The effects of pressure on the visceral branch of the vagus nerve of shallow- and deep-living fishes differed among species (35). Inhibition of the peak amplitude of the action potential for shallow-living fishes was found at pressures of 103–409 atm. No inhibition of peak amplitude was found for two deep-living fishes, *Coryphaenoides armatus* and *Bathysaurus mollis* (Fig. 3).

In summary, the low threshold pressures at which membrane-based functions are perturbed, as well as the magnitudes of these perturbations, indicate that tolerance of the deep sea requires pervasive biochemical adaptations in membrane systems. Even human divers, who may withstand dives to depths of ~300 m, may encounter pressures that are adequate to disrupt a variety of membrane-localized functions.

Pressure Effects on Membrane-Localized Enzymes and Membrane Lipids

The diverse effects of changes in pressure on behavior, osmotic and ionic regulation, and the functions of nervous and contractile systems of animals are not fully understood in biochemical terms. However, because all of these characteristics of animals depend in some measure on the integrity of membrane systems composed primarily of proteins and different classes of lipids (phospholipids, gangliosides, cholesterol, and others), there is strong likelihood that changes induced by pressure in the structures of these two critical membrane components lie at the base of many, and perhaps all, of the pressure effects discussed in the preceding section.

Support for this conjecture comes from studies of the effects of pressure on the physical properties of lipids (23), and studies of the interactions among pressure, temperature, and lipid fluidity on the activity of ion-dependent ATPases such as the Na^+-K^+ ATPase of membranes (24). The Na^+-K^+ ATPase of membranes is a protein that requires an annulus of lipid molecules to be catalytically active. The physical state—the fluidity—of this phospholipid annulus is critically important in establishing the rate of ATPase function. Lipid fluidity is affected by pressure, temperature, and the chemical composition of the lipids.

The pronounced effects of hydrostatic pressure on membrane fluidity have been documented (23, 119). Using synaptic and myelin fractions of goldfish brain, Chong and Cossins (23) showed that membrane fluidity decreased as pressure was raised from 1 to 2000 atm. This effect was observed at all measurement temperatures

TABLE 2. Interactions of Pressure and Temperature on Transition Temperatures and Bilayer Order Parameters for Artificial Membranes, Natural Membranes, and Membrane-Localized Enzymatic Activities[a]

Study System	dT/dP (°C/atm)
I. Phase Transition Temperatures	
Artificial membranes	
Dipalmitoyl phosphatidylcholine	0.022–0.027
Natural membranes	
Acholeplasma laidlawii	0.016–0.03
Rabbit alveolar macrophage membrane	0.027
Enzyme systems: Arrhenius "break" temperatures	
Nitrogenase	0.02
Na^+-K^+-ATPase	0.015
II. Bilayer Order Parameters (determined by fluorescence polarization)	
Goldfish synaptosomal membranes	0.013 (5–20)
	0.018 (20–35)
Acholeplasma laidlawii membrane	0.015 (5–20)
Deep-sea fish liver mitochondria	0.017

[a]Data from summaries of literature in Macdonald (65) and Wann and Macdonald (119).

between 5.6 and 34.3°C. Decreasing measurement temperature also decreased membrane fluidity. An increase in pressure of 1000 atm decreased membrane fluidity by about the same amount as a 13–19°C reduction in temperature.

The relationship between the effects of temperature and pressure seen with the goldfish membrane preparation is quantitatively similar to that found for a variety of other lipid-based systems (Table 2). For natural and artifical membranes, temperature decreases and pressure increases have similar effects on membrane fluidity. These interactions between temperature and pressure on lipid systems can be ex-

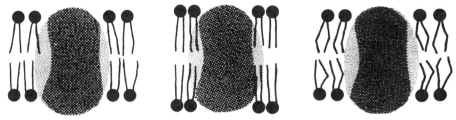

Figure 4. The effect of hydrostatic pressure and fatty acid acyl chain composition on the organization of a bilayer membrane. *Left drawing:* Membrane with predominantly saturated fatty acid acyl chains in the phospholipids, shown at a pressure of 1 atm. The darkly shaded globular structure spanning the bilayer is an enzyme, for example, an ion-dependent ATPase, which must move in the plane of the membrane (indicated by light stippling) in order to carry out its function. Movement in the plane of the membrane will be impeded by a rigid lipid annulus around the protein. *Middle drawing:* The effect of high pressure on the organization of the phospholipids in the membrane of the left frame. Pressure compresses the lipids together, reducing their fluidity, and this lateral compression of the lipids leads to an increase in membrane thickness. Movements of the membrane-spanning enzyme are impeded by the tight packing of acyl chains around the enzyme. *Right drawing:* The effect of homeoviscous adaptation to pressure: The introduction of increased amounts of unsaturated acyl chains—which are kinked and, therefore, pack together less tightly—into the membrane lipids favors restoration of membrane fluidity under pressure. The enzyme can, again, easily undergo movement in the plane of the membrane, and the bilayer thickness is restored to its 1-atm value.

pressed using the Clapeyron relation:

$$dT/dP = \Delta V(T/\Delta H)$$

This equation provides an expression for how a pressure-dependent change in a system, for example, in a lipid phase transition temperature (T)—the temperature at which a lipid-based system changes from an ordered *crystalline* phase to a more fluid *liquid-crystalline* phase—is related to the volume (ΔV) and enthalpy (ΔH) changes that accompany the phase transition. Thus, if a phase transition occurs with a large ΔV, then the effect of a change in pressure on the transition temperature will be large unless there is a correspondingly large enthalpy change during the phase transition. Values of dT/dP can also be used to characterize the interacting effects of temperature and pressure on membrane order parameters, for example, as measured by fluorescence polarization (Table 2).

The similarities in dT/dP seen for phase transitions in artificial lipid bilayers composed of dipalmitoyl phosphatidylcholine (DPPC), for natural membranes, for the "break" temperatures in Arrhenius plots [which some authors have proposed to result from lipid phase changes (67)], and for membrane order parameters suggest a common causal mechanism based on lipid physical state for all of these temperature–pressure effects. This conclusion is also supported by the observation that large values of ΔV are associated with large values of ΔH (119). This correlation would be expected if the "melting" of lipid bilayers is accompanied by a substantial increase in bilayer volume.

The enthalpy and volume changes associated with alterations in the order of a lipid bilayer—artificial or natural—arise largely from changes in the organization of the acyl chains of phospholipids (and gangliosides, see below) (Fig. 4). Low

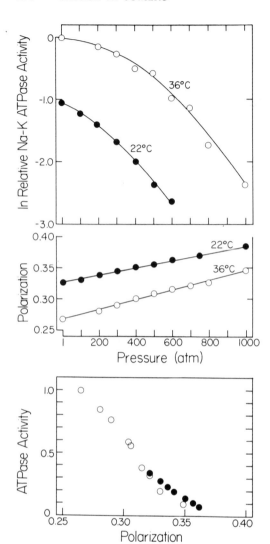

temperature favors a more ordered alignment among these acyl chains because of reduced thermal motion in the membrane. High pressure favors a more ordered state because more tightly aligned acyl chains occupy a smaller volume than less organized—more fluid—acyl side chains. At high pressures membrane phospholipids are laterally compressed, and the membrane increases in thickness (65). When the membrane lipids are in a relatively fluid state, conformational movements of lipoprotein enzymes may require less energy than when the lipid annulus around the protein is relatively rigid, as portrayed in Figure 4.

The effects of pressure- and temperature-induced changes in phospholipid fluidity on the activity of the Na^+-K^+ ATPase of dog kidney are shown in Figure 5 (24). Figure 5 (upper panel) shows how decreased temperature and elevated pressure reduce Na^+-K^+ ATPase activity. The nonlinearity of the rate versus pressure curve indicates that the activation volume of the reaction increases with pressure. Figure 5 (middle panel) shows how the fluorescence polarization signal of this system, due to the probe diphenylhexatriene (DPH), varies with temperature and pressure. Diphenylhexatriene is a hydrophobic probe that partitions into the hydrophobic region of the phospholipid bilayer. The stronger the polarization signal, the greater is the degree of order in the bilayer (23, 24). Increases in pressure and decreases in temperature increase the measured polarization, indicating increases in order of the bilayer. Figure 5 (lower panel) shows

Figure 5. *Upper panel:* The effect of pressure and temperature on the activity of the Na^+-K^+ ATPase isolated from dog kidney. Rates are expressed relative to the rate found at 1 atm. Note the nonlinearity of the effect of pressure on rate. *Middle panel:* The effect of pressure on the polarization signal from diphenylhexatriene (DPH) incorporated into the membrane phospholipids surrounding the Na^+-K^+ ATPase protein. Polarization signal strength is inversely proportional to the fluidity of the membrane phospholipids. *Lower panel:* The correlation between Na^+-K^+ ATPase activity and membrane phospholipid fluidity (as indicated by the strength of the DPH polarization signal). This panel combines the data given in the upper two panels, and shows how Na^+-K^+ ATPase activity depends closely on membrane fluidity, regardless of the temperature or pressure of measurement. Closed symbols refer to a measurement temperature of 22°C; open circles to a temperature of 36°C. [Modified after Chong et al. (24).]

how tightly coupled the activity of the Na$^+$-K$^+$ ATPase is to the fluidity of the membrane. A given level of polarization, whether induced by a change in temperature or pressure, leads to a given level of Na$^+$-K$^+$ ATPase activity. In common with other lipid systems (Table 2), the change in temperature needed to offset a 1500-atm change in pressure was near 20°C (24).

The effects of pressure on Na$^+$-K$^+$ ATPase of fish gills are qualitatively similar to those on the dog kidney enzyme. Figure 6 shows the effects of pressure on Na$^+$-K$^+$ ATPases from gills of marine fishes differing in depths of occurrence (see legend). The Na$^+$-K$^+$ ATPase activity is inhibited by pressure, and this inhibition increases at higher measurement pressures. Although the Na$^+$-K$^+$ ATPases of these fishes are all inhibited by rising pressure, adaptive differences are found between shallow- and deep-living species. The Na$^+$-K$^+$ ATPases of the two deepest-living fishes, the rattails *Coryphaenoides armatus* and *Coryphaenoides leptolepis,* are least inhibited by pressure. The Na$^+$-K$^+$ ATPase of a shallower-living congener, *Coryphaenoides acrolepis,* is distinctly more pressure sensitive and resembles the enzymes of cold-adapted species found at lesser depths. Figure 6b shows that, despite wide interspecific differences in sensitivity to pressure of the Na$^+$-K$^+$ ATPase reactions, the activation volume of the reaction is conserved among species at their *in situ* pressures. The conservation of activation volume at physiological pressures suggests that the physical states of the membranes also are conserved among species at their normal pressures (36a).

The strong effects of pressure and temperature on the activity of a membrane-localized, lipid-dependent enzyme like Na$^+$-K$^+$ ATPase indicate that resistance adaptation to depth requires adjustments in phospholipid composition to maintain proper membrane fluidity. As in the case of temperature adaptation (see Chapter 3), "homeoviscous" adjustments of lipid fluidity serve as an important mechanism for ensuring the preservation of membrane-based functions under pressure. Deep-sea organisms, which encounter very low temperatures as well as high pressures, would seem to require phospholipids with especially high degrees of inherent fluidity.

Evidence for homeoviscous adaptation in deep-living organisms is shown in Figure 7, which plots saturation ratio (the ratio of saturated to unsaturated fatty acids in membrane phospholipids) versus the depth at which these fishes were captured (29). Phospholipid acyl chains having one or more double bonds are kinked in structure and for this reason they align in a less orderly fashion than saturated acyl chains (Fig. 4). Therefore, to a first approximation, the higher the percentage of acyl chains bearing one or more double bonds, the more fluid the membrane. The decrease in fatty acid saturation with depth of occurrence is statistically significant, and indicates a strong trend for conservation of membrane fluidity through the marine water column. The degree of homeoviscous adaptation found in these marine fishes is as high as that noted in temperature acclimation studies (29). The fluorescence polarization of membrane preparations from these same specimens also provided evidence for a strong degree of homeoviscous adaptation (28).

Evidence for acclimation of membrane lipids to pressure has been reported for a marine bacterium captured at great depth and cultured in the laboratory at a series of pressures between 1 and 680 atm (30). The membrane lipids isolated from these differently acclimated populations of this barophilic bacterium showed signifi-

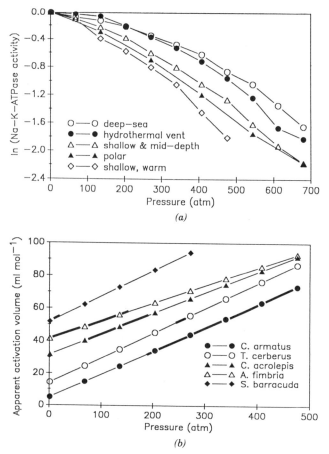

Figure 6. (*a*) The effect of measurement pressure on the logarithm of the maximal velocities of Na⁺K⁺ ATPases isolated from gills of fishes occurring at different depths. The activity at 1 atm was assigned a value of one to allow comparisons among species of the relative effects of pressure on Na⁺-K⁺ ATPase activity. Fishes are grouped according to their environments. Cold deep sea species are *Coryphaenoides armatus* and *Coryphaenoides leptolepis*; hydrothermal vent species are *Thermarces cerberus* and an unnamed fish of the genus *Bythites*; shallow and mid-depth, cold species are *Coryphaenoides acrolepis, Antimora microlepis, Sebastolobus altivelis, Sebastolobus alascanus, Anoplopoma fimbria*, and *Porichthys notatus*; polar species are the Antarctic fishes *Gymnodraco acuticeps, Rhigophila dearborni, Trematomus bernacchii, Trematomus centronotus*, and *Trematomus loenbergi*; shallow-warm species are *Sphyraena barracuda, Sphyraena helleri, Lutjanus kasmira*, and *Mulloidichthyes auriflamma*. [From Gibbs and Somera (36a).] (*b*). The effect of measurement pressure on the activation volume of the Na⁺- K⁺ ATPase reaction of several marine fishes occurring at different depths. The darker portion of each line represents the approximate depth of occurrence of the species. [From Gibbs and Somero (36a).]

cantly decreased acyl chain saturation as a function of elevated culture pressure.

A second type of membrane lipid, gangliosides of brain neurons, also has been shown to alter acyl chain saturation in adaptation to pressure and temperature (5). Gangliosides, which are localized primarily in nerve cell plasma membranes, are complex glycosphingolipids composed of sialic acid (*N*-acetylneuraminic

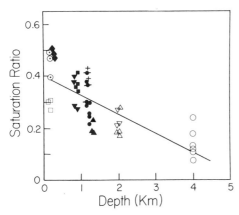

Figure 7. The relationship of membrane phospholipid saturation ratio (ratio of weight percent saturated to unsaturated fatty acids) to capture depth of several species of marine teleost fishes. Species studied were *Coryphaenoides armatus* [○], *Nezumia aequalis* [●], *Antimora rostrata* [△], *Coryphaenoides rupestris* [▲], *Lepidorhombus whiffiagonis* [□], *Conocara murrayi* [▽], *Phycis blennoides* [■], *Lepidion eques* [▼], *Lophius budegassa* [◆], *Alepocephalus bairdii* [+], *Heliocolenus dactylopterus* [⊙], and *Mora mora* [◕]. [Modified after Cossins and Macdonald (28).]

acid), sugars, and a long chain fatty acid (39). The fatty acid composition of brain gangliosides of fishes occurring at different depths, and having different body temperatures, exhibited an even more pronounced homeoviscous adaptation than phospholipids. On average, brain gangliosides of cold-adapted fishes contained 92 more double bonds per 100 fatty acid acyl residues than the gangliosides of warm-water species. Brain gangliosides of the deep-sea fish *Antimora rostrata* had the lowest contents of saturated fatty acids and the highest levels of monoenoic acids. Depth of occurrence had an effect on ganglioside composition above and beyond the effect of low temperature. Because of the high concentrations of gangliosides in neural membranes (about 1 mol of ganglioside per 4–5 moles of phospholipid in the outer layer of the membrane), and because gangliosides are especially prevalent at sites of synaptic contact, adaptation in gangliosides may be especially important in facilitating neural adaptation to depth. For example, adaptations in ganglioside fatty acid composition may be instrumental in overcoming the strong inhibitory effects of pressure on synaptic transmission discussed earlier.

A further illustration of the interactions of pressure and temperature is the finding that animals exposed to different pressures exhibit altered thermal preferenda (12). Exposure of the shallow water blenny fish *Chasmodes bosquianus* to a pressure of 50 atm was accompanied by an increase in preferred temperature of 1.8°C. A similar response was found for a marine crustacean, *Parhyale hawaiiensis*, which increased its preferred temperature by 4.2°C/100 atm rise in pressure (12). Choice of a higher water temperature would lead to an offsetting of the ordering effects of high pressure on lipid-based systems. Acclimation to different temperatures can change the pressure threshold at which HPNS first appear (6). Aquatic animals acclimated to higher temperatures typically showed a lower pressure threshold onset of HPNS symptoms.

Summary: Pressure Effects on Membrane-Based Systems

A wide variety of membrane-linked processes are perturbed by changes in hydrostatic pressure. Pressure thresholds at which perturbation is first observed may be very low relative to the pressures experienced by deep-living marine and freshwater organisms. Thus, adaptation to depth of membrane-linked functions may be critical and widespread among aquatic organisms. These adaptations are not well understood in deep-living, vertically migrating species or in diving animals. Temperature and pressure interact strongly in determining the status of these membrane-linked processes. Increases in temperature typically can offset the effects of increased pressure, and vice versa. For deep-living organisms, the combination of low temperatures and high pressures establishes an especially great stress to membrane-linked processes. Many of the pressure and temperature induced alterations in membrane-localized functions appear to be a result of shifts in the fluidity of the phospholipid bilayer. Modulating this fluidity in pressure- and temperature-adaptive manners, for example, via adjustments in acyl chain saturation, is a critical form of tolerance adaptation in all organisms living under high pressure. The role played by the protein moiety of lipoprotein, membrane-localized enzymes in adaptation to pressure is not known.

Resistance Adaptations: Protein Function and Structure

Studies of diverse proteins that are not dependent on a lipid annulus for activity have shown that pressures in the biological range can severely perturb protein function and structure (50, 51, 103). Many, and perhaps most, of the reactions in which proteins participate occur with significant volume changes due either to water structure effects or changes in the volume of the protein itself (122). Enzymatic catalysis often is strongly affected by pressure, as shown by the sizes of activation volumes listed in Table 1. The assembly of multisubunit proteins, for example, microtubules (tubulin filaments) (91), myosin (53), and actin filaments (113), generally occurs with large volume changes (Table 1), which are thought to result from the displacement of the organized water found at the subunit–subunit interaction sites, and from

the creation of void spaces at the interaction sites when subunits join together (122). It follows that the activities of enzymes, the regulation of enzymatic activity, and the assembly of multisubunit (multimeric) proteins are perturbed by pressures encountered by deep-living organisms. The magnitudes of these perturbations may be large, and the threshold pressures at which these perturbations first appear may be low. Thus, adaptation to pressure of protein-based systems, as well as lipid-based systems, will be seen as critical for allowing tolerance of the deep sea.

The discussion below addressees the following questions concerning these protein adaptations. First, what types of protein-based processes are most pressure sensitive and, therefore, the most critical adaptive sites in the acquisition of tolerance of elevated pressures? Second, what are the threshold pressures at which the disruption of protein function is large enough to favor selection for pressure-resistant protein variants? Are these threshold pressures similar for different types of proteins? How do these threshold pressures compare with those noted for membrane-based processes? Third, what types of modifications in proteins establish tolerance of high pressure? What fraction of the primary structure (the amino acid sequence) of a protein must be changed to reduce the protein's sensitivity to pressure? Are certain amino acids more suited than others for function at elevated pressures? Fourth, do the pressure-adaptive modifications of proteins entail any reduction in protein functional ability at 1 atm? That is, do the adaptations in proteins that permit deep-sea organisms to withstand high pressures limit the performance of these organisms at shallow depths and work to restrict these pressure-adapted organisms to deeper regions of the water column?

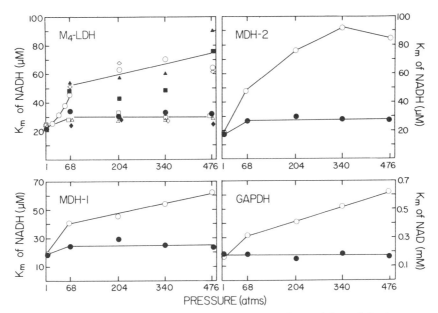

Figure 8. The effects of hydrostatic pressure on homologs of four dehydrogenase enzymes purified from fishes living at different depths. *Upper left:* Lactate dehydrogenase (skeletal muscle, M_4 isozyme). Species studied: *Sebastolobus alascanus* (180–440 m) [○]; *Sebastolobus altivelis* (550–1300 m) [●]; *Antimora rostrata* (800–3500 m) [△]; *Coryphaenoides acrolepis* (475–2825 m) [□]; *Halosauropsis macrochir* (1600–5300 m) [◆]; *Pagothenia borchgrevinki* (surface) [■]; *Scorpaena guttata* (shallow) [▲]; *Sebastes melanops* (shallow) [◇]. Data are from Siebenaller (94), and Siebenaller and Somero (98,99). *Lower left:* Malate dehydrogenase-1 (MDH-1). In this frame, and in both right-hand frames, the points refer to *Sebastolobus alascanus* [○] and *Sebastolobus altivelis* [●]. *Upper right:* malate dehydrogenase-2 (MDH-2). *Lower right:* Glyceraldehyde-3-phosphate dehydrogenase (GAPDH). The MDH and GAPDH are from Seibenaller (95).

Adaptations In Enzymatic Function

Volume changes may accompany each step in the catalytic sequence of an enzymatic reaction: the binding of ligands (substrates, cofactors, and modulators) to the enzyme, the activation of the enzyme–substrate complex, and the release of products at the termination of the catalytic reaction (71). The binding of ligands to proteins appears to be especially sensitive to pressure, probably because ligand binding often entails changes in protein conformation and in the hydration states of the reactants (95, 98, 99, 103, 108). The effects of pressure on the binding of the pyridine nu-

cleotide cofactors NAD^+ and NADH to dehydrogenases of marine fishes, as approximated by the effects of pressure on the apparent Michaelis–Menten constant (K_m) of cofactor, illustrate the sensitivity of ligand binding to pressure and provide at least partial answers to several of the questions raised above.

Figure 8 shows the effects of pressure on the K_m of NADH or NAD^+ for dehydrogenase enzymes purified from muscle tissue of fishes adapted to different depths (depth ranges are given in the figure legend). The four enzymes studied are skeletal muscle type lactate dehydrogenase (M_4-LDH), two isozyme forms of malate dehydrogenase (MDH-1, MDH-

2), and glyceraldehyde-3-phosphate dehydrogenase (GAPDH). These enzymes are all involved in the major catabolic, ATP-generating reaction pathways of cells (see Chapter 8). Their function dictates that the binding of both substrate and cofactor be closely regulated, and that ligand binding be conserved at environmental temperatures and pressures. Large increases in K_m (indicative of decreased binding ability) are antithetical to enzyme function.

Despite the similar values for K_m of cofactor at 1 atm for all of the homologs of each dehydrogenase, large differences among the homologs are found at elevated measurement pressures. For all dehydrogenases of deep-living fishes, pressure has at most a very slight effect on the K_m of NADH or NAD$^+$. For all of the dehydrogenases of the shallow-living fishes, a pressure of only 68 atm or less (e.g., M$_4$-LDH) is adequate to increase significantly the K_m of cofactor.

Several conclusions about the adaptations of enzymes to high pressure are suggested by these findings. First, ligand-binding events are very sensitive to pressure: the low threshold pressures at which perturbation of the K_m is noted are similar to those found for certain membrane-associated processes (see above). Thus, for fishes living at depths greater than a few hundred meters, pressure-adaptive modifications in dehydrogenase enzymes may be essential for ensuring adequate enzyme function under pressure. Pressure-adaptive changes in enzymes thus may play important roles in speciation processes in the deep sea and in establishing species' depth distribution patterns in the marine water column. The differences in the pressure sensitivities of the four dehydrogenases from the two scorpaenid fishes of the genus *Sebastolobus* illustrate this point. *S. alascanus*, the shallower congener, lives at depths of from 180–440 m as an adult; *S. altivelis*

occurs at depths of 550–1300 m as adults (98). The correlation between the depth at which *S. altivelis* replaces *S. alascanus*, ~500 m, and the pressure at which dehydrogenase function is perturbed for enzymes of *S. alascanus*, ~50 atm, (Fig. 8) is suggestive of a causal relationship. The origin of a deep-living congener, therefore, may depend in part on the occurrence in a shallow-living species of mutations that code for pressure-resistant proteins. Individuals with these new protein variants may be able to move into deeper waters, and, if reproduction occurs at depth, allopatric speciation may occur.

A third conclusion based on the data in Figure 8 is that convergent evolution has occurred in the dehydrogenase enzymes of fishes adapted to high pressure. For M$_4$-LDH, the homologs of all four deep-living species, members of four different teleost families, display similar insensitivities to pressure. Because each species arose from a different shallow-living ancestral species, one sees that acquisition of pressure-insensitive NADH binding has occurred repeatedly during colonization of the deep sea. Whether the same amino acid substitutions have led to this convergence in function in the different species is not known.

Adaptations In Protein Structure

The amount of change in primary structure of an enzyme needed to establish pressure insensitivity of function may be small. Comparisons of M$_4$-LDHs of the *Sebastolobus* congeners showed only a single amino acid substitution: The histidine residue at site 115 in the primary sequence of the *S. alascanus* enzyme is replaced by an asparagine residue in the enzyme of *S. altivelis* (96). Site 115 in the primary sequence is not in the active site of the enzyme, but in the "hinge" region of the enzyme involved in the large change in conformation that occurs during cofactor binding. Evidence for the involvement of histidine 115 in the pressure sensitivity of the M_4-LDH of *S. alascanus* comes from a study of the effects of measurement pH on the pressure sensitivities of the M_4-LDHs of the *Sebastolobus* congeners (94). Varying assay pH between 7.5 and 9.0 had no effect on the pressure sensitivity of the enzyme of *S. altivelis*, but for the enzyme of *S. alascanus* this increase in pH eliminated the pressure sensitivity in NADH binding. At pH 9, the imidazole side chain of a histidine residue will be uncharged. These findings suggest that

changes in water structure induced by the reversible hydration of histidine 115 could establish the pressure sensitivity of the K_m of NADH for the M_4-LDH of *S. alascanus* (94, 103).

Structural adaptations in the M_4-LDHs of deep-living fishes also include increased resistance to denaturation by high pressure. Table 3 lists the pressures at which 50% of LDH activity was lost for M_4-LDH homologs of six rattail (macrourid) fishes (43). The species of the genus *Coryphaenoides* display marked differences in sensitivity to pressure that are correlated with their depths of distri-

TABLE 3. Pressure Effects on Protein Structure

I. Inactivation Pressures for M_4-LDHs of Six Macrourid Fishes[a]

Species	Depth of Maximal Abundance (m)	Pressure of 50% Inactivation (atm)
Nezumia bairdii	600	840
Coryphaenoides rupestris	1000	565
C. acrolepis	1200	770
C. carapinus	2000	1265
C. armatus	2900	1715
C. leptolepis	3500	1570

II. Thermal Stabilities of G-actin and Pressure Stabilities of F-actin (113) from Skeletal Muscle of Differently Adapted Vertebrates

Species (Habitat)	Rate Constant of Denaturation[b] ($\times 10^3$/min)	V of Assembly (at 1 atm)^2 (cm^3/mol)
Dipsosaurus dorsalis (desert lizard)	0.00	139
Rabbit	1.76	
Chicken	1.60	107
Sebastolobus alascanus	4.78	56
Pagothenia borchgrevinki (Antarctic fish)	9.20	
Coryphaenoides acrolepis	0.22	63
C. armatus	0.58	9
Halosauropsis macrochir	0.28	58

[a]The pressure at which half-inactivation of LDH activity occurred during 1-h incubations at 3–4°C. [Data from Hennessey and Siebenaller (43).]
[b]The rate constant of denaturation was determined by measuring the amount of native G-actin remaining in solution at various times of incubation at 37°C at 1-atm pressure (111). The volume change accompanying the G-actin to F-actin equilibrium was measured by determining the effect of pressure on the critical concentration of actin, which is reciprocally related to the equilibrium constant of the polymerization reaction (113).

TABLE 4. Enzymatic Activities,[a] Protein Concentration, and Water Content in White Skeletal Muscles of Marine Fishes Living at Different Depths[b]

Species	Depth Range (m)	% Water	Protein (mg/g)	LDH	MDH	PK	CS	CPK
S. alascanus	180–440	81	149	58	5	7	0.35	119
S. altivelis	550–1300	82	152	25	3	4	0.19	75
N. bairdii	260–1965	81	144	7	18	5	0.62	
C. rupestris	550–1960	85	142	16	10	5	0.58	
C. carapinus	1250–2740	85	120	5	7	5	0.50	
C. aramatus	1885–4815	84	177	53	19	7	0.79	
C. leptolepis	2288–4639	82	144	4	7	3	0.41	

[a]Enzyme abbreviations: LDH = lactate dehydrogenase; MDH = Malate dehydrogenase; PK = pyruvate kinase; and CS = citrate synthase; CPK = creatine phosphokinase.
[b]Data on the *Sebastolobus* congeners are from Siebenaller and Somero (100); data on the macrourid species are from Siebenaller et al. (101). See Table 3 for additional data on these species.

bution. As in the case of pressure effects on the Na^+-K^+ ATPase (Fig. 6), the enzymes of *C. acrolepis* and *C. armatus* display pressure-adaptive differences; the enzyme of the deeper-occurring congener is significantly less sensitive to high pressure.

Differences in the structural stabilities of M_4-LDHs of deep- and shallow-living fishes also are shown by their different sensitivities to proteolytic digestion (44, 45). At 10°C and 1 atm pressure, the enzymes of shallow-living fishes were inactivated 3–4 times faster by proteolytic enzymes than the LDHs of the deep-sea forms. At elevated pressures, the enzymes of the shallow-living fishes exhibited a larger increase in rate of inactivation than found for the LDHs of the deep-sea forms. These data, like the pressure denaturation data, support the hypothesis that enzymes of deep-sea animals may have more rigid structures than the homologous enzymes of shallow-living, cold-adapted animals. The enhanced stabilities of enzymes of deep-living species under high pressure, and their resistance to proteolytic digestion suggests that these enzymes not only may retain their native structures well under elevated pressures, but may also be degraded rel-

atively slowly *in vivo* by proteolytic systems (103). The latter possibility remains to be investigated.

Evidence that selection favors stronger protein structures in deep-sea organisms also has come from studies of skeletal muscle actin homologs of vertebrates adapted to different pressures and tempertures (Table 4). It is well established that proteins of high-body-temperature species are more resistant to thermal denaturation than the homologous proteins of cold-adapted species (48, 50). This trend is evident for subunits of skeletal muscle actins: Monomeric actin (G-actin) of the thermophilic desert reptile *Dipsosaurus dorsalis*, whose core body temperature may reach 47°C, is most resistant to heat denaturation; actins of two shallow-living Antarctic fishes (*Pagothenia borchgrevinki* and *Gymnodraco acuticeps*) are most heat sensitive. The actins of three deep-sea fishes, the rattails *C. armatus* and *C. acrolepis*, and the halosaur *Halosauropsis macrochir*, are almost as heat resistant as the actin of the desert iguana, and are more thermally stable than actin from the chicken or the rabbit. Thus, despite having body temperatures of only 2–4°C, G-actins of the deep-sea fishes have heat stabilities comparable to actins of the most thermophilic terrestrial animals (111).

Stability in actin structure of deep-sea animals also was found in the resistance of actin filaments (F-actin) to depolymerization by pressure (Table 3). F-actin of the deep-living fish *C. armatus* was only slightly depolymerized by high pressure—as indicated by the low ΔV of polymerization—relative to the effects noted for actins of other fishes or of birds and mammals (113).

The threshold pressures at which pressure-adaptive modifications in actin self-assembly become important are higher than those noted for adaptations in dehydrogenases. No differences in heat stability or polymerization thermodynamics were found between the actins of the *Sebastolobus* congeners (111). Actin of the fish *C. acrolepis*, which occurs to depths of ~2100 m, polymerizes with a volume change that is similar to that found for a shallow-living fish (*S. alascanus*), and larger than the ΔV for F-actin formation measured for its deeper-living congener, *C. armatus*. Thus, for actin assembly, pressure may not be a selectively important perturbant at depths of less than ~2000 m.

Because the subunit aggregation states of other multisubunit proteins are extremely sensitive to pressure (71, 103), it is probable that adaptations of the sort found for actin are widespread among other proteins of the contractile apparatus, for example, myosins and tropomyosin, among proteins that form the microfilaments, intermediate filaments, and microtubules of the cell, and among multimeric enzymes. In addition to stabilizing the polymerized state of these proteins, these adaptations may minimize the loss of native three-dimensional conformation that occurs for free (depolymerized) subunits in solution, a process known as subunit "conformational drift" (60, 61, 121).

A further concomitant of the increased structural rigidity of enzymatic proteins of deep-living organisms is a reduction in catalytic efficiency, as documented for M_4-LDHs. The catalytic rate constants (k_{cat}'s) of M_4-LDHs from cold-adapted shallow-water fishes, endothermic fishes, deep-sea fishes, and a mammal differed systematically (103, 107). The highest k_{cat} was found for M_4-LDH of the cold-adapted (−1.86°C) Antarctic fish, *P. borchgrevinki,* and the lowest k_{cat} was that of the rabbit LDH. Although the deep-living fishes had body temperatures only slightly above that of *P. borchgrevinki,* their enzymes had k_{cat} values only one half to two thirds as high as those of this species. The M_4-LDH of *S. alascanus* had a k_{cat} essentially the same as that of the enzyme of *P. borchgrevinki,* while the k_{cat} of the LDH of *S. altivelis* was in the low-range typical of deep-sea fishes.

This loss of catalytic efficiency as a consequence of increased protein structural stability may be a reflection of the roles played by reversible conformational changes in enzymatic reactions. If an enzyme changes its conformation reversibly during ligand binding, catalysis, or product release, a more flexible enzyme might be able to function more rapidly than a more rigid enzyme under otherwise identical conditions. In terms of speciation processes and depth distribution patterns, this relationship between flexibility and rate of function suggests that deep-living forms could be at a disadvantage at shallow depths where their less efficient enzymes would support a lower level of metabolic performance on a per enzyme molecule basis. Therefore, protein adaptations that facilitate resistance to high pressures may also tend to limit the performance of pressure-tolerant organisms at shallow depths.

Adaptation to pressure has been reported for hemoglobins (Hb's) of certain deep-sea fishes that possess swimbladders, for example *C. armatus*. Fishes with swimbladders possess hemoglobins with extremely strong Bohr effects (see Chapter 10). The very low oxygen affinity induced in these Hb's by acidification of the blood facilitates the secretion of oxygen into the swimbladder. Secretion of oxygen into the swimbladder is more difficult under high ambient pressure. The Hb's of some deep-sea fishes have especially strong Bohr effects that lead to the lowest oxygen affinities recorded for any Hb (73). Estimated half-saturation (P_{50}) values for oxygen for these Hb's range between 5 and 15 atm.

Summary: Protein Adaptations to Pressure

Although only a few types of proteins have been compared from animals adapted to different pressures, the adaptive differences that have been found permit at least tentative conclusions to be drawn about protein adaptation in the deep sea.

1. These adapations lead to stability in ligand binding, in catalytic rates, and in subunit assembly abilities.

2. The amino acid substitutions required to convert a pressure-sensitive protein to one less sensitive to pressure may represent only minor changes in primary structure.

3. Adaptations to pressure may entail some cost in functional performance in terms of rates of catalysis.

4. Different types of proteins have different pressure thresholds at which perturbation becomes selectively significant.

5. The adaptation of proteins to pressure plays an important role in speciation events in the deep sea and, therefore, in determining depth distribution patterns. The pressure ad-

aptations that allow tolerance of the deep sea may also have a role in restricting the occurrence of deep-sea species to high-pressure environments where the tolerance to high pressure outweighs the reductions in enzymatic activity brought about by these adaptations.

Pressure Effects on Nucleic Acids and on Protein Synthesis

Pressure appears to have only minimal effects on the secondary structures of double-stranded nucleic acids (Table 1). DNA replication is blocked by elevated pressure in *Escherichia coli*, but it is not clear whether this effect is a result of changes in nucleic acid conformation or in the polymerase enzyme (124).

An area of high-pressure biology that remains unexplored in deep-living animals is the effect of pressure on protein synthesis, a process that involves highly complex interactions among a large suite of proteins and nucleic acids (mRNAs, tRNAs, and rRNAs). Studies of pressure effects on protein synthesis in prokaryotes have shown that protein synthesis in whole cells and in *in vitro* systems is inhibited by pressure (62). For *E. coli*, the apparent activation volume for the rate-limiting step in translation was ~100 cm^3/mol; while for another bacterium, *Pseudomonas fluorescens*, $\Delta V\ddagger$ for translation was 50 cm^3/mol. In the *E. coli* system protein synthesis was totally inhibited at 680 atm, but upon release of pressure, synthesis resumed immediately. The pressure sensitivity of protein synthesis was highly dependent on the salt concentration of the experimental medium in the *in vitro* studies; the concentration of magnesium ion was especially critical. The component of the translational apparatus that was most pressure sensitive was the 30S subunit of the ribosome. The assembly of ribosomal subunits may be very sensitive to pressure; the volume change of subunit dissociation for bacterial ribosomes was -242 cm^3/mol (77).

Protein synthesis by rabbit reticulocyte lysates was inhibited by pressure to different extents at different temperatures (93). At 25°C no inhibition was found until the pressure exceeded 100 atm; at 30°C, inhibition occurred by this pressure. Protein synthesis by whole cells was less sensitive to pressure; no effects were found until pressure was increased above 200 atm. In the cell-free system, higher concentrations of magnesium ion reduced the sensitivity to pressure, as seen in the prokaryotic system. The effects of pressure on the rate of protein synthesis were not linear over the full pressure range studied, possibly indicating a change in the rate-limiting step in translation at different pressures.

Rates in the Deep Sea: Capacity Adaptations

Adaptations in lipid and protein systems that confer resistance to high pressures are unlikely to play important roles in establishing rates of metabolic function in deep-sea animals. The modifications in lipid and protein systems discussed above enable deep-living organisms to withstand the effects of high pressure and low temperature, but other environmental factors are of primary importance in determining the rates at which these pressure-adapted membranes, enzymes, contractile proteins, and other macromolecular systems conduct their functions *in situ*.

Minimal Depth of Occurrence and Metabolic Rate

In evaluating the relationship between depth and metabolic rate, it is essential to define what is meant by a "deep-living" species. The criterion for "deep" used in the context of metabolic capacities differs from that used in the context of resistance adaptations to pressure. In the latter context, "deep" referred to depths, that is, to pressures, at which physiological and biochemical functions were perturbed. In the context of metabolic rates, "deep" refers instead to a set of depth-dependent environmental conditions, especially light availability and food supply, that interact to select for appropriate rates of metabolic activity.

A meaningful way to distinguish shallow-living from deep-living marine species is to refer to their minimal depths of occurrence (MDOC), the depth below which 90% of a population lives (17). The concept of minimal depth of occurrence is especially critical for understanding midwater animals, many of which undergo large vertical migrations on a daily basis. Species that co-occur at depths of several hundred meters during the daytime may have very different depths of occurrence at night. The biological and physical characteristics of the near-surface waters, not those of the deeper waters inhabited during the daytime, may play the dominant roles in selecting for the physiological properties of the migrators. For instance, vertical migrators who enter surface waters at night encounter a much more abundant food supply than that present at depth. The abundant energy resources these species experience near the surface may permit a more active life than nonmigrators can afford. Thus, because of the very different energy resources present near the surface and at depth, it is the shallowest depths at which a species commonly occurs, rather than the greatest depths, that may be most instrumental in establishing rates of metabolic function.

For pelagic fishes (Fig. 9) and crustaceans (Fig. 10), the rate of oxygen consumption falls rapidly with increasing MDOC. This decrease in rate with increasing MDOC is much larger than can be explained by the decrease in water temperature with depth, using Q_{10} relationships (see Chapter 3). Rather than an ~50% reduction in rate in going from temperate surface waters to 2–4°C waters, an ~20-fold decrease in oxygen consumption rate is noted for both crustaceans and fishes.

The similar decreases in oxygen consumption rate with increasing MDOC found for mid-water fishes from waters

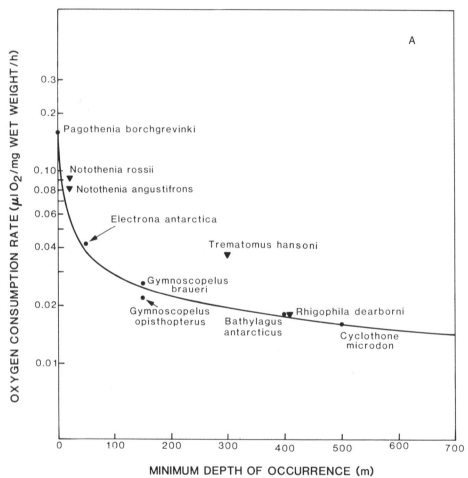

Figure 9. (*A*) The relationship of oxygen consumption rate (μL O₂ consumed/mg of wet weight · h) to minimum depth of occurrence for pelagic and mid-water fishes from Antarctica. Measurement temperatures ranged between −1.86 and 0.5°C. [From Torres and Somero (117).] (*B*) The relationship between minimum depth of occurrence and respiration rate for pelagic and mid-water fishes in coastal Californian waters. Measurement temperatures were either 10°C (MDOC < 100 m) or 5°C (MDOC > 100 m). [Figure from Torres et al. (116).] (*C*) Evidence for metabolic compensation to temperature in mid-water fishes of Antarctic and Californian waters. Data shown in (*A*) and (*B*) have been replotted to show how routine respiration rates as a function of temperature. Rates at *in situ* temperatures are similar for the Antarctic [——] and California fishes [•]. Respiration rates for the California fishes at 0.5°C [– – –] are much lower than the rates measured for the Antarctic fishes at this low temperature [——]. [Figure from Torres and Somero (117).]

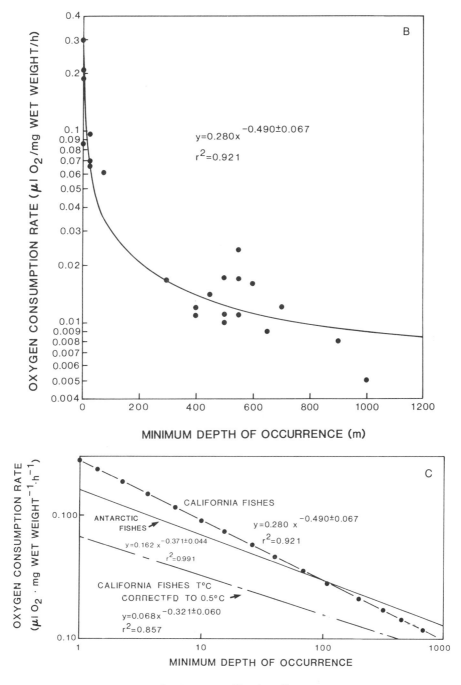

Figure 9. (*Continued*)

off the coast of California (116) and from the Scotia Sea in Antarctica (117) show that depth per se, not temperature, is primarily responsible for the decrease in metabolism with depth. The Californian mid-water fishes occur in waters with high thermal stratification; the temperature of the Scotia Sea varies little throughout the water column. When oxygen consumption rates are compared at am-

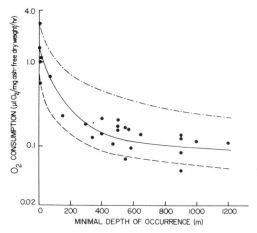

Figure 10. The relationship between maximal respiration rate (-·-·-), respiration measured at an oxygen partial pressure of 30–70 mmHg (——), and minimal respiration (- - -) and species' minimal depth of occurrence for mid-water crustaceans collected off the coast of California. Measurement temperatures were similar to those of the species' habitats [Modified after Childress (15).]

bient temperatures for the two groups of fishes as in the lower panel of Figure 9, the similarities in absolute rates are striking. Estimation of the metabolic rates of the Californian fishes at 0.5°C, the temperature used to measure rates of the Antarctic fishes, yields values significantly lower than those of the Antarctic fishes. These data show that temperature compensation of metabolism is well developed in these two groups of mid-water fishes (see Chapter 3).

Most of the decrease in metabolic rate with depth takes place within the first 200–400 m of the water column. Species with MDOCs below 400 m exhibit little further decrease in metabolic rate with depth, a finding that suggests that hydrostatic pressure is not an important determinant of metabolic rates in the deep sea (21, 69, 70, 100, 103). Where the effects of increased hydrostatic pressure on oxygen consumption rates have been examined for deep-sea animals, only small

effects, relative to those noted in Figures 9 and 10, have been observed (59, 68–70, 85, 114). Therefore, neither increased pressure nor decreased temperature can account for the large decrease in metabolic rate with increasing MDOC observed for pelagic marine animals.

Factors Selecting for Low Metabolic Rates in Deep-Living Pelagic Animals

The evolution of the low capacities for metabolism observed in pelagic fishes and crustaceans with MDOCs greater than ~200–400 m may be a result of the influences of a suite of physical and biological factors that select for a reduced intensity of locomotory activity in deep-living pelagic animals (19, 21, 100, 101, 109). These factors include: the size and distribution of the food supply in deep pelagic waters, the amount and sources of light in the deep sea (virtually all sunlight disappears below 200–300 m), the nature of predator–prey interactions in the deep sea, the dependence of a particular type of animal on the visual sense for locating food and detecting predators, and the selective advantages of rapid growth in deep-living forms. Each of these factors may favor a reduced channeling of available metabolic energy into locomotion, the activity that may consume by far the largest share of metabolic energy in actively swimming ectothermic animals (120).

Evidence in support of a reduced locomotory capacity in deep-living pelagic (bathypelagic) animals has come from visual observations made with manned submersibles and remote cameras, and from biochemical studies of the locomotory musculature of these species.

Biochemical Determinants of Low Metabolic Rates in the Deep Sea

For marine fishes an extensive data set is available showing how the activities (ex-

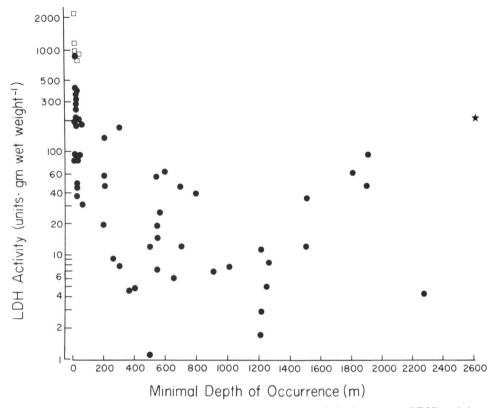

Figure 11. The relationship of white skeletal muscle lactate dehydrogenase (LDH) activity to minimal depth of occurrence for a variety of shallow- and deep-living marine teleost fishes. Each symbol represents a different species. Most points are an average of determinations made with several individuals of a species. All activities are expressed as units per gram of fresh weight of muscle (at a common measurement temperature of 10°C). Warm-bodied tunas are indicated by open squares, and the 21°N hydrothermal vent fish by a star. [Modified after Hand and Somero (41).]

pressed as international units of enzymatic activity per gram wet weight of muscle) of several enzymes important in the supply of ATP to the contractile apparatus decrease with MDOC (Fig. 11). The most striking decrease is found for LDH activity of white skeletal muscle, the predominant type of muscle tissue in most marine fishes, and the type of muscle used to power short-term bursts of swimming, which are important in food capture and escape from predators. Lactate dehydrogenase, functioning as a pyruvate reductase, is critical for the regeneration of oxidized cofactor (NAD⁺) for the glyceraldehyde-3-phosphate reaction (see Chapter 8). The activity of LDH is, therefore, a strong, quantitative index of the capacity of a muscle for sustaining high-speed, anaerobically powered swimming.

The activity of LDH decreases by over 1000-fold between extremely robust swimmers like tunas (the highest five activities shown) and sluggish, deep-living fishes. The true differences are even greater when the normal body temperatures of the species are taken into account (all of the LDH activities shown were determined at 10°C). For a tuna fish, LDH activities range up to ~5000 units/g of fresh weight at a typical body temperature near 25°C. For a sluggish deep-sea fish with a body temperature near 2°C, LDH activity is of the order of one to a few units per gram. The finding that a strong correlation exists between white muscle LDH activity and whole fish oxygen consumption rate shows that LDH activity is a valid index of whole organism metabolic capacity of fishes (19). Therefore, with the exception of the hydrothermal vent zoarcid fish, *Thermarces cerberus,* which occurs at depths near 2500 m and has an LDH activity of 185 units/g of muscle, all deep-sea fishes studies have LDH activities indicative of much lower metabolic rates than most shallow-living fishes. In addition, the white muscle buffering capacity of deep-sea fishes is significantly lower than in shallow-living fishes, a further indication of a reduced capacity for lactic acid production in deep-sea fishes (14).

Two further characterisitcs of the data in Figure 11 merit discussion. First, in agreement with the oxygen consumption data in Figures 9 and 10, the largest fraction of depth-related decrease in enzymatic (metabolic) activity occurs through the first 200–400 m of the water column. Second, and again in agreement with the patterns shown in Figures 9 and 10, at any given MDOC there is substantial variation in enzymatic (metabolic) activity among species. This depth-independent, interspecific variation is characterisitc of both shallow- and deep-living animals, and reflects significant differences in the feeding strategies and locomotory capacities of the animals. For example, relatively sluggish demersal fishes have lower metabolic rates and muscle enzymatic activities than more active pelagic species. Among pelagic species, those with a more active feeding mode exhibit higher metabolic rates than more sluggish animals, for example, sit-and-wait predators. Interspecific differences can be seen among confamilial and congeneric species, as shown by the data in Table 4. The macrourid fish showing the highest enzymatic activities, *Coryphaenoides armatus,* is known to undergo substantial vertical transects in the bathypelagic realm, and is a relatively active swimmer compared to some of its congeners, for example, *C. leptolepis,* a sluggish fish that drifts near the bottom. The differences among these confamilial rattail fishes show that below 200–400 m, lifestyle, rather than depth per se, is the primary determinant of metabolic capacity (101).

Table 4 also shows that enzymatic activities in white muscle of the scorpaenid congeners, *S. alascanus* and *S. altivelis,* differ. For all enzymes studied, activities were ~ 25–50% lower in muscle of *S. altivelis.* Estimates of oxygen consumption rates of the two species (Siebenaller (97). for *S. alascanus,* and Smith and Brown (104) for *S. altivelis*) correlate well with the observed differences in enzymatic activity: for 48-g specimens of the two species at a measurement temperature of 3.5°C, the rates of oxygen consumption for *S. alascanus* and *S. altivelis* were 9.3 and 6.9 µL/h · g of wet weight, respectively.

The depth-related changes in enzymatic acticity found in interspecific comparisons have also been observed in a study of the ontogeny of *S. altivelis* (97). Ontogenetic vertical migration of *S. altivelis,* a species that is pelagic as a larva and young juvenile and demersal as a later juvenile and adult, is marked by reductions in activities of enzymes in muscle.

Further evidence that the low metabolic rates of deep-living pelagic animals result from selection for reduced locomotory capacity, rather than from pervasive selection for reduced levels of cell and organ function, is provided by estimates of enzymatic activity in brain tissue

(100, 109). Unlike the MDOC-related and life-style-related differences in muscle enzymatic activity, brain tissue from teleost fishes with widely different MDOCs and feeding strategies had essentially identical activities for several enzymes of energy metabolism. Therefore, the reduction in metabolic capacity found as a function of MDOC reflects a specific change in the properties of muscle, not organ-wide reductions in metabolic capacity.

The reductions in muscle enzymatic activity with increasing MDOC cannot be explained entirely by increases in water content or decreases in protein content. As indicated by the data in Table 4 as well as data from other studies (100, 102, 109), changes in muscle water content and protein concentration with depth are minor compared to the changes in enzymatic activity. The reduced locomotory capacity of deep-sea fishes appears to result from specific reductions in the levels of enzymes that supply ATP to the contractile apparatus. The amount of contractile machinery does not appear to differ among fishes; the actin content of fish white muscle is approximately the same in tunas, shallow-living ectothermic fishes, and deep-sea fishes (102, 112).

In conclusion, the reduced metabolic expenditures for locomotory activity in bathypelagic animals appear to be the consequence of several factors. Low levels of light make visual detection of food difficult if not impossible. Expending large amounts of energy in the search for food in a dark environment seems a less adaptive response than merely waiting for food to appear in the animal's local vicinity. Visual detection by predators may likewise be difficult, reducing the needs for strong locomotory escape responses. The reduction in locomotory energy expenditures may have an important effect on the channeling of energy into growth. Data on bathypelagic crustaceans

and fishes suggest that the growth rates of bathypelagic animals are not low like metabolic rates (18, 20). Selection may favor rapid growth rates as a means for facilitating early reproduction and escape from the dangers of predation. In terms of growth, then, bathypelagic species may compare favorably to related, shallow-living species, albeit the caloric densities of the deep-sea species are low compared to shallow-living animals.

Metabolic Capacities of Benthic Animals

The reductions in metabolic rates with increasing MDOC for benthic crustaceans, molluscs, and worms are much less than the decreases found for pelagic species (21). The basis for the difference between pelagic and benthic animals is not entirely clear. However, based on the proposed importance of reduced locomotory activity with increasing MDOC in pelagic species, it is possible that the locomotory and predatory strategies of benthic species may determine the effect of MDOC on metabolic rate. The importance of robust locomotory abilities in many benthic species, shallow or deep-living, may be minimal relative to pelagic species. Therefore, the strong depth-dependent selection for reduced locomotory abilities proposed for pelagic animals may not exist for deep-living benthic animals.

The Deep-Sea Hydrothermal Vents

The low biomass of typical deep-sea habitats reflects the small fraction of photosynthetic production that reaches the deeper zones of the water column and the deep-sea floor. Most of the deep sea is food-energy limited. The occurrence of high biomass at the deep-sea hydrothermal vents (Fig. 12) is possible only because a nonsolar energy source is available to drive the net fixation of car-

Figure 12. The deep-sea hydrothermal vent site at the Galapagos Spreading Center. Dense thickets of vestimentiferan tube worms (*Riftia pachyptila*) dominated this particular vent [photograph by R. Hessler].

bon dioxide. The principal energy source at the deep-sea vents appears to be hydrogen sulfide, which is produced geothermally (22, 52). A variety of bacteria are capable of oxidizing sulfide and using a portion of the energy released to drive net carbon dioxide fixation via the Calvin–Benson cycle (22, 52, 106). In this sense, then, sulfide is the "sunlight," and sulfide oxidizing bacteria the "green plants" of the deep-sea vents.

The gutless vestimentiferan tube worm *R. pachyptila* and the vesicomyid clam *Calyptogena magnifica,* major contributors to the vent biomass, contain endosymbiotic sulfur bacteria that are capable of oxidizing sulfide and using the energy released to drive the Calvin–Benson cycle (22, 106). For these symbioses to be possible, adaptations are needed for preventing sulfide from poisoning aerobic respiration. Sulfide concentrations in the micromolar range are sufficient to poison the cytochrome *c* oxidase system and, thereby, inhibit aerobic respiration. The

cytochrome *c* oxidases of the vent animals are as sensitive to sulfide as the homologous enzymes of other animals (41, 82, 83).

Detoxification of Hydrogen Sulfide

Hydrothermal vent animals employ several mechanisms for preventing poisoning of aerobic respiration by sulfide. In *Riftia* sulfide is bound tightly to the hemoglobin (Hb), which is present in high concentrations in the blood and coelomic fluid (3). Unlike most Hb's, Hb of *Riftia* is not poisoned by sulfide. Sulfide binding does not occur at the heme site on the molecule, but appears to take place where disulfide bridges occur in the Hb. Sulfide may add to a disulfide bridge, forming a mixed disulfide (3). When bound to *Riftia's* Hb, sulfide is unable to poison respiration, and the sulfide is protected from spontaneous oxidation, which would occur rapidly if the sulfide were to remain free in solution. Thus, binding sulfide to

Riftia's Hb protects the aerobic respiration of the worm and prevents the energy of sulfide from being wasted through spontaneous oxidation in the blood. The Hb transports sulfide to the bacterial symbionts, which are located in the trophosome. There, sulfide is released and oxidized by the symbionts. Some of the energy of the sulfide is trapped in reducing power and ATP, and a fraction of these energy currencies is used to drive net carbon dioxide fixation via the Calvin–Benson cycle (Fig. 13). Sulfide oxidizing activities also are present in the superficial cell layers of soft tissues such as the body wall musculature (83). These activities may detoxify any sulfide entering the animal cells.

In *Calyptogena* a different anatomical and physiological organization of an animal–sulfur bacteria symbiosis occurs. The bacterial symbionts are localized in the large gills of the bivalve. A sulfide binding protein is present in the serum of *Calyptogena*, but this is not the Hb, which is found within erythrocytes (2). The Hb of *Calyptogena* is poisoned by sulfide. In *Calyptogena* as well as *Riftia* the superficial cell layers of the soft tissues contain sulfide-oxidizing activities which may be important in detoxifying sulfide (83).

In the vent brachyuran crab, *Bythograea thermydron*, which lacks symbiotic sulfur bacteria, sulfide that enters the circulation is oxidized in the hepatopancreas (118). Thiosulfate is the principal product formed during sulfide detoxification, and accumulates in the blood to concentrations in the millimolar range. This detoxification ability allows *Bythograea* to forage actively in the high-sulfide waters bathing the symbiont-containing tube worms and bivalves.

Exploitation of the Energy of Sulfide

Symbiont containing tissues of *Riftia* and *Calyptogena* have high levels of enzymatic activities associated with the oxidation of reduced sulfur compounds (33, 34, 83). These tissues also contain Calvin–Benson cycle enzymes with activities in the range of those found in green plant leaves (33, 34). Oxidation of sulfide and other reduced sulfur compounds, for example, thiosulfate, may provide sufficient energy to synthesize much, if not all, of the reduced carbon compounds needed by symbiont and host. *Riftia* may be largely dependent on the nutrients supplied by its symbionts—the translocation products (Fig. 13)—because the worm lacks any means for ingesting or processing particulate food. *Calyptogena*, too, appears to lack an effective system for feeding on particulate matter, and it may be highly dependent on its symbionts for nutrition.

Although net carbon dioxide fixation driven by the metabolism of reduced sulfur compounds appears restricted to the bacterial symbionts in these organisms, sulfide-driven ATP production may occur in the animal mitochondria. In a gutless protobranch clam (*Solemya reidi*) found in sulfide-rich habitats in shallow waters (pulp mill effluent sites and sewage outfall zones), sulfide taken up into the cells is oxidized by the mitochondria, with a yield of one ATP per sulfide oxidized (84). The mitochondria appear to generate thiosulfate in this process, which is transported to the symbionts for oxidation to sulfate. Thiosulfate oxidation by the symbionts can supply energy for driving the Calvin–Benson cycle (1). Thiosulfate is relatively nontoxic and can be accumulated in *Solemya* to very high levels (1). The mitochondrial oxidation of sulfide to thiosulfate achieves three important functions: The cytochrome *c* oxidase system is protected from inhibition by sulfide, ATP is generated for the animal, and an energy-rich but nonpoisonous reduced sulfur compound is generated to be fed to the bacterial symbionts. It is not known if mitochondrial oxidation of sulfide occurs in the hydrothermal vent organisms.

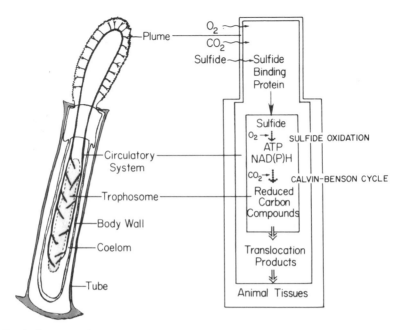

Figure 13. A drawing of the hydrothermal vent vestimentiferan tube worm *R. pachyptila.* The left frame portrays the important anatomical structures of the worm. The right frame illustrates the physiological and biochemical organization of the symbiosis. Much of the worm's coelomic cavity is filled with an organ called the trophosome, which contains large numbers of intracellular bacterial symbionts. The symbionts receive hydrogen sulfide, carbon dioxide, and oxygen *via* the circulatory systems. The obturacular plume is the principal site of gas exchange with the environment. The symbionts oxidize sulfide and use some of the energy released to drive net carbon fixation *via* the Calvin–Benson cycle. Some of the reduced carbon compounds synthesized within the bacterial symbionts are exported to the host (translocation products). [Figure from Somero (106).]

Metabolic Rates at the Deep-Sea Hydrothermal Vents

The metabolic rates of hydrothermal vent organisms are generally similar to those of related species from shallower waters, as indicated by measurements of oxygen consumption rates (21) and by estimates of tissue enzymatic activities (41, 83). For example, the muscle LDH activity of the vent zoarcid fish *Thermarces cerberus* (Fig. 11) is higher than for any other deep-sea fish and is in the range found for many shallow-living demersal fishes.

Conclusions

Two basic classes of physiological and biochemical adaptations permit life in the deep sea. Resistance adaptations confer tolerance of the deep sea by modifying the macromolecular systems and membranes of cells to reduce or eliminate perturbation by high pressures and low temperatures. Capacity adaptations establish rates of metabolic activity that are consistent with the properties of the food resources—food quantity, quality, and distribution—in the deep sea. The total amount of food and the distribution of this food help determine the appropriate levels of metabolic activity. For pelagic fishes and certain invertebrates, there appears to be strong selection for reduced energy expenditure in locomotion. The reduction in locomotory ability noted for deep-living pelagic animals permits 2–3 orders of magnitude reductions in metabolic rate compared to highly active shallow-living organisms. For benthic species, depth-related reductions in metabolism appear to be much less.

The abundant fauna at the deep-sea hydrothermal vents, and the absence of depressed metabolic rates in the vent species, illustrate that neither low temperature nor high pressure necessarily forces a reduction in the amount of life or in the rate of living processes in the deep sea. When a rich source of food is present in the deep sea, the biomass and the rate of metabolic turnover can reach values similar to those characteristic of many nutrient-rich, shallow-water environments.

References

1. Anderson, A. E., J. J. Childress, and J. A. Favuzzi. *J. Exp. Biol.* 133:1–31, 1987. Net uptake of CO_2 driven by sulfide and thiosulfate oxidation in the bacterial symbiont-containing clam *Solemya reidi.*

2. Arp, A. J., J. J. Childress, and C. R. Fisher. *Physiol. Zool.* 57:648–622, 1984. Metabolic and blood gas transport characteristics of the hydrothermal vent bivalve *Calyptogena magnifica.*

3. Arp, A. J., J. J. Childress, and R. D. Vetter. *J. Exp. Biol.* 128:139–158, 1987. The sulphide-binding protein in the blood of the vestimentiferan tube-worm, *Riftia pachyptila* is the extracellular hemoglobin.

4. Ashford, M. L. J., A. G. Macdonald, and K. T. Wann. *J. Physiol. (London)* 333:531–543. The effects of hydrostatic pressure on the spontaneous release of transmitter at the frog neuromuscular junction.

5. Avrova, N. F. *Comp. Biochem. Physiol. B* 78B:903–909, 1984. The effect of natural adaptations of fishes to environmental temperature on brain ganglioside fatty acid and long chain base composition.

6. Beaver, R. W. and R. W. Brauer. *Comp. Biochem. Physiol. A* 69A:665–674, 1981. Pressure-temperature interactions in relation to development of high pressure convulsions in ecotherm vertebrates.

7. Bodansky, A. and W. Kauzmann. *J. Phys. Chem.* 66:177–179, 1962. The apparent molar volume of sodium hydroxide at infinite dilution and the volume change accompanying the ionization of water.

8. Brandts, J. F., R. J. Olivera, and C. Westort. *Biochemistry* 9:1038–1047, 1970. Thermodynamics of protein denaturation. Effect of pressure on the denaturation of ribonuclease.

9. Brauer, R. W. *Philos. Trans. R. Soc. London, B Ser.* 304:17–30, 1984. Hydrostatic pressure effects on the central nervous system–perspective and outlook.

10. Brauer, R. W., M. Y. Bekman, J. B. Keyser, D. L. Nesbitt, G. N. Sidelov, and S. L. Wright. *Comp. Biochem. Physiol. A* 65A:109–117, 1980. Adaptation to high hydrostatic pressures of abyssal gammarids from lake Baikal in eastern Siberia.

11. Brauer, R. W., V. G. Sidelyova, M. B. Dail, G. I. Galazii, and R. D. Roer. *Comp. Biochem. Physiol. A* 77A:699–705, 1984. Physiological adaptation of cottoid fishes of lake Baikal to abyssal depths.

12. Brauer, R. W., M. R. Jordan, C. G. Miller, E. D. Johnson, J. A. Dutcher, and M. E. Sheehan. In *High Pressure Effects on Selected Biological Systems* (A. J. R. Pequeux and R. Gilles, eds.), pp. 3–30. Springer-Verlag, Berlin, 1985.

13. Campenot, R. B. *Comp. Biochem. Physiol.* 52:133–140, 1975. The effects of high hydrostatic pressure on transmission at the crustacean neuromuscular junction.

14. Castellini, M. A. and G. N. Somero. *J. Comp. Physiol.* 143:191–198, 1981. Buffering capacity of vertebrate muscle: correlations with potentials for anaerobic function.

15. Childress, J. J. *Comp. Biochem. Physiol. A* 50A:787–799, 1975. Respiratory rate of mid-water crustaceans as a function of depth of occurrence and the relation to the oxygen minimum layer off Southern California.

16. Childress, J. J. In *Oceangraphy: The Present and Future* (P. Brewer, ed.) pp. 127–136. Springer-Verlag, New York, 1982.

17. Childress, J. J. and M. Nygaard. *Deep-Sea Res.* 20:1093–1109, 1973. The chemical

composition of midwater fishes as a function of depth of occurrence off Southern California.

18. Childress, J. J. and M. Price. *Mar. Biol. (Berlin)* 50:47–62, 1978. The growth rate of a bathypelagic crustacean, *Gnathophausia ingens* (Mysidacea: Lophogastridae). I. Dimensional growth and population structure.

19. Childress, J. J. and G. N. Somero. *Mar. Biol. (Berlin)* 52:273–283, 1979. Depth-related enzymic activities in muscle, brain and heart of deep-living pelagic marine teleosts.

20. Childress, J. J., S. Taylor, G. Calliet, and M. H. Price. *Mar. Biol. (Berlin)* 61:27–40, 1980. Patterns of growth, reproduction and energy usage in some meso- and bathypelagic fishes.

21. Childress, J. J. and T. J. Mickel. In *The Hydrothermal Vents of the Eastern Pacific: An Overview* (M. Jones, ed.), Bull. Biol. Soc. No. 6, pp. 249–260, 1985.

22. Childress, J. J., H. Felbeck, and G. Somero. *Sci. Am.* 256:114–120, 1987. Symbiosis in the deep sea.

23. Chong, P. L.-G. and A. R. Cossins. *Biochemistry* 22:409–415, 1983. A differential polarized phase fluorometric study of the effects of high hydrostatic pressure upon the fluidity of cellular membranes.

24. Chong, P.L.-G, P. A. G. Fortes, and D. M. Jameson. *J. Biol. Chem.* 260:14484–14490, 1985. Mechanisms of inhibition of (Na,K)-ATPase by hydrostatic pressure studied with fluorescent probes.

25. Colton, A. and J. S. Colton. *Physiologist* 25:222, 1982. Depression of inhibitory synaptic transmission by oxygen and pressure.

26. Conti, F., R. Frioavanti, J. R. Segal, and W. Stühmer. *J. Membr. Biol.* 69:33–34, 1982. Pressure dependence of the sodium current of squid giant axon.

27. Conti, F., R. Fioravanti, J. R. Segal, and W. Stühmer. *J. Membr. Biol.* 69:35–40, 1982. Pressure dependence of the potassium currents of squid giant axon.

28. Cossins, A. R. and A. G. Macdonald. *Biochim. Biophys. Acta* 776:144–150, 1984.

Homeoviscous theory under pressure. II. The molecular order of membranes from deep-sea fish.

29. Cossins, A. R. and A. G. Macdonald. *Biochim. Biophys. Acta* 860:325–335, 1986. Homeoviscous theory under pressure. III. The fatty acid composition of liver mitochondrial phospholipids of deep-sea fish.

30. DeLong, E. F. and A. A. Yayanos. *Science* 228:1101–1103, 1985. Adaptation of the membrane lipids of a deep-sea bacterium to changes in hydrostatic pressure.

31. Doubt, T. J. and P. M. Hogan. *J. Appl. Physiol.* 45:24–32, 1978. Effects of hydrostatic pressure on conduction and excitability in rabbit atria.

32. Doubt, T. J. and P. M. Hogan. *J. Appl. Physiol.* 47:1169–1175, 1979. Action potential correlates of pressure-induced changes in cardiac conduction.

33. Felbeck, H. *Science* 213:336–338, 1981. Chemoautotrophic potential of the hydrothermal vent tube worm *Riftia pachyptila*.

34. Felbeck, H., J. J. Childress, and G. N. Somero. *Nature (London)* 293:291–293, 1981. Calvin-Benson cycle and sulphide oxidation enzymes in animals from sulphide-rich habitats.

35. Forbes, A. et al. *J. Physiol. (London)* 372:57P, 1986. Functional fish axons from 4000 m depths, 400 atm pressure: Preliminary results of Challenger cruise 6B/85.

36. Friedmann, M. E. and H. A. Scheraga. *J. Phys. Chem.* 69:3795–3800, 1965. Volume changes in hydrocarbon-water systems. Partial molar volumes of alcohol-water solutions.

36a. Gibbs, A. and G. N. Somero. *J. Exp. Biol.* 143:475–492, 1989. Pressure adaptation of Na^+-K^+-ATPase in gills of marine teleosts.

37. Goldinger, J. M., B. S. Kang, Y. E. Chow, C. V. Paganelli, and S. K. Hong. *J. Appl. Physiol.* 49:224–231, 1980. Effect of hydrostatic pressure on ion transport and metabolism in human erythrocytes.

38. Greaney, G. S. and G. N. Somero. *Biochemistry* 18:5322–5332, 1979. Effects of

anions on the activation thermodynamics and fluorescence emission spectrum of alkaline phosphatase: Evidence for enzyme hydration changes during catalysis.

39. Hadley, N. F. *The Adaptive Role of Lipids in Biological Systems*. Wiley, New York, 1985.

40. Hall, A. C. and A. G. Macdonald. *J. Physiol. (London)* 305:108P, 1980. Hydrostatic pressure alters the sodium content of erythrocytes.

41. Hand, S. C. and G. N. Somero. *Biol. Bull. (Woods Hole, Mass.)* 165:167–181, 1983. Energy metabolism pathways of hydrothermal vent animals: Adaptation to a food-rich and sulfide-rich deep-sea environment.

42. Henderson, J. V., M. T. Lowenhaupt, and D. C. Gilbert. *Undersea Biomed. Res.* 4:19–26, 1977. Helium pressure alteration of function in squid giant synapse.

43. Hennessey, J. P., Jr. and J. F. Siebenaller. *J. Comp. Physiol.* 155:647–652, 1985. Pressure inactivation of tetrameric lactate dehydrogenase homologues of confamilial deep-living fishes.

44. Hennessey, J. P., Jr. and J. F. Siebenaller. *J. Exp. Zool.* 241:9–15, 1987. Pressure-adaptive differences in proteolytic inactivation of M_4-lactate dehydrogenase homologues from marine fishes.

45. Hennessey, J. P., Jr. and J. F. Siebenaller. *Biochim. Biophys. Acta* 913:285–291, 1987. Inactivation of NAD-dependent dehydrogenases from shallow- and deep-living fishes by hydrostatic pressure and proteolysis.

46. Hochachka, P. W. *Biochem. J.* 143:535–539, 1974. Temperature and pressure adaptation of the binding site of acetylcholinesterase.

47. Hochachka, P. W., K. B. Storey, and J. Baldwin. *Comp. Biochem. Physiol. B* 52B:13–18, 1975. Design of acetylcholinesterase for its physical environment.

48. Hochachka, P. W. and G. N. Somero. *Biochemical Adaptation*. Princeton Univ. Press, Princeton, NJ, 1984.

49. Hogan, P. M. In *High Pressure Effects on Selected Biological Systems* (A. J. R. Pequex and R. Gilles, eds.) pp. 94–109. Springer-Verlag, Berlin, 1985.

50. Jaenicke, R. *Annu. Rev. Biophys. Bioeng.* 10:1–67, 1981. Enzymes under extremes of physical conditions.

51. Jaenicke, R. *Naturwissenschaften* 70:332–341, 1983. Biophysical processes under high pressure.

52. Jannasch, H. W. and M. J. Mottl. *Science* 229:717–725, 1985. Geomicrobiology of deep-sea hydrothermal vents.

53. Josephs, R. and W. F. Harrington. *Proc. Natl. Acad. Sci. U.S.A.* 58:1587–1594, 1967. An unusual pressure dependence for a reversibly associating protein system: sedimentation studies on myosin.

54. Kauzmann, W. *Adv. Protein Chem.* 14:1–63, 1959. Factors in interpretation of protein denaturation.

55. Kauzmann, W., A. Bodanszky, and J. Rasper. *J. Am. Chem. Soc.* 84:1777–1788, 1962. Volume changes in protein reactions. II. Comparison of ionization reactions in proteins and small molecules.

56. Kendig, J. J. In *High Pressure Effects on Selected Biological Systems* (A. J. R. Pequex and R. Gilles, eds.), pp. 110–123. Springer-Verlag, Berlin, 1985.

57. Kendig, J. J., J. R. Trudell, and E. N. Cohen. *J. Pharmacol. Exp. Ther.* 195:216–224, 1975. Effects of pressure and anaesthetics on conduction and synaptic transmission.

58. Kendig, J. J. and E. N. Cohen. *Am. J. Physiol.* 230:1244–1249, 1976. Neuromuscular function at hyperbaric pressures: Pressure-anaesthetic interactions.

59. King, F. D. and T. T. Packard. *Deep-Sea Res.* 22:99–105, 1975. The effect of hydrostatic pressure on respiratory election transport system activity in marine zooplankton.

60. King, L. and G. Weber. *Biochemistry* 25:3632–3637, 1986. Conformational drift of dissociated lactate dehydrogenases.

61. King, L. and G. Weber. *Biochemistry* 25:3637–3640, 1986. Conformational drift and cryoinactivation of lactate dehydrogenase.

62. Landau, J. V. and D. H. Pope. *Adv. Aquat. Microbiol.* 2:49–76, 1980. Recent advances in the area of barotolerant protein synthesis in bacteria and implications concerning barotolerant and barophilic growth.

63. Low, P. S. and G. N. Somero. *Comp. Biochem. Physiol. B* 52B:67–74, 1975. Pressure effects on enzyme structure and function *in vitro* and under simulated *in vivo* conditions.

64. Macdonald, A. G. *Physiological Aspects of Deep-Sea Biology.* Cambridge Univ. Press, London, 1975.

65. Macdonald, A. G. *Philos. Trans. R. Soc. London, Ser. B* 304:47–68, 1984. The effects of pressure on the molecular structure and physiological functions of cell membranes.

66. Macdonald, A. G. and I. Gilchrist. *Comp. Biochem. Physiol. A* 71A:349–352, 1982. The pressure tolerance of deep-sea amphipods collected at their ambient high pressure.

67. MacNaughtan, W. and A. G. Macdonald. *Comp. Biochem. Physiol. A* 72A:405–414, 1982. Effects of pressure and pressure antagonists on the growth and membrane-bound ATP-ase of *Acholeplasma laidlawii* B.

68. Meek, R. P. and J. J. Childress. *Deep-Sea Res.* 20:1111–1118, 1973. Respiration and the effect of pressure in the mesopelagic fish *Anoplagaster cornuta* (Beryciformes).

69. Mickel, T. J. and J. J. Childress. *Deep-Sea Res.* 29:1293–1301, 1982. Effects of pressure and pressure acclimation on activity and oxygen consumption in the bathypelagic mysid *Gnathophausia ingens*.

70. Mickel, T. J. and J. J. Childress. *Physiol. Zool.* 55:199–207, 1982. Effects of temperature, pressure and oxygen concentration on the oxygen consumption rate of the hydrothermal vent crab *Bythograea thermydron* (Brachyura).

71. Morild, E. *Adv. Protein Chem.* 34:93–166, 1981. The theory of pressure effects on enzymes.

72. Neumann, R. C., Jr. et al. *J. Phys. Chem.* 77:2687–2691, 1973. Pressure dependence of weak acid ionization in weak buffers.

73. Noble, R. W. et al. *Biochim. Biophys. Acta* 870:552–563, 1986. Functional properties of hemoglobins from deep-sea fish: Correlations with depth distribution and presence of a swimbladder.

74. Noguchi, H. *Biopolymers* 4:1105–1113, 1966. Studies of the helix–coil transition of poly-L-lysine in film and solution.

75. Ornhagen, H. C. L. and P. M. Hogan. *Undersea Biomed. Res.* 4:347–358, 1977. Hydrostatic pressure and mammalian cardiac-pacemaker function.

76. Paladini, A. A., Jr. and G. Weber. *Biochemistry* 20:2587–2593, 1981. Pressure-induced reversible dissociation of enolase.

77. Pande, C. and A. Wishnia. *J. Biol. Chem.* 261:6272–6278, 1986. Pressure dependence of equilibria and kinetics of *Escherichia coli* ribosomal subunit association.

78. Parmentier, J. L., B. B. Shrivastav, and P. B. Bennett. *Undersea Biomed. Res.* 8:175–183, 1981. Hydrostatic pressure reduces synaptic efficiency by inhibiting transmitter release.

79. Penniston, J. T. *Arch. Biochem. Biophys.* 142:322–332, 1971. High hydrostatic pressure and enzymic activity: Inhibition of multimeric enzymes by dissociation.

80. Pequeux, A. J. R. *Symp. Soc. Exp. Biol.* 26:483–484, 1972. Hydrostatic pressure and membrane permeability.

81. Pfeiler, E. *J. Exp. Zool.* 205:393–402, 1978. Effects of hydrostatic pressure on (Na + K)-ATPase and Mg^{2+} ATPase in gills of marine teleost fish.

82. Powell, M. A. and G. N. Somero. *Science* 219:297–299, 1983. Blood components prevent sulfide poisoning of respiration of the hydrothermal vent tube worm *Riftia pachyptila*.

83. Powell, M. A. and G. N. Somero. *Biol. Bull. (Woods Hole, Mass.)* 171:274–290, 1986. Adaptations to sulfide by hydrothermal vent animals: Sites and mechanisms of detoxification and metabolism.

84. Powell, M. A. and G. N. Somero. *Science* 233:563–566, 1986. Hydrogen sulfide ox-

idation is coupled to oxidative phospho-rylation in mitochondria of *Solemya reidi.*

85. Quetin, L. B. and J. J. Childress. *Mar. Biol. (Berlin)* 38:327–334, 1976. Respiratory adaptations of Pleuroncodes planipes to its environment off Baja California.

86. Rasper, J. and W. Kauzmann. *J. Am. Chem. Soc.* 84:1771–1777, 1962. Volume changes in protein reactions. I. Ionization of proteins.

87. Regnard, P. *C. R. Seances Soc. Biol. Ses Fil.* 39:265–269, 1887. Les Phénomènes de la vie sous les hautes pressions—La contraction musculaire.

88. Roer, R. D. and M. G. Shelton. *Comp. Biochem. Physiol. A* 71A:271–276, 1982. Effects of hydrostatic pressure on Na transport in the freshwater crayfish, *Procambarus clarkii.*

89. Roer, R. D., M. Y. Bekman, M. G. Shelton, R. W. Braver, and S. G. Shvetzov. *J. Exp. Zool.* 233:65–72, 1985. The effects of changes in hydrostatic pressure on Na transport in gammarid amphipods from Lake Baikal.

90. Roer, R. D. and A. J. R. Pequeux. In *High Pressure Effects on Selected Biological Systems* (A. J. R. Pequeux and R. Gilles, eds.), pp. 32–51. Springer-Verlag, Berlin, 1985.

91. Salmon, E. D. *Science* 189:884–886, 1975. Pressure-induced depolymerization of brain microtubules *in vitro.*

92. Sauter, J.-F., P. Braswell, P. Wankowicz, and K. W. Miller. In *Underwater Physiology* (A. J. Bachrach and M. M. Matzen, eds.), Vol. 7, pp. 629–638. Undersea Biomedical Society, Bethesda, MD. The effects of high pressure of inert gases on cholinergic receptor binding and function.

93. Scheck, A. C. and J. V. Landau. *Biochim. Biophys. Acta* 698:149–157, 1982. The effect of high hydrostatic pressure on eukaryotic protein synthesis.

94. Siebenaller, J. F. *Mar. Biol. Lett.* 4:233–243, 1983. The pH-dependence of the effects of hydrostatic pressure on the M_4-

lactate dehydrogenase homologs of scorpaenid fishes.

95. Siebenaller, J. F. *J. Comp. Physiol.* 154:443–448, 1984. Pressure-adaptive differences in NAD-dependent dehydrogenases of congeneric marine fishes living at different depths.

96. Siebenaller, J. F. *Biochim. Biophys. Acta* 786:161–169, 1984. Structural comparison of lactate dehydrogenase homologs differing in sensitivity to hydrostatic pressure.

97. Siebenaller, J. F. *Physiol. Zool.* 57:598–608, 1984. Analysis of the biochemical consequences of ontogenetic vertical migration in a deep-living teleost fish.

98. Siebenaller, J. F. and G. N. Somero. *Science* 201:255–257, 1978. Pressure-adaptive differences in lactate dehydrogenases of congeneric fishes living at different depths.

99. Siebenaller, J. F. and G. N. Somero. *J. Comp. Physiol.* 129:295–300, 1979. Pressure-adaptive differences in the binding and catalytic properties of muscle-type (M_4) lactate dehydrogenases of shallow- and deep-living marine fishes.

100. Siebenaller, J. F. and G. N. Somero. *Physiol. Zool.* 55:171–179, 1982. The maintenance of different enzyme activity levels in congeneric fishes living at different depths.

101. Siebenaller, J. F., G. N. Somero, and R. L. Haedrich. *Biol. Bull. (Woods Hole, Mass.)* 163:240–249, 1982. Biochemical characteristics of macrourid fishes differing in their depths of distribution.

102. Siebenaller, J. F. and P. H. Yancey. *Mar. Biol. (Berlin)* 78:129–137, 1984. Protein composition of white skeletal muscle from mesopelagic fishes having different water and protein contents.

103. Siebenaller, J. F. and G. N. Somero. *CRC Crit. Rev. Aquatic Sciences* 1:1–25, 1989. Biochemical adaptation to the deep sea.

104. Smith, K. L., Jr. and N. O. Brown. *Mar. Biol. (Berlin)* 76:325–332, 1983. Oxygen consumption of pelagic juveniles and demersal adults of the deep-sea fish *Sebastolobus altivelis.*

105. Somero, G. N. In *Transport Processes, Iono- and Osmoregulation* (G. Gilles and M. Gilles-Baillien, eds.), pp. 454–468. Springer-Verlag, Berlin, 1985.

106. Somero, G. N. *News Physiol. Sci.* 2:3–6, 1987. Exploitation of hydrogen sulfide by animal–bacteria symbioses.

107. Somero, G. N. and J. F. Siebenaller. *Nature (London)* 282:100–102, 1979. Inefficient lactate dehydrogenases of deep-sea fishes.

108. Somero, G. N., J. F. Siebenaller, and P. W. Hochachka. In *The Sea* (G. T. Rowe, ed.), Vol. 8, pp. 261–330, Wiley (Interscience), New York, 1983.

109. Sullivan, K. M. and G. N. Somero. *Mar. Biol. (Berlin)* 60:91–99, 1980. Enzyme activities of fish skeletal muscle and brain as influenced by depth of occurrence and habits of feeding and locomotion.

110. Suzuki, K. and Y. Taniguchi. *Symp. Soc. Exp. Biol.* 26:103–124, 1972. Effect of pressure on biopolymers and model systems.

111. Swezey, R. R. and G. N. Somero. *Biochemistry* 21:4496–4502, 1982. Polymerization thermodynamics and structural stabilities of skeletal muscle actins from vertebrates adapted to different temperatures and hydrostatic pressures.

112. Swezey, R. R. and G. N. Somero. *Mar. Biol. Lett.* 3:307–315, 1983. Skeletal muscle actin content is strongly conserved in fishes having different depths of distribution and capacities for locomotion.

113. Swezey, R. R. and G. N. Somero. *Biochemistry* 24:852–860, 1985. Pressure effects on actin self-assembly: Interspecific differences in equilibrium and kinetics of the G to F transformation.

114. Teal, J. M. and F. G. Carey. *Deep-Sea Res.* 14:725–733, 1967. Effects of pressure on the respiration of euphausiids.

115. Thompson, R. B. and J. R. Lakowicz. *Biochemistry* 23:3411–3417, 1984. Effect of pressure on the self-association of melittin.

116. Torres, J. J., B. W. Belman, and J. J. Childress. *Deep-Sea Res. Part A* 26:185–197, 1979. Oxygen consumption rates of midwater fishes as a function of depth of occurrence.

117. Torres, J. J. and G. N. Somero. *Comp. Biochem. Physiol. B.* 90:521–528, 1988. Vertical distribution and metabolism in Antarctic mesopelagic fishes.

118. Vetter, R. D., M. E. Wells, A. L. Kurtzman, and G. N. Somero. *Physiol. Zool.* 60:121–137, 1987. Sulfide detoxification by the hydrothermal vent crab *Bythograea thermydron* and other decapod crustaceans.

119. Wann, K. T. and A. G. Macdonald. *Comp. Biochem. Physiol. A* 66A:1–12, 1980. The effects of pressure on excitable cells.

120. Webb, P. W. *Bull. Fish. Res. Board Can.* 190:1–158, 1975. Hydrodynamics and energetics of fish propulsion.

121. Weber, G. *Biochemistry* 25:3626–3631, 1986. Phenomenological description of the association of protein subunits subjected to conformational drift. Effects of dilution and of hydrostatic pressure.

122. Weber, G. and H. G. Drickamer. *Q. Rev. Biophys.* 16:89–112, 1983. The effect of high pressure upon proteins and other biomolecules.

123. Yayanos, A. A. *Science* 200:1056–1059, 1978. Recovery and maintenance of live amphipods at a pressure of 580 bars from an ocean depth of 5700 m.

124. Yayanos, A. A. and E. C. Pollard. *Biophys. J.* 9:1464–1482, 1969. A study of the effects of hydrostatic pressure on macromolecular synthesis in *Escherichia coli*.

125. Zipp, A. and W. Kauzmann. *Biochemistry* 12:4217–4228, 1973. Pressure denaturation of metmyoglobin.

Chapter 5 | Feeding and Digestion

C. Ladd Prosser and Edward J. DeVillez

Most animals devote much time and energy to acquiring and digesting food. Carnivores spend less time eating than do herbivores; rapidly growing herbivores (e.g., caterpillars) eat continuously during daylight, browsing herbivores (e.g., ruminants) eat frequently but rest or ruminate or digest between eating periods. Some carnivores seek and pursue food, others (e.g., spiders) wait for food to approach. Low availability of food is a stress and many types of feeding are adapted to consuming adequate amounts of essential nutrients in different habitats. The amount of food consumed is regulated according to its availability and the physiological state of the consumer. For acquisition of food there are many kinds of structural mechanisms. Animals with high metabolism have high food consumption. Once food is consumed, mechanical and chemical means make products of digestion available for metabolism. All digestive enzymes are hydrolytic and the similarity of enzymes used by different animals for processing various foodstuffs indicates that these enzymes arose early in eukaryotic evolution (56b).

Feeding Mechanisms

Methods of intake differ according to whether nutrients occur dissolved in solution, as particulates, or as large masses of substance either in water or in air.

Dissolved Food

Natural waters contain dissolved organic molecules in concentrations of 0.2 to 20 mg/L; the amount is greater in regions where there is organic detritus and is less in open waters. Dead particulate matter may be present at ~1 mg/L. In freshwater lakes the total organic material is 10 to 15 mg/L of which 1 mg may be colloidal, 1–2 mg of planktonic, and the remainder dissolved organic matter (43c).

Many kinds of aquatic invertebrate animals and some algae are able to extract organic molecules from solution, some of them against concentration gradients as much as several thousandfold (1). Parasitic animals in a rich environment of host tissues do not require special digestive structures. Their body surfaces resemble intestinal epithelia in surface area and transport mechanisms. Many cnidarians, molluscs, echinoderms, annelids, and arthropods extract some organic substances directly from solution. Measurements with radiolabeled amino acids or by high performance liquid chromatography (HPLC) show that the uptake is active and not by exchange diffusion.

Absorption can be from nanomolar concentrations and against concentration gradients as much as 1:3000 (46, 66). Neutral amino acids (aa's) are absorbed, but not acidic or basic aa's. Uptake from natural waters is not via bacteria since it occurs in axenically cultured larvae (echinoderms) and when bacteria are excluded (*Mytilus*, polychaete worms) (74). There are species differences; the ciliate *Tetrahymena* does not absorb dissolved amino acids but feeds by food vacuoles (15). Significant absorption (37–75% extraction) by the gills has been demonstrated in *Mytilus* within a single pass of water (in a few seconds) through the mantle cavity (74). Kinetic analyses show that influx follows Michaelis–Menton kinetics with high V_{max} and low K_m. Active absorption requires the presence of sodium in the medium and apparently a quaternary complex of substrate amino acid (aa), two Na^+ ions, and a carrier molecule function in the transport. The ratio of Na to substrate is 2:1 in sea urchin larvae, 3:1 in Mytilus (16, 65). Uptake is saturable at high concentrations of aa's. The extent to which amino acids are transported for metabolic needs varies according to animal type; in starfish, in several cnidarians, and a polychaete the absorbed amino acids may be stored in the epidermis. Translocation to deep tissues has been demonstrated in *Mytilus* and the polychaete *Glycera*.

Active uptake of D-glucose from a dilute medium has been examined in the annelid *Nereis diversicolor* (29, 30). The uptake is Na dependent and is inhibited by phlorizin and by competing glycosides. The transport mechanism is similar to that for glucose in mammalian intestine and kidney proximal tubules. The transported glucose has been visualized inside epidermal cells and very little is transported to extracellular fluid for metabolism by body cells. Freshwater *Tubifex* absorb acetate and propionate from dilute solutions. Uptake is inhibited by competitive analogs, by ouabain and Na depletion, and by CN and iodoacetate. Labeled acids appear in respired CO_2 and in stored glycogen (35).

The absorbed amino acids and sugars may, in different kinds of animals counteract diffusional loss of solutes, may serve as substrates for epidermal cells, and may provide limited energetic requirements. Whatever the function, a variety of aquatic animals can absorb specific L-amino acids from water at concentrations such as occur naturally.

Particulate Food

Particulate food consists of detritus particles, organic substances adsorbed on inert particles, dead planktonic cells, and living phytoplankton such as diatoms, flagellates and bacteria. In surface and shallow seawater the amount of particulate matter present may vary from 0.2 to 1.8 mg/L, it decreases with depth and varies greatly with season and location. Abundance in inshore waters is greater than in the open ocean or lakes. Several methods are used by animals to ingest particles and plankton from water:

1. *Pseudopodial and Phagocytic Feeding.* Some Protozoa, particularly rhizopods, engulf food particles by pinocytotic food vacuoles. Food is concentrated in the vacuoles as water is extruded from them.

Particles are caught on the reticulate pseudopodia of radiolarians and foraminiferans and then brought to the cell body for ingestion. Pinocytosis (engulfing particles with water) is a widespread cellular phenomenon. Amoebocytic cells engulf detritus particles in tissue spaces of multicellular animals. Ciliate protozoans carry particles by ciliary currents to a "mouth" where food vacuoles are formed. Water currents generated by flagellae in channels of a sponge carry particles, which are taken up phagocytically by cells lining the channels.

2. *Mucoid Feeding.* Some invertebrate animals secrete mucous sheets that filter out particles and then they eat the bolus. The tube-dwelling polychaete *Chaetopterus* and the echiuroid *Urechis* secrete mucous nets, pump water through the net by posterior appendages, and swallow a sackful of food about every 15 min. The nets of *Chaetopterus* retain not only small particles but protein molecules as large as hemocyanin; pore size is estimated to be ~0.004 μm (40 Å). Mucous funnels are used by larvae of the insect *Chironomus*, also some gastropods extrude mucous filters that are swallowed after they are filled.

3. *Ciliary Filter Feeding.* Many animals use cilia arranged in sheets or rows that filter particulate matter that is then carried to the mouth. In some animals ciliary and mucous filtering are combined, in other animals ciliary currents direct food into digestive tissues containing food vacuoles. Ciliary filters of bivalve molluscs have been much studied; the gill filaments bear several sets of cilia, which sort large from small particles and transport food particles to labial palps. The ciliary currents function in respiration as well as in feeding. Gill filaments bear several sets of cilia, which sort large from small particles and transport food to the palps; straining is done between rows of laterofrontal cilia of the gill filaments. Particles

1–2 μm in size are retained. The labial palps discriminate food from inedible particles, which are discarded as pseudofeces (5, 52). The ciliary filter of the mussel *Mytilus* excludes particles 30–40 μm, retains or passes particles of 7 μm, and does not retain particles smaller than 1.5 μm (66). Oysters usually retain flagellates and food particles larger than 2–3 μm. Adult oysters pump some 20–50 mL of water/per hour per mg of N (43a). Intertidal bivalves feed only during high tide, subtidal animals feed more nearly continuously. When in air, *Mytilus* closes its shell and is anaerobic, in water it gapes and metabolizes aerobically. On reimmersion after exposure to air *Mytilus* shows payment of oxygen debt (67a, 72). Snails and bivalves from populations living high in the intertidal zone filter more water than individuals from a low-intertidal population; high-intertidal animals take three times more prey than those from low-intertidal populations during the same immersion and feeding period (51).

In sea anemones, ciliary currents in the pharynx direct food particles into the coelenteron where particulates are taken up by phagocytosis into cells of mesenteric folds. Prefeeding movements of tentacles and column precede discharge of nematocysts into prey and ingestion follows. The presence of polypeptides and amino acids induces the feeding response (71). Some species of sea anemones contain zooxanthellae and zoochlorellae in endodermal cells. These green symbionts fix CO_2, synthesize products, particularly glycerol, glucose, and alanine, which are translocated to tissues of the anemone. Other anemones are commensal on fish or crabs, which transport the anemones into regions of zooplankton. The diet of sessile anemones in general is less restricted than that of mobile populations. Sea anemones from the high intertidal zone consume more prey and absorb more than low intertidal species when tested during the same immersion time (75).

Mechanisms for Taking Food Masses
Some burrowing animals swallow food-containing mud. Some holothurians use their tentacles to force mud into their digestive tracts. Mud dwelling annelids such as *Lumbricus* and *Arenicola* and sipunculids like *Phascolopsis* use pharynxes as pumps; aquatic annelids such as *Nereis* and *Glycera* evert and retract the pharynx. Burrowing crustaceans such as *Upogebia* and *Callianassa* use mouth appendages. The organic content of the mud is utilized and the residue ejected as feces.

Many gastropods (snails) eat food masses. An odontophore (ridged stiff tongue) projects into the buccal cavity and a toothed radula moves by protractor and retractor muscles over the odontophore. These gastropods are grazers, many of them highly selective in diet. Some nudibranch snails feed on a single species of sponge, cnidarian, or ectoproct. Radular structure is used for classification of these gastropods. Cephalopods are a class of molluscs that feed on active prey—mostly crustaceans, fishes, annelids, and molluscs. Cephalopods capture moving prey by means of suckers on arms (8 in octopods, 10 in decapods) and transfer prey to the buccal cavity, which has jaws for breaking the food; usually food is killed by a poison secreted by salivary glands. These toxins are mostly amines or polypeptides—octopamine, tyramine; the glands also secrete proteases. Some molluscs are burrowers; the Teredinidae feed on wood and digest cellulose by action of cellulase-containing bacteria as well as by cellulase in digestive diverticula. A few molluscs are gutless and consume dissolved organic material.

Echinoderms are diverse in feeding habits. Holothurians are mostly mud and detritus feeders. Some starfish are carniv-

orous and the crown of thorns damage to coral reefs has been extensive. Sea urchins are herbivores, widespread in shallow seawater. The preferred diet may be peculiar to a species. In sea urchins, Aristotle's lantern is used in biting and the teeth are replaceable. Echinoids graze on algal reefs, asteroids graze on corals.

Mammals have a variety of feeding habits. Some mammals are periodic feeders, carnivores such as lions eat only occasionally, as after a kill. Others feed more or less continuously, for example, grazing ruminants.

Mechanisms for Taking Fluid Food
Structures for piercing may be associated with sucking, (e.g., hookworms, leeches, mites, and cyclostomes that feed on blood). Sucking structures are diverse—in nematodes, lepidopterans, suckling mammals, and hummingbirds. A hummingbird spends 20% of its time feeding; digestion is rapid (15 min for a food marker to pass through the gut); sugar extraction is as high as 98% (20). Phloem sap is forced into stylets of aphids by plant sap pressure; some aphids suck large quantities of sap to obtain sufficient proteins; excess sugar is regurgitated as honeydew.

A leech feeds on blood only once during its active life and it may ingest the equivalent of 890% of its body mass in 29 min. Body distension beyond a critical point causes cessation of feeding (44b). Frequency of pharyngeal and body wall peristalsis is initially slow. Plasma from a blood meal in excreted (44c). A blood-sucking tick can be induced to feed on saline to which are added either (1) glutathione, ATP, or DPNH or (2) any of several amino acids. Glucose is synergistic (24).

Nutrients from Symbiotic Organisms
Wood-eating termites and the wood roach *Cryptocercus* harbor in the hindgut flag-

ellates that digest cellulose. The flagellates are anaerobic and can be removed by exposing their hosts to oxygen. The most important fermentation products of wood digestion are fatty acids. In ruminant mammals much of the fiber in forage is broken down in the capacious rumen and reticulum by symbiotic bacteria and protozoans. In some nonruminant mammals breakdown of food fiber occurs in the cecum and colon. Conditions in the massive rumen are virtually anaerobic and the redox potential is highly reducing (-0.35 V). Starches, glucose, and other soluble carbohydrates are utilized by the microorganisms and cellulose, hemicellulose, xylan, and lignin are digested. Larger coarser fibers are periodically regurgitated and masticated by the ruminant host. Saliva is secreted in large quantities and contains high concentrations of $NaHCO_3$, which is swallowed to the rumen where the acid pH causes CO_2 liberation from the bicarbonate. Methane is formed by the bacterial fermentation and the large quantities of gas are either absorbed or eliminated by eruction (belching). Several types of bacteria have been cultured from rumen contents; some act on cellulose of plant fiber, others are noncellulolytic. The principal products of rumen digestion are CO_2, methane, acetic, propionic and butyric acids, and lesser amounts of formic, valeric and caproic acids. Absorption of water and of the acids occurs mainly in the abomasum. The rumen microorganisms, in addition to providing oxidizable acids to the host become a source of protein. Rumen microorganisms synthesize many of their nitrogenous constituents from such simple compounds as urea; urea can serve as an important nitrogen source. Polysaccharides other than starch, glycogen, and cellulose do not constitute much of animal diets. Digestion of lichenin, inulin, and chitin may occur in some invertebrate animals.

Other examples of the participation of symbionts in nutrition are the associations of zooxanthellae and zoochlorellae with anthozoans and bivalves (74b). The giant clam *Tridacna* has thick fleshy mantle edges in which there live many zooxanthellae that carry out photosynthesis. Amoebocytes of the clam come and go to the mantle edge, digest the starch-laden algae, and transport the products to body tissues (74). A turbellarian worm *Convoluta* has its body cells packed with zoochlorellae; clumps of worms in intertidal pools are brilliant green. The adult worm has no digestive tract, except a mouth. The extent to which hydroids and corals that contain zoooxanthellae use them for food is not agreed upon. In some coral heads the plant biomass exceeds the animal by three times and under favorable conditions photosynthesis approximately equals respiration. Sea anemones containing zooxanthellae were placed in seawater containing $^{14}CO_2$ and the tagged carbon atom was evident in gastrodermis within 18 h (52a). Corals with zooxanthellae fix four times more carbon and respire two times more in light than in shade. Of the carbon fixed, 57% goes to zoxanthellal growth and 95% is translocated to the hosts. The carbon budget is given in Table 1.

In starved sea anemones the adenylate (metabolic intermediates) ratio declined at the same rate in light as in dark; the host provides carbon compounds for the algae (64).

Cellulose is digested by members of four orders of insects—Thysanura, Isoptera, Coleoptera, and Hymenoptera. A few wood-eating insects (termites, wood roaches, aphids) synthesize a previously unidentified cellulase (C_x), but their principal cellulase (C_1) is produced by symbiotic protozoa, bacteria, or fungi. Some termites and wood wasps culture fungi that produce cellulase (47a).

TABLE 1. Daily Budget in μg of
C/cm² · day (53)

	Light	Shade
Total C fixed	125.6	26.5
Respired by algae	3.1	1.3
Net C fixed		
Growth of algae	1.2	0.7
Translocation in coral	121.4	24.5
Respired by coral	94.8	52.5
Released as dissolved		
organic compounds	8.2	14.4
Assimilated by coral	18.4	?

Food Selection

Animals are omnivorous, herbivorous, or carnivorous. All show food preferences; filter feeders select preferred food particles from an incoming mixture. Feeding is stimulated in many cnidarians—*Hydra*, *Physalia*—by glutathione. The gastropod *Aplysia* distinguishes preferred species of algae by contact chemoreception.

Neural Control of Feeding

Neural mechanisms controlling feeding have been examined in detail in opisthobranchs *Aplysia* and *Pleurobranchaea*, the pulmonate *Helisoma* and the slug *Limax*. Feeding in these gastropods consists of two series of actions. The first is appetitive orientation and seeking in response to chemical signals and assuming a posture for oral contact and eating. Next is a series of stereotyped actions leading to ingestion, movement of the odontophore cartilage, and radula by alternate protraction and retraction of the muscular buccal mass and contraction of radular muscles. In *Helisoma* the initial orientation is mediated by the pedal ganglion, the movements of odontophore and radula by cerebral and buccal ganglia (42). In *Pleurobranchaea* two oscillators have been identified, one in buccal and one in cerebral ganglia. The central pattern is modulated by sensory input from receptors in

the buccal mass (25, 78). In *Helisoma* five pairs of bilateral protractor motorneurons are electrically coupled and these fire 180° out of phase with eight bilateral retractor motorneurons that are also electrically coupled (42). In the snail *Lymnaea* sensory signals from lips go to command neurons in the brain that activate motorneurons in buccal ganglia. Bite cycle frequency is higher on soft food than on hard food; inflation of the crop increases cycle frequency and may terminate feeding (5a). In the carnivore *Pleurobranchaea* the appetitive phase is initiated by stimulation of chemoreceptors on the oral veil, which send excitatory signals to several populations of interneurons and motorneurons that drive protraction and retraction of the proboscis (78). Retractor neurons are connected by reciprocal inhibition to paracerebral interneurons and protractor motorneurons. The feeding oscillator includes components in both buccal and cerebral ganglia. The consummatory phase is stimulated by food in the buccal cavity. Buccal chemoreceptors activate ventral white cells of buccal ganglia that are command neurons for intense biting and swallowing. Chemical stimulation of the oral veil and buccal chemoreceptors activates the same interneuron-motorneuron oscillator, weakly for biting and vigorously (via ventral white cells) for swallowing. When a food mass enters the esophagus, paracerebral interneurons are inhibited and feeding ceases (25). Aversive conditioning occurs when an attractant (shrimp or squid juice) is paired with a shock to the head. After several presentations, head and siphon withdrawal and escape locomotion occur when the food is presented without a shock. Subthreshold tactile stimuli can also be presented as positive conditioning stimuli.

Stomachs of crustaceans are muscular structures, driven to contract rhythmically by oscillatory discharge from the stomatogastric ganglion (78). The number and arrangement of neurons in the ganglion differ with species. In crayfish the anterior stomach has a gastric mill containing teeth moved by 25 muscles driven by 9 motorneurons. The pyloric stomach is driven by 14 motorneurons. The stomatogastric ganglion of the lobster *Panulirus* contains 33 neurons activating a total of 30 muscles. Rhythmic activity of the gastric mill consists of (1) opening and closing by two lateral teeth and (2) power and return strokes of a single medial tooth (61). Movement of the pylorus consists of alternating dilation and constriction. The neurons of each region are coupled and generate separate but interconnected rhythms. Pyloric bursts occur at 0.5-s intervals and gastric bursts at 3-s intervals. Two different mechanisms generate stomatogastric rhythms. The pyloric rhythm is generated by intrinsic cellular bursting and synaptic interactions. The anterior burster is endogenously active but the other cells can be made to burst when supplied with input from extrinsic (central nervous) ganglia. The gastric mill rhythm is due to reciprocal inhibition between neurons that close and open lateral teeth. The gastric and pyloric rhythms are endogenous but are modulated by input from other neural centers. Interneurons of the gastric portion of the stomatogastric ganglion inhibit excitatory cells of the commissural ganglion, which in turn excite gastric mill neurons. Coupling between pyloric and gastric mill systems synchronizes bursts. Command neurons in the central nervous system excite pyloric dilator neurons.

Feeding in insects is initiated by stimulation of chemoreceptors on appendages—mouthparts, antennae, and tarsi. Thresholds of chemoreception vary with physiological state and become higher after feeding. Proboscis extension in a fly or butterfly is a measure of chemical stim-

ulation. Each taste hair of the labella (mouthpart) or tarsus of a blowfly has five sensory cells—two for salt, one for sugar, one for water, and one for movement. Relative sensitivity of sugar receptors is sucrose = 1, glucose = 0.71, fructose = 0.46; salt receptors are stimulated mainly by cations (19). Chemosensory sensilla of a carrion beetle respond to fatty acids, those with chain lengths longer than six carbon atoms are depolarizing while shorter fatty acids are hyperpolarizing (7). Many chemoreceptors of insects are not specific but are sensitive to different degrees to stimulation by several substances. Fifty olfactory cells on antennae of saturnid moths were tested for responses to 13 different odors and no two cells gave the same pattern of sensitivities. A behavioral response depends on integration of input from many receptors of odors or tastes (19, 24, 79).

A dual inhibitory system regulates feeding in blowflies. Excitation from peripheral chemoreceptors is balanced in the nervous system by inhibition from internal stretch receptors. Foregut stretch receptors show delayed, slowly decaying response; stretch of abdominal receptors gives an immediate response and terminates drinking. After nerves to foregut or abdomen are cut, the flies show hyperphagia; normally the two inhibitory systems stop feeding (9).

In fishes and mammals, centers in the hypothalamus regulate feeding. Goldfish can be conditioned to regulate food intake. Stimulation of the ventral medial hypothalamus of bluegill sunfish can elicit biting, swallowing, and exploration of objects at the bottom of an aquarium. Lesions in the hypothalamus of bluegills and goldfish can stop feeding. Stimulation of the olfactory tract and the inferior lobe of the hypothalamus can stop feeding. Stimulation of the olfactory tract and the inferior lobe of the hypothalamus can initiate feeding. Neurons in the hypothalamic feeding area respond when food is placed in the mouth (18).

In mammals, the hypothalamus contains centers for regulation of feeding. Lesion of the ventromedial hypothalamus in rats results in increased food intake and weight gain (33). Stimulation of the ventrolateral hypothalamus increases food intake; lesions cause aphagia and adipsia, a complex of ventral regions of hypothalamus constitutes a satiety center; apparently there is a set point for regulation of body mass (57).

Behavior patterns of various carnivores reflect innate differences in nervous systems. Some carnivores such as cephalopods and dogs are hunters, continually searching for prey; other carnivores use a "sit-and-wait" strategy—spiders, octopus, and cats.

Herbivores may be extreme specialists—feeding on only single or a few plant species; other herbivores are generalists that eat indiscriminately. Diet selection by generalists may be in response to available quantity or to quality—nitrogen content and caloric value. Many animals are both generalists and specialists; they have food *preferences.* Selection may be by sensory or nutritional properties. Selection in the short term may be by taste, but in the long term may result from learning. Rats in a laboratory can be conditioned to avoid a flavor after it has been paired with an emetic compound.

Mussels from two sites differing in quality of available food were compared—mussels from high quality food sites had higher feeding rate and lower absorption efficiency and lower retention time than the mussels from a low-quality food site. Ingested substance containing little organic matter is rejected by bivalves as pseudofeces; gut retention is proportional to absorption efficiency (5). Some crustaceans—shrimp and lobsters—feed

on dead animals. For a shrimp *Leander,* exposed to 28 organic components taurine was the most effective attractant. The most effective stimulus to ingestion by coelenterates *Hydra* and *Physalia* is reduced glutathiones (G-SH). Dead animals are rejected by coelenterates but may be taken if treated with a low concentration of G-SH.

Many insects are highly selective of food plant. Chemoreceptors in taste sensilla may show highest sensitivity to substances absent from the plant on which the insects feed. The degree of intake specificity of plants as food is determined by secondary organic compounds rather than by nutrients. Potato beetles and tobacco hornworms are restricted to Solanaceae; the beetles feed on potato and not on tobacco; potato treated with nicotine is unacceptable. Hornworms feed on tobacco or tomato but not on potato. The evolution of flowering plants during the Cretaceous period was accompanied by a reciprocal development of insects. Repellents in plants may protect them against insects in general but by mutation of an insect species a given substance may become an attractant. A compound (cucurbitacin) in cucumbers is repellent to many kinds of insect but members of one family, cucumber beetles, have evolved a positive response to it (49). Of two groups of the butterfly genus *Papilio,* one feeds on plants of the family Rutaceeae, the other mainly on Umbelliferae; both groups feed on Umbelliferae if the leaves contain methyl chavicol (anise). The agents that attract cabbage butterflies to cruciferous plants are the mustard glycosides sinagran and sinalbin (54). The caterpillars will feed on other plants painted with mustard oil glycosides. Conditioning of larvae can alter the limits of chemobehavior of the emerged adults. Parasitic Hymenoptera have been conditioned by smearing a moth larva that is normally rejected for egg depositing with juice from a plant or from a larva normally used as food. In general, adult insects lay their eggs on the kind of plant or host animal on which they had fed as larvae. The distinction between conditioning and innate behavior has not been made effectively.

Herbivorous mammals vary their diet according to availability and season, to maximize nutrient intake. For several African ruminants, monocotyledenous plants are less digestible than dicotyledenous plants; the monocots contain less fermentable carbohydrates but are rich in cellulose and lignin. Perissodactyla—horses and rhinoceroses—digest cellulose by bacteria in colon and cecum but are less efficient than Artiodactyla—ruminants—and when intake is limited, the ruminants outcompete horses but when intake is ample, horses are more effective feeders. Domesticated ruminants (cows and sheep) have low intake when food is rich in energy yielding compounds and low in fiber; intake is maximum at an energy content of ~600 J/g (62). Yellow-bellied marmots prefer areas of high rather than low plant food biomass; they balance food ingestion with predator avoidance (12); legumes with lower fiber and higher phosphorous and sodium are preferred over graminoids. Populations of rodents (*Microtus*) produce more young in regions where soil content of sodium is high. Seed-eating birds have shorter caecae than browsers; captive grouse have shorter caecae than wild ones (62). Subarctic birds and mammals show preferences based on secondary compounds, as do many insects. Willow ptarmigans feed on willow, rock ptermigans on birch, white-tailed ptarmigans on green alder, all similar in energy content. Ruffed grouse avoid plants high in resins. For several birds and mammals detoxification of resins is more costly than avoidance.

Digestive Enzymes

Digestion of Carbohydrates

Monosaccharides occur as five- or six-membered oxygen-containing rings. The natural sugars are D-sugars that occur in two isomeric forms, the α form is more strongly dextrorotatory than the β form; these differ in the positions of H and O on the same carbon atom. Structural formulas are found in biochemistry textbooks. Monosaccharides include pentoses: ribose, ribulose; and hexoses–aldoses: glucose, galactose; and ketose: fructose. Oligosaccharides are disaccharides, trisaccharides, and substituted sugars; these are classified according to their products of enzymatic hydrolysis. Polysaccharides are of two classes: (1) structural polysaccharides—cellulose, lignin, dextrans, alginic acid, agar, and chitin; and (2) universally digestible polysaccharides—starch, the principal soluble plant polysaccharide, and glycogen the principal animal polysaccharide (56b, 67).

The most important plant storage product is starch, which consists of amylose, a straight chain of α-1,4-glucosidic links, and amylopecten, which is branched and frequently present on the outside of starch grains. Glycogen is an α-1,4,- and an α-1,6-linked structure having ~10 glycosyl subunits between branch points. The most important structural polysaccharide in plants is cellulose, which consists of unbranched glucose in β-1,4 linkage; it is less soluble in aqueous solution than starch and is resistant to boiling and to acid treatment. Numerous other plant polysaccharides are rarely used by animals. These include lichenin, hemicellulose, xylan, lignin, inulin, and pectin.

Chitins are a family of nitrogen-containing linear polymers of *N*-acetylglucosamine in β-1,4 linkage produced by diatoms, fungi, and many animals, often in exoskeletons. Chitin occurs in ciliates, in ectoderm of hydrozoa, corals, and scyphozoans, and in the skeletal elements of molluscs, arthopods, and annelids (38). Many animals have a chitinase that, in combination with chitobiase, splits chitin to acetylglucosamine. *Mytilus* has chitinase in the crystalline style; the enzyme occurs as two isozymes, MW 60,000 and 76,000. A glucosaminidase occurs in hepatopancreas (6). *Crassostrea* has endogenous chitinase and chitobiase in the style (48).

Digestion of Starch

Enzymes for digesting starch are more widely found than those for any other substrate (44a). Animal amylases are α-amylases and are usually activated by chloride ions. In mammals pancreatic amylase is more widely distributed than salivary amylase (71b). In general, amylase is very active in herbivorous and omnivorous animals; in carnivores, amylase or a similar enzyme may act on glycogen that is abundant in tissues of prey. Porcine pancreatic α-amylase contains 496 amino acid residues compared to 497 in human and 493 in rat and mouse enzymes (56). The amino acid compositon of *Drosophila* amylase in similar to but not immunologically reactive with mammalian enzyme (34a). Rat amylase is more structurally and functionally similar to *Drosophila* amylase than to the mosquito fish enzyme (74a).

Insect amylases are secreted by salivary or midgut cells (3b). Crustacean amylases have been purified from the midgut gland (71a). Herbivorous molluscs have an extracellularly acting amylase, either in the midgut gland, crystalline style, or both. Amoebocytes and digestive diverticula of bivalves digest starch, glycogen, and several sugars. The style is formed rapidly after feeding; it is in a sac in which

it is pushed forward by cilia and the tip wears against a gastric shield. As the tide ebbs, the style dissolves and is minimal at low tide; on return of the tides, feeding starts and the style is maximal 3 h after high tide (51). When oysters are kept immersed and fed continuously, the rhythm is lost, hence synthesis is induced by feeding (44). In a sand flat population of a snail, the style disappears at low tide and is regained daily (14). Amylase is widely distributed in invertebrates (71b).

Digestion of Cellulose

Many animals that utilize cellulose do so by symbiotic microorganisms. Three classes of enzymes of cellulase complex are exocellobiohydrolases: C_1, B-1,4-endoglucanases (C_x), and β-glucosidases (cellobiase). C_1 and C_x are not endogenous in vertebrates. Wood eating insects do not make C_1 but some insects and wood boring molluscs appear to make C_x and cellobiase. C_1 cellulase is made by bacteria, some symbiotic protozoans, and fungi. Among insects, silverfish and firebrats make C_1 (75a). Wood-boring insects differ in utilizing the different polysaccharides of wood. Shipworms, *Bankia*, and an intidal bivalve form glucose from cellulose; if lignin is absent from wood, cellulose digestion is enhanced (17). Some beetles use only the stored food in wood—starch and sugars. Bark beetles can digest hemicellulose. Cellulases of cultured fungi are used by termites and siricid woodwasps; acquired fungal enzymes may be responsible for cellulose digestion in xylophagous larvae of cerambycid and other beetles (47a). Primitive termites and wood roaches depend on symbiotic flagellates to hydrolyze cellulose and these insects die if their symbionts are removed (36b). In ruminant mammals much of the fiber in forage is hydrolyzed in the rumen by bacteria. In some nonruminants (horses, rabbits) cellulose is broken down by symbionts in the cecum and capacious colon. Echinoids are not able to digest cellulose but have β-glucosidase (43).

Digestion of Sugars; Disaccharidases

Sugar-digesting enzymes, disaccharidases, are synthesized in brush border epithelia of vertebrate intestine. These enzymes are synthesized as large precursors of >220 kD, which are split to two subunits post-translationally before secretion into the intestinal lumen (61a). Each subunit carries a single catalytic site and is anchored on the cytosolic side of the membrane by a hydrophobic segment at the amino terminus (36c).

Four glycosidases have been isolated from brush border of mammalian intestine (61a):

1. Sucrase–isomaltase complex consists of two subunits, one 120 kD in size hydrolyzes maltose and sucrose, the other of 140 kD has glucopyranosidase action (hydrolyzes isomaltose). The complex accounts for 80% of the maltase activity of the small intestine. Synthesis and membrane localization of sucrose–isomaltase in chicken is similar to that in mammals but antibodies for chicken enzymes do not cross-react with those of goat and rabbit (36d).

2. Maltase–glucoamylase (γ-amylase) lacks a Cl requirement. Its molecular weight is 250 kD; in mammals it consists of two subunits, in pigeon it consists of one. This enzyme accounts for 20% of intestinal maltase action.

3. A β-glycosidase complex is a lactase-hetero-β-glycosidase that accounts for all the lactase activity. This complex has a molecular weight of 320 kD. It hydrolyzes lactose, cellobiose, and phlorizin.

4. Trehalase is a single enzyme, not a complex; it hydrolyzes α-trehalose, a disaccharide occurring in algae, mushrooms, and arthropods.

Neutral lactase hydrolyzes lactose, the principal carbohydrate nutrient in the milk of all land mammals; it is abundant in the intestinal brush border of the offspring at or before birth. A lysosomal acid lactase in enterocytes hydrolyzes milk rich in galactose-rich oligosaccharides occurs in tammar wallaby and kangaroo; a similar enzyme may occur in crabeater seals (13b). Lactase usually disappears from the intestine of mammals after weaning when there is no lactose in the diet. In humans lactase is lost on weaning but in some ethnic groups a genetic change to lactase in adults has occurred, an adaptation to a milk diet since the advent of dairying some 6000 years ago (3a). Mammals develop sucrase and maltase after weaning. Ruminants that metabolize sugars in the rumen have very little sucrase or maltase in the intestine.

Some invertebrate disaccharidases have wider specificities than those of vertebrates. Hydrolysis of sucrose is generally found in the gut of invertebrate animals; it is by an α-glucosidase that is also active on maltose. A number of insects have been shown to have β-fructosidase activity. Midgut cells of a hornworm have a β-glucosidase (cellobiase) of 129,000 D that hydrolyzes β-glucosides, β-galactosides, and β-fucosides but not α-glucosides or α-galactosides (59). The β-glucosidase partially digests hemicellulose; β-fructosidase digests leaf sucrose (67b). Midgut cells of a dipteran *Rhynochosclara* have in peritrophic membrane and midgut caeca a trehalase; a caecal membrane-bound cellobiase acts on three β-sugars (23b).

In *Locusta* the same enzyme has both cellobiase and lactase activities (50). Lactase is absent from most invertebrates but cellobiase activity is common. Sugars are poorly digested by carnivorous insects; however, the blood-sucking tsetse fly *Chrysops* has amylase and sucrase in its posterior midgut; the omnivorous fly *Calliphora* has amylase in salivary glands and amylase, sucrase, and maltase in its midgut. Trehalase is found in the gut of invertebrate animals that feed on fungi (e.g., earthworms). Trehalase activity is found on the hemolymph side of the gut in the silkworm, *Bombyx* (3d). Several echinoderms have some membrane bound enzymes acting on β-fructoside (sucrose), isomaltose, and β-galactose (13a). Homogenates from intestinal epithelium of the polychaete *Neanthes* show sucrase, maltase, lactase, and cellobiase activities (42a).

Digestion of Proteins

Proteins and polypeptides are chains, linear or folded, of amino acids. Digestive endopeptidases attack specific internal peptide bonds in proteins and peptides; exopeptidases attack stepwise from either the amino or carboxyl terminus. Some exopeptidases attack both proteins and peptides while others attack only short chains of peptides. Proteinases and peptidases may be formed as pro-enzymes in zymogen granules from which the protease is cleaved as it is secreted. Other proteinases and peptidases, for example, the cathepsins, are intracellular, frequently associated with particulate lysosomes, which digest nonessential or discarded intracellular proteins or endocytotically obtained proteins.

Acid Proteinases

Acid (aspartic) proteinases act in a medium of low pH (1.8–2.0). Pepsinogen, MW ~39,000, is secreted in the stomach of all vertebrates except stomachless fish and cyclostomes. Pepsinogen is activated autocatalytically below pH 5.0 to yield pepsin. Several molecular species of pepsin may occur in one animal. Pepsin attacks peptide bonds with an adjacent aromatic amino acid; it is particularly active on peptide linkages between dicarboxylic and aromatic amino acids if the second carboxyl of the dicarboxylic acid is free and if there is no free amino group near the peptide link. Chymosin (renin) is a distinct protease, pH optimum 3.7; it occurs only in suckling mammals, particularly ruminants. The pH of the stomach of most vertebrates is acidic, ~pH 2.0, because of the HCl secreted by parietal cells of stomach. Secretion of HCl is stimulated by vagal reflexes and is antagonized by atropine. Acid secretion is also stimulated by histamine and gastrin. Gastrin and histamine are major gastric stimulants in frogs and mammals, acetylcholine is more important in fishes (eel) (2, 22). In fishes pyloric caeca are important sites of secretion of both carbohydrases and proteinases (27).

Acid proteinases with pepsin-like or cathepsin-like properties have been extracted from a few invertebrates (24a). In many cases, definitive information is needed as to whether a particular intracellular proteinase has a digestive function.

Chymotrypsin and Trypsin Proteinases

Chymotrypsinogens and trypsinogens are universally secreted by vertebrate pancreas (9a, 57a, 67, 72a); molecular weights are ~25,000. Proelastases are abundant in the pancreatic secretions of carnivores. All of these proteinases have a common active center composed of serine and histidine residues (serine proteinases). Trypsinogen is activated by intestinal enterokinase (an enteropeptidase that cleaves a unique lysine—isoleucine bond). Trypsin then activates more trypsinogen, also chymotrypsinogen, proelastase, and procarboxypeptidase (67). A proteinase inhibitor is found in the pancreas; it presumably controls the activation process until enterokinase is encountered in the intestine. The activated enzymes are characterized by acting at a mildly alkaline pH 7–8.

Chymotrypsin has a molecular weight of 23,500 and consists of three chains linked by two disulfides. It acts preferentially on peptide bonds on the carboxyl side of the aromatic side chains—tryosine, tryptophan, and phenylalanine, or large hydrophobic residues such as methionine. Trypsin (MW 23,200) requires a lysine or arginine residue; the specificity of elastase is directed toward smaller uncharged side chains. About 40% of the amino acid sequences of these three enzymes are identical (60% interior); different specificities are a result of small structural differences in binding sites (67). These enzymes diverged from a common ancestor. The amino acid sequence of dogfish trypsinogen has 69% homology with that of beef (9a).

Proteinases of Invertebrate Animals
Acid thiol and metal-type digestive endopeptidases have been reported for selected invertebrates but require confirmation as to endogenous source and digestive function. Serine enzymes are widely distributed and well characterized (3b, 14a, 49a, 52, 71b). Crayfish may have multiple but similar forms of trypsin. Forms from *Astacus fluviatilis* and *A. leptodactylus* show antibody cross reactivity but do not cross react with that of the crab, *Carcinus*, or bovine trypsin. There is considerable amino acid sequence homology to that of other trypsins; the carboxy-terminal sequence is similar in beef and crayfish. Beef trypsin has twice as much serine and less glutamic acid than crayfish. Crayfish trypsin, like bacterial, has three disulfide bridges, beef has six (76). Chymotrypsin activity is high in shrimp (69). A hornet has two chymotrypsins and one trypsin. One chymotrypsin from the midgut of hornet larvae has been sequenced. It consists of a single polypeptide chain of 216 residues and has a 37% identity with bovine chymotrypsin,

trypsin, and porcine elastase (37a). There is some evidence for the synthesis of zymogens of serine proteinases in coelenterates, a few insects, and echinoderms; low levels of autodigestion and high stability at pH 4–9 suggest that zymogens may not be required (71b).

Animal tissue collagenases are zinc metallopeptidases that require calcium and function in morphogenesis. Crab collagenase is a digestive serine proteinase with a basic pH optimum and no metal requirement. Collagenases from the dog pancreas and the insect *Hypoderma lineatum* also may be serine proteinases (31).

In a silkworm larva and in a clothes moth, the pH optimum for protein digestion is 9.5. Wool contains the protein keratin that has peptides bonded by disulfides. The gut fluid of a clothes moth *Tineola* contains a substance that maintains a very low oxidation–reduction potential: -200 mV in the midgut, $+20$ mV in a non-keratin feeding *Blatella*. Reduction of the disulfide bonds produces a substrate hydrolyzed by methionine and serine proteinases and peptidases. A clothes moth contains a complex of metal- and proline-proteinases (71c).

A unique proteinase, originally described as a low molecular weight protease (MW 11,000) because of anomalous gel filtration, has been purified from several decapods (3c, 69, 77) and sequenced from the gastric juice of *Astacus fluviatilis* (68). This invertebrate proteinase is not homologous to any established proteinase family; it has an unusual property of hydrolyzing peptide bonds on the amino side of the small uncharged residues, Ala, Thr, Ser, Gly, Val (43b, 76).

Exopeptidases
Carboxypeptidase A is an exopeptidase of 307 amino acid residues that acts on short peptide chains; it is secreted by the pancreas. Mammalian pancreatic carboxypeptidase A acts best on a carboxy-

terminal residue with an aromatic or bulky aliphatic side chain; it is inhibited by a nearby amino group. Pancreatic carboxypeptidase A contains one atom of zinc and has a molecular weight of 34,000 D (67). Digestive metallocarboxypeptidases occur widely in vertebrates and invertebrates and may include carboxy peptidase B, which acts on a terminal basic amino acid.

Aminopeptidases act on a peptide bond next to a free terminal amino acid with a free amino group. Aminopeptidase is inhibited by a free carboxyl nearby in a peptide. Mammalian aminopeptidases are secreted by intestinal epithelial cells.

Di- and tripeptidases are also intestinal enzymes that are relatively specific: glycyl-leucyl dipeptidase and glycylglycine dipeptidases. Exopeptidases have been reported from many invertebrate animals; the di- and tripeptidases usually occur in association with aminopeptidase. In many molluscs, crustaceans, and annelids the final stages of protein hydrolysis occur intracellularly. In the insect *Rhodnius*, a major aminopeptidase is found between inner and outer microvillar membranes (23a).

Digestion of Fats and Nucleic Acids

Neutral fats consist of higher fatty acids linked to the trihydric alcohol glycerol. Fats may be partially hydrolyzed to diglycerides and monoglycerides or fully hydrolyzed to the alcohol and fatty acid. Neutral fat may be absorbed as micelles (colloidal droplets) without hydrolysis or may be digested intracellularly. Lipases hydrolyze the esters of higher fatty acids; esterases hydrolyze esters of shorter acids. In vertebrate animals pancreatic lipase accomplishes most digestion of neutral fats, acting in the intestinal lumen and its substrates are emulsified by bile salts. Bile salts activate pancreatic lipase, increase the total area of oil–water interface

at which enzymes can act, and form micelles of the products of digestion. Absorption of fatty acids and monoglycerides is favored by micelle production.

Cholesterol esters and lecithin are rapidly hydrolyzed by specific esterases.

Absorption of the products of lipid digestion is facilitated by bile salts. Bile acids are shorter in mammals than in fishes—27 carbon atoms in elasmobranchs and 24 in mammals. In mammals the principal acid is cholic, conjugated with the bases of taurine or glycine. There is some diversity of bile acids—deoxycholic in rabbit, nutriacholic in an Indian rat, pythocholic acid in boid snakes. Bile salts have an acidic group on a side chain such that the molecule is a detergent. In most vertebrates this acidic group is provided by taurine or glycine through a terminal carboxyl but in some fishes and amphibians the acidic group is an alcohol sulfate. Bile salts are different according to diet, those in herbivores are different from those in carnivores. Steroid bile salts may be absent from invertebrate animals. In some crustaceans acylsarcosyl-taurines are active.

Several insects are specialized for fat digestion (3b). A salivary lipase occurs in larvae of the meat eating calliphorid fly *Phormia*. Extracts of the midgut of larvae of the waxmoth *Galleria* are active in alkaline medium in the following order: tributyril > methyl butyrate > olive oil. The waxmoth utilizes as much as 38% of the higher fatty acids of beeswax (23c). Bacteria from the gut of *Galleria* act on wax (esters of cerotic acid); the bacteria evidently break down the waxes to lower fatty acids that can then be digested by enzymes of the waxmoth. In crustaceans the midgut gland is rich in esterase and also has lipase activity. In molluscs both intracellular and extracellular lipases and esterases occur in digestive diverticula (71b).

Many marine invertebrates, especially copepods, synthesize and store large quantities of wax esters (that favor buoyancy) and fishes that consume these animals contain wax esters. Oceanic copepods may be as high as 70% wax ester on a dry weight basis. Sea birds that feed on these fishes or invertebrates use wax esters for some 63% of digestible energy. They digest cetyl palmitate and deposit the resulting fatty acids in storage depots. Digestion of the waxes occurs in duodenum (56a). Some birds such as cedar waxwings digest wax of bayberries.

Nucleic acids constitute a significant percentage of ingested nitrogen. Pancreatic nucleases, RNAase and DNAase, split nucleic acids to oligonucleotides and mononucleotides. In ruminants some 20% of the nitrogen entering the stomach from the rumen is polynucleotide nitrogen. In ruminants the excretion of purine products is unusually high. An efficient breakdown (75–85%) of microbial RNA and DNA has been shown to occur in the duodenum of ruminants (63).

Absorption
Absorption of the products of digestion is partly by diffusion and partly by carrier-mediated transport in the brush border. Absorption of sugars and amino acids, whether in mammalian intestine or into parasitic worms, is coupled to active transport of sodium (Chapter 2). The properties of the mucosal and serosal sides of epithelial cells of intestine are different and there is active transport of sodium across the serosal membrane making the gut lumen (mucosal side) electrically negative.

Sugar accumulates in the mucosal epithelial cells of intestine and transport to blood occurs mainly when sodium is present in the lumen. Across the epithelium a potential develops that is proportional to the Na^+ transported. Ouabain, an inhibitor of Na^+-K^+ ATPase, added to the serosal side reduces the potential and the sugar absorption. One model is: A carrier protein molecule on the mucosal side binds to both glucose and sodium, making a ternary complex that crosses the mucosal surface down a gradient and then dissociates. At the basolateral surface Na^+ is extruded by a ouabain-sensitive (Na-K)ATPase and glucose diffuses into the blood. In rabbit duodenum two Na^+-dependent glucose carriers have been identified for a ratio of 3 Na^+ to 1 glucose. In addition, some glucose is absorbed by diffusion (66a, 73). Absorption of glucose is inhibited if K^+ or Li^+ is substituted for Na^+. Galactose may compete with glucose for the carrier site and D-glucose, but not L-glucose, is moved. The carrier protein has been isolated from intestine and has a molecular weight of ~75,000. Energy required for the sugar transport is provided by the Na^+ gradient. Xylose can be bound by the same carrier as glucose but the concentration for half-saturation is 100 mM for xylose and 2 mM for glucose. Lactose absorption is coupled with proton transport. In the intestine of poikilothermic vertebrates the absorption of sugar is similar to that in mammals. In the goldfish, glucose and water are transported from lumen to blood; the serosa is positive to the mucosa. Intestinal parasites—cestodes and acanthocephalans—absorb glucose against a concentration gradient. Acanthocephalan parasites in insects can absorb trehalose, unlike the parasites from other hosts.

Amino acids, like glucose, are absorbed when Na^+ is present on the mucosal side of the brush border. There are multiple amino acid carriers and data obtained with brush border vesicles are not in full agreement with *in vivo* absorption. Some di- and tri-peptides are absorbed from the lumen and after hydrolysis pass to blood as amino acids. Several Na^+-dependent carriers are identified for acidic, basic, neutral, and proline (plus imino acids); also fewer Na^+-independent carriers occur for the different amino acid groups (1a, 36a, 66a). L-Amino acids are absorbed much more than D-amino acids and there may be some competition between D and L forms of the same amino acid. In rabbit ileum mounted to measure the short-circuit current due to Na^+, net transport of alanine and Na^+ was abolished by cyanide, dinitrophenol, or ouabain; these agents did not abolish the uptake across the brush border into epithelial cells. The amount of alanine leaving the cells was proportional to the amount of sodium being pumped (36a). A model for amino acid absorption is like that for glucose absorption in postulating that a complex is formed between an amino acid and a carrier and then a ternary complex with Na^+ is formed. It is postulated that the ternary complex moves across the mucosal membrane, the amino acid and Na^+ are released at the serosal surface, and the Na^+ is then pumped out. Absorption of phenylalanine in midgut of the cockroach *Blabera* (55) and of lepidopteran larvae is by cotransport with K^+ (26). Locust rectum has a proline transport system that is independent of Na^+ (48a).

Fatty acids and glycerides are absorbed differently from sugars and amino acids. In mammals, fat is absorbed as triglycerides into lymphatic vessels and transported as neutral fat into the circulation (37). Pancreatic lipase acts on emulsified fats in the intestine, hydrolyzing them to monoglycerides and fatty acids. These substances aggregate as micelles of 40–50 Å in diameter (7a, 39) with the aid of bile salts. These micelles enter epithelial mucosal cells. Inside the cells the free fatty acids in the presence of ATP, Md^{2+}, and CoA form a CoA–fatty acid complex. Monoglycerides are converted to phosphatidic acids and triglycerides. The triglyceride or neutral fat appears as droplets in endoplasmic reticulum. Chy-

lomicrons are extruded by negative pinocytosis into the lymphatics (39).

Few studies have been made on the absorption of fats from the intestine of invertebrates. Cestodes absorb monoglycerides and fatty acids from micelles of their hosts (4, 13). In a cockroach, after a meal of fat, fat droplets can be seen in epithelial cells of the gut.

Adaptations of Digestion

In all animals much energy is devoted to feeding and digesting. Limited food supply is a stress that effects alterations in behavior and physiology. Feeding is periodic by many animals; secretion of digestive enzymes corresponds to the presence of food in the digestive tract. In continuously feeding animals (grazers) there is continuous flow of food into the stomach and intestine. Some herbivorous animals are very selective; some animals, for example, insects, pandas, some monkeys, feed on only one kind of plant, some on one type of wood; other herbivores are generalists. Some carnivores eat only living prey, some feed on carrion, some are parasites on host animals.

Secretion of digestive enzymes is influenced by the kind and amount of food consumed. Gland cells can be stimulated directly by food in the digestive lumen, by chemical agents via blood (hormones), by nerve reflexes, or by all three methods.

Control of salivary secretion in mammals is entirely nervous; all three pairs of salivary glands receive parasympathetic and sympathetic innervation. The nature of the fluid secreted varies according to the gland secreting and the nerve stimulated. Some salivary secretion is continuous; food in the mouth can stimulate reflex secretion and conditioning by stimuli other than food can elicit salivary secretion (78). Digestive functions of the mouth in vertebrates are selection, intake, and trituration of food. Ruminants masticate regurgitated fiber. Salivary secretions of mammals are slightly alkaline and contain a lubricating mucus. A few mammals—primates and rodents—have salivary amylase but most digestion of starch and glycogen is in the intestine.

Secretion of gastric juice is stimulated by the vagus nerve and can be affected by conditioning and by emotional state. Food or mechanical stimulation in the stomach activates cells of the antral mucosa to liberate a hormone gastrin, that passes into the blood and stimulates parietal cells to secrete HCl and, to a lesser extent, so-called chief cells to secrete pepsinogen. Gastrin occurs in two forms—I and II; it is a polypeptide of MW $\sim$2000. Amines and peptides, particularly aliphatic ones, are potent enhancers of gastric motility and of proliferation of mucosal cells (45).

In teleosts, as in other vertebrates, food in the stomach stimulates gastric secretion, and there is evidence for both vagal and hormonal control.

When the acid chyme from the stomach enters the small intestine of mammals a hormone is liberated from the intestinal mucosa. This substance, secretin, is carried in the blood and stimulates pancreatic secretion. Secretin is a polypeptide of MW $\sim$5000 D. It also stimulates bile secretion by the liver. A hormone from mucosa of upper intestine, cholecystokinin (CCK) has several actions. It causes smooth muscle of the gallbladder to contract; it stimulates secretion of amylase and trypsinogen from pancreas and increases the amylase/trypsinogen ratio threefold (69a). Secretion of CCK is stimulated by food in the intestine. When rats were on a high-protein diet for 12 days the content of trypsinogen, chymotrypsinogen, proelastase, and carboxypeptidase in pancreatic juice was increased. When on a high carbohydrate diet, pancreatic secretion of amylase increased (60). When rats were transferred from low to high starch diets an increase in sucrase and maltase could be detected in 15–45 min (32, 70). The action of a high-carbo-

hydrate diet in increasing amylase secretion is probably mediated by CCK. The rapid change in response to diet results from effects on release from zymogen granules, which constitute 20–25% of pancreatic volume. Cells that stain with antibodies to mammalian gastrin and CCK have been observed in endocrine eyestalk and stomach of crustacea but no digestive function of these peptides has been demonstrated (23).

Diet-induced changes in secreted enzymes occur in intestinal mucosa as well as in pancreatic acinar cells. Rats were fed isocaloric diets of high starch, low fat, and of high fat and low starch; when starch was decreased, disaccharidases decreased but dipeptidases were unaltered in amount. Starvation resulted in decreased sucrase but no change in aminopeptidases (8, 27a). However, aminooligopeptidase activity increased in rat intestine after perfusion with a tetrapeptide. Synthesis in the endoplasmic reticulum and transfer to Golgi apparatus took place in 15 min after a pulse of the peptide and transfer to luminal brush border in 3 h (58).

In addition to affecting pancreatic secretion, CCK has central nervous effects on feeding. Administration of CCK or a mixture of oleate plus protein hydrolysate, which releases CCK results in reduction of feeding by pigs. Cholecystokinin may stimulate a satiety center (3).

In frogs mechanical distension of the stomach leads to reduced feeding and this is not abolished by denervation; vagotomy diminishes pepsin secretion. Low doses of gastrin stimulate acid secretion in frogs. Prostoglandin E_2 inhibits gastric acid secretion in the gastric brooding frog *Rheobatrachus silus* (70a). In teleosts, food in the stomach stimulates gastric secretion. A holocephalan fish *Chimaera* lacks a stomach but its intestinal mucosa produces substances having CCK and secretin actions. Many fishes have pyloric ceca that function like the proximal intestine for digestion and absorption and increase total digestive membrane area. These ceca of fish are unlike those at the ileo–colic junction of many mammals where microbial fermentation occurs (10, 11).

Insects specialize in diet and choice of plant food may be determined more by taste preferences than by nutritional value (Chapter 6). Bloodsucking insects can be stimulated to feed from solutions if they contain some blood components. Mosquitoes suck blood, or a synthetic solution if adenine is present. A tsetse fly is stimulated to artificial feeding by ATP. A tick is stimulated to feed by glutathione, ATP, or amino acids. The production of intestinal protease in the fly *Calliphora* is stimulated by meat in the diet. If neurosecretory cells are removed from the *Calliphora* brain, intestinal protease is much reduced; injection of hormone from corpora cardiaca restores protease production. In the blowfly, *Phormia*, food intake is regulated by distension of the foregut.

Activity of digestive enzymes in the marine copepod *Calanus* is influenced by diet. Digestive enzymes were compared after feeding a flagellate, *Cryptomonas*, rich in starch, then a diatom, *Thalassiorsira*, that lacks starch. Activities of amylase, trypsin, and laminarinase in *Calanus* were higher after feeding on *Chryptomonas* (34). During feeding of a ctenophore *Pleurobranchia* on *Artemia* several hydrolytic enzymes increased in activity: acid phosphatase, aminopeptidase, and carboxypeptidase after 4 h, proteinase, chitinase, and glucosaminidase after 8–31 h (36). Vegetarian sea urchins are rich in α- and β-glucosidases and β-galactosidase; carnivorous starfish have active proteinases (43, 44b).

Intestinal absorption is regulated adaptively in different kinds of animals to feeding habits and acutely in individual animals to diet (10). Adaptations are (1) morphological—intestinal length and absorptive surface and (2) enzymatic—kinetics and specificity of transport. In general, the intestine of herbivorous mammals is longer than of carnivores. Mice were maintained after weaning on high carbohydrate (CHO) with minimum protein or on carbohydrate-free high protein isocaloric diets. Irreversible anatomical effects seen in the adults were higher body weights, longer intestines, and heavier intestines in animals on the high CHO diet. In short-term dietary regimes, reversible changes in the ratio of sugar to amino acid absorption occurred within 1 to 2 days. On a high CHO diet the V_{max} of absorption increased but the K_m for glucose absorption was little changed; this indicates an increase in absorption sites but similar transport kinetics (19a, 28, 40, 41).

Small mammals (wood rats) were compared with iguana lizards of similar size and fed similar alfalfa diet. The rat (at 37°C body temperature) requires nine times more energy, extracts slightly more energy from its food, consumes eight times more food, eliminates waste (feces) in 1–2 days, the iguana in seven days. Small intestine is 1.8 times and colon 6.8 times longer in the rat. The rat has more absorptive sites and transport molecules have higher affinity (41). In tests of absorption of 18 solutes, 12 transporters were identified, all of them regulated by dietary levels. Some transporters such as those for sugars and nonessential amino acids were up-regulated; others such as for two vitamins and three minerals were down-regulated by addition of solute to the diet (20a, 27a).

Among birds, browsing species that utilize much cellulose have longer caeca than do seed eaters. In omnivores (starling), the gut length may be 450% longer in winter than in summer (62).

In herbivorous mammals, food intake is low when food is rich in energy, low in fiber; when food is limited, ruminants outcompete nonruminants (horse and rhinoceros); when food is abundant, the nonruminants are superior in nutrient extraction (62).

It is concluded that digestive efficiency is influenced by kind and quantity of food in all animals and that digestive structure, enzyme secretion, and absorption are influenced by diet either directly on reacting cells or by hormone or nervous reflexes. The cellular mechanisms by which dietary changes are perceived and transduced are poorly known.

Summary

A large fraction of time and energy is used by animals in acquiring and digesting foodstuffs. The stress of shortage of food or deficiency in essential dietary components limits growth and population survival.

Uptake of water and ions was described in Chapter 2. Uptake of organic molecules satisfies all or part of their needed nutrients from substances dissolved in the medium. Modes of absorption from aquatic environments are similar to absorption by digestive epithelia and are often against steep concentration gradients. Uptake of particulates is by a variety of filtration means. Uptake of carbohydrates, proteins, and fats is by various methods of sucking, biting, and chewing.

Feeding generalists consume many kinds of foodstuffs, both plant and animal. Specialists may be carnivores or herbivores and plant eaters may consume many kinds of plant or only a few plants that contain secondary organic attractant substances. Some polysaccharides are taken only by animals with symbiotic bacteria or protozoa, for example, digestion of cellulose or chitin. A few insects culture fungi for cellulose digestion. Rarely are cellulases and chitinases endogenous.

Fatty acids are common products of symbiotic digestion, for example, in ruminants. Behavioral adaptations of carnivores are methods of attack and feeding. Grazers feed more or less continually, most carnivores periodically.

Initiation and cessation of feeding are regulated neurally and hormonally. Fishes and mammals, and probably other vertebrates, have feeding control centers in the hypothalamus. Cholecystokinin regulates inhibition of feeding in vertebrates. Herbivorous larval insects feed until a size for molting is reached. In blowflies, stretch receptors in the foregut signal satiation. A leech may suck blood equivalent to several times its resting body volume in a single meal. In gastropod molluscs, a chain of neurons signal a sequence from chemosensation to food consumption and trituration.

Digestive enzymes are hydrolytic and correlate well with diet. Amylases that break down starch or glycogen are widely distributed. Oligosaccharidases provide for hydrolysis of disaccharides that result from breakdown of polysaccharides and are of several categories—α-glycosidases, β-glycosidases, and galactosidases. Cellulases are commonly made by symbionts, but a few wood-eating insects make one of the three necessary enzymes. Digestion by insects of several relatively insoluble plant polysaccharides such as lignin is by symbionts.

Hydrolyses of proteins are by endopeptidases and exopeptidases, enzymes with a long evolutionary history. Endopeptidases are identified according to pH optimum and specific amino acids adjacent to a peptide bond that is broken. Exopeptidases—aminopeptidases, carboxypeptidases, and dipeptidases attack short chains. Acid proteinases—pepsins or cathepsin—occur in several invertebrates, for example, *keratinase* of clothes moth. A unique alkaline proteinase has been found in crustaceans.

Digestion of fats and phospholipids is by highly specialized enzymes that attack specific parts of the molecules.

Absorption of sugars and amino acids is largely by active processes coupled with transport of sodium, rarely with protons. Carrier molecules for transport have been identified in absorptive epithelia. Absorption of products of fat digestion is usually by different routes from absorption of sugars and amino acids.

An important holistic adaptation of digestion is induction by food of digestive enzymes and absorption mechanisms. Secretion of digestive enzymes is regulated by kind of ingested food. Adaptive changes in synthesis can occur during days after a dietary change. In some vertebrates, hormones such as CCK may be one step in the induction sequence. Adaptations also occur in enzymes of absorptive transport and in the area of the absorptive surface. These commonly vary according to relative amounts of protein and carbohydrate in the diet.

Nutrition, feeding, digestion, and absorption are well coordinated. Neural, hormonal, and biochemical synthetic controls are adaptive to a wide variety of diets, feeding structures, and behavior.

References*

1. Allendorf, P. Kiel. Meeresforsch. 5:557–565, 1981. Absorption of dissolved amino acids in the nutrition of Benthic species.

1a. Alpers, D. H. In Physiology of Gastrointestinal Tract (L. Johnson, ed.), Vol. 2, Chap. 53, pp. 1469–1487. Raven Press, New York, 1987. Digestion and absorption of carbohydrates and proteins.

2. Ando, M. et al. Zool. Sci. 3:429–436, 1986. Regulation of HCl secretion in eel stomach.

*After the completion of this chapter, a useful book appeared: C. E. Stevens. Comparative Physiology of the Vertebrate Digestive System. Cambridge Univ. Press, London and New York, 1988.

3. Anika, S. et al. Physiol. Behav. 19:761–766, 1977. Satiety elicited by CCK in rats.

3a. Aoki, D. Proc. Natl. Acad. Sci. U.S.A. 83:2929–2933, 1986. Evolution of lactose absorption in man.

3b. Applebaum, S. W. in Comprehensive Insect Physiology, Biochemistry and Pharmacology, pp. 279–311. G. A. Kerkut and L. Gilbert, eds. Pergamon Press, Oxford. Biochemistry of digestion.

3c. Armstrong, J. R. and E. J. DeVillez. Can. J. Zool. 56:2225–2229, 1978. Low molecular weight proteinase from hepatopancreas of crustaceans.

3d. Azuma, M. and O. Yamashita. Tissue Cell 17:539–551, 1985. Cellular localization and function of midgut trehalase in silkworm larva.

4. Bailey, H. H. and D. Fairbairn. Comp. Biochem. Physiol. 26:819–836, 1986. Absorption of fatty acids and monoglycerides.

5. Bayne, B. L. and R. C. Newell. In The Molluscs Part 1, pp. 407–515. (K. Wilbur, ed.), Academic Press, Orlando, FL, 1984. Filter feeding by molluscs.

5a. Benjamin, P. R. and R. M. Rose. J. Exp. Biol. 80:93–118, 137–163, 1979. Central pattern in feeding cycle in Lymnaea.

6. Birbeck, T. H. and J. G. McHenry. Comp. Biochem. Physiol. B 77B:861–865, 1984. Chitinase in Mytilus.

7. Boeck, J. J. Comp. Physiol. 90:183–205, 1974; in Olfaction and Taste (D. Denton, ed.), Vol. 5, pp. 239–245. Academic Press, New York, 1975. Coding, odor quality, olfactory pathways, insects.

7a. Borgstrom, B. Biochim. Biophys. Acta 106:171–183, 1965. Johnston, J. M. and B. Borgstrom. Biochim. Biophys. Acta 84:412–423, 1964. Bile salt micelles; intestinal absorption of fats.

8. Bourdell, G. Am. J. Physiol. 244:G125–G130, 1983. Effect of protein and lipid diets on pancreatic adaptation in rat.

9. Bowdan, E. and V. Dethier. J. Comp. Physiol. 158:713–722, 1986. Dual inhibitory feeding in blowfly.

9a. Bradshaw, R. A., H. Neurath, R. W. Tye, K. A. Walsh, and W. P. Winter. Nature

(*London*) 226:237–239, 1970. Dogfish trypsinogen.

10. Buddington, R. K. *J. Comp. Physiol., B* 157:677–688, 1987. Does natural diet influence the intestine's ability to regulate glucose absorption in rainbow trout and carp.

11. Buddington, P. K. and J. M. Diamond. *Am. J. Physiol.* 252:G65–G76, 1987. Pyloric ceca of fish: A new absorptive organ.

12. Carey, H. V. *Holarctic Ecol.* 8:259–264, 1985; *Oikos* 44:273–279, 1985. Foraging and nutritional ecology of yellow-bellied marmots.

13. Chappell, L. H., C. Arme, and C. P. Read. *Biol. Bull. (Woods Hole, Mass.)* 136:313–326, 1969. Absorption of fatty acids by cestodes.

13a. Clifford, C. et al. *Comp. Biochem. Physiol. B* 71B:105–110, 1982. Digestive enzymes and subcellular localization of disaccharidases in some echinoderms.

13b. Crisp, E. H., M. Messer, and D. D. Shaughnessy. *Comp. Bioch. Physiol.* 90B:371–374, 1988. Intestinal lactase and other disaccharides of a suckling crab-eater seal (*Lobodon*).

14. Curtis, L. A. *Mar. Biol. (Berlin)* 59:137–140, 1986. Daily cycling of crystalline style of omnivorous estuarine snail.

14a. Dall, W. and D. J. W. Moriaty. In *Functional Aspects of Nutrition and Digestion of Crustacea*. Vol. 5, pp. 215–261. L. H. Mantel, ed. Academic Press, New York. 1983. Internal anatomy and functional regulation.

15. Davis, J. P. and G. C. Stephens. *J. Comp. Physiol., B* 152:27–33, 1983. Net flux of 14 amino acids in *Tetrahymena*.

16. Davis, J. P., G. Stephens, et al. *J. Comp. Physiol. B* 156:121–127, 1985. Na^+-dependent amino acid transport in bacteria-free sea urchin larvae.

17. Dean, R. C. *Biol. Bull. (Woods Hole, Mass.)* 155:297–316, 1978. Wood digestion by shipworm *Bankia*.

18. Demski, L. S. In *Brain Mechanisms of Behavior in Lower Vertebrates* (P. Laming, ed.), pp. 225–237. Cambridge Univ. Press, London and New York, 1980. Hypothalamic mechanisms of feeding in fishes.

19. Dethier, V. *The Hungry Fly*. Harvard Univ. Press, Cambridge, MA, 1976.

19a. Diamond, J. M. and W. M. Karasov. *J. Physiol. (London)* 349:419–440, 1984. Effect of diet carbohydrate on monosaccharide uptake by mouse intestine *in vitro*.

20. Diamond, J. M. et al. *Nature (London)* 320:62–63, 1986. Digestion and foraging in hummingbirds.

20a. Diamond, J. M. and W. Krasner. *Proc. Natl. Acad. Sci. U.S.A.* 84:2242–2245, 1987. Adaptive regulation of intestinal transporters.

21. D'Souza, M. P. et al. *Proc. Natl. Acad. Sci. U.S.A.* 84:6980–6984, 1987. Lysozomal proton pump.

22. Fange, R. and D. Grove. *Fish Physiol.* 3:161–260, 1979. Digestion in fishes.

23. Farrel, P. et al. *Gen. Comp. Endocrinol.* 65:363–372, 1987. Immunochemical and biological characterization of gastrin/cholecystokinin-like peptides in Palaemon.

23a. Ferreira, C., A. F. Ribeiro, E. S. Garcia, and W. R. Terra. *Insect Biochem.* 18:521–530, 1988. Digestive enzymes between double membranes of posterior midgut cells of *Rhodnius prolixers*.

23b. Ferreira, C. and W. Terra. *Biochem. J.* 213:43–51, 1983. Properties of plasma membrane β-glucosidase from midgut cells of fly Rhynchosciara.

23c. Florkin, M., F. Lorzet, and H. Sarlet. *Arch. Int. Physiol.* 57:71–88, 1949. Digestion of wax by *Galleria* larvae.

24. Galun, R. and S. H. Kindler. *J. Insect Physiol.* 14:1409–1421, 1968. Stimulation of feeding in tick *Ornithodoros*.

24a. Gildberg, A. *Comp. Bioch. Physiol.* 91B:425–435, 1988. Aspartic proteinases in fishes and aquatic invertebrates.

25. Gillette, M. W. and R. J. Gillette. *Neuroscience* 3:1791–1806, 1983. Bursting neurons command feeding behavior in *Pleurobranchaea*.

26. Girrdani, B. et al. *Biochim. Biophys. Acta* 692:81–88, 1982; *J. Exp. Biol.* 113:487–492, 1984. Amino acid absorption in lepideptron larvae and cockroach.

27. Glass, H. J. et al. *Comp. Biochem. Physiol. B* 86B:281–289, 1987. Carbohydrate and protein digestion in halibut.

27a. Goda, T. et al. *Am. J. Physiol.* 245:G418–G423, 1983. Dietary induced decrease in microvillar carbohydrase in rat jejunum.

28. Gomme, J. *J. Membr. Biol.* 62:29–46, 47–52, 1981. D-glucose transport across apical membranes of epithelium in *Nereis*.

29. Gomme, J. *Am. Zool.* 22:691–708, 1982. Epidermal nutrient absorptions in marine invertebrates.

30. Gomme, J. and S. Albrechtsen. *Biochim. Biophys. Acta* 770:47–54, 1984. Specificity of D-glucose transport by apical membrane of *Nereis* epidermis.

31. Grant, G. et al. *Biochemistry* 19:4653–4659, 1980. Amino acid sequence of collagenolytic protease from hepatopancreas of *Uca*.

32. Grossman, M., H. Greengard, and A. Ivy. *Am. J. Physiol.* 138:676–682, 1944. Effect of dietary composition on pancreatic enzymes.

33. Grossman, S. P. *Psychol. Rev.* 82:200–224, 1975. Role of hypothalamus in regulation of food and water intake.

34. Harris, R. P. et al. *Mar. Biol. (Berlin)* 90:353–361, 1986. Effects of algal diet on enzymes of *Calanus*.

34a. Hickey, D. A. and B. F. Binkel. *Critical Rev. Biotech.* 5:229–241, 1987. Regulation of amylase gene expression in *Drosophila*.

35. Hipp, E. et al. *J. Exp. Zool.* 240:289–297, 1986. Integumentary uptake of acetate and propionate by *Tubifex*.

36. Hoeger, U. and T. P. Mommsen. *Mar. Biol. (Berlin)* 81:123–130, 1984. Hydrolytic enzymes in 2 ctenophores: *Pleurobranchia* and *Beroe*.

36a. Hokin, L. E. *Ann. N. Y. Acad. Sci.* 242:12–13, 1974; *Arch. Biochem. Biophys.* 151:453–463, 1972; *Exp. Zool.* 194:197–206, 1975. Energetics of sodium pump.

36b. Hogan, M. E., S. W. Schulz, M. Slator,

T. Czolij, and R. W. O'Brien. *Insect Biochem.* 18:45–51, 1988. Cellulases from termite *Coptotermist lacteus*.

36c. Hopper, U. In *Physiology of Gastrointestinal Tract* (L. Johnson, ed.), Vol. 2, Chap. 55, pp. 1499–1526. Raven Press, New York, 1987. Membrane transport mechanisms for hexoses and amino acids.

36d. Hu, C., M. Spiess, and G. Semenza. *Biochim. Biophys. Acta* 896:275–286, 1987. Mode of anchoring of sucrase-isomaltase and maltase-glucoamylase in chicken intestinal brush border membrane.

37. Isselbacher, K. J. *Gastroenterology* 50:78–82, 1966. Biochemistry of fat absorption.

37a. Jany, K. D., K. Bekelar, B. Pfeiderer, and J. Ishay. *Biochem. Biophys. Res. Commun.* 110:1–7, 1983. Amino acid sequence of an insect chymotripsin.

38. Jeuniaux, C. *Bull. Soc. Zool. Fr.* 107:363–368, 1982. Chitin in animals.

39. Johnston, J. M. In *Handbook of Physiology* (C. F. Code, ed.), Sect. 6, Vol. III, pp. 1353–1375. Physiol. Soc., Washington, DC, 1968. Lipid absorption.

40. Karasov, W. and J. Diamond. *Am. J. Physiol.* 245:G443–G462, 1983; 249:G770–G785, 1985. Adaptive regulation of sugar and amino acid transport by vertebrate intestine.

41. Karasov, W. H. et al. *Proc. Natl. Acad. Sci. U.S.A.* 80:7674–7677, 1983. Regulation of proline and glucose transport in mouse intestine by dietary substrates.

42. Kater, S. B. et al. *Neurophysiology* 36:142–155, 1973; *Brain Res.* 146:1–21, 1978; 159:331–349, 1980; *J. Neurobiol.* 11:73–102, 1980. Central program for feeding in snail *Helisoma*.

42a. Kay, D. G. *Comp. Biochem. Physiol. A* 47A:573–582, 1974. Distribution of digestive enzymes in gut of polychaete *Neanthes*.

43. Klinger, T. *Comp. Biochem. Physiol. A* 78A:597–600, 1984. Activities of alpha and beta glucosidases and beta galactosidases in echinoids.

43a. Korringa, P. *Q. Rev. Biol.* 27:266–308; 339–365, 1952. Filter feeding in oysters.

43b. Kraus, E., R. Zwilling, et al. *Anal. Biochem.* 119:153–157, 1982. Protease of *Astacus* as aid in protein sequencing.

43c. Krogh, A. *Biol. Rev. Cambridge Philos. Soc.* 6:412–442, 1931. Utilization of dissolved foods by aquatic animals.

44. Langton, R. W. and P. A. Gabbott. *Mar. Biol. (Berlin)* 24:181–187, 1974. Tidal rhythm of extracellular digestion in *Ostrea.*

44a. Larner, J. et al. *J. Biol. Chem.* 212:9, 1955; 215:723–736, 1955; 223:709–725, 1956; *Arch. Biochem. Biophys.* 58:252–257, 1955; *Biochim. Biophys. Acta* 20:53–61, 1956. Intestinal enzymes for digesting starch and oligosaccharides.

44b. Lawrence, J. In *Echinodern Nutrition* (M. Jangoux and J. Lawrence, eds.), pp. 283–316. Balkem, Rotterdam, 1982. Digestion.

44c. Lent, *J. Exp. Biol.* 131:1–15, 1987. Feeding behavior and electrophysiology in Leeches.

45. Lichtenberger, L. et al. *Am. J. Physiol.* 243:G341–G347, 1982. Dietary amines induce gastrin release.

46. Manahan, D., J. Davis, and G. C. Stephens. *Science* 220P:204–206, 1983. Axenic cultures sea urchin larval—selective uptake of neutral amino acids.

47. Manahan, D., T. S. Wright, and G. C. Stephens. *Am. J. Physiol.* 244R:832–838, 1983. Net uptake of 16 amino acids by Mytilus.

47a. Kukor, J. J., D. P. Cowan, and M. M. Martin. *Physiol. Zool.* 61:364–371, 1988. Fungal enzymes used in cellulose digestion in larvae of cerambycid beetles.

48. Mayastch, S. M. and R. A. Smucker. *Micro. Ecol.* 14:157–166, 1987. Role of bacteria in chitinase and chitobase activities in crystalline style of *Crassostrea virginica.*

48a. Meredith, J. and J. Phillips. *J. Exp. Biol.* 137:341–360, 1988. Na-independent proline transport in locust rectum.

49. Metcalf, R. L. et al. *Proc. Natl. Acad. Sci. U.S.A.* 77:3769–3772; 78:4007–4010, 1981; *Bull. Entomol. Soc. Am.* 25:30–35, 1979. Molecular parameters and olfaction in fruit fly; secondary plant substances and food selection by beetles.

49a. Michel, C. and E. J. Devillez. In *Physiology of annelids*, pp. 509–553. P. J. Mill, ed. Academic Press, New York, 1978. Digestion.

50. Morgan, M. R. *J. Insect Biochem.* 5:609–617, 1975. Relation between gut cellulobiose, lactase, amyl beta-glucosidase and amyl beta-galactosidase in *Locusta mionatoria.*

51. Morton, B. S. *J. Exp. Mar. Biol. Ecol.* 26:135–151, 1977. Tidal rhythm of feeding and digestion in *Crassostrea gigas.*

52. Morton, B. S. In *The Mollusca* (K. Wilbur, ed.), Vol. 5, Part 2, pp. 65–147. Academic Press, New York, 1983. Feeding and digestion in bivalves.

52a. Muscatine, L. and C. Hand. *Proc. Natl. Acad. Sci. U.S.A.* 44:1259–1263, 1958. Transfer of carbon compounds from symbiotic algae to tissues of a coelenterate.

53. Muscatine, L. et al. *Proc. R. Soc. London, Ser. B* 222:181–202, 1984. Fate of photosynthetically fixed carbon in colonies of larvae.

54. Nayar, J. K. and G. Fraenkel. *J. Insect Physiol.* 8:505–525, 1962. Host plant selection in silk moth *Bombyx mori.*

55. Parenti, P. et al. *J. Comp. Physiol. B* B156:549–556, 1986. Na dependent absorption of alanine in midgut of cockroach *Blabera.*

56. Pasero, L. et al. *Biochim. Biophys. Acta* 869:147–157, 1986. Amino acid sequences in porcine alpha-amylase.

56a. Place, A. and D. Roby. *J. Exp. Zool.* 338:29–41, 1986; 340:149–161, 1986. Assimilation and deposition of wax esters in planktivorous seabirds.

56b. Prosser, C. L. and F. A. Brown, Jr. *Comparative Animal Physiology*, 2nd ed. Saunders, Philadelphia, PA, 1961.

57. Rabin, B. *Brain Res.* 43:317–342, 1972. Ventromedial hypothalamic control of food intake.

57a. Reeck, G. R., W. P. Winter, and H. Neurath. *Biochemistry* 11:503, 1972. Trypsinogen from African lungfishes.

58. Reisenauer, A. and G. Gray. *Science* 227:70–72, 1985. Induction of amino-oligo- peptidase in rat intestine wall.

59. Santos, C. and W. Terra. *Biochim. Biophys. Acta* 831:179–185, 1985. Properties of beta-D-glucosidase (cellobiase) from midgut cells of hornworm cellobiose.

60. Schick, J. et al. *Am. J. Physiol.* 247:G611–G616, 1984. Adaptive responses of exocrine pancreatic enzymes to inverse changes in protein and CHO in the diet.

61. Selverston, A. I. et al. *J. Comp. Physiol.* 91:33–51, 1974; *Prog. Neurobiol.* 7:215–290, 1976; in *Identified Neurons of Arthropods* (G. Hoyle, ed.), pp. 209–225. Plenum, New York, 1977; In *Simpler Networks* (J. Fentress, ed.), pp. 82–98. Sinauer, Sunderland, MA, 1979; *J. Comp. Physiol.* 129:5–17, 1979; *Behav. Brain Sci.* 3:535–575, 1980; *J. Neurophysiol.* 44:1102–1121, 1980; 48:1416–1432, 1982; *J. Comp. Physiol.* 145:191–207, 1981. Network analysis of patterned discharge in somatogastric ganglion of Panulirus.

61a. Semenza, G. *Annu. Rev. Cell Biol.* 2:255–314, 1986. Anchoring and synthesis of stakled brush border membrane proteins, especially oligosaccharidases.

62. Sibly, R. M. In *Physiological, Ecological, and Evolutionary Approach to Resource Use* (C. R. Townsend and P. Calow, eds.), pp. 109–139. Sinauer, Sunderland, MA, 1981. Digestive systems length and efficiency in vertebrates.

63. Smith, R. H. and A. B. McAllan. *Br. J. Nutr.* 25:181–190, 1971. Nucleic acid metabolism in ruminants.

64. Steen, R. G. *J. Exp. Zool.* 240:315–326, 1986. *In vivo* metabolite measures in sea anemone.

65. Stephens, G. C. et al. *Am. J. Physiol.* 244:R832, 1983. Axenic culture of sea-urchin larvae.

66. Stephens, G. C. In *Transport Processes* (R. Gilles and M. Gilles-Baillion, eds.), pp. 280–291. Springer-Verlag, Berlin and New York, 1985.

66a. Stevens, B. R., J. Kaunitz, and E. Wright. *Annu. Rev. Physiol.* 46:417–433, 1984. Intestinal transport of amino acids and sugars.

67. Stryer, L. *Biochemistry*, 3rd ed. Freeman 1988.

67a. Tammes, P. M. L. and A. D. G. Dral. *Arch. Neerl. Zool.* 11:87–112, 1955. Straining of suspensions by mussels.

67b. Terra, W. R., A. Valentia, and C. D. Santos. *Insect Biochem.* 17:1143–1147, 1987. Utilization of sugars, hemicellulose, starch, protein, fats, and minerals by insect *Eninnyis* larvae.

68. Titani, K., H. Neurath, et al. *Biochemistry* 26:222–226, 1987. Amino acid sequence of protease from crayfish *Astacus*.

69. Tsai, I. et al. *Comp. Biochem. Physiol. B* 85B:235–239, 1986. Chymotrypsins in digestive tracts of shrimps.

69a. Tseng, H. C. and S. Rothan. *Am. J. Physiol.* 243:G304–G312, 1982. Stimulation of pancreatic secretion of amylase and trypsinogen by injected CCK-Pz.

70. Tsuboi, K. et al. *Am. J. Physiol.* 249:G510–518, 1985. Elevated intestinal carbohydrase activities after high CHO diet feeding in rats.

70a. Tyler, M. J. et al. *Science* 220:609–610, 1983. Inhibition of gastric acid secretion in the gastric brooding frog *Rheobatrachus*.

71. Van Praeb, M. *Adv. Mar. Biol.* 22:65–99, 1985. Nutrition of Sea Anemones. Food evokes ciliary reversal in throat and lip; pre-feeding movement of tentacles and column.

71a. Van Wormhoudt, A. and D. Favrel. *Comp. Bioch. Physiol.* 89B:201–207, 1988. Alpha amylase in *Palaemon elegans*; polymorphism during moult cycle.

71b. Vonk, H. J. and J. R. H. Western. *Comparative Biochemistry and Physiology of Enzymatic Digestion*. Academic Press, Orlando, FL, 1984.

71c. Ward, G. W. *Austr. J. Biol. Sci.* 28:1–23, 1975. Proteinases in keratinolytic larvae of clothes moth.

72. Widdows, J. and J. M. Shick. *Mar. Biol. (Berlin)* 85:217–232, 1985. Physiological respiration of *Mytilus* and *Cardium* to air exposure.

72a. Wilcox, P. C. *Methods Enzymol.* 19:64, 1970. Chymotrypsins.

73. Wright, J. K. et al. *Annu. Rev. Biochem.* 55:225–248, 1986. Molecular aspects of sugar:ion cotransport.

74. Wright, S. H. and G. C. Stephens. *J. Exp. Zool.* 205:337–352, 1978. Removal of amino acid during single pass across gill of Mytilus.

74a. Yardley, D. G. *J. Comp. Physiol. B.* 159:237–241, 1989. Amylase from mosquitofish, rat, and *Drosophila*.

74b. Yonge, C. M. *Sci. Rep. Great Barrier Reef Exped., Br. Mus.* 1:59–91, 1930; 283–321, 1936; *Biol. Rev. Cambridge Philos. Soc.* 19:68–80; *Mem.—Geol. Soc. Am.* 67:429–442, 1957. Symbiotic relations between zooxanthellae, corals, and *Tridacna*.

75. Zamer, W. E. *Mar. Biol. (Berlin)* 1986. Physiological energetics of intertidal *Anthopleura*.

75a. Zinkler, D. and M. Göetze. *Comp. Bioch. Physiol.* 88B:661–666, 1987. Cellulose digestion in firebrat *Thermobia domestica*.

76. Zwilling, R. and H. Neurath. *Methods Enzymol.* 80:533–564, 1981. Invertebrate proteases.

77. Zwilling, R. et al. *FEBS Lett.* 127:75–78, 1981. Low-molecular weight proteinase from crustacea.

78. Gillette, R. In *Neural and Integrative Animal Physiology* (C. Ladd Prosser, ed.). pp. 574–611. Wiley-Liss, Inc., New York, 1991.

79. Goldsmith, T. H. In *Neural and Integrative Animal Physiology* (C. Ladd Prosser, ed.). pp. 171–245. Wiley-Liss, Inc., New York, 1991.

Chapter 6 | Nutrition

James G. Morris

All living organisms from the simplest unicellular prokaryote to the most complex multicellular mammals, require nutrients or substances that provide nourishment for synthesis of the many compounds necessary for maintenance, growth, and reproduction. Green plants and some green flagellates have the simplest nutrient requirements: a source of energy (light), carbon dioxide, water, and the salts of inorganic elements. These organisms, which use light as a source of energy, are classified as *phototrophs.* Another group of organisms (largely bacteria), which also have simple nutrient requirements, utilize energy released from the oxidation of inorganic compounds such as iron or sulfur. These organisms are classified as *chemotrophs.* Phototrophs and chemotrophs are often collectively referred to as autotrophs to distinguish them from heterotrophic organisms. The majority of animals are *heterotrophs* and obtain energy from the controlled oxidation of organic compounds of varying complexity. These or-

ganisms generally have to be supplied with a number of organic compounds in addition to most of the inorganic elements needed by plants. It appears that during evolution phototrophic organisms became independent of exogenous organic nutrients while heterotrophic organisms remained reliant on many preformed organic compounds. As the great preponderance of animals are heterotrophic, and as these animals require many nutrients in the diet, the major emphasis of this chapter will be on heterotrophic animals.

Most metabolic pathways are common to heterotrophic animals; therefore, organic compounds that are essential in the diet of an individual animal are frequently required by all animals. A need for a nutrient by one organism but not by another generally reflects some modification of a metabolic pathway associated with specialization of that organism. The essential organic compounds for heterotrophic animals can be categorized as belonging to one of four classes of nutrients;

in order of magnitude in the diet they are substances that provide energy, nitrogenous substances, minerals, and vitamins. A fifth class of substances, compounds that stimulate feeding is discussed in Chapter 4 of the companion to this book, *Neural and Integrative Animal Physiology.*

1. *Substances that Provide Energy.* Energy per se is not a nutrient, but compounds that yield available energy by molecular rearrangement (glycolysis) or oxidation are nutrients for animals. Included in this class of compounds are those that on oxidation yield energy, which can be used for maintenance of cellular processes and for the synthesis of new tissue. While the energy released from the oxidation of inorganic compounds such as elemental sulfur can be used by some bacteria, the main compounds supplying energy for animals are organic compounds, particularly carbohydrates and fats. These compounds constitute the bulk of the diet of most mature heterotrophic animals, and represent often in excess of 80% of the dry matter consumed. The daily requirement, in terms of dry matter, for this class of nutrient is usually at about 1.0% of the body mass for large animals and may exceed 10% of body mass in small homeotherms.

2. *Nitrogenous Substances.* Animals are unable to use molecular nitrogen but require fixed nitrogen in compounds such as amino acids; and, for some animals other nitrogenous compounds such as purines are required. The aggregate requirement for these nutrients is of the order of 5 to 40% of the dietary dry matter depending on the animal species and physiological state; highest requirements occur during periods of active growth and reproduction. The requirement for specific nitrogenous compounds (e.g., individual amino acids) rarely exceeds 2% of the dietary dry matter intake.

3. *Minerals.* Mineral elements are es-

sential nutrients for all plants and animals. While qualitatively there is a high degree of uniformity in regard to the essential minerals required by all living organisms, the quantitative requirements are variable. For most animals the requirement of all minerals rarely exceeds 5% of the dry matter in the diet.

4. *Vitamins.* This class of nutrients includes the recognized vitamins and other micronutrients of organic origin such as essential fatty acids, which cannot be synthesized by animals. The requirement for essential fatty acids is generally of the order of 1 to 2% of the dietary dry matter, whereas the aggregate requirement for all the vitamins is no greater than 1% of the dietary dry matter.

5. *Phagostimulatory Compounds.* The recognition of food is essential for the survival of all animals. Animals with complex nervous systems may employ visual, tactile, and chemical (olfactory and gustatory) cues in food recognition. Animals without nervous systems rely on contact chemoreceptors to locate food. For example, protozoa exhibit orientation behavior toward food or away from noxious substances. Larger animals with relatively simple nervous systems also respond to chemical entities. In the terrestrial slug *Ariolimax californicus* starch is a potent feeding stimulant (91). A large number of compounds, particularly sugars, various amino acids, ascorbic acid, thiamin, betaine, oxaloacetic and citric acids, have been shown to stimulate feeding in phytophagous insects (32). Many of the phagostimulatory compounds are nutrients but are not necessarily essential for the metabolism, or for the survival of the animal; one stimulatory compound can be substituted for another. In contrast to phagostimulatory compounds, plants elaborate secondary compounds (allelochemics), which are either toxic or aversive to animals and discourage herbivory by them.

Nutrients that are essential at the cellular level across species are remarkably similar despite the diversity of foods that support animal life. However, the essentiality of the various nutrients in the diet can be modified by the anatomy and physiology of the gastrointestinal tract. Symbiotic microflora often have the capacity to digest organic compounds that are resistant to the host's digestive enzymes and produce end-products of digestion that are assimilated and utilized by the host. In addition to facilitating digestion, these microbes may synthesize essential micronutrients, which make the host animal independent of the dietary sources of these nutrients. Because of the contribution of gut microbes to the nutrition of the host, the demonstration of essentiality of a nutrient may require animals to be grown under axenic conditions.

Origin of the Nutritive Types

While organisms derive utilizable energy from various sources, a *common currency* of most energy transfers is adenosine triphosphate (ATP); this concept was originally proposed by Lipmann (61a) and is one of the major unifying themes of biology. ATP is the link between exergonic or energy yielding reactions and numerous energy requiring (or endergonic) reactions. Various proposals have been made to classify organisms in relation to energy sources that they can utilize and their ability to synthesize essential metabolites. Unfortunately, none of these schemes is universally accepted, especially among microbiologists.

Substances That Provide Energy to Animals

Plants use carbohydrates as storage substances that can be oxidized to provide energy for metabolic processes. These storage carbohydrates are generally sugars and starches that can be used as sources of energy by most animals. Plants also use carbohydrates (cellulose and hemicellulose) as structural components; these carbohydrates are resistant to enzymes secreted by animals but can be degraded by anaerobic bacteria. Oxidation of carbohydrates provides the main source of energy for all herbivorous and many omnivorous animals. A large number of organic compounds, including carbohydrates ranging from simple sugars to complex polysaccharides, fatty acids of variable chain length, alcohols, and amino acids, are metabolized by animals in a series of controlled reactions. These reactions are coupled to permit the capture of some of the chemical energy released in the oxidation reactions.

Virtually every organic compound, which is biologically synthesized, is capable of being degraded by some species of bacteria (49); in contrast, the range of organic compounds that can be digested and assimilated by animals is limited. Some potential sources of chemical energy such as long-chain paraffins cannot be digested, whereas others, such as hemicellulose and cellulose, can only be digested if the animal has symbiotic bacteria. The inability of animals to degrade complex organic molecules into simpler units is frequently a barrier against utilization of a potential food as a source of energy.

Compounds that are oxidized for energy for the general metabolic pool may also fulfill other roles. For example, the oxidation of amino acids provides energy, but amino acids are also the basic building blocks of proteins. Similarly, in a diet where fat is the main substrate oxidized, and protein is restricted to the minimal level to provide nitrogen and essential amino acids, there may be a lack of glucose precursors. A response in growth to the addition of carbohydrate to the diet may follow, as glucose fulfills specific roles in addition to that of an oxidizable substrate for energy.

Protista have been classified into three trophic categories: the photoautotrophs that harness radiant energy by photosynthesis; the photoheterotrophs which, although phototrophic in energy requirements, are unable to use carbon dioxide for cell synthesis and require organic compounds; and lastly the chemoheterotrophs that require chemical energy and organic carbon sources (77). Many of the photoautotrophic flagellates, largely members of the Euglenida, Cryptomonadida, and Volvocida combine autotrophy with heterotrophy in varying degrees, and are often described as the acetate flagellates their preferred carbon source being acetate, simple fatty acids, and alcohols. These flagellates are able to shift from autotrophy in light to heterotrophy in the dark. Within the same genus some species (e.g., *Euglena pisciformis*) cannot be maintained in the dark on organic media and are entirely dependent on autotrophy. Others (e.g., *E. gracilis*) can use organic media for energy when deprived of illumination (61). The majority of free-living protista are heterotrophic, exploiting a range of diets. Some are herbivores and feed on algae, generally of the unicellular variety, some feed on bacteria, whereas others consume microfauna such as rotifers, gastrotrichs, and small crustaceans (61).

Most higher heterotrophs, and many protista, require compounds with three or more carbon atoms, either in the form of sugars or complex carbohydrates. Trypanosomatids have a basically aerobic metabolism and utilize glucose, fructose, and mannose while a few use galactose, sucrose, and raffinose (77). Trichomonadida have a metabolism similar to trypanosomatids in that the hexose monosaccharides, glucose, fructose, and galactose, along with the disaccharide maltose, are utilized. Some species can utilize mannose and sucrose while none uses pentose sugars.

Symbiotic protozoa that inhabit the forestomach of ruminant animals may achieve concentrations as high as 10^6 organisms/mL of rumen fluid. These protozoa are anaerobes and catabolize all the major plant components: soluble sugars, starch, pectin, and complex carbohydrates such as hemicellulose and cellulose, and lipid (15, 16, 51).

Carbohydrates are the major source of energy for helminths, and glycogen is the main carbohydrate reserve. Appreciable quantities of trehalose also occur in parasitic nematodes and acanthocephalans. Parasitic helminths contain lipid, but there is no good evidence that any adult helminth can catabolize lipid (5). *Ascaris lumbricoides* catabolize glucose by glycolysis to phosphoenolpyruvate, which undergoes carbon dioxide fixation to give oxaloacetate, which in turn is reduced to malate. The malate then undergoes a dismutation reaction resulting in the formation of pyruvate and succinate. These products are metabolized to volatile fatty acids. For adult helminths, which fix carbon dioxide (e.g., *Fasciola hepatica*, *Ascaris lumbricoides* and *Moniliformis dubius*), the tricarboxylic cycle is not involved in carbohydrate metabolism. In other helminths such as the plerocercoides of *Schistocephalus solidus* and adult *Haemonchus contortus* there is evidence of a functional tricarboxylic cycle under aerobic conditions that may metabolize up to 75% of the carbohydrate. Functional tricarboxylic acid cycles occur in the developing egg of *A. lumbricoides* and the infective larvae of *H. contortus*. The free-living miracidia and cercariae of trematodes appear to be aerobic in their metabolism (5).

Marine mollusca produce carbohydrases, which degrade the reserve polysaccharides amylose, glycogen, and laminaran (64). Few species contain carbohydrases that attack dextrin or structural carbohydrates with the exception of the polysaccharide chitin which also contains the amino sugar glucosamine (59). Carbohydrases including α-glucosidase, β-glucosidase, laminaranase, and a chitobiase have been isolated from the digestive gland of the scallop *Placopecten magellanicus*. An α-amylase occurs in high concentration in the crystalline style of the same species (106). α-Amylase, cellobiase, and cellulase have been isolated from the stomachs of bivalves indicating that substrates for these enzymes are probably digested by bivalves (80). Some mollusca are capable of digesting native cellulose, but the origins of the cellulases have not been defined in most cases. Both aquatic and terrestrial gastropods contain symbiotic microflora, which may extend the range of carbohydrates capable of digestion beyond those for which carbohydrases are elaborated by the host.

For the great majority of insects studied, carbohydrate is essential or at least required for optimal growth (26). Dietary carbohydrate requirements of some insects are not constant and vary with the stage of development. The sugars utilized best are those commonly found in plant and animal tissues: the hexose monosaccharides glucose and fructose and the disaccharide sucrose. The majority of insects studied utilize with varying efficiency, the hexoses galactose and mannose; the disaccharides maltose, trehalose, and melibiose; and the trisaccharides raffinose and melezitose. The disaccharides lactose and cellobiose as well as the pentose sugars ribose, cylose, rhamnose, and arabinose are generally without nutritive value to insects. The polysaccharides starch and dextrins (α-1,4- and α-1,6-linked glucose polymers) are well utilized by most cereal-feeding insects that live on stored products. However, cellulose (β-1,4-linked glucose polymer) is utilized only by those insects with cellulase-producing symbiotes.

Artificial diets, varying in composition from no carbohydrate to more than 80% carbohydrate have been used for the rearing of insects. For some insects, such as the waxmoth and hidebeetle larvae, carbohydrate can be substituted in the diet for lipid without compromising growth. However, lepidoptera, hymenoptera, and a few species of locusts, have an absolute requirement for carbohydrate (26). The addition of >10% glucose to the diet of shrimp markedly inhibits growth. However, in contrast to glucose, the disaccharides sucrose, maltose, and trehalose and polysaccharides such as dextrin and starch have high nutritive value (55). The reason for this apparent difference between crustaceans and insects probably relates to the high rate of absorption of glucose from the stomach of penaeids that leads to inefficient utilization (55).

Fish can grow on diets devoid of carbohydrate (23). All fish studied, including carnivorous fish, have some ability to digest and utilize hexoses (6). Amylase and maltase are the two carbohydrate-digesting enzymes that occur at the highest activity in the intestine of common carp. Also present at lower activities are sucrase and lactase (74). Omnivorous fish such as carp, channel catfish, and red sea bream utilize higher levels of dietary carbohydrate than carnivorous fish such as yellowtail and salmonids (74).

Adult amphibians (frogs, toads, newts, salamanders, and the limbless amphibians) are mainly carnivorous. However, the immature stages are omnivorous, consuming diverse sources of food including bacteria, algae, and plant material as well as practicing cannibalism (2).

Absorbed carbohydrates are utilized by reptiles. In alligators, injected fructose disappears at the fastest rate. The rates of disappearance (in decreasing order) of other sugars are ribose, mannose, arabinose, talose, xylose, and glucose (21). Alligators, caimans, and at least one turtle (*Pseudemys scripta elegans*) are unable to hydrolyze lactose, maltose, or starch (66).

Birds utilize the common monosaccharides glucose and fructose, as well as the disaccharide sucrose. Lactose is not digested by birds as they lack the enzyme lactase. Starch is digested by most birds, including most carnivorous birds.

All mammals utilize dietary glucose. Felids have a limited ability to catabolize fructose, presumably because of a relative deficiency of fructokinase. Lactose is digested by all young mammals whose mother's milk contains lactose. The ability to digest lactose is generally greatly diminished as the animal matures. The milk of many sea mammals (*Pinnipedia* and *Cetacea*) is low in lactose and the milk from some (e.g., California sea lion, *Otariidae*) is virtually devoid of lactose; the offspring of these mammals have no lactase activity in the brush border of the intestinal mucosa. In contrast to the above mammals, the milk of the great blue whale (*Balaenopteridae*) contains a high concentration of lactose. Sucrose and starch are poorly digested by the young ruminant before development of symbiotic microflora. Cellulose and hemicellulose are digested only by those mammals with symbiotic cellulase-producing bacteria.

Nitrogenous Substances

A few heterotrophic microorganisms (such as *Rhizoblum*) can reduce (fix) atmospheric nitrogen using energy derived from exogenous sources, but the vast majority of plants, and all animals, require nitrogen to be present in a combined state. Autotrophic organisms (most photosynthetic algae and higher plants) use nitrogen in the form of either nitrate or ammonium nitrogen. Organisms that are heterotrophic in respect to energy needs require many amino acids either as free amino acids, or as oligo- or polypeptides. Mesotrophic organisms represent a group between autotrophs and heterotrophs and can exist on a single amino acid or ammonia plus fatty acids.

For higher animals, the most common form of dietary nitrogen is amino nitrogen, present as amino acids in proteins. All animals except green flagellate protista are ultimately dependent on photoautotrophic plants for these amino acids, just as plants are the ultimate source of all carbon compounds that are oxidized by animals to provide chemical energy.

Proteins and Amino Acids

Proteins are major constituents of the bodies of all multicellular animals. All animal proteins are combinations of 20 α-amino acids, the *standard* amino acids, that are coded for in protein synthesis (102). About 120 other amino acids occur by post-translational modification of amino acids. The amino acid selenocysteine, in glutathione peroxidase, has recently been shown to be specified by the TGA codon, which is usually a termination codon (13). Accordingly, selenocysteine has been proposed as the 21st standard amino acid. The proportion and sequence of amino acids in the protein provides the basis for individual differences between proteins. Animal proteins, with few exceptions (hair, nails, hooves, and skin), are in a dynamic state and are continually being degraded and resynthesized or "turned over". Some of the body proteins (e.g., scleroproteins—collagen and elastin) turn over slowly, whereas others (e.g., enzymes, proteins of the gut wall and liver) turn over rapidly. Adult animals require a regular supply of amino acids to replace those that cannot be recovered from degraded tissue proteins for the synthesis of new replacement protein. In growing animals, amino acids, in addition to those required for protein turnover, are needed for the net synthesis of protein associated with the accretion of new tissue.

Photoautotropic plants synthesize all 20 amino acids from simple nitrogenous compounds. Animals are unable to synthesize about 10 of these amino acids from constituents normally present in the diet at a rate commensurate with optimal performance. These 10 amino acids, which cannot be synthesized, are known as *essential* amino acids. The other dietary amino acids are designated as *dispensable* amino acids, as they can be synthesized by the body (38). Sometimes the term "nonessential" is used rather than dispensable; but this terminology is not strictly correct because dispensable amino acids are required in the diet to supply nitrogen. A diet that contains only the minimal level of essential amino acids will result in poor performance of the animal as the diet lacks nitrogen for the synthesis of the dispensable amino acids and other compounds such as purines, pyrimidines, and so on. The dietary essential

amino acids for animals have been identified by either the systematic omission of each amino acid from a diet adequate in all other nutrients, or by measuring whether ^{14}C from glucose or other compounds such as acetate can be incorporated into the particular amino acid. For mammals, and probably other animals, essential amino acids can be divided into two groups: those in which the entire amino acid has to be present in the diet (lysine and threonine), and those for which only the carbon skeleton is essential. For these latter amino acids the corresponding α-keto acid can substitute, at least partially, for the essential amino acid. As α-keto acids of the essential amino acids are not significant dietary components, these amino acids are designated as essential.

The concept of essential amino acids was first developed from studies on amino acid requirements of rats, and consequently the term "essential amino acids for the rat" is often used to designate these amino acids. The 10 essential amino acids required in the diet of the growing rat (10 EAA-rat) are arginine, histidine,

Examples of α-Keto Acids that Can Substitute for Essential α-Amino Acids

Essential Amino Acid	α-Keto Acid	Remarks
L-Arginine	2-Keto-5-guanidinovalerate	Citrulline will substitute for arginine
L-Histidine	Imidazolpyruvate	
L-Methionine	2-Keto-4-methiolbutyrate	
L-Phenylalanine	Phenylpyruvate	Normally a trivial route of metabolism
L-Tryptophan	Indolepyruvate	
L-Leucine	2-Ketoisocaproic	As the branched-chain amino acid transaminases occur mainly in muscle (not liver), this leads to less efficient use of these α-keto acids for amino acid synthesis.
L-Isoleucine	2-Keto-3-methylvalerate	
L-Valine	2-Ketoisovalerate	
L-Lysine	No α-keto acid	These two amino acids cannot be substituted for by the corresponding α-keto acid and must be supplied per se.
L-Threonine	No α-keto acid	

isoleucine, leucine, lysine, methionine, phenylalanine, threonine, tryptophan, and valine. The requirement for two of these essential amino acids, methionine and phenylalanine, can be reduced to about one half if the diet contains adequate levels of the dispensable amino acids cystine and tyrosine, respectively. These amino acids spare methionine and phenylalanine, because they can be derived from these amino acids and are constituents of proteins and other substances. When cystine and tyrosine are absent from the diet the requirement for methionine and phenylalanine is increased to supply the former amino acids. The dietary requirement for arginine can also be reduced in some animals if proline is present in the diet because a common intermediate is shared by both amino acids.

Some amino acids are required only by the young to achieve maximal growth. For example, the growth rate of very young rats can be stimulated by the addition of asparagine to the diet, whereas older rats do not show a growth response to asparagine.

Excessive amounts of a single amino acid in the diet can produce an amino acid toxicity. For mammals, methionine, tryptophan, and tyrosine are among the most toxic amino acids. Another condition known as amino acid imbalance is produced when the diet is limiting in one essential amino acid and excessive levels of other amino acids are present. The imbalance can be relieved by increasing the level of the limiting amino acid in the diet, or reducing the level of those in excess. Amino acid imbalances are readily demonstrated with the branched chain amino acids, leucine, isoleucine, and valine, which are metabolized in mammals in their peripheral tissues, by a common enzyme (39).

Nineteen of the 20 amino acids that commonly occur in animal proteins are L-amino acids (the exception is glycine, which does not have an asymmetric carbon atom). D-amino acids are produced by bacteria and some insects. Worms and marine invertebrates may contain significant quantities of free D-amino acids. These D-amino acids can substitute for L-amino acids in the diet to a variable degree, depending on the amino acid and animal species. Some D-amino acids cannot be utilized by mammals, for example, D-lysine, D-threonine, whereas others, such as D-methionine are well utilized by most species. The activity of the enzyme D-amino acid oxidase in the animal is one of the main factors determining whether a D-amino acid can be utilized.

Essential Amino Acid Requirements of Animals

No cases of nitrogen fixation are known in marine cryptomonad and chrysomonad phytoplankton, but these organisms are able to utilize inorganic nitrogen sources (ammonia, nitrite, and nitrate) for the synthesis of all 20 amino acids. Simple organic nitrogenous sources can be utilized by some marine chrysomonads and cryptomonads as the sole source of nitrogen. These include urea, glycine, arginine, ornithine, lysine, and a number of other amino acids (1).

Many heterotrophic animals have been able to achieve total or partial independence of the amino acid profile in the diet by virtue of symbiotic microorganisms. Symbiosis has developed in animals ranging in size from ciliate protozoa with endosymbiotic algae and bacteria, to large mammals such as macropods, ruminants, manatees, at least one species of whale (45), and leaf eating primates with symbiotic bacteria and protozoa (10, 15, 58). The hoatzin (*Opisthocomus hoazin*) one of the world's few obligate leaf eating birds, also has a foregut fermentation and is one of the smallest endotherms with this type of digestive system (36a).

The contribution of microbial synthesized amino acids to the evolution of sym-

TABLE 1. Essential Amino Acid Requirements of Ciliate Protozoa[a]

Amino Acid	Tetrahymena pyriformis	T. pyriformis (some European strains)	Paramecium aurelia	P. caudatum P. multimicro nucleatum	Glaucoma chattoni A G. scintillan A
Arginine	+	+	+	+	+
Histidine	+	+	+	+	+
Isoleucine	+	+	+	+	+
Leucine	+	+	+	+	+
Lysine	+	+	+	+	+
Methionine	+	+	+	+	+
Phenylalanine	+	+	+	+	+
Serine	+[b]	+	−[c]	+	+[d]
Threonine	+	+	+	+	+
Tryptophan	+	+	+	+	+
Valine	+	+	+	+	+
Proline	−	+	+	−	+
Tyrosine	−	−	+	+	−
Glycine	−	+	+[c]	−	−[d]
Cysteine	−	−	−	−	−
Glutamic acid	−	−	−	−	−
Aspartic acid	−	+	−	−	−
Alanine	−	+	−	−	−

[a]From van Wagtendonk and Soldo (100).
[b]Only one strain absolutely dependent on serine.
[c]Serine spares glycine.
[d]Glycine can replace serine.

biotic microflora is not readily separated from the other contributions of symbiosis. Besides microbial synthesized amino acids, symbiosis generally supplies end products that on metabolism yield energy as well as vitamins and accessory factors. *Euglena gracilis*, one of the first protista to be grown on a chemically defined medium (20), has been shown to require the 10 EAA-rat. Eleven amino acids (10 EAA-rat plus serine) are required for most of the heterotrophic protista (77). For 16 strains of *Tetrahymena* examined, all were able to grow on a medium containing the 10 EAA-rat plus serine and folic acid at physiological levels. No growth occurred in any strains if glycine was substituted for serine unless folic acid levels were elevated; these high levels of folic acid permitted 14 strains of *Tetrahymena* to synthesize serine from glycine. All strains save two could use threonine in place of

serine provided folic acid was elevated, indicating that threonine aldolase was present (57). In *Tetrahymena* tyrosine spares, but cannot replace, phenylalanine, cystine spares methionine, and both citrulline and ornithine spare arginine. The essential amino acids for ciliate protozoa are shown in Table 1.

The essential amino acid requirements for both free-living and parasitic platyhelminths are not known. Nematodes appear to require tyrosine in addition to the 10 essential amino acids. It has been suggested that parasitic helminths probably require the 10 EAA-rat and possibly others for normal growth (5).

Mollusca have been grown on chemically defined diets containing amino acids as the sole source of nitrogen. However, as these have not been cultured axenically, it is difficult to assess the requirement for essential amino acids (11). The

Isopod *Ligia pallasii* has been grown on a diet containing the 10 EAA-rat, plus nine dispensable amino acids (12) in the presence of antibiotics (streptomycin and penicillin).

Many insects form symbiotic associations with microorganisms that provide essential nutrients for the host, and without their symbiotes many insects exhibit poor growth or cannot survive (10, 58). It has been hypothesized that many insect types originally had microbial associations in their digestive tract that permitted them to digest their food more completely without the production of a broad range of digestive enzymes. The majority of these associations have evolved to where these bacteria have penetrated the intestinal wall and now reside in fatty tissue or special organs in the host (10). These associations may obscure the true essential amino acid requirements of the host insect. For example, when nymphs of the aphid *Myzus persicae* are rendered aposymbiotic, deletion of any one of the 10 EAA-rat from a synthetic diet causes severe depression of growth. For normal aphids, growth retardation occurs when only histidine, isoleucine, or methionine are deleted from the diet (67). This observation indicates that the other 7 essential amino acids can be synthesized by the symbiotes. In a study of 13 species of insects (one of which normally carries symbiotes but was rendered aposymbiotic) the 10 EAA-rat were also found to be essential for these insects (26). Isotopic measurements suggest that the same 10 EAA-rat are required by a further 6 species of insects that could not be reared on a synthetic diet.

Omission of any of the 10 EAA-rat from the diet of insects typically causes growth of newly hatched individuals to cease entirely within the first or early instars. Omission of proline causes a more gradual curtailment of growth and de-

velopment that may proceed through several larval stages with a proportion of the individuals reaching a delayed adulthood (26). In some species, such as the silkworm, the requirement for proline is not absolute, but can be partially supplied by arginine or ornithine. For *Argyrotaenia velutinana* and *Heliothis zea*, proline is a dispensable amino acid determined by dietary deletion, but no labeled proline was derived from ^{14}C glucose indicating that the arginine to proline pathway in these insects is operable.

Some species of insects have been shown to have a requirement for amino acids in addition to the 10 EAA-rat, the most common one being proline but also glycine, glutamic, or aspartic acids. The requirement for glycine is probably related to uric acid synthesis. Birds and insects are uricotelic and in birds glycine is required for uric acid synthesis. Glutamic or aspartic acid is required by the silkworm and flesh-fly (*Phormia regina*); and, for the screwworm (*Cochliomyia hominivorax*) glutamic acid is essential. Caution needs to be exercised in interpreting experiments with insects in which only the 10 EAA-rat have been included in the diet and a response has been obtained to the addition of a dispensable amino acid. Dietary amino acids are required to supply nitrogen for synthesis of dispensable amino acids and other nitrogenous compounds as well as carbon skeletons. For rats, a diet containing a mixture of dispensable amino acids yields superior growth in comparison to one in which all the dispensable nitrogen is derived from one amino acid, or where all nitrogen is in the form of essential aminio acids. For several species of insects, (*Tribolium confusum*, *Drosophila melanogaster*, and *Blattella germanica*), supplementation of the diet with only glutamate gives nearly optimal growth. For others, such as the mosquito (*Culex pipiens*) and aphid (*Myzus*

TABLE 2. Proportions of L-Amino Acids in Mixtures Modified to Approach Optimality for Various Insects[a]

Amino Acid	Bombyx mori	Argyrotaenia velutinana	Pectinophora gossypiella	Anthonomus grandis	Oryzaephilus surinamensis	Drosophila melanogaster
Essential (Rat)						
Arginine	6	6	15	9	5	3
Histidine	3	5	4	3	3	4
Isoleucine	5	4	6	4	3	12
Leucine	8	8	9	10	8	8
Lysine	6	9	6	8	5	8
Methionine	3	4	4	5	3	3
Phenylalanine	6	5	7	7	3	5
Threonine	5	5	5	5	5	8
Tryptophan	3	4	4	1	1	2
Valine	6	5	7	5	3	11
Dispensable (Rat)						
Alanine	7	6	(7)	9	0	0
Aspartic acid	(13)	9	5	6	6	0
Cysteine/cys	2	3	2	3	1	0
Glutamic acid	12	(10)	6	7	17	(34)
Glycine	4	5	(7)	(9)	(30)	0
Proline	4	5	2	3	0	0
Serine	5	2	2	3	0	0
Tyrosine	2	5	2	3	7	0

[a]Values are rounded percentages of the diet. Major dispensable amino acids are shown in parentheses. From Dadd (26).

persicae), only glycine or serine is required in addition to the 10 essential amino acids for the rat for nearly optimal growth.

Asparagine appears to be essential for certain species of mosquito (e.g., *C. pipiens* and *Culiseta incidens*), and this requirement cannot be satisfied with aspartic acid. However, for other mosquito species (e.g., *Aedes aegypi*) neither asparagine nor aspartic acid are essential. The essential nature of asparagine may be the result of a lack of the enzyme asparagine synthetase, which effects the transamination of amide nitrogen from glutamine to aspartic acid to form asparagine. Certain aphids with intracellular symbiotes can use inorganic sulfate to replace or spare dietary cystine or methionine. Insect parasitoids appear to have similar amino acid requirements to those of nonparasitic insects (96).

The proportions of essential and dispensable amino acids in mixtures that were found to give growth approaching optimal in insects are shown in Table 2.

The dietary crude protein (N × 6.25) requirement of most fish for optimal growth is higher than for most mammals (Table 3). Fry require a diet in which nearly one half of the digestible ingredients consists of protein, and this requirement declines only modestly with age to ~35% dietary protein in the case of yearling salmonids (73). Carnivorous fish, such as plaice (*Pleuronectes platessa*), are not able to decrease the activities of the

TABLE 3. Estimated Dietary Protein Requirement of Certain Fish[a]

Common Name	Species	Crude Protein Level for Optimal Growth (g/kg of diet)
Rainbow trout	*Salmo gairdneri*	400–600
Carp	*Cyprinus carpio*	380
Eel	*Oncorhynchus tschawytscha*	400
Plaice	*Pleuronectes platessa*	500
Gilthead bream	*Chrysophrys aurata*	400
Grass carp	*Ctenopharyngodon idella*	410–430
Brycon	*Brycon* spp.	356
Red sea bream	*Crysophrys major*	550
Yellowtail	*Seriola quinqueradiata*	550

[a]From Cowey (23a).

aminotransferases for dispensible amino acids in response to diets low in protein (22). These fish require a high protein diet to supply amino acids, which are continually being deaminated.

A number of finfish have been shown to require all 10 EAA-rat (Table 4), and it is probably safe to conclude all fish need a source (from the diet or from microbial synthesis) of the 10 EAA-rat.

The requirement of most fish for almost all essential amino acids is higher than for mammals (104) (Table 5A and B) and this results in a higher requirement for protein as shown in Table 3.

Birds require the 10 essential amino acids of the rat, but in addition, also re-

quire glycine or serine for uric acid synthesis, which is the end product of catabolism of nitrogenous products (75). Cystine and tyrosine are able to spare methionine and phenylalanine in birds, as in other animals.

A number of mammals have evolved anatomical modifications of the stomach, which permit the maintenance of symbiotic microflora. While ruminants are a prime example of this evolution (51), many other mammals including macropods (e.g., kangaroos and wallabies) (50), peccaries, some primates (e.g., langur and colobus monkeys), manatees, and baleen whales have microbial digestion in the stomach. Ruminants and possibly

TABLE 4. Finfish Known to Require the 10 Essential Amino Acids for the Rat

Common Name	Species	Kind of Studies
Channel catfish	*Ictalurus punctatus*	Growth
Chinook salmon	*Oncorhynchus tshawytscha*	Growth
Common carp	*Cyprinus carpio*	Growth
European eel	*Anguilla anguilla*	Growth
Japanese eel	*Anguilla japonica*	Growth
Plaice	*Pleuronectes platessa*	^{14}C-labeling
Rainbow trout	*Salmo gairdneri*	Growth
Red sea bream	*Chrysophrys major*	Growth
Sea bass	*Dicentrarchus labrax*	^{14}C-labeling
Sockeye salmon	*Oncorhynchus nerka*	Growth
Sole	*Solea solea*	^{14}C-labeling
Tilapia	*Tilapia zillii*	Growth

TABLE 5. Part A. Essential Amino Acid Requirements of Fish and Eels[a]

Fish	Arginine	Histidine	Isoleucine	Leucine	Valine
Chinook salmon	2.4(40)[b]	0.7(40)	0.9(41)	1.6(41)	1.3(40)
Coho salmon	2.3(40)	0.7(40)			
Common carp	1.6(38.5)	0.8(38.5)	0.9(38.5)	1.3(38.5)	1.4(38.5)
Japanese eel	1.7(37.7)	0.8(37.7)	1.5(37.7)	2.0(37.7)	1.5(37.7)
Channel catfish	1.03(24)	0.37(24)	0.62(24)	0.84(24)	0.71(24)
Lake trout			0.72(27.6)	0.96(27.6)	0.78(23.7)
Rainbow trout	1.2(36)				
	1.4(35)				
	2.8(47)				
Gilthead bream	1.7(34)				
Tilapia	1.59(40)				

other mammals with symbiotic microflora in their stomach can grow, albeit slowly, when given diets devoid of preformed amino acids but containing nonprotein nitrogen. Amino acids synthesized by the bacteria and protozoa from nonprotein nitrogen (and from food amino acids nitrogen) are digested and absorbed when these organisms pass out of the forestomach into the intestine. While protozoa are the most conspicuous of the microflora of ruminants, their contribution to the amino acid requirements of the hosts is less than that of the bacteria.

TABLE 5 (continued). Part B

Fish	Lysine	Phenylalanine[c]	Methionine[d]	Threonine	Tryptophan
Chinook	2.0(40)[b]	2.1(41)[3]	1.6(40)[5]	0.9(40)	0.2(40)
Coho salmon					0.2(40)
Sockeye salmon					0.2(40)
Common carp	2.2(38.5)	2.5(38.5)[1]	1.2(38.5)[1]	1.5(38.5)	0.3(38.5)
					0.13(42.5)
Japanese eel	2.0(37.7)	2.2(37.7)[1]	1.2(37.7)[1]	1.5(37.7)	0.4(37.7)
Channel catfish	1.23(24)	1.2(24)[1]	0.56(24)[1]	0.53(24)	0.12(24)
	1.5(30)				
Rainbow trout	1.3(35)		1.0(46.4)[1]		0.25(55)
	1.9(45)				0.58(42)
	2.9(47)		1.1(35)[2]		
			1.0(35)[3]		
Gilthead bream	1.7(34)		1.4(34)[6]		0.2(34)
Tilapia	1.62(40)		1.27(40)[4]		
Sea bass			1.0(50)[6]		

[a] Adapted from Wilson and Halver (104).
[b] Requirements are expressed as percentage of amino acid in the diet. The values in parentheses are the percentage of total protein in the test diet.
[c] Superscripts on phenylalanine values 1, 2, and 3 = presence of tyrosine at 0.0, 0.3, and 0.4%, respectively.
[d] Superscripts on methionine values 1, 2, 3, 4, 5, and 6 = presence of cystine at 0.0, 0.3, 0.5, 0.7, 1.0% or not stated, respectively.

A high concentration of bacteria exists in the large intestine of most mammals, which is favored by the slower rate of passage of food residues than in the small intestine. Microbial protein, synthesized in the large intestine, is probably unavailable to the host animal unless it practices coprophagy (65). Among mammals, largomorphs, rodents in general, a primate (*Lepilemur*), a marsupial (koala, *Phasolarctos cinereus*) (50), and horses (under certain conditions) practice coprophagy, and the microbial synthesized amino acids may contribute to their amino acid economy. Rabbits produce two types of feces: special soft or night feces and normal pelleted feces. The soft feces, which are consumed directly from the anus, have a higher bacterial and lower fiber concentration (hence a higher protein content) than normal feces.

Arginine is a dispensable amino acid for many adult mammals (humans, rat, pregnant swine, etc.) and some growing animals (e.g., children). Adult rats synthesize citrulline in the mucosa of the small intestine, and this citrulline is converted to arginine in the kidney (105). Adult cats and ferrets cannot synthesize citrulline at a rate commensurate to the rate of degradation of arginine. For cats a single meal of an amino acid diet devoid of arginine can result in death because of hyperammonemia (68). Lack of citrulline synthesis by cats is a result of a low activity of the enzyme pyrroline-5-carboxylate synthase (87), and also ornithine aminotransferase, and possibly carbamoyl phosphate synthase (69). This nutritional peculiarity does not pose a problem to cats as long as they are maintained on a carnivorous (meat) diet, which is high in arginine. In starvation arginine is made available from tissue catabolism, and as the urea cycle enzymes do not show adaptation to low protein diets, cats are always prepared to catabolize a high protein (meat) meal. The requirement of most animals for arginine can be substituted by citrulline. However, ornithine cannot substitute for arginine in any of the species tested (chickens, rats, and cats).

Protein is required in the diet of animals to supply essential amino acids, and as a source of nitrogen for synthesis of dispensable amino acids and other nitrogenous compounds. The quantitative protein needs are the sum of the amino acids required for maintenance of the animal's tissues and that required for growth and reproduction. In omnivorous animals such as rats, the first enzyme of an amino acid that irrevocably commits the amino acid to degradation (generally an aminotransferase) is regulated by the intake of protein. When the diet is high in protein, the activity of the degradative enzyme increases; similarly when the diet is low in protein, the activities of these enzymes decrease (89). The activity of enzymes associated with the disposal of the nitrogen from deamination of amino acids (in mammals the urea cycle enzymes) also increases or decreases when the dietary intake of protein increases or decreases. The high protein requirement of truly carnivorous animals (e.g., felids, but not canids) is a result of the transaminases for dispensable amino acids not being adaptive to changes in the intake of protein (88, 92). These transaminases and urea cycle enzymes are set at a relatively high level at all times. This relative lack of regulation has two consequences for the carnivorous animal: first, the animal is always prepared to metabolize a high protein (meat) meal (i.e., an advantageous modification); second, the lack of regulation has the apparent shortcoming of a reduced ability to conserve nitrogen when the animal ingests a low protein diet. However, a carnivorous diet by definition is high in protein, so this shortcoming is only realized when a felid (cat) consumes a noncarnivorous diet. As arginine synthesis in the kidney of cats is extremely low, incoming dietary arginine provides intermediates for the urea cycle and a further stimulus for ureogenesis and nitrogen disposal.

The total free amino acid pool in animal tissues is small compared to the amino acids present in proteins. However, free amino acids in blood vary with dietary intake and can be used to assess the adequacy of an amino acid in the diet. The free amino acid and tissue amino acid patterns in vertebrates resemble one another. However, in invertebrates the free amino acids are often dominated by one or two amino acids. Free amino acids range from 10 to 50 mg/100 g of fresh tissue in birds and mammals. However, in invertebrates the sum of the free amino acids is much higher (300–2000 mg/100 g of fresh tissue) and in some cases as high as 3000 mg/100 g of fresh weight for marine invertebrates. The concentration of total free amino acids in helminths is variable from 100 to 400 mg/100 g in cestodes, and 112 up to 2600 mg/100 g in marine species of trematodes. The main free amino acids in helminths are proline, alanine, and glutamate. In marine species, the free amino acid in highest concentration is frequently taurine (5).

Taurine

Taurine (2-amino ethanesulfonic acid) is a β-amino acid that is present as a free amino acid in most animal tissues (54). Mollusca such as *Mytilus*, *Pecten*, and *Sepia* contain high concentratiuns of taurine, which act in osmoregulation in these species. In *Mytilus edulis* free taurine accounts for ~1% of body weight (60). Taurine concentration in the tissues of some marine mollusca, crustaceans, and echinoderms varies directly with the salinity of the surrounding water. Taurine is virtually absent in plants, and the taurine present in most animals is derived by de novo synthesis from methionine and

cystine. Taurine is an essential dietary nutrient for the domestic cat and presumably all felids as endogenous synthesis is limited and the bile acids are exclusively conjugated with taurine. Taurine deficiency produces a wide range of clinical conditions in cats: adult females have a high incidence of abortions, fetal development is abnormal, post-natal growth of kittens is reduced, adult cats and kittens develop feline central retinal degeneration (41) and dilated cardiomyopathy, which is normalized by dietary taurine supplementation (82).

Purines and Pyrimidines

These nitrogen-containing substances are the constituent bases in RNA and DNA. In mammals they are synthesized from dispensable amino acids glycine, aspartate, and glutamine. Purines, but not pyrimidines, stimulate phototrophic growth of the cryptomonad *Hemiselmis virescens* (1), which may indicate synthesis is limiting. All ciliate protozoa examined have an absolute requirement for a purine and a pyrimidine in their culture medium (53). The purine and pyrimidine requirements can generally be satisfied, respectively, by guanine (which is spared by adenine or adenine nucleotides or nucleosides) and uracil (46).

The free living nematode *C. briggsae* and the plant parasitic nematode *Aphelenchoides rutgirsi* seem to be able to synthesize pyrimidines and purines de novo. However, a number of other protozoa when grown under symbiote-free conditions have been shown to require a purine or a purine and pyrimidine in the culture medium (94, 94a).

There is no evidence that helminths synthesize purine or pyrimidine bases de novo, but they can incorporate free purines (e.g., adenine) and purine nucleosides (e.g., adenosine) into nucleic acids. Similarly, exogenous pyrimidimes (e.g., uracil) and pyrimidine nucleosides (e.g., uridine and thymidine) can be incorporated into nucleic acids by the salvage pathway (5).

Although most insects have no dietary requirement for nucleic acids, optimal larval diets for Diptera must contain nucleic acids or certain constituents of them (26). In these dipteran species, nucleic acids are not essential for complete development to the adult stage, but the growth of the larvae is retarded, adult size reduced, and mortality increased. These observations suggest that the pathways for nucleic acid synthesis are rate limiting. In contrast to the general ability of insects to synthesize nucleic acids, a number of Diptera including the flesh-inhabiting screwworm *Cochliomyia hominivorax*, certain strains of *Drosophila melanogaster*, and mosquitoes have an absolute dietary requirement for nucleic acid or certain nucleotides for development.

Fish, birds, and mammals do not require a dietary source of nucleic acids, although dietary nucleic acid bases may be utilized for nucleotide synthesis.

Choline

Choline is the most abundant N-methyl compound in the animal body and has diverse functional roles: a neurotransmitter (acetylcholine), a constituent of membranes (phosphatidylcholine), neural tissue (sphingomyelins), and muscles (carnitine). Most of the choline in the body is present as phospholipids: phosphatidylcholine and sphingomyelins. These phospholipids have a polar head (from choline) and a hydrophobic tail (from the lipid) making them able to have one part of the molecule in water and the other part in a lipid. Phospholipids are major constituents of lipoproteins, which transport lipid around the body. Phosphatidylcholine is also known as lecithin.

$$HO-CH_2-CH_2-\overset{+}{\underset{|}{\underset{CH_3}{\overset{CH_3}{N}}}}-CH_3$$

Choline

Choline is not a dietary essential for *Tetrahymena pyriformis* except for one clone (46), nor for the photoflagellates, *Cryptomonads* and *Chrysomonads* (1). A dietary deficiency of choline by the host of the rat tapeworm *H. diminuta* has no effect on the tapeworm, though a host dietary deficiency of vitamin B6 inhibits the establishment, growth, and development of this parasite (5).

Choline appears to be a dietary requirement of Arthropods. The majority of insects have been shown to require choline in the diet, the apparent exceptions may be a result of dietary impurities, or choline supplied by associated microorganisms (7). Insects probably cannot synthesize choline from labile methyl group donation to ethanolamine as occurs in mammals, or if synthesis occurs, as in the case of *Pieris brassicae* (8), it is inadequate to supply physiological needs. Phosphatidylcholine is a dietary essential for the crustacean *Penaeus japonicus*.

Choline deficiency in fish produces similar deficiency signs as for higher animals. Trout can synthesize sufficient choline from dietary methylaminoethanol and from dimethylaminoethanol but not from aminoethanol or betaine (56).

Birds and mammals are able to synthesize choline from ethanolamine and methyl group donors, especially methionine. When methionine is present in the diet in amounts only sufficient to meet the amino acid needs, a choline deficiency can be demonstrated. In mammals and birds, a deficiency of choline leads to an accumulation of fat in the liver. The accumulated fat is presumably a result of lack of phosphatidylcholine, which limits formation of liver lipoproteins and their secretion into blood. Other clinical signs of choline deficiency in rats include hemorrhages and renal damage. In mammals, choline is constantly degraded in the mitochondria by oxidation of the alcohol group to a carboxylic acid with the resultant production of glycine betaine. One of the methyl groups from betaine can contribute to the labile methyl pool by being donated to homocysteine to form methionine.

Carnitine

The immediate precursor of carnitine is ε-*N*-trimethyllysine, which is produced by the methylation of lysine. The methyl groups are donated by *S*-adenosyl methionine for the enzymatic trimethylation of peptide-linked lysine. In the turn-over of proteins containing lysine, which have been modified by the methylation of the terminal group ε-amino group (e.g., histones, cytochrome C, and myosin), trimethyllysine is released and can participate in carnitine synthesis. In mammals, carnitine acts as an acyl carrier for the transport of fatty acid into the mitochondrion for oxidation and phospholipid synthesis and in the modulation of intramitochondrial acyl-CoA/CoA ratios.

$$(CH_3)_3 \overset{+}{\equiv} N - CH_2 - CHOH - CH_2 - COO^-$$

Carnitine

Carnitine was shown to be an essential vitamin for the larval stage of the mealworm *Tenebrio molitor* (31). All members of the family *Tenebrionidae* studied to date have been shown to have a dietary requirement for carnitine. There does not appear to be any clear evidence for a carnitine requirement in any other insect,

nor for that matter any other genetically normal organism with the exception of two microorganisms (*Pediococcus soyae* and a mutant of the yeast *Torulopsis* (*Candida*) *bovina*) (31). However, growth of *D. melanogaster* and indices of development of *Oryzaephilus surinamensis* are improved by the addition of carnitine to the diet. In both species, full development can occur without dietary carnitine so it appears that biosynthesis is limiting.

Carnitine has not been shown to be a dietary requirement for higher animals (85) and is synthesized in large amounts by them. Some animal tissues contain high concentrations of carnitine (1 mg/g of dry weight), yet the requirement for carnitine by *Tenebrionidae* is only ~1 µg/g of diet (31).

Other nitrogenous compounds that occur in high concentrations in the muscles of mammals are carnosine (a peptide of β-alanine and histidine), anserine, and *N*-acetylaspartate, but these have not been shown to be required in the diet.

Non Nitrogenous Substances

Sterols

Sterols constitute a large class of compounds derived from the complex heterocyclic hydrocarbon perhydrocyclopentanophenanthrene. Cholesterol is the most common animal sterol and is the parent substance for a large array of compounds with important physiological functions. Cholesterol is a constituent of cell membranes; it participates in the transport of fatty acid in the blood; in its oxidized form, it is a constituent of bile salts important in the digestion of fat; and as a precursor for vitamin D and steroid hormones. Apparently all animals use cholesterol in lipid metabolism, but some animals cannot synthesize it de novo.

CH_3
CH_3
$CH-CH_2-CH_2-CH_2-CH$
CH_3
CH_3
CH_3
HO

Cholesterol

Mammals synthesize cholesterol from acetate. The process can be divided into three phases: (1) the synthesis of mevalonate, which is the basic building block for the synthesis of isoprenoids; (2) the conversion of mevalonate to the long chain hydrocarbon squalene; (3) cyclization of squalene to lanosterol, the precursor of cholesterol.

Most species of *Tetrahymena* do not appear to have a nutritional requirement for sterol, although three species of *Tetrahymena* have a requirement that can be satisfied by any one of a large number of sterols (46). Growth of *T. setosa* is stimulated by as little as 0.1 µg of ergosterol/mL of medium (37). *Paramecium aurelia* requires both a sterol and a fatty acid for growth (93). All parasitic helminths appear to be unable to synthesize sterol de novo; however, many are able to synthesize isoprenoids of one sort or another (5). The rat tapeworm *H. diminuta* cannot synthesize cholesterol from acetate or mevalonic acid but can synthesize farnesol, a sesquiterpenoid (three isoprenoid units). Among the nematodes, *Ascaris lumbricoides* does not synthesize cholesterol from acetate or mevalonate nor can it dealkylate plant sterol to give cholesterol as occurs in phytophagous insects. Helminths readily incorporate exogenous fatty acids into sterol esters.

Arthropods are incapable of de novo synthesis of sterol from acetate and require a dietary or exogenous source of sterol for their normal growth and development. The sterol requirement of insects in most cases can be satisfied by cholesterol, which is the principal sterol of insect bodies. Cholesterol is required by insects as a structural component of cell membranes and as a precursor of the molting hormone *ecdysone*. Although cholesterol synthesis cannot be accomplished by insects, many isoprenoid chemicals (juvenile hormones, ubiquinones, defence terpenoids, and squalene) that share a common pathway with cholesterol synthesis in vertebrates can be synthesized. The barrier to synthesis in insects appears to be at squalene cyclization. Some insects can satisfy their requirement for sterol by modification of phytosterols (e.g., sitosterol, stigmasterol, and campesterol) and by conversion of these C_{24} alkylated sterols to cholesterol. Many zoophagous species are incapable of undertaking these conversions. At one time the ability to utilize phytosterols was proposed as a distinction between phytophagous insects and insects that are omnivorous and carnivorous. While this distinction is not absolute, many insects fall into these respective groupings (25).

In mammals cholesterol derivatives are parent substances of vitamins D_3 (7-dehydrocholesterol) and of many steroid hormones that have profound effects on metabolism and reproduction.

Essential Fatty Acids

In plants and animals, the synthesis of long chain fatty acids proceeds by the successive additions of two carbon acyl units (chain elongation) to the carboxyl end of a primer molecule. Unsaturated fatty acids are produced by desaturase enzymes, which are mixed function oxidases and act on specific C—C bonds and abstract hydrogen to produce a double bond of the cis configuration. Desaturases are denoted by Δ and a numeral; the numeral indicates the carbon bond from the *carboxyl* end of the fatty acid where the desaturase acts. Plants have the ability to insert cis-double bonds at C-9 and three carbon intervals (C-12,C-15) as well as introducing a double bond at C-6. Animals can also introduce double bonds by the desaturases Δ5, Δ6, and Δ9 but not longer than C-9. For example, stearic acid (18:0) can be converted to oleic acid 18:1 (9) by a Δ9 desaturase, but not to linoleate 18:2 (9, 12) or linolenic acid 18:3 (9, 12, 15).

Mammals and many other animals do not possess desaturases capable of double-bond insertion at either C-3 or C-6 from the *methyl* end of the long chain fatty acid. These fatty acids with 18 carbon atoms are designated *essential fatty acids*. The parent fatty acid with double bonds at C-6 and C-9 from the methyl group of the fatty acid (*n6* family) is linoleic acid, the parent with double bonds at the C-3, C-6 and C-9 positions (*n3* family) is linolenic acid. Both these fatty acids or their elongation products are metabolically essential for most animals.

Essential fatty acids are components of the phospholipids of cell membranes and are precursors for several series of biologically active substances known collectively as prostaglandins, leukotrienes, and thromboxanes. Essential fatty acids are dietary requirements of almost all animals, with the exception of protozoa and some insects.

A fatty acid requirement for oleate (but not linoleate or linolenate) has been reported for *Paramecium aurelia* (93, 94). Linolenic and γ-linolenic are taken up and incorporated into complex lipids by *Tetrahymena* species but do not stimulate growth.

Little is known about the essential fatty acid requirements of helminths but they do not appear to possess the ability to synthesize long chain polyunsaturated fatty acids. The occurrence of a wide range of mono- and polyunsaturated fatty acids in the lipids of parasitic trematodes and cestodes has been used as presumptive evidence for an essential fatty acid requirement (5). Free living nematodes can synthesize polyunsaturated fatty acids, and there are claims that *Ascaris lumbricoides* can convert linoleic into arachidonic acid. The free-living nematodes, along with the free-living protozoa, appear to be unique in that they are the only animals that can synthesize polyunsaturated fatty acids (5).

Although a universal dietary requirement of insects for polyunsaturated fatty acids has not been demonstrated, the majority of ~50 species from five orders studied showed an essential fatty acid requirement (26). There is considerable variation between species in the time taken to induce a deficiency: Diptera (including mosquitoes) do not exhibit a deficiency until after one generation (24); lepidoptera tend to show essential fatty acid deficiency dramatically by failure of the pupal-adult ecdysis. Aphids with intracellular symbiotes can be reared for successive generations on lipid-free diets. Essential fatty acids in insects are apparently critical at metamorphosis, and a deficiency leads to failure at the pupal-adult stage of development as well as in preventing deformities in adults and slow larvae growth.

The essential fatty acid requirement of most insects can be satisfied by either or both linoleic (18:2*n6*) or linolenic (18:3*n3*) acids. These fatty acids are normal constituents of plants. Of 18 lepidopteran species examined, three had no apparent fatty acid requirement, nine utilized either linoleic or linolenic acid (for three linolenic was more potent), five utilized only linolenic acid, and one required both fatty acids (26). Possibly linolenic acid may be a general requirement for all lipidopteran species. The mosquito *Culex pipiens* also requires arachidonic acid (20:4*n*6), which is only found in animal lipids.

Despite the general lack of essential fatty acid synthesis by insects and the possibility of lipogenesis by symbiotic microorganisms, there are examples (especially the cricket *Acheta domesticus*) that support the possibility of substantial de novo synthesis of polyunsaturated fatty acids in insects.

Crustaceans, including lobster *Homarus americanus* and shrimp (*Penaeus japonicus, P. indicus, P. aztecus,* and *Palaemon serratus*) and crab *Carcinus maenas,* appear to require 18:3*n*3 alone or in combination with an *n*6 fatty acid (74).

Investigations into the essential fatty acid requirement of cold-water teleosts (salmon) showed that linolenic was the essential fatty acid rather than linoleic. It was suggested that linolenic allowed a greater degree of unsaturation of phospholipids in membranes as well as maintenance of permeability and flexibility under cold conditions. In support of this argument the fatty acid composition of human triglycerides tend to approach 50% unsaturated with 16:0 the predominant fatty acid. A triglyceride of two 16:0 and one 18:0 residue has a melting point at 35–36°C just below that of body temperature. Some fish require essential fatty acids of both the *n*3 and *n*6 series and the tropical fish *Tilapia zilli* apparently requires only *n*6 fatty acid.

In mammals and birds, the major requirements for essential fatty acids are those of the *n*6 series of which linoleate (9,12-C18:2) is the precursor. A deficiency of dietary linoleate results in an increased proportion of unsaturated fatty acid from oleic acid with the eventual production of the "triene" ecosatrienoic acid (5,8,11-C20:3 or C20:3*n*9) in place of arachidonate the "tetraene" (5,8,11,14-

C20:4 or C20:4*n*6) derived from linoleate. The activity of the Δ6 desaturase in felids is very low, restricting the conversion of linoleate to arachidonate, so rendering arachidonic acid and linoleate essential dietary constituents.

A dietary requirement for essential fatty acids was first demonstrated in rats given a fat-extracted diet. These rats exhibited severely depressed growth, scaly skin, and reproductive failure when the diet did not include linoleic acid (18:2*n*6). Subsequently a dietary requirement for essential fatty acids has been demonstrated in almost every homoiotherm investigated with the possible exception of ruminants and other animals with symbiotes. Supplementation of a deficient diet with linolenic acid (18:3*n*3) reversed some of the clinical signs of essential fatty acid deficiency (e.g., restored growth), but linoleic or arachidonic acid (20:4*n*6) was necessary for restoration of reproduction.

Because of the high proportion of polyunsaturated fatty acids present in membrane phospholipids, the structural role of polyunsaturated fatty acids was one of the first functions identified. In mammals, polyunsaturated fatty acids, and in particular arachidonic acid, are the precursor of a group of extremely potent oxygenated derivatives: prostaglandins, (Figure 1), thromboxanes, and leukotrienes. These products have profound metabolic effects on blood flow, aggregation of blood platelets, tissue reaction to stimulants, host defense mechanisms, and antibody production. The oxidation of arachidonic acid by the cyclooxygenase-catalyzed pathway produces the prostaglandins from which are derived the thromboxanes, whereas the lipoxygenase-catalyzed pathway produces the leukotrienes and lipoxins. When cells are stimulated, arachidonic acid is released from membrane phospholipids and con-

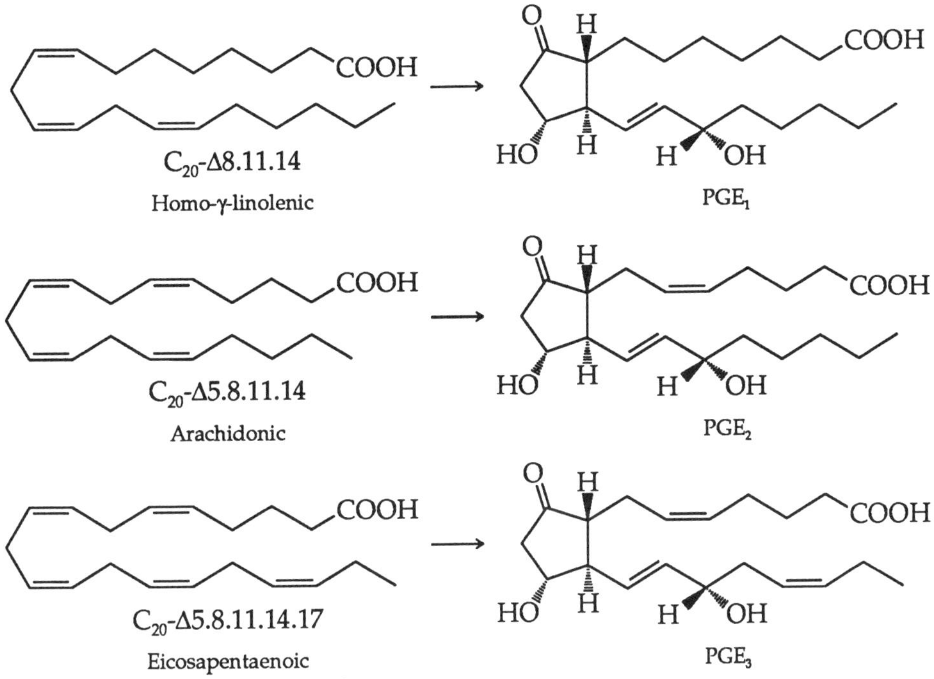

Figure 1. Formation of prostaglandins from unsaturated fatty acids.

verted to leukotrienes by a 5-lipoxygenase. Leukotrienes then participate in host defence reactions and pathophysiological conditions such as immediate hypersensitivity and inflammation.

Vitamins

A vitamin is an organic substance that cannot be synthesized by an animal but is required in the diet in low concentrations for the normal function of the animal. Vitamins may be supplied by symbiotic organisms or from the diet in amounts that are generally <0.1% of the dietary dry matter. By convention, vitamins are grouped on the basis of whether they are water soluble or fat soluble. This arbitrary subdivision has little to do with their metabolic function although the water soluble vitamins frequently act as coenzymes, whereas the fat soluble vitamins occupy more diverse roles.

Water Soluble Vitamins

Many of the vitamins of the B complex are virtually universal requirements of animals. This is probably a reflection of the participation of these vitamins in common metabolic pathways. For example, thiamin is required by all animals ranging from acetate flagellates (62) to primates; similarly, vitamin B_{12} is also required by a wide spectrum of animals. In contrast to the water soluble vitamins, the fat soluble vitamins are generally required only by vertebrate animals. Where a requirement occurs in invertebrate animals it relates to a specialized function (e.g., vitamin A in vision or vitamin E in reproduction).

Thiamin
Thiamin was the first B vitamin to be discovered when it was shown to be the active principle that prevented beri-beri in

humans and polyneuritis in birds. Of all nutrients a deficiency of thiamin has the most marked effect on appetite (90). Thiamin as thiamin pyrophosphate is a coenzyme for the oxidative decarboxylation of α-keto acids, particularly pyruvate and α-ketoglutarate to acetyl CoA and succinyl CoA, respectively. Thiamin pyrophosphate is also a coenzyme, for the transketolase reaction of the phosphogluconate pathway. Thiamin pyrophosphate is therefore involved in linking glycolysis to the tricarboxylic acid. In microorganisms, thiamin also participates as a coenzyme for the nonoxidative decarboxylation of α-keto acids and the phosphoketolase reaction. The latter reaction involves the cleavage of a ketopentose phosphate to a triose phosphate and acetyl phosphate.

Thiamin contains a pyrimidine and a thiazole ring linked by a methylene bridge. The complete molecule is synthesized by green plants, many bacteria, molds, and some flagellates. Thiamin is present, albeit in low concentrations, in natural waters, both fresh and marine, particularly in regions where there are high concentrations of bacteria. Laboratory studies on axenic cultures of algae have shown that thiamin in the culture medium is required by almost one third of all strains studied (29, 84). Only three B vitamins (thiamin, vitamin B_{12}, and biotin) have been found to be of general importance for algae. Some microorganisms require the complete thiamin molecule; others can synthesize thiamin from either the pyrimidine or the thiazole ring.

Probably all multicellular animals have a metabolic requirement for thiamin; however, many microorganisms including those in the digestive tract or associated with the animal may synthesize sufficient thiamin to make the host independent of a dietary source. All insects require in their diet (or from their symbiotes) a source of thiamin. *Phormia* require 3 mg/kg of diet, which is approximately the same as required by rats and fish (72, 73).

A large number of fresh-water fish as well as some salt-water fish, mollusca, and crustaceans (74) contain an enzyme (thiaminase), which cleaves the thiamin molecule at the quaternary nitrogen of the methylene bridge rendering it biologically inactive. Thiaminases are produced by some plants and are present in rice polishings, beans, and mustard as well as in certain bacteria. The function of these thiaminases is not apparent; they may have a role comparable to secondary compounds in plants in protecting animals from predation.

Riboflavin (Vitamin B_2)

Riboflavin serves as a constituent coenzyme of flavoproteins which are involved in oxidation–reduction reactions. The phosphorylated form of riboflavin (riboflavin 5'-phosphate), erroneously called flavin mononucleotide, is a component of NADH dehydrogenase located on the inner surface of the mitochondrion, where it transfers electrons from NADH to other flavin carriers. The other important riboflavin coenzyme, flavin adenine dinu-

Thiamine

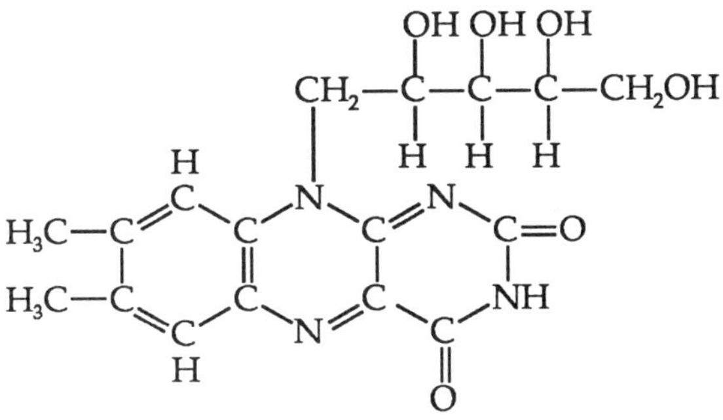

$$\text{Riboflavin}$$

Riboflavin structure:

OH OH OH
CH$_2$—C—C—C—CH$_2$OH
    H H H

H$_3$C—C=C   C   N   C   N   C=O
H$_3$C—C—C   C   N   C   C   NH
              H   C
                            O

Riboflavin

cleotide (FAD) is the prosthetic group in dehydrogenases including succinate and acetyl CoA dehydrogenase as well as D-amino and monoamine oxidases. Some of the flavoproteins have nonheme iron (an iron-sulfur center), or molybdenum involved in the catalytic mechanism.

Riboflavin is synthesized by many microorganisms—bacteria, yeasts, and possibly flagellates. All insects require either riboflavin in the diet or to be supplied by symbiotes (26). The requirement for riboflavin for most mammals is ~3 to 4 mg/kg of diet dry matter. In contrast, some insects (e.g., *Musca domestica*) and cold-water fish are reported to require ~20 mg/kg of diet dry matter. The riboflavin requirement of warm-water fish has been reported as being only about one half that of cold-water fish (74). Mammals that practice coprophagy (e.g., rabbits) or have fermentative pregastric digestion (e.g., ruminants and macropods) do not require a dietary source of riboflavin.

Niacin

Niacin in the form of nicotinamide adenine dinucleotides (NAD$^+$ and NADP$^+$) is present in both the mitochondrion and cytosol of cells and participate in oxidation–reduction reactions involved in H$^+$ and electron transfer. The reduction of NAD$^+$ (to NADH) is accompanied by oxidation of the substrate. The interconversion of pyruvate and lactate in the cytosol of cells is one example of reversible oxidation and reduction involving NAD$^+$.

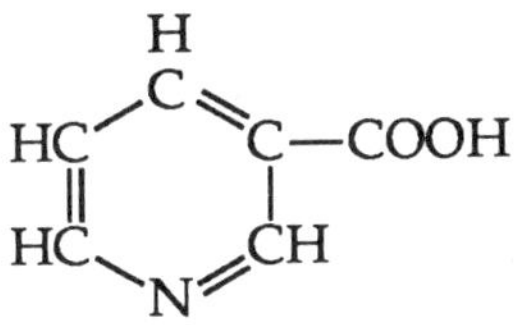

Niacin structure:

        H
        C
HC = C—COOH
HC   CH
    N

Niacin

An example of a similar reaction occurring in the mitochondrion is the reduction of oxaloacetate to malate. Malate can cross the mitochondrion to the cytosol where it is oxidized and decarboxylated by NADP-malate dehydrogenase to give pyruvate and NADPH. The main reductant of acetyl groups for fatty acid synthesis is NADPH. The NADH generated in the TCA cycle can also be oxidized by ubiquinones and in the process, high energy phosphate bonds are generated.

Quinolinate appears to be the immediate precursor for the de novo synthesis of niacin in all living forms that have been studied (43). Plant and some microorganisms synthesize quinolinate from aspar-

$$\begin{array}{ccc} \text{COO}^- & & \text{COO}^- \\ | & & | \\ \text{C}{=}\text{O} + \text{NADH} + \text{H}^+ \xrightleftharpoons{\text{lactic dehydrogenase}} & \text{HO}{-}\text{C}{-}\text{H} + \text{NAD}^+ \\ | & & | \\ \text{CH}_3 & & \text{CH}_3 \\ \text{Pyruvate} & & \text{L-Lactate} \end{array}$$

tate and a C_3 compound, whereas many mammals obtain the quinolinate for niacin synthesis from the degradation of tryptophan.

Most species of *Tetrahymena* require niacin in the culture medium. Niacin is required by snails reared under axenic conditions (98) and in the diet of all insects with the possible exception of those with symbiotes (26). Although niacin can be synthesized from tryptophan by bacteria and many animals, large variation occurs in the efficiency of conversion. Deficiencies of riboflavin and pyridoxine reduce the efficiency of niacin synthesis from tryptophan. For humans, 1 mg of niacin is produced from the catabolism of 60 mg of tryptophan. When tryptophan is catabolized by cats, virtually no niacin is produced. Felids have all the enzymes of the niacin pathway, but the activity of the competing enzyme leading to the production of glutarate is much higher than that of the enzyme for niacin synthesis. Niacin is therefore an essential dietary requirement of cats, whereas most other mammals can synthesize adequate amounts from tryptophan.

A deficiency of niacin in humans results in the disease known as *pellagra*, which was an important nutritional disease in the United States when corn constituted a major component of the diet. In corn, much of the niacin is in a bound form, and corn is also low in tryptophan.

Pyridoxine

Pyridoxal phosphate is frequently the prosthetic group of enzymes for the attachment of free amino acids. In amino-transferases, which transfer amino groups from amino acids to α-keto acids, pyridoxal phosphate is the coenzyme. Pyridoxal phosphate is a major coenzyme for the catabolism of amino acids and the synthesis of dispensable amino acids.

Vitamin B_6 (pyridoxine)

Vitamin B_6 (pyridoxine) activity is present in vitamers in which R is substituted by CH_2OH (pyridoxol), CHO (pyridoxal), or CH_2NH_2 (pyridoxamine).

Pyridoxine stimulates growth, but is not essential for *Chilomonas*. Some species and strains of species of *Tetrahymena* require pyridoxine and can use pyridoxamine, pyridoxal, pyridoxal-5-phosphate, and pyridoxamine-5-phosphate almost equally but are practically unresponsive to pyridoxol (3). Axenically cultured snails (98), and insects require a dietary source of pyridoxine. Even in the case of insects which have intracellular gut symbiotes (e.g., *Sitophilus oryzae*), pyridoxine is a dietary requirement (4). Pyridoxine is required in the diet of birds and all mammals with the exception of those with fermentative pregastric digestive systems.

The requirement of mammals and birds for pyridoxine is of the order of 3 mg/kg of diet dry matter. Warm-water and cold-water fish require from 3 to 6 mg/kg of diet and 10 mg/kg of diet dry matter, respectively (73, 74). Crustaceans appear to have a higher pyridoxine requirement than fish, birds, or mammals. Shrimp (*Penaeus* spp) is reported to require 120 mg/kg of diet and insect requirements range from ~1 to 18 mg/kg of diet dry matter.

Pantothenic Acid
Pantothenic acid is a component of coenzyme A (CoA), which functions in the transfer of acyl groups. Acetyl CoA introduces acetate to the TCA cycle where the acyl group combines with oxaloacetate to form citrate. Pantothenic acid is not required by phytoflagellates but is required by most species of *Tetrahymena* (46). The slug *Arion ater* requires pantothenic acid in the diet, and when it is removed, there is a rapid depression of growth rate and survival is poor (86). Insects, fish, birds, and mammals require a dietary source of pantothenic acid, except where this is supplied by symbiotic bacteria. The dietary requirement of mammals and birds is of the order of 10–15 mg/kg of diet dry matter, whereas fish appear to have a higher requirement, up to 50 mg/kg of diet. The insect dietary requirements are within the range of mammal and fish requirements.

Folic Acid
The parent compound of the folate vitamins is pteroylglutamic acid, also known as folic acid. The major subunits are a pteridine moiety linked by a methylene bridge to *p*-amino benzoic acid, which in turn is joined by a peptide bond to glutamic acid. In the animal body, folate is present in the reduced state either as 7,8-dihydrofolate or 5,6,7,8-tetrahydrofolate (THFA) with various C_1 moieties linked to the N^5, N^{10}, or $N^{5,10}$ positions. The metabolic role of folic acid is the transfer of these single carbon groups.

Most species of *Tetrahymena* require folic acid for purine synthesis. Thymine can spare folic acid in *Tetrahymena* but not in *Glaucoma*. A requirement for folic acid has been demonstrated in snails reared under axenic conditions (98). Folic acid has been shown to be a dietary requirement of some insects, but for those in which a requirement has not been demonstrated, some doubt remains whether the test diet was completely free of folic acid. Tetrahydrofolate is necessary for nucleic acid synthesis in *D. melanogaster* and dietary nucleic acid spares the folic acid requirement. Complete larval development can occur in *M. domestica* without folic acid in the diet if nucleic acids are supplied (81). Folic acid appears to have

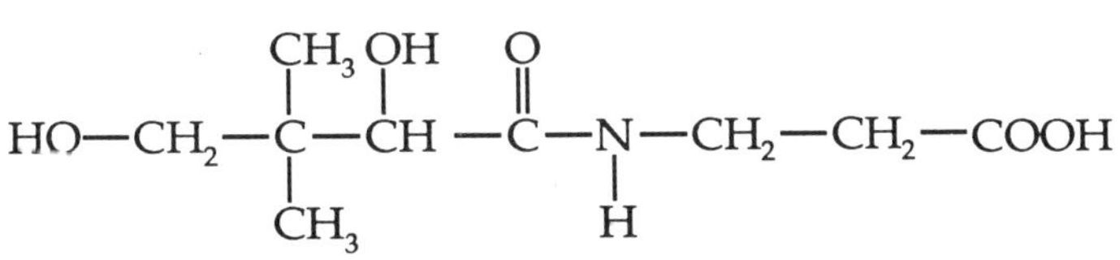

Pantothenic acid

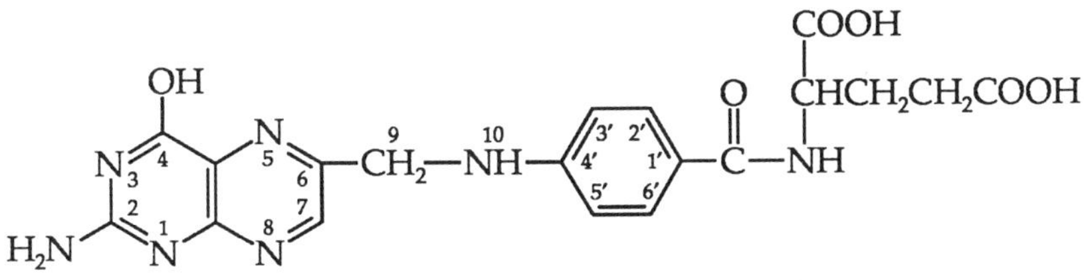

Folic acid

additional functions in *M. domestica* besides nucleic acid synthesis, because feeding an inhibitor of folic acid (aminopterin) in the presence of nucleic acids did not sustain normal growth.

Folic acid is an essential nutrient for all vertebrate animals. A deficiency of folic acid is usually expressed by clinical signs that include anemia. Folic acid has been shown to be required by salmonid fish. Some species of warm water teleosts show a reduction in growth rate and erythrocyte numbers when subjected to a folic acid deficient diet. The intestinal bacteria in the common carp can synthesize folic acid (74). Birds and all mammals have a metabolic requirement for folic acid. Folates are quite ubiquitous in nature and are present in both plant and animal tissues.

Vitamin B₁₂
Vitamin B_{12} or cobalamin is a coenzyme for methylmalonyl CoA mutase and methyltransferase. Cobalamin contains a large tetrapyrrole ring surrounding a single atom of cobalt. The cobalt is linked to the four nitrogen atoms of the corrin ring and also forms a bond with a carbon of an adenosyl group and a nitrogen of a benzimidazole group. The cobalt—carbon bond of cobalamin is the only known biological example of a carbon—metal bond.

The metabolism of folic acid and vitamin B_{12} is closely linked in mammals and birds, as both vitamins function in the transfer of single carbon units and are involved in the synthesis of thymidylate and therefore DNA. A deficiency of either vitamin produces a similar megaloblastic anemia. The interaction of the two vitamins is explained by the "methyl trap" hypothesis. Folate is active in thymidylate synthesis in the form of 5,10-methylene-THFA. Methyltransferase (which has cobalamin as a coenzyme) removes the methyl group from methyl-THFA converting it to the "active" THFA and transfers the methyl group to homocysteine

Vitamin B_{12}

(which in turn is converted to methionine). The THFA gains a methylene group donated by serine (which in turn is converted to glycine) and transfers this methylene group to deoxyuridylate to form thymidylate.

Vitamin B_{12} in the stable cyano form in which it is prepared chemically. In the active coenzyme form, the CN is replaced by a 5'-deoxyadenosyl group joined to the cobalt by a methylene bridge.

All the vitamin B_{12} in the world is synthesized by bacteria; thus, animals ultimately depend on bacteria for their vitamin B_{12}. Protozoa in general do not require vitamin B_{12} (53); however, there are many exceptions among the green and colorless flagellates. Both *Ochromonas malhamensis* and *Euglena gracillis* have been used as assay organisms for vitamin B_{12}. Natural fresh and marine waters contain vitamin B_{12}, especially along the shore line where bacterial activity is highest. Aquatic organisms obtain vitamin B_{12} either from the water or directly from bacteria. A number of algal species require vitamin B_{12} (29, 40). Bivalves such as clams and oysters, which siphon large quantities of B_{12}-synthesizing organisms from the sea, are a rich natural source of vitamin B_{12}. Vitamin B_{12} is essential for the snail *Biomphalaria glabrata* when grown in axenic culture (98).

Vitamin B_{12} is a metabolic requirement of all teleost fish, birds, and mammals. Microbial synthesis in the gut of some species of fish (e.g., *Tilapia*) and some mammals (e.g., ruminants) supplies the vitamin B_{12} requirement, provided that the diet contains an adequate level of cobalt. In animals that do not synthesize vitamin B_{12} in their gut, cobalt (except in the form of vitamin B_{12}) is not an essential dietary requirement.

Biotin
Biotin is a prosthetic group for many carboxylation reactions, the most notable being pyruvate carboxylase (needed for the synthesis of oxaloacetate in gluconeogenesis and replacement of TCA cycle intermediates), acetyl CoA carboxylase (fatty acid synthesis), and propionyl CoA carboxylase (for the entry of propionate and those amino acids that yield propionate) into the TCA cycle.

Axenic cultures of certain marine protista have shown that biotin, thiamin, and vitamin B_{12} are essential vitamins. Biotin concentrations of surface seawater ranges from <0.1 to 57.9 ng/L, the concentration being generally higher in coastal waters and lower in deep ocean waters (78).

Biotin

Among the protista, some species of *Tetrahymena*, *Crithidia*, and *Ochromonas* require biotin and have been used as microbiological assay organisms for biotin. Many other species do not appear to require a dietary source of biotin (46). Biotin has been demonstrated to be an essential nutrient for some insects, but in others the low requirement and the difficulty of achieving an axenic culture may preclude demonstration of a requirement (26). Synthesis of biotin by symbiotic organisms in the gastrointestinal tract renders most vertebrate animals independent of a dietary source of this vitamin. A metabolic requirement for biotin can be demonstrated if the biotin in the gut is rendered unavailable. The albumin of raw avian eggs contains a protein, avidin, which binds four molecules of biotin per molecule of avidin. This avidin–biotin complex cannot be digested and assimilated from the gut. Under these condi-

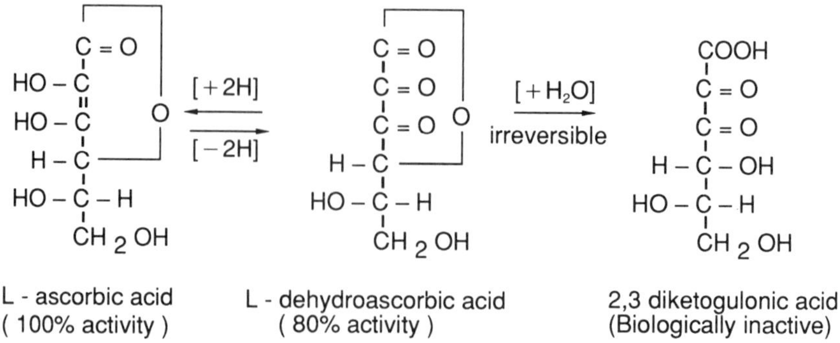

```
     C = O                      C = O                    COOH
HO – C                     C = O                     C = O
HO – C      O   [+2H]      C = O   O   [+H₂O]        C = O
H – C          ⇌           H – C      irreversible   H – C – OH
            [−2H]
HO – C – H                 HO – C – H                HO – C – H
     CH₂OH                      CH₂OH                    CH₂OH

L - ascorbic acid      L - dehydroascorbic acid     2,3 diketogulonic acid
( 100% activity )          ( 80% activity )          (Biologically inactive)
```

Figure 2. Oxidation of ascorbate.

tions a requirement for biotin has been demonstrated for all species tested.

The requirement for biotin for most species appears to be in the range of 0.1 to 0.3 mg of biotin/kg of diet dry matter.

Vitamin C

Ascorbate, unlike the other water-soluble vitamins, is not an established component of any enzyme and its exact functional role is not known. It is involved as a reductant in the hydroxylation of prolyl and lysyl residues of collagen and prolyl residues in elastin, in p-hydroxyphenylpyruvate oxidase and noradrenalin formation. Some of these functions relate to the ability of ascorbate to be reversibly oxidized and reduced (30). Ascorbate is capable of undergoing reversible oxidation to L-dehydroascorbic acid, which has ~80% of the biological activity of L-ascorbic acid. However, L-dehydroascorbic acid is easily oxidized further to 2,3-diketogulonic acid, which is biologically inactive. In species that do not synthesize ascorbate, a dietary deficiency of vitamin C results in the condition known as scurvy. Scurvy was a major disease that plagued sailors on prolonged sea voyages.

Most (but not all) invertebrates do not synthesize ascorbate, whereas most (but not all) vertebrates synthesize ascorbate from glucose. The synthetic pathway may have evolved when animals moved from an aquatic to a terrestrial environment. One theory of the evolution of the distri-

bution of ascorbate synthesis in animals, proposes that it originated in the kidney of amphibians, as synthesis is not present in earlier evolved animals such as invertebrates (particularly insects), and most fish (14). Reptiles, which are descendants of amphibians, are also able to synthesize ascorbate in the kidney. In mammals this synthetic ability moved from the kidney to the liver. Subsequently, some vertebrate species, including primates, guinea pigs, fruit-eating bats, some birds (e.g., mynah and passerines) lost the ability to synthesize ascorbate and for these species ascorbate is a vitamin. The failure of guinea pigs, fruit eating bats, and primates to synthesize ascorbate results from the low activity or absence of the last enzyme of the pathway, L-gulonolactone oxidase (GLO).

The above theory of the evolution of ascorbate synthesis has many anomalies. The Australian lungfish (*Ceratodus*), which evolved prior to the emergence of amphibians synthesizes ascorbate in the kidney. Monotremes have GLO activity only in the kidneys, and while most marsupials have activity only in the liver, some macropods (e.g., *Macropus rufogriseus*) and bandicoots (*Peramelina* spp.) have activity in both liver and kidney.

Ascorbate is not a dietary essential for the snail *Biomphalaria glabrata* (90a), but is required by most insects whose natural diet is fresh plant tissue. However, some insects such as the plant-boring lepidopteran *Pectinophthora gossypiella* can be reared on a synthetic diet devoid of ascorbate and accumulate ascorbate in their tissues. In insects there is a clear association between phytophagy and a requirement for ascorbate, but ascorbate is not required by all insects that feed on plant tissue. However, no example is known of an insect with a dietary requirement for ascorbate whose natural diet excludes live plant tissue. As feeding studies of ascorbate-requiring insects (with L-gulonolactone or precursors of gulonolactone along the putative pathway from glucose) did not demonstrate an ability to substitute for ascorbate, it has been assumed that insects lack L-gulonolactone oxidase.

Penaeid shrimp, Japanese eels (*Anguilla japonica*), channel catfish (*Ictalurus punctatus*), and all salmonids require dietary L-ascorbate. Deficiency signs include structural deformities, scoliosis, lordosis, and abnormal support cartilage of the eyes, gills and fins, and internal hemorrhage. Ascorbate deficiency in mammals is also associated with capillary fragility and hemorrhage as well as defective collagen and elastin. Lizards and snakes also show similar clinical signs from ascorbate deficiency.

Myo-*inositol* (*Inositol*)

Myo-inositol is one of nine isomers of inositol, a cyclic alcohol (hexitol), and is the only one that is biologically active. Biosynthesis of inositol appears to be widespread in microorganisms, plants, and animals. Despite its ubiquitous nature, the dietary significance of inositol is not clear. In plants inositol exists as the mono-, di- and triphosphate esters, and the hexaphosphate ester known as phytic acid. Phytic acid binds divalent ions, particularly calcium and magnesium, forming mixed salts of low availability to most animals. In animal tissue *myo*-inositol occurs both in the free form and in phospholipids such as phosphatidylinositol, diphosphoinositide, and triphosphoinositide. Glycosylated forms of phosphati-

dylinositol (termed glycosylphosphatidylinositols) play important roles in biological membrane function and serve as the means by which cell surface proteins are anchored to the membrane (63). Molecules with a similar structure have been implicated in the transmembrane signaling function of the hormone insulin. *Myo*-inositol is an essential dietary nutrient as a lipogenic growth factor for a number of diverse species of insects with no distinct taxonomic association. Most insects that require dietary inositol normally feed on living plant tissue or seeds, both of which are rich sources of combined inositol (26). *Myo*-inositol is also an essential dietary nutrient for salmonid fish (73); and a deficiency causes anorexia, anemia, and reduced growth rate. Inositol is not normally required by mammals. *Myo*-inositol has been shown to be a lipotrophic factor in the alleviation of fatty liver of rats fed a particular diet (33). Gerbils (*Meriones unguiculatus*) require inositol when the diet contains saturated fatty acids of moderate chain length (capric, lauric, and myristic acids) (42). It appears that under certain dietary and physiological conditions synthesis of inositol may be inadequate. The amount, degree of saturation, and fatty acid content of the fat in the diet, as well as the level of choline, are factors that affect the need for *myo*-inositol.

Fat Soluble Vitamins

The fat-soluble vitamins include vitamins A, D, E, and K. Vertebrates have a metabolic requirement for all four vitamins, although only vitamin A (or the pro-vitamin carotene) and vitamin E are usually essential in the diet. Among invertebrate animals, a metabolic requirement for these vitamins can be demonstrated in only a limited number of species. A dietary requirement for fat soluble vitamins has not been demonstrated in platyhelminths or in nematodes. Responses to vitamins A, D, and E have been reported in mollusca, but no response to fat-soluble vitamins was found for the marine snail *Ligia pallasii* (12).

Insects in general are insensitive to a lack of dietary fat-soluble vitamins over one larval cycle, but responses to fat soluble vitamins have been obtained in certain isolated species (26).

Vitamin A

Vitamin A is a term that is used to designate biologically active compounds including retinol, 3-dehydroretinol, and their esters. Vitamin A naturally exists only in animals and is a product of the enzymatic oxidation of carotenoids that are synthesized by all green plants. Among the carotenoids, β-carotene has the highest potency as the molecule is symmetrical and yields on oxidation two molecules of retinol. Other carotenoids such as cryptoxanthin, α-carotene, and γ-carotene are less potent.

The land snail *Helix pomatia* is reported to require vitamin A or carotene (48), but no evidence of a vitamin A requirement was found for the sea slater *Ligia pallasii* over a 40-week period (12). Similarly, the snail *Biomphalaria glabrata* has been cultured successfully in the absence of vitamin A or carotene (90a). In vertebrate animals, retinol (vitamin A alcohol) fulfills many functions including that of being a component of the visual pigment rhodopsin (108). Insect eye pigments are also rhodopsins; and a severe dietary deprivation of vitamin A or carotene has been shown to lead to impaired light response in several species of insects. Deletion of carotene from the diet of two locusts (*Locusta migratoria* and *Schistocerca gregarai*) leads to growth retardation in the second generation and anomalous color and behavior patterns (26, 32). Larvae of the moth *Plodia interpunctella* died in the first instar, apparently without feeding, when given semisynthetic diets devoid of carotene.

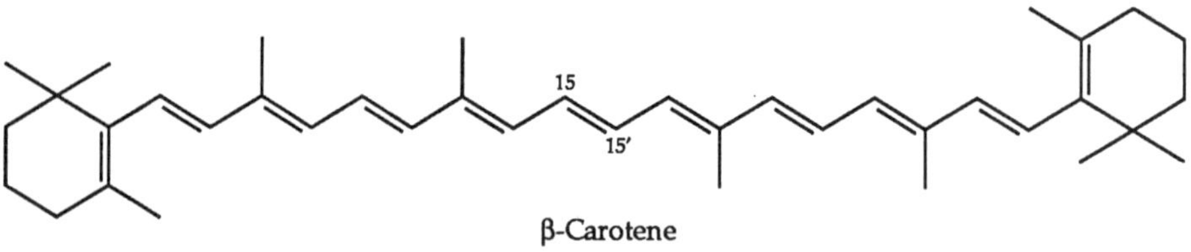

β-Carotene

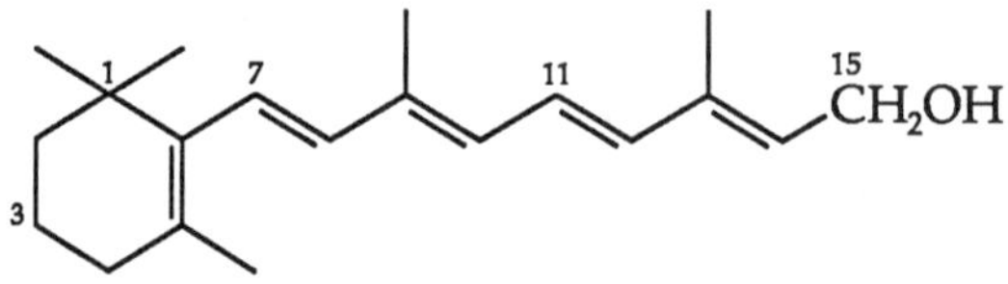

All-*trans*-retinol

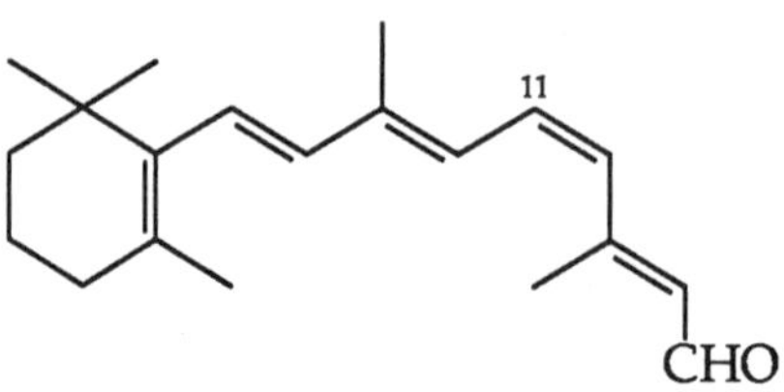

11-*Cis*-retinal

Vitamin A

Similarly, larvae of the silkworm moth, when reared aseptically on a carotene or vitamin A-free diet, went through progressively delayed molts compared to larvae given diets containing carotene. The adults from larvae reared on carotene-free diets also produced fewer viable eggs. While the silkworm seems to require carotene or vitamin A for optimal growth, a requirement for carotene by insects in general, other than for vision, is equivocal (32).

In crustaceans, vitamin A is concentrated in the eye and probably has a role in the crustacean visual cycle. Turtles when fed lettuce in captivity develop vitamin A deficiency (21).

Fish require either carotene or vitamin A in their diet. Cold-water fish can utilize β-carotene as a precursor for vitamin A at temperatures of 12.4 to 14°C, but not at 9°C (83). Channel catfish utilize β-carotene if it is provided in the diet at over 2000 IU/kg of diet (74). The vitamin A present in freshwater fish is predominately vitamin A_2 (all-*trans*-3-dehydroretinol), whereas saltwater fish have mainly vitamin A_1 (all-*trans* retinol), the same form as occurs in mammals. The 11-*cis*-3-dehydroretinal form of naturally occurring vitamin A_2 is active in fish but not mammals or birds.

Amphibians, in general, and tadpoles of *Rana catesbiana* contain only vitamin A_2. In these tadpoles the vitamin A_2 is replaced by vitamin A_1 in the adults (101). Tadpoles of other species of frogs (*R. esculenta* and *R. temporaria*) have only vitamin A_1 (18). Adult *Rana temporaria* frogs convert carotenoids to vitamin A that is stored, predominately in the liver (85%), but also in the eye and kidney in about equal amounts (70). The amount of vitamin A stored in the kidney of this frog is much greater than that in most mammals except in cats. Another amphibian, the axolot *Amblyostoma tigrinum* is able to assimilate vitamin A and store it in its liver, but it is unable to utilize β-carotene and

apparently growth is not compromised in the absence of vitamin A in the diet (17).

All birds require either a dietary source of either carotene or vitamin A for normal growth and reproduction. The color of some birds (e.g., the pink of the flamingo) can develop only if the diet contains carotenoid type pigments such as astaxanthin.

All mammals, including those with symbiotic microflora in the gut, require a dietary source of vitamin A or carotene. The main site of conversion of dietary carotenoids to vitamin A is the intestinal mucosa, although some conversion may also occur in the liver. The efficiency of the conversion of β-carotene to vitamin A ranges from zero in cats (which lack the 15,15′ dioxygenase), to a stoichiometric conversion of 1 mol of β-carotene into 2 mol of retinal by the soluble enzyme of the rat intestinal mucosa or liver (36).

Vitamin A deficiency in mammals affects a number of organ systems because of the role of vitamin A in RNA and DNA synthesis and cell differentiation. Only ~0.01% of the body's vitamin A is in the eye, but one of the early signs of vitamin A deficiency is defective vision caused by loss of the visual pigment rhodopsin. As the rods of the retina are affected before the cones, the initial loss of vision occurs most acutely in low light intensities (nyctalopia). Vitamin A is required for normal bone development in young animals and a deficiency leads to compression and stretching of nerves, especially the cranial nerves. Reproduction is compromised in vitamin A deficiency both in the male and in the female. Spermatogenesis in males requires vitamin A: In females, vitamin A deficiency leads to keratinization of the mucus secreting cells of the reproductive tract, which leads to failure in implantation or failure to maintain pregnancy or, if pregnancy is sustained, teratogenic defects in the fetus. Vitamin A is required for normal epithelial growth and differentiation, and a deficiency of vitamin A is associated with an increase in the proportion of squamous keratinized cells and a decrease in the proportion of mucus secreting columnar cells in the respiratory, alimentary, urinary, and reproductive tracts. These cellular changes render epithelial and mucosal tissues more susceptible to bacterial infections.

In young animals, vitamin A deficiency leads to impairment of growth. Retinoic acid can sustain growth, but it cannot substitute for retinol or retinal to support vision or reproduction.

Vitamin D

If vitamin D were to be discovered today, it would be classified as a hormone rather than a vitamin. Vitamin D is unique among the vitamins as it is the only vitamin that is synthesized in one tissue (skin), and that undergoes sequential hydroxylations in two tissues (liver and kidney) to become the active compound; a hormone that modulates the activity of bone and the intestinal mucosa (28, 44). Evidence for an essential role for vitamin D in other than vertebrate animals is questionable. Sterols aid in calcium absorption in the snail *Helix aspersa*, but cholesterol and ergosterol appear to be more stimulatory than vitamins D_2 or D_3 (99). Addition of cod liver oil (a source of vitamins A and D) to a purified diet for axenic snails was without demonstrable effect on growth or reproduction (98). The evidence suggesting a role for vitamin D in crustaceans is also equivocal (19, 56). There is no substantive evidence suggesting that vitamin D has a role in insects (26, 32).

Vitamin D is required by all vertebrate animals. Fingerling trout require vitamin D in the diet, and in its absence grow slowly and have impaired calcium homeostasis as evidenced by clinical signs of tetany of the white skeletal muscles (73). No hypocalcemia or changes in bone ash were found in trout deprived of vi-

tamin D (74). South African clawed toads (*Xenopus laevis*) are very susceptible to rickets and osteoporosis (similar to that which occurs in mammals), when deprived of vitamin D (9). Reptiles also develop rickets and osteoporosis in the absence of vitamin D.

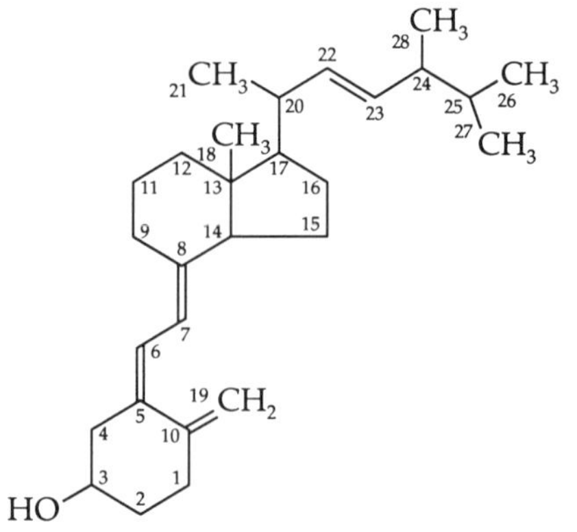

Vitamin D₂
(ergocalciferol)

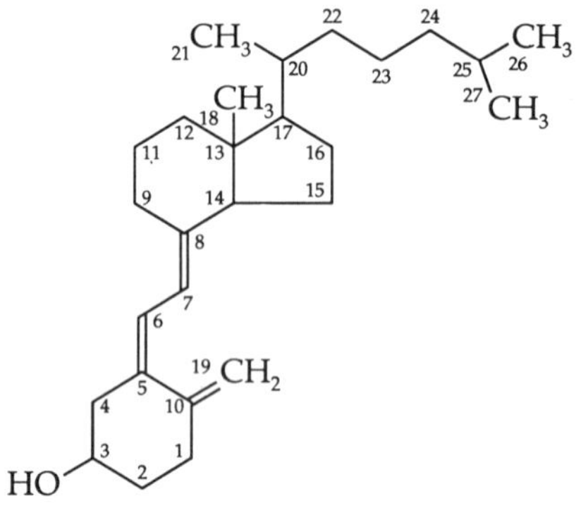

Vitamin D₃
(cholecalciferol)

Vitamin D can be supplied either from the diet or from ultraviolet (UV) irradiation of the skin. A source of vitamin D activity in the diet of animals is vitamin D_2 (ergocalciferol), which is the product of UV irradiation of the plant sterol ergosterol. Vitamin D_3 (cholecalciferol) is formed when 7-dehydrocholesterol in the skin of animals is irradiated with UV light. These two forms of vitamin D are assumed to possess equal potency in most mammals, as studies in rats have shown that low doses of ergocalciferol produce the same degree of intestinal calcium transport stimulation, bone calcium mobilization, and calcification of rachitic cartilage as cholecalciferol (71). However, in certain other animals, large differences exist between the potency of the two forms of the vitamin. In new world primates (e.g., marmosets, tamarinds, etc.), vitamin D_2 has much lower activity than vitamin D_3 (52). Cholecalciferol has greater potency than ergocalciferol in domesticated pigs but not in rats (47), and it has been suggested that this specific difference may be a reflection of the natural source of vitamin D for these two species. As rats are nocturnal animals, they consume mainly ergocalciferol from vegetable matter, whereas pigs would obtain most of their vitamin D from exposure to sunlight. These findings caution against the general assumption of equal potency of these two prohormones in mammals. For birds, the activity of vitamin D_2 is only $\sim\frac{1}{30}$ to $\frac{1}{40}$ that of vitamin D_3 (75). For trout, vitamin D_3 is at least three times as effective as vitamin D_2 in satisfying the requirement for vitamin D.

The metabolic pathway of vitamin D is shown in Figure 3. Vitamin D (either D_2 or D_3), whether derived from the diet or synthesized in skin, is metabolized by the liver resulting in an hydroxylation to give 25-OH-D. This hydroxylation reaction is inhibited by its product, which prevents an excessive amount of the more biologically active form of vitamin D being produced. The second hydroxylation reaction occurs in the kidney which results in the production of 1,25-$(OH)_2$-D, the most active form of vitamin D. This second hydroxylation is mediated by the parathyroid hormone (PTH), which is elevated when the concentration of calcium declines in the blood.

Metabolically, 1,25-$(OH)_2$-D, is inti- mately concerned in calcium and phosphorus homeostasis. In the short-term regulation of plasma calcium, 1,25-$(OH)_2$-D acts with PTH causing a resorption of bone and the release of calcium and phosphorus. Parathyroid hormone also acts on the kidney causing enhanced excretion of phosphorus, the net effect being a normalization of plasma calcium. Long-term regulation of calcium homeostasis requires that the gut increase calcium uptake. The active metabolite of vitamin D binds with a specific receptor in the enterocyte nucleus of the intestinal mucosa and initiates events leading to the formation of calcium-binding protein and a stimulation of calcium and phosphorus absorption by the intestine.

A large number of metabolites of vitamin D ($\sim$33) have been isolated and identified, but their biological function is uncertain.

A deficiency of vitamin D results in defective mineralization of the cartilaginous matrix of growing bones, and those bones that are weight bearing become deformed and assume the characteristic curvature of rickets. In adult animals, a deficiency of vitamin D results in a reduced uptake of calcium from the gut and demineralization of bones.

An excessive intake of any nutrient can be deleterious. The intake of the nutrient, in terms of multiples of the minimal requirement of a nutrient that are tolerated before toxicity is encountered, is one measure used to compare the relative toxicities of nutrients. Using this method of comparison, vitamin D is among the most toxic of all nutrients, and vitamin D_3 is more toxic than vitamin D_2 (76). The potential toxicity of vitamin D is the basis for the inclusion of cholecalciferol as the active ingredient in a common rodenticide. Excessive exposure to sunlight does not result in vitamin D toxicity. In the photolysis of provitamin D in the epidermis, a thermally labile precursor known as previtamin D_3 is produced. Once pro-

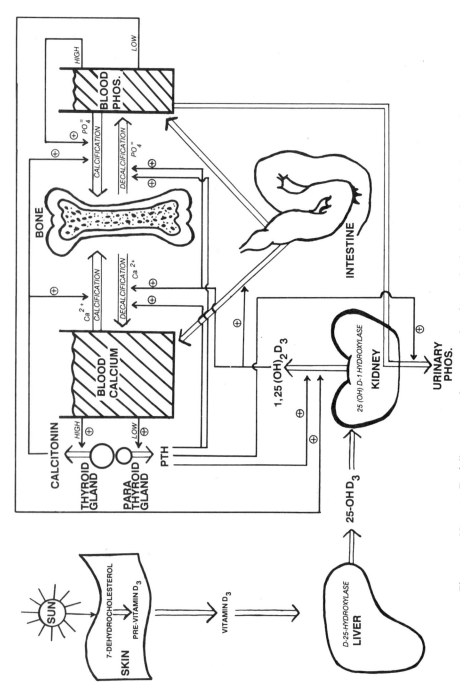

Figure 3. Vitamin D_3 following production in the skin is hydroxylated in the liver and kidneys to the active form of the vitamin, $1,25(OH_2)D_3$. $1,25(OH_2)D_3$ in concert with the hormones of the parathyroid and thyroid glands (parathyroid hormone (PTH) and calcitonin) play the major role in the regulation of the concentrations of calcium and phosphorus in the blood, and intestinal absorption. Double lines indicate pathways and single lines indicate control. (From DeLuca (27). Used with permission.)

duced, previtamin D$_3$, in the absence of sunlight is converted over a period of 2 to 3 days to vitamin D$_3$. Prolonged exposure to sunlight results in photodegradation of previtamin D$_3$ to biologically inert photoproducts (101a).

Vitamin E

Vitamin E is the generic term for a group of lipid-soluble tocol and tocotrienol derivatives that possess vitamin activity, the most active form of vitamin E being D-α-tocopherol. The word "toco" in tocopherol comes from the Greek "*tokos*" meaning childbirth, because it restores fertility to deficient rats. Tocopherols protect polyunsaturated lipids in cell membranes against oxidative damage. Factors that alter the vitamin E requirement of animals are oxidizing agents, ozone, peroxidizing fatty acids, and catalytic trace elements such as copper.

The only identified role of vitamin E in invertebrates is its effect on reproduction. In rotifers of the genus *Asplanchna* (*A. brightwelli, A. intermedia,* and *A. sieboldi*), the size and shape of the female is controlled by the vitamin E in the source of food. In the absence of tocopherol, females are relatively small, sacculate, and amictic (female producing). When tocopherol is present they are 50–200% larger in length, and may possess characteristic body-wall outgrowth, giving a cruciform or campanulare shape, and are often mictic (male producing) (34, 35).

Vitamin E is required for reproduction in mollusca. Axenically cultured snails (*Biomphalaria glabrata*), given a defined basal diet showed no difference in growth rate regardless of whether the basal diet was or was not fortified with α-tocopherol. However, there was no oviposition in those snails given the basal diet. When this diet was fortified with α-tocopherol, oviposition commenced 2–4 weeks later (97).

The main effect of vitamin E deficiency in insects is not expressed until the reproductive stage is reached. Tocopherol addition to the diet improves the fecundity of some moths and beetles and in at least one case, that of the cricket (*Acheta domesticus*), is necessary for viable sperm in the male (32).

Although no requirement for vitamin E has been demonstrated in crustaceans, the essentiality of a dietary source of polyunsaturated fatty acids suggests that vitamin E may be required for their protection against peroxidation.

Teleost fish require a dietary source of vitamin E and deficiencies have been described in trout, salmon, carp, and channel catfish (73, 74). In these species there is impaired erythropoiesis, which is characterized by many immature irregularly shaped erythrocytes, fragility and fragmentation of erythrocytes, extreme anemia, high mortality, accumulation of serous fluid in the body cavity, and accumulation of body water (exudative diathesis).

Crocodilians develop steatitis (accu-

α-Tocotrienol

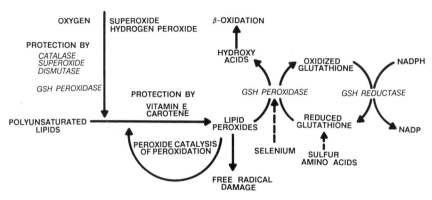

Figure 4. Relation between vitamin E and selenium in the glutathione peroxidase interaction. The net effect is to reduce accumulation of toxic peroxides (95a).

mulation of ceroid pigment in the adipose tissue) when given diets low in vitamin E. Vitamin E deficiency in birds may result in either encephalomalacia, exudative diathesis, or nutritional muscular dystrophy (75). Birds with encephalomalacia move with an uncoordinated gait; the head is often retracted, the legs outstretched, and the toes flexed. Exudative diathesis is an edematous condition where fluid accumulates in the tissues, particularly those of the breast. In nutritional muscular dystrophy there is degeneration of muscle, particularly those of the breast. The addition of vitamin E or selenium to the diet will prevent the occurrence of exudative diathesis and muscular dystrophy, but encephalomalacia requires vitamin E supplementation.

In mammals, a number of clinical conditions responsive to either vitamin E or selenium have been recorded. These include depressed growth in young animals, nutritional muscular dystrophy or "white muscle disease", reproductive disorders including the birth of weak young. In dogs, hamsters, pigs, rabbits, and rats a lack of vitamin E results in male infertility. Vitamin E is required to prevent fetal resorption in female rats. Selenium addition to the diet will prevent

muscular dystrophy in goats and lambs but not in rabbits or guinea pigs, while vitamin E is effective in rabbits and guinea pigs, but little effect on sheep and goats in the absence of an adequate level of dietary selenium (90).

The relationships between vitamins which protect against oxidative damage to cells and enzymes involved in oxidative protection (glutathione peroxidase which contains selenium, catalase and superoxide dismutase) are shown in Figure 4.

Vitamin E prevents the formation of peroxides especially in the cell membrane. Once peroxide formation occurs they act autocatalytically producing further peroxides. Selenium is an integral part of the enzyme glutathione peroxidase which reduces lipid peroxides to hydroxy acids which can be metabolized by β-oxidation. In the absence of selenium, the enzyme glutathione peroxidase is depleted which permits accumulation of lipid peroxides and breakdown of cellular membranes.

Vitamin K

Hemostasis or the cessation of bleeding in animals is achieved by formation of a fibrin clot, which prevents loss of blood

Vitamin K_1 (green leafy vegetables)

Vitamin K_2 (intestinal bacteria)

from the site of injury. Production of the fibrin clot is the end product of a cascade of reactions that produces thrombin (an enzyme), which acts on the protein fibrinogen to produce fibrin. The classical defect of vitamin K deficiency in animals produces a lowered concentration of factors in the cascade: factor II, factor IX, and factor X. These factors are often referred to as "vitamin K dependent factors" as they require activation, a step which involves the participation of vitamin K.

Vitamin K is used as the generic descriptor of 2-methyl-1,4-naphthoquinone and of all derivatives that exhibit an antihemorrhagic activity in animals fed a vitamin K deficient diet. The compound 2-methyl-3-phytyl-1,4-naphthoquinone is generally referred to as vitamin K_1 or phylloquinone. A compound first iso lated from putrefied fish meal and called vitamin K_2 is one of a series of vitamin K's with unsaturated side chains that are found in animal tissues and bacteria (95). A number of trans isomers of substituted 2-methyl-1,4-naphthoquinones synthesized by green plants and bacteria have vitamin K activity. Bacteria, which are normal inhabitants of the gastrointestinal tract, render most animals independent of a dietary source of vitamin K.

The major metabolic function of vitamin K is its participation as a cofactor for a microsomal enzyme system that carboxylates peptide bound glutamyl residues on proteins to γ-carboxyglutamyl residues. This post-translational modification of the precursor clotting factors allows these proteins to participate in specific protein-Ca^{2+}-phospholipid reactions that are necessary for their biological function. In addition to the four classical vitamin K dependent coagulation proenzymes (factors II, VII, IX, and X), four additional carboxyl coagulation factors (proteins C, S, M, and Z) are recognized. All four proteins have amino acid sequences homologous to that of prothrombin, and all require calcium for activation (79). Other proteins, such as a bone protein osteocalcin, undergo the same post-translational modification as the vitamin K dependent clotting factors.

Vitamin K deficiency occurs infrequently in animals, except for chickens that have a high requirement for the vitamin and are often fed antibiotics that depress intestinal synthesis (95). The low incidence in other species is attributed to synthesis of vitamin K by bacteria in the gastrointestinal tract. Only when the diet contains an antagonist of vitamin K (e.g., dicumarol) does a deficiency become evident.

Vitamin K does not appear to be required by insects, though brood rearing of worker honey bees is substantially increased by the addition of menadione to an amino acid-based synthetic diet; however, as this was accompanied by a large increase of food intake, the effect may be a result of either a phagostimulatory or a nutritional response (26). Trout and channel catfish require vitamin K for normal blood clotting (73, 74). Gingival bleeding attributed to vitamin K deficiency has been reported in crocodilians in captivity (21). Vitamin K is a metabolic requirement of all mammals, but those with microbial digestive systems (e.g., ruminants and many others that harbor a significant bacteria in their gastrointestinal tracts) do not require a dietary source of the vitamin.

Minerals

In addition to compounds containing carbon, hydrogen, and oxygen, plants and animals need ~20 to 30 inorganic elements (see Chapter 2 Water and Ions). The nutritional requirements for these essential inorganic elements are remarkably similar for all living cells, which possibly reflects evolution from a common organism and the use of similar metabolic reactions in life processes. Many enzymes require divalent metal ions such as magnesium and manganese for their activation, while in other enzymes, metal ions, particularly those of the transition series of the periodic table, participate in oxidation–reduction reactions.

One of the most interesting elements from an evolutionary viewpoint is phosphorus, which occupies a pivotal role in the genetic transference of information (DNA and RNA diesters of phosphoric acid), in energy transference (ATP, creatine phosphate, and phosphoenolpyruvate—anhydrides, amides, and enol esters of phosphoric acid), and as intermediates of metabolism (e.g., glucose-6-phosphate and fructose-6-phosphate, which are esters of phosphoric acid). A major reason which permits phosphates to occupy this unique place among the elements used by living systems is that they are ionized at a physiological pH (103). Electrically neutral molecules have appreciable solubility in lipid membranes, whereas ionized molecules are retained in cells.

The minerals essential for vertebrate animals are frequently divided into two groups: the macro (or major) minerals and the minor (or trace) minerals. The macro minerals are calcium, phosphorus, magnesium, potassium, sodium, chlorine, and sulfur. The minor minerals are zinc, iron, copper, manganese, molybdenum, cobalt, iodine, fluorine, selenium, chromium, vanadium, silicon, tin, and nickel. There have not been any experiments where the essentiality of all inorganic elements has been demonstrated in any one species of animal. Nevertheless, it is fairly safe to assume that if the requisite experiments were undertaken, a virtual universal requirement would be found.

Demonstration of the essentiality of a dietary element generally requires three steps (1) the preparation of a purified diet deficient in the element, (2) feeding this diet to animals to induce the deficiency, (3) showing that the deficiency is reversed by addition of the element under investigation. Even at high degrees of purification (e.g., using crystalline amino acid in place of protein and purified carbohydrates), a diet may still contain many atoms of an element. Avogadro's number states that one gram mole of an element contains 6×10^{23} molecules of the element. A diet containing only one part per billion (1×10^{-9} by weight), for example, of aluminum, contains a large number (2×10^{13}) atoms Al/kg. Therefore, demonstration that an element is essential in the diet can more readily be shown than the reverse case that it is not essential.

Most animals obtain their mineral requirements from their normal diet, without seeking out specific foods rich in the mineral or salts of the mineral element. There are some notable exceptions to this generalization. Herbivorous animals, particularly ruminants and rabbits, actively seek salts of sodium and have a specific appetite for sodium. The pasture plants on which these animals graze do not require sodium as a nutrient (or require very low concentrations). This imbalance between the animal's requirement for a nutrient and the ability of the forage to supply it may have caused the evolution of a specific appetite for sodium. Another example of a specific appetite is that of the chicken for calcium. The laying hen actively seeks out sources of calcium such as shells, which she eats in the latter part of the day to supply calcium for the shell gland during the night.

Vertebrate animals have a high requirement for calcium, especially during the period of active growth of the skeleton and during lactation and pregnancy. Similarly invertebrate animals with a calcified shell or exoskeleton also have a high requirement for calcium when the exoskeleton is shed at molting. This calcium may be partially supplied by calcium stores laid down before molting or from dissolved calcium in the water of aquatic environments.

Sodium and potassium are the major extracellular and intracellular cations of animal cells and are important in the maintenance of osmotic and water balance, in nerve conduction, as well as in many other cellular functions such as transport of nutrients into the cell. The major portion of the body's chloride is in the extracellular fluid acting as the anion to balance sodium. Most of the magnesium in the body of vertebrate animals is associated with the skeleton, but magnesium fulfills an important role in maintenance of the irritability of nerves and as a necessary divalent metal ion for the activation of enzymes.

Deficiencies of many of the trace elements in animals can be related directly to known functional roles of the element in essential enzymes or in metabolic pathways. For example, iron deficiency results in anemia because iron is an essential element of heme, a component of hemoglobin in the erythrocyte. Copper deficiency can also result in anemia as copper is required for the mobilization of iron. Copper deficiency has other clinical signs directly related to a deficits of copper-containing enzymes: defective collagen and elastin synthesis (lysyl oxidase), depigmentation of normally dark hair because of inadequate melanin synthesis (tyrosine oxidase), and defective myelin synthesis (cytochrome oxidase).

Deficiencies of other trace elements can be related to the participation of the trace element as a necessary component of an enzyme (e.g., selenium in glutathione peroxidase) or a vitamin that acts as a coenzyme (e.g., cobalt in vitamin B_{12}). However, there are other trace elements such as zinc that is required for cell growth, development, and differentiation in all living species. Although zinc is an essential component of >200 enzymes, alteration in the function of these enzymes alone cannot account for the manifestation of zinc deficiency. Studies on the function of zinc in *Euglena gracilis* have demonstrated that zinc is required in replication, transcription, and translation; a deficiency of zinc leads to alterations in the structure of DNA. Zinc has been shown to be an integral part of many DNA and RNA polymerases. The intrinsic Zn ions in RNA polymerase play (a) a regulatory role in specific promoter recognition and RNA chain initiation, (b) a role in the recognition and orientation of the initiation nucleotide for catalysis, and (c) a structural role for maintaining the proper configuration necessary for its function (107).

Conclusion

The transition from prokaryotic to eukaryotic cells was accompanied by marked changes in the nutritional requirements of organisms. Many transitional eukaryotes were photosynthetic and also capable of heterotrophic metabolism. Heterotrophs evolved metabolically in the Precambrian era, and some of the major nutritional requirements of today's animals were apparently established at that time; examples are the universal need, from protozoa to mammals, for 10 essential amino acids and vitamins such as thiamin and vitamin B_{12}. The ability to synthesize an essential compound requires production of enzymes for the synthetic pathway, the more steps in the pathway the greater the number of enzymes involved. The pathways for the synthesis of the essential amino acids contain a greater number of steps than those for the synthesis of the dispensable amino acids. It appears that in animals either the competitive advantage of being able to synthesize the essential amino acids was less than the costs involved, or, the evolution of animals towards greater degrees of cellular organization and specialization, which in itself required the production of new organic compounds, was more readily accomplished without the incorporation of additional metabolic pathways for the synthesis of nutrients.

Metabolism in all animal cells uses many common pathways to derive energy from food and to synthesize organic molecules for maintenance, growth, and reproduction. Consequently, the nutrient requirements of animals exhibit a high degree of parallelism across species. However, variations in nutrient requirements of animals are present and appear to have originated for one or more of the following reasons.

First, some animals evolved pathways for the synthesis of essential compounds, whereas others did not. Specialized animals, such as protozoa and insects, did not develop the pathways for the synthesis of purines, pyrimidine, and cholesterol, whereas vertebrate animals evolved these pathways. Cholesterol synthesis in turn permitted the evolution of a wide range of steroid hormones. Second, some animals developed a metabolic need for a constituent that must have been normally present in the diet and that became a precursor for an essential metabolite. For example, the fat soluble vitamins are not required by most invertebrates (with the possible exception of vitamin A for vision in some species) but are metabolic requirements for all vertebrates. Third, other animals in the course of evolution lost key enzymes or total pathways for the synthesis of nutrients normally present in the diet. Among mammals, cats provide examples of such metabolic changes. The absence (or low activity) of key enzymes has resulted in their having a dietary requirement for nutrients not needed by other mammals; two examples are the absence of the 15,15′ dioxygenase (inability to synthesize vitamin A from carotene), and the low activity of the Δ5 desaturase (limited ability to synthesize arachidonic acid from linoleic acid). Another example of defective synthesis of an essential nutrient occurs in cats where there has been an increase in the activity of an enzyme that takes substrate along a pathway that competes with the synthetic pathway. This has apparently occurred in the tryptophan to niacin pathway rendering niacin a dietary requirement for cats.

Apparently evolution of animals proceeded both by the development of new synthetic pathways and the deletion of existing synthetic pathways. The evolution of new synthetic pathways involves an additional expenditure of energy for the synthesis of new proteins, whereas the deletion of synthetic pathways probably followed where the nutrient synthesized by the pathway was always present in the diet. The maintenance of redundant pathways (enzymes) for the synthesis of nutrients that are supplied in the diet is energetically inefficient. Deletion of a synthetic pathway extracts a price, because it restricts the animal's ability to utilize new sources of food which are deficient in the nutrient elaborated by the deleted pathway.

References

1. Antia, N. J. In *Biochemistry and Physiology of Protozoa* (M. Levandowsky and S. H. Hutner, eds.), 2nd ed., Vol. 3, pp. 67–115. Academic Press, New York, 1980. Nutritional physiology and biochemistry of marine cryptomonads and chrysomonads.

2. Baker. D. K. In *CRC Handbook Series in Nutrition and Food, Section D: Nutritional Requirements* (M. Rechcigl, Jr., ed.), Vol. 1, pp. 425–428. CRC Press, Cleveland, OH, 1977. Feeding behavior.

3. Baker H. and O. Frank, *Clinical Vitaminology: Methods and Interpretation*. Wiley (Interscience), New York, 1968.

4. Baker, J. E. *J. Insect Physiol.* 21:1337–1342, 1975. Vitamin requirements of larvae of *Sitophilus oryzae*.

5. Barrett, J. *Biochemistry of Parasitic Helminths*. University Park Press, Baltimore, MD, 1981.

6. Black, E. C., Robertson, A. C. and R. R. Parker. In *Comparative Physiology of Carbohydrate Metabolism in Heterothermic Animals* (A. W. Martin, ed.), pp. 89–124. Univ. of Washington Press, Seattle, 1961. Some aspects of carbohydrate metabolism in fish.

7. Bridges, R. G. *Adv. Insect Physiol.* 9:51–110, 1972. Choline metabolism in insects.

8. Bridges, R. G. In *Metabolic Aspects of Lipid Nutrition in Insects* (T. E. Mittler and R. H. Dadd, eds.), pp. 159–181. Westview Press, Boulder, CO, 1983. Insect phospholipids.

9. Bruce, H. M. and A. S. Parkes. *J. Endocrinol.* 7:64–80, 1950. Rickets and osteoporosis in *Xenopus laevis*.

10. Buchner, P. *Endosymbiosis of Animals with Plant Microorganisms*. Wiley (Interscience), New York, 1965.

11. Campbell, J. W. and S. H. Bishop. In *Comparative Biochemistry of Nitrogen Metabolism* (J. W. Campbell, ed.), Vol. 1, pp. 103–206. Academic Press, London, 1970. Nitrogen metabolism in molluscs.

12. Carefoot, T. H. *Comp. Biochem. Physiol. A* 79A:655–665, 1984. Studies on the nutrition of the supralittoral isopod *Ligia pallasii* using chemically defined artificial diets: Assessment of vitamin, carbohydrate, fatty acid, cholesterol and mineral requirements.

13. Chambers, I., Frampton, J., Goldfarb, P., Affara, N., McBain, W., and P. R. Harrison. *EMBO J.* 5:1221–1227, 1986. The structure of mouse glutathione peroxidase gene: the selenocysteine in the active site is encoded by the 'termination' codon, TGA.

14. Chatterjee, I. B. *Science* 182:1271–1272, 1973. Evolution and the biosynthesis of ascorbic acid

15. Clarke, R. T. J. In *Microbial Ecology of the Gut* (R. T. J. Clarke and T. Bauchop, eds.), pp. 251–275. Academic Press, London, 1977. Protozoa in the rumen ecosystem.

16. Coleman, G. S. In *Biochemistry and Physiology of Protozoa* (M. Levandowsky and S. H. Hutner, eds.), Vol. 2, pp. 381–408. Academic Press, New York, 1979. Rumen ciliate protozoa.

17. Collins, F. D., Love, R. M. and R. A. Morton. *Biochem. J.* 53:626–629, 1953. Studies in vitamin A; vitamin A and its occurrence in *Amblystoma tigrinum*.

18. Collins, F. D., Love, R. M. and R. A. Morton. *Biochem. J.* 53:632–636, 1953. Studies in vitamin A; visual pigments in tadpoles and adult frogs.

19. Conklin, D. E. In *Proceedings of the Second International Conference on Aquaculture Nutrition: Biochemical and Physiological Approaches to Shellfish Nutrition* (G. D. Pruder, C. J. Langdon, and D. E. Conklin, eds.), Spec. Publ. No. 2, pp. 146–165. Louisiana State University Division of Continuing Education, Baton Rouge, 1983. The role of micronutrients in the biosynthesis of the crustacean exoskeleton.

20. Cook, J. R. In *The Biology of Euglena* (D. E. Buetow, ed.), Vol. 1, pp. 243–314. Academic Press, New York, 1968. The cultivation and growth of Euglena.

21. Coulson, R. A. In *CRC Handbook Series in Nutrition and Food, Section D: Nutritional Requirements* (M. Rechcigl, Jr., ed.), Vol. 1, pp. 457–466. CRC Press, Cleveland, OH, 1977. Qualitative requirements and utilization of nutrients: Reptiles.

22. Cowey, C. B., Brown, D. A., Adron, J. W. and A. M. Shanks. *Mar. Biol. (Berlin)* 28:207–213, 1974. Studies on the nutrition of marine flatfish. The effect of dietary protein content on certain cell components and enzymes in the liver of *Pleuronectes platessa*.

23. Cowey, C. B. and J. R. Sargent. *Adv. Mar. Biol.* 10:383–492, 1972. Fish nutrition.

23a. Cowey, C. B. *Proc. World Symp. Finfish Nutr. Fishfeed Technol.* Hamburg, 1978, 1:3–16, 1979. Protein and amino acid requirements of finfish.

24. Dadd, R. H. In *Current Topics in Insect Endocrinology and Nutrition* (G. Bhaskaran, S. Friedman, and J. G. Rodriguez, eds.), pp. 189–214. Plenum, New York, 1981. Essential fatty acids for mosquitoes, other insects and vertebrates.

25. Dadd, R. H. In *Metabolic Aspects of Lipid Nutrition in Insects* (T. E. Mittler and R. H. Dadd, eds.), pp. 107–147. Westview Press, Boulder, CO, 1983. Essential fatty acids: Insects and vertebrates compared.

26. Dadd, R. H. In *Comprehensive Insect Physiology Biochemistry and Pharmacology* (G. A. Kerkut and L. I. Gilbert, eds.), 4: pp. 313–390. Pergamon, New York, 1985. Nutrition: organisms.

27. DeLuca, H. F. *Vitamin D: Metabolism and Function.* Springer-Verlag Berlin, 1979.

28. DeLuca, H. F. *FASEB J.* 2:224–236, 1988. The vitamin D story: A collaborative effort of basic science and clinical medicine.

29. Droop, M. R. In *Physiology and Biochemistry of Algae* (R. A. Lewin, ed.), pp. 141–159. Academic Press, New York, 1962. Organic micronutrients.

30. Englard, S. and S. Seifter. *Annu. Rev. Nutr.* 6:365–406, 1986. The biochemical functions of ascorbic acid.

31. Fraenkel, G. In *Carnitine Biosynthesis, Metabolism, and Functions* (R. A. Frenkel and J. D. McGarry, eds.), pp. 1–6. Academic Press, New York, 1980. The proposed vitamin role of carnitine.

32. Friend, W. G. and R. H. Dadd. *Adv. Nutr. Res.* 4:205–247, 1982. Insect nutrition: a comparative perspective.

33. Gavin, G. and E. W. McHenry. *J. Biol. Chem.* 139:485, 1941. Inositol: A lipotropic factor.

34. Gilbert, J. J. *Am. J. Clin. Nutr.* 27:1005–1016, 1974. Effect of tocopherol on the growth and development of rotifers.

35. Gilbert, J. J. In *Nutrition in the Lower Metazoa* (D. C. Smith and Y. Tiffon, eds.), pp. 57–71. Pergamon, Oxford, 1980. Some effects of diet on the biology of the rotifers Asplanchna and Brachionus.

36. Goodman, D. S., Huang, H. S. and T. Shiratori. *J. Lipid Res.* 6:390–396, 1965. Tissue distribution and metabolism of newly absorbed vitamin A in the rat.

37. Guyer, W. and K. Bloch. *J. Protozool.*

30:625–629, 1983. The sterol requirement of *Tetrahymena setosa* HZ-1.

38. Harper, A. E. *J. Nutr.* 104:965–967, 1974. "Non-essential" amino acids.

39. Harper, A. E., Miller, R. H. and K. P. Block. *Annu. Rev. Nutr.* 4:409–454. 1984. Branched-chain amino acid metabolism.

40. Hastings, J. L. and W. H. Thomas. In *CRC Handbook Series in Nutrition and Food, Section D: Nutritional Requirements* (M. Rechcigl, Jr., ed.), Vol. 1, pp. 87–163, CRC Press, Cleveland, OH, 1977. Qualitative requirements and utilization of nutrients: Algae.

41. Hayes, K. C. and J. A. Sturman. *Annu. Rev. Nutr.* 1:401–425, 1981. Taurine in metabolism.

42. Hegsted, D. M., Gallagher, A. and H. Hanford. *J. Nutr.* 104:588–592, 1974. Inositol requirement of the gerbil.

43. Henderson, L. M. *Annu. Rev. Nutr.* 3:289–307, 1983. Niacin.

44. Henry, H. L. and A. W. Norman. *Annu. Rev. Nutr.* 4:493–520, 1984. Vitamin D: Metabolism and biological actions.

45. Herwig, R. P., Staley, J. T., Nerini, M. K. and H. W. Braham. *Appl. Environ. Microbiol.* 47:421–423, 1984. Baleen whales: Preliminary evidence for forestomach microbial fermentation.

46. Holz, G. G., Jr. In *Biology of Tetrahymena* (A. M. Elliot, ed.), pp. 89–98. Dowden Hutchinson & Ross, Stroudsburg, PA, 1973. The nutrition of tetrahymena: Essential nutrients, feeding, and digestion.

47. Horst, R. L., Napoli, J. L. and E. T. Littledike. *Biochem. J.* 204:185–189, 1982. Discrimination in the metabolism of orally dosed ergocalciferol and cholecalciferol by the pig, rat and chick.

48. Howes, N. H. *Biochem. J.* 31:1489–1498, 1937. A semi-synthetic diet for *Helix pomatia*.

49. Hughes, D. E. and A. G. Callely. In *Biology of Nutrition* (R. N. T.-W.-Fiennes, ed.), pp. 187–204. Pergamon, Oxford, 1972. Energy production by microorganisms.

50. Hume, I. D. *Digestive Physiology and Nutrition of Marsupials. Monographs in Marsupial Biology.* Cambridge Univ. Press, London, 1982.

51. Hungate, R. E. *The Rumen and Its Microbes.* Academic Press, New York, 1966.

52. Hunt, R. D., Garcia, F. G. and D. M. Hegsted. *Lab. Anim. Care* 17:222–234, 1967. A comparison of vitamin D_2 and D_3 in new world primates 1. Production and regression of osteodystrophia fibrosa.

53. Hutner, S. H., Baker, H., Frank, O. and D. Cox. In *Biology of Nutrition* (R. N. T.-W.-Fiennes, ed.), pp. 85–177. Pergamon, Oxford, 1972. Nutrition and metabolism in protozoa.

54. Jacobsen, J. G. and L. H. Smith, Jr. *Physiol. Rev.* 48:424–511, 1968. Biochemistry and physiology of taurine and taurine derivatives.

55. Kanazawa, A. In *Proceedings of the Second International Conference on Aquaculture Nutrition: Biochemical and Physiological Approaches to Shellfish Nutrition* (G. D. Pruder, C. J. Langdon, and D. E. Conklin, eds.), Spec. Publ. No. 2, pp. 87–105. Louisiana State University Division of Continuing Education, Baton Rouge, 1983. Penaeid nutrition.

56. Ketola, H. G. *J. Anim. Sci.* 43:474–477, 1976. Choline metabolism and nutritional requirements of lake trout, *Salvelinus namaycush*.

57. Kidder, G. W. *Chem. Zool.* 1:93–159, 1967. Nitrogen: Distribution, nutrition, and metabolism.

58. Koch, A. In *Symbiosis* (S. M. Henry, ed.), Vol. 2, pp. 1–106. Academic Press, London and New York, 1967. Insects and their Endosymbionts.

59. Kristensen, J. H. *Mar. Biol.* (Berlin) 14:130–142, 1972. Carbohydrases of some marine invertebrates with notes on their food and on the natural occurrence of the carbohydrates studied.

60. Lange, R. *Comp. Biochem. Physiol.* 10:173–179, 1963. The osmotic function of amino acids and taurine in the mussel, *Mytilus edulis*.

61. Laybourn-Parry, J. *A Functional Biology of Free-Living Protozoa.* Univ. of California Press, Los Angeles, 1984.

61a. Lipmann, F. *Adv. Enzymol.* 1:99–162, 1941. Metabolic generation and utilization of phosphate bond energy.

62. Lloyd, D. and M. H. Cantor. In *Biochemistry and Physiology of Protozoa* (M. Levandowsky and S. H. Hutner, eds.), Vol. 2, pp. 9–65. Academic Press, New York, 1979. Subcellular structure and function in acetate flagellates.

63. Low, M. G. and A. R. Saltiel. *Science* 239:268–275, 1988. Structural and functional roles of glycosylphospatidylinositol in membranes.

64. Martin, A. W. In *Comparative Physiology of Carbohydrate Metabolism in Heterothermic Animals* (A. W. Martin, ed.), pp. 35–64. Univ. of Washington Press, Seattle, 1961. The carbohydrate metabolism of the mollusca.

65. McBee, R. H. In *Microbial Ecology of the Gut* (R. T. J. Clarke and T. Bauchop, eds.), pp. 185–222. Academic Press, London, 1977. Fermentation in the Hindgut.

66. Miller, M. R. In *Comparative Physiology of Carbohydrate Metabolism in Heterothermic Animals,* (A. W. Martin, ed.), pp. 125–144. Univ. of Washington Press, Seattle, 1961. Carbohydrate metabolism in amphibians and reptiles.

67. Mittler, T. E. *J. Nutr.* 101:1023–1028, 1971. Dietary amino acid requirements of the aphid *Myzus persicae* affected by antibiotic uptake.

68. Morris, J. G. and Q. R. Rogers. *Science* 199:431–432, 1978. Ammonia intoxication in the near-adult cat as a result of a dietary deficiency of arginine.

69. Morris, J. G. *J. Nutr.* 115:524–531, 1985. Nutritional and metabolic responses to arginine deficiency in carnivores.

70. Morton, R. A. and D. G. Rosen. *Biochem. J.* 45:612–627, 1949. Carotenoids, vitamin A and 7-dehydrosteroid in the frog (*Rana temporaria*).

71. Napoli, J. L., Fivizzani, M. A., Schnoes, H. K. and H. F. Deluca. *Arch. Biochem.* *Biophys.* 197:119–125, 1979. Synthesis of vitamin D_5: Its biological activity relative to vitamins D_3 and D_2.

72. National Academy of Sciences-National Research Council. *Nutrient Requirements of Laboratory Animals,* 3rd rev. ed. National Academy Press, Washington, D.C., 1978.

73. National Academy of Sciences-National Research Council. *Nutrient Requirements of Coldwater Fishes,* No. 16. National Academy Press, Washington, D.C., 1981.

74. National Academy of Sciences-National Research Council. *Nutrient Requirements of Warmwater Fishes and Shellfishes,* rev. ed. National Academy Press, Washington, D.C., 1983.

75. National Academy of Sciences-National Research Council. *Nutrient Requirements of Poultry,* 8th rev. ed. National Academy Press, Washington, D.C., 1984.

76. National Academy of Sciences-National Research Council. *Vitamin Tolerance of Animals.* National Academy Press, Washington, D.C., 1987.

77. Nisbet, B. *Nutrition and Feeding Strategies in Protozoa.* Croom Helm, London, 1984.

78. Ohwada, K. *Mar. Biol. (Berlin)* 14:10–17, 1972. Bioassay of biotin and its distribution in the sea.

79. Olson, R. E. *Annu. Rev. Nutr.* 4:281–337, 1984. The function and metabolism of Vitamin K.

80. Owen, G. *Adv. Comp. Physiol. Biochem.* 5:1–35, 1974. Feeding and digestion in the bivalvia.

81. Perry, A. S. and S. Miller, *J. Insect Physiol.* 11:1277–1287, 1965. The essential role of folic acid and the effect of antimetabolites on growth and metamorphosis of housefly larvae *Musca domestica*.

82. Pion, P. D., Kittleson, M. D., Rogers, Q. R. and J. G. Morris. *Science* 237:764–768, 1987. Myocardial failure in cats associated with low plasma taurine: a reversible cardiomyopathy.

83. Poston, H. A., Riis, R. C., Rumsey, G. L. and H. G. Ketola. *Cornell Vet.*

67:472–509, 1977. The effect of supplemental dietary amino acids, minerals, and vitamins on salmonids fed cataractogenic diets.

84. Provasoli, L. and A. F. Carlucci. *Bot. Monogr.* (*Oxford*) 10:741–787, 1974. Vitamins and growth regulators.

85. Rebouche, C. J. In *Carnitine Biosynthesis, Metabolism, and Functions* (R. A. Frenkel and J. D. McGarry, eds.), pp. 57–72. Academic Press, New York, 1980. Comparative aspects of carnitine biosynthesis in microorganisms and mammals with attention to carnitine biosynthesis in man.

86. Ridgway J. W. and A. A. Wright. *Comp. Biochem. Physiol. A* 51A:727–732, 1975. The effect of deficiencies of B vitamins on the growth of *Arion ater L.*

87. Rogers, Q. R. and J. M. Phang. *J. Nutr.* 115:146–150, 1985. Deficiency of pyrroline-5-carboxylate synthase in the intestinal mucosa of the cat.

88. Rogers, Q. R., Morris, J. G. and R. A. Freedland. *Enzyme* 22:348–356, 1977. Lack of hepatic enzymatic adaptation to low and high levels of dietary protein in the adult cat.

89. Schimke, R. T. *Sci. Pub.—Pan Am. Health Organ.* 222:21–30, 1971. Enzyme synthesis and degradation: Effects of nutritional status.

90. Scott, M. L. *Nutrition of Humans and Selected Animal Species.* Wiley, New York, 1986.

90a. Senna, I. A. and E. C. Vieira. *Am. J. Trop. Med. Hyg.* 19:568–570, 1970. Culture of the snail *Biomphalaria glabrata* on a simplified medium.

91. Senseman, D. M. *J. Chem. Ecol.* 3:707–715, 1977. Starch: A potent feeding stimulant for the terrestrial slug *Ariolimax californicus.*

92. Silva, S. V. P. S. and J. R. Mercer. *Comp. Biochem. Physiol. B* 80B:603–607, 1985. Effect of protein intake on amino acid catabolism and gluconeogenesis by isolated hepatocytes from the cat (*Felis domestica*).

93. Soldo, A. T. and W. J. van Wagtendonk.

J. Gen. Microbiol. 53:341–348, 1968. Lipid interrelationships in the growth of *Paramecium aurelia,* Stock 299.

94. Soldo, A. T. and W. J. van Wagtendonk. *J. Protozool.* 16:500–506, 1969. The nutrition of *Paramecium aurelia,* Stock 299.

94a. Soldo, A. T. and E. J. Merlin. *J. Protozol.* 24:556–562, 1977. The nutrition of *Parauronema acutum.*

95. Suttie, J. W. *Handb. Lipid Res.* 2:211–277, 1978. Vitamin K.

95a. Omaye, S. T. and A. L. Tappel. *J. Nutr.* 104:747–753, 1974. Effect of dietary selenium on glutathione peroxidase in the chick.

96. Thompson, S. N. *Annu. Rev. Entomol.* 31:197–219, 1986. Nutrition and in vitro culture of insect parasitoids.

97. Vieira, E. C. *Am. J. Trop. Med. Hyg.* 16:792–796, 1967. Influence of vitamin E on reproduction of *Biomphalaria glabrata* under axenic conditions.

98. Vieira, E. C., Senna, I. A., Rogana, S. M. G. and M. L. V. C. Tupynamba. In *Germfree Research,* (J. B. Heneghan, ed.), pp. 657–660. Academic Press, New York, 1973. Studies on the nutrition of the snail *Biomphalaria glabrata* under axenic conditions.

99. Wagge, L. E. *J. Exp. Zool.* 120:311–342, 1952. Quantitative studies of calcium metabolism in *Helix aspera.*

100. Wagtendonk, W. J. van and A. T. Soldo. In *Comparative Biochemistry of Nitrogen Metabolism* (J. W. Campbell, ed.), Vol. 1, pp. 1–56. Academic Press, London, 1970. Nitrogen metabolism in protozoa.

101. Wald, G. *Harvey Lect.* 41:117, 1946. The chemical evolution of vision.

101a. Webb, A. R. and M. F. Holick. *Annu. Rev. Nutr.* 8:375–399, 1988. The role of sunlight in the cutaneous production of vitamin D_3.

102. Weber, A. L. and S. L. Miller. *J. Mol. Evol.* 17:273–284, 1981. Reasons for the occurrence of the twenty coded protein amino acids.

103. Westheimer, F. H. *Science* 235:1173–1178, 1987. Why nature chose phosphates.

104. Wilson, R. P. and J. E. Halver. *Annu. Rev. Nutr.* 6:225–244, 1986. Protein and amino acid requirements of fishes.

105. Windmueller, H. G. In *Glutamine: Metabolism, Enzymology and Regulation* (J. Mora and R. Palacios, eds.), pp. 235–257. Academic Press, New York, 1980. Enterohepatic aspects of glutamine metabolism.

106. Wojtowicz, M. B. *Comp. Biochem. Physiol. A* 43A:131–141, 1972, Carbohydrases of the digestive gland and the crystalline style of the Atlantic deep-seascallop (*Placopecten magellanicus, gmelin*).

107. Wu, F. Y.-H. and C.-W. Wu. *Annu. Rev. Nutr.* 7:251–272, 1987. Zinc in DNA replication and transcription.

108. Feng, A. S. and J. C. Hall. In *Neural and Integrative Animal Physiology* (C. Ladd Prosser, ed.). pp. 247. Wiley-Liss, Inc., New York, 1991.

Chapter 7

Excretory Nitrogen Metabolism

James W. Campbell

Animals excrete three main nitrogenous end-products: ammonia, urea, or uric acid. Most of the nitrogen in these compounds represents the α-amino-*N* removed from amino acids during catabolism. Dietary amino acids in excess of the amounts needed for growth and maintenance of protein turnover are preferentially degraded over carbohydrates and lipids. This is because animals cannot store excess amino acids, unlike carbohydrates and lipids, which can be stored as glycogen and triglycerides (184). Amino acid degradation releases ammonia which, because of its toxicity, must be disposed of or detoxified. Aquatic species can excrete ammonia directly, whereas semiaquatic and terrestrial species convert it to urea or uric acid for excretion. Because most tissues depend on glucose as an energy source, a major disposition of the carbon skeleton resulting from the degradation of excess dietary amino acids is conversion to glucose, that is, amino acid gluconeogenesis. Liver is considered to be the "glucostat" of the body (169),

and is therefore the main site of amino acid gluconeogenesis in mammals and most other vertebrates. Other tissues, especially the kidney (107), are also capable of gluconeogenesis. Liver is also the main site of formation of nitrogenous end-products because their synthesis is intimately coupled with gluconeogenesis when amino acids are the source of the carbon skeleton (140). In addition to metabolizing excess dietary amino acids, the liver also metabolizes amino acids formed by extrahepatic tissues. Muscle, for example, releases alanine and glutamine (120) as well as ammonia itself (199). These two amino acids can act as gluconeogenic substrates whereas ammonia must be excreted directly or detoxified by conversion to an end-product for excretion. During protein catabolic states, amino acids from the breakdown of muscle protein are also major gluconeogenic substrates for liver. An understanding of the excretory nitrogen metabolism of animals therefore requires an understanding of hepatic amino acid gluconeogenesis as well as the interrelationship between liver and the extrahepatic tissues. These two aspects of nitrogenous end-product metabolism in animals will be considered in this chapter.

Nitrogenous End-Products Are Formed during Catabolism of Excess Amino Acids

The general cellular mechanisms involved in hepatic amino acid degradation and in amino acid gluconeogenesis in mammals are considered in this section along with comparative data on these processes.

Gluconeogenesis

Gluconeogenesis is metabolically a relatively expensive process in that the synthesis of 1 mol of glucose requires 4 mol of ATP, 2 mol of GTP, and 2 mol of NADH (195). As shown in Figure 1, it is not the simple reversal of glycolysis but has four distinct steps. These are catalyzed by the enzymes pyruvate carboxylase, phosphoenol pyruvate (PEP)-carboxykinase, fructose-1, 6-bisphosphatase, and glucose-6-phosphatase. The latter three steps are unique to gluconeogenesis. PEP-carboxykinase, being the first of these, is a main point of regulation of gluconeogenesis (192, 246). When there is a need for glucose, there is an increase in cytosolic PEP-carboxykinase. This may be mediated by glucagon (152), glucocorticoids (311, 370), or other hormones such as thyroid hormone and epinephrine (198). In protein catabolic states, pyruvate carboxylase may also be a point of regulation of gluconeogenesis (132).

Amino Acid Substrates for Gluconeogenesis

The majority of protein amino acids are gluconeogenic in that they can either be degraded to pyruvate or tricarboxylic acid

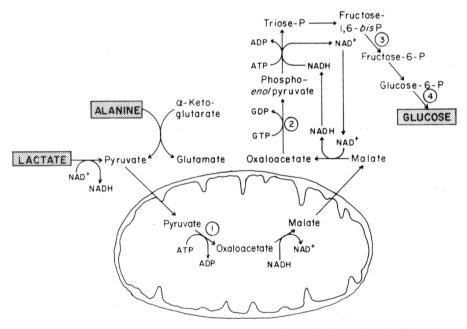

Figure 1. Gluconeogenesis from lactate or alanine. The circled numbers indicate important gluconeogenic reactions: (1) pyruvate carboxylase, (2) phosphoenolpyruvate carboxykinase, (3) fructose-1,6-bisphosphatase, and (4) glucose-6-phosphatase.

cycle intermediates for eventual conversion to oxaloacetate. In carnivorous species, they are a major source of energy. In trout, for example, 90% of the calories utilized during sustained swimming is from protein (350). Cats, whose natural diet also consists mainly of protein, illustrate some unique adaptations of mammals to such diets. They, for example, maintain essentially maximal rates of hepatic gluconeogensis irrespective of protein intake (308). This is unlike omnivorous mammals in which increases or decreases in dietary protein intake cause corresponding changes in rates of hepatic gluconeogenesis (192). Cats also exhibit other unique aspects of amino acid metabolism, including an absolute dietary requirement for arginine due to a high demand for ornithine to sustain a high rate of hepatic ureagenesis (14, 309, 320).

The liver also acts on amino acids formed by extrahepatic tissues. Alanine and glutamine, for example, are major products of muscle metabolism in mammals, accounting for 50% or more of the amino acids released by this tissue (292, 341). Alanine and glutamine are also released by adipose tissue (342). These two amino acids are formed during the catabolism of other amino acids, especially the branched-chain amino acids (leucine, isoleucine, and valine), which are rapidly transaminated in both cardiac and skeletal muscle (120, 139). The resulting α-keto acids may be returned to the liver, or may be oxidized directly by muscle as an alternate energy source (121, 250). The synthesis of alanine requires the action of alanine aminotransferase in muscle and the synthesis of glutamine, glutamate dehydrogenase, and glutamine synthetase (reactions 1 and 2); (178, 292).

$$\text{L-glutamate} + NAD(P)^+ + H_2O \rightleftharpoons$$
$$\alpha\text{-ketoglutarate} + NH_4^+ \quad (1)$$
$$+ NAD(P)H + H^+$$

$$\text{L-glutamate} + \text{MgATP}$$
$$+ NH_4^+ \xrightarrow{Mg^{2+}} \text{L-glutamine} \quad (2)$$
$$+ \text{MgADP} + P_i$$

The latter enzyme thus functions to "detoxify" part of the ammonia formed in muscle. Glucocorticoids, which cause increased muscle protein catabolism, also cause the induction of glutamine synthetase in this tissue (178).

The main fate of alanine formed in extrahepatic tissues is conversion to glucose in liver (96, 136). This has been referred to as the "glucose–alanine cycle" (95). Glutamine may also serve as a gluconeogenic substrate for liver but its main fate is uptake by either intestine or kidney (123, 360, 365). It is a major energy source for intestinal tissues. Alanine formed during glutamine catabolism in intestinal tissues may be released to be taken up by liver for glucose synthesis. In fact, as much as 50% of the alanine utilized by liver may come from the intestine. In kidney, glutamine serves as a source of ammonia for acid–base balance. Glutamine taken up by mammalian liver is converted to glutamate and ammonia through the action of a glutaminase (215). Ammonia formed by the glutaminase reaction is rapidly converted to urea for excretion.

Amino Acid Catabolism–Transdeamination
The main pathway for amino acid catabolism requires an initial transfer of the α-amino function to α-ketoglutarate by a transaminase (aminotransferase is the preferred name) to form glutamate and the corresponding α-keto acid (reaction 3).

$$\text{L-}\alpha\text{-amino acid}$$
$$+ \alpha\text{-ketoglutarate} \rightleftharpoons \quad (3)$$
$$\text{L-glutamate} + \alpha\text{-keto acid}$$

Glutamate is then taken up by mitochondria, where it is oxidatively deaminated by glutamate dehydrogenase. The coupled amino acid aminotransferase–glutamate dehydrogenase reactions are referred to as "transdeamination," a process first proposed by Braunstein (32, 343).

The aminotransferases utilize pyridoxal phosphate (vitamin B_6 derivative) as a cofactor. The two most common ones, aspartate and alanine aminotransferases, occur as isozymes in the mitochondrial and cytosolic compartments of vertebrate liver. The compartmental isozymes of aspartate aminotransferase, and probably those of alanine aminotransferase, are separate gene products (81, 128, 227). The hormonal regulation of hepatic aminotransferases during protein catabolic states is much the same as for PEP-carboxykinase (189a, 243a). Aminotransferase reactions are at or near equilibrium so any increase or decrease in plasma concentrations of amino acids causes a corresponding increase or decrease in their rate of degradation (184, 343). There are several other major aminotransferases [see (59)], but all show broad substrate specificities so probably more than 100 different amino acids undergo enzyme-catalyzed transamination reactions (71). However, there is an almost ubiquitous involvement of α-ketoglutarate as an amino acceptor in these reactions. Because of this, it has been suggested that the aminotransferases are "living fossil" enzymes in that the α-ketoglutarate–glutamate system may have served to disseminate chirality to other amino acids during the origin of life. Glutamate crystallizes as a mixture of D and L crystals so it could have undergone spontaneous resolution (166).

Mammalian liver glutamate dehydrogenase is localized exclusively in the mitochondrial matrix (213). With but few exceptions, one, a parasitic flatworm (233), the enzyme is also mitochondrial in other animal tissues. Vertebrate glutamate dehydrogenases utilize either

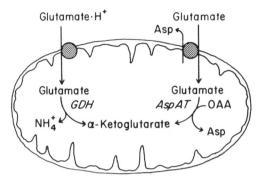

Figure 2. Glutamate transport systems in mammalian liver mitochondria and fate of imported glutamate. The abbreviations are Asp, aspartate; AspAT, aspartate aminotransferase; GDH, glutamate dehydrogenase; and OAA, oxaloacetate.

NAD^+ or $NADP^+$ as cofactors. This is also true of some invertebrate enzymes (204) although others are either $NADP^+$ (23, 154, 203) or NAD^+ specific (280, 290). The regulation of glutamate dehydrogenase is complex (106). A major allosteric activator is ADP (156) and activation by ADP is a property shared by all animal glutamate dehydrogenases, whether they are NAD^+ or $NADP^+$ specific.

Coupling of the various aminotransferases with glutamate dehydrogenase serves as an amino acid oxidase system for most of the protein amino acids and involves the formation of glutamate in the cytosolic compartment of liver. This is usually followed by its deamination within the mitochondrial compartment (133). There are two systems for glutamate translocation into mammalian liver mitochondria. One is the electrogenic glutamate–aspartate system (190, 191). The other is an electroneutral glutamate-H^+ symport system (242). Glutamate entering via exchange with aspartate is selectively transaminated with oxaloacetate to form aspartate (Fig. 2). Glutamate entering mitochondria by the glutamate-H^+ (glutamic acid) symport system is oxidatively deaminated by glutamate dehydrogenase. Under normal conditions, glutamate entry into liver mitochondria via this system is not rate limiting in amino acid catabolism. However, when deamination is maximally stimulated, it can become so (242). The channeling of glutamate to either aspartate aminotransferase or glutamate dehydrogenase in mammalian liver mitochondria may be due to a unique spatial relationship of the mitochondrial membrane translocators with a multifunctional complex consisting of aspartate aminotransferase, glutamate dehydrogenase, and other enzymes in the mitochondrial matrix (93, 295).

Alternate Routes of Amino Acid Catabolism
A second general mechanism for amino acid catabolism in specific mammalian tissues has been proposed (199, 200). This mechanism requires coupling of transamination reactions with the purine nucleotide cycle, which is made up of the enzymes adenylosuccinate synthetase and lyase and AMP deaminase (Fig. 3). Ammonia is released by the purine nucleotide cycle in the cytosolic compartment. In addition to mammalian muscle,

in which it was first described, the cycle has been shown to be present in mammalian brain (301, 302) and kidney tissues (27). These tissues are characterized by high aspartate aminotransferase and AMP deaminase activities and low glutamate dehydrogenase. In working muscle, the enzymes of the cycle function as a unit and are a major source of the ammonia formed by this tissue. Deamination of AMP serves to "pull" the myokinase reaction (2ADP $\rightarrow$ ATP + AMP) toward ATP formation. An increase in ATP turnover and a decrease in creatine phosphate levels stimulates the purine nucleotide cycle (172, 206). If the cycle is inhibited experimentally, muscle becomes dysfunctional (101). The purine nucleotide cycle is also a major source of ammonia produced by brain tissue (301). The cycle operates in mammalian kidney tissues, but to what extent is not agreed upon (240, 325, 344). Transdeamination via the purine nucleotide cycle was also proposed as a major alternative to transdeamination via glutamate dehydrogenase for mammalian liver (199). However, this has been ruled out for

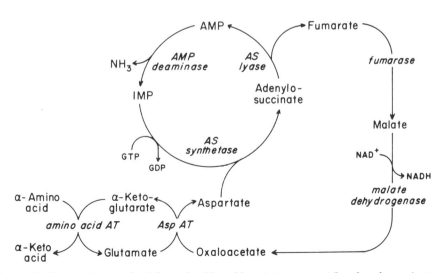

Figure 3. The purine nucleotide cycle. The abbreviations are AS, adenylosuccinate and AT, aminotransferase.

this tissue (185, 285, 350a) as well as for fish and avian liver (47, 51).

Although transdeamination via the purine nucleotide cycle does not appear to be a major mechanism for amino acid catabolism in vertebrates, it may nevertheless operate in this capacity in some invertebrates. Gastropod and bivalve mollusc tissues have high aspartate aminotransferase and AMP deaminase activities, but low glutamate dehydrogenase activity (21, 22, 115, 144, 210, 316), and the other two enzymes of the cycle, adenylosuccinate synthetase and lyase, are present in gastropod hepatopancreas (43).

Amino acids may also be catabolized without undergoing an initial transamination. For example, L-amino acid oxidase, a flavoprotein, converts amino acids directly to their corresponding α-keto acids and ammonia. L-amino acid oxidase activity is high in vertebrate kidney (238). It is probably localized in peroxisomes along with D-amino acid oxidase (209) since both are H_2O_2-forming enzymes (34). Despite their very broad substrate specificities, L-amino acid oxidases are generally felt to be of minor significance in amino acid catabolism in vertebrates. On the other hand, recent work with L-amino acid oxidase in bivalves suggests the enzymes may nevertheless be important in these and other invertebrates (21, 37).

Some amino acids are also acted upon by specific enzymes. Proline, serine, and theronine are examples. Proline undergoes an initial conversion to Δ-pyrroline-5-carboxylate by proline oxidase, also a flavoprotein. This product is then converted to glutamate via an NAD^+-dependent dehydrogenase (153). The utilization of proline as a direct energy source is especially important in tissues with high oxidative rates such as insect flight muscle (9, 294) and cephalopod mantle tissue (322). An NAD^+-specific L-threonine dehydrogenase is present in both avian (5) and mammalian liver (18, 275). Glycine can be produced via this reaction so the enzyme may be important in birds which require a source of glycine for hepatic urate synthesis. Threonine and serine can also be deaminated directly by serine dehydratase, a pyridoxal enzyme showing a high degree of sequence similarity with other pyridoxal enzymes that form a family of enzymes thought to have evolved from a common ancestral protein (60, 256). Serine dehydratase is induced in mammalian liver by protein catabolic states (243) and is present almost exclusively in the cytosolic compartment (330). Although present in several invertebrates (144), serine dehydratase is low or absent in nonmammalian vertebrates (288, 372).

Comparative Aspects of Amino Acid Catabolism and Gluconeogenesis

Although gluconeogenesis has been extensively studied in some classes of vertebrates, there is relatively little information on the process in invertebrates. Earthworms show an increased rate of urea synthesis during starvation (20), which is typical of gluconeogenic states in mammals. However, they lack glucose-6-phosphatase and thus appear incapable of glucose synthesis (268). The aquatic annelid *Tubifex* can convert aspartate to glycogen but the preferred substrates for glycogen synthesis are fatty acids (234). Insects can convert alanine to glucose and appear to regulate this hormonally (127). A mitochondrial pyruvate carboxylase and a cytosolic PEP-carboxykinase are present in bivalve tissues (375), and some gastropods have been reported to synthesize glucose from alanine as well as from aspartate, glutamate, and other substrates (374). The main focus of research on the physiological function of the aminotransferases and glutamate dehydrogenase in invertebrates has been their role in amino acid turnover during

TABLE 1. Amino Acid Metabolism in Fish Hepatocytes

SUBSTRATE	Substrate converted to:	
	GLUCOSE	CO_2
	($\mu mole \cdot g^{-1} \cdot hr^{-1}$ at 15°C)[a]	
Rainbow trout –		
lactate	5.4	3.7
alanine	0.8	0.8
asparagine	1.8	4.3
glycine	0.9	1.2
histidine	0.2	0.1
proline	<0.1	0.2
serine	4.0	4.2
valine	<0.1	<0.1
Coho salmon –		
lactate	7.6	6.8
alanine	2.4	2.5
Sockeye salmon –		
lactate	2.2	1.5
alanine	0.4	0.6

[a]Recalculated from refs. 105,223

cell volume regulation (19, 258), but it seems clear that these enzymes may function in many invertebrates for the utilization of amino acids as energy sources and in the conversion of amino acids to carbohydrates and lipids for storage.

Stored glycogen and especially lipid are the preferred substrates of fish but when these are depleted, muscle protein becomes the sole energy source (24). Fish on high protein diets show elevated blood amino acids (75), which can be rapidly cleared from the circulation (17). Fish liver can utilize amino acids directly as energy sources as can muscle tissue. Blood leucine is rapidly taken up by fish muscle (350), which contains the branched-chain amino acid aminotransferases and corresponding α-keto dehydrogenases necessary for its utilization (160, 255, 335). As shown in Table 1, the carbon skeletons of

several amino acids are oxidized to CO_2 by fish liver cells. The main pathway for this is transdeamination (356). Aminotransferases are present in fish liver and fish liver mitochondria contain relatively high levels of glutamate dehydrogenase (51, 105). The rate of glutamate oxidation by ray liver mitochondria is comparable to that of fatty acids, ketone bodies, and tricarboxylic acid cycle intermediates and is tightly coupled to ATP formation (232). Other amino acids utilized by fish liver cells include serine, asparagine, and glutamine (46, 51). Because serine aminotransferase is localized in fish liver mitochondria (373), serine is probably degraded in the cytosol by serine dehydratase, despite the reported absence of this enzyme from fish liver (288). Fish liver contains both asparaginase- and phosphtate-dependent glutaminase activities necessary for the initial hydrolysis of asparagine and glutamine.

Several amino acids, including alanine, are converted to glucose by fish hepatocytes (Table 1) and this process is regulated hormonally in much the same way as it is in mammals (151). Alanine, but apparently not glutamine, is produced by fish muscle (193) and again may serve as a major substrate for hepatic gluconeogenesis. In all fish species studied, and in the same species under different physiological states including starvation (282), lactate is a more effective precursor of glucose than is alanine (105, 143, 223, 281, 329). In the American eel, gluconeogenesis from lactate, but not from alanine, responds to glucagon and cortisol (281). On the other hand, glucagon (or cAMP) also stimulates alanine gluconeogenesis by trout liver cells (224). Like higher vertebrates (315), fish show a high degree of species specificity with respect to the subcellular localization of PEP-carboxykinase in liver cells (143, 223). In trout, it is almost exclusively mitochondrial. Because the lactate dehydrogenase reaction generates the NADH required for the conversion of triose phosphates to glucose in the cytosol, lactate is a much better gluconeogenic substrate than either pyruvate or alanine. With the latter two precursors, cytosolic reducing equivalents must be generated intramitochondrially and translocated to the cytosol.

Compared with fish, mammals, and birds, there have been relatively few studies on amino acid gluconeogenesis in amphibians and reptiles. Most adult amphibia are carnivorous (insectivorous) and rely on dietary proteins as a major source of energy. Amphibian metamorphosis usually involves the transition from an aquatic, herbivorous tadpole to a semiterrestrial, carnivorous adult. During this transition, there is an increase in the activities of the hepatic aminotransferases (54), glutamate dehydrogenase

(10, 175), and cytosolic PEP-carboxykinase (235). These changes are directed toward an increased gluconeogenic capability from amino acids due to the alimentation of a high protein diet in the adult. There is a concomitant increase in the ammonia-detoxifying capacity of liver (66). The hormonal regulation of gluconeogenesis in amphibians differs from that in mammals. Glucocorticoid receptors are present in liver (162), but glucocorticoids do not induce aminotransferases in this tissue (55). With thyroid hormone-induced metamorphosis, mitochondrial PEP-carboxykinase, but not the cytosolic isozyme, is induced (253). On the other hand, thyroxine has been reported to decrease PEP-carboxykinase mRNA in frog liver (229). The two compartmental isozymes of PEP-carboxykinase are immunologically identical (253), so the nature of the enzyme(s) as well as its regulation in amphibian liver appear to be quite different from that in higher vertebrates. Amphibian gastric mucosa contains the four enzymes unique to the gluconeogenic pathway, PEP carboxykinase being mainly mitochondrial, and can convert alanine and glutamate, as well as other substrates, to glucose (97).

Reptiles are very resistant to long-term starvation and catabolize protein only after other reserves, which can be maintained for long periods of time, are depleted. Tissue glycogen levels show little change in turtles even after 154 days of starvation (260) and they decrease in tortoise liver no more than 5% after 80 days (157). Lipid is the preferred substate during hibernation of lizards (257) and hepatic aminotransferases show an actual overall decrease during this time in snakes although they increase somewhat during the early stages of hibernation (304). Nor do other gluconeogenic enzymes increase in turtle liver during starvation (260). Lizards rapidly resynthesize glycogen from nondietary sources after exercise (119), so they are clearly capable of gluconeogensis. Also, after prolonged starvation, there is an increased rate of ureagenesis in the torotoise, presumably reflecting an increased rate of amino acid catabolism (157). Reptiles thus appear capable of converting amino acids to glucose. Gluconeogenic enzymes reported in reptilian liver include the aminotransferases (304), glutamate dehydrogenase (44), a cytosolic PEP-carboxykinase (259), fructose-1,6-bisphosphatase, and glucose-6-phosphatase (260).

Blood glucose concentrations are higher in birds than in mammals (296). PEP-carboxykinase is exclusively mitochondrial in the liver of some bird species and, because of this, there has been some question as to the role of the liver in avian gluconeogenesis (251, 315, 340). The exclusive localization of the enzyme in the mitochondrial compartment imposes a requirement for a source of cytosolic reducing equivalents other than malate. This is because oxaloacetate formed by pyruvate carboxylation within mitochondria is converted to PEP for efflux to the cytosol (Fig. 1). Avian kidney has a hormonally inducible PEP-carboxykinase and has been considered to be the major gluconeogenic organ in birds (358). Avian kidney is definitely gluconeogenic and utilizes alanine, glutamine, and glutamate as well as other substrates for glucose synthesis (82, 254). However, there is now evidence that the liver is a major, if not the main, gluconeogenic organ in birds. In intact, fasted birds, liver takes up alanine, glutamine, and glycine and produces glucose and urate (339). So, despite alanine and other amino acids being relatively poor glucose precursors for isolated hepatocytes *in vitro* (11, 30), avian hepatic amino acid gluconeogenesis is nevertheless an important process *in vivo*. As discussed below, the coupling of gluconeogenesis with urate formation in avian liver (99) may provide the necessary cytosolic reducing equivalents for glucose synthesis. Gluconeogenesis is stimulated by glucagon (or cAMP) and epinephrine in isolated avian hepatocytes as well as in perfused liver (241, 248), so its regulation appears to be similar to that in mammals.

Ammonia Generated during Amino Acid Catabolism Is Toxic

Ammonia is released during hepatic amino acid catabolism. Because the major pathway for amino acid catabolism is transdeamination via glutamate dehydrogenase, the main site of release is within liver mitochondria. Release of ammonia via the purine nucleotide cycle, which occurs in muscle and brain, takes place in the cytosol (230). As discussed above, specific enzymes such as serine dehydratase also release ammonia in the cytosol. In addition to differences in the compartments in which ammonia is released, there is also a very fundamental difference in the form in which it is released. The glutamate dehydrogenase reaction (reaction 1) produces NH_4^+, whereas AMP deaminase (reaction 4), produces NH_3.

$$AMP + H_2O \longrightarrow IMP + NH_3 \quad (4)$$

The former is a very weak acid while the latter is a very strong base and rapidly binds protons. Reactions releasing NH_3 can be easily monitored by following the ΔpH of the medium. "Ammonia is toxic" according to dogma, and some possible mechanisms for this toxicity are considered in this section.

Comparative Ammonia Tolerance

Except for catastrophic events, environmental ammonia levels sufficiently high to cause harm usually occur only under circumstances where animals are crowded. The sensitivity of animals to environmental ammonia varies markedly from species to species. Respiratory epithelium is one of the first tissues affected. Chickens exposed to 20 ppm of ammonia show behavior indicating it is an irritant at this concentration and long-term exposure to these levels leads to an increased susceptibility to viral infection (3). Exposure to 60–70 ppm causes respiratory distress in chickens (348). Concentrations of 280 ppm are lethal for pigs (321). Fish can tolerate relatively high concentrations of NH_4^+ but are adversely affected by micromolar (μM) concentrations of NH_3 (283). Exposure of fish to 6 μM NH_3 results in marked increases in free amino acids, and especially glutamate, serine, and theonine, in brain and other tissues (77). The increase in glutamate probably reflects glutamine formation, especially in brain (194). Most fish species are killed by NH_3 concentrations of ~20 μM (0.3–0.4 ppm). Fish are relatively more tolerant of injected ammonium salts than are birds or mammals. When ammonium acetate is injected intraperitoneally, fish show an average LD_{50} of 24 mmol/kg of body wt whereas for both birds and mammals the LD_{50} is 10–11 mmol/kg (362, 363).

The most remarkable adaptations to high atmospheric ammonia levels are those shown by Mexican free-tailed or guano cave bats. These bats inhabit caves in which ammonia levels can reach 2000 ppm (222). It is only at levels of 3000 ppm or above that they are affected (221). Bats exposed to high levels of atmospheric ammonia apparently retain carbon dioxide to neutralize blood ammonia so that blood pHs of 7.6–7.7 or below are maintained. Both carbon dioxide and ammonia are blown off when the bats return to normal atmospheres (326–328).

Substrate Depletion
Hepatic coma is the terminal stage of ammonia intoxication in man so the brain has been a major focus of studies on ammonia toxicity (72, 98). One of the first molecular mechanisms for ammonia toxicity, the Krebs cycle substate depletion hypothesis, was proposed by Bessman (16). According to this hypothesis, increased blood ammonia levels cause an excessive uptake of ammonia by brain. Increased ammonia concentrations in brain tissue then cause an increased reductive amination of α-ketoglutarate to form glutamate which, in turn, is converted to glutamine by the high levels of glutamine synthetase present in this tissue (73). Glutamate formation diverts NADH from the electron-transport chain as well as α-ketoglutarate from the TCA cycle (reaction 1). Glutamine formation depletes ATP itself (reaction 2). Decreased oxygen consumption by brain during ammonia intoxication (16), the kinetics of the glutamate dehydrogenase reaction (291), and the accumulation of glutamine in brain under these circumstances (261), all support the Bessman hypothesis as a major cause of hepatic coma. It also seems likely that glutamate, in its role as a neurotransmitter, as well as possible changes in γ-aminobutyric acid derived from it, may also be involved (28, 72).

Intracellular pH Effects
The more general effects of ammonia are due to changes in pH that affect rates of enzyme reactions. Ammonia also affects other cellular processes. In some cases, the effect is due to changes in pH, but in other cases, it appears to be due to the unique properties of ammonia or the ammonium ion itself. Changes in intracellular pH are due mainly to the generation of ammonia in the cytosol by enzymatic reactions. Changes also occur when cells are exposed to ammonium salts, as in experimental ammonia intoxication. If the penetrating species is NH_3, which is the usual case, there is an alkalization of the

cell's interior; if NH_4^+ is the permeant, there is acidification (286). Ammonium salts are known to cause alkalization of cellular organelles (279) and this results in impaired function. For example, lysosomal proteases have acidic pH optima and are inhibited by exposure to ammonia and other amines (15). An acidic pH in several organelles including the Golgi vesicles is required for certain protein recycling processes and ammonium salts specifically interfere with these processes (138, 337). Ammonium ions specifically activate phosphofructokinase, thereby stimulating glycolysis (331). Ammonium ions also inhibit fatty acid oxidation by hepatic mitochondria, which is the cause of fatty livers in diseased states characterized by hyperammonemia (202). These are some of the immediate effects of increased ammonia levels. Chronic exposure to elevated ammonia levels may also result in longer-term effects. For example, ammonia interferes with iron uptake by reticulocytes (118), eventually leading to a decreased oxygen carrying capacity of blood. Because ammonia is now known to affect several developing systems, it also seems likely that normal cell growth and division might be affected by chronic exposure to elevated ammonia levels. Ammonium ions mimic fertilization in sea urchin eggs and cause the onset of active protein and DNA synthesis (78, 86). It is also a regulator of ribosomal RNA synthesis in *Xenopus* eggs (306) and may act as a differentiation factor in *Dictyostelium* (135).

Membrane Effects

The NH_3 molecule was once considered to be the only permeant form of ammonia, despite the fact that the molecule is quite lipophobic (125). It has now been shown that NH_4^+ is also a permeant and transport systems for this cation are common among prokaryotes, fungi, and plants as well as animals (180). Although there may be specific animal transport systems for the ammonium ion, for example, in the gut of *Sarcophaga* larvae (270), the main mechanism for its translocation across cell membranes is by its substitution for K^+ on the Na^+-K^+ activated ATPase system (61, 92, 187). High levels of NH_4^+ can therefore deplete cells of K^+, which may account for the early cardiopathy of ammonia intoxication (363).

When NH_3 is the permeant, its translocation results in H^+ binding on the opposite side of the membrane. In the chemiosmotic theory for mitochondrial ATP synthesis, ADP phosphorylation is driven by a H^+ gradient across the inner mitochondrial membrane. This gradient is the result of outward H^+ pumping during respiration. In green plants, the gradient across the thylakoid membrane is inward because of light-dependent H^+ uptake. The uncoupling effects of NH_3 were first discovered in plants because of the ease of changing NH_3 concentrations by simple adjustment of the pH of media containing chloroplasts. Ammonia was shown to uncouple photosynthetic phosphorylation by penetrating thylakoid membranes and binding H^+, thereby eliminating the gradient necessary for ATP synthesis (76). Ammonia- and amine-sensitive phosphorylation is now used experimentally as a criterion for light-dependent phosphorylation in chloroplasts (158). Ammonia also uncouples phosphorylation in animal mitochondria (33), but this is more difficult to demonstrate experimentally. The approach used to demonstrate uncoupling of beef heart mitochondria is shown in Figure 4. Additional support for the role of NH_3 as an uncoupler in mitochondria comes from work with inside-out submitochondrial particles that have H^+ gradients in the same direction as chloroplasts. Ammonia

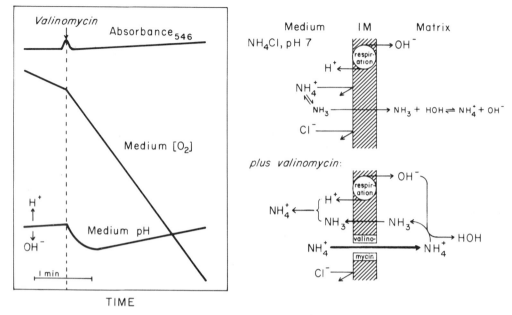

Figure 4. Uncoupling of oxidative phosphorylation by ammonia. Valinomycin, which causes the mitochondrial membrane to become permeable to NH_4^+, was added to heart mitochondria respiring in ammonium chloride. The NH_4^+ entering the mitochondrial matrix under these conditions is converted to NH_3 by the relative alkalinity of the matrix. NH_3, which is permeable to the mitochondrial membrane, then exits to bind H^+, thereby abolishing the H^+ gradient and preventing H^+ flux. This is indicated in the experimental data on the left, which show a transient alkalinization of the medium following addition of valinomycin. Uncoupling is indicated by the marked increase in O_2 uptake (depletion of medium O_2) after this addition. That there is little or no accumulation of NH_4^+ within the matrix, which would cause swelling because of osmotic water uptake, is shown by the absence of change in the optical properties (absorbancy) of the mitochondrial suspension. [Redrawn from (33).]

and other amines inhibit ATP synthesis by these particles (6).* It is thus possible that ammonia formed intramitochondrially during hepatic amino acid catabolism could uncouple phosphorylation as does OH^- (303) were it allowed to efflux in an unmodified form. As described below, in both the ureotelic and uricotelic hepatic ammonia-detoxifying systems, this is prevented by converting it to a form that does not bind protons.

Despite the specific effects of ammonia described here, the general toxicity of ammonia remains a relatively vague concept. In most instances, toxicity is probably due not to a single effect but to a combination of several effects at the organellar or cellular level, and these may vary in relative importance from tissue to tissue as well as from species to species.

*The authors actually conclude that NH_4Cl is not an effective uncoupler in comparison with certain amines. However, these comparisons were made at pH 7 and when the concentrations of NH_3 are estimated at this pH, the inhibition of ATP synthesis by NH_4Cl was equivalent to that of the most effective amine tested.

There Are Specialized Molecular and Cellular Mechanisms for Handling Ammonia

As discussed in the above section, increasing ammonia concentrations affect several cellular processes, leading to their malfunction. Animals must therefore have molecular and cellular mechanisms to maintain low concentrations of ammonia in organelles, cells, and body fluids. In ureoteles and uricoteles, ammonia generated intramitochondrially during hepatic amino acid catabolism is converted to a form that will not bind H^+ and therefore not interfere during exit with the normal proton gradient necessary for oxidative phosphorylation. There is thus an interaction between the mitochondrial and cytosolic compartments during the formation of these end-products. The end-products themselves are relatively nontoxic so their accumulation in body fluids is not a problem. Ammonotelic species prevent the buildup of ammonia in body fluids by efficient excretion systems. However, how mitochondrially generated ammonia is handled in cells of ammonoteles, especially vertebrate hepatocytes, to prevent its possible action as an uncoupler, is not known.

Ureotelic System

In ureotelic amphibians, turtles, mammals, and a few invertebrates, ammonia generated intramitochondrially is converted to urea for excretion. In mammals, urea has a special role in the kidney's ability to form a hypertonic urine (79); in many other vertebrates, it functions as an osmotically active solute in water regulation (177). This may also be a major function of urea in some invertebrates (48).

The pathway for urea biosynthesis is shown in Figure 5. Both carbamyl phosphate synthetase (62) and ornithine trans-carbamylase (272) are exclusively mito-chondrial in mammalian liver and their combined activities result in citrulline formation by these organelles. There are three carbamyl phosphate synthetases in animal tissues. The one in ureotelic liver is designated carbamyl phosphate synthetase (CPS)-I. It utilizes ammonia as a substrate, requires N-acetyl-L-glutamate for activity, and functions in ammonia detoxication (273). CPS-II, which utilizes glutamine as a substrate and does not require N-acetylglutamate, is part of a multifunctional cytosolic protein involved in pyrimidine synthesis (129). CPS-III, which also utilizes glutamine as a substrate but shows a requirement for N-acetylglutamate, is a mitochondrial enzyme found in invertebrates and fish, and functions especially in the synthesis of urea for osmotic purposes (4, 48, 49, 345, 346). CPS-I in mammalian liver mitochondria accounts for 15–20% of the total mitochondrial protein (62). Argininosuccinate (ASA) synthetase and lyase are exclusively cytosolic in ureotelic liver (74). In mammalian liver, arginase is also predominantly cytosolic although a small fraction appears to be associated with the outer mitochondrial membrane (58). In chicken (130, 170) and elasmobranch (50, 176) liver and kidney tissues, arginase is mainly mitochondrial.

Glutamate, which is formed in the cytosol mainly from alanine via transamination (133), is taken up by ureotelic liver mitochondria and is either transaminated with oxaloacetate to form aspartate or is oxidatively deaminated by glutamate dehydrogenase (183). As discussed above (see Fig. 2), which of these routes taken is dictated by the mode of entry of glutamate into the mitochondrion. As shown in Figure 5, aspartate formed by glutamate transamination exits to the cytosol where it is utilized in the ASA synthetase reaction (217, 220). Ammonia formed via

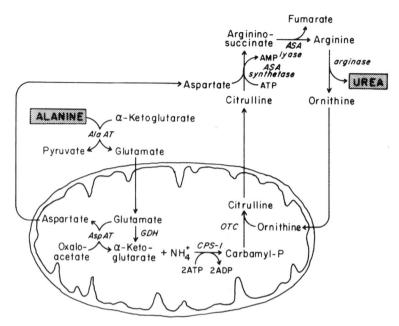

Figure 5. Ureagenesis during alanine catabolism in mammalian liver. The abbreviations are AlaAT, alanine aminotransferase; ASA, argininosuccinate; AspAT, aspartate aminotransferase; CPS-I, carbamyl phosphate synthetase-I; GDH, glutamate dehydrogenase; and OTC, ornithine transcarbamylase.

glutamate dehydrogenase serves as a substrate for CPS-I. Carbamyl phosphate, the product of CPS-I, is converted to citrulline by ornithine transcarbamylase in the presence of L-ornithine. Citrulline then exits to the cytosol by an energy-independent translocation (112). In the cytosol, ASA synthetase converts aspartate and citrulline to argininosuccinate, which is in turn cleaved by ASA lyase to arginine and fumarate. The latter can be converted to malate by fumarase present in the cytosol (252). The former is hydrolyzed to urea and ornithine by arginase. Urea enters the hepatic venous system for excretion, whereas ornithine is translocated into mitochondria by an exchange with either citrulline or H$^+$ (214, 217) to complete the *urea cycle*.

Tissue Distribution of Urea Cycle Enzymes
Only the liver of ureotelic vertebrates has the functional capacity for urea synthesis

de novo. Brain, and especially kidney, have significant levels of ASA synthetase and lyase and are therefore capable of converting citrulline to arginine (274). Arginase occurs in most, if not all, extrahepatic tissues in mammals (150) and other vertebrates (7, 39). Its probable function in these tissues is in the pathway for the conversion of arginine, formed during intracellular protein degradation, to proline or glutamate for oxidation (371).

The intestine is the only extrahepatic tissue capable of significant citrulline synthesis (272). CPS-I and ornithine transcarbamylase in intestine are similar to their counterparts in liver but are regulated differently (293). Citrulline is a major product of intestinal glutamine utilization (29, 366). It is released by the intestine and taken up from the circulation by kidney and to a lesser extent, by brain where it can be converted to arginine. Because of the very high levels of arginase in liver, little or no arginine is released from that organ, so arginine produced in kidney is a main source of this amino acid for muscle protein synthesis (94).

Detoxication of Extrahepatic Ammonia by the Ureotelic System
Muscle, brain (88), and intestine are major sources of systemic ammonia, although it is also formed by other extrahepatic tissues. As shown in the classical experiments of Duda and Handler (87), the principal fate of systemic ammonia is incorporation into the amide function of glutamine. The total body capacity for glutamine synthesis in the rat is equal to the total capacity for urea synthesis. Glutamine is thus the major form of ammonia shunted to the liver for detoxication. Liver also extracts ammonia directly from the systemic circulation. Glutamine synthetase is present in mammalian liver, but there is little glutamine formation from circulating ammonia (201). This is because of the metabolic zonation of glutamine synthetase in mammalian liver. The enzyme is restricted to a one- to two-cell layer surrounding the terminal venules (Fig. 9B) and is thought to constitute a "fail-safe" mechanism that operates only if the liver's capacity for urea synthesis, concentrated in the portal part of liver, is exceeded (140). Glutamine and ammonia entering the liver from the general circulation serve as substrates for urea synthesis. Glutamine is taken up by hepatocytes via the Na^+-dependent N amino acid transport system (173, 307) and there is also a transport system for its uptake by mitochondria (182). There are two systems for the release of the amide-N of glutamine as ammonia in mammalian liver. One is direct hydrolysis of the amide by glutaminase [reaction 5; (215)] and the other, an initial transamination followed by hydrolysis [reactions 6 and 7; (70)]. While transamination of

glutamine definitely occurs in liver (142), its physiological role is not clear. The trans-

$$\text{L-glutamine} + H_2O \longrightarrow$$
$$\text{L-glutamate} + NH_4^+ \quad (5)$$

$$\text{L-glutamine} + \alpha\text{-keto acid}$$
$$\xrightleftharpoons{\text{transaminase}} \alpha\text{-ketoglutaramate}$$
$$+ \alpha\text{-amino acid} \quad (6)$$

$$\alpha\text{-ketoglutaramate} \xrightarrow{\Omega\text{-amidase}}$$
$$\alpha\text{-ketoglutarate} + NH_4^+ \quad (7)$$

amination system occurs in both cytosolic and mitochondrial compartments (70), whereas glutaminase is restricted to the latter (183, 215). Hepatic glutaminase is regulated in a feed-forward manner by NH_3. It is therefore activated at alkaline pHs and inhibited at acid pHs (141). This accounts for the shunting of glutamine away from the liver during metabolic acidosis for use by the kidney (359). Ammonia released intramitochondrially from the amide function of glutamine is channeled directly to CPS-I (218).

Regulation of Urea Synthesis
Schimke (297–299) showed that hepatic levels of urea cycle enzymes, as well as other enzymes of amino acid catabolism, are markedly affected by the availability of dietary protein: They increase on high protein diets and after prolonged starvation. Although the end results are essentially the same under both physiological conditions, the mechanisms underlying the changes may differ. High protein diets cause an increase in CPS-I and ornithine transcarbamylase mRNA, whereas these mRNAs actually decrease during starvation (226). Changes in mRNAs for the other enzymes have not been investigated. In perfused liver, ureagenesis, as well as gluconeogenesis, is a linear function of amino acid concentration in the perfusion fluid (205). As we

will see below, the availability of amino acids, whether from the diet or from the breakdown of muscle protein, is a major factor in the regulation of hepatic ureagenesis (314).

The short-term regulation of hepatic ureagenesis in mammals has been the focus of numerous studies. These include studies on (a) the availability of N-acetylglutamate for activation of CPS-I, (b) the availability of ornithine for the ornithine transcarbamylase reaction or its possible direct activation of CPS-I (c) the availability of ATP for the CPS-I reaction, and (d) the kinetics of the individual enzymes relative to the concentrations of their substrates and products. Based on amounts of enzyme activity, ASA synthetase would appear to be the rate-limiting reaction. This is true experimentally when the concentrations of ammonia, ornithine, and aspartate are saturating. Under conditions *in vivo*, where the concentrations of these substrates are not saturating, mitochondrial citrulline formation is limiting (65, 219). Either the CPS-I or the ornithine transcarbamylase reaction could therefore be the point of regulation. As indicated above, CPS-I is affected by the availability of N-acetylglutamate, which converts inactive enzyme to the active form. N-Acetylglutamate is synthesized within mammalian liver mitochondria from glutamate and acetyl CoA by a synthetase that is activated by low levels of arginine. The effect of arginine on this system was therefore one of the first short-term regulatory mechanisms proposed for the urea cycle (334). However, the concentration of arginine that normally exists in liver mitochondria *in vivo* is saturating for N-acetylglutamate synthetase so changes in its concentration would appear not to affect levels of N-acetylglutamate (103, 104). Because CPS-I controls the intramitochondrial concentration of NH_3, its regulation by N-acetylglutamate may be directed more to the

availability of this substrate than to regulation of the cycle *per se* (57). There is an increase in both the concentration of N-acetylglutamate in liver and its rate of synthesis concomitant with an increased rate of ureagenesis in animals on high proteins diets, which nevertheless indicates that N-acetylglutamate is involved in the longer-term regulation of the cycle (228). Ornithine has been known to stimulate the urea cycle since the latter's discovery by Krebs. The mechanism for this was unclear for some time. It is now known that, under physiological conditions, external ornithine supplied to mitochondria is channeled directly to ornithine transcarbamylase in the mitochondrial matrix. This is most likely due to association of CPS-I and ornithine transcarbamylase with the inner mitochondrial membrane (267). Under these conditions, the measured rates of citrulline synthesis by mitochnondria *in vitro* can account for the amount of urea excreted by the rat (65).

From these observations, it can be concluded that the two most important factors for the short-term regulation of ureagenesis in mammalian liver are the availability of ornithine and of ammonia (57, 65, 219). A key role for ornithine availability as a major regulator of the cycle is consistent with physiological observations on the protective effect of arginine against ammonia intoxication (126, 320). Because neither the transaminases not glutamate dehydrogenase are rate limiting in amino acid catabolism (183), another factor determining the rate of urea synthesis is simply the availability of amino acids to the liver for deamination to provide ammonia for CPS-I. Hepatic amino acid transport systems are limiting in this process (133) and glucagon, which stimulates both ureagenesis and gluconeogenesis, also stimulates the major hepatic amino acid transport systems (173).

The hormonal regulation of hepatic

ureagenesis in mammals is much the same as that for gluconeogensis. Glucagon, for example, induces PEP-carboxykinase (370) and also the enzymes of the urea cycle (161, 161a, 289). Glucocorticoids, which cause muscle protein degradation, thereby increasing the availability of amino acids to the liver, are also effective inducers of urea cycle enzymes (299). Both glucagon and glucocorticoids regulate glucose availability. The decrease in urea synthesis caused by the addition of glucose during perfusion of liver with amino acids (N-sparing) (165a) is indicative of a role for glucose in regulating hepatic ureagenesis, once again illustrating the close coupling of nitrogenous end-product formation with gluconeogenesis. Glucocorticoids stimulate mitochondrial citrulline synthesis (208) and elevate CPS-I mRNA levels (134) in much the same manner as do high protein diets. In the developing rat, the urea cycle enzymes appear in liver around day 17.5 and increase slowly until birth. Their development is paralleled by the liver's capacity for urea synthesis (271). The induction of the enzymes in fetal liver is glucocorticoid dependent (161). Shortly after birth, there is a rapid increase in enzyme activity to the adult level due to an increased rate of gene transcription (293). The development of the citrulline-forming system in fetal and neonatal liver is quite different from its development in the intestine. In the latter tissue, there is a marked decline in CPS-I and ornithine transcarbamylase mRNAs after birth and these mRNAs are not affected by either glucagon or glucocorticoid treatment (293).

Uricotelic System

In birds, crocodilians, and reptiles, which represent over 75% of the known species of terrestrial vertebrates (209a), ammonia formed during amino acid catabolism is converted to uric acid for excretion. Uric acid excretion represents a major water-conserving mechanism and its development during vertebrate evolution may have been responsible for the rapid radiation of the dinosaurs and their kin during the Triassic Period (45). Uric acid and other purines are formed by several invertebrates, but whether this is analogous to the coupled synthesis of urate with gluconeogenesis in higher vertebrate liver is not clear. Uric acid is accumulated in tissues of several invertebrates but, in many cases, is recycled by symbionts (64, 266). Mutations in *Drosophila* that prevent uric acid synthesis are not lethal (145, 238a), indicating its formation is not essential even though insects are generally considered to be uricotelic.

Just as the urea cycle is an adaptation of the basic arginine biosynthetic pathway, the uric acid pathway is an adaptation of the basic biosynthetic pathway for purines. As elucidated by Buchanan and co-workers (35), the precursors of the purine ring are glutamine, glycine, formate (as folate derivatives), carbon dioxide, and aspartate. The soluble fraction of pigeon liver is capable of converting all of these substrates, but not ammonium salts, to purine (317). The purine pathway is therefore localized mainly in the cytosol. However, the addition of mitochondria stimulates purine synthesis by the soluble fraction (372). The discovery that glutamine synthetase is localized in avian and reptilian liver mitochondria, in contrast with its cytosolic localization in mammalian liver (352), served to explain these observations and allowed for the formulation of the uricotelic system for hepatic ammonia detoxication. This system is shown in Figure 6. Ammonia released intramitochondrially is converted to glutamine for exit to the cytosol. Glutamate taken up by uricotelic liver mitochondria serves as a substrate for both glutamine synthetase and glutamate dehydrogenase in about equal amounts (42).

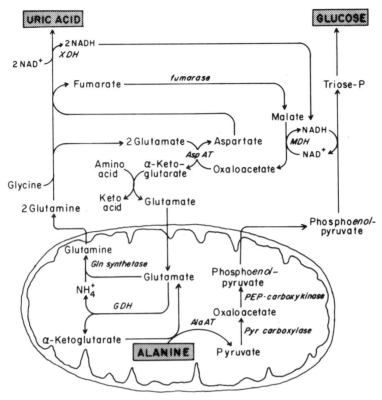

Figure 6. Uricogenesis and gluconeogenesis during alanine catabolism in avian liver. The abbreviations are AlaAT, alanine aminotransferase; AspAT, aspartate aminotransferase; Gln, glutamine; GDH, glutamate dehydrogenase; MDH, malate dehydrogenase; PEP, phosphoenolpyruvate; Pyr, pyruvate; and XDH, xanthine dehydrogenase.

This, plus the localization of aspartate aminotransferase on the outer side of the inner mitochondrial membrane (117), indicates the formation of aspartate used in purine biosynthesis in avian liver occurs outside the mitochondrial matrix. This is consistent with the presence of an "aspartate–malate cycle" for NADH generation in the cytosol as depicted in Figure 6. The remainder of the enzymes of the *de novo* purine biosynthetic pathway are localized in the cytosol, possibly as a loosely associated complex (287) or multifunctional protein (300). As shown in Figure 7, glutamine contributes N-3 and N-9 of the purine ring; glycine, N-7; and aspartate, N-1. Of the three amino acid precursors, only glycine contributes car-

bon atoms to the purine ring. The detailed enzymatic steps of the pathway may be found in Henderson and Paterson (146) and elsewhere. Briefly, the initial reactions of purine synthesis lead to the formation of an aminoimidazole ring consisting of C-4, C-5, C-8, N-3, N-7, and N-9. The first step unique to purine synthesis is catalyzed by phosphoribosylpyrophosphate (PRPP)-amidotransferase (reaction 8), which utilizes glutamine to form phosphoribosylamine. This reaction is a major point of regulation of the purine

$$\text{gluiamine} + \text{PRPP} \xrightarrow{\text{Mg}^{2+}}$$

$$\text{phosphoribosylamine} \quad \quad (8)$$

$$+ \text{ glutamate} + \text{PP}_i$$

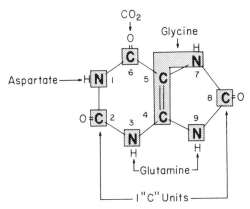

Figure 7. Metabolic origins of the atoms of uric acid.

pathway. Ammonia-dependent phophoribosylamine synthesis has been reported in avian liver (277), but this only occurs at high ammonia and PRPP concentrations and alkaline pHs (287). It may

therefore be due to nonenzymatic synthesis (36). Three of the first five steps leading to formation of the aminoimidazole ring are catalyzed by individual catalytic domains of a multifunctional protein. The two enzymes that utilize glutamine in this sequence are separate proteins (80, 300). In the next series of reactions, aspartate is added to form N-1 and the aspartate carbon atoms are then removed as fumarate. As shown in Figure 6, fumarate can be converted to malate to become a source of reducing equivalents for glucose synthesis (68, 84). The product of the purine biosynthetic pathway is IMP, which occupies a position equivalent to arginine in the urea cycle. However, rather than a simple one-step hydrolysis to the excretory end-product that occurs with arginine, processing of IMP to uric

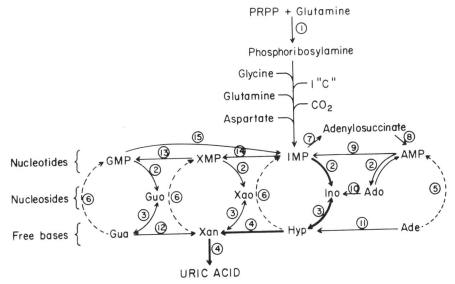

Figure 8. Synthesis and interconversions of the purines and purine nucleosides and nucleotides in animal tissues. The enzymatic reactions shown are (1) phosphoribosyl-pyrophosphate (PRPP)-glutamine amidotransferase; (2) 5′-nucleotidase; (3) nucleoside phosphorylase; (4) xanthine dehydrogenase; (5) adenine phosphoribosyltransferase; (6) hypoxanthine-guanine phosphoribosyltransferase; (7) adenylosuccinate synthetase; (8) adenylosuccinate lyase; (9) AMP deaminase; (10) adenosine deaminase; (11) adenine deaminase (adenase); (12) guanine deaminase (guanase); (13) XMP aminase; (14) IMP dehydrogenase; and (15) GMP reductase. The "shunt pathway" for IMP to urate, which is present in uricotelic liver, is shown by bold arrows. The "salvage pathway" for the utilization of preformed purines by tissues is shown by dashed lines.

acid requires multiple steps (Figure 8). The IMP is first converted to inosine by 5'-nucleotidases, which are present in most eukaryotic cells (85). There are three 5'-nucleotidases: a plasma membrane integral protein, a lysosomal enzyme, and a cytosolic enzyme (349). Only the cytosolic enzyme is involved in the conversion of IMP to uric acid. It is therefore quite high in birds compared with mammals (165, 237). The enzyme is mainly membrane bound in mammals and is active on AMP. Inosine is then converted to hypoxanthine and ribose-1-phosphate by purine nucleoside phosphorylase (225). The conversion of hypoxanthine to xanthine and of xanthine to uric acid is catalyzed by xanthine dehydrogenase. The dehydrogenase utilizes NAD^+ as an electron acceptor, whereas xanthine oxidase utilizes oxygen. Avian xanthine dehydrogenase is a nonheme iron flavoprotein containing molybdenum (63). Surveys of several animal species (186, 369) have shown the dehydrogenase to be the predominant form in most uricotelic vertebrates and also in invertebrates that excrete or store uric acid or its degradation products [see also (284)].

Because of the localization of PEP-carboxykinase in the mitochondrial compartment of chicken liver, the role of the liver in this species and in birds in general with respect to amino acid gluconeogenesis has been questioned. Also, alanine alone is usually a poor substrate for glucose formation by isolated avian hepatocytes (30, 207). The localization of PEP-carboxykinase in mitochondria means that oxaloacetate formed in this compartment is not available for malate or aspartate formation for the export of reducing equivalents to the cytosol where they are required for glucose synthesis (Figs. 1 and 6). Avian kidney contains an inducible cytosolic PEP-carboxykinase and was felt for some time to be the major organ for glucose synthesis from amino acids (251,

358). However, in intact chickens, liver would appear to be the main site of gluconeogenesis, with kidney forming a lesser amount of glucose (38, 339). The conversion of hypoxanthine to uric acid by xanthine dehydrogenase in the cytosol generates two NADH (Fig. 6), so this reaction may be the main source of reducing equivalents for hepatic glucose synthesis in birds and other uricotelic species (68). Another source is from what may be called the asparate–malate cycle as shown in Fig. 6. Inhibition of this cycle by the transaminase inhibitor aminooxyacetate, and not the malate–asparate shuttle, may be the reason this compound decreases gluconeogenesis from alanine by avian hepatocytes (249). A source of glycine is also required for uric acid synthesis. The failure to provide glycine to either isolated hepatocytes or perfused liver limits uric acid synthesis and therefore glucose synthesis.

Tissue Distribution of the Uric Acid Pathway

A supply of purines is required of all cells to sustain nucleic acid turnover, as well as to maintain adequate adenine and guanine nucleotide pools. Because 5'-nucleotidases, nucleoside phosphorylases, and hypoxanthine oxidizing enzymes have broad tissue distributions, uric acid may be formed by any tissue that possesses the *de novo* pathway. In mammals, the liver was once thought to be the main site of purine synthesis, exporting the oxypurines for "salvage" by the extrahepatic tissues (269). The salvage enzymes are hypoxanthine-guanine phosphoribosyl transferase (HGRPT; reaction 9) and adenine phosphoribosyl transferase (APRT). However, several extrahepatic tissues, in-

$$\text{hypoxanthine (guanine)} + \text{PRPP} \xrightarrow{Mg^{2+}}$$
$$\text{IMP (GMP)} + \text{PP}_i \quad (9)$$

cluding chick kidney and lymphoid tissues (8) and mammalian brain (159) and muscle (305), possess the complete *de novo* pathway. The salvage of exogenous purines by many cells is now known to be very low. Less than 2–5% of circulating hypoxanthine is salvaged by most mammalian tissues (231, 265, 351). In fact, only 2–5% of adenine present in ingested RNA is utilized for nucleic acid synthesis and this is mainly by intestinal tissues (318). This low rate of utilization of exogenous purines suggests the *de novo* pathway may be the main source of purines in all but a few tissues. Exogenous oxypurines are rapidly converted to uric acid or allantoin by most mammalian tissues (231, 265, 312, 351), indicating a broad tissue distribution for xanthine oxidase and uricase. Avian muscle releases uric acid *in vivo* (339). This could be due either to synthesis *de novo* or to the action of xanthine dehydrogenase on exogenous oxypurines. Xanthine dehydrogenase is not present in pigeon liver (90, 236) so, at least in this species, liver appears to export oxypurines for conversion to uric acid by extrahepatic tissues.

Detoxication of Extrahepatic Ammonia by the Uricotelic System
The main fate of systemic ammonia in birds is conversion to glutamine, as it is in mammals (364). Avian muscle forms alanine and glutamine as well as glycine and lysine during starvation (339) and, again, the former two amino acids constitute more than 50% of the amino acids released by this tissue. Some of the glycine released by muscle is taken up by kidney and converted to serine. Serine released by kidney can be reconverted to glycine for uric acid synthesis in liver (372). The liver also takes up alanine, glutamine, and glycine from the circulation and releases glucose and urate. Thus, as in mammals, glutamine is the major form of extrahepatic ammonia shunted to avian

liver for detoxication. Both ammonium salts and glutamine stimulate uric acid formation by perfused avian liver (13). However, hepatic metabolism of glutamine in birds is different from that in mammals because incoming glutamine can be converted directly to uric acid in the cytosol rather than be converted back to ammonia by mitochondrial glutaminase as occurs in mammalian liver. Avian liver mitochondria contain only barely detectable levels of glutaminase and their utilization of glutamine is extremely low compared with their utilization of glutamate (353). The level of hepatic glutaminase has been reported to increase 10–17% during protein catabolic states (361), but the function of this is not known. Ammonia removed from circulation by avian liver is presumably taken up by mitochondria for conversion to glutamine.

Regulation of Uricogenesis
The purine pathway operates at a much higher rate in chicken liver than in rat liver (196), reflecting its role in ammonia detoxication. PRPP-glutamine amidotransferase, the first committed step in the purine biosynthetic pathway is a major point for its regulation. In other tissues, this is through negative feedback inhibition by nucleotides, especially by the monophosphates (52, 148, 155, 368). The avian enzyme is also sensitive to this feedback inhibition, which is somewhat inconsistent with the detoxication function of the pathway in avian liver. However, the enzyme disaggregates into a small molecular weight form that is insensitive to inhibition by nucleotides (164). High PRPP levels as well as other effectors can cause this disaggreation. One of the major differences between avian and mammalian liver is their PRPP content. The concentration of PRPP in chicken liver is over 200 times that in rat liver (196). This high level of PRPP in avian liver therefore maintains PRPP

amidotransferase in a form that is insensitive to inhibition by IMP and other nucleotide products of the purine pathway. Despite the differing contents of PRPP in chicken and rat liver, the activity of PRPP synthetase in the two tissues is about the same. One reason for the high level of PRPP in bird liver may be the low level of HGPRT and APRT. These enzymes compete with the amidotransferase for PRPP (reactions 8 and 9). The level of HGPRT in avian liver is 3–9% that in rat liver while the level of APRT is 4% (163, 196). Suppression of the level of salvage enzymes is critical in uricotelic liver in order to prevent their competing with the amidotransferase for PRPP and also to prevent their converting hypoxanthine, formed by nucleoside phosphorylase, back to IMP.

Another adaptation shown by avian liver for a uricotelic-type metabolism is the low affinity of cytosolic 5'-nucleotidase for AMP relative to its affinity for IMP (237). This also serves to shunt IMP toward uric acid formation (Fig. 8). Because of this, there is far more flux through the purine pathway than through the adenine nucleotide pool and it has been suggested that xanthine dehydrogenase, more than any other reaction, is limiting in uric acid synthesis (319). Because glycine protects against ammonia intoxication in birds (25), in much the same manner as arginine (ornithine) does in mammals (126, 320), glycine (or serine) may be a limiting substate for uricogenesis. Compared with the operation of the *de novo* purine pathway strictly for purine nucleotide biosynthesis, there has thus been a general "deregulation" of the pathway in uricotelic liver so that it may respond to increasing ammonia concentrations during periods of high amino acid catabolism.

During starvation or high dietary protein intake, there is an increase in the activities of the enzymes of the purine pathway in avian liver as well as in those involved in amino acid breakdown via transdeamination (31, 171, 236, 264, 361). In some cases, high protein diets induce increased hepatic levels of mRNA for the enzymes (357). Of the purine pathway enzymes examined, there is a very major increase in xanthine dehydrogenase under conditions of increased protein catabolism. Xanthine dehydrogenase also develops in avian liver after hatching along with the gluconeogenic capacity of this organ (324), and this can be induced prematurely by glucocorticoids (367). These observations are consistent with xanthine dehydrogenase being limiting in urate formation. During development of the chicken, the kidney is the major gluconeogenic organ and the main substrates are lactate, pyruvate, and glycerol. Glutamine, alanine, and other amino acids can also be utilized, but to a much lesser extent. Isolated embryonic chick kidney tubules form ammonia from amino acids and this is inhibited by lactate, pyruvate and glycerol indicating that, *in vivo*, amino acid catabolism is minimized (82). The gluconeogenic capacity of avian liver starts developing a few days prior to hatching and, at hatching, is ~25% the adult level (167). There is then a relatively rapid development to the adult level. There is a similar change in hepatic glutamine synthetase (149) and other enzymes of amino acid catabolism (263). Thus in birds, as well as in mammals, there is a concomitant development of the capacity for nitrogenous end-product formation in liver along with development of the capacity for amino acid gluconeogenesis. Both coincide with or, in some instances, slightly anticipate, the alimentation of a protein-containing diet.

Similarities between the Ureo- and Uricotelic Systems

In addition to their responses to dietary and hormonal stimuli, the ureotelic and uricotelic systems for ammonia detoxication share several other properties. The mitochondrial components of both systems are located in the mitochondrial matrix along with glutamate dehydrogenase (353). The kinetics of glutamine synthetase and CPS-I, and especially their affinities for ammonia, are similar. Both are also inhibited by Ca^{2+} (53, 354). Glutamine synthetase is generally distributed to most parenchyma cells in avian (45) as well as crocodilian (313) liver. This is distinct from the metabolic zonation of the enzyme in mammalian liver where it is restricted to cells directly surrounding the perivenules (113, 114). These cells do not contain CPS-I (224a) so the expression of CPS-I or glutamine synthetase in adult mammalian hepatocytes is mutally exclusive (313a). This mutually exclusive expression of one or the other of the enzymes is the result of differentiation of adult mammalian liver since all fetal hepatocytes express both enzymes (186a, 224a). The expression of glutamine synthetase but not CPS-I by avian hepatocytes suggests that the mechanism underlying the differentiation of the few glutamine synthetase containing perivenule hepatocytes in adult mammalian liver may also have been involved in the transition to uricoteley during divergence of the mammalian and avian lines of descent (45). The mutually exclusive expression of one or the other of the two enzymes does not occur in lower vertebrates (313a). In any event, the distribution of glutamine synthetase in avian liver is nevertheless similar to the distribution of CPS-I and ornithine transcarbamylase in mammalian liver (45, 111) (Fig. 9).

An important similarity between the two detoxication systems with respect to the mitochondrial handling of ammonia is their conversion of ammonia to a form that does not bind H^+ at physiological pH values. As shown in Figure 10, both citrulline and glutamine titrate the same as

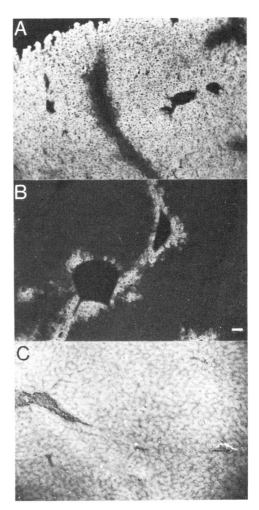

Figure 9. Demonstration of metabolic zonation of glutamine synthetase and carbamyl phosphate synthetase(CPS)-I in liver tissue by immunofluorescent staining. (*A*) Distribution of CPS-I in mammalian (hamster) liver. (*B*) Distribution of glutamine synthetase in mammalian liver. (*C*) Distribution of glutamine synthetase in avian (chicken) liver. Lightness indicates positive staining. The white bar (in *B*) equals 40 μm. (Courtesy of Dr. Darwin D. Smith, Jr.)

glycine, which has no ionizable R chain. They differ from histidine, which shows an additional pK^a ~pH 6 due to the ionization of the imidazole function. Both citrulline (112) and glutamine (42) rapidly efflux from their respective mitochondria

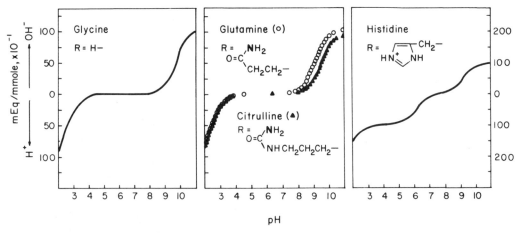

Figure 10. Ionization of the R group of amino acids as illustrated by the acid–base tritration curves of glycine, glutamine, citrulline, and histidine. (Courtesy of Dr. Jean E. Vorhaben).

by energy-independent processes and, unlike NH_3, do not equilibrate H^+ across the inner mitochondrial membrane. They therefore do not uncouple phosphorylation.

The Ammonotelic System
Most aquatic animals excrete ammonia as the major nitrogenous end-product. Many invertebrates are ammonotelic and, of these, crustaceans are among the best studied (278). Ammonia is produced in crustacean tissues mainly via transdeamination, diffuses into the body fluids, and is cleared from these fluids by the gills. This is also true for teleost fish (92). Although conversion of ammonia to glutamine has been considered as a mechanism for its detoxication in ammonoteles, the evidence for this is, at best, indirect (174). Blood glutamine levels are generally low in ammonoteles, especially fish (46), so there is no evidence that this compound acts as an interorgan transporter of ammonia as it does in higher vertebrates (193). The high rates of hepatic amino acid catabolism in carnivorous fish make them ideal systems for the study of the cellular mechanisms

for handling ammonia, if other than simple diffusion gradients of NH_4^+ are involved. The rate of alanine deamination by catfish hepatocytes can account for 50% of the total ammonia excreted by live fish and the rate with glutamine, 85%. The rate of glutamate deamination by intact catfish liver mitochondria can account for 160% of the rate of ammonia excretion (46), so liver is a main site of ammonia formation in fish. There should then be a relatively rapid efflux of ammonia from fish liver mitochondria and, if there are special adaptations for an ammonotelic-type hepatic metabolism, they should be evident since neither the citrulline- nor the glutamine-forming systems are present in these mitochondria. Most studies on ammonia efflux from mitochondria have been done with those from the mammalian kidney because of their special role in providing ammonia for NH_4^+ excretion during metabolic acidosis (123). Three mechanisms for ammonia efflux from these mitochondria have been proposed (190). In all three, NH_3 has been considered to be the permeant. In two of the mechanisms, H^+ for NH_4^+ formation following efflux is from the electrogenic proton pump. In the third mechanism, H^+ effluxes as an undissociated carboxylate of glutamate, implying a stoichiometric appearance of ammonia and glutamate in the medium. However, there is no evidence for any of these mechanisms in catfish mitochondria and NH_4^+ itself appears to be the exiting species (46). Thus, one possible adaptation of ammonoteles for the cellular handling of ammonia generated during amino acid catabolism is a hepatic mitochondrial ammonium ion translocation system.

Another adaptation of ammonotelic liver mitochondria is their ability to sustain glutamate deamination for long periods because of the "uncoupling" of glutamate transamination. Unlike either mammalian liver or kidney mitochondria, there is little aspartate formation by fish mitochondria from glutamate or glutamine even after relatively long incubation periods with these substrates *in vitro* (46, 355). The synthesis of both urea and uric acid as end-products requires the utilization of an aspartate-*N* in the cytosol. Since this is not required in ammonotelic liver, mitochondrial transamination of glutamate can be suppressed thereby increasing the availability of oxaloacetate for malate formation for the export of reducing equivalents used in the cytosol for glucose synthesis.

The Nitrogenous End-Products Can Arise from Pathways Not Involved in Ammonia Detoxication

The predominant nitrogenous end-product in urine or excreta has been used as a general means of classifying animals as to being ammonotelic, ureotelic, or uricotelic (purinotelic). While this may be appropriate physiologically, basing phylogenetic or ontogenetic generalizations on such data has serious limitations. As emphasized in the foregoing sections, nitrogenous end-product formation reflects a kind of hepatic metabolism for the detoxication of ammonia. This metabolism has become established during evolution in a particular group of animals as a result of genetic changes leding to modifications in enzymes and transport systems as well as in their localization in different cellular compartments in liver. However, ammonia, urea, and uric acid or its degradation products may appear in excreta because they are produced by pathways other than those involved in ammonia detoxication. Some of the alternate routes for the formation of the main end-products, as well as alternate reasons for their appearance in urine or other body fluids, are discussed in this section. Examples of the limitations of making generalizations based on the simple analyses of urine or other body fluids are also considered.

Urinary Ammonia

The vertebrate kidney has a major role in regulating pH and osmotic composition of body fluids in addition to its role in the elimination of nitrogenous and other end-products. In invertebrates, specialized structures, such as the Malpighian tubule–rectal system in insects (262), may assume these functions. In essentially all classes of vertebrates studied, metabolic acidosis results in increased NH_4^+ excretion (177). Because the kidney can extract amino acids, especially glutamine, from the systemic circulation and deaminate and/or deamidate them, the presence of ammonia in urine or excreta does not necessarily reflect an ammonotelic-type hepatic metabolism. It may simply reflect a physiological response to metabolic acidosis which, for example, occurs in female birds during the laying cycle (310). Crocodilians are also an example of animals in which the urine composition does not reflect the kind of hepatic metabolism they possess. Crocodiles and alligators excrete both uric acid and ammonia, with the latter normally constituting the highest percentage of urinary-N. They are therefore "ammonotelic" by the classical definition. However, they have the same hepatic ammonia-deoxifying system that is present in birds and squamate reptiles, so are constitutively uricotelic (313). The excretion of large quantities of ammonia by crocodilians, which is probably produced by the kidney from circulating precursors, thus represents a secondary adaptation to a semiaquatic existence. When in a water-limited environment such as seawater, crocodilians switch to uric acid excretion (131, 333a).

Alternate Routes for Urea Synthesis

Arginase, which has an almost ubiquitous species and tissue distribution, has a general catabolic role in amino acid degradation in addition to its specific role in ureagenesis. Ornithine, formed by the action of arginase on dietary arginine or on arginine derived from protein breakdown, is converted to glutamyl semialdehyde by ornithine aminotransferase. This can be converted to either glutamate or proline for oxidation (147, 168). An example of this function of arginase is in insects. Insect mitochondria are impermeable to glutamate and arginase plays a major role in the formation of proline, which is permeable (2). Insects are incapable of arginine biosynthesis (276), so any urea appearing in their excreta arises mainly through arginine degradation and not through the detoxication of ammonia.

The report by Needham (239) that the developing chick passes through a "ureotelic" stage apparently represents an esthetically pleasing concept because it has persisted in the literature despite acknowledgment of its probable invalidity by Needham and associates shortly after the initial report (67). Urea that appears in the developing chick is most likely from the action of arginase on arginine derived from yolk protein (89), since a complete

urea cycle is not present in either liver or kidney tissues (91). With the exception of CPS-I, the remaining urea cycle enzymes have been detected in avian tissues (333, 347). Genomic probes are now available for the mammalian urea cycle enzymes (26, 83, 189, 216, 247), so questions of the possible presence of the genes for these enzymes and of the reasons for their failure to be expressed as an integrated pathway in birds, crocodilians, and squamate reptiles can now be asked.

A very major source of urea in animal excreta is from the breakdown of dietary purines. In lower vertebrates, this pathway consists of the enzymes uricase, allantoinase, and allantoicase which are localized mainly in liver peroxisomes [Fig. 11; (109, 179, 244, 332)]. In at least one instance, the latter two enzymes have also been reported to be cytosolic in fish liver (122). The three enzymes are not present in fish intestinal peroxisomes (336). In fish liver peroxisomes, uricase is in soluble form in the matrix whereas in amphibia and mammals, it is present as a crystalline structure called the uricosome (1, 109, 110). Uricase, as well as the remaining enzymes of the purine degradative pathway, are absent from squamate reptiles, crocodilians, and birds. Uricase in some fish species is immunochemically similar to the mammalian enzyme (100). In other cases, fish, amphibian, and mammalian enzymes have been found not to be cross-reactive (109). Allantoinase and allantoicase occur as separate catalytic domains on a multifunctional protein in amphibians whereas in invertebrates and fish, they are separate peptides (332). Glyoxylate formed from purine degradation may be transaminated to form glycine (245), whereas urea formed from purine breakdown appears in the excreta. In fish, the purine degradative pathway is the main source of excreted urea (124).

Urease is present in members of sev-

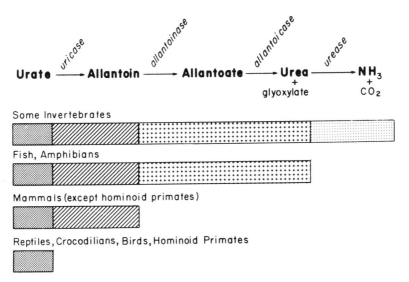

Figure 11. Degradation of uric acid by animals.

eral invertebrate phyla (12, 69, 137, 197) and, in Florkin's classical formulation of the purine degradative pathway (102), represents the final step in certain species (Fig. 11). However, invertebrate urease is not necessarily always associated with the other enzymes of purine degradation (323). In some terrestrial gastropods, urea is synthesized *de novo* via a modified urea cycle in hepatopancreas tissue, a tissue that also contains a unique constitutive urease (40, 211, 212). It has been suggested that ammonia released by the urease reaction serves to precipitate calcium carbonate in an essentially isohydric manner according to reaction 10 (41). Florkin (102) suggested that there have been systematic "deletions" of the purine

$$NH_3 + Ca^{2+} + HCO_3^- \longrightarrow$$
$$CaCO_3 + NH_4^+ \quad (10)$$

degradative enzymes starting with urease during evolution. While the presence or absence of specific enzymes in specific groups of invertebrates has subsequently proven not to necessarily correspond with the original formulation of this hy-

pothesis, the overall view of the evolutionary deletions in the pathway is correct. Urease was deleted in the invertebrate–vertebrate transition; allantoinase and allantoicase, in the amphibian–cotylosaurian transition; and uricase, in the sauropsid line leading to the higher reptiles, crocodillians, and birds (45). Uricase is present in most mammals but has also been deleted in hominoid primates (108). All or part of the uricase gene is present in man, so its "deletion" appears to be at the transcriptional level (192a).

Urinary Uric Acid
The Dalmatian dog excretes large amounts of uric acid and this led to the suggestion that this particular breed of dog was unique from other nonhominoid mammals in lacking hepatic uricase. The absence of hepatic uricase in "man, the great apes, and the Dalmatian dog," became another of the somehow esthetically pleasing, but incorrect concepts in excretory nitrogen metabolism. In fact, the excretion of uric acid by the Dalmatian has been cited as an example of the facile evolution of uricotelism (56). Uricase is, however, present in Dalmatian liver at a

level of activity comparable to that in other dogs (181). High uric acid excretion is therefore due to other aspects of the Dalmatian's physiology. These appear to be mainly alterations in transport systems. In most mammals, uric acid is reabsorbed by the kidney whereas in the Dalmatian, there is a net renal excretion (177). It is also clear that the liver is involved, possibly also because of altered transport systems (116, 188, 338). Whatever the ultimate reason for the high rate of uric acid excretion by the Dalmatian, it is probably not due to a uricotelic-type hepatic metabolism.

Synopsis

A major objective of this chapter has been to emphasize the coupling of nitrogenous end-product formation with gluconeogenesis. Most animals take in amino acids in excess of what is needed to sustain growth and protein turnover. Unlike carbohydrates and lipids, amino acids are not stored to any great extent. The carbon skeletons of amino acids can be utilized directly, especially in carnivores, but they are mainly converted to glucose. Glucose can then serve as an energy substrate for different tissues or can be stored as glycogen. In addition to diet, amino acids are also formed as metabolic end-products by muscle and other tissues, especially during protein catabolic states. Amino acid gluconeogenesis is therefore an on-going process in animals. In vertebrates, liver is the "glucostat" of the body. Amino acids coming to this organ either via the hepatic portal system from the intestine or via the systemic circulation from the extrahepatic tissues serve as major gluconeogenic substrates. The first step in their utilization is removal of the α-amino function as ammonia. Although deamination can take place directly because of the action of specific enzymes, it usually involves the process of transdeamination. This requires an initial transamination with α-ketoglutarate in the cytosol of liver cells to form glutamate. Glutamate is taken up by mitochondria and then oxidatively deaminated in the matrix by glutamate dehydrogenase. It is the ammonia formed by this deamination reaction that must be eliminated or detoxified by conversion to a precursor of one of the other main nitrogenous end-products, urea or uric acid. Oxidative phosphorylation depends on a proton (H^+) gradient across the inner mitochondrial membrane and were the ammonia generated intramitochondrially allowed to efflux in an unprotonated form (NH_3), it would uncouple phosphorylation by interfering with this proton gradient because of NH_4^+ formation. How this is prevented in ammonotelic liver is not well known. In ureotelic liver, intramitochondrially generated ammonia is converted to the carbamoyl function of citrulline. Citrulline then effluxes to the cytosol where it is converted to urea for excretion. In uricotelic liver, intramitochondrially generated ammonia is converted to the amide function of glutamine. Glutamine then effluxes to the cytosol for conversion to uric acid for excretion. Neither the carbamoyl function of citrulline nor the amide function of glutamine binds protons at physiological pH values so the efflux of these two compounds does not uncouple phosphorylation. The amino function of aspartate is also required for both urea synthesis from citrulline and uric acid synthesis from glutamine; in ureotelic liver this aspartate is formed in the mitochondrial matrix via the transamination of glutamate with oxaloacetate, whereas in uricotelic liver, this reaction takes place on the cytosolic side of the mitochondrial inner membrane. Ammonia formed by extrahepatic tissues in both ureoteles and uricoteles is returned to the liver, either as ammonia *per se* or as glutamine, for detoxification by the appropriate mitochondrial ammonia-detoxifying system. The two metabolic pathways utilized for end-product synthesis in ureoteles and uricoteles were originally biosynthetic pathways for arginine and purines, respectively, and were usurped during evolution for their excretory function. This required changes in the cellular compartmentation of specific enzymes as well as changes in the basic mode of regulation of the pathways. In their biosynthetic function, they were regulated by feed-back inhibition based on the availability of end-product. Their function in an excretory capacity required that they be regulated so that flux through the pathways is limited only by substrate (ammonia or amino acids) availability. Both systems are sensitive to dietary protein and their developmental regulation is directed toward the alimentation of a protein-containing diet. There are alternate ways in which the major nitrogenous end-products can be formed, especially by extrahepatic tissues, so their appearance in animal excreta does not necessarily imply a specific kind of excretory nitrogen metabolism. The functioning of the specific hepatic ammonia-handling systems requires fundamental genetic events, so it is the cellular and molecular aspects of these systems in liver tissue that must be considered in drawing ontogenetic or phylogenetic conclusions with respect to the major nitrogenous end-products.

References

1. Afzelius, B. A. *J. Cell Biol.* 26:835–843, 1965. The occurrence and structure of microbodies. A comparative study.
2. Aigahi, T. and M. Osanai. *J. Comp. Physiol. B* 155B:653–657, 1985. Arginase activity in the silkworm, *Bombyx mori*: Developmental profiles, tissue distribution and physiological roles.
3. Anderson, D. P., C. W. Beard, and R. P. Hanson. *Avian Dis.* 8:369–379, 1964.

The adverse effects of ammonia on chickens including resistance to infection with Newcastle disease virus.
4. Anderson, P. M. *Science* 208:291–293, 1980. Glutamine- and N-acetylglutamate-dependent carbamoyl phosphate synthetase in elasmobranchs.
5. Aoyama, Y. and Y. Motokawa. *J. Biol. Chem.* 256:12367–12373, 1981. L-Threonine dehydrogenase in chicken liver.
6. Azzone, G. F., H. Gutweniger, E. Viola, E. Strinna, S. Massari, and R. Colonna. *Eur. J. Biochem.* 62:77–86, 1976. Anion and amine uptake and uncoupling in submitochondrial particles.
7. Baby, T. G., S. C. Goel, and S. R. R. Reddy. *Physiol. Zool.* 49:286–291, 1976. A comparative study of arginase activity in lizards.
8. Badenoch-Jones, P. and P. J. Buttery. *Biochem. J.* 158:549–556, 1976. The effects of added purines on urate and purine synthesis *de novo* by isolated chick liver, kidney and lymphoid cells.
9. Balboni, E. *Biochem. Biophys. Res. Commun.* 85:1090–1096, 1978. A proline shuttle in insect flight muscle.
10. Balinsky, J. B., G. E. Shambaugh, Ill, and P. P. Cohen. *J. Biol. Chem.* 245:128–137, 1970. Glutamate dehydrogenase biosynthesis in amphibian liver preparations.
11. Bannister, D. W. and I. O'Neill. *Int. J. Biochem.* 13:437–444, 1981. Control of gluconeogenesis in chick (*Gallus domesticus*) isolated hepatocytes: Effect of redox state and phosphenolpyruvate carboxykinase (EC 4.1.1.32).
12. Barnes, D. J. and C. J. Crossland. *Comp. Biochem. Physiol. B* 55B:371–376, 1976. Urease activity in the staghorn coral, *Acropora acuminata*.
13. Barratt, E., P. J. Buttery, and K. N. Boorman. *Biochem. J.* 144:189–198, 1974. Urate synthesis in the perfused chick liver.
14. Beliveau, G. P. and R. A. Freedland. *Comp. Biochem. Physiol. B* 71B:13–18, 1982. Metabolism of serine, glycine and threonine in isolated cat hepatocytes *Felis domestica*.

15. Benynon, R. J. and J. S. Bond. *Am. J. Physiol.* 251:C141–C152, 1986. Catabolism of intracellular protein: molecular aspects.

16. Bessman, S. P. and N. Pal. In *The Urea Cycle* (S. Grisolia, R. Báguena, and F. Mayor, eds.), pp. 83–89. Wiley (Interscience), New York, 1976. The Krebs cycle depletion theory of hepatic comma.

17. Bever, K., M. Chenoweth, and A. Dunn. *Am. J. Physiol.* 240:R246–R252, 1981. Amino acid gluconeogenesis and glucose turnover in kelp bass (*Paralabrax sp.*).

18. Bird, M. I. and P. B. Nunn. *Biochem. J.* 214:687–694, 1983. Metabolic homeostasis of L-threonine in the normally-fed rat. Importance of liver threonine dehydrogenase activity.

19. Bishop, S. H. In *Estuarine Processes* (M. Wiley, ed.), Vol. 1, pp. 414–431. Academic Press, New York, 1976. Nitrogen metabolism and excretion: Regulation of intracellular amino acid concentrations.

20. Bishop, S. H. and J. W. Campbell. *Comp. Biochem. Physiol.* 15:51–71, 1965. Arginine and urea biosynthesis in the earthwork, *Lumbricus terrestris*.

21. Bishop, S. H., L. L. Ellis, and J. M. Burcham, In *The Mollusca* (P. W. Hochachka, ed.), Vol. 1, pp. 243–327. Academic Press, New York, 1983. Amino acid metabolism in molluscs.

22. Bishop, S. H., D. E. Greenwalt, and J. M. Burcham. *J. Exp. Zool.* 215:277–287, 1981. Amino acid cycling in ribbed mussel tissues subjected to osmotic shock.

23. Bishop, S. H., A. Klotz, L. L. Drolet, D. H. Smullin, and R. J. Hoffman. *Comp. Biochem. Physiol. B* 61B:185–187, 1978. NADP-specific glutamate dehydrogenase in *Metridium senile* (L.).

24. Black, D. and R. M. Love. *J. Comp. Physiol. B* 156B:469–479, 1986. The sequential metabolism of and restoration of energy reserves in tissues of Atlantic cod during starvation and refeeding.

25. Bloomfield, R. A., A. A. Letter, and R. P. Wilson. *Arch. Biochem. Biophys.* 129:196–201, 1969. The effect of glycine on ammonia intoxication and uric acid biosynthesis in the avian species.

26. Bock, H.-G. O., T.-S. Su, W. E. O'Brien, and A. L. Beudet. *Nucleic Acids Res.* 11:6505–6512, 1983. Sequence for human argininosuccinate synthetase cDNA.

27. Bogusky, R. T., L. M. Lowenstein, and J. M. Lowenstein. *J. Clin. Invest.* 58:326–335, 1976. The purine nucleotide cycle. A pathway for ammonia production in the rat kidney.

28. Bradford, H. K., H. K. Ward, and C. M. Thanki. In *Glutamine, Glutamate, and GABA in the Central Nervous System* (L. Hertz, E. Kvamme, E. G. McGeer, and A. Schousboe, eds.), pp. 249–260. Alan R. Liss, New York, 1983. Glutamine as a neurotransmitter precursor: Complementary studies *in vivo* and *in vitro* on the synthesis and release of transmitter glutamate and GABA.

29. Bradford, N. M. and J. D. McGivan. *Biochim. Biophys. Acta* 689:55–62, 1982. The transport of alanine and glutamine into isolated intestinal cells.

30. Brady, L. J., D. R. Romsos, and G. A. Leveille. *Comp. Biochem. Physiol. B* 63B:193–198, 1979. Gluconeogenesis in isolated chicken (*Gallus domesticus*) liver cells.

31. Brand, L. M. and J. M. Lowenstein. *J. Biol. Chem.* 253:6872–6878, 1978. Effect of diet on adenylosuccinase activity in various organs of rat and chicken.

32. Braunstein, A. E. In *Transaminases* (P. Christen and D. E. Metzler, eds.), pp. 2–19. Wiley (Interscience), New York, 1985. Transamination and transaminases.

33. Brierley, G. P. and C. D. Stoner. *Biochemistry* 9:708–713, 1970. Swelling and contraction of heart mitochondria suspended in ammonium chloride.

34. Bright, H. J. and D. J. T. Porter. In *The Enzymes* (P. D. Boyer, ed.), 3rd ed., pp. 421–505, Academic Press, New York, 1975. Flavoprotein oxidases.

35. Buchanan, J. M. In *The Nucleic Acids* (E. Chargaff and J. N. Davidson, eds.), Vol. 3, pp. 303–322, Academic Press, New York, 1960. Biosynthesis of purine nucleotides.

36. Buchanan, J. M. *Adv. Enzymol. Relat. Areas Mol. Biol.* 39:91–183, 1973. The amidotransferases.

37. Burcham, J. M., D. E. Greenwalt, and S. H. Bishop. *Mar. Biol. Lett.* 1:329–340, 1980. Amino acid metabolism in euryhaline bivalves: The L-amino acid oxidase from ribbed mussel gill tissue.

38. Burns, R. A. and P. J. Buttery. *Arch. Biochem. Biophys.* 233:507–514, 1984. Purine metabolism and urate biosynthesis in isolated chicken hepatocytes.

39. Campbell, J. W. *Arch. Biochem. Biophys.* 93:448–455, 1961. Studies on tissue arginase and ureogenesis in the elasmobranch, *Mustelus canis.*

40. Campbell, J. W. In *Urea and the Kidney* (B. Schmidt-Nielsen, ed.), pp. 48–80. Excerpta Medica, Amsterdam, 1970. Comparative biochemistry of arginine and urea metabolism in invertebrates.

41. Campbell, J. W. and B. D. Boyan. In *The Mechanisms of Mineralization in the Invertebrates and Plants* (N. Watabe and K. M. Wilbur, eds.), pp. 109–133. Univ. of South Carolina Press, Columbia, 1976. On the acid-base balance of gastropod molluscs.

42. Campbell, J. W. and J. E. Vorhaben. *J. Biol. Chem.* 251:781–786, 1976. Avian mitochondrial glutamine metabolism.

43. Campbell, J. W. and J. E. Vorhaben. *J. Comp. Physiol.* 129:137–144, 1979. The purine nucleotide cycle in *Helix* hepatopancreas.

44. Campbell, J. W., D. D. Smith, Jr., and J. E. Vorhaben. *Science* 228:349–351, 1985. Avian and mammalian mitochondrial ammonia-detoxifying system in tortoise liver.

45. Campbell, J. W., J. E. Vorhaben, and D. D. Smith, Jr. *J. Exp. Zool.* 243:349–363, 1987. Uricoteley: Its nature and origin during the evolution of tetrapod vertebrates.

46. Campbell, J W., P. L. Aster, and J. E. Vorhaben. *Am. J. Physiol.* 244:R709–R717, 1983. Mitochondrial ammoniagenesis in liver of the channel catfish (*Icalurus punctatus*).

47. Campbell, J. W., P. L. Aster, C. A. Casey, and J. E. Vorhaben. In *Isolation, Characterization, and Use of Hepatocytes* (R. A. Harris and N. W. Cornell, eds.), pp. 31–40. Elsevier, Amsterdam, 1983. Preparation and use of fish hepatocytes.

48. Campbell, J. W., R. B. Drotman, J. A. McDonald, and P. R. Tramell. In *Nitrogen Metabolism and the Environment* (J. W. Campbell and L. Goldstein, eds.), pp. 1–54. Academic Press, London, 1972. Nitrogen metabolism in terrestrial invertebrates.

49. Casey, C. A. and P. M. Anderson. *J. Biol. Chem.* 258:8723–8732, 1983. Glutamine- and N-acetyl-L-glutamate-dependent carbamoyl phosphate synthetase from *Micropterus salmoides.* Purification, properties, and inhibition by glutamine analogs.

50. Casey, C. A. and P. M. Anderson. *Comp. Biochem. Physiol. B* 82B:307–315, 1985. Submitochondrial localization of arginase and other enzymes associated with urea synthesis and nitrogen metabolism in liver of *Squalus acanthias.*

51. Casey, C. A., D. F. Perlman, J. E. Vorhaben, and J. W. Campbell. *Mol. Physiol.* 3:107–126, 1983. Hepatic ammoniagenesis in the channel catfish, *Ictalurus punctatus.*

52. Caskey, C. T., D. M. Ashton, and J. B. Wyngaarden. *J. Biol. Chem.* 239:2570–2579, 1964. The enzymology of feedback inhibition of glutamine phosphoribosylpyrophosphate amidotransferase by purine ribonucleotides.

53. Cerdan, S., C. J. Lusty, K. N. Davis, J. A. Jacobsohn, and J. R. Williamson. *J. Biol. Chem.* 259:323–331, 1984. Role of calcium as an inhibitor of rat liver carbamyl phosphate synthetase I.

54. Chan, S.-K. and P. P. Cohen. *Arch. Biochem. Biophys.* 104:325–330, 1964. Activity of three transaminases during

spontaneous and induced metamorphosis in *Rana catesbeiana*.

55. Chan, S.-K., and P. P. Cohen. *Arch. Biochem. Biophys.* 104:335–337, 1964. A comparative study of the effect of hydrocortisone injection on tyrosine transaminase activity of different vertebrates.

56. Charig, A. J. *Symp. Zool. Soc. London* 52:597–628, 1984. Competition between therapsids and archosaurs during the Triassic Period: A review and synthesis of current theories.

57. Cheung, C.-W. and L. Raijman. *J. Biol. Chem.* 255:5051–5057, 1980. The regulation of carbamyl phosphate synthetase (ammonia) in rat liver mitochondria. Effects of acetylglutamate and ATP translocation.

58. Cheung, C.-W. and L. Raijman. *Arch. Biochem. Biophys.* 209:643–649, 1981. Arginine, mitochondrial arginase, and the control of carbamyl phosphate synthesis.

59. Christen, P. and D. E. Metzler, eds. *Transaminases.* Chapter 6, pp. 365–451. Wiley (Interscience), New York, 1985. Sections by R. J. Ireland and K. W. Joy, Y. Morino and S. Tanase, I. K. Smith, A. L. J. Cooper, and A. Meister, R. A. John and L. J. Fowler, A. Ichihara, and M. Mason and D. R. Deshmukh.

60. Christen, P., U. Graf-Hausner, F. Bossa, and S. Doonan. In *Transaminases* (P. Christen and D. E. Metzler, Eds.), pp. 173–185. Wiley (Interscience), New York, 1985. Comparison of covalent structures of the isoenzymes of aspartate aminotransferase.

61. Claiborne, J. B., D. H. Evans, and L. Goldstein. *J. Exp. Biol.* 96:431–434, 1982. Fish branchial $Na^+NH_4^+$ exchange is via basolateral Na^+-K^+ activated ATPase.

62. Clarke, S. *J. Biol. Chem.* 251:950–961, 1976. A major polypeptide component of rat liver mitochondria: Carbamyl phosphate synthetase.

63. Cleere, W. F. and M. P. Coughlan. *Comp. Biochem. Physiol. B* 50B:311–322, 1975. Avian xanthine dehydrogenase.

I. Isolation and characterization of the turkey liver enzyme.

64. Cochran, D. G. *Annu. Rev. Entomol.* 30:29–49, 1985. Nitrogen excretion in cockroaches.

65. Cohen, N. S., C.-W. Cheung, and L. Raijman. *J. Biol. Chem.* 262:203–208, 1987. Channeling of extramitochondrial ornithine to matrix ornithine transcarbamylase.

66. Cohen, P. P. *Science* 168:533–543, 1970. Biochemical differentiation during amphibian metamorphosis.

67. Cohen, P. P. and G. W. Brown, Jr. In *Comparative Biochemistry* (M. Florkin and H. S. Mason, eds.), Vol. 3, pp. 161–244. Academic Press, New York, 1960. Ammonia metabolism and urea biosynthesis.

68. Coolbear, K. P., G. R. Herzbery, and J. T. Brosnan. *Biochem. J.* 202:555–558, 1982. Subcellular localization of chicken liver xanthine dehydrogease. A possible source of cytoplasmic reducing equivalents.

69. Cooley, L., D. R. Crawford, and S. H. Bishop. *Biol. Bull.* (*Woods Hole, Mass.*) 151:96–107, 1976. Urease from the lugworm, *Arenicola cristata*.

70. Cooper, A. J. L. and A. Meister. *Comp. Biochem. Physiol. B* 69B:137–145, 1981. Comparative studies of glutamine transaminases from rat tissues.

71. Cooper, A. J. L. and A. Meister. In *Transaminases* (P. Christen and D. E. Metzler, eds.), pp. 534–563. Wiley (Interscience), New York, 1985. Metabolic significance of transamination.

72. Cooper, A. J. L. and F. Plum. *Physiol. Rev.* 67:440–519, 1987. Biochemistry and physiology of brain ammonia.

73. Cooper, A. J. L., F. Vergara, and T. E. Duffy. In *Glutamine, Glutamate, and GABA in the Central Nervous System* (L. Hertz, E. Kvamme, E. G. McGeer, and A. Schousboe, eds.), pp. 77–93. Alan R. Liss, New York, 1983. Cerebral glutamine synthetase.

74. Cornell, N. W., A. M. Janski, and A.

Rendon. *Fed. Proc., Fed. Am. Soc. Exp. Biol.* 44:2448–2452, 1985. Compartmentation of enzymes: ATP citrate lyase in hepatocytes from fed or fasted rats.

75. Cowey, C. B., D. Knox, M. J. Walton, and J. D. Adron. *Br. J. Nutr.* 38:463–470, 1977. The regulation of gluconeogenesis by diet and insulin in rainbow trout (*Salmo gairdneri*).

76. Crofts, A. R. *J. Biol. Chem.* 242:3352–3359, 1967. Amine uncoupling of energy transfer in chloroplasts. I. Relation to ammonium ion uptake.

77. Dabrowska, H. and T. Wlasow. *Comp. Biochem. Physiol. C* 83C:179–184, 1986. Sublethal effect of ammonia on certain biochemical and haematological indicators in common carp (*Cyprinus carpio* L.).

78. Danilchik, M. V., Z. Yablonka-Reuveni, R. T. Moon, S. K. Reed, and M. B. Hille. *Biochemistry* 25:3696–3702, 1986. Separate ribosomal pool in sea urchin embryos: Ammonia activates a movement between pools.

79. Dantzler, W. H. *BioScience* 32:108–113, 1982. Renal adaptation of desert vertebrates.

80. Daubner, S. C., M. Young, R. D. Sammons, L. F. Courtney, and S. J. Benkovic. *Biochemistry* 25:2951–2957, 1986. Structural and mechanistic studies on the HeLa and chicken liver proteins that catalyze glycinamide ribonucleotide synthesis and formylation and aminoimidazole ribonucleotide synthesis.

81. DeRosa, G. and R. W. Swick. *J. Biol. Chem.* 250:7961–7967, 1975. Metabolic implications of the alanine aminotransferase isoenzymes.

82. Dickson, A. J. and A. J. Bate. *Comp. Biochem. Physiol. B* 86B:185–190, 1987. Kidney gluconeogenesis: its importance to net glucose synthesis during the development of chick embryos.

83. Dizikes, G. J., E. B. Spector, and S. D. Cederbaum. *Somatic Cell Mol. Genet.* 12:375–384, 1986. Cloning of rat liver arginase cDNA and elucidation of regulation of arginase gene expression in H4 rat hepatoma cells.

84. Domenech, C., A. Mazo, R. Artigas, A. Cortes, and J. Bozal. *Biol. Chem. Hoppe-Seyler* 367:1069–1074, 1986. Malate dehydrogenase in the cytosolic fraction of chicken liver.

85. Drummond, G. I. and M. Yamamoto. In *The Enzymes* (P. D. Boyer, ed.), 3rd ed., Vol. 4, pp. 337–354. Academic Press, New York, 1971. Nucleotide phosphomonoesterases.

86. Dubé, F. and D. Epel. *Exp. Cell Res.* 162:191–204, 1986. The relation between intracellular pH and rate of protein synthesis in sea urchin eggs and the existence of a pH-independent event triggered by ammonia.

87. Duda, G. D. and P. Handler. *J. Biol. Chem.* 232:303–314, 1958. Kinetics of ammonia metabolism *in vivo*.

88. Duffy, T. E., F. Plum, and A. J. L. Cooper. In *Glutamine, Glutamate, and GABA in the Central Nervous System* (L. Hertz, E. Kvamme, E. G. McGeer, and A. Schousboe, eds.), pp. 371–388. Alan R. Liss, New York, 1983. Cerebral ammonia metabolism *in vivo*.

89. Eakin, R. E. and J. R. Fisher. In *The Chemical Basis of Development* (W. D. McElroy and B. Glass, eds.), pp. 514–522. Johns Hopkins Univ. Press, Baltimore, MD, 1958. Patterns of nitrogen excretion in developing chick embryos.

90. Edson, N. L., H. A. Krebs, and A. Model. *Biochem J.* 30:1380–1385, 1936. The synthesis of uric acid in the avian organism: hypoxanthine as an intermediary metabolite.

91. Emmanuel, B. and H. Gilanpour. *Comp. Biochem. Physiol. B* 61B:287–289, 1978. Studies on enzymes of nitrogen metabolism, and nitrogen end products in the developing chick (*Gallus domesticus*) embryo.

92. Evans, D. H. and J. N. Cameron. *J. Exp. Zool.* 239:17–23, 1986. Gill ammonia transport.

93. Fahien, L. A., E. H. Kmiotek, G. Wold-

egiorgis, M. Evenso, E. Shrago, and M. Marshall. *J. Biol. Chem.* 260:6069–6079, 1985. Regulation of aminotransferase-glutamate dehydrogenase interactions by carbamyl phosphate synthetase-I, Mg^{2+} plus leucine *versus* citrate and malate.

94. Featherston, W. R., Q. R. Rogers, and R. A. Freedland, *Am J. Physiol.* 224:127–129, 1973. Relative importance of kidney and liver in synthesis of arginine by the rat.

95. Felig, P. *Metab., Clin. Exp.* 22:179–207, 1973. The glucose-alanine cycle.

96. Felig, P., T. Pozefsky, E. Marliss, and G. F. Cahill, Jr. *Science* 167:1003–1004, 1970. Alanine: Key role in gluconeogenesis.

97. Finol, J. and J. Chacín. *Comp. Biochem. Physiol. B* 65B:651–656, 1980. Gluconeogenesis in toad gastric mucosa.

98. Fischer, J. E. In *The Liver: Normal and Abnormal Functions* (F. F. Becker, ed.), Part B, pp. 725–753, Dekker, New York, 1975. Acute hepatic failure: Hepatic coma and the hepatorenal syndrome.

99. Fister, P., E. Eigenbrodt, and W. Schoner. *FEBS Lett.* 139:27–31, 1982. Simultaneous stimulation of uric acid synthesis and gluconeogenesis in chicken hepatocytes by α-adrenergic action of epinephrine and calcium.

100. Fitzpatrick, D. A. and K. F. McGeeney. *Comp. Biochem. Physiol. B* 51B:37–39, 1975. Comparative immunology of vertebrate urate oxidases.

101. Flanagan, W. F., E. W. Holmes, R. L. Sabina, and J. L. Swain. *Am. J. Physiol.* 251:C795–C802, 1986. Importance of purine nucleotide cycle to energy production in skeletal muscle.

102. Florkin, M. *Biochemical Evolution.* Academic Press, New York, 1949.

103. Freedland, R. A., G. L. Crozier, B. L. Hicks, and A. J. Meijer. *Biochim. Biophys. Acta* 802:407–412, 1984. Arginine uptake by isolated mitochondria.

104. Freedland, R. A., A. J. Meijer, and J. M. Tager. *Fed. Proc., Fed. Am. Soc. Exp. Biol.* 44:2453–2457, 1985. Nutritional influences on the distribution of the urea cycle: Intermediates in isolated hepatocytes.

105. French, J., T. P. Mommsen, and P. W. Hochachka. *Eur. J. Biochem.* 113:311–317, 1981. Amino acid utilization in isolated hepatocytes from rainbow trout.

106. Frieden, C. In *The Urea Cycle* (S. Grisolia, R. Báguena, and F. Mayor, eds.), pp. 59–71. Wiley (Interscience), New York, 1976. The regulation of glutamate dehydrogenase.

107. Friedman, P. A. and J. Torretti, *Am. J. Physiol.* 234:F415–F423, 1978. Regional glucose metabolism in the cat kidney *in vivo*.

108. Friedman, T. B., G. E. Polanco, J. A. Appold, and J. E. Mayle. *Comp. Biochem. Physiol. B* 81B:653–659, 1985. On the loss of uricolytic activity during primate evolution. I. Silencing of urate oxidase in a hominoid ancestor.

109. Fujiwara, S., H. Ohashi, and T. Noguchi. *Comp. Biochem. Physiol. B* 86B:23–26, 1987. Comparison of intraperioxisomal localization form and properties of amphibian (*Rana catasbeiana*) uricase with those of other animal uricases.

110. Fujiwara, S., K. Nakashima, and T. Noguchi. *Comp. Biochem. Physiol. B* 88B:467–469, 1987. Insoluble uricase in liver peroxisomes of Old World monkeys.

111. Gaasbeek Janzen, J. W., W. H. Lamers, A. F. M. Moorman, A. de Graaf, J. Lós, and R. Charles. *J. Histochem. Cytochem.* 32:557–564, 1984. Immunohistochemical localization of carbamoylphosphate synthetase (ammonia) in adult rat liver. Evidence for a heterogeneous distribution.

112. Gamble, J. G. and A. L. Lehninger. *J. Biol. Chem.* 248:610–618, 1972. Transport of ornithine and citrulline across the mitochondrial membrane.

113. Gebhardt, R. and E. Mecke. *EMBO J.* 2:567–570, 1983. Heterogeneous distribution of glutamine synthetase among rat liver parenchymal cells *in situ* and in primary culture.

114. Gebhardt, R. and E. Mecke. In *Glutamine Metabolism in Mammalian Tissues* (D. Häussinger and H. Sies, eds.), pp. 98–121. Springer-Verlag, Berlin, 1984. Cellular distribution and regulation of glutamine synthetase in liver.

115. Gibbs, K. L. and S. H. Bishop. *Biochem. J.* 163:511–516, 1977. Adenosine triphosphate-activated adenylate deaminase from marine invertebrate animals. Properties of the enzyme from lugworm (*Arenicola cristata*) body-wall muscle.

116. Giesecke, D. and W. Tiemeyer. *Experientia* 40:1415–1416, 1984. Defect of uric acid uptake in Dalmatian dog liver.

117. Gil, M., M. Cascante, A. Cortés, and J. Bozal. *Int. J. Biochem.* 19:355–363, 1987. Intramitochondrial location and some characteristics of chicken liver asparatate aminotransferase.

118. Glass, J. and M. T. Nunez, *J. Biol. Chem.* 261:8298–8302, 1986. Amines as inhibitors of iron transport in rabbit reticulocytes.

119. Gleeson, T. T. *J. Comp. Physiol. B* 147B:79–84, 1982. Lactate and glycogen metabolism during and after exercise in the lizard *Sceloporus occidentalis*.

120. Goldberg, A. L. and T. W. Chang. *Fed. Proc., Fed. Am. Soc. Exp. Biol.* 37:2301–2307, 1978. Regulation and significance of amino acid metabolism in skeletal muscle.

121. Goldberg, A. L. and R. Odessey. *Am. J. Physiol.* 223:1384–1391, 1972. Oxidation of amino acids by diaphragms from fed and fasted rats.

122. Goldberg, H. *Mol. Cell. Biochem.* 16:17–21, 1977. Organization of purine degradation in the liver of a teleost (carp; *Cyprinus carpio* L.). A study of its subcellular distribution.

123. Goldstein, L. *Int. Rev. Physiol.* 11:283–316, 1976. Ammonia production and excretion in the mammalian kidney.

124. Goldstein, L. and R. P. Forster. *Comp. Biochem. Physiol.* 14:567–576, 1965. Role of uricolysis in the production of urea by fishes and other aquatic vertebrates.

125. Goldstein, L., J. B. Clairborne, and D.

E. Evans. *J. Exp. Zool.* 219:395–397, 1982. Ammonia excretion by the gills of the two marine teleost fish: The importance of NH_4^+ permeance.

126. Goodman, M. W., L. Zieve, F. N. Konstantinides, and F. B. Cerra. *Am. J. Physiol.* 247:G290–G295, 1984. Mechanism of arginine protection against ammonia intoxication in the rat.

127. Gourdoux, L., Y. Lequellec, R. Moreau, and J. Dutrieu. *Comp. Biochem. Physiol. B* 74B:273–276, 1983. Gluconeogenesis from some amino acids and its endocrine modification in *Tenebrio molitor* L. (Coleoptera).

128. Graf-Hausner, U., J. K. Wilson, and P. Christen. *J. Biol. Chem.* 258:8813–8826, 1983. The covalent structure of mitochondrial aspartate aminotransferase from chicken. Identification of segments of the polypeptide chain invariant specifically in the mitochondrial isoenzyme.

129. Grayson, E. R., L. Lee, and S. E. Evans. *J. Biol. Chem.* 260:15840–15849, 1985. Immunochemical analysis of the domain structure of CAD, the multifunctional protein that initiates pyrimidine biosynthesis in mammalian cells.

130. Grazi, E., E. Magri, and G. Balboni. *Eur. J. Biochem.* 60:431–436, 1975. On the control of arginine metabolism in chicken kidney and liver.

131. Grigg, G. C. *J. Comp. Physiol. B* 144B:261–270, 1981. Plasma homeostasis and cloacal urine composition in *Crocodylus porosus* caught along a salinity gradient.

132. Groen, A. K., C. W. J. van Roermund, R. C. Vervoorn, and J. M. Tager. *Biochem. J.* 237:379–389, 1986. Control of gluconeogenesis in rat liver cells.

133. Groen, A. K., H. J. Sips, R. C. Vervoorn, and J. M. Tager. *Eur. J. Biochem.* 122:87–93, 1982. Intracellular compartmentation and control of alanine metabolism in rat liver parenchymal cells.

134. Groot, C. J. de, A. J. van Zonneveld, P. G. Mooren, D. Zonneveld, A. van den Dool, A. J. W. van den Bogaert, W. H.

Lamars, A. F. M. Moosman, and R. Charles. *Biochem. Biophys. Res. Commun.* 124:882–888, 1984. Regulation of mRNA levels of rat liver carbamoylphosphate synthetase by glucocorticosteroids and cyclic AMP as estimated with a specific cDNA.

135. Gross, J. D., J. Bradburn, R. R. Kay, and M. J. Peacey. *Nature (London)* 303:244–245. 1983. Intracellular pH and the control of cell differentiation in *Dictyostelium discoideum.*

136. Hall, S. E. H., R. Goebel, I. Barnes, G. Hetenyi, Jr., and M. Berman. *Fed. Proc., Fed. Am. Soc. Exp. Biol.* 36:239–244, 1977. The turnover and conversion to glucose of alanine in newborn and grown dogs.

137. Hanlon, D. P. *Comp. Biochem. Physiol. B* 52B:261–264, 1975. The distribution of arginase and urease in marine invertebrates.

138. Harper, J. R., V. Quaranta, and R. A. Reisfeld. *J. Biol. Chem.* 261:3600–3606, 1986. Ammonium chloride interferes with a distinct step in the biosynthesis and cell surface expression of human melanoma-type chondroitin sulfate proteoglycan.

139. Harris, R. A., R. Paxton, S. M. Powel, G. W. Goodwin, M. J. Kuntz, and A. C. Han. *Adv. Enzyme Regul.* 25:219–237, 1986. Regulation of branched-chain α-keto acid dehydrogenase complex by covalent modification.

140. Häussinger, D. and W. Gerok. In *Regulation of Hepatic Metabolism* (R. G. Thurman, F. C. Kauffman, and K. Jungermann, eds.), pp. 253–291. Plenum, New York, 1986. Metabolism of amino acids and ammonia.

141. Häussinger, D. and H. Sies. *Eur. J. Biochem.* 101:179–184, 1979. Hepatic glutamine metabolism under the influence of the portal ammonia concentration in the perfused rat liver.

142. Häussinger, D., T. Stehel, and W. Gerok. *Biol. Chem. Hoppe-Seyler* 366:527–536, 1985. Glutamine metabolism in isolated perfused liver. The transamination pathway.

143. Hayashi, S. and Z. Ooshiro. *J. Comp. Physiol. B* 132B:343–350, 1979. Gluconeogenesis in isolated liver cells of the eel *Anguilla japonica.*

144. Hayashi, Y. S., O. Matsushima. H. Katayama, and K. Yamada. *Comp. Biochem. Physiol. B* 83B:721–724, 1986. Activities of three ammonia-forming enzymes in the tissues of the brackish-water bivalve *Corbicula japonica.*

145. Heinkoff, S., D. Nash, R. Hards, J. Bleskan, J. F. Woolford, F. Naquib, and D. Patterson. *Proc. Natl. Acad. Sci. U.S.A.* 83:3919–3923, 1986. Two *Drosophila melanogaster* mutations block successive steps of *de novo* purine biosynthesis.

146. Henderson, J. F. and A. R. P. Paterson. *Nucleotide Metabolism: An Introduction.* Academic Press, New York, 1973.

147. Hensgens, H. E. S. J., A. J. Meijer, J. R. Williamson, J. A. Gimpel, and J. M. Tager. *Biochem. J.* 170:699–707, 1978. Proline metabolism in isolated rat liver cells.

148. Hershfield, M. S. and J. E. Seegmiller. *J. Biol. Chem.* 252:6002–6010, 1977. Regulation of *de novo* purine synthesis in human lymphoblasts.

149. Herzfeld, A. and S. M. Raper. *Biol. Neonate* 27:163–176, 1975. Glutamine synthetase and glutamyltransferase in developing chick and rat.

150. Herzfeld, A. and S. M. Raper. *Biochem. J.* 153:469–478, 1976. The heterogenity of arginases in rat tissues.

151. Higuera, M. de la and P. Cardenas. *Comp. Biochem. Physiol. B* 85B:517–521, 1986. Hormonal effects on gluconeogenesis from (U-^{14}C) glutamate in rainbow trout (*Salmo gairdneri*).

152. Hod, Y., S. M. Morris, and R. W. Hanson. *J. Biol. Chem.* 259:15603–15608, 1984. Induction by cAMP of the mRNA encoding the cytosolic form of phosphoenolypyruvate carboxykinase (GTP) from the chicken. Identification and characterization of a cDNA clone for the enzyme.

153. Hoek, J. B. and R. M. Njogu. *J. Biol. Chem.* 255:8711–8718, 1980. The role of

glutamate transport in the regulation pathway of proline oxidation in rat liver mitochondia.

154. Hoffmann, R. J., S. H. Bishop, and C. Sassaman. *J. Exp. Zool.* 203:165–170, 1978. Glutamate dehydrogenase from coelenterates NADP-specific.

155. Holmes, E. W., J. A. McDonald, J. M. McCord, J. B. Wyngaarden, and W. N. Kelley. *J. Biol. Chem.* 248:144–150, 1973. Human glutamine phosphoribosylpyrophosphate amidotransferase.

156. Hornby, D. P., M. J. Aitchison, and P. E. Engel. *Biochem. J.* 223:161–168, 1984. The kinetic mechanism of ox liver glutamate dehydrogenase in the presence of the allosteric effector ADP. The oxidative deamination of L-glutamate.

157. Horne, F. and C. Findeisen. *Comp. Biochem. Physiol. B* 58B:21–26, 1977. Aspects of fasting metabolism in the desert tortoise *Gopherus berlandieri.*

158. Horner, R. D. and E. N. Moudrianakis. *J. Biol. Chem.* 261:13408–13414, 1986. Effect of permeant buffers on the initiation of photosynthetic phosphorylation and postillumination phosphorylation in chloroplasts.

159. Howard, W. J., L. A. Kerson, and S. H. Appel. *J. Neurochem.* 17:121–123, 1970. Synthesis *de novo* of purine in slices of brain and liver.

160. Hughes, S. G., G. L. Rumsey, and M. C. Neshiem. *Comp. Biochem. Physiol. B* 76B: 429–431, 1983. Branched-chain amino acid aminotransferase activity in the tissues of lake trout.

161. Husson, A., M. Bouazza, C. Buquet, and R. Vaillant. *Biochem. J.* 216:281–285. 1983. Hormonal regulation of two urea-cycle enzymes in cultured feotal hepatocytes.

161a. Husson, A., C. Buquet, and R. Vaillant. *Differentiation* 35:212–218, 1987. Induction of the five urea-cycle enzymes by glucagon in cultured foetal rat hepatocytes.

162. Incerpi, P., P. Luly, and S. Scapin. *Comp. Biochem. Physiol. B* 75B:645–648,

1983. Glucocorticoid receptor of frog (*Rana esculenta*) liver.

163. Itoh, R. and K. Tsushima. *Comp. Biochem. Physiol. B* 54B:43–46, 1976. A comparative study of hypoxanthine phosphoribosyltransferase activity in birds and mammals.

164. Itoh, R., E. W. Holmes, and J. B. Wyngaarden. *J. Biol. Chem.* 251:2234–2240, 1976. Pigeon liver amidophosphoribosyltransferase. Ligand-induced alterations in molecular and kinetic properties.

165. Itoh, R., C. Usami, T. Nishino, and K. Tsushima. *Biochim. Biophys. Acta* 526:154–162, 1978. Kinetic properties of cytosol 5'-nucleotidase from chicken liver.

165a. Jahoor, F. and R. R. Wolfe. *Am. J. Physiol.* 253:E543–E550, 1987. Regulation of urea production by glucose infusion in vivo.

166. Jenkins, W. T. In *Transaminases* (P. Christen and D. E. Metzler, eds.), pp. 365–373. Wiley (Interscience), New York, 1985. Introductory remarks.

167. Jo, J.-S., N. Ishihara, and G. Kikuchi. *Arch Biochem. Biophys.* 160:246–254, 1974. Occurrence and properties of four forms of phosphoenolpyruvate carboxykinase in chicken liver.

168. Jones, M. E. *J. Nutr.* 115:509–515, 1985. Conversion of glutamate to ornithine and proline: Pyrroline-5-carboxylate, a possible modulator of arginine requirements.

169. Jungermann, K. and N. Katz. In *Regulation of Hepatic Metabolism* (R. G. Thurman, F. C. Kauffman, and K. Jungermann, eds.), pp. 211–252. Plenum, New York, 1986. Metabolism of carbohydrates.

170. Kadowaki, H., H. W. Israel, and M. C. Neshein. *Biochim. Biophys. Acta* 437:158–165, 1976. Intracellular localization of arginase in chicken kidney.

171. Katunuma, N., Y. Matsuda, and Y. Kuroda. *Adv. Enzyme Regul.* 8:73–81, 1970. Phylogenic aspects of different regula-

tory mechanisms of glutamine metabolism.

172. Katz, A., K. Sahlin, and J. Henriksson. *Am. J. Physiol.* 250:C834–C840, 1986. Muscle ammonia metabolism during isometric contraction in humans.

173. Kilberg, M. S., M. E. Handlogten, and H. N. Christensen. *J. Biol. Chem.* 255:4011–4019, 1980. Characteristics of an amino acid transport system in rat liver for glutamine, asparagine, histidine, and closely related analogs.

174. King, F. D., T. L. Cucci, and R. R. Bridgare. *Comp. Biochem. Physiol. B* 80B:401–403, 1985. A pathway of nitrogen metabolism in marine decapod crabs.

175. King, K. S. and P. P. Cohen. *Comp. Biochem. Physiol. B* 51B:113–117, 1975. Purification and comparative properties of glutamate dehydrogenase from frog and tadpole liver.

176. King, P. A. and L. Goldstein. *Am. J. Physiol.* 245:R581–R589, 1983. Renal ammoniagenesis and acid excretion in the dogfish, *Squalus acanthias*.

177. King, P. A. and L. Goldstein. *Renal Physiol. (Basel)* 8:261–278, 1985. Renal excretion of nitrogenous compounds in vertebrates.

178. King, P. A., L. Goldstein, and E. A. Newsholme. *Biochem. J.* 216:523–525, 1983. Glutamine synthetase activity of muscle in acidosis.

179. Kinsella, J. E., B. German, and J. Shetty. *Comp. Biochem. Physiol. B* 82B:621–624, 1985. Uricase from fish liver: isolation and some properties.

180. Kleiner, D. *Biochim. Biophys. Acta* 639:41–52, 1981. The transport of NH_3 and NH_4^+ across biological membranes.

181. Klemperer, F. W., H. C. Trimble, and A. B. Hastings. *J. Biol. Chem.* 125:445–449, 1938. The uricase of dogs, including the Dalmatian.

182. Kovacevic, Z. and K. Bajin. *Biochim. Biophys. Acta* 687:291–295, 1982. Kinetics of glutamine efflux from liver mitochondria loaded with ^{14}C-labeled substrate.

183. Kovacevic, Z. and J. D. McGivan. *Physiol. Rev.* 63:547–605, 1983. Mitochondrial metabolism of glutamine and glutamate and its physiological significance.

184. Krebs, H. A. *Adv. Enzyme Regul.* 10:397–420, 1972. Some aspects of the regulation of fuel supply in omnivorous animals.

185. Krebs, H. A., R. Hems, P. Lund, D. Halliday, and W. W. C. Read. *Biochem. J.* 176:733–737, 1978. Sources of ammonia for mammalian urea synthesis.

186. Krenitsky, T. A., J. V. Tuttle, E. L. Cattau, Jr., and P. Wang. *Comp. Biochem. Physiol. B* 49B:687–703, 1974. A comparison of the distribution and electron acceptor specificities of xanthine oxidase and aldehyde oxidase.

186a. Kuo, C. F., K. E. Paulson, and J. E. Darnell, Jr. *Molec. Cell. Biol.* 8:4966–4971, 1988. Positional and developmental regulation of glutamine synthetase expression in mouse liver.

187. Kurtz, I. and R. S. Balaban. *Am. J. Physiol.* 250:F497–F502, 1986. Ammonium as a substrate for Na^+-K^+-ATPase in rabbit proximal tubles.

188. Kuster, G., R. G. Shorter, B. Dawson, and G. A. Hallenbeck. *Arch. Intern. Med.* 129:492–496, 1972. Uric acid metabolism in Dalmatians and other dogs.

189. Lambert, M. A., L. R. Simnard, P. N. Ray, and R. R. McInnes. *Mol. Cell. Biol.* 6:1722–1728, 1986. Molecular cloning of cDNA for rat argininosuccinate lyase and its expression in rat hepatoma cell lines.

189a. Lamers, W. H., R. W. Hanson, and H. M. Meisner. *Proc. Natl. Acad. Sci. U.S.A.* 79:5137–5141, 1981. cAMP stimulates transcription of the gene for cytosolic phosphenol pyruvate carboxykinase in rat liver nuclei.

190. LaNoue, K. F. and A. C. Schoolwerth. *Annu. Rev. Biochem.* 48:871–922, 1979. Metabolite transport in mitochondria.

191. LaNoue, K. F. and M. E. Tischler. *J. Biol. Chem.* 249:7522–7528, 1974. Electrogenic characteristics of the mitochondrial glutamate-aspartate antiporter.

192. Lardy, H. and P. E. Hughes. *Curr. Top.*

Cell. Regul. 24:171–179, 1984. Regulation of gluconeogenesis at phosphoenolpyruvate carboxykinase.

192a. Lee, C. L., X. Wu, R. A. Gibbs, R. G. Cook, D. M. Munzy, and C. T. Caskey. *Science* 239:1288–1291, 1988. Generation of cDNA probes directed by amino acid sequence: Cloning of urate oxidase.

193. Leech, A. R., L. Goldstein, C.-J. Cha, and J. M. Goldstein. *J. Exp. Zool.* 207:73–80, 1979. Alanine biosynthesis during starvation in skeletal muscle of the spiny dogfish, *Squalus acanthias.*

194. Levi, G., G. Morisi, A. Coletti, and R. Catanzaro. *Comp. Biochem. Physiol. A* 49A:623–636, 1974. Free amino acids in fish brain: Normal levels and changes upon exposure to high ammonia concentrations *in vivo,* and upon incubation of brain slices.

195. Linder, M. J. and P. H. Weigel. *Trends Biochem. Sci.* 11:200–201, 1986. A balanced equation for gluconeogenesis: Are the textbooks all wet?

196. Lipstein, B., P. Boer, and O. Sperling. *Biochim. Biophys. Acta* 521:45–54, 1978. Regulation of *de novo* purine synthesis in chicken liver slices. Role of phosphoribosylpyrophosphate availability and of salvage purine nucleotide synthesis.

197. Loest, R. A. *Comp. Biochem. Physiol. B* 63B:103–107, 1979. Urease from a sea urchin *Lytechinus variegatus:* Partial purification and kinetics.

198. Loose, D. S., D. K. Cameron, H. P. Short, and R. W. Hanson. *Biochemistry* 24:4509–4512, 1985. Thyroid hormone regulates transcription of the gene for cytosolic phosphoenolpyruvate carboxykinase (GTP) in rat liver.

199. Lowenstein, J. M. *Physiol. Rev.* 52:382–414, 1972. Ammonia production in muscle and other tissues: The purine nucleotide cycle.

200. Lowenstein, J. and K. Tornheim. *Science* 171:397–400, 1971. Ammonia production in muscle: The purine nucleotide cycle.

201. Lund, P. and M. Watford, In *The Urea Cycle* (S. Grisolia, R. Báguena, and F. Mayor, eds.), pp. 479–488. Wiley (Interscience), New York, 1976. Glutamine as a precursor of urea.

202. Maddaiah, V. T. *Biochem. Biophys. Res. Commun.* 127:565–569, 1985. Ammonium inhibition of fatty acid oxidation in rat liver mitochondria: a possible cause of fatty liver in Reye's syndrome and urea cycle defects.

203. Male, K. B. and K. B. Storey. *Comp. Biochem. Physiol. B* 76B:823–829, 1983. Kinetic characterization of NADP-specific glutamate dehydrogenase in the sea anemone, *Anthopleura xanthogrammica:* Control of amino acid biosynthesis during osmotic stress.

204. Male, K. B. and K. B. Storey. *J. Comp. Physiol. B* 151B:199–205, 1983. Tissue specific isozymes of glutamate dehydrogenase from the Japanese beetle *Popilla japonica:* Catabolic vs. anabolic GDH's.

205. Mallette, L. E., J. H. Exton, and C. R. Park. *J. Biol. Chem.* 244:5713–5723, 1969. Control of gluconeogenesis from amino acids in the perfused rat liver.

206. Manfredi, J. P. and E. W. Holmes. *Arch. Biochem. Biophys.* 233:515–529, 1984. Control of purine nucleotide cycle in extracts of rat muscle: Effects of energy state and concentrations of cycle intermediates.

207. Mapes, J. P. and H. A. Krebs. *Biochem J.* 172:193–203, 1978. Rate-limiting factors and gluconeogenesis in avian liver.

208. Martin, A. D. and M. A. Titheradge. *Biochem. J.* 222:379–387, 1984. Stimulation of mitochondrial pyruvate metabolism and citrulline synthesis by dexamethasone. Effect of isolation and incubation medium.

209. Masters, C. and R. Holmes. *Physiol. Rev.* 57:816–882, 1977. Peroxisomes: New aspects of cell physiology and biochemistry.

209a. May, R. M. *Science* 241:1441–1449, 1988. How many species are there on Earth?

210. McCormick, A., K. T. Paynter, M. M. Brodey, and S. H. Bishop. *Comp.*

Biochem. Physiol. B 84B:163–166, 1986. Aspartate aminotransferase from ribbed mussel gill tissues: Reactivity with β-L-cysteinesulfinic and other properties.

211. McDonald, J. A. and J. W. Campbell. *Comp. Biochem. Physiol. B* 66B:215–222, 1980. Invertebrate urease: Effect of metals and sulfhydryl reagents on the *Otala* enzyme.

212. McDonald, J. A., J. E. Vorhaben, and J. W. Campbell. *Comp. Biochem. Physiol. B* 66B:223–231, 1980. Invertebrate urease: Purification and properties of the enzyme from the land snail *Otala lactea.*

213. McGivan, J. D. and J. B. Chappell. *FEBS Lett.* 52:1–7, 1975. On the metabolic function of glutamate dehydrogenase in rat liver.

214. McGivan, J. D., N. M. Bradford, and A. D. Beavis. *Biochem. J.* 162:147–156, 1977. Factors influencing the activity of ornithine aminotransferase in isolated rat liver mitochondria.

215. McGivan, J. D., N. M. Bradford, A. J. Verhoeven, and A. J. Meijer. In *Glutamine Metabolism in Mammalian Tissues* (D. Häussinger and H. Sies, eds.), pp. 122–137. Springer-Verlag, Berlin, 1984. Liver glutaminase.

216. McIntyre, P., J. F. B. Mercer, M. G. Peterson, P. Hudson, and N. Hoogenraad. *Eur. J. Biochem.* 143:183–187, 1984. Selection of cDNA clone which contains the complete coding sequence for the mature form of ornithine transcarbamylase from rat liver: Expression of the cloned protein in *Escherichia coli.*

217. Meijer, A. J. and H. E. S. J. Hensgens. In *Metabolic Compartmentation* (II. Sies, ed.), pp. 259–286. Academic Press, London, 1982. Ureogenesis.

218. Meijer, A. J. and A. J. Verhoeven. *Biochem. Soc. Trans.* 14:1001–1004, 1986. Regulation of hepatic glutamine metabolism.

219. Meijer, A. J., C. Lof, I. C. Ramos, and J. Verhoeven. *Eur. J. Biochem.* 148:189–196, 1985. Control of ureogenesis.

220. Meijer, A. J., J. A. Gimpel, G. Deleeuw,

M. E. Tischler, J. M. Tager, and J. R. Williamson. *J. Biol. Chem.* 253:2308–2320, 1978. Interrelationships between gluconeogenesis and ureogenesis in isolated hepatocytes.

221. Mitchell, H. A., *J. Mammol.* 44:543–551, 1963. Ammonia tolerance of the California leaf-nosed bat.

222. Mitchell, H. A. *J. Mammol.* 45:568–577, 1964. Investigations of the cave atmosphere of a Mexican bat colony.

223. Mommsen, T. P. *Can. J. Zool.* 64:1110–1115, 1986. Comparative gluconeogenesis in hepatocytes from salmonid fishes.

224. Mommsen, T. P. and R. K. Suarez. *Mol. Physiol.* 6:9–18, 1984. Control of gluconeogenesis in rainbow trout hepatocytes: Role of pyruvate branchpoint and phosphoenolpyruvate-pyruvate cycle.

224a. Moorman, A. F. M., P. A. J. de Boer, W. J. C. Geerts, H. V. Zande, W. H. Lamars, and R. Charles. *J. Histochem. Cytochem.* 36:751–755, 1988. Complementary distributions of carbamoylphosphate synthetase (ammonia) and glutamine synthetase in rat liver acinus is regulated at pretranslational level.

225. Mora, M., J. C. Manzanero, and J. Boxal. *Comp. Biochem. Physiol. B* 88B:1143–1149, 1987. Pigeon liver nucleoside phosphorylase: Kinetic behavior and sulfhydryl groups.

226. Mori, M, S. Miura, M. Tatibana, and P. P. Cohen. *J. Biol. Chem.* 256:4127–4132, 1981. Cell-free translation of carbamoylphosphate synthetase-I and ornithine transcarbamylase messenger RNA. Effect of dietary protein and fasting on translatable mRNA levels.

227. Morino, Y. and S. Tanase. In *Transaminases* (P. Christen and D. E. Metzler, eds.), pp. 384–389. Wiley (Interscience), New York, 1985. Aminotransferases utilizing pyruvate.

228. Morita, T., M. Mori, and M. Tatibana. *J. Biochem. (Tokyo)* 91:563–569, 1982. Regulation of N-acetyl-L-glutamate degradation in mammalian liver.

229. Morris, S. M., Jr. *Arch Biochem. Biophys.*

259:144–148, 1987. Thyroxine elicits divergent changes in mRNA levels for two urea cycle enzymes and one gluconeogenic enzyme in tadpole liver.

230. Moss, K. M. and J. D. McGivan. *Biochem. J.* 150:275–283, 1975. Characteristics of asparate deamination by the purine nucleotide cycle in the cytosol fraction of rat liver.

231. Moyer, J. D. and J. F. Henderson. *Can. J. Biochem. Cell Biol.* 61:1153–1157, 1983. Salvage of circulating hypoxanthine by tissues of mouse.

232. Moyes, C. D., T. W. Moon, and J. S. Ballantyne. *J. Exp. Zool.* 237:119–128, 1986. Oxidation of amino acids, Krebs cycle intermediates, fatty acids and ketone bodies by *Raja erinacea* liver mitochondria.

233. Mustafa, T., R. Komuniecki, and D. F. Mettrick. *Comp. Biochem. Physiol. B* 61B:219–222, 1978. Cytosolic glutamate dehydrogenase in adult *Hymenolepis diminuta*.

234. Mustafa, T., J. Seuss, J. B. Jørgensen, and K. H. Hoffman. *J. Comp. Physiol. B* 149B:477–483, 1983. Gluconeogenesis in facultative anaerobic invertebrates: Evidence of oxalacetate decarboxylation and anaerobic end product incorporation into glycogen from the tissues of *Tubifex* sp.

235. Nagano, H., H. Itoh, and R. Shukuya. *FEBS Lett.* 33:125–128, 1973. Changes in hepatic phosphoenolpyruvate carboxykinase and its intracellular distribution during bullfrog metamorphosis.

236. Nahahara, A., T. Nishino, M. Kanisawa, and K. Tsushima. *Comp. Biochem. Physiol. B* 88B:589–593, 1987. Effect of dietary protein on purine nucleoside phosphorylase and xanthine dehydrogenase activities of liver and kidney in chicken and pigeon.

237. Naito, Y., R. Itoh, and K. Tsushima. *Int. J. Biochem.* 5:807–810, 1974. 5'-Nucleotidase of chicken liver: Comparison of soluble 5'-nucleotidase activities in chicken and rat liver.

238. Nakano, M. and T. S. Danocoski. *J. Biol.*

Chem. 241:2075–2083, 1966. Crystalline mammalian L-amino acid oxidase from rat kidney mitochondria.

238a. Nash, D. and J. F. Henderson. *Adv. Comp. Physiol. Biochem.* 8:1–51, 1982. The biochemistry and genetics of purine metabolism in *Drosophila melanogaster*.

239. Needham, J. *Chemical Embryology*, Vols. 2 and 3. Cambridge Univ. Press, London and New York, 1931.

240. Nissim, I., M. Yudkoff, and S. Segal. *J. Biol. Chem.* 261:6509–6514, 1986. Effect of 5-amino-4-imidazolecarboxamide riboside on renal ammoniagenesis. Study with ^{15}N aspartate.

241. Niwa, H., T. Yamano, T. Sugano, and R. A. Harris. *Comp. Biochem. Physiol. B* 85B:739–745, 1986. Hormonal effects and the control of gluconeogenesis from sorbitol, xylitol and glycerol in perfused chicken liver.

242. Njogu, R. M. and J. B. Hoek. *FEBS Lett.* 152:222–226, 1983. The effect of inhibitors of glutamate transport on the pathway of glutamate oxidation in rat liver mitochondria.

243. Noda, C., T. Nakamura, and A. Ichihara. *J. Biol. Chem.* 258:1520–1525, 1983. α-Adrenergic regulation of enzymes of amino acid metabolism in primary cultures of adult rat hepatocytes.

243a. Noguchi, T., M. Diesterhaft, and D. Granner. *J. Biol. Chem.* 257:2386–2390, 1982. Evidence for a dual effect of dibutyryl cyclic AMP on the synthesis of tyrosine aminotransferase in rat liver.

244. Noguchi, T., S. Fujiwara, and S. Hayashi. *J. Biol. Chem.* 261:4221–4223, 1986. Evolution of allantoinase and allantoicase involved in urate degradation in liver peroxisomes. A rapid purification of amphibian allantoinase and allantoicase complex, its subunit locations of the two enzymes, and its comparison with fish allantoinase and allantoicase.

245. Noguchi, T., S. Fujiwara, Y. Takada, T. Mori, M. Nagano, N. Hanada, E. Saeki, and O. Yasuo. *Comp. Biochem. Physiol. B* 77B:279–283, 1984. Enzymatic and immunological comparison of alanine:

Glyoxylate aminotransferase from different fish and mammalian livers.

246. Nordlie, R. C. and H. A. Lardy. *J. Biol. Chem.* 238:2259–2263, 1963. Mammalian liver phosphoenolpyruvate carboxykinase activities.

247. Nyunoya, H., K. E. Broglie, E. E. Widgen, and C. J. Lusty. *J. Biol. Chem.* 260:9346–9356, 1985. Characterization and derivation of the gene coding from mitochondrial carbamyl phosphate synthetase I of rat.

248. Ochs, R. S. and R. A. Harris. *Arch. Biochem. Biophys.* 190:193–201, 1978. Studies on the relationship between glycolysis, lipogenesis, gluconeogenesis, and pyruvate kinase activity of rat and chicken hepatocytes.

249. Ochs, R. S. and R. A. Harris. *Biochim. Biophys. Acta* 632:260–269, 1980. Aminooxyacetate inhibits gluconeogenesis by isolated chicken hepatocytes.

250. Odessey, R. and A. L. Goldberg. *Am. J. Physiol.* 223:1376–1383, 1972. Oxidation of leucine by rat skeletal muscle.

251. Ogata, K., M. Watford, L. J. Brady, and R. W. Hanson. *J. Biol. Chem.* 257:5385–5391, 1982. Mitochondrial phosphoenolpyruvate carboxykinase (GTP) and the regulation of gluconeogenesis and ketogenesis in avian liver.

252. O'Hare, M. C. and S. Doonan. *Biochim. Biophys. Acta* 827:127–134, 1985. Purification and structural comparisons of the cytosolic and mitochondrial isoenzymes of fumarase from pig liver.

253. Ohki, Y., Y. Goto, and R. Shukuya. *Biochim. Biophys. Acta* 661:230–234, 1981. Induction of mitochondrial phosphoenolpyruvate carboxykinase in liver of *Rana catesbeiana* tadpole treated with 3,5,3'-triiodothyronine and undergoing natural metamorphosis.

254. O'Neill, I. E. and D. W. Bannister. *Comp. Biochem. Physiol. B* 83B:595–598, 1986. Gluconeogenic and lipogenic properties of isolated kidney tubules from the domestic fowl (*Gallus domesticus*).

255. Ottesen, I. and L. Klungsøyr. *Comp.* *Biochem. Physiol. B* 79B:489–491, 1984. Branched chain α-keto acid dehydrogenase in tissues from rainbow trout (*Salmo gairdnerii*).

256. Parsot, C. *EMBO J.* 5:3013–3019, 1986. Evolution of biosynthetic pathways: a common ancestor for threonine synthase, threonine dehydratase and D-serine dehydratase.

257. Patterson, J. W., P. M C. Davies, D. A. Veasey, and J. R. Griffiths. *Comp. Biochem. Physiol. B* 60B:491–493, 1978. The influence of season on glycogen levels in the lizard *Lacerta vivipara* Jacquin.

258. Paynter, K. T., G. A. Karam, L. L. Ellis, and S. H. Bishop. *Comp. Biochem. Physiol. B* 82B:129–132, 1985. Subcellular distribution of aminotransferases, and pyruvate branch point enzymes in gill tissue from four bivalves.

259. Penney, D. G. and E. H. Kornecki. *Comp. Biochem. Physiol. B* 46B:405–415, 1973. Activities, intracellular localization and kinetic properties of phosphoenolpyruvate carboxykinase, pyruvate kinase and malate dehydrogenase in turtle (*Pseudemys scripta elegans*) liver, heart, and skeletal muscle.

260. Penney, D. and K. Papademas. *Comp. Biochem. Physiol. B* 50B:137–143, 1975. Effect of starvation on fructose diphosphatase, glucose-6-phosphatase and phosphoglucomutase activities in organs of *Pseudemys* (*Chrysemys*) *scripta elegans*.

261. Perry, T. L. In *Glutamine, Glutamate, and GABA in the Central Nervous System* (L. Hertz, E. Kvamme, E. G. McGeer, and A. Schousboe, eds.), pp. 581–594. Alan R. Liss, New York, 1983. Levels of glutamine, glutamate and GABA in CSF and brain under pathological conditions.

262. Phillips, J. E. *Fed. Proc., Fed. Am. Soc. Exp. Biol.* 36:2480–2486, 1977. Excretion in insects: Function of gut and rectum in concentrating and diluting the urine.

263. Pons, A., F. J. Garciá, A. Palou, and M. Alemany. *Arch. Int. Physiol. Biochim.*

94:219–226, 1986. Amino-acid metabolism enzyme activities in the liver, intestine and yolk sack membrane of developing domestic fowl.

264. Pons, A., F. J. Garciá, A. Palou, and M. Alemany. *Comp. Biochem. Physiol. B* 85B:275–278, 1986. Effect of starvation and a protein diet on the amino acid metabolism enzyme activities of the organs of domestic fowl hatchlings.

265. Post, G. R. and A. G. Fischer. *Int. J. Biochem.* 18:63–66, 1986. Hypoxanthine and adenine metabolism in bovine thyroid tissue.

266. Potrikus, C. J. and J. A. Breznak. *Proc. Natl. Acad. Sci, U.S.A.* 78:4601–4605, 1981. Gut bacteria recycle uric acid nitrogen in termites: A strategy for nutrient conservation.

267. Powers-Lee, S. G., R. A. Matisco, and M. Bendayan. *J. Biol. Chem.* 262:15683–15688, 1987. The interaction of rat liver carbamoyl phosphate synthetase and ornithine transcarbamoylase with inner mitochondrial membrane.

268. Prentø, P. *Comp. Biochem. Physiol. B* 86B:333–341, 1987. Blood sugar, sugar metabolism and related enzymes in the earthworm *Lumbricus terrestris*. L.

269. Pritchard, J. B., F. Chavez-Peon, and R. D. Berlin. *Am. J. Physiol.* 219:1263–1267, 1970. Purines: Supply by liver to tissues.

270. Prusch, R. D. *J. Exp. Biol.* 64:89–100, 1976. Unidirectional ion movements in the hindgut of larval *Sarchophaga bullata* (Diptera: Sarcophagidae).

271. Räihä, N. C. R. In *The Urea Cycle* (S. Grisolia, R. Báguena, and F. Mayor, eds.), pp. 261–272. Wiley (Interscience), New York, 1976. Developmental changes of urea cycle enzymes in mammalian liver.

272. Raijman, L. *Biochem. J.* 138:225–232, 1974. Citrulline synthesis in rat tissues and liver content of carbamoyl phosphate and ornithine.

273. Raijman, L. and M. E. Jones. *Arch. Biochem. Biophyhs.* 175:270–278, 1976. Purification, composition, and some

properties of rat liver carbamyl phosphate synthetase (ammonia).

274. Ratner, S. *Adv. Enzymol. Relat. Areas Mol. Biol.* 39:1–90, 1973. Enzymes of arginine and urea synthesis.

275. Ray, M. and S. Ray. *J. Biol. Chem.* 260:5913–5918, 1985. L-Threonine dehydrogenase from goat liver.

276. Reddy, S. R. R. and J. W. Campbell. *Experientia* 33:160–161, 1977. Enzymic basis for the nutritional requirement of arginine in insects.

277. Reem, G. H. *J. Biol. Chem.* 243:5695–5701, 1968. Enzymatic synthesis of 5'-phosphoribosylamine from ribose-5-phosphate and ammonia, an alternate first step in purine biosynthesis.

278. Regnault, M. *Biol. Rev. Cambridge Philos. Soc.* 62:1–24, 1987. Nitrogen excretion in marine and fresh-water Crustacea.

279. Reijngoud, D.-J., P. S. Oud, J. Kás, and J. M. Tager. *Biochim. Biophys. Acta* 448:290–302, 1976. Relationship between medium pH and that of lysosomal matrix studied by two independent methods.

280. Reiss, P. M., S. K. Pierce, and S. H. Bishop. *J. Exp. Zool.* 202:253–258, 1977. Glutamate dehydrogenase from tissues of the ribbed mussel *Modiolus demissus*: ADP activation and possible physiological significance.

281. Renaud, J. M. and T. W. Moon, *J. Comp. Physiol. B* 135B:115–125, 1980. Characterization of gluconeogenesis in hepatocytes isolated from the American eel, *Anguilla rostrata* LeSueur.

282. Renaud, J. M. and T. W. Moon. *J. Comp. Physiol. B* 135B:127–137, 1980. Starvation and the metabolism of hepatocytes isolated from the American eel, *Anguilla rostrata* LeSueur.

283. Rice, S. D. and R. M. Stokes. *Fish. Bull.* 73:207–211, 1975. Acute toxicity of ammonia to several developmental stages of rainbow trout, *Salmo gairdneri*.

284. Riemke, E., E. Mitidieri, O. E. Alfonso, and L. P. Ribiero. *Comp. Biochem. Physiol. B* 61B:53–57, 1978. Comparative as-

pects of xanthine dehydrogenase activity of *Panstrongylus megistus*.

285. Rognstad, R. *Biochim. Biophys. Acta* 496:249–254, 1977. Sources of ammonia for urea synthesis in isolated rat liver cells.

286. Roos, A. and W. F. Boron. *Physiol. Rev.* 61:296–434, 1981. Intracellular pH.

287. Rowe, P. B., E. McCairns, G. Madsen, D. Sauer, and H. Elliott. *J. Biol. Chem.* 253:7711–7721, 1978. *De novo* purine synthesis in avian liver. Co-purification of the enzymes and properties of the pathway.

288. Rowsell, E. V., J. A. Carnie, S. D. Wahbi, and A. H. Al-Tai. *Comp. Biochem. Physiol. B* 63B:543–555, 1979. L-Serine dehydratase and L-serine pyruvate aminotransferase activities in different animal species.

289. Rozen, R., C. Noel, and G. C. Shore. *Biochim. Biophys. Acta* 741:47–54, 1983. Effects of glucagon on biosynthesis of the mitochondrial enzyme, carbamoyl-phosphate synthetase I, in primary hepatocytes and Morris hepatoma 5123D.

290. Ruano, A. R., J. L. A. Riaño, M. R. Amil, and M. J. H. Santos. *Comp. Biochem. Physiol. B* 82B:197–202, 1985. Some enzymatic properties of NAD$^+$-dependent glutamate dehydrogenase of mussel hepatopancreas (*Mytilus edulis* L.)—requirement of ADP.

291. Rubio, V. and S. Grisolia. In *The Urea Cycle* (S. Grisolia, R. Báguena, and F. Mayor, eds.), pp. 91–93. Wiley (Interscience), New York, 1976. Some considerations regarding the feasibility of Bessman's hypothesis for ammonia toxicity.

292. Ruderman, N B. and M. Berger. *J. Biol. Chem.* 249:5500–5506, 1974. The formation of glutamine and alanine in skeletal muscle.

293. Ryall, J C., M. A. Quantz, and G. C. Shore. *Eur. J. Biochem.* 156:453–458, 1986. Rat liver and intestinal mucosa differ in the developmental pattern and hormonal regulation of carbamoyl-phosphate synthetase I and ornithine carbamoyltransferase gene expression.

294. Sacktor, B. In *Insect Biochemistry and Function* (D. J. Candy and B. A. Kilby, eds.), pp. 1–88. Chapman & Hall, London, 1976. Biochemistry of insect flight.

295. Salerno, C., P. Fasella, and L. A. Fahien, In *Transaminases* (P. Christen and D. E. Metzler, eds.), pp. 192–208. Wiley (Interscience), New York, 1985. Interaction of aminotransferases with other metabolically linked enzymes.

296. Sarkar, N. K. *Int. J. Biochem.* 8:427–432, 1977. Implication of alanine and aspartate aminotransferases, glutamic dehydrogenase and phosphoenolpyruvate carboxykinase in glucose production in rats, chickens and pigs.

297. Schimke, R. T. *J. Biol. Chem.* 237:459–468, 1962. Adaptive characteristics of urea cycle enzymes in the rat.

298. Schimke, R. T. *J. Biol. Chem.* 237:1921–1924, 1962. Differential effects of fasting and protein-free diets on levels of urea cycle enzymes in rat liver.

299. Schimke, R. T. *J. Biol. Chem.* 238:1012–1018, 1963. Studies on factors affecting the levels of urea cycle enzymes in rat liver.

300. Schrimsher, J. L., F. J. Schendel, and J. Stubbe. *Biochemistry* 25:4356–4365, 1986. Isolation of a multifunctional protein with aminoimidazole ribonucleotide synthetase, glycinamide ribonucleotide synthetase, and glycineamide ribonucleotide transformylase activities. Characterization of aminoimidazole ribonucleotide synthetase.

301. Schultz, V. and J. M. Lowenstein. *J. Biol. Chem.* 251:485–492, 1976. Purine nucleotide cycle: Evidence for the occurrence of the cycle in brain.

302. Schultz, V. and J. M. Lowenstein. *J. Biol. Chem.* 253:1938–1943, 1978. The purine nucleotide cycle. Studies of ammonia production and interconversions of adenine and hypoxanthine nucleosides by rat brain *in situ*.

303. Selwyn, M. J. *Nature (London)* 330:424–

425, 1987. Holes in mitochondrial inner membranes.

304. Seniów, A. *Comp. Biochem. Physiol. B* 54B:69–76, 1976. The influence of sex differences, skin-shedding cycle and hibernation on the regulation of asparate and alanine aminotransferase activities in the grass snake, *Natrix n. natrix.*

305. Sheenan, T. G. and E. R. Tully. *Biochem. J.* 216:605–610, 1983. Purine biosynthesis *de novo* in rat skeletal muscle.

306. Shiokawa, K., Y. Kawazoe, H. Nomura, T. Miura, N. Nakakura, T. Horiuchi, and K. Yamana. *Dev. Biol.* 115:380–391, 1986. Ammonium ion as a possible regulator of the commencement of rRNA synthesis in *Xenopus laevis* embryogenesis.

307. Shotwell, M A., M S. Kilberg, and D. L. Oxender. *Biochim. Biophys. Acta* 737:267–284, 1983. The regulation of neutral amino acid transport in mammalian cells.

308. Silva, S. V. P. S. and J. R. Mercer. *Comp. Biochem. Physiol. B* 80B:603–607, 1986. Effect of protein intake on amino acid catabolism and gluconeogenesis by isolated hepatocytes from the cat (*Felis domestica*).

309. Silva, S. V. P. S. and J. R. Mercer. *Biochem. J.* 240:843–846, 1986. Protein degradation in cat.

310. Simkiss, K. *Comp Biochem. Physiol.* 34:777–788, 1970. Sex differences in the acid-base balance of adult and immature fowl.

311. Sistare, F. D. and R. C. Haynes, Jr. *J. Biol. Chem.* 260:12754–12760, 1985. Acute stimulation by glucocorticoids of gluconeogenesis from lactate/pyruvate in isolated hepatocytes from normal and adrenalectomized rats.

312. Smith, C. M., L. M. Ravaneo, M. P. Kekomake, and K. O. Raivio. *Can. J. Biochem.* 55:1134–1139, 1977. Purine metabolism in isolated rat hepatocytes.

313. Smith, D. D., Jr. and J. W. Campbell. *Comp. Biochem. Physiol. B* 86B:755–762, 1987. Glutamine synthetase in liver of the American alligator, *Alligator mississippiensis.*

313a. Smith, D. D., Jr. and J. W. Campbell. *Proc. Nat. Acad. Sci. U.S.A.* 85:160–164, 1988. Distribution of glutamine synthetase and carbamoyl-phosphate synthetase I in vertebrate liver.

314. Snodgrass, P. J. and R. C. Kin. *J. Nutr.* 111:586–601, 1981. Induction of urea cycle enzymes of rat liver by amino acids.

315. Söling, H.-D. and J. Kleineke. In *Gluconeogenesis: Its Regulation in Mammalian Species* (R. W. Hanson and M. A. Mehlman, eds.), pp. 369–462. Wiley (Interscience), New York, 1976. Species dependent regulation of hepatic gluconeogenesis in higher animals.

316. Sollock, R. L., J. E. Vorhaben, and J. W. Campbell. *J. Comp. Physiol.* 129:129–135, 1979. Transaminase reactions and glutamate dehydrogenase in gastropod hepatopancreas.

317. Sonne, J. C., I. Lin, and J. M. Buchanan. *J. Biol. Chem.* 220:369–378, 1956. Biosynthesis of the purines. IX. Precursors of the nitrogen atoms of the purine ring.

318. Sonoda, T. and M. Tatibana. *Biochim. Biophys. Acta* 521:55–66, 1978. Metabolic fate of pyrimidines and purines in dietary nucleic acids ingested by mice.

319. Spychala, J. and G. Van den Berghe. *Biochem. J.* 242:551–558, 1987. Adenine nucleotide metabolism in isolated chicken hepatocytes.

320. Stewart, P. M., M. Batshaw, D. Valle, and M. Walser. *Am. J. Physiol.* 241:E310–E315, 1981. Effects of arginine-free meals on ureagenesis in cats.

321. Stombaugh, D. P., H. S. Teague, and W. L. Roller. *J Anim. Sci.* 28:844–847, 1969. Effects of atmospheric ammonia on the pig.

322. Storey, K. B., J. H. A. Fields, and P. W. Hochachka. *J. Exp. Zool.* 205:111–118, 1978. Purification and properties of glutamate dehydrogenase from the mantle muscle of the squid, *Loligo pealeii.*

323. Streamer, M. *Comp. Biochem. Physiol. B*

65B:669–674, 1980. Urea and arginine metabolism in the hard coral, *Acropora acuminata*.

324. Strittmatter, C. F. *J. Biol. Chem.* 240:2557–2564, 1965. Studies on avian xanthine dehydrogenase. Properties and patterns of appearance during development.

325. Strzelecki, T., J. Rogulski, and S. Angielski. *Biochem. J.* 212:705–711, 1983. The purine nucleotide cycle and ammonia formation from glutamine by rat kidney slices.

326. Studier, E. H. *J. Exp. Zool.* 163:79–85, 1966. Studies on the mechanisms of ammonia tolerance of the guano bat.

327. Studier, E. H. *J. Exp. Zool.* 170:253–258, 1969. Respiratory ammonia filtration and ammonia tolerance in bats.

328. Studier, E. H. and A. A. Fresquez. *Ecology* 50:492–494, 1969. Carbon dioxide retention: A mechanism of ammonia tolerance in mammals.

329. Suarez, R. K. and T. P. Mommsen. *Can. J. Zool.* 65:1869–1882, 1987. Gluconeogenesis in teleost fish.

330. Suda, M. and H. Nakagawa. *Methods Enzymol.* 17B:346–351, 1971. L-Serine dehydratase (rat liver).

331. Sugden, P. H. and E. A. Newsholme. *Biochem. J.* 150:113–122, 1975. The effects of ammonium, inorganic phosphate and potassium ions on the activity of phosphofructokinase from muscle and nervous tissues of vertebrates and invertebrates.

332. Takada, Y. and T. Noguchi. *J. Biol. Chem.* 258:4762–4764, 1983. The degradation of urate in liver peroxisomes. Association of allantoinase with allantoicase in amphibian liver but not in fish and invertebrate liver.

333. Tamir, H. and S. Ratner. *Arch. Biochem. Biophys.* 102:249–258, 1963. Enzymes of arginine metabolism in chicks.

333a. Taplin, L. E. *Biol. Rev.* 63:333–377, 1988. Osmoregulation in crocodilians.

334. Tatibana, M. and K. Shigesada. In *The Urea Cycle* (S. Grisolia, R. Báguena, and

F. Mayor, eds.), pp. 301–313. Wiley (Interscience), New York, 1976. Regulation of urea biosynthesis by the acetylglutamate-arginine system.

335. Teigland, M. and L. Klungsøyr. *Comp. Biochem. Physiol. B* 75B:703–705, 1983. Accumulation of α-ketoisocaproate from leucine in homogenates of tissue from rainbow trout (*Salmo gaidneri*) and rat. An improved method for determination of branched chain keto acids.

336. Temple, N. J., P. A. Martin, and M. J. Connock. *Comp. Biochem. Physiol. B* 64B:57–63, 1979. Intestinal peroxisomes of goldfish (*Carassius auratus*)—Examination for hydrolase, dehydrogenase and carnitine acetyltransferase activities.

337. Thorens, B. and P. Vassalli. *Nature (London)* 321:618–620, 1986. Chloroquine and ammonium chloride prevent terminal glycosylation of immunoglobulins in plasma cells without affecting secretion.

338. Tiemeyer, W., K. Hoferer, and D. Giesecke. *Comp. Biochem. Physiol. A* 85A:417–421, 1986. Uric acid uptake in erythorcytes of beagle and Dalmatian dogs.

339. Tinker, D. A., J. T. Brosan, and G. R. Hertzberg. *Biochem. J.* 240:829–836, 1986. Interorgan metabolism of amino acids, glucose, lactate, glycerol and uric acid in the domestic fowl (*Gallus domesticus*).

340. Tinker, D. A., M. Kung, J. B. Brosan, and G. R. Gerzberg. *Int. J. Biochem.* 15:1225–1230, 1983. Avian phosphoenolpyruvate carboxykinase: Effect of age, starvation and photoperiod.

341. Tischler, M. E. and A. L. Goldberg. *Am. J. Physiol.* 238:E487-E493, 1980. Production of alanine and glutamine by atrial muscle from fed and fasted rats.

342. Tischler, M. E. and A. L. Goldberg. *J. Biol. Chem.* 255:8074–8081, 1980. Leucine degradation and release of glutamine and alanine by adipose tissue.

343. Torchinsky, Y. M. *Trends Biochem. Sci.*

12:115–117, 1987. Transamination: Its discovery, biological and chemical aspects (1937–1987).

344. Tornheim, K., H. Pang, and C. E. Costello. *J. Biol. Chem.* 261:10157–10162, 1986. The purine nucleotide cycle and ammoniagenesis in rat kidney tubules.

345. Tramell, P. R. and J. W. Campbell. *J. Biol. Chem.* 245:6634–6641, 1970. Carbamyl phosphate synthesis in a land snail, *Strophocheilus oblongus*.

346. Trammell, P. R. and J. W. Campbell. *Comp. Biochem. Physiol. B* 40B:395–406, 1971. Carbamyl phosphate synthesis in invertebrates.

347. Tsuji, S. *J. Biochem. (Tokyo)* 94:1307–1315, 1983. Chicken ornithine transcarbamylase: Purification and some properties.

348. Valentine, H. A. *Br. Poult. Sci.* 5:149–159, 1964. A study of the effect of different ventilation rates on the ammonia concentrations in the atmosphere of broiler houses.

349. Van den Berghe, G., C. Van Pottlesberghe, and H.-G. Hers. *Biochem. J.* 162:611–616, 1977. A kinetic study of the soluble 5′-nucleotidase of rat liver.

350. Van den Thillart, G. *J. Comp. Physiol. B* 156B:511–520, 1986. Energy metabolism of swimming trout (*Salmo gairdneri*). Oxidation rates of palmitate, glucose, alanine, leucine and glutamate.

350a. Van Waarde, A. *Biol. Rev.* 63, 259–298, 1988. Operation of the purine nucleotide cycle in animal tissues.

351. Vincent, M.-F., G. Van den Berghe, and H.-G. Hers. *Biochem. J.* 222:145–155, 1984 Metabolism of hypoxanthine in isolated rat hepatocytes.

352. Vorhaben, J. E. and J. W. Campbell. *J. Biol. Chem.* 247:2763–2767, 1972. Glutamine synthetase: A mitochondrial enzyme in uricotelic species.

353. Vorhaben, J. E. and J. W. Campbell. *J. Cell Biol.* 73:300–310, 1977. Submitochondrial localization and function of enzymes of glutamine metabolism in avian liver.

354. Vorhaben, J. E., D. D. Smith, and J. W. Campbell. *Int. J. Biochem.* 14:747–756, 1982. Characterization of glutamine synthetase from avian liver mitochondria.

355. Walton, M. J. and C. B. Cowey. *Comp. Biochem. Physiol. B* 57B:143–149, 1977. Aspects of ammoniagenesis in rainbow trout, *Salmo gairdneri*.

356. Walton, M J. and C. B. Cowey. *Comp. Biochem. Physiol. B* 73B:59–79, 1982. Aspects of intermediary metabolism in salmonid fish.

357. Wasserman, G. F., W. T. Mueller, S. J. Benkovic, W. S. L. Liao, and J. Taylor. *Biochemistry* 23:6704–6710, 1984. Evidence that the folate-requiring enzymes of *de novo* purine biosynthesis are encoded by individual mRNAs.

358. Watford, M., Y. Hod, Y.-B. Chiao, M. F. Utter, and R. W. Hanson. *J. Biol. Chem.* 256:10023–10027, 1981. The unique role of the kidney in gluconeogenesis in the chicken. The significance of a cytosolic form of phosphoenolpyruvate carboxykinase.

359. Welbourne, T. C. *Biol. Chem. Hoppe-Seyler* 367:301–305, 1986. Hepatic glutaminase flux regulation of glutamine homeostasis: Studies *in vivo*.

360. Welbourne, T. C. and V. Phromphetcharat. In *Glutamine Metabolism in Mammalian Tissues* (D. Häussinger and H. Sies, eds.), pp. 161–177. Springer-Verlag, Berlin, 1984. Renal glutamine metabolism and hydrogen ion homeostasis.

361. Wiggins, D., P. Lund, and H. A. Krebs. *Comp. Biochem. Physiol. B* 72B:565–568, 1982. Adaptation of urate synthesis in chicken liver.

362. Wilson, R. P., R. O. Anderson, and R. A. Bloomfield. *Comp. Biochem. Physiol.* 28:107–118, 1969. Ammonia toxicity in selected fish.

363. Wilson, R. P., M. E. Muhrer, and R. A. Bloomfield. *Comp. Biochem. Physiol.* 25:295–301, 1968. Comparative ammonia toxicity.

364. Wilson, R. P., A. A. Letter, L. E. Davis, R. A. Bloomfield, and M. E. Mucher. *J. Anim. Sci.* 25:1274–1275, 1966. Comparative ammonia metabolism utilizing ^{15}N-ammonia.

365. Windmueller, H. G. *Adv. Enzymol.* 53:201–237, 1982. Glutamine utilization by the small intestine.

366. Windmueller, H. G. In *Glutamine Metabolism in Mammalian Tissues* (D. Häussinger and H. Sies, eds.), pp. 61–77. Springer-Verlag, Berlin, 1984. Metabolism of vascular and luminal glutamine by intestinal mucosa *in vivo*.

367. Wittman, J., A. Mengi, and M. Goldberg. *Enzyme* 35:70–76, 1986. Activation of hepatic and inactivation of renal xanthine dehydrogenase activity by dexamethasone during the prenatal period in chick eggs.

368. Wood, A. W. and J. E. Seegmiller. *J. Biol. Chem.* 248:138–143, 1973. Properties of 5-phosphoribosyl-1-pyrophosphate amidotransferase from human lymphoblasts.

369. Wurzinger, K.-H. and R. Hartenstein. *Comp. Biochem. Physiol. B* 49B:171–185, 1974. Phylogeny and correlations of aldehyde oxidase, xanthine oxidase, xanthine dehydrogenase and peroxidase in animal tissues.

370. Wynshaw-Boris, A., J. M. Short, D. S. Loose, and R. W. Hanson. *J. Biol. Chem.* 261:9714–9720, 1986. Characterization of the phosphoenolpyruvate carboxykinase (GTP) promoter-regulatory region. I. Multiple hormone regulatory elements and effects of enhancers.

371. Yip, M. C. M. and W. E. Knox. *Biochem. J.* 127:893–899, 1972. Function of arginase in lactating mammary gland.

372. Yoshida, T. and G. Kikuchi. *Arch. Biochem. Biophys.* 145:658–668, 1971. Significance of glycine cleavage system in glycine and serine catabolism in avian liver.

373. Youngson, A., C. B. Cowey, and M. J. Walton. *Comp. Biochem. Physiol. B* 73B:393–398, 1982. Some properties of serine pyruvate aminotransferase purified from liver of rainbow trout *Salmo gairdneri*.

374. Zwaan, A. de In *The Mollusca* (P. W. Hochachka, ed.), Vol. 1, pp. 137–242. Academic Press, New York, 1983. Carbohydrate catabolism in bivalves.

375. Zwaan, A. de and W. J. A. van Marrewijk. *Comp. Biochem. Physio. B* 44B:1057–1066, 1973. Intracellular localization of pyruvate carboxylase, phosphoenolpyruvate carboxykinase and "malic enzyme" and the absence of glyoxylate cycle enzymes in the sea mussel (*Mytilus edulis* L.).

Chapter 8 | Design of Energy Metabolism

Peter W. Hochachka

When most students of comparative physiology think about energy metabolism, two processes come to mind: energy production and energy utilization, which are usually assumed to be in balance. Although it is widely appreciated that electrochemical potential energy is interconvertible with the cell's energy currency [adenosine triphosphate (ATP); see abbreviations list at the end of this chapter] for practical purposes the coupling between energy production and energy utilization in most cells means a coupling between metabolic pathways producing ATP and metabolic processes [often adenosine triphosphatases (ATPases)] utilizing it. The pathways producing ATP as a utilizable energy source are of two types, anaerobic (thus O_2 independent) and aerobic (thus O_2 dependent). These are described in great detail in standard biochemical textbooks (e.g., 52). Here it is sufficient to briefly overview the main pathways utilized by animal tissues.

The simplest mechanism for generating ATP is phosphagen mobilization. In vertebrate tissues containing creatine phosphate (PCr) this mobilization is catalyzed by CPK, a process obviously independent of O_2 and which can be written as follows:

$$PCr + ADP + H^+ \rightleftharpoons ATP + creatine$$

Fermentation, or the partial (O_2 independent) catabolism of substrates to anaerobic end-products, is another means of forming ATP. In animals, the commonest fermentative pathway is that of anaerobic glycolysis. At high pH (>8.0), the summed reaction can be written as follows:

$$glucose + 2ADP^{3-} + 2HPO_4^{2-} \longrightarrow$$
$$2\ lactate^- + HATP^{4-}$$

At low pH (<6.0), the adenylates are more protonated and the reaction can be written

$$glucose + 2HADP^{2-} + 2H_2PO_4^- \longrightarrow$$
$$2\ lactate^- + 2H^+ + 2HATP^{3-}$$

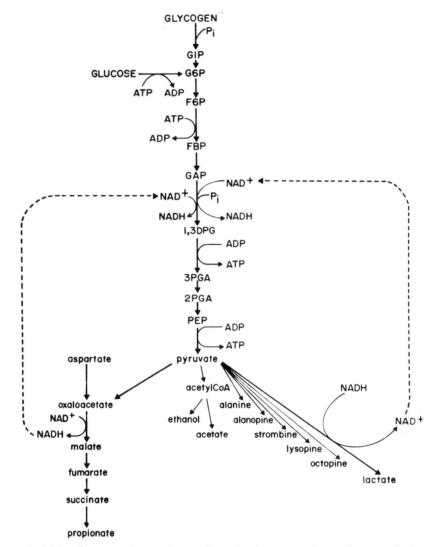

Figure 1. Major fermentation pathways in animal tissues. Anaerobic metabolism currently is viewed as a series of linear, and loosely linked, pathways. The most important of these, aside from classical glucose → lactate fermentation (yielding 2 mol of ATP/mol of glucose) are

1. Glucose → octopine, lysopine, alanopine, or strombine, energy yield 2 mol of ATP/mol of glucose.
2. Glucose → succinate, energy yield 4 mol of ATP/mol of glucose.
3. Glucose → propionate, energy yield 6 mol of ATP/mol of glucose.
4. Aspartate → succinate, energy yield 1 mol of ATP/mol of aspartate.
5. Aspartate → propionate, energy yield 2 mol of ATP/mol of aspartate.
6. Glutamate → succinate, energy yield 1 mol of ATP/mol of glutamate.
7. Glutamate → propionate, energy yield 2 mol of ATP/mol of glutamate.
8. Branched-chain amino acids → volatile fatty acids, energy yield 1 mol of ATP/mol of substrate.
9. Glucose → acetate, energy yield 4 mol of ATP/mol of glucose.

The pathway of glycolysis is phylogenetically ancient and many of its features are strongly conserved (6). However, as illustrated in Figures 1 and 2, the terminal dehydrogenases are subject to adaptive change; in many invertebrate organisms, lactate dehydrogenase is replaced by functionally analogous imino acid dehydrogenases. In such cases, an imino acid (octopine, alanopine, strombine, or tauropine) replaces lactate as the anaerobic glycolytic end-product. However, neither the proton stoichiometries nor the ATP yields are changed (36). If glycogen is the substrate fermented instead of glucose, the yield of ATP is 3 mol of ATP formed/mol of glucosyl unit. In addition to carbohydrates, some amino acids can also be fermented. Aspartate, for example, can be fermented to succinate or propionate, a process that in molluscs is stoichiometrically coupled to glucose (or glycogen) fermentation (Fig. 2). A third major category of potential fuel, the fats, are so reduced that they are not fermentable by animal cells.

In all fermentations in animals, an organic molecule (e.g., pyruvate) serves as a terminal proton and electron acceptor, forming an organic end-product (e.g., lactate). In contrast, O_2 is required as a terminal acceptor for the complete oxidation of substrates such as glucose, glycogen, fatty acids, or amino acids. The pathways by which such complete oxidations are achieved are much more complex than most fermentation pathways. In the case of glucose, the first phase of complete oxidation (namely, the conversion of glucose to pyruvate) is the same as in gly-

colysis. However, instead of being reduced to lactate, pyruvate is converted to acetylCoA, which serves as the entry substrate into the Krebs cycle (Fig. 3), a process occurring in the mitochondrial matrix. The two carbon atoms of acetylCoA appear as CO_2 with the simultaneous formation of reducing equivalents in the form of NADH. Under anaerobic conditions, NADH is reoxidized to NAD^+ by an organic substrate that is reduced in the process (pyruvate and fumarate, e.g., being reduced to lactate and succinate, respectively). Under aerobic conditions, NADH is reoxidized to NAD^+ via the electron-transport system (ETS) located in the mitochondrial inner membrane or cristae (Fig. 4). During the electron-transfer process, H^+ ions are believed to be pumped out of the mitochondria across the inner mitochondrial membrane.

The resulting H^+ ion concentration gradient plus an electric potential across the membrane are thought to supply the driving force for ATP synthesis from ADP and P_i (inorganic orthophosphate), a reaction catalyzed by ATP synthase. The latter is a mitochondrial enzyme located on, and spanning, the inner mitochondrial membrane (Fig. 4). The net reaction for glucose oxidation (via glycolysis, the Krebs cycle, the ETS, and ATP synthase) can be written as follows:

$$glucose + 36(ADP + P_i) + 6O_2 \longrightarrow$$
$$36ATP + 6CO_2 + 6H_2O$$

When fatty acids are the fuel being combusted, the pathway of oxidation is

---

10. $CH_2O + SO_4^{2-} + H^+ \rightarrow H_2S + HS^- + H_2O + CO_2$, energy yield 6 mol of ATP/mol of glucose.

Pathways 6, 7, 8, and 10 are not shown in the diagram. Pathways 1–4 are known in various bivalve molluscs; 2, 3, 8, and 9 are often utilized by helminths; 5, 6, and 7, while theoretically possible in bivalve molluscs, do not appear to be utilized to any significant extent (19). [From (40)].

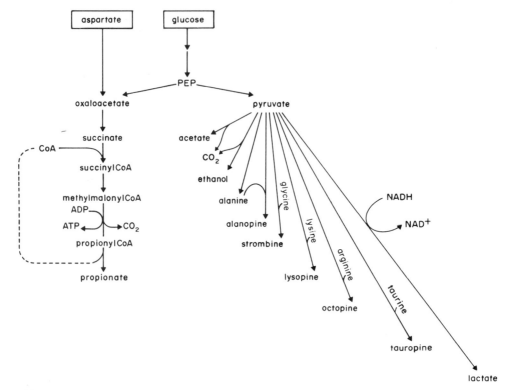

Figure 2. The PEP branchpoint in bivalve molluscs and other hypoxia tolerant vertebrates. The ratio of PEP carboxykinase (PEPCK)/pyruvate kinase (PK) varies in different species and tissues. It is usually high in tissues most tolerant to O_2 lack. In tissues or species with low PEPCK activities, the main precursor of succinate or propionate is aspartate; in some tissues, such as oyster heart, the fermentation of aspartate and glucose (or glycogen) are redox coupled (NAD^+ required for glycolysis is formed primarily by malate dehydrogenase catalyzed NADH oxidation, a step in aspartate fermentation). [From (40).]

termed the β-oxidation spiral which, following fatty acylCoA formation, involves four basic steps:

1. acylCoA + FAD^+ → α,β-unsaturated acylCoA + $FADH_2$
2. α,β-unsaturated acylCoA + H_2O → β-hydroxyacylCoA
3. β-hydroxyacylCoA + NAD^+ → β-ketoacylCoA + NADH + H^+
4. β-ketoacylCoA + CoASH → acylCoA (−2 C) + acetylCoA

This pathway, also located in the mitochondria, generates acetylCoA, which is

then completely metabolized by the Krebs cycle, the ETS, and ATP synthase. The overall equation for fatty acid oxidation, using palmitate as an example, can be written as

$$palmitate + 23O_2 + 129(ADP + P_i) \longrightarrow$$
$$129ATP + 16CO_2 + 145H_2O$$

Finally, if amino acids are the fuels being combusted, they are metabolized by pathways that all ultimately feed into the Krebs cycle (Fig. 5), where the intermediates can be fully metabolized. Being at about the same oxidation state as car-

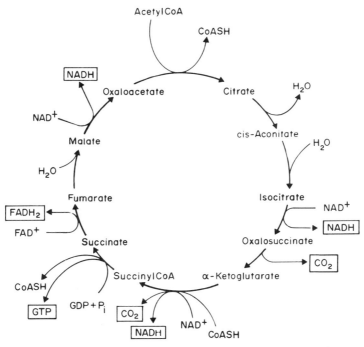

Figure 3. The Krebs cycle, initiated by oxaloacetate condensation with acetylCoA and ending with oxaloacetate formation from malate, is the hub of cellular energy metabolism.

bohydrates, the ATP yields of amino acids during oxidation are also similar. For example, alanine oxidation yields 15 ATPs per alanine, exactly the same value as for pyruvate. Oxidation of glutamate yields 27 ATPs while the oxidation of proline could in theory be coupled to the synthesis of 30 ATPs (40).

For practical purposes, then, the above relatively small number of metabolic pathways are the main means by which cells can balance energy production with energy utilization rates. That we are able today to summarize so complex a set of processes so easily and so simply is a measure of the great achievements of researchers in this area over the last half-century. While a great deal of detail remains to be filled in, for students of physiology it is the properties of these metabolic processes that are of greater inherent interest. Like all biological systems, these pathways are the outcome of

cycles of mutation and selection, and they are thus in a proper sense "designed" systems, analogous to products of engineering design (2). Although organisms and metabolic pathways are designed by natural selection while machines are designed by engineers, both design systems share the fundamental attributes of function and purpose; both also involve trial of variants with the selection of those that work best. Our goal in this chapter is to consider the properties of fuels and pathways that are under such selection. What properties are advantageous and why? What properties set limits and constraints and why? To set the stage, let us consider ATP itself.

ATP as Utilizable Energy

In most tissues, the rates of ATP turnover are low because most metabolic pathways operate rather sluggishly. Muscle tissues,

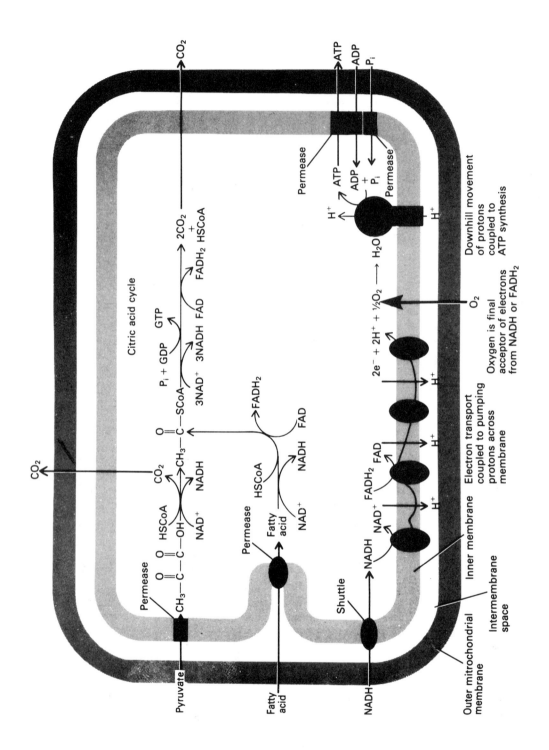

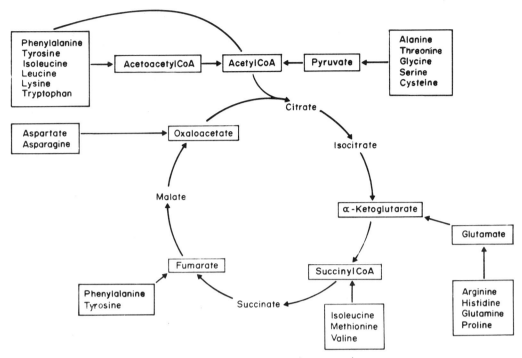

Figure 5. Sites of entry of free amino acids into the Krebs cycle are requisite steps for the complete catabolism of these potential carbon and energy sources. [From (40).]

however, are exceptional: During transition from rest to work, ATP cycling or turnover (in flux units defined as μmol of ATP/g · min) can increase by orders of magnitude, rather than by small percentage changes. Highest rates of ATP turnover for a range of animals are given in Tables 1 and 2 and vary systematically between large and small animals. Insect flight muscles probably have the highest ATP turnover rates, >5000 μmol of ATP/ g · min (Table 2). Interestingly, despite

such large differences in ATP fluxes, ATP concentrations do not vary much. In fact, fast-twitch or white-type muscle (WM) usually contains only ~5–8 μmol ATP/g while slow-twitch oxidative or red muscle (RM) and insect flight muscles usually contain even less (27, 43). Adenosine triphosphate contents in brain and liver are also less than in skeletal muscle. If only these endogenous ATP supplies were available, they would be quickly exhausted during high work rates. Instead,

Figure 4. An abbreviated metabolic map of the major metabolic reactions in mitochondria. The substrates of oxidative phosphorylation (pyruvate, fatty acids, ADP, and P_i) are transported into the matrix from the cytosol by transporters or exchangers (termed permeases in the diagram); O_2 simply diffuses into the matrix. NADH, generated in the cytoplasm during glycolysis, is not transferred directly into the matrix, because the inner membrane is impermeable to NAD^+ or NADH. Rather, a "shuttle" system is used to transport the electrons on cytosolic NADH into the electron-transport chain. The products of metabolism, such as ATP, are transported into the cytoplasm, and CO_2 diffuses into the cytoplasm across the mitochondrial membranes. [After (16).]

TABLE 1. Estimated Maximum Power Output for Skeletal Muscle (of Humans) Utilizing Different Substrates and Metabolic Pathways

Process	Metabolic Power Output (μmol of ATP/g of wet weight · min)
Aerobic Metabolism	
Fatty acid oxidation	20.4
Glycogen oxidation	30.0
Anaerobic Metabolism	
Glycogen fermentation	60.0
PCr and ATP hydrolysis	96–360

ATP levels are refractory to increased flux in a variety of tissues and species (21, 25, 59, 67). The only way high ATPase rates can be sustained with minimal change in ATP level is through balanced ATP replenishment. The simplest metabolic pathway for achieving this balance involves phosphagen mobilization. This is of major importance in some tissues (skeletal, heart muscles and smooth muscles; nervous tissues) but only of minimal significance, in others (such as liver, gills, lungs, and mantle). In the former tissues, a minimum set of criteria must be met for a phosphagen to be a useful endogenous fuel and thus to have been selected for this function.

Criteria for "Good" Phosphagen-Type Fuels

Phosphagens Should Be Storable at High Concentrations

All currently known phosphagens are substituted phosphoguanidinum compounds (Fig. 6). The two best studied systems are PCr and PArg (formed from Cr and Arg by CPK and APK, respectively). Creatine itself is formed from methionine, glycine, and arginine; in humans, this occurs in liver and pancreas, while in some species the first step occurs in the kidney and the second in the liver (72). Creatine is accumulated in skeletal muscles, myocardium, and brain, but apparently mechanisms controlling storage amounts are unknown, although feedback repression of synthesis in the liver may limit total availability (72). Nevertheless, by usual standards, the total pool sizes of PCr + Cr are high (usually ~30 μmol/g in WM (white muscle) and somewhat lower in RM (red muscle), heart, and brain). On a short- and long-term basis, the amount of PCr available for high power output (burst work) seems to be controlled by hypertrophy, not by concentration adjustments (41). This is not the case for PArg. It too can be stored at high con-

TABLE 2. Calculated Rate of ATP Cycling at Maximum Sustainable Exercise in Various Animals and Humans

Species	Weight (kg)	Estimated Muscle Metabolic Rate[a] (μmol of ATP/g · min)
Locust[b]	0.0010	5400
Hummingbird[c]	0.0025	600
Pygmy mouse	0.0072	227
Chipmunk	0.09	214
Ferret	0.5	82
Suni	3.5	86
Tammar wallaby	4.8	57
Dik–dik	4.5	49
African goat	20	48
Man	70	30
Eland	213	32
Zebu cattle	254	27

[a]From (51) with modification.
[b]From (1).
[c]From (66a). All other data from (39).

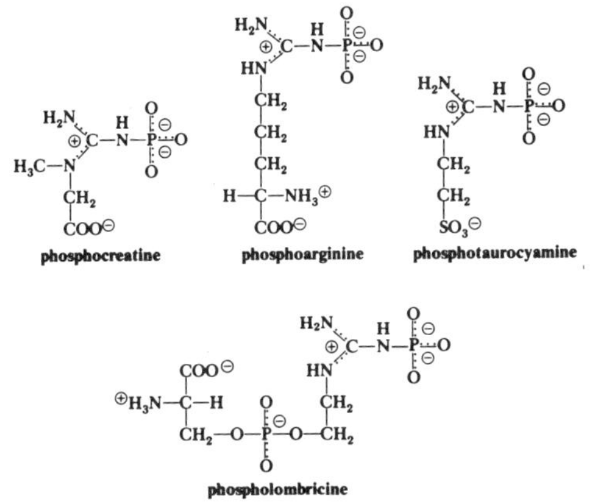

Figure 6. Phosphagens are phosphorylated derivatives of guanidinio compounds used to store high-energy phosphate, especially in muscles. The commonest phosphagens are phosphocreatine (PCr) in vertebrates and phosphoarginine (PArg) in most invertebrates.

centrations (up to 50 μmol/g in some species) but in this case there seems to be a reasonable correlation between burst work capacities and the amount of PArg stored (19).

Phosphagens Should Be Rapidly Mobilizable

For a phosphagen to be a useful ATP source, its ~P group must be transferable to ADP at high rates and at correct times. For PCr and PArg, this capacity is based upon (a) large amounts of the cytosolic isozyme of CPK or APK, (b) suitable kinetic properties (33), and (c) correct cellular localization (in the case of muscle, bound to myosin ATPase). Within any vertebrate species, CPK activities tend to be highest in WM, lower in RM, myocardium, and brain; between-species comparisons indicate that relative to other ATP-yielding enzymes, highest CPK activities occur in animals capable of high burst speeds; sluggish species have lower activities. A similar relationship appears to hold for PArg kinases (19). The process

of ATP formation from phosphagen is further facilitated by the kinetic properties of these enzymes.

During high rates of ATP turnover, the direction of net flux is

$$PCr + ADP \longrightarrow ATP + Cr$$

$$PArg + ADP \longrightarrow ATP + Arg$$

Even if CPK keeps this reaction close to equilibrium at all times, it is instructive that for the human muscle cytosolic isozymes of CPK, the K_d values for PCr are 72 and 32 mmol/l for the binary and ternary complexes, respectively, while for ADP they are 0.2 and 0.06 mmol/l (45). This means that under most *in vivo* concentrations PCr is not saturating and the enzyme responds sensitively to changes in [PCr]. On the other hand, CPK affinity for ADP is relatively high, making it so competitive for ADP that at least in the early phases of work, rising [ADP] drives the CPK reaction to the right (25). It is not known with certainty for later stages of work whether or not CPK functions under ADP saturation. However, in invertebrate muscles there is no doubt that during hard work ADP levels gradually rise high enough to saturate APK fully (4). From this point on its catalytic activity must be mainly determined by PArg supplies or by metabolic modulators including reaction products (23). Thus both enzyme content and enzyme kinetic adaptations of CPK and APK favor ~P transfer to ADP. However, because phosphagen supplies are nonsaturating *and* diminish during burst of work, these rates must decline rapidly with time.

Phosphagen Mobilization Should Minimally Perturb ATP Pools

Another requirement of a useful phosphagen is that its utilization should proceed with minimal effect on ATP levels.

In this regard, compounds such as PCr can be viewed as being fitted for the job because the equilibrium constant (K_{eq}) for ATP formation by the CPK reaction is large: At pH 7.0, the K_{eq} for the reaction is ~2×10^9 (18, 25). As a result PCr can be almost completely used up with minimal change in ATP concentrations. It would not do to use another compound (such as a nucleotide triphosphate, NTP), with a K_{eq} for hydrolysis similar to that of ATP because the concentration ratio of ATP/ADP would change along with that of NTP/NDP. With this kind of phosphagen, muscle metabolic transitions would sustain much larger fluctuations in ATP levels; using up 90% of such a hypothetical ~P donor, for example, would cause a 90% decline in ATP concentration. Biologists think that is why compounds such as PCr and PArg, rather than NTPs, have evolved as muscle phosphagens and why the seemingly extreme equilibria of their ~P group transfer reactions are, in this role, advantageous.

Phosphagens Should Not Generate Metabolically Deleterious End-Products

None of the end-products of phosphagen mobilization appear to be deleterious, with one exception: Arginine (formed from PArg), which may have relatively nonspecific but undesirable effects on enzymes. This led Somero and his students (7) to suggest that an important function of the octopine dehydrogenase reaction, an LDH analog found in many invertebrates, is to serve as a sink for arginine (Fig. 2). Thus, even here, the end-product, directly phosphagen derived, does not seem to be a problem. However, CPK or APK *in vivo* are coupled to cell ATPases:

$$PCr + ADP \longrightarrow ATP + Cr$$

$$\text{ATPase} \quad H_2O$$
$$P_i$$

In a closed system, such as muscle at high

intensity work (56), the end-products that accumulate are creatine plus P_i. One reason why the former is not deleterious is because it is involved in no other important metabolic sequences; its main metabolic fate is reconversion to the phosphagen form. Similar arguments can be made for arginine, although it can also be metabolized by other pathways (7). However, P_i is another matter. It is a highly reactive metabolite involved in numerous enzyme reactions in intermediary metabolism. Just as in the case of ATP, where unduly high concentrations must be avoided because of generalized perturbing effects, the need to control the accumulation of P_i may place limits upon how much phosphagen can be stored as an ATP buffer.

Phosphagen Mobilization Should Not Cause Osmotic, Ionic, or Charge Disturbances

When fully depleting the phosphagen supply, the CPK or APK reaction might be expected to lead to ionic and charge perturbations. For example, in the resting state, 30 μmol PCr^{2-}/g would require similar amounts of a divalent cation. Creatine is of course uncharged and its accumulation would appear to leave the cell with an anion gap. This problem is alleviated by CPK coupling to ATPase, since in terms of charge, P_i^{2-} is equivalent to PCr^{2-}; moreover, the former is a better buffer at near-neutral pH values than all known phosphagens. So the stoichiometry not only avoids serious osmotic or charge imbalances, but simultaneously minimizes perturbing effects of any $[H^+]$ changes.

A Good Phosphagen Can Take on Ancillary Roles During Aerobic Metabolism

The value of phosphagens as fuels would be greatly amplified if they were also usable in aerobic metabolism. For PCr, at least three such aerobic functions have been suggested; namely, in the shuttling of P_i between sites of ADP formation and utilization (5), in the preferential channelling of ATP to SR-bound ATPase (49), and in the facilitated diffusion of ATP (53). One or all of these roles (or a combination of them) presumably explains why hummingbird muscle, which is overloaded with mitochondria and displays a minimal anaerobic metabolism, nevertheless contains ~2000 *units of CPK* (66a).

Whereas similar aerobic roles may be played by PArg in molluscan muscles (65), they seem to be minimized in insects. Insect flight muscle contains unexpectedly low levels of PArg and of the enzyme mobilizing it. A tracheal-based O_2 delivery system, which can achieve a rapid and efficient rest → work transition is probably the reason insect flight muscle does not require significant ATP buffering by PArg. Mammalian and avian O_2 delivery, on the other hand, is based on pumps and pipes (hearts and vessels), which is a comparatively sluggish system, requiring seconds or even minutes to fully match the maximum aerobic energy demands of the organism (14, 32). A phosphagen system for buffering ATP concentrations would seem to be particularly advantageous during these early stages of activated ATPase function when demands for ATP momentarily outstrip the aerobic ATP synthesis capacity.

Phosphagens: Advantages and Constraints

From this discussion it is evident that two potentially important advantages can arise from using phosphagens in support of cell work: *high power output and effective buffering of ATP levels*. We can assert with confidence that the first is more important than the second, because numerous invertebrates (23) and at least one teleost (20) are known to tolerate very large

drops in [ATP] during intense muscle work. Therefore, the *maintenance of [ATP] during very high ATP turnover rates is neither universal nor absolutely requisite*. On the other hand, in all currently known systems the power output obtainable by using phosphagen as fuel is higher than obtainable from any other pathway (Table 1). Perhaps this is why phosphagens and phosphagen kinases were selected in the first place and why these pathways of ATP generation are of limited utility in most nonmuscular tissues, where high power outputs are never required.

Even in this special context, however, phosphagens show some critical limitations, the most obvious being the 30–50 µmol/g of maximum storage level (placing a modest ceiling on the total amount of work supportable by this fuel). Again, we can make this assertion with confidence, since some muscles may store *less*, but none are known that store more, than this amount. While this analysis cannot specify the exact basis for limits to phosphagen storage, the most likely reasons seem to be either: (a) that phosphagens are strongly anionic (and if levels were too high they would begin to create uncontrollable side reactions), or (b) that the amount of phosphagen-derived P_i could lead to significant metabolic perturbations. Whatever the underlying reasons, the modest limit put on storage amount means that phosphagen-based work is only possible for short times (e.g., 5–10 s in mammals); any further work therefore requires the more universal back-up pathways for ATP replenishment that are also used by cells containing little or no phosphagen. These fall into two categories, anaerobic pathways, which form ATP in the absence of O_2, and aerobic ones, which require O_2.

Fermentative Metabolism

Under O_2-limiting conditions, anaerobic glycogenolysis or glycolysis appears to be the main back-up mechanism for ATP replenishment after phosphagen supplies are depleted. The pathway is phylogenetically very old and its component enzymes are considered to be highly conservative. Their relatively constant properties are thought to be maintained by rigorous natural selection (6). The only sites where major adaptational changes are known are at the terminal step, normally catalyzed by lactate dehydrogenase (LDH), and at the PEP branchpoint. Thus many invertebrate groups have evolved different terminal dehydrogenases (Fig. 1); the energy yields of these modified glycolytic pathways are, however, unchanged from the classical process of fermentation to lactate. At the level of PEP, some animals possess high PEP carboxykinase activities, which allow a large flow of carbon away from mainstream glycolysis and towards succinate or propionate, as anaerobic end-products (Fig. 2). These branching pathways may or may not be coupled with simultaneous aspartate fermentation, but in all cases they are energetically more efficient than classical glycolysis (29). As with phosphagens, minimal conditions must be met in order for a compound to be a useful fermentable fuel for muscle work.

Fermentable Fuels Should Be Storable at High and Adjustable Concentrations

This storage criterion is well met by tissue glycogen. In vertebrate WM, glycogen can be stored at ~100 μmol of glycosyl unit/g, a level that may be modified according to need. In the liver of most vertebrates, concentrations may be two or three times higher. In some invertebrates with greater relative dependence on this pathway, as in bivalve adductor muscle, glycogen content is so high it constitutes a significant fraction of tissue weight (19). To increase efficiency of storage, some muscles rely upon glycogen–membrane

associations or upon large intracellular granules (29). Some gastropods store galactogens (galactose-based polymers) along with glycogen but the former do not appear to be used for muscle work (19). It is unclear why this should be so. In many marine invertebrates, the free amino acid pool is expanded and several of these (aspartate and the branched-chain amino acids in particular) can be fermented at low rates to supplement glycolysis (13).

Fermentable Fuels Should Amplify the Molar Yield of ATP

The ATP amplification of glycogen → lactate conversion is three; that is, 3 mol of ATP/mol of glycogen-derived glucosyl units (Fig. 1). Thus, the complete fermentation of 100 μmol/g generates at least 300 μmol of ATP, and the cell work possible on this fuel is ~10 times that supportable by PCr (because the latter is stored at only 30 μmol/g and only generates maximally equimolar amounts of ATP). The amplification in the aspartate → propionate path is only twofold, as it is in the glucose → lactate path; that is one reason why glycogen is a better fuel for anaerobic muscle work than either aspartate or glucose. On the other hand, ATP molar amplification in glycogen → propionate fermentation can be as high as 7; that may be why this kind of metabolism is favored in metabolically arrested, O_2-limited states in invertebrates (35).

Fermentable Fuels Should Be Rapidly Mobilizable at Appropriate Times and Rates

As far as we know to this time, the main reason why sequences such as aspartate → propionate or branched-chain amino acids → volatile fatty acids are not

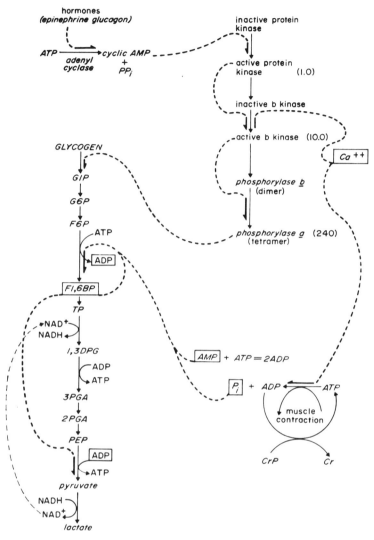

Figure 7. Regulatory properties of glycogenolysis in vertebrate muscles. Heavy arrows indicate activation of a regulated enzyme by a signalling metabolite. Molar concentration ratios of glycogen phosphorylase cascade are given in parentheses. [From (40).]

favored for anaerobic muscle work is because they cannot supply sufficient power (19). In contrast, the high power output of the glycogen → lactate pathway is probably the chief reason why it is almost universally used in support of extended anaerobic work (Table 1). The "flare up" characteristics of this pathway (Fig. 7) are based upon the regulatory properties of glycolytic enzymes and upon the occur-

rence, at virtually every step in the glycogen → lactate conversion, of a tissue-specific isozyme form or forms. Most of the control of flux through glycolysis is determined by the catalytic and regulatory properties of isozymes of phosphorylase, PFK and PK, although other isozymes may also play a role (33).

In addition to the kind of isozymes present, the properties of the glycolytic

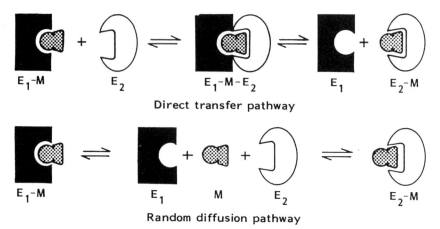

Figure 8. Schematic representation of the mode of transfers of metabolite (M) between two sequential enzyme sites (E_1 and E_2). In direct transfer mechanism, metabolite (M) does not dissociate from E_1 to reach at the E_2 site. This is in contrast to the random diffusion mechanism in which metabolite (M) dissociates from E_1 into aqueous solvent before forming a complex with E_2. [Modified from (63).]

pathway are strongly determined by the amount or concentration of component enzymes. In fact, the concentrations of enzyme sites in cytosol fluid are surprisingly high and may well exceed the concentrations of many intermediates in the pathway. Close examination indicates that glycolytic intermediates fall into two categories: (a) precursors and end-products, whose concentrations exceed those of enzymes acting upon them, and (b) intermediary metabolites, whose concentrations are substantially less than the enzymes acting upon them. Under these conditions, strong selective advantage would arise from adapting the pathway for direct enzyme-to-enzyme metabolite transfer of intermediates [i.e., for channelling, rather than for diffusion from enzyme to enzyme in aqueous solution (Fig. 8)]. Experimental evidence for this kind of behavior is available for GAPDH, PGK, LDH, and other dehydrogenases (62, 63). In the case of the above three enzyme series, the two target enzymes for GAPDH function are PGK and LDH: 1,3-DPG is directly transferred to PGK while NADH is handed off directly to LDH. The

LDH function in turn transfers NAD^+ back to GAPDH, so that the catalytic cycle can continue (Fig. 9).

An important feature of this system is that the distribution of reactive substrate and product complexes at an individual enzyme site (K_{eq}^{int}) is near unity; this allows transfer in either direction and in effect smooths out free energy profiles for the overall metabolic pathway. With equal energy partitioning among enzyme-bound intermediates, the unidirectional drive or flux of glycolysis arises because of *the removal of final end-products (ATP and lactate) by other enzyme pathways (ATPases in the case of ATP) or, in some cases, by export to a segregated location (efflux out of the cell in the case of lactate or other anaerobic end-products)*. Srivastava and Bernhard (62, 63) argue that selective forces are acting less upon each enzyme component (to increase its k_{cat} or turnover number, so as to maximize flux through the reaction as assayed in aqueous solution) than upon the direct transfer capacities of pathway segments or of the pathway as a whole. Evolutionary drive is towards optimization of the facility for metabolite transfer,

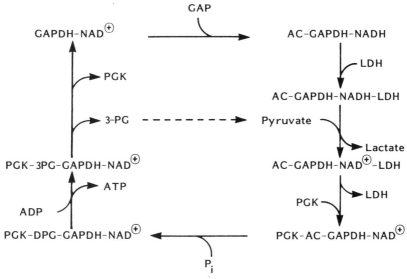

Figure 9. Schematic model for GAPDH function when acylated (acyl-GAPDH) by LDH and by PGK during forward glycolytic flux. Note the specific interaction of acyl-GAPDH-NADH and LDH and the specific interaction of acyl-GAPDH-NAD⁺ and PGK. [Modified from (62).]

for enzyme–enzyme interactions favorable for direct transfer of intermediates. In this view, pathway segments such as pyruvate dehydrogenase (PDH), in which the three enzyme components are covalently linked and intermediates remain bound throughout the catalytic transformation of substrates to products, represent a kind of evolutionary optimization of direct metabolite transfer. Srivastava and Bernhard (62, 63) in fact argue that this is an extremely widespread feature of metabolism, not only a characteristic of the glycolytic pathway. In their view, diffusion necessarily plays a major role in metabolite transfer between functionally distinct compartments, but a more minor role within each such compartment.

While stimulation of muscle glycolysis is triggered by a variety of signals (hormones, ions, and metabolites), the mechanisms ensuring that the pathway is correctly phased in with other ATP-generating pathways is frequently underemphasized. Nonetheless, during an-

aerobic work it is widely accepted that phosphagen hydrolysis usually precedes full glycolytic activation; PCr utilization, for example, approaches completion before glycolysis is phased in. This is because, as an ATP synthesizing process, glycolysis is competing with CPK (APK in invertebrate systems) for limiting amounts of ADP. Three enzymes are involved in the competition (Fig. 7): CPK (or APK), PGK, and PK. These enzymes are present in the 10^{-4} M range (48, 51). During early phases of work, both the high activity of CPK and its affinity for ADP ensure that this pathway dominates (see above). On the other hand, the kinetic properties of PGK and PK are well designed for back-up function. In the case of PK, for example, the K_d for PEP is lowered by an order of magnitude from a nonphysiological range to values between 0.01 and 1 mM on addition of ADP. Phosphoenolpyruvate has the same effect on binding of ADP; that is, the binding of one substrate leads to about an order-of-magnitude increase in enzyme affinity for the cosubstrate (15). This means that as anaerobic work continues and [ADP] rises, PK becomes ever more competitive and serves to complete the coupling of myosin ATPase and glycolysis (33).

Although the rate of work supportable by this system is related in some complex way to the catalytic potentials of the enzyme pathway *per se* (22), it is clear that the enzyme content of a pathway cannot be elevated without limit because of potentially debilitating effects of the end-products. It is the need to minimize such problems that defines another criterion of useful fermentable fuels.

Fermentable Fuels Should Not Generate Metabolically Deleterious End-Products

As in phosphagen hydrolysis, the end-products of concern are not strictly those formed by glycolysis *per se*, but from the glycolysis–myosin ATPase couple. In this system, the net reaction and the end-products are

$$\text{glycogen (glucosyl unit)} \longrightarrow 2 \text{ lactate}^-$$

with

$$2P_i + 2ADP \longrightarrow 2ATP + 2H_2O$$
$$2H^+$$

At steady state, two end-products accumulate: lactate anions (formed by glycolysis per se) and protons (formed mainly in ATP hydrolysis). Very similar proton stoichiometry prevails for other fermentation pathways in molluscs or other invertebrates (36), the main difference being that n (the number of moles of ATP per mole of glucosyl unit) varies from 2 (in glucose fermentation) to ~7 in the glycogen → propionate pathway:

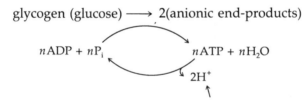

$$\text{glycogen (glucose)} \longrightarrow 2(\text{anionic end-products})$$
$$nADP + nP_i \longrightarrow nATP + nH_2O$$
$$2H^+$$

Thus, during anaerobic work two classes of end-products accumulate: organic anions (commonly lactate, octopine, alanopine, strombine and tauropine) (24), and protons. In principle, both could have deleterious effects. However, compounds such as lactate (which accumulate in mammalian muscle to 20–40 μmol/g) can rise to 200 μmol/g in animals such as the diving turtle. So apart from modest osmotic and cation gap problems, glycolytic end-products present few difficulties for metabolism *per se* (30). Protons, however, are another matter. Many workers consider H^+ to be a critical limiting factor in anaerobic systems, so it is useful to place this metabolic intermediate in context.

In the first place, just how much H^+ is cycled through normoxic metabolism is usually overlooked. A simple calculation can show that in a 70-kg human (with a resting metabolic rate of ~700 mmol of O_2/h) over 100 mol of protons are produced per day. The main sink for these H^+ ions is oxidative metabolism, and a close balance between rates of H^+ production and removal ensures stable pH. It is well known that the balance of this system breaks down during anaerobiosis, but it is usually also overlooked that *pathways such as glycogen → lactate represent net*

H⁺-consuming reactions at metabolically relevant pH values. At pH 7.4, for example, this pathway consumes 0.4 H^+/glycosyl unit, while the pathway glycogen → propionate at this pH consumes an order of magnitude more H^+/glucosyl unit (36); this means that both H^+ stoichiometry and ATP yield are pathway specific. In both cases, it is the coupling with ATPase that leads to net H^+ production. The metabolically relevant relationship therefore is *between H⁺ and ATP, not between H⁺ and glucosyl units*. For the (glucose → lactate) ATPase couple, 1 μmol of ATP is cycled/μmol of H^+ produced, while for the (glycogen → lactate) ATPase couple, 1.5 μmol of ATP are cycled through/μmol of H^+ produced. In succinate fermentation, 2 μmol of ATP are cycled/μmol of H^+, whereas in glycogen → propionate, up to 3.5 μmol of ATP are cycled/μmol of H^+ generated (30). The strategy of maximizing the amount of ATP turned over per mole of H^+ formed in muscle working anaerobically *at its limit means a balance between H⁺ formed and H⁺ consumed during ATP cycling, as in aerobic metabolism* (70).

Interestingly, such a balance could be achieved by alcohol fermentations, but such pathways typically do not occur in animal tissues for they generate their own set of problems (30). Thus when animal tissues are O_2 limited, they necessarily rely upon metabolic pathways that generate H^+ ions, a process that cannot proceed indefinitely. That may explain why *glycogen is stored in muscle far below possible levels*; in some tissues of some species (e.g., liver of turtles or mantle of bivalves) (29), glycogen may be stored at over 1 M levels (as glucosyl units). Here, of course, the main end-product of glycogen metabolism is glucose, not lactate and H^+, as in muscles. That is also an important reason why a number of properties—such as (a) stored [glycogen], (b) catalytic capacities of glycolytic enzymes, (c) myosin or other cell ATPases, and (d) cell buffering power—all coadapt. Upward adjustments in any one of these parameters usually mean upward adjustments in all of them. Even the best adapted glycolytic systems, however, still are critically time limited, so anaerobic glycolysis must in turn be backed up by aerobic, ATP-replenishing pathways.

Oxidative Metabolism

The transition from rest to maximum sustained work in mammals requires the ATP turnover rates of the active muscles to increase by an order of magnitude or more. The requisite ATP is derived from the complete oxidation of carbohydrates, fats and proteins, or amino acids. Interestingly, most of the enzymes of β-oxidation, the Krebs cycle, and the ETS do not appear to occur in isozymic form. Perhaps for this reason the dependence on enzyme–enzyme direct transfer mechanisms appears even greater than in anaerobic metabolism. The most obvious situation concerns the enzymes of the ETS, which have long been considered to operate as functional complexes. When NADH is oxidized to NAD^+, two electrons and one H^+ are released. Protons, of course, are soluble in aqueous solution as hydronium ions (H_3O^+). Free electrons, in contrast, cannot exist in aqueous solution. Thus to reduce O_2, electrons are passed from NADH or from $FADH_2$ to O_2 along a chain of electron carriers, all of which are components of the inner mitochondrial membrane. At least 12, and probably more, electron carriers are grouped into four multiprotein, intramembranous or transmembranous particles or complexes; even if the *in vivo* structural order of these complexes may not yet be known with certainty, the *in vivo* functional order can be determined with certainty with a variety of approaches. For example, *the order of the electron-transfer components in the ETS is the order in which they become oxidized after the addition of O_2 to anoxic mitochondrial preparations*. In addition to serving as electron carriers, three of these complexes serve as proton pumps, in essence setting up H^+ concentration gradients, which are thought to supply some of the energy for ATP synthesis (the chemiosmotic theory). How these two primary functions (of electron transfer and of proton pumping) are integrated is not yet well understood; current concepts are available in numerous reviews and books (e.g., 16).

The importance of structural organization to function is also widely appreciated in studies of the β-oxidation pathway (8). The important point is that β-oxidation is designed to receive long chain fatty acylCoA substrates and release acetyl CoA; in the case of palmityl CoA, for example, *none of the 54 intermediates between starting substrate and pathway end-product are found in vivo*, which is a classic example of optimized metabolite channelling (63).

Unlike β-oxidation and the ETS, the Krebs cycle reactions, until recently, were widely considered as essentially functional in solution in the mitochondrial matrix. Recently, however, this position has been gradually changing. For example, gently disrupted mitochondria can be prepared that show exposed and readily sedimentable Krebs cycle enzymes. When either fumarate oxidation or the MDH-CS (malate dehydrogenase-citrate synthase) coupled enzyme system are analyzed, relative kinetic advantages are observed over completely solubilized enzyme preparations. These kinds of data led Srere and his co-workers to propose that the Krebs cycle *in situ exists as a sequential complex of enzymes, a metabolon favoring channelling of metabolites along the metabolic pathway and electrons directly to the ETS* (as in the coupling between MDH and NADH · ubiquinone oxidoreductase, Complex I in the ETS). The kinetic advantages observed are more sensitive to disruption than are the binding interactions with the overall particle; that is, enzyme–enzyme interactions favoring metabolite hand-offs along the Krebs cycle are more fragile than are enzyme-inner mitochondrial membrane interactions (57). Perhaps for this reason experimental support for the metabolon concept has been difficult to obtain and only now is the concept receiving broader acceptance.

Finally, another functionally important design feature of oxidative metabolism is that the complete combustion of most substrates and the coupling to ATPases take place in at least two separate compartments, the cytosol and the mitochondria. This introduces problems of exchange between these compartments of (a) carbon metabolites *per se*, (b) reducing equivalents, and (c) adenylates *per se*—all of which may influence flux rates. [The interested reader is referred to (42, 68)]. Because no single fuel feeds oxidative metabolism and because aerobically working tissues remain perfused, relatively open systems, the criteria for useful substrates depend on whether the fuel is stored internally (endogenous) or in other depots (exogenous).

Criteria for Endogenous Fuels for Aerobic Metabolism

Storage Quantities Should Be Adjustable and High

The main endogenous fuels in mammalian muscles are glycogen and triglycerides. Both pools are expandable during short-term training or in long-term adaptation (41). Similar fuel preferences are observed in insects such as locusts and moths (66). In salmon and squid, expendable proteins are dominant energy sources during migration, although there is uncertainty as to pathways of mobilization of the resultant amino acids (55). The amino acid pool is also utilizable for aerobic metabolism in many marine invertebrates (37, 65) as well as insects (66).

Fuels For Aerobic Metabolism Should Markedly Amplify Molar ATP Yields

Probably the most striking difference between phosphagens and glycogen in anaerobic ATP generation and fuels for aerobic metabolism *is the immense increase in yield of ATP/mol of the starting substrate* achievable by the complete oxidation of fuels. For triglyceride (calculated for glycerol + 3 palmitate), free fatty acids (calculated for palmitate), glycogen, glucose, proline and lactate, the respective molar yields of ATP are 403, 129, 38, 36, 21, and 18, assuming complete combustion; these values are 18–400 times higher than the molar yield of ATP from phosphagen hydrolysis, and 6–130 times the yield from anaerobic glycogenolysis. The advantages for sustained work are self-evident.

Good Fuels for Aerobic Metabolism Should Not Generate Undesirable End-Products

It has been emphasized by Atkinson and Camien (3) that the main end-products of triglyceride and glycogen oxidation are only molecular CO_2 and H_2O, while the oxidations of proteins, amino acids, and carboxylates yield CO_2, H_2O, and HCO_3^- as well as NH_4^+. Fortunately, none of these are very deleterious as metabolic end-products. Even if NH_4^+ and HCO_3^- can perturb metabolism, such problems can be minimized by their incorporation into urea and by urea excretion. Similarly, CO_2 can be removed at the lungs or gills while H_2O is essentially harmless.

Two other end products of aerobic metabolism which are of concern are alanine and IMP. The former may accumulate during augmentation of Krebs cycle intermediates, but the levels are normally too modest to present any serious problems (40). Also, under working conditions in mammalian muscle, AMP may be deaminated to IMP leading to adenylate depletion via the sequence:

$$ATP + H_2O \longrightarrow ADP + P_i$$
$$2ADP \longrightarrow ATP + AMP$$
$$AMP \longrightarrow IMP + NH_3 \begin{cases} NH_4^+ \\ H^+ \end{cases}$$

However, not all animals have this system. In many molluscs and crustaceans AMP deaminase is absent and [ATP] may drop drastically (below 1 μmol/g) with a concomitant rise in [ADP] and [AMP] *but with no net change in the adenylate pool* (4, 23). In exhausted trout, muscle [ATP] is similarly allowed to drop (to as low as 0.5 μmol/g), but in this case, because AMP deaminase is present, there is also a large reduction in the adenylate pool (20). This implies that the work obtainable from ATP hydrolysis remains adequate at quite widely varying adenylate levels, with or without AMP deaminase function, with or without defending [ATP] at a threshold of 2–3 μmol/g, with or without ADP and AMP accumulation (18).

If this is the case, one wonders what are the functions of AMP deaminase in species that possess it? The most likely answer is that by supplying IMP for adenylsuccinate synthetase, AMP deaminase sets the stage for fumarate formation from aspartate and, thus, for augmenting Krebs cycle intermediates when they are needed. This would explain the occurrence of this enzyme in vertebrate muscles, as well as its absence in invertebrates, where proline and glutamate are abundant and serve anapleurotic roles (19, 65). It also explains why the pathway of AMP → IMP may have arisen in the first place.

Criteria for Useful Exogenous Fuels of Aerobic Muscle Metabolism

Good Exogenous Fuels Should Be Storable at Sites Other Than Working Muscles

This criterion in vertebrates is met for all major classes of carbon substrates. For glucose this requirement is met by liver glycogen depots; for free fatty acids (FFA), by adipose fat. When lactate and amino acids are used as fuels they are mobilized mainly from glycogen and protein stores in nonworking muscles (9, 31, 35, 40). In ruminants, the situation is complicated by the rumen of the digestive system in which short chain volative fatty acids, especially 2-, 4-, 5-, and 6-carbon chains, are the dominant carbon and energy sources entering the circulation. The metabolic complications occur mainly in the liver where these fuels are partitioned between glucogenesis, lipogenesis, and oxidation (28).

In contrast to carbon substrates, O_2 is an exogenous substrate that cannot be stored in any appreciable amounts and for this reason mechanisms for assuring high rates of transport must be provided in active animals.

Exogenous Fuels Should Be Able to Sustain High Flux Rates

Because aerobically working muscles are perfused open systems, a critical feature of any exogenous substrate is that it must be capable of sustaining adequate flux rates. In units defined as $\mu mol/g \cdot min$, *the flux rates required depend on the molar yield of ATP*. Thus an ATP turnover rate in human muscle of 20 $\mu mol/g \cdot min$ could be supported by a palmitate flux of 0.15 units, a glucose flux of ~0.6 units, a lactate flux of ~1.2 units, and an O_2 flux of ~3.3 units. The more energetically efficient a fuel is (in terms of ATP/substrate burned) the less rapidly it need be fluxed from depot site to working muscle. For this reason, enzymes mobilizing efficient fuels (e.g., acylCoA synthetases) can occur at lower activities per gram of tissue than those mobilizing relatively inefficient fuels or those operating in anaerobic metabolism.

The highest flux is required by O_2, which is another important reason why its transfer from lungs or gills to tissue mitochondria is facilitated by hemoglobins (or their analogs) and by muscle myoglobins. Despite such transfer-facilitating mechanisms, many physiologists and metabolic biochemists conclude that in vertebrates generally O_2 availability becomes limiting at maximum aerobic metabolic rates (e.g., during maximum sustainable exercise). Some believe the limitation is because the cardiac pump is limiting. This argument emphasizes that, per unit of muscle mass, higher rates of O_2-based metabolism are attainable when only a small muscle mass is involved than when the whole organism is exercising. In species where this is so, humans, for example, maximum whole-organism work would therefore not be limited by O_2 uptake capacities at the muscle but by the heart's capacity to circulate O_2 loaded blood (60). In some species, dog, for example, the pump capacity appears to be relatively expanded and the available evidence, suggests that O_2 uptake capacities at the working muscles set the highest whole-organism sustainable work rate. Recent analyses indicate that O_2 limitation (either delivery or diffusional based) is probably the rule during maximum whole-organism work in organisms with lungs or gills but may not be so in insects. In the latter, O_2 delivery by tracheoles is so efficient that the *in vivo* cytochrome *c* oxidase turnover numbers are reasonably close to the theoretical maximum rates, in contrast to vertebrates, where cytochrome *c* oxidase *in vivo* works at only some 10% of its theoretical maximum (32).

Flux of Exogenous Fuels Should Be Adjustable According to Needs

In general, uptake at the working muscle varies with delivery rate (perfusion) and with plasma [substrate]. This feature requires provision for plasma substrate availability varying with working needs. Plasma glucose often does not satisfy this provision at low work rates, but may do so at high rates (47). Plasma FFAs, on the other hand, increase in availability during modest exercise, making them a quantitatively important fuel (50). In humans, their contribution decreases during heavy exercise (47), while in llamas, their contribution increases with prolonged work (38). Similarly, insects such as locusts begin long-term flight on carbohydrate, but as work continues, they rely more and more on fat (66).

Most earlier analyses of fat metabolism have overlooked the potential role of peroxisomes in the overall mobilization process. According to current concepts, one function of peroxisomes is to partially degrade long chain fatty acids (C_{16} and longer) to medium and short chain fatty acids. The latter are then either transferred to working tissues or are fully oxidized by mitochondrial metabolism *in situ*. The advantage of this arrangement seems to arise from (a) higher water solubility of shorter chain fatty acids and (b) higher rates of mitochondrial oxidation than sustainable by long chain analogs. This function of peroxisomes seems to be a feature of many tissues. In the rat, the relative abundances of peroxisomes are: heart > liver > diaphragm > kidney > skeletal muscle (71).

A final oxidative fuel to be considered is lactate *per se*. Although lactate has been emphasized as a muscle fuel from time to time (44, 64), all but a handful of workers have overlooked the magnitude of its contribution. In most recent analyses (9, 31, 38), it is supposed that lactate is formed in sites such as fast-twitch glycolytic (FG) fibers at about the same rates as it is utilized in more oxidative-type fibers, and that, at the upper limit, plasma lactate replacement rates equal oxidation rates in aerobically working muscles. In rats, as much as 50% of the carbon flux to CO_2 (during maximum sustained muscle work or at high lactate levels) accords with such a lactate "shuttling" model, while in dogs and tammar wallabies, the maximum value is 20–30% (40).

These lactate flux rates are only high by the standards of aerobic pathways; anaerobic glycolytic fluxes can be two orders of magnitude higher. Because of its energetic inefficiency, however, anaerobic glycolysis hardly affects total ATP turnover. In wallabies, for example, glycogenolysis in FG fibers (adequate to account for the maximum flux of lactate to aerobically working muscles) would contribute only 1–2% to the total ATP cycling rate. These rates of glycolysis [which may be anaerobic or aerobic (9)] are similar during aerobic exercise to those in other animals (10, 69) and lead to an interesting conclusion; namely, that during sustained exercise *the function of glycogenolysis is less to make ATP at lactate production sites than it is to mobilize glycogen (via lactate) as a fuel for aerobically working muscles* (38, 40, 73). If, in a 70-kg human with 30 kg of muscle, 10 kg were used for steady work rates, while 10 kg were used intermittently, it is easy to show that enough glycogen is available (100 μmol/g) to sustain observed lactate turnover rates of 500 μmol/kg · min for ~3 h, while with training the work time could increase (because of higher amounts of stored glycogen). Hence, the supplies of precursor are theoretically adequate to meet a significant energy need, which may be why during exercise in humans, lactate is useful as a fuel (9) while the use of plasma FFA seems to slow down (47). However, this sequential change in fuel use is not seen in all animals. In llamas, lactate flux can contribute significantly to short-term aerobic work, but after 30 min even at modest exercise (<50% of $\dot{V}_{O_2(max)}$) lactate availability drops to less than at rest, and so makes a minimal contribution to muscle energy demands. At this time, [FFA] in the plasma and FFA fluxes both rise and largely account for muscle ATP turnover rates (38).

The shuttling role of lactate between sites of formation and sites of utilization also is minimized or totally lacking in fishes and insects. In fishes, lactate is formed in large amounts in white muscle during swimming exercise. However, unlike the situation in mammals, lactate seems to be preferentially retained in white muscle and lactate clearance takes many hours of recovery when perfusion of muscle is well above normal (54). While this pattern has been known for over three decades, the molecular basis for it is not fully appreciated even today. One popular theory is that the glycogen → lactate pathway is arranged to discharge during burst work, while during recovery most of the lactate is merely reconverted to glycogen *in situ*, a kind of discharging–recharging spring coil view of glycogenolysis (54).

In flight muscles of insects such as the locust, LDH for practical purposes is deleted (58, 66), so that muscle function here is obligately aerobic, and there is no reliance on the lactate shuttling role found in mammals. Muscles in very small mammals (shrews) and very small birds (hum-

mingbirds) likewise display relatively reduced activities of LDH, which is apparently retained as a mechanism for buffering cytosolic redox particularly during rest → work transitions in a manner analogous to CPK buffering of the ATP pool under similar conditions (34). Presumably this function can be abandoned in insect flight muscle because the tracheole-based O_2 delivery can match the square-wave muscle work transition from rest to flight, a situation that cannot be achieved in organisms that circulate O_2 from lungs or gills to working tissues (32).

Integrating Aerobic and Glycolytic Pathways

The final design criterion of oxidative metabolism which we need to consider is a means for integrating it appropriately with anaerobic metabolism. In this regard, it is widely observed *in vivo* that as exercise intensity increases, $\dot{V}_{O_2}$ rises to a maximum level at which time further muscle work depends on anaerobic glycolysis. The problem thus arises of how this integration of aerobic and glycolytic pathways is achieved. Although many regulatory signals may be involved (40), in mammals, at least during high intensity sustained exercise, O_2 delivery to working muscles may represent an important limitation and thus signal when anaerobic metabolism should be phased in. The argument for this conclusion depends on the observation that muscle ATP turnover rates can be higher in muscles working in isolation than *in vivo*. For example, in one-legged exercise in humans, work rates ranging up to 180 μmol of ATP/g · min can be sustained by a fully oxidative metabolism; work at this very high rate for the entire muscle mass of a human would require a cardiac output some threefold higher than possible. Thus, many exercise biochemists (60) argue that *in vivo* the cardiac pump, and thus O_2 delivery to muscle mitochondria, is a major determinant of control. In this situation, O_2 concentration in the muscle would signal when anaerobic glycolysis need be strongly activated.

These studies are instructive because they predict that during intense work involving only a relatively small muscle mass, O_2 delivery to muscle mitochondria should not be limiting. Experimentally, this prediction in fact seems to be realized. During heavy (but not maximal) work in isolated dog gracilis, for example, recent studies show that autoregulation mechanisms can maintain O_2 delivery to muscle at high enough rates to saturate mitochondrial cytochrome *c* oxidase (14, 26). In this situation, as in insect flight muscle (32), O_2 concentration presumably has minimal control influence. If O_2 concentration is not the signal for phasing in glycolysis, what is?

Most workers consider that another important control mechanism revolves around acceptor control of muscle mitochondrial function. According to current evidence (11), mitochondria in nonworking muscle are in state 4: Flux rates through the ETS are low because of limiting [ADP] or [P_i]. During muscle work, mitochondrial respiration and phosphorylation rates increase; the mitochondria are said to enter state 3 as [ADP] and/or [P_i] increase. While the overall response *in vitro* is hyperbolic, at K_m substrate levels respiration increases essentially directly with [ADP] or [P_i] (46), which is presumably also why [with *in vitro* K_m values for ADP of ~22–28 μM (12)] work rates vary in a linear fashion with [P_i]/[PCr], also quantified *in vivo* (11). Aside from being consistent with studies of isolated mitochondria (46), these *in vivo* studies also satisfactorily explain the advantage, with endurance training, of increasing the capacities of the ETS and oxidative phosphorylation (17); for any given submaximal work rate in trained muscle, *flux through the ETS/g will be lower, because the amount of ETS/g is up to twofold higher*. Muscle mitochondria from endurance-trained individuals, therefore, almost necessarily operate at lower [ADP] or [P_i]; that is on the steep parts of the ADP or P_i saturation curves. In addition, because *in vivo* operation appears to occur at lower concentrations of acceptors and because the maximum capacity for state 3 flux is elevated, the overall metabolic scope for activity is effectively expanded. Finally, when the state 3 capacity is surpassed in either trained or untrained muscle, [ADP] and [P_i] may rise even further, creating conditions favoring glycolytic competition for both. This ensures that in all species, despite great variation in speed capacity or size (61), anaerobic glycolysis is phased in at the appropriate time: When the ATP replenishing capacity of oxidative phosphorylation is being surpassed.

Abbreviations Used

acetylCoA–acetyl coenzyme A

ADP–adenosine diphosphate

AMP–adenosine monophosphate

APK–arginine phosphokinase

Arg–arginine

ATP–adenosine triphosphate

CoASH–reduced form of coenzyme A

CPK–creatine phosphokinase

Cr–creatine

ETS–electron-transfer system

FAD/FADH–oxidized and reduced flavin adenine dinucleotide

GAPDH–glycosaldehyde-3-P-dehydrogenase

IMP–inosine monophosphate

LDH–lactate dehydrogenase

NAD$^+$/NADH–oxidized and reduced nicotinamide adenine dinucleotide

NDP/NTP–nucleoside diphosphate/triphosphate

PArg–phosphoarginine

PCr–phosphocreatine

PEP–phosphoenolpyruvate

PGK–phosphoglycerate kinase

PK–pyruvate kinase

RM–red (slow oxidative) muscle

WM–white (fast-twitch glycolytic) muscle

References

1. Armstrong, G. and W. Mordue. *Physiol. Entomol.* 10:353–358, 1985. O_2 consumption in flying locusts.
2. Atkinson, D. E. *Cellular Energy Metabolism and its Regulation.* Academic Press, New York, 1977.
3. Atkinson, D. E. and M. N. Camien. *Curr. Top. Cell. Regul.* 21:261–302, 1982. The role of urea synthesis in the removal of bicarbonate and the regulation of blood pH.
4. Baldwin, J. and P. W. Hochachka. *Mol. Physiol.* 7:29–40, 1985. A glycolytic paradox in *Limaria* muscle.
5. Bessman, S. P. and P. J. Geiger. *Science* 211:448–452, 1981. Transport of energy in muscle: The phosphorylcreatine shuttle.
6. Boiteux, A. and B. Hess. *Philos. Trans. R. Soc. London, Ser. B* 293:5–22, 1981. Design of glycolysis.
7. Bowlus, R. D. and G. N. Somero. *J. Exp. Zool.* 208:37–152, 1979. Solute compatibility with enzyme function and structure: Rationales for the selection of somotic agents and end products of anaerobic metabolism in marine invertebrates.
8. Bremer, J. and H. Osmundsen. In *Fatty Acid Metabolism and Its Regulation* (S. Numa, ed.), pp. 113–154. Elsevier, Amsterdam, 1984. Fatty acid oxidation and its regulation.
9. Brooks, G. A. *Fed. Proc., Fed. Am. Soc. Exp. Biol.* 45:2924–2929, 1986. Lactate production under fully aerobic conditions: The lactate shuttle during rest and exercise.
10. Brooks, G. A., D. M. Donovan, and T. P. White. *J. Appl. Physiol.* 56:20–525, 1984. Estimation of anaerobic energy production and efficiency in rats during exercise.
11. Chance, B., S. Eleff, J. S., Leigh, Jr., D. Sokolow, and A. Sapega. *Proc. Natl. Acad. Sci. U.S.A.* 78:6714–6718, 1981. Mitochondrial regulation of phosphocreatine/inorganic phosphate ratios in exercising human muscle: A gated ^{31}P NMR study.
12. Chance, B., Leigh, J. S., Jr., J. Kent, and K. McCully. *Fed. Proc., Fed. Am. Soc. Exp. Biol.* 45:2915–2920, 1986. Metabolic control principles and ^{31}P NMR.
13. Collicutt, J. M. and P. W. Hochachka. *J. Comp. Physiol.* 115:147–157, 1977. The anaerobic oyster heart: Coupling of glucose and aspartate fermentation.
14. Connett, R. J., T. E. Gayeski, and C. R. Honig. *Am. J. Physiol.* 248:H922–H929, 1985. Energy sources in fully aerobic rest-work transitions: A new role for glycolysis.
15. Dann, L. G. and H. G. Britton. *Biochem. J.* 169:39–54, 1978. Kinetics and mechanism of action of muscle pyruvate kinase.
16. Darnell, J., H. Lodish, and D. Baltimore. *Molecular Cell Biology.* Scientific American, New York, 1986.
17. Davies, K. J. A., L. Packer, and G. A. Brooks. *Arch. Biochem. Biophys.* 209:539–554, 1981. Biochemical adaptation of mitochondria, muscle, and whole-animal respiration to endurance training.
18. Dawson, M. J., D. G. Gadian, and D. R. Wilkie. *Nature (London)* 274:861–866, 1978. Muscular fatigue investigated by phosphorous nuclear magnetic resonance.
19. De Zwaan, A. In *The Mollusca* (P. W. Hochachka, ed.), Vol. 1, pp. 137–175. Academic Press, New York, 1983. Carbohydrate catabolism in bivalves.
20. Dobson, G. P. and P. W. Hochachka. *J. Exp. Biol.* 129:125–140, 1987. Role of glycolysis in adenylate depletion and repletion during work and recovery in teleost white muscle.
21. Driedzic, W. R. and P. W. Hochachka.

Am. J. Physiol. 230:579–582, 1976. Control of energy metabolism in fish white muscle.

22. Emmett, B. and P. W. Hochachka. *Respir. Physiol.* 45:261–267, 1981. Scaling of oxidative and glycolytic enzymes in mammals.

23. England, W. R. and J. Baldwin. *Physiol. Zool.* 56:614–622, 1983. Anaerobic energy metabolism in the tail musculature of the Australian yappy, *Cherax destructor*: Role of phosphagens and anaerobic glycolysis during escape behaviour.

24. Gade, G. *Eur. J. Biochem.* 160:311–318, 1986. Purification and properties of tauropine dehydrogenase from the shell adductor muscle of the Ormer, *Haliotis lamellosa*.

25. Gadian, D. G., G. K. Radda, T. K. Brown, E. M. Chance, M. J. Dawson, and D. R. Wilkie. *Biochem. J.* 194:215–228, 1981. The activity of creatine kinase in frog skeletal muscle studied by saturation-transfer nuclear magnetic resourance.

26. Gayeski, T. E. J. and C. R. Honig. *Am. J. Physiol.* 251:789–H799, 1986. O_2 gradients from sarcolemma to cell interior in red muscle at maximal $\dot{V}_{O_2}$.

27. Guppy, M., W. C. Hulbert, and P. W. Hochachka. *J. Exp. Biol.* 82:303–320, 1979. Metabolic sources of heat and power in tuna muscles. II. Enzyme and metabolite profiles.

28. Hochachka, P. W. In *Comparative Animal Physiology* (C. L. Prosser, ed.). pp. 212–278. Saunders, Philadelphia, PA, 1973. Metabolism.

29. Hochachka, P. W. *Living Without Oxygen*, pp. 1–181, Harvard University Press, Cambridge, MA, 1980.

30. Hochachka, P. W. *Adv. Shock Res.* 9:49–65, 1983. Protons and glucose metabolism in shock.

31. Hochachka, P. W. *Fed. Proc., Fed. Am. Soc. Exp. Biol.* 45:2948–2952, 1986. Balancing conflicting metabolic demands of exercise and diving.

32. Hochachka, P. W. *Adv. Exp. Med. Biol.* 222:143–149, 1988. Patterns of O_2-dependence of metabolism.

33. Hochachka, P. W., G. P. Dobson, and T. P. Mommsen. *Isozymes: Curr. Top. Biol. Med. Res.* 8:91–113, 1983. Role of isozymes in metabolic regulation during exercise: Insights from comparative studies.

34. Hochachka, P. W., B. Emmett, and R. K. Suarez. *Can. J. Zool.* 66:1128–1138, 1988. Limits and constraints in the scaling of oxidative and glycolytic enzymes in homeotherms.

35. Hochachka, P. W. and M. Guppy. *Metabolic Arrest and the Control of Biological Time.* Harvard Univ. Press, Cambridge, MA, 1987.

36. Hochachka, P. W. and T. P. Mommsen. *Science* 219:1391–1397, 1983. Protons and anaerobiosis.

37. Hochachka, P. W., T. P. Mommsen, J. M. Storey, K. B. Storey, K. Johansen, and C. J. French. *Mar. Biol. Lett.* 4:1–21, 1983. The relationship between arginine and proline metabolism in cephalopods.

38. Hochachka, P. W., T. P. Mommsen, J. H. Jones, and C. R. Taylor. *Am. J. Physiol.* 253:R298–R305, 1987. Substrate and O_2 fluxes during rest and exercise in a high altitude adapted animal, the Llama.

39. Hochachka, P. W., W. B. Runciman, and R. V. Baudinette. *Mol. Physiol.* 7:7–28, 1985. Why exercising tammar wallabies turn over lactate rapidly: Implications for models of mammalian exercise metabolism.

40. Hochachka, P. W. and G. N. Somero. *Biochemical Adaptation*, Princeton Univ. Press, Princeton, NJ, 1984.

41. Holloszy, J. O. and F. W. Booth. *Annu. Rev. Physiol.* 38:273–291, 1976. Biochemical adaptations to endurance exercise in muscle.

42. Holloszy, J. O. and E. F. Coyle. *J. Appl. Physiol.* 56:831–838, 1984. Adaptations of skeletal muscle to endurance exercise and their metabolic consequences.

43. Howald, H., G. von Glutz, and R. Billeter. *Energy Stores and Substrate Utilization in Muscle During Exercise*, 3rd. Int. Symp., pp. 75–86. Symposia Specialists, Inc., Quebec City, Quebec, Canada 1978.

44. Issekutz, B., Jr., W. A. S. Shaw, and A. C. Issekutz. *J. Appl. Physiol.* 40:312–319, 1976. Lactate metabolism in resting and exercising dogs.

45. Jacobs, H. K. and S. A. Kuby. *J. Biol. Chem.* 255:8477–8482, 1980. Studies on muscular dystrophy. A comparison of the steady-state kinetics of the normal human ATP-creatine transphosphorylase isoenzymes (creatine kinases) with those from tissues of Duchenne muscular dystrophy.

46. Jacobus, W. E., R. W. Moreadith, and K. M. Vandegaer. *J. Biol. Chem.* 257:2397–2402, 1982. Mitochondrial respiratory control. Evidence against the regulation of respiration by extramitochondrial phosphorylation potentials or by [ATP]/[ADP] ratios.

47. Jones, N. L., G. J. F. Heigenhauser, A. Kuksis, C. G. Matsos, J. R. Sutton, and C. J. Toews. *Clin. Sci.* 59:469–478, 1980. Fat metabolism in heavy exercise.

48. Lebherz, H. G., J. K. Petell, J. E. Shackelford, and M. J. Sardo. *Arch. Biochem. Biophys.* 214:642–656, 1982. Regulation of concentrations of glycolytic enzymes and creatine phosphokinase in "fast-twitch" and "slow-twitch" skeletal muscles of the chicken.

49. Levitskii, D. O., T. S. Levchenko, V. A. Saks, V. G. Sharov, and V. N. Smirnov. *Biokhimiya (Moscow)* 42:1389–1395, 1977. Functional coupling between Ca^{++}-ATPase and creatine phosphokinase in sarcoplasmic reticulum of myocardium.

50. Maughan, R. J., C. Williams, D. M. Campbell, and D. Hepburn. *Eur. J. Appl. Physiol. Occup. Physiol.* 39:7–16, 1978. Fat and carbohydrate metabolism during low intensity exercise: Effects of the availability of muscle glycogen.

51. McGilvery, R. W. In *Metabolic Adaptation to Prolonged Physical Exercise* (H. Howald and J. R. Poortmans, eds.), pp. 12–30. Birkhaeuser, Basel, 1975. The use of fuels for muscular work.

52. McGilvery, R. W. *Biochemistry: A Functional Approach*, Saunders, Philadelphia, PA, 1983.

53. Meyer, R. A., H. L. Sweeney, and M. J. Kushmerick. *Am. J. Physiol.* 246:C365–C377, 1984. A simple analysis of the "phosphocreatine shuttle."

54. Milligan, C. L. and C. M. Wood. *J. Exp. Biol.* 123:23–144, 1986. Tissue intracellular acid-base status and the fats of lactate after exhaustive exercise in the rainbow trout.

55. Mommsen, T. P., C. J. French, and P. W. Hochachka. *Can. J. Zool.* 58:785–1799, 1980. Sites and patterns of protein and amino acid utilization during the spawning migration of salmon.

56. Petrofsky, J. S., C. A. Phillips, M. N. Sawka, D. Hanpeter, and D. Stafford. *J. Appl. Physiol.* 50:493–502, 1981. Blood flow and metabolism during isometric contractions in cat skeletal muscle.

57. Robinson, J. B., Jr., L. Inman, B. Sumegi, and P. A. Srere. *J. Biol. Chem.* 262:1786–1790, 1987. Further characterization of the Krebs tricarboxylic acid cycle metabolon.

58. Sacktor, B. *Biochem. Soc. Symp.* 41:111–131, 1976. Biochemical adaptations for flight in the insects.

59. Sahlin, K., G. Palmskog, and E. Hultman. *Pfluegers Arch. Gesamte Physiol. Menschen Tiere* 374:193–198, 1978. Adenine nucleotide and IMP contents of the quadriceps muscle in man after exercise.

60. Saltin, B. *J. Exp. Biol.* 115:45–354, 1985. Malleability of the system in overcoming limitations: Functional elements.

61. Seeherman, H. J., C. R. Taylor, G. M. O. Maloiy, and R. B. Armstrong. *Respir. Physiol.* 44:11–23, 1981. Design of the mammalian respiratory system. II. Measuring maximum aerobic capacity.

62. Srivastava, D. K. and S. A. Bernhard. *Curr. Top. Cell. Regul.* 28:1–68, 1986. Enzyme-enzyme interactions and the regulation of metabolic reaction pathways.

63. Srivastava, D. K. and S. A. Bernhard. *Science* 234:1080–1086, 1986. Metabolite transfer via enzyme-enzyme complexes.

64. Stainsby, W. N. and H. G. Welch. *Am. J. Physiol.* 211:177–183, 1966. Lactate me-

tabolism of contracting dog skeletal muscle *in situ*.

65. Storey, K. B. and J. M. Storey. In *The Mollusca* (P. W. Hochachka, ed.), Vol. 1, pp. 91–136. Academic Press, New York, 1983. Carbohydrate metabolism in cephalopods.

66. Storey, K. B. In *Circulation, Respiration, and Metabolism* (R. Gilles, ed.), pp. 193–207. Springer-Verlag, Berlin, 1985. Metabolic biochemistry of insect flight.

66a. Suarez, R. K., G. S. Brown, and P. W. Hochachka. *Am. J. Physiol.* 251:R537–R542, 1986. Metabolic sources of energy for hummingbird flight.

67. Sutton, J. R., N. L. Jones, and C. J. Toews. *Clin. Sci.* 61:331–337, 1981. The effect of pH on muscle glycolysis during exercise.

68. Tager, J. M., A. K. Groen, R. J. W. Wanders, J. Duszynski, H. V. Westerhoff, and R. C. Vervoorn. *Biochem. Soc. Trans.* 11:40–43, 1983. Control of mitochondrial respiration.

69. Taylor, C. R., G. M. O. Maloiy, E. R. Weibel, V. A. Langman, J. M. Z. Kamau, H. J. Seeherman, and N. C. Heglund. *Respir. Physiol.* 44:25–37, 1981. Design of the mammalian respiratory system. III. Scaling maximum aerobic capacity to body mass: Wild and domestic mammals.

70. Vaghy, P. L. *J. Mol. Cell. Cardiol.* 11:33–940, 1979. Role of mitochondrial oxidative phosphorylation in the maintenance of intracellular pH.

71. Veerkamp, J. H. and H. T. B. Moerkerk. *Biochim. Biophys. Acta* 875:01–310, 1986. Peroxisomal fatty acid oxidation in rat and human tissues.

72. Walker, J. B. *Adv. Enzymol.* 50:177–242, 1979. Creatine: Biosynthesis, regulation, and function.

73. Weber, J.-M., W. S. Parkhouse, G. P. Dobson, J. C. Harman, D. H. Snow, and P. W. Hochachka. *Am. J. Physiol.* 253:R896–R903, 1987. Lactate kinetics in exercising thoroughbred horses. II. Regulation of metabolite turnover rate in plasma.

Chapter 9 | *Respiration and Metabolism*

Warren Burggren and John Roberts

Introduction

The terms respiration and metabolism are defined in several ways, influenced in part by the research discipline. Thus, a physiological ecologist might view respiration in terms of whole animal energy consumption, a systems level physiologist might define this term as the movement of O_2 and CO_2 between environment and gas exchange organs, while a cellular physiologist might regard respiration as the movement of electrons between a series of carrier molecules within the mitochondria. For the purposes of this chapter we will define respiration as the sum of the processes by which the respiratory gases are transferred between environment and tissues. We define metabolism as the intracellular process that consumes substrates and produces by-products in the course of generating chemically stored energy for anabolism. Metabolism and respiration directly interact in that respiratory processes are a proximate limiting factor in metabolism.

Respiratory Gases: Physical, Environmental, and Physiological Considerations

Two physicochemical phenomena dominate any discussion of respiration. First, O_2 and CO_2 have very different properties with respect to their rates of diffusion through and storage in both respiratory media and body fluids. Thus, an understanding of the elimination of CO_2 is not derived directly from reversing how O_2 uptake occurs. Second, the two respiratory media—air and water–have vastly different physicochemical properties. Consequently, breathing with water presents a very different set of limitations from breathing with air.

Units in Respiratory and Metabolic Physiology

Respiratory physiologists typically are interested in several variables relating to a respiratory gas. The concentration of gas has traditionally been expressed as a volume per volume of solution or gas mixture, at a specified temperature and humidity. More recently, the preferred expression of concentration has been in units of quantity of gas. (Typically, these units are micromoles per volume of solution or gas mixture. Parts per million (ppm) and similar expressions are rarely used in respiratory physiology.) Units involving quantity rather than volume of solution are preferable because they are independent of temperature and humidity.

The partial pressure of a gas is a second important variable. The partial pressure of a particular gas is that proportion of the total atmospheric pressure of a gas mixture that is contributed by that gas. That is, the partial pressure exerted by a gas is directly proportional to its fractional concentration in the overall gas mixture. In a hypothetical gas mixture that is 10% O_2, 90% N_2, devoid of water vapor, and at 1 atm (760 mmHg), the O_2 partial pressure (P_{O_2}) is $760 \times 0.10 = 76$ mmHg. Importantly, the partial pressure of a particular gas depends on its fractional concentration of the overall gas mixture and the barometric pressure, but not on the absolute quantity of gas present. Thus, if water is at gaseous equilibrium with a 10% O_2/90% N_2 gas mixture overlying it, the P_{O_2} of the gas mixture dissolved in the water is also 76 mmHg, even though the concentration of O_2 molecules dissolved in the water is clearly far lower than the concentration of O_2 molecules in the gas. Only a gas physically dissolved in fluid contributes to the partial pressure of that gas—gases chemically bound (e.g., O_2 bound to hemoglobin, CO_2 in the form of the bicarbonate ion) make no contribution to the partial pressure. Thus, if there is considerable chemical binding of a gas in a fluid, the gas partial pressure can be quite low in a fluid even while the gas concentration is very high.

A third important variable to consider for aquatic environments is the solubility of the gas in the respiratory medium or body fluid. The solubility coefficient (alpha, α) describes the change in quantity of gas that will dissolve physically in a given volume of fluid at a specified temperature in response to an incremental change in gas partial pressure. Typically, the units of the solubility coefficient are mmol/mmHg L^{-1} (or ml/mmHg L^{-1}). Importantly, solubility coefficients vary with temperature, the nature of the solvent, and the specific gas.

Finally, it is important to distinguish between α, the solubility coefficient and β, the capacitance coefficient. The capacitance coefficient includes not only the gas physically dissolved in solution (as described by α), but also chemically bound gas (38). In water, for example, where there are only insignificant amounts of chemically bound molecular O_2, $\alpha = \beta$ for O_2. In that same water, however, CO_2 can be present as dissolved molecular CO_2, as the bicarbonate ion, and as carbonate compounds; thus, β_{CO_2} will be $>\alpha$. In body fluids such as blood, where both O_2 and CO_2 may be chemically bound in large quantities, β is considerably higher than α (see Chapter 10).

Diffusion of Respiratory Gases

The rate of diffusion of O_2 and CO_2 through respiratory media and biological tissues is governed by a variety of physicochemical factors. Fick (49) in 1870 first interrelated these factors in what has become known as Fick's law of diffusion:

$$\dot{M}_x = \frac{D_x A (C_{ext,x} - C_{int,x})}{L}$$

where $\dot{M}_x$ is the mass of gas x transferred per unit time, A is the surface area available for diffusion, $(C_{ext,x} - C_{int,x})$ is the concentration difference across the diffusion barrier, and L is the thickness of the diffusion barrier. The term D_x is the diffusion coefficient for gas x, and reflects the "diffusibility" of a gas (the rate of diffusion of a gas is in inverse proportion to its molecular mass). However, the solubility of the gas in the material through which it is diffusing has a major effect on the overall rate of gas movement by affecting local concentration gradients. This factor was recognized by August Krogh (99), who introduced the concept of a biological diffusion constant, which is the product of the diffusion coefficient (D) and the solubility coefficient α. Krogh's diffusion constant, K, has been determined empirically for O_2 diffusing through a variety of biological tissues (Table 1).

Unfortunately, measurement of gas concentrations in biological tissues is difficult at best when compared with measurement of gas partial pressure. Also, the solubilities of O_2 and CO_2 in biological fluids are very different. Fick's equation is:

$$\dot{M}_x = \frac{K_x A \cdot (P_{ext,x} - P_{int,x})}{L}$$

where $P_{ext,x} - P_{int,x}$ is the partial pressure gradient of x across the diffusion pathway and K_x is Krogh's diffusion constant. The term K for CO_2 emerges as ~25–30 times higher than for O_2 in water (Table 2). As a consequence, the partial pressure gradient for O_2 along a diffusion pathway must be ~25–30 times greater than the gradient for CO_2 to move equivalent amounts of these two gases. This phenomenon takes on practical tones in gas exchange organs, where generally the partial pressure gradient for O_2 from environment to tissues is far larger than the partial pressure gradient for CO_2 in the opposite direction.

Like all physicochemical processes, diffusion is temperature dependent.

TABLE 1. Krogh's O_2 Diffusion Constant (K_{O_2}) for a Variety of Biological Materials and Other Substances

Material	Reference	K_{O_2}[a]
Air	(99)	11
Water	(99)	3.4
Water	(15)	3.2
Water	(176)	3.6
9% Serum albumin	(98)	2.6
15% Gelatin	(99)	2.8
Frog egg jelly	(15)	2.4
9% Serum albumin	(98)	2.6
Frog skeletal muscle	(99)	1.4
Frog connective tissue	(99)	1.2
Rat myocardium	(65)	1.9
Human erythrocytes	(65)	1.3
Chitin	(99)	0.13
Chitin	(108)	0.13

[a]Units for K are $cm^2/atm \cdot min \times 10^{-6}$

However, the net effect of temperature on diffusion is relatively small, with the absolute value of Krogh's diffusion constant increasing by only ~1.4% for every degree Celsius increase in the medium. This is considerably less than the 5–15%/°C increase that occurs in metabolic rate, indicating that processes in diffusion can become increasingly limiting to gas exchange as temperature increases.

A final factor to consider in diffusion is time. Diffusion time increases with the square of the diffusion path length. Thus, whereas an O_2 molecule in water diffuses 1 μm in ~10^{-4} s, it would require ~100 s to diffuse 1000 μm, or 1 mm. Clearly, diffusion alone rapidly becomes inadequate as a method for gas transport as the diffusion distance increases.

Water and Air as Respiratory Environments

Air and water are very different with respect to serving as a respiratory media. The specific respiratory medium used by a particular species has acted as a major

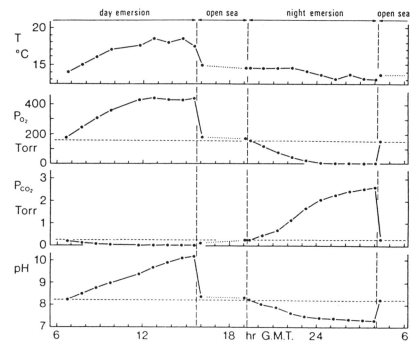

Figure 1. Daily variations in temperature, pH, P_O and P_{CO} of seawater in a tidal rock pool (coast of Brittany, France) during a 22 h period in June. The dashed lines correspond to P_{O_2}, P_{CO_2} and pH of air-equilibrated water at 15°C (38).

selection pressure in the evolution of respiratory structures and processes. Let us first consider the general properties of water and air as respiratory media.

Water

When water is at equilibrium with the air above it, then the P_{O_2} and P_{CO_2} of the water will be identical to those of the air. In a biological context, however, natural bodies of water are not always at equilibrium with air. The prevailing balance between respiration and photosynthesis can significantly affect environmental P_{O_2} and P_{CO_2}. In shallow lakes or tidepools with a large floral community and intense solar

TABLE 2. A Comparison of the Physicochemical Properties Related to Gas Exchange of Air and Water at 20°C

Physiochemical Properties	Units	Water	Air (1 atm)	Water/Air
O_2 Diffusion coefficient	cm²/s	0.000025	0.198	1/7920
CO_2 Diffusion coefficient	cm²/s	0.000018	0.155	1/8600
O_2 Capacitance coefficient	nmol/mL · mmHg	1.82	54.7	1/30
CO_2 Capacitance coefficient	nmol/mL · mmHg	51.4	54.7	1/1.06
Krogh's O_2 diffusion constant	nmol/cm · s · mmHg	0.000046	10.9	1/237,000
Krogh's CO_2 diffusion constant	nmol/cm · s · mmHg	0.00093	8.5	1/9140
Viscosity	centipoise, cP	0.101	0.0018	56
Density	kg/L	1.001	0.00121	827
Heat capacity	cal/L · °C	1000	0.295	3400
Heat conductivity	μcal/s · °C · cm	1440	60.3	24

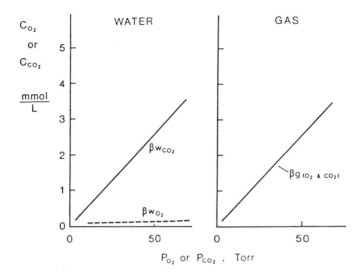

Figure 2. Concentrations of O_2 and CO_2 in distilled water and in air (15°C) as a function of P_{O_2} and P_{CO_2}.

radiation, high rates of photosynthesis during daylight hours can result in an elevation of P_{O_2} to above 400 mmHg (Fig. 1)! During darkness, when respiration prevails, however, P_{O_2} may fall to as low as a few millimeters of mercury. Aquatic habitats characterized by such large diurnal swings are probably more common than generally perceived, and present substantial challenges to respiratory systems of animals living in them (110). Even large bodies of water (i.e., lakes and oceans) show significant O_2 stratification. Most oceans, for example, show reduced O_2 at depths below 50–100 m, because of the dominance of respiration over photosynthesis in the poorly illuminated deeper waters (169). The sediments at the bottom of bodies of water—both deep and shallow—are notorious for near anoxic or anoxic conditions unless vertical currents mix surface and deeper water. Although hydrostatic pressure increases with increasing depth in the water column, in air-equilibrated water the dissolved gases remain at essentially the same concentrations and partial pressures as surface water.

The changes in O_2 and CO_2 concentration in water as a function of gas partial

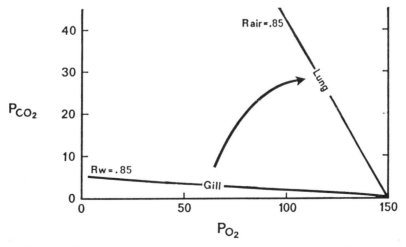

Figure 3. Relationship between P_{O_2} and P_{CO_2} in the medium leaving gills and lungs (136). R is ratio of CO_2 produced to O_2 consumed.

pressure at 15°C are shown in Figure 2. The solubility coefficient for CO_2 in distilled water is ~20–35 times greater than that for O_2, depending on temperature (Table 1). This has important consequences to aquatic respiration because, in essence, this means that water is much more effective as a CO_2 *sink* than an O_2 source. To illustrate this, a change in P_{O_2} of 100 mmHg in 1 L of distilled water at 15°C will change the total O_2 content by ~200 μM. A similar change in total CO_2 concentration of ~200 μM would be produced by a change in water P_{CO_2} of only a little over 3 mmHg! Thus, an animal breathing water must produce a far larger reduction in the P_{O_2} in water flowing over its respiratory membranes than the required rise in P_{CO_2} needed for adequate CO_2 elimination. This is illustrated graphically in Figure 3, which for both water and air breathers indicates the general relationship between P_{O_2} and P_{CO_2} in the expired respiratory medium. The respiratory quotient, which is the ratio of CO_2 elimination (M_{O_2}) to O_2 uptake (M_{O_2}), (M_{CO_2}/M_{O_2}), is assumed to be 0.85, which is representative of the metabolism of a normal balance of carbohydrates and protein (see below). Because β for O_2 and CO_2 in a gas mixture (but not in water) are the same, the P_{CO_2} of expired air will rise by 0.85 times the rate at which the P_{O_2} will fall. In water breathers with the same respiratory quotient, however, the fall in P_{O_2} greatly exceeds the small rise in P_{CO_2}.

Water not only has a low O_2 capacitance, but also has a high absolute viscosity and density. The viscosity and density of water contribute to the establishment of unstirred boundary layers of water immediately adjacent to respiratory membranes—these boundary layers can constitute the major diffusion resistance in an aquatic gas exchange organ. Also, the combination of high density and viscosity makes water very expensive in

metabolic terms to pump through a respiratory organ. This is reflected in a much higher *metabolic cost of ventilation* for aquatic animals compared with air breathing animals. However, the high density of water and the accompanying buoying effect provides postural support, unlike air.

Water has a very high thermal capacitance (Table 2). That is, water is a highly effective heat sink, as evident from its use in many industrial heat transfer systems. In aquatic animals, heat produced through tissue metabolism is very rapidly lost to the surrounding water, particularly if blood perfuses an aquatic gas exchange organ with a large surface area exposed to the surrounding water. As a direct consequence, almost all water-breathing animals are necessarily ectotherms. A limited form of endothermy has evolved in some very large fishes, where specialized counter-current vascular structures help to conserve heat produced in the core tissues (Chapter 3).

Finally, water-breathing animals may experience problems with hydromineral balance because of the use of water as a respiratory medium. Because the respiratory membranes must be very thin to allow rapid gas diffusion, the respiratory structures of water-breathing animals also represent the major sites for uncontrolled water and ion fluxes between environment and body fluids, as dictated by ion concentrations and osmotic pressure differentials (Chapter 2).

Air

The inert gas N_2 constitutes over 78% of dry atmospheric air, with the respiratory gases O_2 and CO_2 constituting 20.95 and 0.03%, respectively. At sea level (atmospheric pressure = 760 mmHg) the P_{O_2} of dry air is ~159 mmHg. Water vapor exerts a partial pressure proportional to relative humidity and temperature. The presence of water vapor in air slightly reduces the

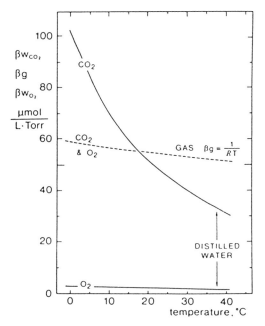

Figure 4. Capacitance coefficients for O_2 and CO_2 in distilled water and in gas as functions of temperature.

fractional concentration of the other gases. Thus, in fully water-saturated air at 25°C and a barometric pressure of 760 mmHg, the P_{O_2} falls to ~154 mmHg. Industrial pollution—specifically the burning of fossil fuels—is another factor acting to alter the composition of air. For example, molecular CO_2 is increasing at a rate of ~0.5% of its present composition per year.

Barometric pressure falls ~90 mmHg for every 1000 m increase in altitude, a phenomena termed the barometric slope. Consequently, the P_{O_2} of ambient dry air falls by ~20 mmHg for the same altitude increase. Recent data from mountain climbing expeditions indicates that the barometric slope underestimates total barometric pressure over land masses. The slightly higher than predicted barometric pressure at high altitude provides for a slightly higher P_{O_2}, an apparently critical factor in permitting humans to as-

cend Mt. Everest without compressed gases.

The changes in capacitance for O_2 and CO_2 as a function of partial pressure in air (or any other gas mixture) at 15°C are indicated in Figure 4. Note that for the same incremental increase in P_{O_2} or P_{CO_2} the concentration of either gas in a gas mixture increases linearly and at the same rate. The practical consequence of this is that, for equivalent partial pressure changes, air as a respiratory medium is equally effective as an O_2 source and CO_2 sink. This is readily evident from examination of capacitance coefficients in Table 2.

Other important features of air that influence its suitability as a respiratory medium include density, viscosity, and thermal characteristics. In absolute terms, air has both very low density and viscosity relative to water. (However, relative to water, the kinematic viscosity—viscosity/density—is very high, which takes on important overtones in locomotion. The low density and viscosity of air combined with air's high O_2 and CO_2 capacitance means that a relatively small amount of metabolic energy need be expended in generating an adequate convective flow of air over a respiratory surface. In fact, estimates of the *metabolic cost of ventilation* in air breathing animals indicate that only ~1–2% (humans) or 10–30% [turtles, *Pseudemys floridana* (93)] of metabolism is devoted to ventilating the respiratory structures, compared with up to 50% of metabolism in fish. However, the low density of air also means that terrestrial animals must exert energy in maintaining posture against the effects of gravity, a metabolic cost not evident in aquatic animals.

Air also acts as a relatively low capacity heat sink compared to water (Table 2). While endothermy has evolved to a very limited extent in aquatic vertebrates (see above), the low thermal capacitance of air has been a critical factor in the wide-spread evolution of endothermic terrestrial vertebrates.

A final consideration of air as a respiratory medium is that of water balance. With the exception of those comparatively few environments in which air ventilating the respiratory surfaces is fully saturated with water vapor, breathing air will result in a variable amount of loss of body water. The loss of water vapor from respiratory surfaces is usually small compared to water loss in feces and urine, but amounts to a significant component of an animal's overall water budget. A wide variety of structures and processes have evolved that minimize respiratory water loss and allow colonization of arid environments.

Mechanisms and Structures for Gas Exchange

Design Considerations

From an engineering perspective, respiratory structures must satisfy some relatively simple and straightforward design constraints.

1. A respiratory medium (air or water) must be delivered to the environmental side of a thin respiratory membrane. Ideally, body fluids simultaneously are conveyed past the internal side of this membrane. As will be seen below, a wide variety of animals subsist without either ventilating or perfusing specialized respiratory structures; however, this seriously limits body size and metabolic rate.

2. To allow rapid diffusion of O_2 and CO_2, the respiratory membranes should be as thin as possible while remaining structurally sound and providing a selective barrier to water and ion passage. The medium-to-blood diffusion distance varies greatly in specialized respiratory struc-

tures, ranging from 0.2 to 1.0 μm when considering most vertebrate lungs (183), the labyrinth organs of some air breathing fishes (80), and the modified branchial chambers of air breathing crabs (23). Tissue diffusion distances in the gills of aquatic fishes (78) can be as low as 0.5 μm in pelagic teleosts or exceed 10 μm in some elasmobranchs and in air breathing fishes with branchial specializations for air breathing. The gas diffusion distances in the gills of aquatic decapod crustaceans show a similar range to that of fishes (114). Animals that exchange respiratory gases partially or wholly through their integument must contend with substantially larger gas diffusion distances of up to 50 μm (47).

3. Most effective gas transport occurs when the quantity of respiratory medium and body fluids flowing past the respiratory membranes are carefully matched. The ventilation/perfusion ratio (also called the V/Q ratio, although more current physiological terminology favors $\dot{V}_{water}/\dot{V}_{blood}$ or $\dot{V}_{air}/\dot{V}_{blood}$) describes the ratio of water or air flow to blood flow in a respiratory organ. There is no single correct ratio that applies to all gas exchange organs, since what is being optimized is not the ratio of volumes of respiratory medium and blood, but rather the ratio of *gas capacitances* of these two components. Thus, for O_2 transfer, ventilation and perfusion tend to be regulated so that the O_2 capacitance of the air or water delivered to the respiratory membranes closely matches the O_2 capacitance of the blood perfusing the respiratory structures (154). In other words, it is energetically wasteful either to deliver more O_2 to the respiratory structure than can be taken up and transported to the underlying body tissues, or to circulate more blood through the respiratory structure than can possibly be oxygenated. Similar arguments apply for CO_2 elimination from blood.

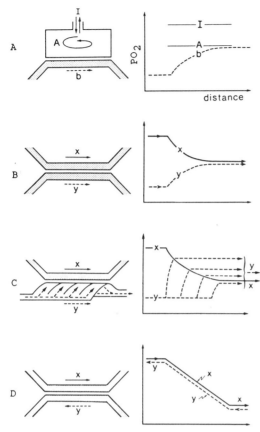

Figure 5. Models of the flow of blood and respiratory medium through gas exchange organs, and the exchange of O_2 achieved as a consequence. To the left of the diagram is a schematic representation of flows within the gas exchange organ. To the right is a diagram that relates the O_2 partial pressures in blood and respiratory medium as a function of passage through the gas exchange organ. (*A*) A ventilated pool arrangement. I, inspired gas; A, alveolar gas; b, blood. (*B*) A concurrent arrangement, in which the respiratory medium (x) flows in the same direction as the blood (y). (*C*) A cross-current arrangement, in which several parallel streams of blood (y) flow obliquely over the stream of respiratory medium (x). (*D*) A countercurrent arrangement, in which the respiratory medium (x) and blood (y) flow in opposite directions through the gas exchanger. (38).

Because it is the gas capacitances of respiratory medium and blood that tend to be optimally matched, and because gas capacitances are affected by the properties of the respiratory medium (air or water), temperature, and by the type and concentration of respiratory pigment (see Chapter 10), the optimal V/Q ratio varies greatly between species. Among air breathing animals, the V/Q ratio is approximately $1:1$ in mammals (187) and $2:1–5:1$ in chelonian reptiles (20). Because of the considerably lower O_2 capacitance of water compared to air (but not greatly dissimilar from the O_2 capacitance of the blood in water and air breathers), ventilation perfusion ratios are $9:1–35:1$ in fishes (139) and decapod crustaceans (114).

4. A final design consideration is the specific pattern of flow of water/air on one side of the respiratory membrane and blood on the other. Four different arrangements are depicted schematically in Figure 5. In a *uniform pool* arrangement, capillary blood flows against an evenly ventilated pool of respiratory medium in which the air or water does not flow in any particular direction immediately adjacent to the respiratory membranes (Fig. 5*A*). In such a system, capillary blood can come to gaseous equilibrium with the respiratory medium. In a concurrent arrangement, the respiratory medium and the capillary blood flow in the same direction on opposite sides of the respiratory membrane (Fig. 5*B*). In this arrangement, respiratory medium and blood have come to gaseous equilibrium at an intermediate partial pressure as they leave the gas exchange organ. In cross-current systems, capillary blood flows at an angle to the main flow of respiratory medium (Fig. 5*C*). This system is more efficient than a concurrent system, in that the partial pressure of O_2 in blood leaving the gas exchange organ can be higher

than the P_{O_2} of respiratory medium leaving the organ. Finally, in a cross-current arrangement (Fig. 5D), the flow of respiratory medium and capillary blood are in opposing directions, producing a nearly complete transfer of gases between medium and blood. In such a system, the blood leaving the gas exchange organ will be near gaseous equilibrium with the air or water entering it.

Diversity of Respiratory Structures

From an evolutionary perspective, the simple design requirements outlined above have been achieved in a variety of ways, in some instances with highly intricate respiratory structures dependent on complex physiological processes and controls, in other instances by surprisingly simple, multifunctional structures. Figure 6 depicts common arrangements of the respiratory surfaces in a variety of animals. The two most commonly studied gas exchange organs—gills and lungs—represent only a small proportion of the many respiratory structures that have evolved in complex metazoans.

Integumentary Gas Exchange in Water and Air

The respiratory challenges facing the first organisms that evolved in ancient marine environments were essentially no different from those currently faced by all single cell organisms (e.g., protozoa) and by aquatic metazoans lacking specialized respiratory surfaces and systems for internal convection of body fluids. Oxygen and CO_2 must be exchanged between environment and tissues entirely by diffusion. While virtually all cell surfaces are permeable to respiratory gases (even when those same cells have low water permeability), the diffusion of gases through aqueous pathways is very slow, as has already been discussed. Thus, the uptake of O_2 is normally limited by the

length of the diffusion path for O_2 (68). The maximum thickness consistent with typical metabolic rates for simple metazoans lacking circulatory systems is given by the equation:

$$\text{limiting thickness} = 8[O_2]\frac{K}{\dot{M}_{O_2}}$$

where $[O_2]$ is the O_2 concentration in the medium, K is the diffusion constant for O_2, and $\dot{M}_{O_2}$ is the rate of O_2 uptake. The slow rate of O_2 diffusion through tissue limits the tissue thickness to a maximum of ~1 mm—that is, metabolizing cells must lie within 1 mm of the respiratory medium. Much larger body sizes occur in some simple invertebrates lacking both respiratory and cardiovascular systems (e.g., coelenterates), in apparent contradiction to the above statements. However, in a jellyfish, for example, only ~1% of the body mass is organic material, the vast bulk being primarily a nonmetabolizing salt solution, and the cells with the highest metabolic rates are usually located close to the body surface.

Traditionally, analyses of integumentary gas exchange (or that across specialized surfaces) have regarded tissue lying between environment and site of metabolism as the primary, if not only, site that limits gas diffusion. Increasingly, however, the significance to gas exchange of a stagnant, unstirred boundary layer of respiratory medium immediately adjacent to the integument has been recognized (18, 47, 75). In fish gills, for example, the boundary layer of unstirred water about the secondary lamellae of fish gills may account for 80–90% of the total resistance to gas diffusion in the gills (75). The thickness of unstirred boundary layers depends on the kinematic viscosity and velocity of the respiratory medium flowing across the animal's integument, the diffusivity (D) of the respiratory gas, and the linear dimension of the animal.

animal body	ambient medium	examples and remarks
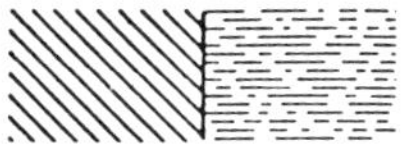 integuments	water	eventually the surrounding medium may be air
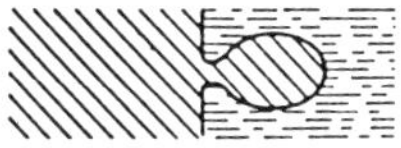 gill	water	crustaceans fishes molluscs tadpoles also-book gill of limulus -podia of urchins
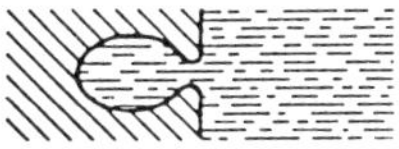 water lung	water	holothurians (sea cucumber) also-buccopharyngeal respiration -cloacal respiration
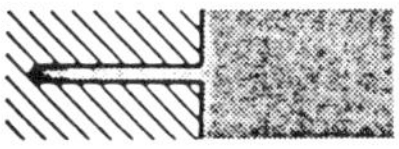 trachea	air	spiders insects also-book lung of scorpions -tracheal lung of spiders
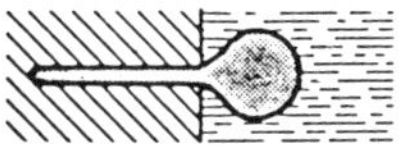 tracheal gill	water	aquatic larvae of insects *e.g.* mayflies
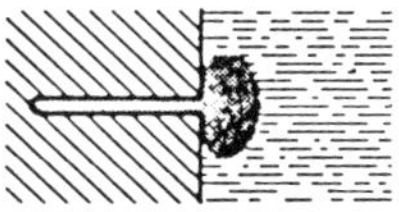 compressible gas gill	water	aquatic insects *e.g. Notonecta*
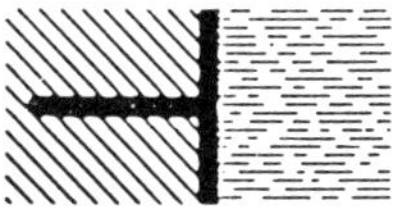 incompressible gas gill	water	aquatic insects *e.g Aphelocheirus* syn.: plastron
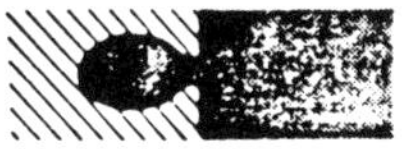 air lung	air	- vascularized cavities (land snail) - air chamber of some air breathing fish - simple air sac (sphenodon) - alveolar lungs (reptiles and mammals) - parabronchial lung of birds

Figure 6. A schematic, highly simplified representation of possible arrangements of gas exchange surfaces separating respiratory medium and animal tissue. (38).

All other conditions equal, the thickness of the diffusion boundary layer is inversely proportional to the square root of the velocity of the respiratory medium across the skin. In practical terms, (1) unstirred boundary layers in air are very thin and rarely influence integumentary gas exchange, (2) water boundary layers for O_2 diffusion frequently are limiting to gas exchange, (3) the larger the animal, the higher the velocity of the water current required to dissipate the boundary layer limiting O_2 diffusion, and (4) even in water, boundary layers have comparatively little influence upon CO_2 movement because of this gas' relatively high

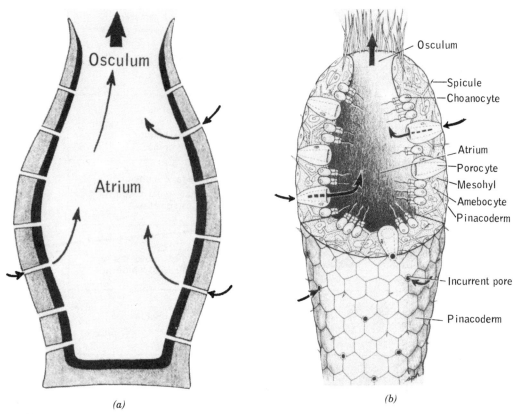

(a)   (b)

Figure 7. Ventilation of internal surfaces in asconoid sponges. (*A*) Schematic diagram of internal water flow. (*B*) A partially section asconoid sponge, showing details of cell types responsible for pore formation and water current generation. (4).

diffusion coefficient. Given these considerations of boundary layers, it seems that the evolution of large body size may have been favored in metazoans either located in moving water (e.g., in tidal currents) or capable of generating water flow over their body surfaces, either by locomotion or actual pumping structures.

Some Porifera, Coelenterata, many Platyhelminthes, Nematoda, Rotifera, a few Annelida, and Bryozoa depend strictly on diffusion across the integument for gas exchange. In some groups there may be channels through which water is circulated internally, greatly reducing the apparent diffusion distances. In sponges, for example, the outer body surface is perforated by numerous small pores (Pori-

fera = pore bearer) formed by specialized porocytes (Fig. 7). These pores connect with an interior, water-filled atrium (spongocoel). The internal surface of the atrium is lined with choanocytes, specialized cells each bearing a flagellum. Noncoordinated beating of these flagella (sponges lack a nervous system) generates a current of water flowing in through the pores to the atrium. The atrium also connects to the external environment by a single large osculum, through which water leaves the atrium. A leuconoid sponge [*Leuconia* (*Leucandra*)] 10-cm tall and 1 cm in diameter generates a water velocity of 8.5 cm/s through the osculum, pumping an impressive 22.5 L of water each day (144). From this copious water stream the sponge both exchanges O_2 and CO_2 and captures particulate food particles. The internal ventilation of sponges provides not only an excellent example of a rudimentary ventilatory system, but also a system in which feeding and respiration are inextricably linked.

In more complex metazoans, which lack specialized respiratory structures, integumentary gas exchange can be greatly facilitated by the presence of a circulatory system. When a hemolymph or blood is circulated through or just beneath an otherwise nonspecialized integument, the distance between environment and metabolizing cells across which gases must diffuse is greatly reduced, and represented by the sum of the distance between environment and the integumentary vasculature plus the distance between tissue blood spaces and the interior of the metabolizing cells. In essence, the evolution of a mechanism for the internal circulation of body fluids removes gas diffusion through the entire thickness of underlying tissues as the primary restriction to increasing body size. Thus, in many annelids (e.g., oligochaetes, hirundineans, and some polychaetes), sipunculids, echiuroids, and a few aquatic insects the highly vascular integument provides the principal route for O_2 and CO_2 exchange, and has allowed relatively large body size in the absence of specialized respiratory structures.

The evolution of the specialized respiratory structures depicted in Figure 6 does not necessarily diminish the importance of integumentary respiration. The integument of nearly every animal has measurable permeability to O_2 and CO_2, even when integumentary specializations have evolved for limiting the flow of heat or water, for locomotion or for camouflage (47). Among invertebrates, integumentary gas exchange is prevalent even in the relatively few phyla with highly specialized respiratory structures, often

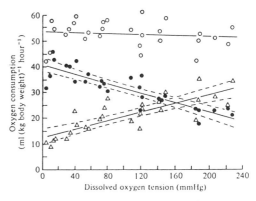

Figure 8. Oxygen consumption in the aquatic pulmonate snail *Lymnaea stagnalis* as a function of ambient P_{O_2}. Total O_2 uptake (open circles), pulmonary O_2 uptake (closed circles), and cutaneous O_2 uptake (triangles) are indicated, along with regression lines and confidence limits. (90).

accounting for 20–50% of total gas exchange (66). The aquatic pulmonate snail *Lymnaea stagnalis* depends heavily on O_2 uptake from the lung when water P_{O_2} is low, but at air saturation the majority of O_2 uptake occurs across the comparatively nonspecialized body wall (Fig. 8) (90). Gill ligation in burrowing polychaetes reduces O_2 consumption by only 50% (180). Similarly, when a gas impermeable covering is temporarily placed over the specialized respiratory surface (primarily the tube feet and respiratory papulae on the oral surface) of the asteroid *Asterias forbsei*, the aboral surface can still provide ~60% of total O_2 exchange (31).

Integumentary gas exchange in vertebrates, though generally perceived to occur primarily in amphibians, in fact is widespread in all classes (Fig. 9). Among the fishes, elasmobranchs depend relatively little on the skin for gas exchange, while some teleost fishes achieve up to one half of their O_2 uptake across the skin (47). The gross morphology of the skin appears to have little influence on the extent of integumentary gas exchange, with

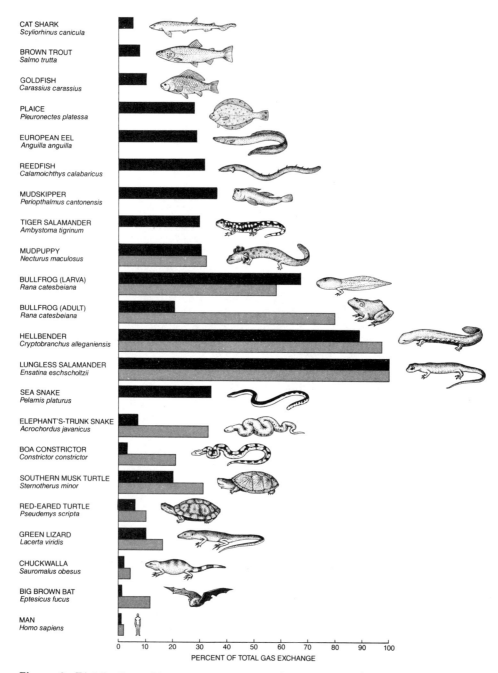

Figure 9. Distribution of integumentary gas exchange in vertebrates. Carbon dioxide elimination is indicated by the hatched bars, while O_2 uptake is indicated by the solid bars. (48).

the skin accounting for ~30% of the total O_2 uptake in both the heavily scaled reedfish *Erpetoichthys* (151) and the scaleless plaice (166). The skin of some fishes is the site of considerable active transport associated with hydromineral regulation. Reflecting the high metabolic rate of the skin under these circumstances, the O_2 consumption of the skin itself is equal to or exceeds O_2 uptake across the skin in a diverse group of fish including eel, rainbow trout, tench, crucian carp, yellow perch, northern pike, brook trout, brown trout, butterfish, cod, and rockling (123, 124, 94). Thus, while the skin of fishes can serve as a gas exchange organ, in many species it apparently does not provide an important route for the exchange of O_2 and CO_2 from underlying tissues.

Virtually all amphibians, even those with extensively developed gills and/or lungs, depend to a greater or lesser extent on integumentary gas exchange (Fig. 9). Within single amphibian species the relative magnitude of integumentary exchange varies greatly with temperature (19, 175), ambient gas partial pressures (175), presence of boundary layers (18), activity levels (133), developmental state (28) and body size (174). Salamanders of the genera *Chioglossa*, *Salamandrina*, *Rhyacotrition*, and the entire family Plethodontidae lack gills or lungs as adults—all O_2 and CO_2 exchange occurs across the highly vascularized skin. Testifying to the effectiveness of integumentary gas exchange, plethodontids can reach relatively large sizes, occupy very warm tropical habitats (thus experiencing the attendant increase in metabolism) and have a high metabolic scope (48). In other amphibians using integumentary exchange to supplement pulmonary gas exchange, specializations of the skin have been developed to increase its surface area and thus facilitate integumentary gas exchange (47). The hairy frog *Astylosternus robustus* has vascularized papillae covering significant amounts of its skin,

while extra skin folds are evident in the mountain frogs *Batrachophrynus* and *Telmatobius*, and the salamander *Cryptobranchus*. Several urodeles and the larvae of anurans develop expanded tail fins, which increase skin surface area. At least in anuran larvae, the skin thickness and in particular the amount of tissue lying between skin capillaries and ambient water can become reduced within a few weeks of exposure to hypoxic water, presumably facilitating cutaneous gas exchange (24).

Many reptiles depend on the skin for gas exchange (Fig. 9). As for fishes, the gross morphology of the skin (e.g., thickness of scales) appears to have little direct bearing on the extent of integumentary gas exchange (47). Even species that have developed integumentary specializations for limiting water flow across the skin (104, 160) may still show surprisingly large gas fluxes across the skin.

The comparatively high metabolic rate of endotherms (see the section on Homeothermy, Poikilothermy) demands a rapid and large exchange of O_2 and CO_2. Integumentary gas exchange generally represents a relatively insignificant component of total gas exchange (Fig. 9). However, flying bats eliminate ~12% of metabolically produced CO_2 across the surface of their wings (70). While <2% of the total O_2 and CO_2 exchange of humans occurs across the skin, the skin is thought to provide all of the O_2 it requires and actually eliminate not only its own CO_2, but an undetermined proportion of CO_2 produced in underlying tissues (50).

Aquatic Gas Exchange—Gills

Expansion of the body surface for increased diffusive gas exchange occurred early in the evolution of motile aquatic animals, with the result that higher levels of metabolism became possible. As the remaining variable in the Fick equation (see, p. 355), expansion of respiratory surface area has resulted in a variety of means for solving the dual problems of maintaining structural integrity and limiting the thickness of gas-exchange surfaces. One means developed was the evagination of integumentary surfaces to form gills (Fig. 5). The morphology of gills can be as simple as outward expansions of the body wall. For example, the dermal papillae of echinoderms provide thin surfaces of large total area for gas exchange between coelomic fluids and seawater (38). Gills may be as complicated as the lamellar gills of fishes where the internal flow of blood is largely counter to the external water flow, thus assuring nearly complete oxygen saturation of the blood. Gills may be ciliated or be waved about by muscles as ways to generate surface flow. Gills are often enclosed within external body cavities, or may be located directly on the body surface in animals that are burrow dwellers, so that they benefit from respiratory currents generated by auxiliary ventilatory structures and by locomotory and feeding activities.

Functionally, gills can be rated in several ways relative to O_2 taken up from the respiratory stream. Most directly, the utilization coefficient, E_w, or the gill extraction efficiency is given as the percentage O_2 withdrawal from water flowing through the gill chambers. It can be calculated also as a variant form of the Fick equation as:

$$E_w = \frac{Ci_{O_2} - Co_{O_2}}{Ci_{O_2}} \, 100$$

when the incurrent and excurrent O_2 concentrations are known (Ci_{O_2} and Co_{O_2}, respectively). The utilization coefficient is regulated strongly by amplitude changes in gill ventilatory flow and blood flow, especially in those animals that use the incurrent water flow for both respiration and for feeding as do lamellibranch mollusks. Additionally, the apparent utilization is modulated by the capacitance coefficient β of body fluids containing respiratory pigments because of their effect

in maintaining steep gradients for O_2 diffusion. Consequently, the transfer factor is a more meaningful estimate.

$$\text{transfer factor} = \frac{\dot{M}_{O_2}}{\dot{V}b\alpha_{O_2} \, (Pi_{O_2} - Pv_{O_2})} \, 100$$

Transfer coefficients can be calculated from the partial pressure of O_2 in incurrent water (Pi_{O_2}) and in venous blood entering gills (Pv_{O_2}), when $\dot{V}b$ = the volume of blood passing through the gills in a given time and, αO_2 = the solubility of oxygen in the blood. Maximum uptake occurs if blood leaving the gills has the same P_{O_2} as the inspired water.

1. *Invertebrate Gills*. Polychaetes have gill-like structures such as parapodia (nereids), gills (*Arenicola*), and branchial tufts (terebellids and most sabellids), which supplement integumentary oxygen uptake. The large marine worm *Arenicola* lives in a U-shaped tube and withdraws 30–35% of the O_2 over a wide range of P_{O_2} values (181). Intermittent ventilatory movements resembling peristalsis pass along the body wall, pumping water through the tube in a pattern dependent on the concentration of dissolved gases. During periods when oxygen is low, the pumping tends to shift from intermittent to continuous (107). Utilization by terebellid worms is higher (50–60%) when pumping is only for respiration than when feeding is also occurring (5–10% O_2 utilization). An active tube dwelling polychaete can remove 60% of the dissolved O_2 in the tube; an inactive one removes only 14%. When the tide level drops below the tube openings, the burrow O_2 drops to 10% of the initial concentration in ~1 h; after such a drop the worm switches to anaerobiosis (107).

Although sparse in some, the retractile dermal papillae of asteroids and echinoids shown in Figure 10 bring the slowly moving coelomic fluid into close association with the cilia-stirred, external water.

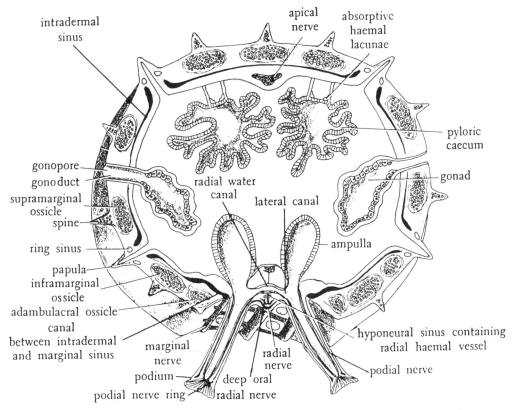

Figure 10. Schematized section through an asteroid ray. Dermal papulae or papillae act with the tube feet to increase the respiratory surface area for gas exchange. (116).

All echinoderms also have tube feet (podia), which function in a similar way, and like the thin-walled papillae, are lined with a ciliated epithelium (Fig. 10). They receive additional ventilatory flow through circulation of ambulacral fluid as the podial ampullae expand and contract. Gas exchange between the radial canal system and other coelomic spaces which are microcirculatory units must occur by ordinary diffusion or with fluid movement driven by ciliary action (167). Some holothurians, such as *Cucumaria*, have coelomic respiratory trees, which open at the cloaca and account for 60% of the total O_2 uptake. As dissolved O_2 diminishes, the rate of cloacal pumping first increases, but below 60–70% air saturation of the respiratory stream, pumping stops and

the animal becomes inactive (121).

Most molluscs have gills called ctenidia. These consist of broad, flattened filaments that are extensively ciliated marginally and laterally (Fig. 11). Pelecypod gills often are more complex and derived from the protobranch form as stacks of broad, thin filamentous plates on a stiffened primary filament much like the gills of decapods and fish. They may be subdivided (Filibranchia and Eulamellibranchia) as narrow, secondary filaments with tissue bridges forming a multilayered, netlike structure of huge surface areas as in those pelecypods whose complex ctenidia have the dual functions of filter feeding and respiration.

In all gill breathers among the molluscs, water flow over and through the

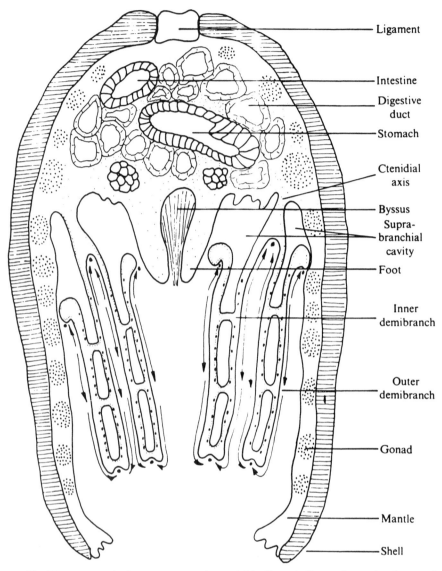

Ligament

Intestine

Digestive
duct

Stomach

Ctenidial
axis

Byssus

Supra-
branchial
cavity

Foot

Inner
demibranch

Outer
demibranch

Gonad

Mantle

Shell

Figure 11. Diagrammatic transverse section of *Mytilus edulis* to show the form of the demibranch gill pairs and the current directions (arrows) generated by the gill-surface cilia. (7).

gills is powered by ciliary action and controlled in volume output by water P_{O_2}. In lamellibranchs, the $\dot{V}_w$ is dominated by "appetite," but is responsive to water P_{O_2} as well. For example, a medium sized *Mytilus californianus* (1 g of dry flesh wt) takes up O_2 at 13°C nearly 5 times faster when pumping water at 3–4 L/h than when pumping at less than 1 L/h (7).

Thus when filter feeders are taking food from the respiratory stream, the $\dot{V}_w$ soars, as is reflected in the increased ventilatory metabolism. But at the same time, utilization coefficients generally fall to about one fifth of that found in nonfilter feeding animals (38, 7). Below ~80 mmHg, another regulatory effect of P_{O_2} becomes marked in *Mytilus edulis*, which is seen as a drop in $\dot{V}_w$, but with a concomitant rise in E_w. Thus when the P_{O_2} of the incurrent respiratory stream drops, a compensatory drop in the respiratory flow ($\dot{V}_w$) occurs that reduces the pumping cost, but sustains oxygen uptake. This relationship has been formulated by Bayne (7) as:

$$\frac{\dot{V}_w}{\dot{V}_{O_2}} E_w Ci_{O_2} = 1$$

Thus the response of *Mytilus edulis* to hypoxia is to increase utilization, by reducing ventilatory flow or metabolism, or both. When starving, the ventilatory flow of bivalves rises with a consequent rise in metabolism to support the additional work of ventilation, but with a sacrifice in utilization efficiency. Filtration rates are also strongly influenced by temperature, but rates acclimate quite rapidly in *Mytilus edulis*. Animals transferred to 5°C have filtration rates of only one third of those transferred to 15°C, but by the 14th day rates are about the same, and are very nearly equal on the 28th day of acclimation to the initial rates of the animals held at 15°C (7).

Utilization rises sharply—greater than 20% in *Mercenaria*—following long periods of anaerobiosis when pumping has stopped. Such a response might be expected to result from tidal exposure. With reinitiation of pumping, utilization rises and remains elevated at levels of 5–10% after several hours. Many bivalves show periodicity in the opening and closing of their shells, and in anoxia the valves remain shut (in *Mytilus*, *Mercenaria*, and *Ostrea*, but not in *Cardium* or *Pecten*). Oxygen withdrawal by several marine gastropods is higher (40–80%) than it is by pelecypods. In cephalopods (*Octopus*), as in pelecypods, exposure to low O_2 increases pumping. Unlike most molluscs, *Aplysia* and the cephalopod, *Octopus* are also sensitive to increased CO_2 levels. In *Aplysia* the frequency of respiratory pumping in-

creases with hypercapnia or with a fall in water pH, but not in response to bubbling air, O_2 or N_2 through the bathing seawater. Because the sensitivity to low pH values resides in the mantle of *Aplysia*, the chemoreceptive osphradium is the likely receptor organ for regulating gill ventilation (34). Oxygen withdrawal by *Octopus*, like that of other nonfilter feeding, aquatic molluscs is high and may reach 80%.

Aquatic crustaceans have gills that are usually ventilated by paddlelike movements of special appendages such as scaphognathites (Fig. 12, Sc). These are derived as flattened exopodites of the second maxilliped. Crabs that live in the low intertidal zone have more gills than do beach dwellers; active aquatic crabs have larger total gill areas than intertidal crabs, and these have larger gill areas than land crabs. Some land crabs not only have reduced gills but accessory respiratory tufts that project from the branchial epithelium that only need to be moistened occasionally to permit oxygen uptake from air. Table 3 lists a number of decapod species in order of decreasing dependence on gill breathing in seawater along with their habits and comments about modifications for aerial respiration. When compared to similar data for fish species given also in Table 3, crustacean gills seem to have about the same surface area and operational efficiency per unit of body weight (114, 145). In *Carcinus*, water flows in the gills in a direction opposite to blood flow and can be shunted along the gill edges or past the lamellae; counterflow is not so evident in crayfish filamentous gills, yet extraction efficiency is not notably less. Rhythmic sweeping of the scaphognathites can generate branchial pressures of up to 45 mmH$_2$O in large grapsoid crabs like *Cancer*, but pressures required to maintain an adequate respiratory flow, ordinarily do not exceed 10–15 mmH$_2$O. Many crustaceans periodically reverse branchial water flow. In some, this may

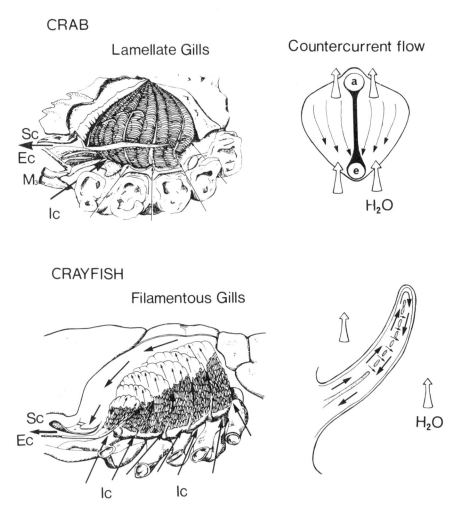

Figure 12. Water and hemolymph flow patterns across gills of a crab and a crayfish. In both diagrams, the overlying carapace has been removed on one side to show waterflow during forward pumping by scaphognathites. Ec, excurrent channel; Ic, incurrent channels; M3, third maxilliped; Sc, scaphognathite blade. Figures at right diagram water flow (open arrows) and hemolymph flow (closed arrows). (114).

be done to clear interlamellar passages or the gills of particulates with a kind of "cough" when supra-ambient pressures may develop (114). Periodic or one-sided pumping seems to be usual in resting crustaceans and it is likely one of several accommodation mechanisms that match volume flow to locomotory activity level (114). At low P_{O_2} levels, the volume of water pumped increases and utilization may rise to 40%. In the lobster *Homarus*, as the water P_{O_2} dropped from $\sim$115–$\sim$50 mmHg, the rate of pumping remained constant (9.5 L/h), but withdrawal increased from 31 to 55% (171).

Caution must be exercised in estimating the average scaphognathite stroke volume (S_v) from the measurement of ventilation volume multiplied by mean saphognathite frequency in the usual way. Sizeable variations in S_v can occur that may relate to changes in branchial

TABLE 3. Habitats and Gill Surface Areas of Various Crabs and Fish (145)

Species (Family, and Common Name)	Gill Area (cm^2/g)	Habits	Respiratory Modifications
		Crabs (Decapoda)	
Callinectes sapidus (Portunidae) blue crab	13.67	Aquatic burrowing, open sea, estuarine, very active	
Libinia emarginata (Inchidae) spider crab	5.66	Aquatic, benthic open sea, sluggish	
Panopeus herbstii (Xanthidae) mud crab	8.74	Amphibious, burrowing, oyster beds, crevices, moderately active	
Uca pugilator (Ocypodidae) sand fiddler crab	6.24	Amphibious, burrowing, sandy beaches, very active	Branchial troughs for water retention; branchial wall vascularization
Uca minax (Ocypodidae) fiddler crab	5.13	Amphibious, burrowing, salt marsh banks, very active	Branchial troughs for water retention; branchial wall vascularization
Ocypode albicans (Ocypodidae) ghost crab	3.25	Terrestrial, burrows in dunes, night, active	Spongy masses for water retention in branchial wall; branchial vascularization
		Fishes	
Euthynnus pelamis (Scombridae) skipjack tuna	13.20	Aquatic, open ocean very active, pelagic	Ram ventilator and obligate swimmer
Coryphaena sp. (Corpyhaenidae) dolphin	3.70	Aquatic, open ocean, very active, pelagic	
Stenotomus crysos (Sparidae) scup	5.06	Aquatic, midwater, open ocean, active	Ram ventilator
Pleuronectes platessa (Pleuronectidae) plaice	4.33	Aquatic, benthic, sandy bottom, active	
Callionomus lyra (Callionymidae) dragonet	2.92	Aquatic, benthic, open ocean, moderately active	Opercular tube, directs excurrent respiratory stream upward
Opsanus tau (Batrachoididae) oyster toadfish	2.14	Aquatic, benthic in oyster beds, sluggish	Withstands low P_{O_2} habitats at low tide exposure, respiratory conformer
Anabas testudineus (Anabantidae) climbing perch	1.50	Amphibious, wet season migratory, moderately active	Air labyrinth, gulps air CO_2 eliminated via gills

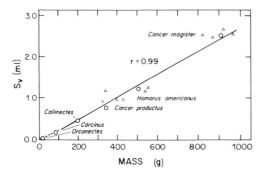

Figure 13. Scaling of scaphognathite stroke volume (Sv) and body mass for *Cancer magister* (open triangles) and five other species (open circles). The regression calculated for the data is, Sv = 2.94B − 0.0756. (114).

flow resistance. Yet, as Figure 13 shows, S_v does bear a consistent correlation with body mass in a variety of decapods under normoxic conditions (114).

Stimulation of ventilation during hypoxia has been observed in a number of decapods. When available O_2 dropped from above 140 to 50 mmHg; individual crayfish, *Astacus,* tripled their breathing rates and doubled ventilation volumes. In crayfish, as in lobsters, O_2 consumption decreases linearly with P_{O_2}, but the utilization percent shows only a slight drop. While pumping in crayfish increases with hypoxia, it declines with hypercapnia; heart rates also decrease with hypoxia. Somehow the lack of O_2 in the respiratory stream of crayfish activates receptors, probably located in the circulatory system, to bring about bradycardia and increased ventilation. Application of CO_2 to periopods or at gill apertures elicits bradycardia in crayfish (1).

External oxygen receptors localized in the cuticle between the coxae of the walking legs and within the lamellae of the book gills have been identified in *Limulus*. The intercoxal units are inhibited in hypoxia while three groups of differing sensitivities are found in the gills; those inhibited by hypoxia, some that are ex-

cited by increased P_{O_2}, and some mechanoreceptors whose tactile sensitivity declines with hypoxia. Proportional dependence in beat frequency and amplitude of the book gill movements in *Limulus* has been found; ventilation ceases completely in anoxic seawater (193).

Numerous aquatic insects have gills through which O_2 passes directly into the profusely branched tracheae. Tracheal gills occur in nymphs of Odonata, Plecoptera, Tricoptera, and Ephmeridae (see also the section on the Tracheal System of Arthropods). Dependence of aquatic Hemiptera on tracheal gills is shown as follows:

	O_2 Uptake with Gills (ml of O_2/kgh)	O_2 Uptake without Gills (ml of O_2/kgh)
Naucoris	241	24
Notonecta	225	29
Corixa	305	47

2. *Vertebrate Gills.* Vertebrate gills originate as six or more lateral pharyngeal evaginations of the branchial endoderm between visceral arches opening to the outside via pores or gill slits. Each arch contains branchial vessels that conduct blood from the ventral to the dorsal arteries. Different degrees of developmental loss of whole or parts of the visceral arches and branchial vessels has occurred. Often these changes can be related to adoption of different feeding mechanisms, and to cardiovascular changes that very likely have occurred with evolution of vertebrates onto land. Some changes have involved the reworking of respiratory and circulatory systems to accommodate various modes of breathing air. Development of branchial filaments (primary lamellae), with either extensive secondary branching or the out-

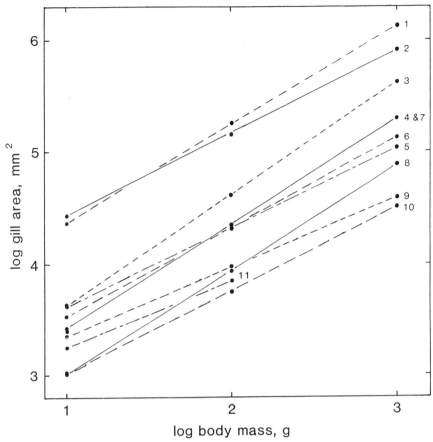

Figure 14. Scaling of the logarithm of gill surface area and body mass of sluggish to active fishes at body weights of 10, 100, and 1000 g. Species represented are: (1) *Thynnus albacres, Thynnus thynnus*, (2) *Coryphaena hippurus*, (3) *Scomber scomber*, (4) *Salmo gairdneri*, (5) *Tinca tinca*, (6) *Opsanus tau*, (7) *Scyliorhinus canicula*, (8) *Torpedo marmorata*, (9) *Anabas testudineus*, (10) *Saccobranchus fossilis*, (11) *Channa punctata*. (80).

ward growth of closely spaced thin secondary lamellae on the filaments occurred early in the evolution of the aquatic classes. This sort of branching and re-branching—to form the secondary lamellae that collectively make up the enormous respiratory exchange surface of the gills—is a familiar means for surface area expansion in animals for various purposes. It is seen again with the expansion of lung surface in amphibious and terrestrial vertebrates (73). In teleosts a reasonable average of gill area is 10 cm^2/g body weight (80) (Fig. 14). Figure 15 illustrates

the functional morphology of these features as seen in many teleosts (73).

Scanning electron micrographs of the surface of secondary lamellae on gill filaments show an extensive surface area expansion resulting in the formation of micro-ridges (Fig. 16) (101, 126). The form of the micro-surface has not yet been correlated with O_2 transfer other than by diffusion, although micro-folding, which may greatly increase the respiratory surface area, is often found on epithelial surfaces of gas-exchange organs (80).

In fishes, variations on a basic pattern

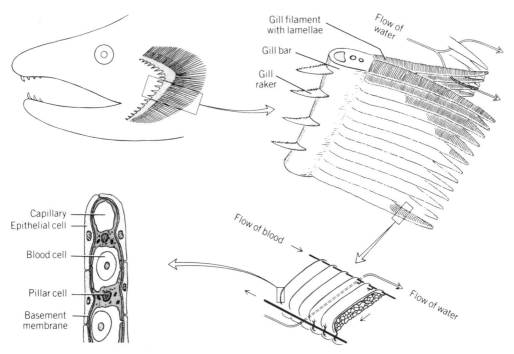

Figure 15. A fish gill bar, constituent gill filaments, and filament lamellae showing the structural basis for the large area, and the thin water-to-blood, counterflow pattern of the respiratory stream and the blood. (73).

of respiration are adaptive for different levels of activity and responsiveness to environmental factors. In very active fish, the total gill area is larger than in more sluggish fish (Table 3, Fig. 14); this same relationship applies to decapod crustaceans (Table 3) (145). Fast-swimming, marine fish are known to keep the mouth open continually and thus provide continuous water flow over the gills, powered by the pressure generated at the mouth by forward motion. Because drag and inertial losses related to the "braking" effects of cyclic opening and closing of the mouth and opercula are eliminated in open mouth swimmers, it is likely that the transition to ram gill ventilation, that is, ventilation via the pressure flow induced at the mouth by swimming, allows acceleration to somewhat higher swimming speeds without an increase in the cost of transport (Fig. 17) (51, 146).

In general, the gills of fishes lie between two pumps, (1) a buccal or force pump powered during mouth closing by the adductor mandibulae complex and several other sets of mandibular adductors, and (2) a branchial aspiration pump powered by opercular abductor muscles (Fig. 18) (3). As the mouth of most fishes begins to open, a small negative pressure develops first in the buccal or prebranchial chamber (flow reversal possible) that is soon followed by opercular abduction and a drop in postbranchial pressure. The postbranchial pressure in most pelagic and mid-water fishes is thought to drop more and faster than the mouth pressure. Thus when the cyclic pattern reaches the reversal phase, with mouth closure followed by opercular adduction, the respiratory flow continues because a pre- to postbranchial pressure gradient is maintained (82). When the differential

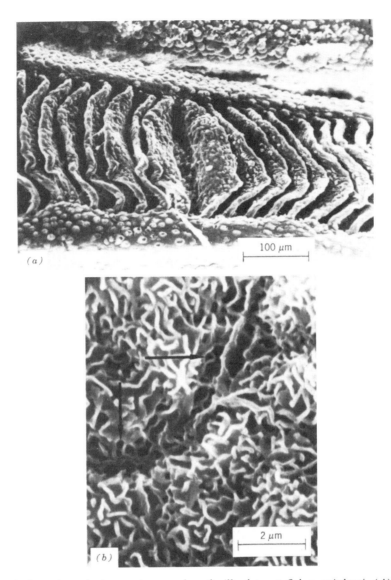

100 μm

(a)

2 μm

(b)

Figure 16. Scanning electron micrographs of gill of trout *Salmo gairdneri*. (A) Several secondary lamellae and body of a filament which is covered by mucus secreting cells. (B) Section of surface of a secondary lamellae. Epithelial cell surface consists of extensions of the cell as microridges. Arrows indicate regions that appear as pores. (Courtesy of K. R. Olson and P. O. Fromm.)

pressure across the gills is measured experimentally in a tethered rainbow trout, an apparent flow-reversal phase can be detected, but its duration usually is <10% of a respiratory cycle, thus attesting to the ability of linked buccal-op-

ercular, cyclic pumps to sustain almost continuous respiratory flow over the gills of fish (Fig. 19) (82).

Apneic and hyperpneic periods that occur with either rhythmic or arrhythmic patterns are common among bony fishes

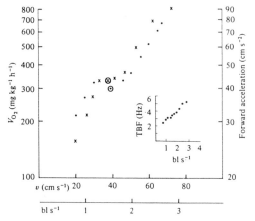

Figure 17. Relationship between O_2 consumption (●), forward acceleration amplitude (x), swimming speed, and tailbeat frequency (inset) of a striped bass, *Morone saxatilis*, at 15°C over the transition velocity between active branchial and ram-gill ventilation. Circles around data points at 38 cm/s represent the transition velocity for adoption of ram ventilation. (51).

that breathe only water or are bimodal in shifting between air and water as the respiratory medium. Irregular, short bouts of apnea occur during feeding, while long apneic periods are routine for epipelagic fishes utilizing ram ventilation during cruise swimming. These variant breathing patterns serve maintenance needs according to the niche of a species. Sculpins and angler fishes spend most of their time resting on the bottom. They are opportunistic suction feeders, usually active only during prey capture. Their respiratory rates are slower than their heart rates in marked contrast to mid-water and pelagic species in which these rates are about equal at low activity levels. Very likely, slow respiratory cycles, which are typical of demersal species with large-volume, elaborate branchiostegal baskets, are energetically efficient because respiratory cycles are mostly spent slowly aspirating water through the gills. The water ejection or opercular return phase in most cases is short (148).

In resting elasmobranchs, ventilation is by rhythmic contraction of the branchial muscles and elastic recoil of the skeleton. Swimming movements may be coupled with breathing in dogfish when they are active (153). Whether or not this coupling between cyclic gill ventilation and tail beat frequency is common among fishes is not known. On the other hand, cardiorespiratory synchrony has been reported to be routine in dogfish and probably is reflexly mediated (153). It also occurs as a transient phenomenon in bony fish, but one that is reliably detected only during hypoxia (tench, 79, 143).

Because ventilatory flow across gills is unidirectional in contrast to the tidal flow of most lungs, the E_w or extraction efficiency generally exceeds 10% in fishes in well-aerated waters while transfer effec-

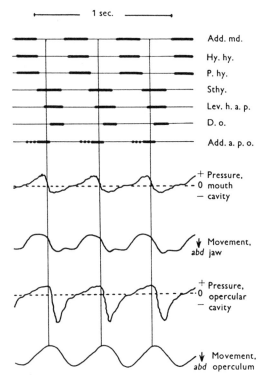

Figure 18. Relation between activity in several muscles and the movements and pressures of the respiratory pumps in trout. (3).

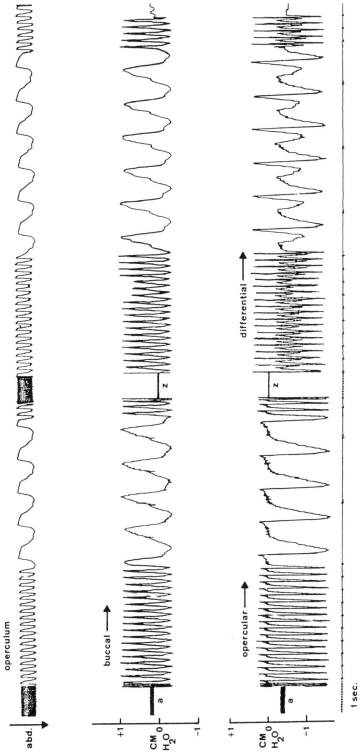

Figure 19. Simultaneous recordings of opercular movements and pressure changes in the branchial cavities of a rainbow trout. The differential pressures wave forms shown on right are inverted (+ for −) relative to buccal and opercular pressure scales at the left. The convention adopted is for the differential pressure to be positive when the buccal pressure exceeds that in the opercular cavity. Symbols are, (a) average pressure; (z) bath zero pressure. (82).

379

tiveness approaches 100% (142). Yet in most fish, water flow over the lamellar surface of gills is laminar so that 80 to 90% of the resistance to O_2 transfer from water to blood is caused by the slow rate of diffusion of O_2 in this thin, unstirred layer surrounding the gill lamellae (80). A high transfer effectiveness is an indicator that blood flow over gills mainly is countercurrent to water flow because the mean arterial oxygen, P_{aO_2}, often is a bit less than the mean O_2 in inflow and outflowing water but greater than that of the excurrent stream, as has been demonstrated in dogfish and trout (75, 132). There are vascular shunts in the margins and middle of fish gills that permit changes in flow pattern with need (see also Chapter 11). In bimodal breathers, posterior branchial arches may be so modified that branchial vessels pass directly to the dorsal aorta as shunts (101).

Nomograms for trout, carp, and dogfish show the relation between arterial oxygen (P_{aO_2}), oxygen consumption (M_{O_2}), and both water and blood flow (83); the P_{aO_2} increases as the ratio of ventilation volume to cardiac output increases (Fig. 20) (83). When the P_{O_2} in water decreases, the ventilatory volume increases markedly with a corresponding slight decrease in breathing rate, stroke volume of the heart increases considerably, also with a decrease in heart rate. When P_{CO_2} increases in the branchial flow of some fishes, the respiratory amplitude increases, the respiratory frequency decreases, and the heart rate increases in much the same way as the response to hypoxia. In about one third of fishes tested, hypercapnia exerts only a slight excitatory effect while in another one third no effects related to hypercapnia or to a drop in pH is detectable. The control by P_{O_2} is much more important than that by P_{CO_2} and pH. Respiratory and circulatory responses combine to maintain relatively constant oxygen consumption

over a wide range in concentration of dissolved O_2 (139, 142).

In hypoxia the effectiveness of O_2 uptake is reduced. In many fish, O_2 consumption increases as P_{O_2} drops, because of the greater ventilatory effort; hence, a double hazard is created. If temperature is raised, ventilation volume increases first and rate later; but at a critical temperature, heart rate slows with inadequate oxygenation and the two respiratory pumps may uncouple (82).

Respiratory movements originate with output from motor nuclei that passes from the medulla oblongata mainly via the trigeminal (V) and facial (VII) cranial nerves. The activity of these motor cells is triggered by networks of interneurons of a central pattern generator that are seen to fire rhythmically in phase with buccal and opercular movements (2, 148). The location of O_2 and CO_2 receptors is uncertain. Probably there are both peripheral and central receptors. Activity of receptors sensitive to O_2 level have been detected in teased nerve branches in the isolated first gill arches of yellowfin tunas; these enter the CNS via the tenth nerve. Their activity showed an increasing discharge frequency with decreasing P_{O_2} in a manner similar to the activity of mammalian carotid and aortic bodies (118).

In a choice gradient, salmonid and centrarchid fishes avoid low O_2 concentrations; chinook salmon avoided 4.5 mg/L. These behavioral regulations suggest a sensitivity greater than is needed for respiratory control and avoidance of concentrations well above hypoxic levels (189).

3. Gills and Bimodal Respiration. The external gills of a few fish and larval amphibians, unlike those of most fish, are ectodermal in origin and presumably are later invaded by lateral projections of the arch arteries. They may actually appear to be internal as in anuran tadpoles where they are located under a fleshy operculum. There is reason to believe that ex-

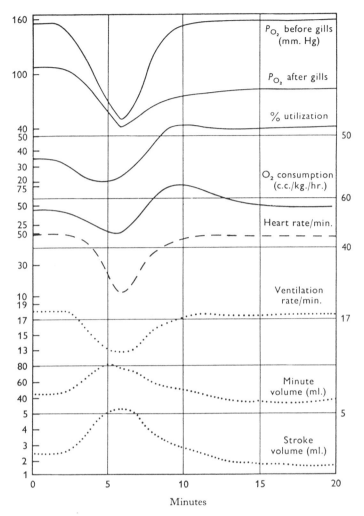

Figure 20. Summary of effect of lowering P_{O_2} of water for 6 min, as shown by values of P_{O_2} before the gills. (83).

ternal gills are antecedent to pharyngeal gills; in embryonic teleosts they develop before the pharyngeal gill slits are formed (73). Larvae of the frog, *Rana*, have four gills consisting of highly branched filaments containing blood channels (24). As is characteristic of most amphibians, larval gas exchange is bimodal. Strictly aquatic, early larval *Rana* (stages IV–V) at 20°C, gain ~41% of their total M_{O_2} via gill respiration and the balance by exchange through the skin surface. Air breathing tadpoles with functional lungs (stages XVI–XIX) gain only ~17% of their total O_2 via gills, the rest entering the skin and lungs. However, CO_2 exchange, as is usual for most amphibian adults in water, occurs largely via cutaneous diffusion (gills and skin) directly into the water. As tadpoles mature and the gills are lost, the pathway for CO_2 excretion shifts mostly to the skin, <20% being lost via lung exhalation (28). Thus, tadpole metamorphosis illustrates well the supposition that efficient CO_2 exchange with lungs was an evolutionary difficulty in the transition from aquatic to aerial respiration—probably only resolved with development of the bicarbonate buffering system as in present-day reptiles (136).

Air Breathing

THE TRACHEAL SYSTEM OF ARTHROPODS. In insects, some spiders, onchyphorans, centipedes, and millipedes gas exchange between metabolizing tissues and the environment occurs directly via a microscopic network of arborizing gas-filled tubes termed the tracheal system (Fig. 21). The circulation of body fluids is not directly involved in gas exchange, distinguishing these invertebrates from other animals with well-developed cardiovascular systems. Great variation is evident among the air-breathing arthropods with tracheal systems (hardly surprising given the nearly 1 million animal species with tracheal systems), and the following is necessarily a general description (4, 106, 122, 170).

In their more primitive arrangement (e.g., in wingless insect), tracheae open to the environment as simple holes in the cuticle. In most insects, however, the tracheal openings are located in a small pit and are guarded by a spiracle. The spiracles are muscular gates under direct neural control, and act to regulate the flow of gas through the tracheal system (122). In most species, there are three pairs of spiracles above the legs and on the first seven or eight abdominal segments (Fig. 21B). The tracheae originating from spiracles usually interconnect to form a pair of longitudinal trunks with numerous cross connections, particularly in the thorax. The tracheae soon divide into progressively finer branches, and terminate at tracheal cells, which embryonically give rise to several tracheoles. The

tracheoles, terminating in sometimes fluid-filled branches as small as 0.2 μm in diameter, represent the smallest functional unit of the tracheal system, and are the sites of O_2 and CO_2 exchange.

The tracheal system forms a vast branching network permeating all tissues (Fig. 21B). Generally, gas within the tracheal system comes within 10 μm to mitochondria within metabolizing cells (117A). Interestingly, individual flight muscle fibers, which have among the highest metabolic rates of any animal tissue (170), may even be penetrated by branches of a single tracheole, potentially delivering gas to within one tenth of a micrometer of individual mitochondria. The volume of the tracheal system is surprisingly large, occupying 41 and 48% of the total body volume of female and male cicada (*Fidicina mannifera*). Approximately one-half of this volume in cicadas is accounted for by large abdominal air sacs.

In relatively sedentary insects or insects smaller than ~1 g, diffusion of O_2 and CO_2 along partial pressure gradients within the tracheal system generally appears adequate for gas exchange. In larger, more active insects, however, diffusion is insufficient. In many large flying insects, an unidirectional flow of gas through the tracheal system is generated by rhythmic longitudinal telescoping of the abdomen, raising and lowering of the abdominal sterna, and by protraction and retraction of the head and prothorax. Such movements are clearly evident in resting hymenopterans (especially wasps), particularly when high body temperatures result in elevated metabolic rates. During flight, unidirectional ventilation of the tracheal system is further facilitated by the rhythmic distortion and compression of the thorax associated with the action of the flight muscles (184). In the desert locust, abdominal pumping generating air flow through the tracheal system increases from 0.7 ml/g · min up

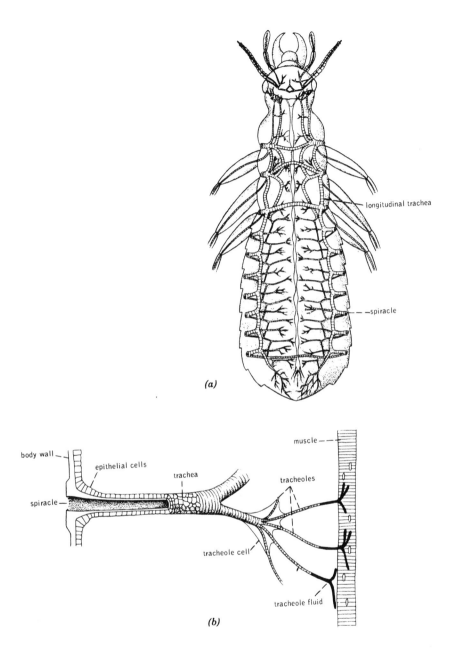

(a)

(b)

Figure 21. The tracheal system of an insect. (*A*) Gross anatomy of the tracheal system, showing opening of trachea at the spiracles, interconnection of largest tracheae, and finner division into tracheoles. (*B*) Details of tracheal anatomy, indicating arborization into tracheoles, which may penetrate individual cells. (4).

to 3.1 ml/min · kg (6). Measurements of tracheal gas concentrations in the cicada *Fidicina mannifera* when motionless or walking indicates relatively effective ventilation, with the fractional gas concentration in the thoracic air sacs for O_2 (F_{O_2}) and CO_2 (F_{CO_2}) remaining at ~17 and 3%, respectively (6). These cicadas are endothermic, raising thoracic temperature prior to flight by nonflapping contraction of wing muscles. During nonflapping warm-up, F_{O_2} falls to as low as 1%, while F_{CO_2} rises to as high as 21% (6). Once wing flapping begins, both F_{O_2} and F_{CO_2} change rapidly back towards resting levels. This not only indicates the efficacy of wing flapping on tracheal ventilation, but also suggests that inadequate tracheal ventilation during nonflapping warm-up may limit metabolic heat production.

Tracheal ventilation in terrestrial insects is under precise neural control (117). Motor output, probably originating from the metathoracic ganglia, stimulates closer muscles guarding the opening of each spiracle. By controlling the diameter of the opening, the resistance to gas flow and the total ventilation volume is regulated. Adjustment of tone of the closer muscles results from reflex chemoreceptive input resulting from both high CO_2 levels and low O_2 levels. In addition, proprioceptors respond to flexion of the abdomen and thorax (as would be produced during locomotion) and reflexly stimulate opening of the closer muscles to facilitate tracheal ventilation.

Aquatic insects secondarily have returned to water from land, and show a wide variety of modifications of tracheal systems for aquatic exchange. Several aquatic insects have lost spiracles in all but the most posterior abdominal segments. Larvae of the mosquito (e.g., *Culex*) have a posterior spiracle modified into a tube that can pierce the water surface. The larvae hang suspended in the surface film, allowing gases to move by

diffusion into the internal tracheal system. The tracheae of such insects are often of disproportionately large volume, allowing a considerable O_2 store and CO_2 sink to be taken below the surface for extended underwater forays. Following a similar concept of introducing the tracheal system into a gas source, some dipteran insect larvae (e.g., *Chrysogaster*, *Notiphila*, *Mansonia*, and *Taeniorhyncus*) actually use a spiracular tube to pierce the stems and roots of aquatic plants, which may contain relatively O_2 rich gas bubbles produced by photosynthesis.

In other aquatic insects, air trapped on the body surface at the air water interface is taken down during submersion. The trachea open into the surface covered by the gas bubble, and thus provide for diffusive gas exchange. A variety of specialized structures, generally termed gas gills (38, 137), are used (Fig. 6). In some species, the gas gill consists of a relatively large, compressible bubble. As O_2 is consumed from the bubble, more O_2 diffuses in from the surrounding water, allowing the bubble to provide many times its initial volume of stored O_2. Eventually, however, the bubble decreases in volume due to hydrostatic compression and the resulting diffusion of N_2 outward into the surrounding water; the bubble must then be renewed. For example, the water beetle *Dytiscus* makes brief surface contact to renew the gas bubble residing beneath the elytra. Spiracles open into this gas filled cavity, allowing diffusive gas exchange with the underlying body tissues. The carnivorous bug *Notonecta*, the water boatman, traps a gas bubble in a dense mat of hairs on the abdomen.

Other aquatic insects have a modified outer cuticle termed the plastron (172). Tiny hairs form dense, hydrophobic mats that permanently trap a thin gas film against the cuticle. In the bug *Aphelocheirus*, for example, the 5-μm long hairs occur at a density of $2 \times 10^6/mm^2$ over most of the cuticle (173). Physical factors such as surface tension and the nonwettable surface of the hairs exert major influences in such small structures, preventing hydrostatic compression of the gas bubble and the associated loss to surrounding water of N_2 (137). As a consequence, the gas film remains permanently in place, allowing O_2 to diffuse outward. Unlike insects that use compressible gas gills, plastron breathers can range over a water depth of several meters and can stay under water indefinitely without need to renew the water bubble at the surface.

In some aquatic insects (usually larvae) the gas-filled tracheal system is completely internally contained, making no connection with the cuticle. In larvae of the midge *Chironomus*, tracheoles lie very close to the outer cuticle of the body surface, and sufficient gas diffusion occurs across the integument. In other insects, abdominal appendages termed tracheal gills may contain a dense network of tracheoles. Movement of tracheal gills helps to disrupt stagnant boundary layers surrounding these structures. Larvae of many of the dragonflies and damselflies (Odonata) have tracheal gills lying within the rectum, which is ventilated by water pumped in and out of the anus.

The tracheal system in arachnids represents a case of convergent evolution, structurally and functionally resembling that of insects but having arisen independently. In most spiders the tracheal system appears to be derived from the book lungs (Fig. 22). In insects the tracheae repeatedly branch into finer and finer tracheoles; in most spiders, ricinuleids and pseudoscorpions, each spiracle opens onto a single atrium from which a tuft of individual tracheoles emerge.

DIFFUSION LUNGS AND BLADDERS. At least three invertebrate phyla and all classes of the vertebrates contain genera in which gas exchange occurs across moist respiratory membranes of a gas-filled chamber enclosed within the body cavity. In some forms the chamber is not ventilated. Simple diffusion from the surrounding atmosphere is apparently adequate for gas exchange, and these structures are termed *diffusion lungs*. In other forms the chamber is actively ventilated, and these structures are termed *ventilation lungs*.

Book Lungs. Many arachnids respire via so-called *book lungs*, primitive paired structures thought to be derived from book gills (4). Scorpions, whip scorpions, amblypygids, and some spiders have no tracheal system, depending entirely on book lungs for gas exchange (106). Multiple lamellae held apart by "spacer bars" are arranged within an internal invagination of the ventral abdominal wall (Figure 22). Hemolymph is circulated within the lamellae, while air flows between the lamellar sheets. Although most gas exchange in the book lung is thought to occur by simple diffusion, contraction of the body wall results in some ventilation of the book lung. Most spiders have spiracles connected to the tracheal system opening into the book lungs (see above).

Diffusion "Lungs" of Pulmonate Gastropods. Pulmonate snails (both freshwater and terrestrial), slugs, along with a few of the more primitive limpets (e.g., *Siphonaria*) achieve gas exchange using both the general body wall and a specialized, gas-filled *lung*. The gills have generally been lost (though they secondarily reevolved in some species) and the mantle has been curled inwards to form the internal lung, connected to the aerial environment by a regulated opening termed the pneumostome. The inner mantle is highly vascularized, and ventilation of the surface is achieved by contraction and relaxation of the mantle floor. Aquatic pulmonate

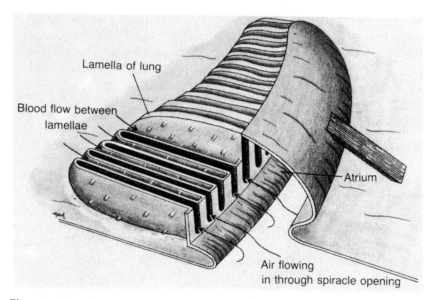

Figure 22. The internal book lung used for respiration in some arachnids. (4).

snails rise to the surface to ventilate the lung when the P_{O_2} of the gas within it falls below ~7–30 mmHg in *Planorbis* and 45–95 mmHg in *Lymnaea*. The opening of the pneumostome is thought to be regulated by P_{CO_2} in snails such as *Limax*, *Helix*, and *Arion* (89). A few deep-water forms of aquatic pulmonate snails never surface, and the lung is water filled (4).

Branchial Chamber Specializations. Several anomuran and brachyuran crabs adapted to varying degrees for terrestrial life show specializations of the branchial chamber associated with aerial respiration (113). In most semiterrestrial and terrestrial species, the gills are greatly reduced in size (e.g., *Coenobita*, *Birgus* and *Cardisoma*) and appear to function primarily in ion and water regulation (63). The branchiostegites (the region of the carapace covering the branchial chambers) are generally greatly expanded in volume and the inner branchiostegal lining is highly convoluted and heavily vascularized, providing a large surface area for gas exchange. While the gas diffusion distance between res-

piratory medium and hemolymph in the reduced gills averages ~4–10 μm in a wide variety of terrestrial crabs, the diffusion distance in the branchiostegal lung can be as small as 0.22 μm, which is comparable to that in mammalian lungs. Ventilation of the branchial chambers in *Cardisoma guanhumi*, for example, is generated by the same pumping action of the scaphognathites that moves water through the branchial chamber (25). In the anomuran crab *Coenobita clypeatus*, however, the convective flow of gas through the branchial chamber produced by the scaphognathites is aided by contraction and relaxation of epimeral muscles, which cause changes in branchial chamber volume (111).

A variety of branchial chamber modifications also occur in air breathing fishes. The family Anabantidae (e.g., *Clarias*, *Anabas*, *Amphipnous*, *Ophiocephalus* and *Trichogaster*) possess paired labyrinth organs, which take the form of highly elaborate, heavily vascularized plates or tufts, which are derived from branchial tissue and lie within a dorsally expanded opercular cavity (35, 85). Variants movements in normal cycles of continuous buccal pumping are used to push air bubbles into the expanded upper regions of the branchial chambers or labyrinths, where gas exchange with blood perfusing the labyrinth organs then occurs (130). In the gourami *Trichogaster trichopterus* an average of 70% of the O_2 in the gas bubble in the opercular chamber is extracted within 5 min of a single air breath (13), which is comparable to rates of pulmonary O_2 depletion in lungfishes (85, 130).

Some air-breathing fishes use a a combination of the existing gills and a heavily vascularized buccal cavity to extract O_2 from air held in the buccal–branchial cavity. Panamanian swamp eels, *Synbranchus marmoratus*, for example, switch to aerial respiration during terrestrial forays out onto stream banks searching for prey, and also while dwelling in the hypoxic water of their flooded burrows (10, 60). The air-breathing organ consists of minute papillae widespread through much of the buccopharyngeal mucosa, resembling a mat of "red velvet" because of the high capillary density. The threshold for switching between breathing modes is size dependent so that a 6-g eel withstands a P_{O_2} as low as 27 mmHg while a large eel of 850-g weight makes the transition to aerial respiration when the P_{O_2} of the burrow water drops to about 67 mmHg (60).

Even fish generally regarded as strictly aquatic can use air bubbles inspired into the branchial chamber to facilitate gas exchange. When in severely hypoxic water the goldfish *Carassius auratus* moves to the surface and gulps air bubbles at a frequency that increases as water P_{O_2} falls (14). The air bubble trapped in the branchial chamber increases the P_{O_2} of the inspired water passing over the gills. At very low and potentially life-threatening aquatic P_{O_2} levels (~18 mmHg), the net effect of air gulping on arterial P_{O_2} is an

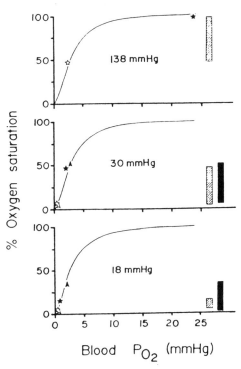

Figure 23. Effects of air gulping at three levels of aquatic P_{O_2} on blood O_2 transport in the goldfish *C. auratus*. Arterial P_{O_2} in the absence of air gulping is shown by solid stars, while the corresponding venous P_{O_2} indicated by an open star. Arterial P_{O_2} during air gulping is indicated by a solid triangle, with an open triangle indicating the associated venous P_{O_2}. The vertical bars at the side of the figure portray the arterial − venous difference in O_2 saturation at each level of aquatic P_{O_2} in the absence (striped bars) and presence (solid bars) of air gulping (14).

increase of only a few millimeters of mercury (Fig. 23). However, the low blood O_2 affinity and the shape of the O_2 equilibrium curve of goldfish blood is such that the amount of O_2 transported to the tissues can be maintained even in near anoxic water that would kill aquatic fishes unable to gulp air.

Gas Bladders and Lungs in Air-Breathing Fishes. The evolution in fishes of a gas-filled chamber connected to the alimentary canal has been heralded as a major evolutionary event in the evolution of structures and mechanisms for air breathing (85, 106, 141). Paleobiologists have argued long and loud about whether such a chamber first arose to regulate buoyancy or to exchange respiratory gases (149). Yet, it is difficult to see how only one of these functions in isolation from the other could have arisen. Certainly, important interactions and conflicts occur between regulation of buoyancy and aerial gas exchange in extant air-breathing fishes (57) and larval amphibians (46). Whatever the selection pressure for gas bladders in fishes, they have apparently evolved independently numerous times in both primitive and advanced fishes. The gas bladder of phystostome fishes, and the true lungs evident in the Polypteridae (*Polypterus* and *Erpetoichythys*), the Dipnoi (*Neoceratodus, Propterus* and *Lepidosiren*) and tetrapod vertebrates, are distinguished on the basis of their point of origin from the embryonic esophagus (Fig. 24). While the gas bladder of many phyletically ancient fishes (e.g., *Acipenser, Lepisosteus* and *Amia*) and almost all teleost fishes has a dorsal or lateral connection with the esophagus, vertebrates with true lungs (e.g., *Polypterus,* the lungfishes *Polypterus, Neoceratodus, Lepidosiren,* and tetrapods) have single or paired lungs arising from a ventral evagination of the esophagus.

Fishes that use the swim-bladder primarily for buoyancy have a gas gland that secretes a gas that may be 80 to 95 percent O_2. The gland is glycolytic and produces lactate that lowers the pH of blood leaving the gland. Acidification causes release of O_2 from HbO_2. A rete of capillaries provides for cycling of O_2 so that high pressures of O_2 can be maintained in the bladder (see pp. 204–206, 226–227 in Ref. 134).

Gas bladders and lungs in air-breathing fishes vary greatly in their structural complexity. In some species (e.g., *Amia*

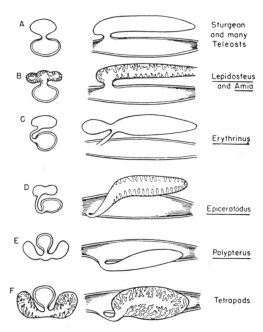

Figure 24. Diagrammatic cross-sections and longitudinal section of the air bladders and lungs of various fish and tetrapods. Note the variations in the position of the opening, number of lobes, and the extent of internal development (149).

and *Erpetoichythys*) the gas bladder or lung remains essentially saccular, with the major evolutionary modifications over strictly aquatic fishes being primarily an increase in vascularization of the gas bladder's inner wall. However, several species (e.g., *Arapaima, Hoplyerythrinus* and *Lepisosteus*) have gas bladders in which the inner wall has become relatively septate (Fig. 25). The greatest degree of septation occurs in the lungs of Dipnoi (21, 22, 106). Internal septa, ridges, and pillars effectively partition the lung into smaller lateral compartments opening into a central cavity in an arrangement quite similar to that in amphibians. Even within the lungfish there is much variation, with the single lung of Australian lungfish (*Neoceratodus*) being less septate than that of the paired lungs of the African (*Protopterus*) or South American (*Lepidosiren*) lungfishes.

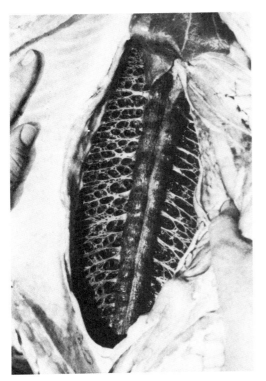

Figure 25. A ventral dissection of *Arapaima*, showing the complex structure of the swim bladder (77a).

In most fishes that use a swim bladder or lung for respiration, surfacing and protrusion of the mouth through the air–water interface is immediately followed by exhalation. The use of X-ray cinematography and gas pressure and flow analysis has revealed that many species almost completely empty the bladder or lung. The bladder is then inflated with a single air breath (or at most a few breaths) and the fish then resubmerges. Ventilation of the gas bladder is achieved by only slight modification of the buccal cycles normally used for propelling water through the gills (103, 141). During inhalation, the buccal floor drops, drawing air into the mouth. The mouth is closed, and elevation of the buccal floor compresses the air bubble, driving it through the pneumatic duct into the gas bladder. Expiration and emptying of the bladder

is due largely to passive recoil of the ribs and body wall, aided by external hydrostatic compression of the body wall transmitted to the gas bladder. In a few air-breathing fish [e.g., *Hoplyerythrinus* (45), and *Piabucina* (62)] the more typical pattern of exhalation followed by inhalation has become reversed (Fig. 26). The gas bladder, which even in most air breathing fishes is a single-lobe saccular organ, has become modified in *Hoplyerythrinus* into a muscular, contractile anterior portion, and a highly vascularized, relatively septate posterior portion. Inspired air is preferentially directed into the anterior portion, helping to prevent mixing with stagnant gas resident in the posterior portion. Both passive and active compression of the body wall results in emptying of the posterior chamber into the buccal cavity and out the mouth, finally followed by release of inspired gas into the posterior portion of the bladder for subsequent gas exchange.

Gas bladders in air-breathing fishes vary greatly not only in their structural complexity, but also in their location in the cardiovascular circuit (21, 85, 141) (Chapter 11). Critical to gas exchange is the anatomical origin of blood perfusing the gas bladder or lung, since the particular P_{O_2} and P_{CO_2} of afferent blood will critically influence the gas diffusion gradients in the bladder or lung. In the gar pike *Lepisosteus*, for example, the bladder is perfused by numerous small segmental arteries derived from the dorsal aorta (Fig. 27*A*). Thus, the afferent blood supply to the gas bladder has first passed through the branchial circulation, and has essentially the same gas composition as arterial blood perfusing other systematic tissues. In the teleost *Hoplerythrinus*, the afferent circulation shows a more specialized condition (44). The gas bladder is perfused with arterial blood via the coeliac artery, which is derived from the efferent branchial circulation of the two most posterior gill arches (Fig. 27*B*). A

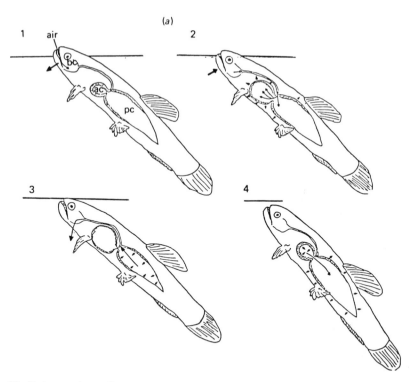

Figure 26. Pattern of gas flow in the gas bladder of the air breathing fish *Hoplerythrinus unitaeniatus*. (1) Air is drawn into the buccal cavity (bc). (2) Buccal cavity elevation occurs, compressing the inspired air the buccal cavity and producing a flow of gas preferentially into the muscular anterior chamber (ac) of the gas bladder. (3) The anterior and posterior chambers become separated by constriction of the narrow region of the gas bladder, and exhalation of gas occurs. (4) Gas flows from the anterior chamber into the posterior chamber, completing the ventilatory cycle (141).

narrow, high impedance blood vessel connects the origin of the coeliac artery with the dorsal aorta. Blood flow to the gas bladder can be regulated so as to optimize the ratio of gas ventilation and blood perfusion. During and immediately following an air breath, ~20–40% of the cardiac output perfuses the posterior arches and most of this passes into the coeliac artery rather than into the dorsal aorta. During prolonged breath holding, however, the impedance of the coeliac artery rises, shunting blood from the posterior arches into the dorsal aorta rather than to the gas bladder.

All fishes using a gas bladder for aerial gas exchange have a more or less common pattern of venous drainage from the bladder, in which oxygenated blood from the bladder flows into the central systemic veins where it completely mixes with systemic venous blood (21). Thus, the contribution of the swim bladder to gas exchange lies primarily in elevating the P_{O_2} of venous blood. At first this sounds intuitively inefficient, given that after passage through the heart venous blood will pass through yet another respiratory organ (i.e., the gills) before eventually reaching the systemic tissues. However, it should be emphasized that in many fishes, both aquatic and air breathing, the gills have numerous shunt vessels by which blood can traverse the branchial circulation without actually being exposed to ambient water (140). This is crit-

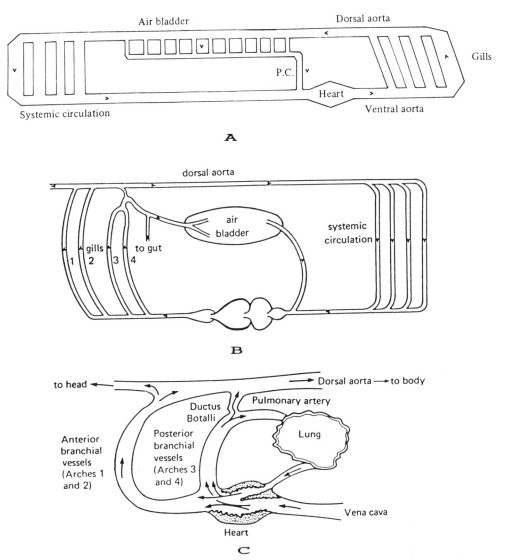

Figure 27. Arrangement of the air breathing organ with respect to the afferent and efferent of circulation in (*A*) the holostian *Lepisosteus osseus* (138), (*B*) the teleost *Hoplyerythrinus uniteniatus* (44), and (*C*) the Dipnoan *Protopterus aethiopicus* (87).

ical, since air-breathing fishes in hypoxic water could lose O_2 gained via the bladder during transit to the gills unless some mechanism existed for preventing exposure of branchial blood to the water.

Of the air-breathing fishes, the lungfishes (Dipnoi) show the most complex and efficient cardiovascular specializations associated with evolution of an air-breathing organ (21, 22, 85) (Chapter 11).

The lungs are perfused via a pulmonary artery derived from the sixth branchial arch (Fig. 27C). As in many air breathing fishes, the proportion of cardiac output that perfuses the lungs is carefully regulated, constantly changing to achieve an optimal match of lung perfusion to lung ventilation. In the Dipnoi a discrete pulmonary vein has evolved, returning oxygenated blood directly to the left side of the heart rather than allowing this blood to mix into the general systemic venous return. Selection of these observed morphological and physiological adaptations for maintenance of separate intracardiac blood streams (e.g., right and left circuits) in bimodal breathers among fish and amphibians very likely had to occur in parallel with separation of the pulmonary and systemic venous return pathways. Intracardiac separation allows preferential perfusion of the systemic body tissues with blood oxygenated in the lungs, and perfusion of the lungs with deoxygenated blood derived from the systemic tissues (see Chapter 11).

The overall contribution of the gas bladder or lung to total gas exchange varies greatly between various species of air-breathing fishes (Table 4). Typically, however, the gas bladder or lung serves primarily as an organ for O_2 uptake, with the great majority of CO_2 elimination usually occurring via the gills and skin.

Not only are there anatomical considerations that dictate the balance between aquatic and aerial respiration (e.g., origin of arterial blood and path of venous drainage of the bladder or lung, respiratory surface area and diffusion distances, effectiveness of V_{air}/V_{blood} matching, etc.), but environmental and physiological factors can alter the dependence on aerial gas exchange within individuals. Some species (e.g., *Neoceratodus*, *Clarias*, *Monopterus* and *Heteropneustes fossilis*) are termed *facultative* air breathers, in that aerial respiration can be used to supplement aquatic respiration, but is not necessary in that the fish will not drown if denied access to air. Other species are *obligate* air breathers, normally requiring aerial respiration for survival (e.g., *Protopterus*, *Trichogaster*, *Erpetoichythys* and *Betta*). It should be emphasized that these functional categories are not rigid, in that a normally facultative air-breathing fish must become an obligate air breather under certain conditions. A major condition influencing the extent of dependence on

TABLE 4. Gas Exchange Partitioning between Water-Breathing Organs (Gills and Skin) and Air-Breathing Organs of Selected Air-Breathing Fishes at Rest and in Air-Saturated Water[a]

Species	T (°C)	% $\dot{M}_{O_2}$ Water	% $\dot{M}_{O_2}$ Air	% $\dot{M}_{O_2}$ Water	% $\dot{M}_{O_2}$ Air
Amia calva	20	63	37	75	25
Heteropneustes fossilis	25	59	41	94	6
Anabas testudineus	25	46	54	91	9
Clarias batrachus	25	42	58	94	6
Lepisosteus osseus	22	27	73	92	8
Monopterus albus	25	25	75		
Electrophorus electricus	26	22	78	81	19
Protopterus aethiopicus	20	11	89	70	30
Protopterus aethiopicus	24	10	90	68	32
Lepidosiren paradoxa (juvenile)	20	64	36	76	24
Lepidosiren paradoxa (adult)	20	4	96	41	59
Misgurnus anguillicaudatus	20	80	20	3	97
Erpetoicthys calabaricus	27	60	40		
Trichogaster trichopterus	27	60	40	15	85

[a]From (13, 102, 112, 151, 163)

aerial respiration is body temperature. As water temperature rises, the O_2 capacitance of the water falls, yet a fish's metabolic rate rises. At some critical temperature, aquatic respiration may be inadequate to support total metabolic demand. This situation is well exemplified by temperate air-breathing fishes such as the bowfin *Amia* (88) or the garpike *Lepisosteous* (138). Amia at 10°C rarely breathes air and can survive entirely on aquatic respiration (Fig. 28). As temperature increases, total M_{O_2} and M_{CO_2} increase. Branchial gas exchange also increases, but not as rapidly as total metabolism, with the balance of gas exchange being achieved at higher temperatures by an increase in aerial gas exchange via the bladder.

Another critical factor influencing the degree of dependence on aerial gas exchange is the P_{O_2} and P_{CO_2} of the aquatic environment. For example, the Australian lungfish *Neoceratodus*, which in air-saturated water at 20–24°C rarely breathes air, will greatly increase both gill and lung ventilation frequency in hypoxic water (86). Prolonged activity produces a similar increase in lung ventilation in this species (61). While an increase in gill ventilation is a typical response to hypoxia in strictly aquatic fish (139), severe aquatic hypoxia or hypercapnia actually results in a sharp decrease or even cessation of branchial ventilation in many air-breathing fishes such as *Amia* (88) or *Hoplyerythrinus* (191). While this shifts the burden of gas exchange to the air-breathing organ, it also helps to reduce the potential loss to the ambient water of O_2 from blood entering the gills.

LUNG STRUCTURE AND FUNCTION
IN TETRAPOD VERTEBRATES

Amphibians. Many living amphibians have become highly modified from their ancestral forms, showing adaptations for a wide variety of semiterrestrial and terrestrial habitats.

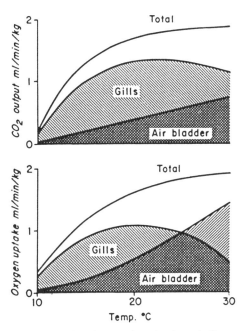

Figure 28. CO_2 elimination (top) and O_2 uptake (bottom) by gills and swimbladder of the bowfin *Amia calva* over the temperature range 10–20°C (88).

The general form of amphibian lungs has probably departed little from that of ancestral tetrapods. In most species the lungs are relatively simple, sac-like structures (17). The separation of the inner walls is usually not extensive and alveoli, when they occur, are large compared with those in higher vertebrates. In anurans, the lungs are relatively globular (Fig. 29), while in species with elongate body shape (e.g., apodans and most urodeles) the lungs are long tubular organs. Weight-specific lung volume in amphibians is large compared with other tetrapods (Table 5), ranging from ~0.1 ml/g of body wt in the bullfrog *Rana catesbeiana* to over 1 ml/g of body wt in the toad *Bufo marinus* (17).

The lungs of many amphibians show structural and physiological adaptations related to buoyancy regulation. Aquatic larvae of the salamander *Ambystoma tigrinum*, often float at neutral buoyancy in

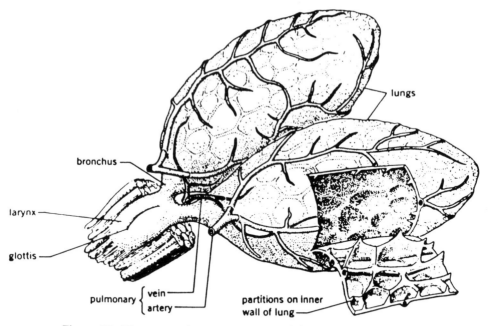

Figure 29. Diagrammatic representation of the lungs of a ranid frog.

the water column to feed on zooplankton. Their air breathing frequency and lung volume at a particular time depend on a complex interplay between gas exchange, the risk of surface predation, and the relative availability of pelagic versus benthic zooplankton (100).

Mechanisms of lung ventilation have been investigated primarily in anuran am-

phibians (37, 92, 188). Lung ventilation is powered primarily by a buccal force pump that is homologous with that used in fishes (see the section on Aquatic Gas Exchange). The mechanical events and movement of gases associated with respiratory movements in the toad *Bufo marinus* are shown in Figure 30. The buccal cavity is in nearly constant motion,

TABLE 5. Pulmonary Morphometrics of a Frog (*R. catesbeiana*), Turtle (*P. scripta*), Chicken (*Gallus domesticus*), and Rabbit (*Sylvilagus*)

	Frog	Turtle	Chicken	Rabbit
Body Weight (g)	300	990	1600	1600
Lung volume/body weight (mL/g)	0.2	0.28	0.01	0.075
Lung surface area/body weight (cm²/g)	2.5	2.5	18	18
Lung surface area/lung volume (cm²/mL)	12	18	1920	240
Thickness of blood–air barrier (μm)	2	0.5	0.1	0.2–10

[a]Where appropriate data are expressed on a weight-specific basis to facilitate comparison. Data compiled from (38), (127) and (17).

rhythmically moving up and down. When the glottis is closed, the lungs are isolated from the rest of the respiratory tract, and the nares (nostrils) are open; these buccal movements have the effect of producing a tidal ventilation of the buccal cavity. The functions of constant ventilation of the buccal cavity have yet to be established. Does it mainly facilitate olfaction, gas exchange across buccal membranes; or perhaps it remains as a physiological vestige of buccal movement that is necessary to gill ventilation in aquatic larval stages.

Ventilation of the lungs occurs when the normal buccal cycle is modified by a complex, and often variable, sequence of events. Lung inflation begins with narrowing or closure of the nares. When the buccal floor raises, gas in the buccal cavity is compressed rather than expelled out the nares. As the glottis opens, gas in the buccal cavity is forced under positive pressure into the lungs. Often several successive lung inflation cycles are required to fill the lung with gas. Expiration occurs when the glottis opens and the elastic recoil of the body wall (and hydrostatic pressure, if in water) compresses the lungs.

Reptiles: The Origin of Aspiration Breathing. The lungs of reptiles show a degree of anatomical complexity that is intermediate between that of amphibians and mammals (128). Some species retain lungs that are essentially saccular, with an inner surface that appears like the face of a honeycomb. In snakes (among which the colubrid snakes have only one lung), the lungs may be sharply divided into an anterior, highly partitioned, and vascularized region where the majority of gas exchange occurs, and a posterior, saccular section that appears to serve a gas storage function. Reptilian lungs reach their most complex form in the pelagic sea turtles such as the green sea turtle, *Chelonia my-*

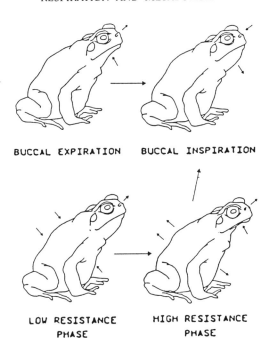

BUCCAL EXPIRATION   BUCCAL INSPIRATION

LOW RESISTANCE   HIGH RESISTANCE
PHASE   PHASE

Figure 30. Mechanical and gas flow events associated with buccal ventilation and lung ventilation in the toad *B. marinus* (92). Resistance refers to position of nares.

das (128). Well-formed highly branching airways and very extensive alveolarization lend the spongy appearance of mammalian lungs.

Lung volume in reptiles varies widely (128). The tegu lizard *Tubinambis nigropunctatus* has a lung volume of only 0.15 ml/g of body wt compared with 0.6 ml/g in the monitor lizard *Varanus exanthematicus* and 1.25 ml/g in the chameleon *Chamaeleo chamaeleon* (129). The evolution of reptilian lungs has been strongly shaped by nonrespiratory selection pressures such as buoyancy and behavioral displays (127, 128, 129), as has been emphasized for amphibians (17). Morphometric data in Table 5 indicate that freshwater turtles have only slightly larger weight-specific lung volumes, and roughly equivalent weight-specific surface areas when compared with bullfrogs. In turtles, however, a significant proportion of total body weight is the weight of the calcified shell, and doubtless the ratio of lung surface area/body weight would be higher if expressed only on the basis of the weight of actively metabolizing tissues.

Lung ventilation in reptiles is achieved by aspiration (suction) rather than by buccal force pumping. Aspiration into the lungs develops when expansion of the body wall and/or internal movements of the viscera generate a subambient gas pressure within the lung cavity. If the glottis opens, gas will flow along the pressure gradient from surrounding air into the lungs. Aspiration breathing probably evolved as a supplement to buccal force pumping, but its precise evolutionary origins are not clear (141).

Given the extreme variation in body morphology of reptiles, it is not surprising to find that generation of subambient pulmonary pressure and the aspiration of gas into the lungs can be generated in several different ways (141). Snakes, which lack a diaphragm, have well-developed ribs and intercostal muscles. Inspiration is generated by expansion of the rib cage through intercostal muscle contraction, and expiration is a result of both intercostal muscle action and passive recoil of the body wall. Crocodilians, which have well-developed ribs and intercostals and lack a diaphragm, show one of the more curious liver functions in a vertebrate, putting this organ to use in lung ventilation (Fig. 31A). Aspiration in *Caiman crocodilis* is generated by the action of diaphragmatic muscle, which is attached to the liver anteriorly and the pelvic girdle posteriorly (53). The liver, in turn, is firmly attached to the posterior regions of the paired lungs. Contraction of the diaphragmatic muscle pulls on the liver, which in turn expands the lungs, generating the subambient pulmonary pressure necessary for aspiration. Contraction of abdominal muscles pushes the liver anteriorly, causing lung expiration. The in-

tercostal muscles appear to act primarily to maintain a constant thoracic volume, allowing the liver to slide back and forth like a piston in a cylinder. In chelonian reptiles (turtles, tortoises) the ribs have become fused with the underside of the carapace, and the diaphragm is nonmuscular and does not make an active contribution to lung ventilation (54, 56). Inspiration is achieved when an outward extension of the pelvic and pectoral girdles draws out the viscera, which in turn are attached to the diaphragm and cause downward expansion of the lungs (Fig. 31B). Expiration occurs when inward movement of the limbs compresses the lungs. Opening and closing of the glottis plays a large role in controlling the flow of gas in and out of the lungs; behavior such as walking or neck extension causes changes in intrapulmonary pressures that are not related to the need for lung ventilation. Hydrostatic pressure when in water, and gravity when in air, are important factors in lung ventilation. The importance of hydrostatic compression of the body cavity varies with the depth of the body in water at the time of inspiration and expiration. The deeper the body, the more muscular activity is required for inspiration compared to expiration (56).

Most reptiles are highly periodic breathers, commonly exhibiting breath holding (apnea) (58). Apnea typically begins following an inspiration, so that the lungs are inflated throughout the period of breathholding. Some species characteristically terminate apnea with only one or a few breaths (e.g., the sea snake *Acrochordus*), while other species tend to breath in episodic bursts of several breaths at a time [e.g., the freshwater turtle *Pseudemys scripta* (12)]. Apnea length appears to be related to lifestyle. Because air is constantly available for lung ventilation, most terrestrial snakes, lizards, and tortoises tend to show more rhythmic breathing with only short periods of

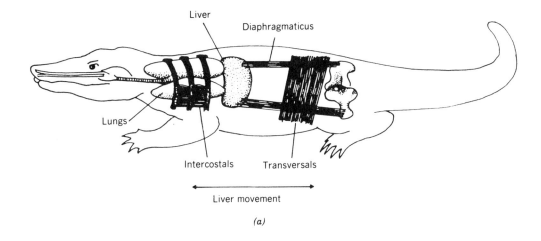

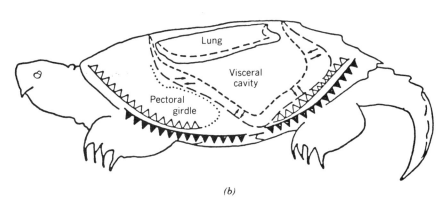

Figure 31. Mechanisms of lung ventilation in reptiles. (*A*) Lung ventilation in *Caiman* (53). (*B*) Pulmonary and visceral cavities and the forces influencing lung ventilation in the snapping turtle *Chelydra serpentina*. The small arrows indicate the effects produced by muscular activity of the pectoral and pelvic girdles. The open triangles indicate the effects of hydrostatic pressure, while the solid triangles indicate the effects of gravity (56).

breatholding than aquatic ones (e.g., *Testudo graeca* in Fig. 32). Diving aquatic species do not have constant access to air. Surfacing to breathe may have high energetic costs as well as risks of predation (46, 96), and aquatic turtles, snakes, lizards, and the Crocodilia typically show long periods of apnea of several minutes to >1 h. Although apnea lengths may be long relative to diving birds or mammals, the large O_2 stores in lung gas (16) (and

to a lesser extent in blood), combined with a comparatively low metabolic rate allow aerobic respiration to continue during all but the longest dives. The intermittent nature of lung ventilation in most reptiles requires intermittent perfusion of the lungs if optimal ventilation–perfusion matching is to be achieved. Lung perfusion during intermittent breathing in reptiles is regulated by adjustments in heart rate (Fig. 32), stroke volume, and the de-

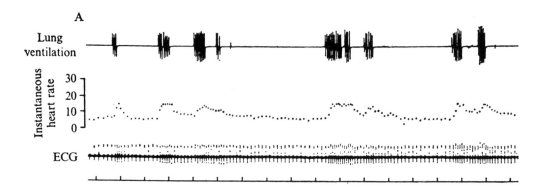

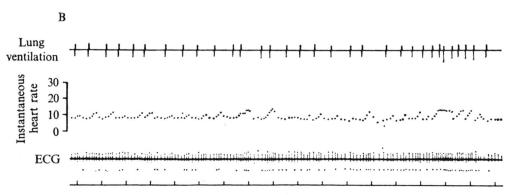

Figure 32. Patterns of lung ventilation and heart rate during undisturbed, intermittent breathing in the freshwater turtles *P. scripta* (A) and the tortoise *T. graeca* (B) (12).

gree of central cardiovascular shunting between pulmonary and systemic circuits (Chapter 11). Indeed, pulmonary blood flow during periods of active lung ventilation in the turtle *P. scripta* may be as much as 40 times higher than towards the end of a long period of apnea (162).

Mammalian Lungs. The lungs of mammals represent a maximization of respiratory surface area in a relatively compact organ of small volume. The surface area/volume ratio in a rabbit is ~20 times greater than in an anuran amphibian (Table 5). The major structural specialization of mammalian lungs is the change from the "saccular" arrangement in amphibians and reptiles to a lung which is composed

of small (25–50 μm in diameter) gas sacs or *alveoli* (Fig. 33). Alveoli, which represent the functional unit of gas exchange in mammalian lungs, are heavily vascularized and provide a structure by which alveolar gas and capillary blood are brought within less than a micrometer of each other. Diffusion of O_2 and CO_2 occur extremely rapidly over such a small diffusion distance and, consequently, pulmonary venous blood has usually attained equilibrium with alveolar gas (183). Near-astronomic numbers of alveoli occur in the lungs of mammals, for example >1 million in each human lung. This confers a large surface area to mammalian lungs relative to those of lower vertebrates (Table 5). As with other tetra-

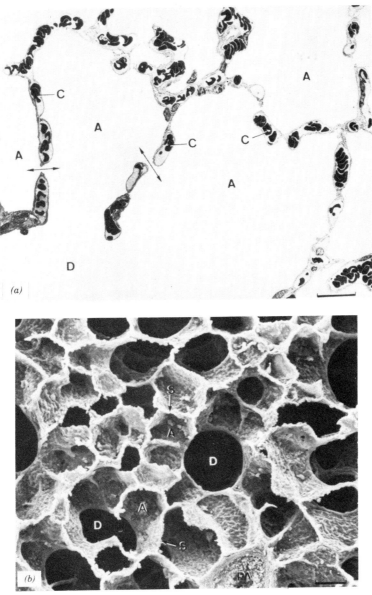

Figure 33. Fine structure of the mammalian lung. (*A*) Scanning electron micrograph of the alveoli of a human lung. Clusters of alveoli (A) branch from alveolar ducts (D). A small branch of the pulmonary artery (PA) is evident. Scale marker = 100 μm (From 184). (*B*) Scanning electron micrograph of the lung of a dog, showing individual alveoli (A) opening onto an alveolar duct (D). Note the high density of alveolar capillaries (C) containing red blood cells. Scale marker = 20 μm (183).

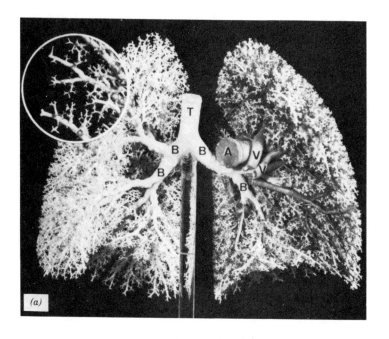

(a)

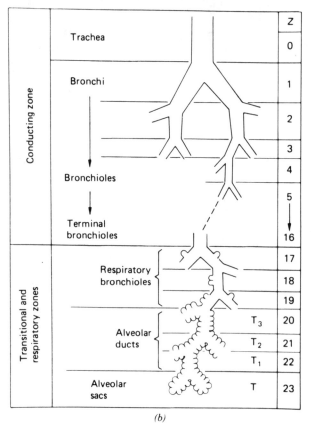

(b)

pod lungs, the flow of gas in alveoli and the flow of blood around them, resembles the "ventilated pool" concept as indicated in Figures 5 and 39.

Alveolar lungs are characterized by a dense, spongelike appearance resulting from the clusters of alveoli surrounding the terminal gas conducting tubes. The major airways form a highly branching delivery system for gas during expiration and inspiration (Fig. 34A). Although there is some variation between species, >20 successive branchings occur between the trachea and the termination of the alveolar ducts (Fig. 34B). The conducting airways are not emptied during expiration, and they do not contribute to gas exchange. Thus, gas contained within them following inspiration is exhaled unaltered. This so-called *anatomical dead space* represents a small but significant proportion of the total gas volume of the lung (Fig. 35). Because exhaled gas lying in these airways mixes with inhaled air during the following inspiration, the gas that eventually enters the alveolar ducts represents a mixture of air and exhaled gas. Thus, the alveoli do not contain air, but rather a gas with a lower P_{O_2} and higher P_{CO_2} than air.

Reflecting the comparatively high metabolic rate and respiratory requirements of mammals, lung ventilation necessarily is a continuous process in most species. Diving mammals that have secondarily evolved anatomical and physiological adaptations for a marine existence (e.g., pinnipeds and cetaceans) are notable exceptions. Voluntary apnea length may range from only several minutes in the sea lion *Zalophus californianus* to over 1 h in the Weddell seal (*Leptonychotes weddelli*) (41). The primary structural adaptation of diving mammals involves strengthening of the nonrespiratory airways. During diving to depth, hydrostatic pressure on the chest wall is transmitted to the lungs. Gas in the respiratory portions of the lung is compressed, and driven into the armored airways. Because gas resident in these airways cannot exchange with pulmonary blood, there is no opportunity for large quantities of nitrogen gas to enter the blood stream during the dive. It is thought that this allows whales and seals to avoid *the bends*, a disease experienced in human divers breathing compressed gas in which bubbles of N_2 form in the blood stream upon surfacing and cause potentially lethal blockages of blood flow to the tissues. Diving mammals generally do not dive with fully inflated lungs. Instead, the major O_2 stores in diving mammals reside within the blood in the form of oxyhemoglobin (16, 41).

Lung inspiration in mammals is achieved through aspiration provided by a highly variable combination of contraction of the diaphragm and of intercostal muscles (Fig. 36). In humans, expiration at rest is often passive, utilizing recoil of the rib cage, but can be active using intercostals and abdominal muscles (e.g., during exercise).

Parabronchial Lungs of Birds. The respiratory system of birds represents a sharp break from an evolutionary trend in tetrapods towards increasing alveolarization of a saccular lung ventilated by

Figure 34. The airways of the lung. (*A*) Ventral view of a plastic resin cast of the human lungs. Note the rapid and extensive branching of the bronchi (B) from the trachea (T). The pulmonary vein (V) and artery (A) show clearly in the left lung (on the right of the photograph) (185). (*B*) Functional organization of the pulmonary airways of the lung in relation to the generations (z) of dichotomous branching (182).

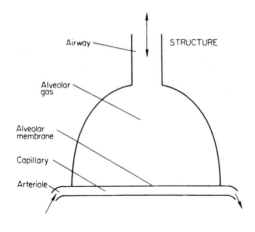

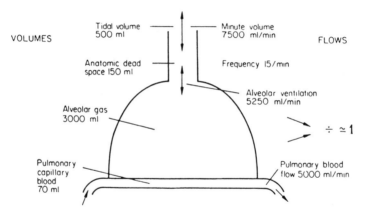

Figure 35. Schematic representation of the human lungs, based on the concept of the alveolus as the functional lung unit. Top panel: Structural elements of the lung. Bottom panel: Gas volumes and blood flows through the human lung (volumes and flows represent combined values for both lungs) (187).

aspiration. Although there are diverse anatomical patterns among birds, several general features distinguish the avian respiratory system. The lungs themselves, which are relatively rigid and show little volume change during the ventilatory cycle, represent only ~20% of the total respiratory volume. The remaining volume resides in nine nonrespiratory gas sacs separated from the viscera by an oblique septum. The anterior sac group is the largest, and consists of the anterior thoracic (also called prethoracic) sac, the single interclavicular sac, and the small cervical sacs (Fig. 37). The posterior sacs consist of the abdominal sac and postthoracic sacs. The abdominal and posterior thoracic sacs show the largest changes in volume during the ventilatory cycle. Both physiological and anatomical studies have attempted to determine the patterns of gas flow through this rather

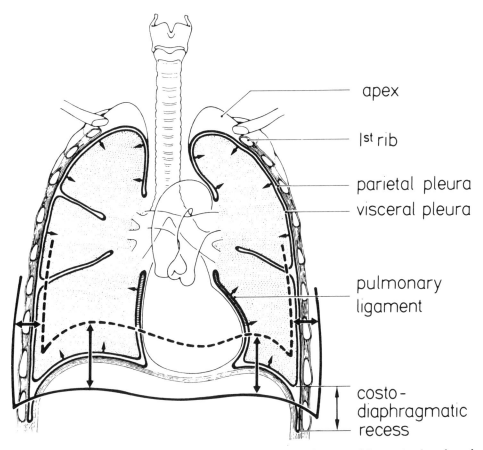

apex

1st rib

parietal pleura
visceral pleura

pulmonary
ligament

costo-
diaphragmatic
recess

Figure 36. Frontal section of the human chest showing the paired lungs in the pleural cavity. Retractile forces in the lungs are indicated by the single arrows, while the excursion of the base and periphery of the lungs is indicated by the double arrows. The space below the dome-shaped diaphragm (the costodiaphragmatic recess), serves as a reserve space for lung expansion (184).

complex respiratory network (91, 154, 155, 170). During inspiration, air is aspirated into the trachea by a simultaneous bellows like expansion of primarily the abdominal and posterior thoracic sacs, produced by expansion of the ribcage and an anterioventral pivoting of the sternum (Fig. 38). Expiration is also normally active, produced primarily by contraction of abdominal muscles surrounding the air sacs. Air flow at the level of the trachea is tidal in birds, as in other vertebrates. However, in a triumph of design, gas within the sacs travels in a circular pathway such that both inspiration and expiration contribute to the continuous unidirectional flow of gas anteriorly through the parabronchi.

The pulmonary structure, like the airway and airsac system, is also quite distinctive in birds (Figs. 37, 39). The lung consists of a series of parabronchial tubes, open at both ends. Blindended gas capillaries, which branch radially from the parabronchi, are intertwined with blood-filled capillaries derived from pulmonary

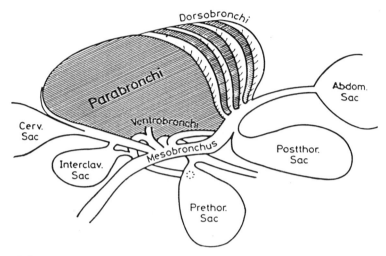

Figure 37. Schematic diagram of the bronchial tree and air sacs of a representative bird (155).

arterial arterioles. Oxygen uptake and CO_2 elimination occurs across the 0.1-μm thick cell layer separating parabronchial gas and capillary blood. Physiological and morphological analyses (154) indicate that the flow of blood and air occur in a cross-current pattern (Fig. 39; see Chapter 11). The continuous unidirectional flow, combined with the intrinsically higher gas transfer efficiency of a cross-current system (Fig. 5) provides for a substantially greater oxygen transfer efficiency than occurs in mammalian lungs.

Respiration and Metabolism: Modifying Agents

Principles and Concepts

Measuring Metabolism

The term *metabolism* means different things to different physiologists. In Chapter 8, metabolism is discussed primarily in the context of cells and the biochemical processes within. Metabolism will be considered in the broader context of the organismal responses to changes in cellular metabolism, and emphasize factors affecting the *rate* of O_2 consumption and CO_2 production of whole animals interacting with their environments.

While it is easy to conceptualize the process of consuming nutrients and substrates, generating energy from them, and expelling waste products, it is less easy to quantify the process of metabolism. *Metabolic rate,* the rate at which chemical energy is consumed by an animal in growth and maintenance, is the amount of energy consumed per unit time. Metabolic rate changes with body size. Thus, metabolic *intensity,* or mass-specific metabolic rate, is often reported rather than the total metabolic rate of the organism to facilitate comparison between dissimilarly sized animals.

Metabolic rate can be measured in any of several ways. For example, the difference in energy value of all ingested food and all excreted waste products unequivocally indicates the metabolic rate of an animal. But in practice, this assumes that there is no net storage or loss of previously stored chemical energy as fat or other substances, and that there is no net growth during the observation period. A more useful approach to the determination of metabolism consists of measuring the total heat production resulting from cellular metabolism. An organism is placed in a calorimeter, which is a highly insulated chamber fitted with thermocouples for quantifying heat production by individual animals in the chamber. This method is highly accurate, but the necessity of completely enclosing an animal in a sealed chamber imposes limits on both animal size and the experimental manipulation that a researcher can perform.

The method of measuring metabolic rate in most common use is the determination of *oxygen consumption*. As a technique, this approach has both great strengths and considerable weakness. One major advantage is the relative ease of measurement. Oxygen partial pressure and/or O_2 concentration can be determined accurately in water or air. By measuring the O_2 depletion from the respiratory medium surrounding an animal in a closed vessel (closed respirometry), or from the respiratory medium flowing past an animal (open or flow-through respirometry), the total O_2 consumed/unit time ($\dot{M}_{O_2}$) can readily be calculated. Another advantage of using $\dot{M}_{O_2}$ as an index of metabolic rate is the fact that the relation between the amount of O_2 consumed and the resultant heat energy produced is largely independent of the metabolic substrate (protein, fat, and carbohydrate); thus facilitating conversion between the various commonly adopted units for measurement of metabolism. A disadvantage of using $\dot{M}_{O_2}$ to express metabolic rate is that cellular metabolism in various forms can proceed at a limited rate in the absence of molecu-

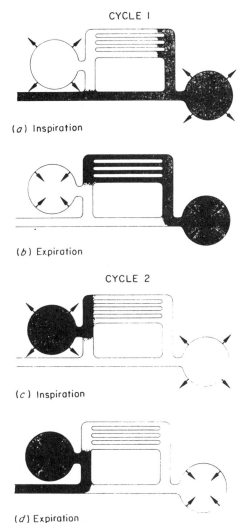

CYCLE 1

(a) Inspiration

(b) Expiration

CYCLE 2

(c) Inspiration

(d) Expiration

Figure 38. Pattern of movement of a single inhalation through the avian respiratory cycle (11).

lar O_2—so-called *anaerobic* metabolism (Chapter 8). While anaerobic metabolism yields 2ATP molecules for every glucose molecule consumed, compared with the yield of 36 molecules of ATP for aerobic metabolism (77), many animals do use anaerobic metabolism under a wide variety of circumstances when either environmental or internal O_2 availability is limiting. Thus, $\dot{M}_{O_2}$ may underestimate total

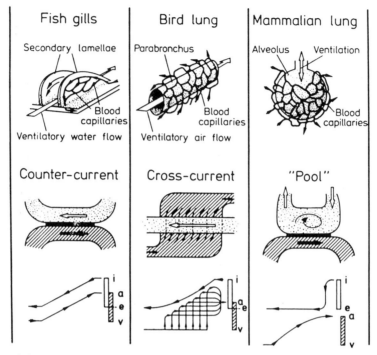

Figure 39. Schematic comparison of the anatomy and flow patterns of blood and respiratory medium in the fish gill, bird lung, and mammalian lung (154).

metabolic rate under some conditions. Unfortunately, in the comparative physiological literature $\dot{M}_{O_2}$ is often implicitly equated with metabolic rate. It is now commonly measured for purposes of understanding processes in O_2-based exchange, transport, and metabolism rather than for the study of overall metabolism.

Respiratory Quotient and
Gas Exchange Ratio

Carbon dioxide production ($\dot{M}_{CO_2}$) is another process in aerobic metabolism and, in theory, could be used as an indicator of metabolic rate. In practice, this approach is rarely used. The accurate measurement of total CO_2 in water, in particular, is relatively difficult because of the complex chemistry of CO_2 in water making respirometry particularly difficult. Even more importantly, CO_2 production must be measured under *steady-state* conditions if it is to be used as an

indicator of metabolic rate. There is a large pool of CO_2 stored in animal tissues as molecular CO_2, bicarbonate ions and as carbamino compounds, and this CO_2 store is very easily mobilized. A sharp increase in muscle activity, for example, not only causes the elimination of molecular CO_2 as a waste product of anaerobic metabolism, but may also cause the production of metabolic acids (e.g., lactic acid) if some component of the metabolic increase is anaerobic. The release of these acids into the body tissues shifts the equilibrium between the various forms of stored CO_2 such that additional molecular CO_2 is produced and is subsequently eliminated by ventilation. Eventually, the period of recovery from exercise is accompanied by a reestablishment of the normal CO_2 equilibrium of the tissues, which is achieved by a transient sequestering rather than elimination of metabolically produced CO_2.

Similar large shifts in CO_2 equilibrium accompany intermittent periods of lung ventilation and breath holding, as has been well documented for amphibians and reptiles (161). While some CO_2 resulting from continued aerobic metabolism can be eliminated across the skin during breath holding (see section on Integumental Gas Exchange), some proportion of CO_2 will temporarily be stored in tissues pending the resumption of air breathing. When lung ventilation occurs, this stored CO_2 quickly leaves the tissues and is eliminated in a brief burst before the resumption of breath holding. Thus, even if CO_2 production at the tissue level is constant during intermittent breathing and breath holding, the actual elimination of CO_2 at the body surface can fluctuate over a wide range. Thus, the measurement of CO_2 elimination provides information on precisely that process, but is a rather misleading metabolic indicator in the presence of labile CO_2 equilibrium.

A frequently used concept in metabolism and its measurement is the *respiratory quotient* (R_Q), which is the ratio of CO_2 production to O_2 consumption ($\dot{M}_{CO_2}/\dot{M}_{O_2}$) *at the cellular level.* The R_Q is usually <1.0 for two reasons. Of minor importance in most animals is the potential for CO_2 fixation as oxaloacetate (76). Critically affecting R_Q, however, is the metabolic substrate. The metabolism of carbohydrate consumed produces equivalent amounts of O_2 and CO_2, respectively (i.e., R_Q = 1.0), while metabolism of protein produces only 0.71 mol of CO_2 for every mol of O_2 consumed. Protein metabolism results in a R_Q of 0.81 if the metabolic end-product is urea and 0.74 if it is uric acid. Following prolonged starvation, fat and carbohydrate stores are exhausted and protein is the only available metabolic substrate, leading to R_Q values of 0.7–0.8. In most animals a mixture of metabolic substrates are used, and the R_Q usually is between 0.8 and 0.9.

It must be emphasized that the respiratory quotient reflects $\dot{M}_{O_2}$ consumption and $\dot{M}_{CO_2}$ at the cellular level. In fact, R_Q cannot be measured directly in intact animals. The actual ratio of O_2 and CO_2 flux between the animal and surrounding environment is termed the *Gas Exchange Ratio* (R_E). Under steady-state conditions, R_E = R_Q. However, if CO_2 is in disequilibrium, R_E can differ considerably from R_Q. In intermittent breathing as mentioned above, R_E can range from as low as 0.3 during prolonged breath holding to as high as 2.0 during a short bout of breathing (27, 161), even though at the cellular level $\dot{M}_{O_2}$ and $\dot{M}_{CO_2}$ remain constant. As an example, Figure 40 shows plots of pulmonary P_{O_2} and P_{CO_2} of lung gas from the turtle *P. scripta* and the tortoise *T. graeca*. At high P_{O_2} values, which prevail during the brief episodes of air breathing, R_E exceeds 1.5, while at the lower P_{O_2} values occurring during diving R_E can fall below 0.3.

Acclimatization, Acclimation, and Adaptation

Like most physiological processes, the overall metabolic rate of an organism is influenced by that organism's history, both in a short-term sense and in genetic, evolutionary terms (134, 135). The broad category of terms used to describe patterns of change in metabolism are *acclimation, acclimatization,* and *adaptation,* and have been discussed extensively in Chapter 1.

Metabolic acclimation has been used to refer to compensatory changes in metabolism that occur in response to an individual animal's chronic exposure to a new set of conditions (e.g., temperature, diet, and activity level). Acclimation is usually used to refer to changes that are experimentally imposed in the laboratory or a modified field setting. Metabolic accli-

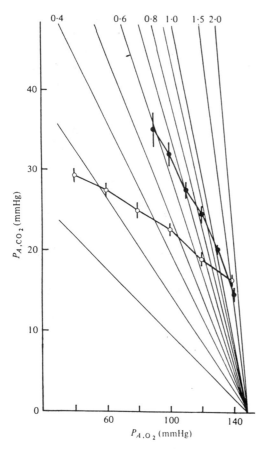

Figure 40. Relationship between P_{O_2} and P_{CO_2} in alveolar gas during voluntary intermittent breathing in the freshwater turtle *P. scripta* (open symbols) and the terrestrial tortoise *T. graeca* (closed symbols). Also indicated are the values of the pulmonary gas exchange ratio from 0.2 to 2.0 (27).

lower or higher metabolic rates as a consequence of some highly persistent environmental factor in a particular geographic region. Because the metabolic phenotype of the organism is genetically determined, the results of adaptation are not permanently reversible, although they can be tempered by processes of acclimation or acclimatization.

A discussion of metabolism of any particular animal necessarily must take into account not only modifying factors, but also the environmental context of these factors. In particular, it is important to differentiate, where possible, the separate effects of acclimation, acclimatization and adaptation.

Metabolism and Modifying Factors

Biological literature includes many records of oxygen consumption by various animals, yet metabolic values are meaningful only for the particular conditions of measurement, and tabulated values are subject to necessary qualification. Rates of metabolism, $\dot{M}_{O_2}$, are influenced by the thermal status of an animal (poikilothermal $\approx$ ectothermal; homeothermal $\approx$ endothermal), activity, ambient temperature, body size, life-cycle stage, season, and time of day, as well by prior conditions met during development and growth (ontogenetic experiences), and genetic background.

Homeothermy and Poikilothermy

A third classification of thermal status often is used when comparing the energetics of poikilothermal and homeothermal animals. The former generally have low or slow metabolism ($\dot{M}_{O_2}$) and are identified as bradymetabolic, while the latter with their more rapid $\dot{M}_{O_2}$ are called tachymetabolic (32). Unfortunately none of the classifying terms is universally applicable for variant patterns of metabolism such as periodic endothermy and

mation is usually reversible by a return to original conditions. *Metabolic acclimatization,* on the other hand, refers to metabolic compensations that result from seasonal and climatic changes that occur under field conditions. Like acclimation, the effects of acclimatization often are reversible once the environmental determinant is removed. *Metabolic adaptation* has been used to identify natural selection acting over generations of organisms, that results in altered phenotypes showing periodic torpor, and the category to be considered later—behavioral thermoregulation. Sometimes the periodic variants are lumped into still another category called heterothermy.

A kind of shivering thermogenesis (periodic thermoregulation) is used by some large insects like dung beetles, honey bees, and hawk moths, and by lamnid sharks and tunas in order to keep locomotory muscles warm. They seem to be homeothermic when active for they also successfully avoid run-away heat gain, which is taken as evidence of thermoregulation, a feature ordinarily implied with the use of the term homeothermy. When a flight or swimming bout ends, the warm parts of the body cool down. Converse situations occur when animals temporarily abandon homeothermy with daily or longer periods of torpor; that is allowing their body temperature to fall and drift with ambient temperature. This periodic torpor minimizes costs of sustained tachymetabolism which is very high for small hummingbirds, bats, and rodents, and verifies the truism that being small and endothermic bears a very high metabolic cost. For most, dropping the metabolic rate nearly to ambient, greatly extends the number of food-free days that can be tolerated during seasons when food is scarce or foraging difficult (5). In a sense, all organisms are endothermic for they generate heat as a metabolic by-product. In current usage, however, endothermy is taken to mean that rates of heat production must exceed rates of heat loss so that body temperature is not only above ambient but that both rate processes are regulated to a stable, above-ambient temperature.

Many interspecific comparisons of poikilotherms and homeotherms appear in the literature that have been made with metabolic rates *corrected* for body temperature differences in animals of about the same body mass. The results suggest rate differentials of at least three times (32, 72, 158, 168).

Oxygen Availability

Conformity and Regulation. Oxygen availability varies in both space and time in many environments, as has been emphasized earlier in this chapter and in Chapter 10. Freshwater aquatic habitats, in particular, may impose limitations to the rate of $\dot{M}_{O_2}$ that can be achieved by an organism. In part this is a result of the lower O_2 capacitance of water compared with air but also because of potential effects produced by floral and faunal respiration. At high altitude, by virtue of the lower barometric pressure and thus lower P_{O_2} at air saturation, hypoxia (relative to sea level) is imposed on both aquatic and terrestrial animals.

The metabolic responses of animals to O_2 limitations in the environment are complex and varied, although metabolic responses generally can be placed in one of two categories. Animals that are able to maintain their $\dot{M}_{O_2}$ at ambient P_{O_2} values below air saturation are termed *oxygen regulators*. Metabolism of O_2 regulators is maintained down to a critical, species-specific P_{O_2} (termed the P_c), below which O_2 uptake begins to fall (Fig. 41). Oxygen regulators include Protozoa, freshwater and terrestrial annelids, echinoderm eggs, many molluscs and crustaceans, some aquatic and probably all terrestrial insects, and almost all vertebrates (134).

In some animals the P_c is effectively at air saturation; the $\dot{M}_{O_2}$ decreases in proportion to decreasing ambient P_{O_2} below air saturation (Fig. 41). Such animals are termed *oxygen conformers*, and include a few Protozoa, numerous coelenterates, free-living annelids and marine worms, most parasitic worms, some molluscs and crustaceans, adult echnioderms, and a few aquatic insects. Generally, O_2 con-

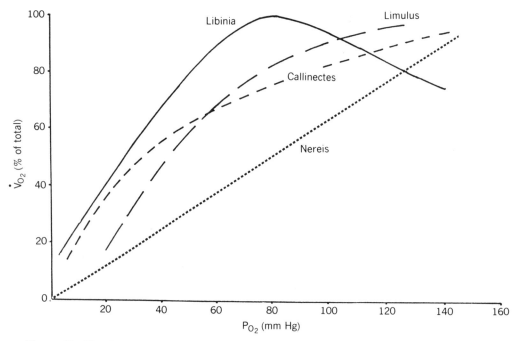

Figure 41. The O_2 consumption (expressed as 100% of total O_2 consumption) as a function of ambient P_{O_2} in four species of invertebrates. *Libinia* is an O_2 uptake regulator, while *Nereis, Calinectes* and *Limulus* are O_2 uptake conformers. Data replotted from (36).

formity is common in marine invertebrates (40, 77, 134, 156), particularly large inactive forms lacking highly specialized respiratory organs and circulatory systems. The inability rapidly to transport O_2 to the metabolizing tissues appears to be the limiting factor of $\dot{M}_{O_2}$ in conformers. This limitation is evidenced by the fact that thin tissue slices may have higher rates of M_{O_2} than whole animals, and that smaller animals, even with their higher mass-specific metabolic rate (see section on Body Size below) show a lower P_c than larger animals of the same species.

The distinction between oxygen conformity and oxygen regulation is not a sharp one, particularly when the P_c is relatively high. Moreover, both invertebrates and ectothermic vertebrates may be O_2 regulators at low temperatures when $\dot{M}_{O_2}$ requirements are low, but they may be O_2 conformers at high tempera-

tures and at the attendant higher metabolic rate. Commonly, P_c decreases with previous exposure to hypoxia, reflecting acclimation to low O_2 by one or more of the physiological systems involved in supplying O_2 to the metabolizing tissues (e.g., cardiovascular system, respiratory system, and blood and its respiratory properties). It is also likely that the metabolic response to O_2 availability is not absolutely fixed, but can range from conformity to regulation to low P_c values depending upon the physiological state of the animal. This possibility is emphasized by contradictory findings of O_2 conformity and regulation in vertebrates. The salamander *Desmognathus fuscus* has been reported to be both an O_2 conformer (55) and a regulator (9) under similar conditions; other examples of conflicting findings on amphibians abound (175). Similarly, the toad fish *Opsanus tau*, long

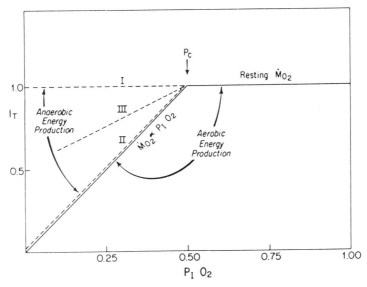

Figure 42. Patterns of contribution of aerobic and anaerobic energy production to total metabolism (M_T) as a function of reduced ambient P_{O_2}. Pattern I = homeometabolic response. Pattern II = poikilometabolic response. Pattern III = hetermetabolic response. See text for further details (69).

considered one of the few examples of an O_2 conformer among vertebrates, has been shown to be an O_2 regulator under certain circumstances (178). A number of other fishes are reported to be O_2 conformers, or at least have very high P_c values (26). Until a definitive study addressing the extent and causes of interspecific variation in P_c is performed, the absolute categorization of aquatic vertebrates as O_2 regulators or conformers should be avoided.

Anaerobic Metabolism and Oxygen Debt. A reduction in $\dot{M}_{O_2}$ with decreasing ambient P_{O_2} does not necessarily result in a similar decline in overall metabolism if the animal can supplement its energy requirements using anaerobic metabolism. Possible interactions between aerobic and anaerobic metabolism as a function of ambient P_{O_2} are indicated in Figure 42. A homeometabolic response, depicted in Pattern I, requires a large anaerobic energy production to maintain total metab-

olism below the P_c for oxygen uptake. Pattern II indicates a poikilometabolic response, in which there is no significant anaerobic contribution below the P_C, and total metabolism falls with O_2 uptake. Pattern III, which actually reflects a very wide group of the most common responses, is a heterometabolic response in which anaerobic metabolism partially compensates for the declining O_2 uptake below the P_c.

Of critical importance to the contribution of anaerobic pathways to overall metabolism is the time course of limited O_2 availability combined with the period of time for which anaerobiosis can be sustained. Some animals, termed *facultative anaerobic organisms*, can so effectively replace aerobic metabolism using anaerobic pathways that they can survive for days or weeks in the complete absence of free O_2, a phenomenon quite common among invertebrates and particular forms that are parasitic, burrowing (159), or intertidal (40, 77, 156). However, some verte-

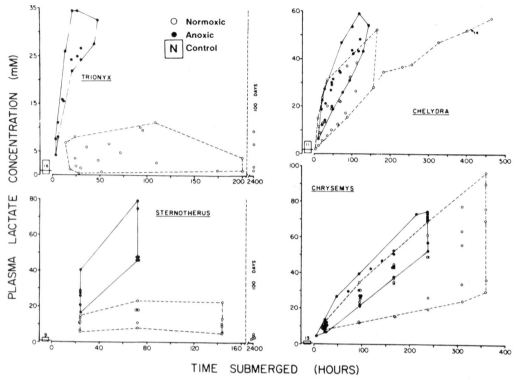

Figure 43. A comparison in four species of freshwater turtles of plasma lactate levels resulting from prolonged submergence in normoxic or anoxic water at 10°C. *Chelydra* and *Chrysemys* cannot maintain aerobic metabolisms without access to air, and utilize anaerobic metabolism as evidenced by lactate accumulation. *Trionyx* and *Sternotherus* utilize anaerobic metabolism, and thus accumulate lactate, primarily when submerged in anoxic water (177).

brates, especially those that might be expected to encounter anoxic conditions at low temperatures during overwintering, show surprising anaerobic capacities. Goldfish (*Carassius auratus*) and carp (*Cyprinus carpio*), for example, survive anoxia at 4°C by utilizing biochemical pathways that are highly unusual for a vertebrate. Carbohydrate is fermented to lactate, producing CO_2 and ethanol (77). The lactate is then converted to pyruvate and then to acetyl CoA and CO_2. Many freshwater turtles (e.g., *Trionyx, Sternotherus, Chelydra* and *Chrysemys*) can survive anoxia for days or even months, apparently using more conventional vertebrate anaerobic pathways that result in the production of

remarkably high plasma concentrations (Fig. 43) (177). The ability of freshwater turtles to utilize anaerobiosis as the sole energy source while surviving the acid–base disturbances resulting from lactic acid as the end-product appears to be an adaptation for cold-weather overwintering under water rather than to warm-water diving.

Anaerobic metabolism, besides having a far lower energy yield in terms of ATP, in addition, produces metabolites that (a) are potentially disruptive to coexisting metabolic pathways or to tissue acid–base balance and (b) represent a potential energy loss if excreted rather than reconverted into energy-rich compounds

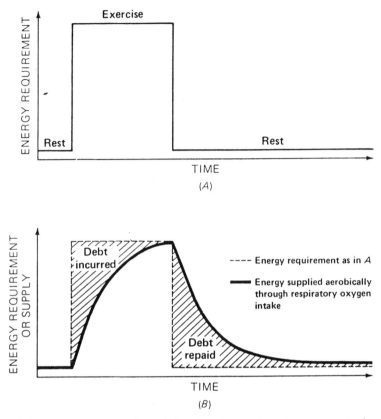

ENERGY REQUIREMENT

Exercise

Rest

Rest

TIME

(A)

ENERGY REQUIREMENT
OR SUPPLY

Debt
incurred

- - - - Energy requirement as in *A*

——— Energy supplied aerobically
through respiratory oxygen
intake

Debt
repaid

TIME

(B)

Figure 44. Schematic representation of the development and repayment of an oxygen debt resulting from exercise (74).

suitable for further metabolism (see Chapter 8). The capacity to accumulate the end-products of anaerobic metabolism represents an important adaptation to animals exposed to anoxia or hypoxia because of either environmental limitations or as a result of sustained activity (135). Invertebrates show a comparatively wide variety of anaerobic end-products that vary in the physiological disruption they cause and in the ease with which they can be further metabolized (40, 78, 156) (see Chapter 8). As already alluded to, lactic acid is the primary end-product of anaerobic metabolism in vertebrate muscle. With the resumption of aerobic metabolism, lactate is carried to the liver where up to 80% is reconverted by glu-

coneogenesis to carbohydrate. The amount of oxygen consumption attributed to the aerobic metabolism of anaerobic end-products is termed the "oxygen debt" (Fig. 44). The size of the posthypoxic M_{O_2} is usually proportional to the amount of anaerobic end-product accumulated. The rate of *repayment* of an oxygen *debt* varies greatly between organisms, and this directly reflects glu coneogenic capacities and the extent (usually low) to which lactate is excreted rather than remetabolized. Oxygen debt may also be defined as an "ATP debt" or a "phosphagen debt" (e.g., phosphoargenine, phosphocreatine), especially in muscle where the ultimate source of energy to recharge contractile processes is

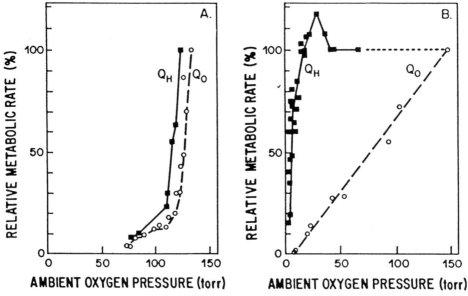

Figure 45. The O_2 consumption (Q_O) and total energy production (Q_H) measured as heat production, as a function of ambient P_{O_2} in the marine invertebrates *Lingula achatina* (A) and (B) *C. virginica* (110).

ATP, which in turn is regenerated at mitochondria by both oxidative and transferase rephosphorylation.

Finally, it should be emphasized that a successful alternative way of dealing with severe hypoxia or completely anoxic conditions can be the reduction or complete suspension of all metabolism, aerobic or anaerobic. Figure 45 indicates optional responses to decreasing O_2 availability in marine invertebrates (67). In *Lingula achatina*, an O_2 conformer, total energy production decreases in parallel with $\dot{M}_{O_2}$, indicating little dependence on anaerobic metabolism during hypoxic or anoxic conditions. In *Crassostrea virginica*, also an O_2 conformer, the total energy production is maintained until very low levels through anaerobiosis. Each animal survives hypoxic exposure—*Crassostrea* responds *actively* by utilizing anaerobic pathways to maintain energy production, *Lingula* responding *passively* by allowing total metabolism to fall and not having to deal with the buildup of potentially disruptive metabolic end-products. Examples of a passive response to severe hypoxia can also be found among vertebrates. In the sturgeon *Acipenser transmontanus*, an apparent O_2 conformer, the declining $\dot{M}_{O_2}$ is accompanied by no change in plasma pH, which might be anticipated if lactic acid were being produced in quantity (26). Moreover, upon return to normoxic conditions, *Acipenser* shows no increased $\dot{M}_{O_2}$ suggestive of the repayment of an O_2 debt.

Diving animals, in addition to turning to anaerobic metabolism during especially long dives (41, 76, 177), may also show a reduction in total metabolism during diving. This response, which has been termed a "strategic retreat," has been observed in diving ducks, seals, and alligators (41). In the gray seal *Halichoerus gryphus*, for example, the increase in post-dive $\dot{M}_{O_2}$, representing the repayment of the O_2 debt incurred during the dive and presumably proportional to the magnitude of anaerobic metabolism during the

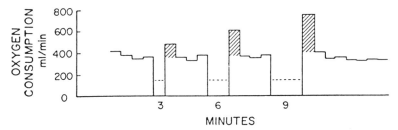

Figure 46. The pattern of oxygen consumption of a gray seal *H. gryphus* during intermittent diving. The shaded area is the post-dive increase in V_{O_2} above pre-dive levels, and is equivalent in area to that below the dashed line in the preceding dive. Note that the increase in $\dot{V}_{O_2}$ over pre-dive levels in $\dot{V}_{O_2}$ measured after diving was smaller than the actual reduction in $\dot{V}_{O_2}$ during the dive (41).

dive, is considerably smaller than the actual reduction in M_{O_2} during the dive (Fig. 46).

Activity

One of the intrinsic modifiers of O_2 consumption most difficult to control is muscular and other activity. By definition, "basal" metabolism represents the oxygen consumption for maintenance only. In humans this is measured in a postabsorptive state, in a relaxed but awake subject, usually in mid-morning. For measurements on animals, movements are minimized by darkness, quiet, and habituation to the metabolism chamber. Measurements may be made on curarized or anesthetized animals, but muscular tone varies with level of anesthesia. Routine metabolism refers to O_2 consumption measured with uncontrolled but minimum activity. Measurements may be made during enforced activity as with treadmills or swimming tunnels; maximum levels being termed "active" metabolism. When different levels of activity are used, O_2 consumption may be extrapolated to zero activity; this extrapolated "lowest" resting value is "standard" metabolism (52). Table 6 lists standard and active M_{O_2} for size ranges of various poikilotherms and homeotherms and factorial differences between the two levels of activity. With a few exceptions, for example, moths and butterflies, the multiple of the increase in energetic cost of activity between standard and active metabolic levels of animals in general is surprisingly similar despite marked differences in body temperature. As Figure 47 shows, differences in the metabolic costs of locomotion per unit distance travelled are highly dependent upon body mass—running being most costly, swimming cheapest, and flying intermediate. The incremental cost for terrestrial runners also is remarkably similar for animals of about the same size whether they are mammals, reptiles, or birds (Table 7) (157, 174).

When standard and activity $\dot{M}_{O_2}$ are measured as a function of some environmental parameter, such as temperature (Fig. 48A), the difference between the two curves is the scope for activity (Fig. 48B). When standard and active $\dot{M}_{O_2}$ measurements are made at different oxygen partial pressures, the activity metabolism remains constant down to an incipient limiting level; the standard metabolism is constant to a lower P_{O_2}, which is the level at which there can be no excess activity and below which, the needs of maintenance cannot be met.

Increased O_2 consumption need not be proportional to the increased activity, and

TABLE 6. Standard and Active Metabolism of Various Animals

Animal	Body Weight (g)	$\dot{M}_{O_2}$ (mL O_2/g · h) Standard	Active	Factorial Difference
Fruit fly *Drosophila*	1.0×10^{-1}	2.3	30	13
Butterfly *Vanessa*	0.3	0.60	100	170
Sunfish *Lepomis*	50	0.031	0.29	9
Salmon, young *Onchorhynchus*	63	0.084	0.60	8
Slider turtle *Pseudemys*	305	0.031	0.64	20
Desert iguana *Dipsosaurus*	35	0.052	0.89	17
Monitor lizard *Varanus*	714	0.080	0.37	5
Hummingbird *Calypte*	3	2.80	42.0	15
Budgerigar *Melopsitacus*	35	1.70	18.9	11
House mouse *Mus*	20	2.50	20.0	8
Dog *Canis*	2.6×10^3	0.33	4.0	12
Lion *Felis*	5.0×10^4	0.23	3.0	13
Human *Homo*	7.5×10^4	0.23	3.2	14
Horse *Equus*	1.0×10^5	0.21	1.6	8
Eland *Taurotragus*	2.4×10^5	0.17	2.2	13
Elephant *Elephas*	3.7×10^6	0.113	1.25[a]	11
Blue whale *Sibbaldus*	1.0×10^9	0.038[a]	0.62[a]	16

[a]$\dot{M}_{O_2}$ estimated from allometric equations for metabolism and body size based on: $\dot{M}_{O_2} = aWb$; a = y intercept, b = slope.

animals differ significantly in their ability to use energy from anaerobic routes and to incur oxygen debts. In Iguana, the anaerobic production of lactic acid provides 75% of the energy of activity (119). The contribution of muscle to total metabolism increases in humans from 20% under basal conditions to 80% during exercise.

Nonmuscular activity is also reflected in metabolic increase, as in reproductive organs and related structures during a reproductive season. *Mytilus* consumes ~60% more oxygen in winter months when gametogenesis is actively occurring

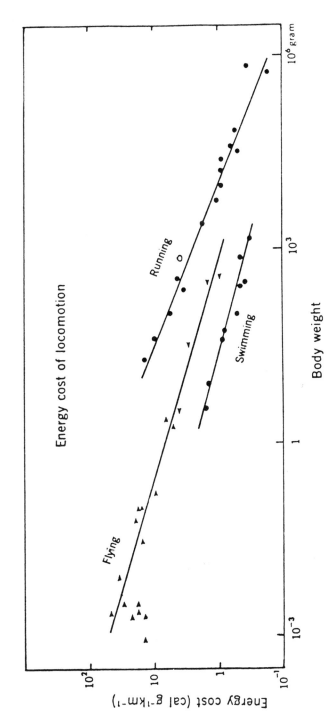

Figure 47. Energy cost of transport for swimming, flying, and running animals related to body size. The single open circle is the cost of locomotion for a swimming duck (157).

417

TABLE 7. Equations for the Cost of Running Related to Body Size (M_b, in g) as Obtained by Various Investigators; the Mean of the Body Mass Exponents for All These Equations Is -0.32. [From (158).]

Number of Species	Equation for Cost (mL of O_2/g · km)	Cost for $M_b = 1$ kg (L of O_2/km)
7 mammals	cost = $8.46\ M_b^{-0.40}$	0.53
8 reptiles	cost = $5.9\ M_b^{-0.33}$	0.60
7 birds	cost = $2.45\ M_b^{-0.20}$	0.62
22 mammals	cost = $6.35\ M_b^{-0.34}$	0.61
52 mammals, birds, reptiles	cost = $5.01\ M_b^{-0.32}$	0.55
69 birds, mammals, reptiles	cost = $3.89\ M_b^{-0.28}$	0.56
72 mammals, birds, reptiles, ants	cost = $8.61\ M_b^{-0.352}$	0.76
62 mammals, birds	cost = $4.73\ M_b^{-0.316}$	0.53

than in the summer when gamete production is suspended (7).

Temperature and Seasons

Metabolism of inactive or resting homeothermal animals is minimal over a seasonally labile range of temperature (zone of thermal neutrality), above and below which their O_2 consumption rises. The metabolism of most poikilotherms rises or falls about two and one-half times per 10°C change (i.e., Q_{10}) within their thermal limits. The standard metabolism of poikilotherms, defined as near basal or resting (52), increases continuously with temperature up to lethal levels. The active or routine (spontaneous) (52) metabolism of fishes (8) (Fig. 48) and amphipods, *Gammarus oceanus* (67) (Fig. 49), may either increase to a plateau or pass through a peak at a temperature above which the animals are incapable of sustaining higher levels of metabolism. The difference between active and standard metabolism (i.e., the energy available for work), which is identified as the metabolic scope for activity (52), also changes with temperature. Peaks in the activity scopes for brook trout occur at 15°C and for carp at 25°C (Fig. 48B), but no such peak occurs for bullheads (8). In some cases (Figs. 48B), peak metabolic scopes

for swimming were obtained during forced or "paced" activity. Thus, the metabolic rates shown in Figure 48 are not necessarily comparable with the acclimated metabolic rates that would be obtained during spontaneous, undisturbed activity when an individual fish or a school of fishes "selects" a particular swimming velocity (125, 147). Metabolic rate comparisons that are free of the experimental bias or operant conditioning that is introduced by such techniques are enforced swimming, have been a great aid in a few cases for the derivation of reasonable judgments about the ecological value of seasonal metabolic compensations to poikilothermal animals. Use of thermal gradients of the shuttlebox type (Fig. 50) (120) and very large tanks that allow free swimming (125) (both minimize operant conditioning), and the controlled use of operant conditioning (150) are ways to study routine thermoregulatory behavior independently of the loading effects of activity upon metabolism. These approaches have verified the view that a thermal preferendum or eccritic temperature for a species is a narrow range of temperatures within which a poikilothermal animal ceases to receive aversive thermal feedback (125, 151) and within which it is least active (147).

Aquatic poikilotherms are often sea-

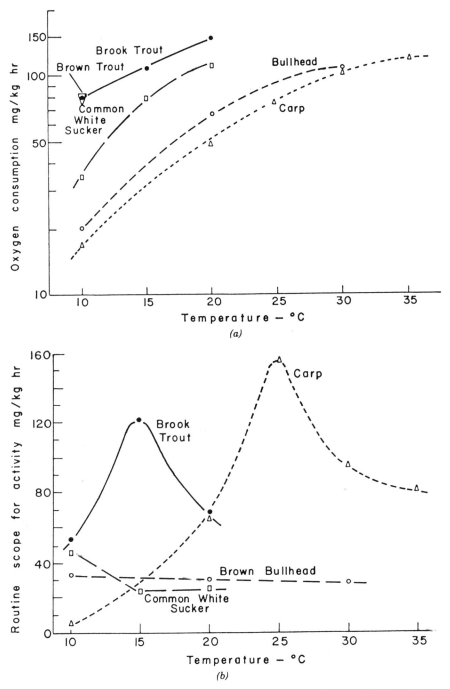

Figure 48. (*A*) Standard O_2 consumption by 100-g specimens of several fish as a function of temperature. (*B*) Influence of temperature on scope for activity (difference between active and standard metabolism) in freshwater fish (8).

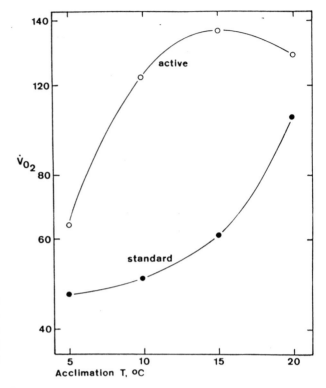

Figure 49. Acclimated V_{O_2} levels as mL/g · h, wet weight, of spontaneously swimming (routine metabolism) and inactive (standard metabolism) amphipods, G. oceanicus (67).

sonally subjected to wide ranges in temperature and thus show evidence of both physiological and behavioral acclimation. Individuals acclimated to cold tend to have higher metabolic rates at a given temperature than ones acclimated to high temperatures; although at low acclimation temperatures their metabolism or-

dinarily is less than that observed during warmer seasons. The net effect is to compensate for environmental changes related to season and latitude (29) (Fig. 51). Much evidence indicates that such seasonal compensations are mediated in animals, not just by temperature alone, but by integrative mechanisms that detect

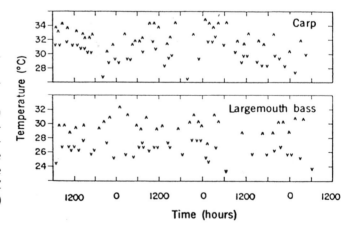

Figure 50. Self-programmed temperature selection by carp, and largemouth bass, during 3-day test periods in an electronic, shuttlebox aquarium. Upper (∧) and lower (∨) turn-around temperatures are shown for movements of fish through the connecting tube between the two tanks of the shuttlebox (2°C differential, right tank higher) (120).

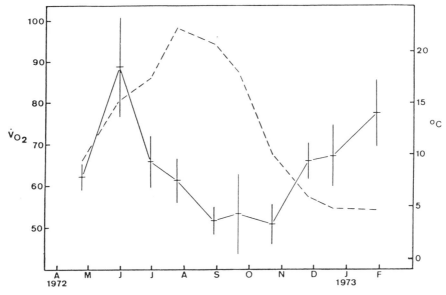

Figure 51. Seasonal changes in the routine V_{O_2} as mL/kg · h of sunfish, *Lepomis gibbosus,* measured at 17.5°C after 20–23 h at 17.5°C (solid line). The dashed line gives temperatures at times of seasonal collections (2 days previous). Vertical bars ± 2 standard errors of means (29).

and measure seasonally appropriate photoperiods. Specific effects of temperatures on metabolism are discussed in Chapter 3.

Photoperiods and Rhythms
Many animals show daily, lunar, annual, and other rhythms of oxygen consumption, which persist in the laboratory under conditions that are constant with respect to light, temperature, and pressure. A widely held view is that these rhythmic responses usually are mediated via the actions of hormones and neurohemal agents.

Receptor transduction systems, originally based upon light detection and photoperiodic time measurement, also must include the ability to record and summate changes in the seasonal precession of daily photoperiods. Consequently, measurement of photoperiod length somehow provides a highly precise calendar index—perhaps the best cue available for

the physiological "anticipation" of seasonal change. However, some assumptions are required. The first is that a high light intensity is not a detection requirement; and second, that the detection sensitivity of the receptor should roughly obey the Weber–Fechner rule as it applies to light detection so that a detectable change in response approximates the $\log_{10}$ of light intensity. Both assumptions are supportable on the basis of current knowledge about photic detection in animals and plants. On a heavily overcast day (1% full sunlight) versus a day with full sunlight, the error in day length measurement is likely to be <4 min (192). Additional assumptions follow from the facts that daily periodicities of animals activities often are endogenous or circadian (usually 24 h). That is, these activity cycles free-run for a number of days under constant, low-light levels, and that somehow these endogenous circadian rhythms (194) are translated by as yet unknown mechanisms into periodicities of longer duration (weekly, monthly, and annual).

Photoperiodic time measurement probably evolved as a means for coordinating or fine tuning reproductive activity to seasons in the year when the probability of offspring survival is best (29, 59, 147). The release of gametes and or fertilized eggs is preceded by increased metabolic demands for support of gamete production, courtship behavior, nest building, and often is followed by other, metabolically demanding activities such as the guarding and feeding of the young.

Superposition of several rhythms such as solar and lunar is known. In the fiddler crab *Uca*, oxygen uptake is 3–50% higher at 0600 than at 1800 h, and in a lunar cycle $\dot{M}_{O_2}$ is 30–50% higher when the moon is at either its highest or lowest transit than when it is at the horizon (179). Locomotor activity of an intertidal isopod, *Excirolana chiltoni*, shows a similar dependence on lunar cycles, but one that also seems tied to the complex interactions of tidal height and frequency with solar and lunar cycles (43).

A two-part question remains: (1) how much of an observed change in oxygen consumption of an individual, found to be relatable to circadian or other rhythms of differing periodicity, is based on increases in oxygen consumption of most cells; and, (2) how much is a result of behavioral modifications that alter routine activity and muscle tone. Probably the answer is that most seasonal and latitudinal compensations are of the second sort, but identification of the true proportioning of effects is seldom attainable (147).

Body Size

The analysis of variation in metabolic rate resulting in differences in body mass is a very well developed field, and draws heavily on so-called allometric or scaling equations (30, 72, 158). The relation between total metabolism (*M*) and body mass (*W*) is given by the equation $M = KW^b$. Most commonly, *M* is expressed as O_2 consumed per unit time. Of course, $\dot{M}_{O_2}$ does not include any anaerobic contributions to metabolism, and thus expressing metabolic rate as heat production (in kcal) more accurately represents total metabolism. The constant *K* represents the absolute level of metabolism, and describes the vertical intercept on a plot of log *M* as a function of log *W*. Larger animals have a higher total body metabolism, and thus produce more heat, than do small animals (Fig. 52*A*). Of great biological significance is the fact that the total metabolism of animals does not increase in direct proportion to increasing body mass. The exponent *b* in the allometric equation, which is the slope of the line in the plot, relates the rate of change of metabolism (*M*) with changing body mass. If metabolism changes in direct proportion with body mass, *b* = 1. If metabolism does not increase as rapidly as body mass, however, then *b* < 1. The solid line in Figure 52*A* represents a line with an exponent *b* = 0.75, while the dashed line represents a line with an exponent *b* = 0.67. Thus, metabolism in this grouping of mammals does not increase at the two-thirds proportionality with increasing body mass as does surface area (158).

The relationship between metabolic rate and body mass is more readily visualized by plotting *mass-specific* metabolic rate (also called *metabolic intensity*) against body mass (Fig. 52*B*). It can be derived by dividing the equation $M = KW^b$ by mass so that it simplifies to $\dot{M}_{O_2} = KW^b/W$ or $\dot{M}_{O_2} = KW^{b-1}$. Very small increases in absolute body mass cause very large decreases in metabolic rate in small animals in particular. Analyses of literally hundreds of species of organisms have repeatedly confirmed this relationship

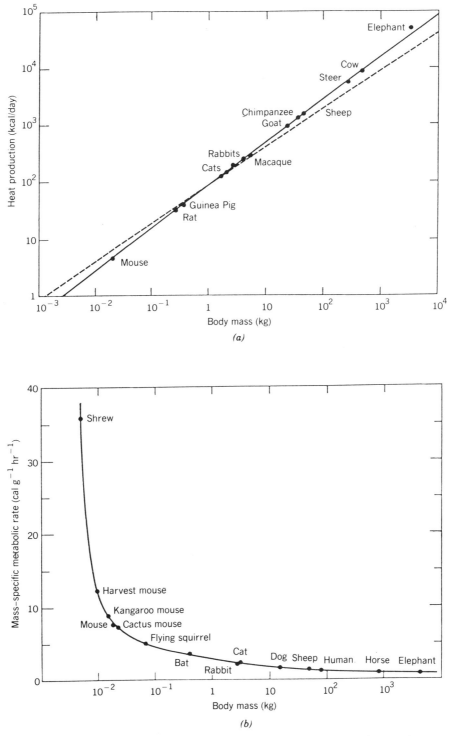

Figure 52. Relationship between body mass and metabolic heat production in a range of mammals with widely differing body mass. In the top panel the absolute heat production is plotted. The solid line has a slope of 0.75, while the dashed line has a slope of 0.67. In the bottom panel the data are reexpressed as mass-specific metabolic rate (115).

(30, 131, 158), which was first appreciated over 150 years ago (152) but only formalized ~5 decades ago by Kleiber (95). The relationship between size and mass-specific metabolic rate appears to be one of the fundamental principles in physiology, applying equally to large ungulates, insect eggs, protozoa, and bacteria.

Importantly, mass-specific metabolism generally decreases with increasing body mass (and with age, especially among endotherms) both for arrays of differently sized individuals within a species as well as between species. The precise relationship as described by the slope, b from the allometic equation does seem to vary depending on whether the comparison is intra or interspecific. Allometric analysis of metabolic rate versus body mass of adults of a variety of different-sized species routinely yields a different slope (30, 131, 158). Yet, intraspecific analysis of total body metabolism versus body mass often yields a slope of 0.75 (71). An additional complication when determining metabolic scaling effects within a species is that smaller individuals may well be at a younger stage of development than large individuals of that species. Thus, allometric parameters may well be influenced by developmental factors that in fact are unrelated to body mass.

The metabolism of an animal, as measured by either heat production or $\dot{M}_{O_2}$, is the outward manifestation of many important physiological processes. That metabolism scales negatively with increasing body mass strongly suggests that metabolic support processes—for example, blood flow, ventilation of gas exchange organs and nutrient assimilation—will also show similar scaling effects, and indeed this is the case (30, 131, 158). Table 8 presents selected data from mammals on the allometry of a wide range of physiological processes intimately related to metabolic support. While some processes scale to body mass with an exponent sim-

TABLE 8. Average Slope (b) of the Allometric Relationship Describing Exchange and Transport Parameters in a Comparison of Different Species of Adult Mammals

Variable	Slope
Metabolic rate	0.75
Body surface area	0.67
Lung volume	1.02
Tidal volume	1.02
Breathing frequency	−0.26
Air–blood barrier	0.04
Oxygen diffusing capacity	1.12
Breathing power	0.74
Heart mass	0.96
Heart volume	0.98
Stroke volume	1.04
Heart rate	−0.26
Cardiac output	0.86
Blood O_2 capacity	1.0
P_{50}	−0.05
Blood carbonic anhydrase	0.89

[a]Published data from tables compiled by Peters (157).

ilar to that of metabolism, others scale directly with body mass or appear to be independent of body mass. Analysis of cardiovascular systems makes this point clear. Stroke volume is proportional to heart mass, which scales in direct proportion with body mass. Heart rate, however, scales negatively with body mass. Since cardiac output is the product of heart rate and stroke volume (Chapter 11) the net effect is that the changes in cardiac output associated with body mass change closely parallel the change in total metabolism.

Why is the relationship between metabolism and body mass so pervasive, and what are the mechanisms that underlie it? An early explanation for the relationship between body mass and metabolism involved the so-called "surface rule." The surface area of objects tends to increase at about two-thirds the rate of their volume (and thus mass). Because this is similar to the rate of growth of metabolism as mass increases, explanations were sought to link the two factors. Since metabolism is associated with heat production, and since the rate of heat loss is directly proportional to the rate of heat loss, it was proposed that the lower mass-specific metabolic rate of smaller animals was an adaptation to prevent excessive heat buildup in large animals. This explanation, apart from providing no mechanism for what limits metabolic rate, began to lose favor as increasing amounts of experimental data from ectotherms (including very small invertebrates) indicated that mass-specific metabolic rate scaled to body mass with the same exponent ($\sim$0.75) as for endotherms.

Several alternative hypotheses have been suggested to account for the scaling between mass-specific metabolic rate and body mass. Essentially, the arguments revolve around scaling of the metabolic ability of tissues and cells themselves, and scaling in the ability of the transport systems to supply the tissues. If the metabolic ability of the tissues dictates scaling effects of metabolism, then this should be reflected in the metabolism of the tissues *in vitro*. Early investigations of individual tissue slices of different organs from different sized animals suggested that cell metabolism was independent of the mass of the animal from which they were acquired, but these studies were seriously flawed due to experimental conditions that resulted in apparent diffusion-limitations to O_2 and nutrient transport. However, experiments on five major tissues from nine species of mammals, performed using adequate oxygenation and appropriate media for metabolic measurements of each tissue, have revealed that the metabolism of individual tissues decreases as body mass increases, though not to as great an extent as the metabolic rate of intact animals (97).

While the results of *in vitro* studies of tissue metabolism depend considerably on the preparation of the tissue (i.e., slice or cell suspension), degree of oxygenation, and media composition (158), *in vitro* studies confirm that the *intrinsic* metabolic rate of tissues in smaller animals is higher. Corroborating these findings, increasing body mass is also accompanied by a mass-specific reduction in the concentration of cytochrome c (39), cytochrome oxidase (84), and citrate synthase, β-hydroxybutyl CoA dehydrogenase, and malate dehydrogenase (42). Mitochondrial density has been shown generally to decrease with increasing body mass in mammalian liver tissue (165) and skeletal muscle (109). The mitochondrial density of heart muscle of the Etruscan shrew (*Suncus etruscus*), one of the smallest living mammals, is not higher than found in much larger mammals, but the density of the cristae in shrew mitochondria is higher (185).

Many of the studies that reveal biochemical–structural limitations to metabolism in the tissues of larger animals report by values <1 but still higher than the 0.75 exponent that applies to the metabolism of intact animals. This has led some investigators to propose that the ability to transport O_2 and metabolic substrates to the tissues may also be a limiting factor. Many hematological studies have shown that some of the important blood properties (e.g., Hb concentration and O_2 capacity) are generally independent of body mass, while variables such as blood O_2 affinity usually (but not always) decrease with increasing body mass (30, 131, 155) (Chapter 10). Of course, the total transport capabilities of the cardiovascular system depend not only on blood properties but also on cardiac output, which as previously noted scales negatively with body mass. One important variable is blood transport time or circulation time, the time required for blood to make a complete transit of the circulation. Like cardiac output, circulation time scales to body mass with the exponent of ~0.75, which is the same as that for metabolic rate, in a wide range of vertebrates (33, 158). Thus, calculated circulation times range from 140s in a 4000-kg elephant to 4 s in a 3-g shrew (158).

Observations on metabolic rates of human fetuses and infants suggest that limitations to metabolism may be related to transport capabilities as well as the properties of the tissues themselves. The metabolic rate of a fetus, which is dependent upon transport by the maternal cardiovascular system and diffusion across the placenta, can as predicted by allometric analysis be regarded as just another maternal organ (190). Within 36 h of birth (a period far too brief to assume changes in cell structure), the metabolic rate of a newborn infant has doubled to a level predictable by allometric analysis for that of a much smaller, free-living organism.

A search for a single limiting factor to account for the change in metabolism as body mass increases may be misguided. The increasingly popular concept of *synmorphosis* (105) suggests that interdependent processes and structures are optimally designed. From this it follows that any single factor critical to metabolism could be the limiting factor (even a single enzyme in the Kreb's cycle). That the exponent 0.75 shows up repeatedly in allometric analysis of factors important to metabolism supports the concept of an integrated network of processes and structures limiting the growth of metabolism with increasing body mass.

Summary

Oxygen is potentially an important limiting factor in many environments. For air-breathing animals, environmental O_2 is limiting at high altitudes and while diving. For water breathers, the available O_2 is much less (low O_2 solubility, low diffusion rate) and much more variable in relation to temperature and stagnation than in air. In some animals, adaptations occur which enhance O_2 uptake or provide for better distribution and utilization by the respiratory organs; some adaptations relate to the permanent life habit of the species, while others are effective only for brief exposures to reduced O_2 supply, as in exercise.

In all organisms O_2 enters cells by diffusion, and O_2 diffusivity may determine cell size in both free-living cells and multicellular organisms. Oxygen limits the attainable size of an animal that does not have a circulatory system, and it limits the distance between capillaries in those that do. At the metabolic rates of most and at usual P_{O_2} levels, a diffusion distance of 0.5–1 mm is maximal. Intracellular P_{O_2} levels probably are normally one or two orders of magnitude lower than extracellular P_{O_2}.

Multicellular animals have evolved numerous mechanisms for acquiring O_2. In only a few animals is O_2 carried directly to tissues as it is in sponges and coelenterates, and in tracheates (mostly insects). Control of ventilation in insects is by breathing movements and selective spiracle action. In all other animals some sort of fluid—coelomic fluid or blood, separated from the water or air by a thin, moist epithelial membrane—is interposed between the O_2 source and the tissues. Ventilation of membranes where exchange occurs may be accomplished by elaborate muscular mechanisms that are reflexly controlled. These reflexes are initiated mainly by reduced O_2 or by elevated CO_2, the former more commonly in aquatic animals and the latter more in terrestrial animals. Receptors sensitive to low O_2, and others sensitive to high CO_2, may be external as on gill arches of several aquatic tetrapod vertebrates, but most depend on chemoreception in the blood stream (e.g., aortic and carotid receptors). The reflex responses are usually increased ventilation rate (rate or amplitude of breathing or both), or they may be changed blood flow. The net effect is a tendency to maintain a constant supply of O_2 to the organism. Sometimes increased ventilation has secondary effects, such as the alkalosis caused by *blowing off* CO_2, that complicate the attempted increase in O_2 uptake. Also, the extra energy consumed in increased ventilation may make the increased supply of O_2 less useful, as in some fish. There are many auxiliary respiratory organs, such as the oral membranes of turtles, the swim bladders of physostome fish, and the skin of some fishes and reptiles and all amphibians.

Metabolic rate varies considerably according to kind of animal, body size, environmental conditions, and locomotor activity. An important modifier of metabolism is O_2 availability. Many animals are metabolic conformers; they increase O_2 utilization in proportion to available O_2, and at normal atmospheric levels they are not metabolizing to capacity. Other animals are metabolic regulators and have constant metabolism down to a critical P_{O_2} level, which is related to their life habits. Most of the regulation in animals is by variation in external respiration (ventilation, and circulation to respiratory organs).

There is incomplete evidence that activity metabolism may proceed by enzyme pathways slightly different from those of rest metabolism. One useful way to separate rest or standard metabolism from activity metabolism is to drive the animal to different speeds of locomotion and then to extrapolate to zero activity; for some animals it is necessary to measure O_2 consumption during periods when there is no perceptible movement as standard metabolism. The difference between rest and activity metabolism may be many orders of magnitude depending on how near maximal the paced activity is for the individual; discordance in published metabolic data is often caused by absence of activity control. Environment, body temperature, hormonal state, nutrition, and the nature of the foodstuffs being oxidized all cause considerable variations in metabolic rate.

For a given kind of animal, small individuals and species have higher metabolic rates than large individuals. Metabolism is proportional to body weight in a few small animals, but metabolic rates in most (unicellular to large multicellular animals) is proportional to body weight to a power (often 0.75), which is between the power function for body surface (0.67) and unity. There is incomplete evidence that oxidative activity, on a protein or weight basis, is higher at the cellular (perhaps mitochondrial) level in small animals. How this is controlled is unknown. Apart from similar metabolism-size relations, the basal metabolism (for unit body weight) shows genetically determined variations related to life habits—sluggish versus highly active, and capable or not of entry into torpor or hibernation.

Some animals shift from aerobic to anaerobic metabolism for brief periods (as do air breathers during diving, or as in severe exercise or during periods of anoxia). Some, after periods of hypoxia, show increased O_2 consumption and pay back the metabolic debt. Others live for long times on glycolytic energy and excrete the resulting acids (e.g., intestinal parasites and oysters). The balance between aerobic and anaerobic pathways and their coupling represent a complex integrated system. Regulation of O_2 supply by central nervous reflexes are on the level of control by the whole organism. How enzyme levels are set at the cellular level to reflect body size, environmental stresses, and available O_2 is only partly known. Some of these control mechanisms, particularly for hormonal and neural control, are discussed in subsequent chapters.

References

1. Ashby, E. A. and J. L. Larimer. *J. Cell. Comp. Physiol.* 65:373–380, 1965. Modification of cardiac and respiratory rhythms in crayfish following carbohydrate chemoreception.

2. Ballintijn, C. M. In *Exogenous and Endogenous Influences on Metabolic and Neural Control* (A. D. F. Addink and N. Spronk, eds.), pp. 127–140. Pergamon Press, Oxford, 1982. Neural control of respiration in fishes and mammals.

3. Ballintijn, C. M. and G. M. Hughes. *J. Exp. Biol.* 43:349–362, 1965. The muscular basis of the respiratory pumps in trout.

4. Barnes, R. D. *Invertebrate Zoology.* Saunders, Philadelphia, PA, 1974.

5. Bartholomew, G. A. In *Animal Physiology: Principles and Adaptations*, (M. S. Gordon, ed.), pp. 333–406. Macmillan, New York, 1982. Body temperature and energy metabolism.

6. Bartholomew, G. A. and M. C. Barnhart. *J. Exp. Biol.* 111:131–144, 1984. Tracheal gases, respiratory gas exchange, body temperature and flight in some tropical cicadas.

7. Bayne, B. L. *Marine Mussels, their Ecology and Physiology.* Cambridge Univ. Press, London, 1976.

8. Beamish, F. W. H. and P. S. Mookherjii. *Can. J. Zool.* 42:177–188, 1964. Standard activity metabolism of brook trout and goldfish in relation to temperature and body size.

9. Bechenbach, A. T. *Physiol. Zool.* 43:338–347, 1975. Influence of body size and temperature on the critical oxygen tension of some plethodontid salamanders.

10. Bicudo, J. E. P. W. and K. Johansen. *Environ. Biol. Fish.* 4(1):55–64, 1979. Respiratory gas exchange in the airbreathing fish, *Synbranchus marmoratus.*

11. Bretz, W. L. and K. Schmidt-Nielsen, *J.*

Exp. Biol. 56:57–65, 1972. Movement of gas in the respiratory system of the duck.

12. Burggren, W. W. *J. Exp. Biol.* 63:367–380, 1975. A quantitative analysis of ventilation tachycardia and its control in two chelonians, *Pseudemys scripta* and *Testudo graeca.*

13. Burggren, W. W. *J. Exp. Biol.* 82:197–213, 1979. Bimodal gas exchange during variation in environmental oxygen and carbon dioxide in the air breathing fish *Trichogaster trichopterus.*

14. Burggren, W. W. *Physiol. Zool.* 55(4):327–334, 1982. "Air gulping" improves blood oxygen transport during aquatic hypoxia in the goldfish *Carassius auratus.*

15. Burggren, W. W. *Physiol. Zool.* 58(5):503–514, 1985. Gas exchange, metabolism, and "ventilation" in gelatinous frog egg masses.

16. Burggren, W. W. *Can. J. Zool.* 66:20–28, 1988. Cardiovascular responses to diving and their relation to lung and blood oxygen stores in vertebrates.

17. Burggren, W. W. In *Comparative Pulmonary Physiology: Current Concepts* (S. C. Wood and C. Lenfant, eds.), pp. 153–192. Dekker, New York, 1988. The structure and function of amphibian lungs.

18. Burggren, W. W. and M. E. Feder, *J. Exp. Biol.* 121:445–450, 1986. Effect of experimental ventilation of the skin on cutaneous gas exchange in the bullfrog.

19. Burggren, W. W., M. E. Feder, and A. W. Pinder. *Physiol. Zool.* 56:263–273, 1983. Temperature and the balance between serial and aquatic respiration in larva of *Rana berlandieri* and *Rana catesbeiana.*

20. Burggren, W. W., M. L. Glass, and K. Johansen. *Can. J. Zool.* 55:2024–2034, 1977. Pulmonary ventilation/perfusion relationships in terrestrial and aquatic chelonian reptiles.

21. Burggren, W. W. and K. Johansen. In *The Biology and Evolution of Lungfishes* (W. Bemis and W. W. Burggren, eds.), pp. 217–236. Alan R. Liss, New York, 1987. Circulation and respiration in lungfishes (Dipnoi).

22. Burggren, W. W., K. Johansen and B. R. McMahon. In *Evolutionary Biology of Primitive Fishes* (R. E. Foremen, A. Gorbman, J. M. Dodd, and R. Olsson, eds.), pp. 217–252. Plenum, New York, 1986. Respiration in phyletically ancient fishes.

23. Burggren, W. W. and B. R. McMahon. In *Biology of the Land Crabs* (Burggren, W. W. and B. R. McMahon, eds.), pp. 298–332. Cambridge Univ. Press, New York, 1988. Circulation.

24. Burggren, W. W. and A. Mwalukoma. *J. Exp. Biol.* 105:205–213, 1982. Respiration during chronic hypoxia and hyperoxia in larval and adult bullfrogs (*Rana catesbeiana*). I. Morphological responses of lungs, skin and gills.

25. Burggren, W. W., A W. Pinder, B. McMahon, M. Wheatly and M. Doyle. *J. Exp. Biol.* 117:133–154, 1985. Ventilation, circulation and their interactions in the land crab, *Cardisoma guanhumi*.

26. Burggren, W. W. and D. J. Randall. *Respir. Physiol.* 34:171–183, 1978. Oxygen uptake and transport during hypoxic exposure in the sturgeon *Acipenser transmontanus*.

27. Burggren, W. W. and G. Shelton. *J. Exp. Biol.* 82:75–92, 1979. Gas exchange and transport during intermittent breathing in chelonian reptiles.

28. Burggren, W. W. and N. H. West. *Respir. Physiol.* 47:141–164, 1982. Changing respiratory importance of the gills, skin and lungs during metamorphosis in the bullfrog, *Rana catesbeiana*.

29. Burns, J. R. *Physiol. Zool.* 48:142–149, 1975. Seasonal changes in the respiration of pumpkinseed, *Lepomis gibbosus*, correlated with temperature, day length, and stage of reproductive development.

30. Calder, W. A. III. *Size, Function and Life History.* Harvard Univ. Press, Cambridge, MA, 1984.

31. Cole, R. N. and W. W. Burggren. *Mar. Biol. Lett.* 2:279–287, 1981. The contribution of respiratory papulae and tube feet to oxygen uptake in the sea star *Asterias forbsei* (Desor).

32. Cossins, A. R. and K. Bowler. Temperature Biology of Animals. Chapman & Hall, New York, 1987.

33. Coulson, R. A. and T. Hernandez, *Comp. Biochem. Physiol.* 69:1–13, 1981. The relationship between metabolic rate and various physiological and biochemical parameters. A comparison of alligator, man and shrew.

34. Croll, R. P. *J. Exp. Biol.* 117:15–27, 1985. Sensory control of respiratory pumping in *Aplysia californica*.

35. Datta Munshi, J. S. In *Respiration of Amphibious Vertebrates* (G. M. Hughes, ed.), pp. 73–104. Academic Press, New York. 1976. Gross and fine structure of the respiratory organs of air-breathing fishes.

36. deFur, P. L. and C. P. Mangum. *Comp. Biochem. Physiol.* 62A:288, 1979.

37. De Jongh, H. J. and C. Gans. *J. Morphol.* 127:259–290, 1969. On the mechanism of respiration in the bullfrog, *Rana catesbeiana*: A reassessment.

38. Dejours, P. *Principles of Comparative Respiratory Physiology.* Elsevier/North-Holland, New York, 1981.

39. Drabkin, D. L. *J. Biol. Chem.* 182:317–333, 1950. The distribution of the chromoproteins, hemoglobin, myoglobin, and cytochrome c, in the tissues of different species, and the relationship of the total content of each chromoprotein to body mass.

40. Ellington, W. R. *J. Exp. Zool.* 228:431–444, 1983. The recovery from anaerobic metabolism in invertebrates.

41. Elsner, R. and B. Gooden. *Diving and Asphyxia.* Cambridge Univ. Press, New York, 1983.

42. Emmett, B. and P. W. Hochachka. *Respir. Physiol.* 45:261–272, 1981. Scaling of oxidative and glycolytic enzymes in mammals.

43. Enright, J. T. *J. Comp. Physiol.* 77:141–162, 1972. A virtuoso isopod. Circa-lunar rhythms and their tidal structure.

44. Farrell, A. P. *Can. J. Zool.* 56:953–958, 1978. Cardiovascular event associated with air breathing in two teleosts, *Arapaima gigas* and *Hoplerythrinus unitaeniatus*.

45. Farrell, A.P. and D. J. Randall. *Can. J. Zool.* 56:939–945, 1978. Air-breathing mechanics in two Amazonian teleosts, *Arapaima gigas* and *Hoplerythrinus unitaeniatus.*

46. Feder, M. E. In *Respiration and Metabolism of Embryonic Vertebrates* (R. Seymour, ed.), pp. 71–86. Junk, Dordrecht, The Netherlands, 1984. Consequences of aerial respiration for amphibian larvae.

47. Feder, M. E. and W. W. Burggren. *Biol. Rev. Cambridge Philos. Soc.* 60:1–45, 1985. Cutaneous gas exchange in vertebrates: Design, patterns, control and implications.

48. Feder, M. E. and W. W. Burggren. *Sci. Am.* 253(5):126–142, 1985. Skin breathing in vertebrates.

49. Fick, A. *Sitzungsber. Phys. Med. Ges. Wurzburg* 16, 1870. Ueber die Messung des Blutquantums in den Herzventrikeln.

50. Fitzgerald, L. R. *Physiol. Rev.* 37:325–336, 1957. Cutaneous respiration in man.

51. Freadman, M. A. *J. Exp. Biol.* 90:253–265, 1981. Swimming energetics of striped bass (*Morone saxatilis*) and bluefish (*Pomatomus saltatrix*): Hydrodynamic correlates of locomotion and gill ventilation.

52. Fry, F. E. J. In *The Physiology of Fishes* (M. E. Brown, ed.), pp. 1–63. Academic Press, New York, 1957. The aquatic respiration of fish.

53. Gans, C. and B. Clark. *Respir. Physiol.* 26:285–301, 1976. Studies on ventilation of *Caiman crocodilis* (Crocodilia: Reptilia).

54. Gans, C. and G. M. Hughes. *J. Exp. Biol.* 47:1–20, 1967. The mechanism of lung ventilation in the tortoise *Testudo graeca* Linne.

55. Gatz, R. N., E. C. Crawford and J. Piiper. *Respir. Physiol.* 20:43–49, 1974. Metabolic and heart rate response of the plethodontid salamander *Desmognathus fuscus* to hypoxia.

56. Gaunt, A. S. and G. Gans. *J. Morphol.* 128:195–227, 1969. Mechanics of respiration in the snapping turtle *Chelydra serpentina* (Linne).

57. Gee, J. H. and J. B. Graham. *J. Exp. Biol.* 74:1–16, 1978. Respiratory and hydrostatic functions of the intestine of catfishes *Hoplosternum thoracatum* and *Brochis splendens* (Callichthyidae).

58. Glass, M. L. and S. C. Wood. *Physiol. Rev.* 63(1):232–260, 1983. Gas exchange and control of breathing in reptiles.

59. Gorbman, A., W. W. Dickhoff, S. T. Vigna, N. B. Clark, and C. L. Ralph. *Comparative Endocrinology.* Wiley, New York, 1983.

60. Graham, J. B. *Am. Zool.* 28:1031–1045, 1988. Ecological and evolutionary aspects of integumentary respiration: Body size, diffusion and the invertebrata.

61. Graham, J. B. and T. A. Baird. *J. Exp. Biol.* 108:357–376, 1984. The transition to air breathing in fishes. III. Effects of body size and aquatic hypoxia on the aerial gas exchange of the swamp eel *Synbranchus marmoratus.*

62. Graham, J. B., D. L. Kramer and E. Pineada. *J. Comp. Physiol.* 122B:295–310, 1977. Respiration of the air breathing fish *Piabucina festae.*

63. Greenaway, P. In *Biology of the Land Crabs.* (W. W. Burggren and B. McMahon, eds.), pp. 211–248. Cambridge Univ. Press, New York, 1988. Ion and water balance.

64. Grigg, G. *Aust. J. Zool.* 13:243–253, 1965. Studies on the Queensland lungfish *Neoceratodus forsteri* (Krefft). I. Anatomy, histology, and functioning of the lung.

65. Grote, J. and G. Thews. *Pfluegers Arch. Gesamte Physiol. Menschen Tiere* 276:142–165, 1962. Die Bedingungen fur die Sauerstoffversorgung des Herzmuskelgewebes.

66. Guimond, G. R. and V. H. Hutchison. In *Respiration of Amphibious Vertebrates* (G. M. Hughes, ed.), pp. 313–338. Academic Press, New York, 1976. Gas exchange of the giant salamanders of North America.

67. Halcrow, K. and C. M. Boyd. *Comp. Biochem. Physiol.* 23:233–242, 1967. The oxygen consumption and swimming activity of the amphipod *Gammarus oceanicus* at different temperatures.

68. Harvey, E. N. *J. Gen. Physiol.* 11:469–475, 1928. The oxygen consumption of luminous bacteria.

69. Herreid, C. F., II. *Comp. Biochem. Physiol. A* 67A:311–320, 1980. Hypoxia in invertebrates.

70. Herreid, C. F., II, W. L. Bretz, and K. Schmidt-Nielsen. *Am. J. Physiol.* 215:506–508, 1968. Cutaneous gas exchange in bats.

71. Heusner, A. A. *Respir. Physiol.* 48:1–12, 1982. Energy metabolism and body size. I. Is the 0.75 mass exponent of Kleiber's equation a statistical artifact?

72. Heusner, A. A. *Ann. Rev. Physiol.* 49:121–133, 1987. What does the power function reveal about structure and function in animals of different size?

73. Hildebrand, M. Analysis of Vertebrate Structure. Wiley, New York, 1982.

74. Hill, R. W. and G. A. Wyse. *Animal Physiology.* Harper & Row, New York, 1989.

75. Hills, B. A. and G. M. Hughes. *Respir. Physiol.* 9:126–140, 1970. A dimensional analysis of oxygen transfer in the fish gill.

76. Hochachka, P. W. and G. N. Somero. *Biochemical Adaptation.* Princeton Univ. Press, Princeton, NJ, 1984.

77. Hochachka, P. W. *Living Without Oxygen.* Harvard Univ. Press, Cambridge, MA, 1980.

77a. Hochachka, P., T. Moon, J. Bailey and C. Hulbert. *Can. J. Zool.* 56:820–832, 1978. The osteoglossid kidney; correlations of structure, function and metabolism with transition to air breathing.

78. Hughes, G. M. *Folia Morphol. (Prague),* 18:78–95, 1970. Morphological measurements on the gills of fishes in relation to their respiratory function.

79. Hughes, G. M. *Am. Zool.* 13:475–489, 1973. Respiratory responses to hypoxia in fish.

80. Hughes, G. M. In *Fish Physiology* (W. S. Hoar and D. J. Randall, eds.), Vol. 10A, pp. 1–72. Academic Press, Orlando, Florida, 1984. General anatomy of the gills.

81. Hughes, G. M. and M. Morgan. *Biol. Rev. Cambridge Philos. Soc.* 48:419–475, 1973. The structure of fish gills in relation to their respiratory function.

82. Hughes, G. M. and J. L. Roberts. *J. Exp. Biol.* 52:177–192, 1970. A study of the effect of temperature changes on the respiratory pumps of the rainbow trout.

83. Hughes, G. M. and S. I. Umezawa. *J. Exp. Biol.* 49:565–582, 1968. On respiration in the dragonet, *Callionomus lyra* L.

84. Jansky, L. *Can. J. Biochem. Physiol.* 41:1847–1854, 1963. Body organ cytochrome oxidase activity in cold- and warm-acclimated rats.

85. Johansen, K. In *Fish Physiology.* (W. S. Hoar and D. J. Randall, eds.), Vol. 4, pp. 361–413. Academic Press, New York, 1970. Air breathing in fishes.

86. Johansen, K., C. Lenfant and G. Grigg. *Comp. Biochem. Physiol.* 20:835–854, 1967. Respiratory control in the lungfish, *Neoceratodus forsteri.*

87. Johansen, K., C. Lenfant and D. Hanson. *Z. Vergl. Physiol.* 59:157–186, 1968. Cardiovascular dynamics in the lungfish.

88. Johansen, K., D. Hanson and C. Lenfant. *Respir. Physiol.* 9:162–174, 1970. Respiration in a primitive air breather, *Amia calva.*

89. Jones, J. D. *Comp. Biochem. Physiol.* 12:297–310, 1964. Respiratory gas exchange in the aquatic pulmonate, *Biomphalaria sudanica.*

90. Jones, J. D. *Comp. Biochem. Physiol.* 4:1–29, 1961. Aspects of respiration in *Planorbis corneus* L. and *Lymnaea stagnalis* L. (Gastropoda: Pulmonata).

91. Jones, J. H., E. L. Effmenn and K. Schmidt-Nielsen. *Respir. Physiol.* 45:121–131, 1981. Control of air flow in bird lungs: Radiographic studies.

92. Jones, R. M. *Respir. Physiol.* 49:251–265, 1982. How toads breathe: Control of air flow to and from the lungs by the nares in *Bufo marinus.*

93. Kinney, J. L. and F. N. White. *Respir. Physiol.* 31:327–332, 1977. Oxidative cost of ventilation in a turtle, *Pseudemys floridana.*

94. Kirsch, R. and G. Nonnotte. *Respir. Physiol.* 29:339–354, 1977. Cutaneous respiration in three freshwater teleosts.

95. Kleiber, M. *Hilgardia* 6:315–353, 1932. Body size and metabolism.

96. Kramer, D. *Can. J. Zool.* 66:89–94, 1988. The behavioral ecology of air breathing by aquatic animals.

97. Krebs, H. A. *Biochim. Biophys. Acta* 4:249–269, 1950. Body size and tissue metabolism.

98. Kreuzer, F. *Helv. Physiol. Pharamacol. Acta* 8:505–516, 1950. Uber die Diffusion von Sauerstoff in Serumeiweisslosungen verschiedener Konzentration.

99. Krogh, A. *J. Physiol. (London).* 52:391–608, 1919. The rate of diffusion of gases through animal tissues, with some remarks on the coefficient of invasion.

100. Lannoo, M. J. and M. D. Backman. *Can. J. Zool.* 62:15–18, 1984. On flotation and air breathing in *Ambystoma tigrinum* larvae: Stimuli for and the relationship between these behaviors.

101. Laurent, P. In *Fish Physiology* (W. S. Hoar & D. J. Randall, eds.), Vol. 10A, pp. 73–183. Academic Press, Orlando, Florida, 1984. Gill Internal Morphology.

102. Liem, K. F. *Copeia* 2:375–380, 1967. Functional morphology of the integumentary, respiratory and digestive systems of the synbranchoid fish *Monopterus albus*.

103. Liem, K. F. In *Environmental Physiology of Fishes* (M. A. Ali, ed.), pp. 57–92. Plenum, New York, 1979. Air ventilation in advanced teleosts: biomechanical and evolutionary aspects.

104. Lillywhite, H. B. and P. F. A. Maderson. In *Biology of the Reptilia.* (C. Gans and F. H. Pough, eds.), Vol. 12, pp. 397–442. Academic Press, New York, 1982. Skin structure and permeability.

105. Lindstedt, S. L. and J. H. Jones. In *New Directions in Physiological Ecology* (M. E. Feder, A. F. Bennett, W. W. Burggren, and R. E. Huey, eds.), pp. 289–304. Cambridge Univ. Press, New York, 1987. Synmorphosis: The concept of optimal design.

106. Little, C. *The Colonization of Land.* Cambridge Univ. Press, London, 1983.

107. Mangum, C. P. *Am. Sci.* 58:641–647, 1970. Respiratory physiology in annelids.

108. Mangum, C. P., B. R. McMahon, P. L. deFur and M. G. Wheatly. *J. Crustacean Biol.* 5:188–206, 1985. Gas exchange, acid-base balance, and the oxygen supply to the tissues during a molt of the blue crab *Callinectes sapidus*.

109. Mathieu, O., R. Krauer, H. Hoppeler, P. Gehr, S. L. Lindstedt, R. McN. Alexander, C. R. Taylor and E R. Weibel. *Respir. Physiol.* 44:113–128, 1981. Design of the mammalian respiratory system. VI. Scaling mitochondrial volume in skeletal muscle to body mass.

110. McMahon, B. R. *Am. Zool.* 28:39–53, 1988. Physiological responses to oxygen depletion in intertidal animals.

111. McMahon, B. R. and W. W. Burggren. *J. Exp. Biol.* 79:265–281, 1979. Respiration and adaptation to the terrestrial habitat in the land hermit crab *Coenobita clypeatus*.

112. McMahon, B. R. and W. W. Burggren. *J. Exp. Biol.* 133:371–394, 1987. Respiratory physiology of intestinal air breathing in the teleost fish *Misgurnus anguillicaudatus*.

113. McMahon, B. R. and W. W. Burggren. In *Biology of the Land Crabs* (W. W. Burggren and B. R. McMahon, eds.), pp. 249–297. Cambridge Univ. Press, New York, 1988. Respiration.

114. McMahon, B. R. and J. L. Wilkens. In *The Biology of Crustacea* (L. Mantel, ed.), Vol. 5, pp. 289–372. Academic Press, New York, 1983. Ventilation, perfusion and oxygen uptake.

115. McMahon, T. A. and J. T. Bonner. *On Size and Life.* Freeman, New York, 1983.

116. Meglitsch, P. A. *Invertebrate Zoology.* Oxford Univ. Press, London, 1967.

117. Mill, P. J. In *Comprehensive Insect Physiology, Biochemistry and Pharmacology,* (G. A. Kerkut and L. I. Gilbert, eds.), pp. 517–593. Pergamon Press, Oxford, 1985. Structure and physiology of the respiratory system.

118. Milsom, W. K. and R. W. Brill. *Respir. Physiol.* 66:193–203, 1986. Oxygen sensitive afferent information arising from the first gill arch of yellowfin.

119. Moberly, W. R. *Comp. Biochem. Physiol.* 27:1–20, 1968. Oxygen consumption of Iguana.

120. Neill, W. H., J. J. Magnuson, and G. C. Chipman. *Science* 176:1443–1445, 1972. Behavioral thermoregulation by fishes: A new experimental approach.

121. Newell, R. C. and W. A. M. Courtney. *J. Exp. Biol.* 42:45–57, 1965. Respiratory movements in Holothuria.

122. Nikam, T. B. and V. V. Khole. *Insect Spiracular Systems.* Wiley, New York, 1989.

123. Nonotte, G. *Comp. Biochem. Physiol.* 70:541–543, 1981. Cutaneous respiration in six freshwater teleosts.

124. Nonotte, G. and R. Kirsch. *Respir. Physiol.* 35:111–118, 1978. Cutaneous respiration in seven sea-water teleosts.

125. Olla, B. L., and A. L. Studholme. 1976. *Jt. U.S./USSR Symp. Compr. Anal. Environ. [Proc.], 2nd, 1975,* pp. 25–31, EPA, No. 600/.9-76/024, 1976. Environmental stress and behavior: Response capabilities of marine fish.

126. Olsen, K. R. and P. O. Fromme. *Z. Zellforsch. Mikrosk. Anat.* 143:439–449, 1973. A scanning electron microscopic study of secondary lamellae and chloride cells of rainbow trout (*Salmo gairdneri*).

127. Perry, S. F. *Respir. Physiol.* 35:245–262, 1978. Quantitative anatomy of the lungs of the red-eared turtle, *Pseudemys scripta elegans.*

128. Perry, S. F. In *Comparative Pulmonary Physiology: Current Concepts* (S. C. Wood and C. Lenfant, eds.), pp. 193–236. Dekker, New York, 1988. Structure and function of reptile lungs.

129. Perry, S. F. and H. R. Duncker. *Respir. Physiol.* 34:61–81, 1978. Lung architecture, volume and static mechanics in five species of lizards.

130. Peters, H. M. *Zoomorphologie* 89:93–123, 1978. On the mechanism of air ventilation in anabantoids (Pisces: Teleostei).

131. Peters, R. H. *The Ecological Implications of Body Size.* Cambridge Univ. Press, London and New York, 1983.

132. Piiper, J. and P. Scheid: In *Fish Physiology* (W. S. Hoar and D. J. Randall, eds.), Vol. 10A, pp. 229–262. Academic Press, Orlando, Florida, 1984. Model analysis of gas transfer in fish gills.

133. Pinder, A. W. and W. W. Burggren. *J. Exp. Biol.* 126:453–468, 1986. Ventilation and partitioning of oxygen uptake in the frog *Rana pipiens:* effects of hypoxia and activity.

134. Prosser C. L. *Comparative Animal Physiology.* 3rd ed. Saunders, Philadelphia, PA 1973.

135. Prosser, C. L. *Adaptational Biology.* Wiley, New York, 1986.

136. Rahn, H. and B. J. Howell. In *Respiration of Amphibious Vertebrates,* (G. M. Hughes, ed.), pp. 271–285. Academic Press, 1976. Bimodal gas exchange.

137. Rahn, H. and C. V. Paganelli. *Respir. Physiol.* 5:145–164, 1968. Gas exchange in gas gills of diving insects.

138. Rahn, H., K. B. Rahn, B. J. Howell, C. Gans and S. M. Tenney. *Respir. Physiol.* 11:285–307, 1971. Air breathing of the garfish (*Lepisosteus osseus*).

139. Randall, D. J. In *Fish Physiology* (W. S Hoar and D. J. Randall, eds.), Vol. 4, pp. 253–292. Academic Press, New York, 1970. Gas exchange in fish.

140. Randall, D. J. In *Cardiovascular Shunts: Phylogenetic, Ontogenetic and Clinical Aspects,* (K. Johansen and W. Burggren, eds.), pp. 71–82. Munksgaard, Copenhagen, 1985. Shunts in fish gills.

141. Randall, D. J., W. W. Burggren, A. P. Farrell and M. S. Haswell. *The Evolution of Air Breathing in Vertebrates.* Cambridge Univ. Press, New York, 1981.

142. Randall, D. J., G. F. Holeton, and E. D. Stevens. *J. Exp. Biol.* 46:339–348, 1967. The exchange of oxygen and carbon dioxide across the gills of rainbow trout.

143. Randall, D. J. and J. C. Smith. *Physiol. Zool.* 40:104–113, 1967. The regulation of cardiac activity on fish in a hypoxic environment.

144. Reiswig, H. M. *Mar. Biol.* 9(1):38–50, 1971. *In situ* pumping activities of tropical Demospongiae.

145. Roberts, J. L. In *Physiological Adaptations to the environment* (F. J. Vernberg, ed.), pp. 395–414. Intext Educational Publ., New York, 1975. Respiratory adaptations of aquatic animals.

146. Roberts J. L. *Biol. Bull.* (Woods Hole, Mass.) 148:85–105, 1975. Active branchial and ram gill ventilation in fishes.

147. Roberts, J. L. In *Marine Pollution: Functional Responses*, (W. B. Vernberg, F. P. Thurberg, A. Calabrese, and F. J. Vernberg, eds.), pp. 365–388. Academic Press, New York, 1979. Seasonal modulation of thermal acclimation and behavioral thermoregulation in aquatic animals.

148. Roberts, J. L., and D. M. Rowell. *Can. J. Zool.* 66:182–190, 1988. Periodic respiration of gill-breathing fish.

149. Romer, A. S. *The Vertebrate Body*, 4th ed. Saunders, Philadelphia, PA, 1970.

150. Rozin, P. N. and J. Mayer. *Science* 134:942–943, 1961. Thermal reinforcement and thermoregulatory behavior in the goldfish. *Carassius auratus.*

151. Sacca, R. and W. W. Burggren. *J. Exp. Biol.* 97:179–186, 1982. Oxygen uptake in air and water in the air-breathing reedfish *Calamoichthys calabaricus:* Role of skin, gills and lungs.

152. Sarrus and Rameaux. *Bull. Acad. Med.* (*Paris*) 3:1094–1100, 1838, Rapport sur un mémoire adressé à l'Académie Royale de Médecine.

153. Satchell, G. H. *Comp. Biochem. Physiol.* 27:835–841, 1968. A neurophysiological basis for the coordination of swimming with respiration in fish.

154. Scheid, P. In *A Companion to Animal Physiology* (C. R. Taylor, K. Johansen and L. Bolis, eds.), pp. 3–16. Cambridge Univ. Press, New York, 1982. A model for comparing gas-exchange systems in vertebrates.

155. Scheid, P., H. Slama, and J. Piiper. *Respir. Physiol.* 14:83–89, 1972. Mechanisms of unidirectional flow in parabronchi of avian lungs: measurement in duck lung preparations.

156. Schick, J. M., J. Widdows, and E. Gnaiger. *Amer. Zool.* 28:161–181, 1988. Calorimetric studies of behavior, metabolism and energetics of sessile intertidal animals.

157. Schmidt-Nielsen, K. *Science* 177:222–228, 1972. Locomotion: Energy cost of swimming, flying, and running.

158. Schmidt-Nielsen, K. *Scaling: Why is Animal Size So Important?* Cambridge Univ. Press, London and New York, 1984.

159. Schroff, G. and U. Schottler. *J. Comp. Physiol.* 138:35–41, 1980. Anaerobic metabolism in *Arenicola.*

160. Seymour, R. S. In *Biology of the Reptilia,* (C. Gans and F. H. Pough, eds.), Vol. 13, pp. 1–51. Academic Press, New York. 1982. Physiological adapations to aquatic life.

161. Shelton, G. and R. G. Boutilier. *J. Exp. Biol.* 100:245–273, 1982. Apnoea in amphibians and reptiles.

162. Shelton, G. and W. W. Burggren. *J. Exp. Biol.* 64:323–343, 1976. Cardiovascular dynamics of the Chelonia during apnoea and lung ventilation.

163. Singh, B. N. In *Respiration of Amphibious Vertebrates,* (G. M. Hughes, eds.), pp. 125–164. Academic Press, New York, 1976. Balance between aquatic and aerial respiration.

164. Smatresk, N. J. and J. N. Cameron. *J. Exp. Biol.* 96:263–280, 1982. Respiration and acid-base physiology of the spotted gar, a bimodal breather. I. Normal values and the response to severe hypoxia.

165. Smith, R. E. *Ann. N. Y. Acad. Sci.* 62:403–422, 1956. Quantitative relations between liver mitochondria metabolism and total body weight in mammals.

166. Steffensen, J. F., J. P. Lomholt and K. Johansen. *Respir. Physiol.* 44:269–272, 1981. The relative importance of skin oxygen uptake in the naturally buried plaice, *Pleuronectes platessa,* exposed to graded hypoxia.

167. Steen, J. B. *Comparative Physiology of Respiratory Mechanisms.* Academic Press, New York, 1971.

168. Taylor, C. R. *Annu. Rev. Physiol.* 49:135–146, 1987. Structural and functional limits

to oxidative metabolism: Insights from scaling.

169. Teal, J. M. and F. G. Carey. *Limnol. Oceanogr.* 12:548–550, 1967. Profiles of temperature and oxygen at different depths in the southern Pacific Ocean.

170. Tenney, S. M. In *Evolution of Respiratory Processes.* (S. C. Wood and C. Lenfant, eds.), pp. 51–106. Dekker, New York, 1979. A synopsis of breathing mechanisms.

171. Thomas, H. J. *J. Exp. Biol.* 31:228–251, 1954. Oxygen uptake of *Homarus.*

172. Thorpe, W. H. *Biol. Rev. (Cambridge Philos. Soc.)* 25:344–390, 1950. Plastron respiration in insects.

173. Thorpe, W. H. and D. J. Crisp. *J. Exp. Biol.* 24:227–269, 1947. Studies on plastron respiration. 1. The biology of *Aphelocheirus* [Hemiptera Aphelocheiridae (Naucoridae)] and the mechanism of plastron retention.

174. Tucker, V. A. *Am. Sci.* 63:413–419, 1975. The energetic cost of moving about.

175. Ultsch, G. R. In *Respiration of Amphibious Vertebrates,* (G. M. Hughes, ed.), pp. 287–312. Academic Press, New York, 1976. Eco-physiological studies of some metabolic and respiratory adaptations of sirenid salamanders.

176. Ultsch, G. R. and G. Gros. *Comp. Biochem. Physiol. A* 62A:685–689, 1979. Mucus as a diffusion barrier to oxygen: possible role in O_2 uptake at low pH in carp (*Cyrpinus carpio*) gills.

177. Ultsch, G. R., C. Herbert, and D. C. Jackson. *Physiol. Zool.* 57(6):620–631, 1984. The comparative physiology of diving in North American freshwater turtles. I. Submergence tolerance, gas exchange and acid-base balance.

178. Ultsch, G. R., D. C. Jackson, and R. J. Moalli. *Comp. Physiol.* 142:439–443, 1981. Metabolic oxygen conformity among lower vertebrates: The toadfish revisited.

179. Webb, H. M. and F. A. Brown. *Physiol. Rev.* 39:127–161, 1959. Biological rhythms.

180. Weber, R. E. In *Physiology of Annelids,* (P. J. Mill, ed.), pp. 369–392. Academic Press, New York, 1978. Respiratory pigments.

181. Wells, G. P. *J. Mar. Biol. Assoc. U. K.* 28:447–464, 1949. Respiratory movements of *Arenicola marina* L.: intermittent irrigation of the tube, and intermittent aerial respiration.

182. Weibel, R. E. *Morphometry of the Human Lung.* Springer-Verlag, Heidelberg, 1963.

183. Weibel, E. R. *Physiol. Rev.* 53:419–495, 1973. Morphological basis of alveolar-capillary gas exchange.

184. Weibel, E. R. In *Pulmonary Diseases and Disorders,* (A. P. Fishman, ed.), pp. 224–271. McGraw-Hill, New York 1980. Design and structure of the human lung.

185. Weible, E. R. *The Pathway for Oxygen: Structure and Function in the Mammalian Respiratory System.* Harvard Univ. Press, Cambridge, 1984.

186. Weis-Fogh, T. *J. Exp. Biol.* 47:561–587, 1987. Respiration and tracheal ventilation in locusts and other flying insects.

187. West, J. B. *Ventilation/Blood Flow and Gas Exchange,* 3rd ed. Blackwell, Oxford, 1977.

188. West, N. H. and D. R. Jones. *Can. J. Zool.* 53:332–344, 1975. Breathing movements in the frog *Rana pipiens.* I. The mechanical events associated with lung and buccal ventilation.

189. Whitemore, C. M., C. E. Warden and P. Doudoroff. *Trans. Am. Fish. Soc.* 89:17–26, 1960. Behavior of fish in a oxygen gradient.

190. Wilkie, D. R. In *Scale Effects in Animal Locomotion* (T. J. Predley, ed.), pp. 23–36, Academic Press, New York, 1977. Metabolism and body size.

191. Willmer, E. N. *J. Exp. Biol.* 11:283–306, 1934. Some observations on the respiration of certain tropical fresh water fish.

192. Withrow, R. B. In *Photoperiodism* (R. B. Withrow, ed.), pp. 439–471. Publ. 55, Amer. Assoc. Adv. Sci., Washington, DC. 1959.

193. Wyse, G. A. and C. H. Page. *Fed. Am. Soc. Exp. Biol. Proc.* 35:2007–2012, 1976. Sensory and central nervous control of gill ventilation in *Limulus.*

194. Hasting, J. W., B. Rusak, and Ziad Boulos. In *Neural and Integrative Animal Physiology* (C. Ladd Prosser, ed.). pp. 435–545. Wiley-Liss, Inc., New York, 1991.

Chapter 10 | *Respiratory Functions of Blood*

Warren Burggren, Brian McMahon, and Dennis Powers

As animals evolved with increased body size and higher metabolic rate, simple diffusion across the body wall became increasingly inadequate as a means of delivering O_2 to the tissues. Consequently, in most animals that have metabolizing tissues more than a few millimeters from respiratory surfaces, there is a specialized cardiovascular system for circulating body fluids to aid in the internal distribution of O_2 (the Insecta are a notable exception, using their tracheal system for gas transport). However, simply creating a convection of a water-based body fluid in itself does not guarantee internal transport of sufficient quantities of O_2, because of the low solubility of molecular O_2 in water-based body fluids. Thus, blood-borne proteins specialized for reversibly binding large quantities of O_2 evolved in most phyla. These so-called transport or respiratory pigments greatly increase the O_2 content of blood. In most mammals, for example, 95% of the O_2 in arterial blood is bound to hemoglobin. The remaining 5% of O_2 in arterial blood is dissolved in plasma, the extracellular fluid component of blood. As will be evident later in this chapter, not all respiratory pigments bind such large quantities of O_2. In some decapod Crustacea, for example, the respiratory pigment may only double or triple the O_2 capacity of the blood (normally termed hemolymph) over that provided by O_2 dissolved in solution. However, even this relatively modest increase in blood O_2 content may still make an important contribution toward supporting aerobic metabolism during physical activity or environmental hypoxia.

In addition to O_2 transport, respiratory pigments may also feature prominently in short-term O_2 storage. Some pigments play a role in CO_2 transport both by binding CO_2 directly to form carbamino compounds and by buffering hydrogen ions released in association with the formation of bicarbonate ions. Finally, those respiratory pigments conveyed in plasma rather than within blood cells also make an important contribution to the overall colloid osmotic pressure of the blood.

Phyletic Distribution of Blood Respiratory Pigments

Hemoglobins

Hemoglobin (Hb), reddish in color, consists of an iron-porphyrin or "heme" group coupled to a protein (globin). It is the most widely distributed of the respiratory pigments, occurring in at least 10 animal phyla and in a few plants (Table 1). Vertebrate hemoglobins are always found in blood cells (commonly termed erythrocytes, red blood cells, or rbcs). The monomeric form, myoglobin, is found in red muscle cells, where it functions as a short-term O_2 storage pigment. Invertebrate Hb may be found in plasma, muscle cells, coelomic or vascular corpuscles, in oocytes, nerve cells (e.g., annelids and molluscs) and others cells (Table 1). Hemoglobins are distributed among one half of the 30 invertebrate phyla (Table 1), ranging from unicellular organisms like *Paramecium* (45) to flat worms (*Platyhelminthes*), nemertine worms (*Nemertina*), roundworms (*Nematoda*), sea cucumbers (*Ecinodemata*), clams (*Mollusca*), some of the segmented worms (Annelids) and a few species of insects (185).

Chlorocruorins (Greenish Hemoglobins)

The term chlorocruorin has historically been used to distinguish a greenish oxygen-carrying pigment found in some polychaete worms from the reddish hemoglobin transport pigments alluded to above. While this color difference is obvious to the eye, it is the result of a rather minor alteration of the porphyrin and should not be distinguished as a class from other hemoglobins. On the other hand, there is a significant literature that

uses the historical term chlorocruorin, and therefore students should be aware of its existance.

Heme pigments are common in the hemolymph of at least four families of polychaete worms, including over 20 species of the Serpulimorpha. For example, *Potamilla* has red Hb in muscle cells and greenish Hb (i.e., chlorocruorin) in hemolymph. *Serpula* has both protoheme and chlorocruoroheme and has both the red hemoglobin and greenish hemoglobin (i.e., chlorocruorin) in its blood (61). In the genus *Spirobis*, *S. borealis* has blood chlorocruorin, *S. corrugatus* has blood Hb, and *S. militaris* has neither.

Hemerythrins

Hemerythrin is another iron-based respiratory pigment. This pigment, rosy or pink in color, is found in nucleated cells rather than free in hemolymph.

Hemerythrin is found in sipunculids [e.g., *Sipunculus*, *Themiste*, and *Golfingia* (*Phascolosoma*), priapulids, in a brachiopod and at least one family of polychaetes (*Magelonidae*)]. The hemerythrins found in the blood and coelomic cells have been found to differ in oxygen affinity and myohemerythrin is sometimes present in the muscle (103).

Hemocyanins

Unlike iron-based Hb, chlorocruorin, and hemerythrin, hemocyanin (Hc) is copper-based. Hemocyanin lacks porphyrin groups, and always occurs dissolved in hemolymph. After Hb, Hc is the next most widely distributed respiratory pigment. Among the molluscs it is found in prosobranchs, pulmonates (excluding the freshwater Planorbidae), amphineurans, cephalopods, some gastropods and in lamellibranchs (149). Many molluscs (e.g., the snail *Busycon* and the amphineuran

TABLE 1. Phyletic Distribution of Hemoglobin

Chordates

Vertebrates—Hb in corpuscles (erythrocytes) in all classes; red muscles contain myoglobin.
   No respiratory pigments in leptocephalan eel larvae and three genera of Antarctic ice fish.
Prochordates—Hb absent in Branchiostoma or most prochordates.

Echinoderms

Holothurians—*Thyone, Cucumaria, Molpadia,* and *Caudina* have Hb in corpuscles.

Annelids

Oligochaetes—*Lumbricus, Tubifex,* and the hirudineans *Hirudo* and *Analastoma* have Hb
   dissolved in solution in blood; *Lumbricus* has myoglobin in muscle.
Polychaetes—Capitellidae, Glyceridae, *Polycirrus hematodes,* and *P. aurantiacus* have Hb in
   coelomic fluid cells. *Terebella* and *Travisia* have Hb in both coelomic fluid cells and in blood
   plasma; Nereidae, *Arenicola, Amphitrite,* Cirratulidae, Eunicidae, and others have Hb
   dissolved in blood plasma; some Serpulids have both Hb and chlorocruorin dissolved in
   plasma; no respiratory pigment in blood or coelomic fluid of Syllidae, Phyllodice,
   Aphroditidae, Chaetoperidae, and Lepidonotus.

Echiuroids

Urechis and *Thalassema* have Hb in coelomic corpuscles. Hb in body wall muscles of *Urechis*
   and *Arhynchite.*

Phoronids

Phoronis and *Phoronopsis* have Hb in corpuscles.

Arthropods

Crustacea—Hb unknown in Malacostraca, common in Entomostraca; Artemia (Anostraca),
   Daphnia (Cladocera); *Triops* and *Apus* (Notostraca).
Insecta—Hb in chironomid larvae and in tracheal and other cells of *Gastrophilus,* a dipteran
   parasite.

Molluscs

Hb in corpuscles in a few pelecypods—*Solen, Arca, Petunculus.* Hb dissolved in plasma in the
   gastropod *Planorbis.* Myoglobin in the radular muscles of many prosobranch gastropods and
   chiton.

Nemerteans

Hb found in dissolved plasma in some species, in corpuscles in others. Hb in ganglion cells
   of Polia.

Platyhelminthes

Hb in a few parasitic trematodes and in rhabdocoeles Derestoma, Syndesmis, and Telorchis.

TABLE 1. (*Continued*)

Nemathelminthes

Hb in pseudocoelic fluid and in hypodermal cells of body wall in *Ascaris, Nippostrongylus, Eustronglyides,* and *Camallanus.*

Protozoa

Hb in *Paramecium* and in certain strains of *Tetrahymena.*
Plants—Hb is found in root nodules of some legumes when plant and bacteria are symbiotic.

Cryptochiton) have Hc in the hemolymph but have myoglobin in some of their muscle fibers (particularly the radular muscles). Hemocyanin is found in the Crustacea, particularly in the Malacostraca where it is almost ubiquitous. The horseshoe crab, *Limulus*, a few arachnids and centipedes also use Hc as a respiratory pigment.

The evolutionary history of Hc is unclear. Some ancient enzymes (e.g., cytochrome oxidases) contain both iron and copper binding domains. Possibly such compounds were protopigments from which a variety of oxygen-carrying molecules evolved. Molluscan and arthropod hemocyanins have some crude similarities at the O_2-binding site but they are very different in their gross structures, suggesting that the pigments from these two groups diverged from the ancestral tyrosinsase-type molecule very early in evolution.

Structure and Evolution of Respiratory Pigments

Hemoglobin

The primary function of hemoglobin is to carry oxygen from the gas exchange organs to the metabolizing tissues. This physiological function depends on the ability of hemoglobin to form a reversible complex between oxygen and the ferrous iron in a hematoporphyrin (heme) pros-

thetic group (Fig. 1*a*) that is buried in a hydrophobic pocket of a protein called globin. The globin, with its associated hematoporphyrin, constitutes a Hb subunit (Fig. 1*b*). Vertebrate Hbs generally have four subunits each ~16,000 D (Fig. 1*c*), whereas invertebrate Hbs have from 1 to >250 subunits.

Structure of Vertebrate Hemoglobins

Primary Structure and Evolution. Although some species have Hbs that are composed of identical subunits, most vertebrates have tetrameric Hbs with at least two different types of subunits. In adult vertebrates these polypeptides are referred to as the alpha (α) and beta (β) chains. In humans both α chains are identical and consist of 141 amino acids each. The heme group is covalently bound between the iron and the imidazole side chain of histidine at position 87, which is called the proximal histidine of the α chain. The β chains are 146 amino acids each with the proximal histidine at position 92.

Other human globin chains have been identified and sequenced, including δ, τ, ε, and ζ chains. Comparisons of their respective amino acid sequences (primary structure) clearly indicate homology. Moreover, analysis of these genes encoding the globin polypeptides indicates linkage homology (6, 120). Although the α and β genes are located on different chromosomes, sequence comparisons clearly

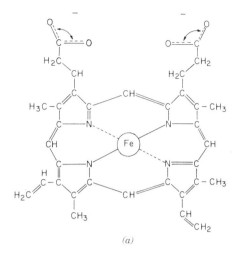

(a)

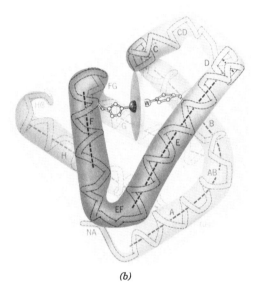

(b)

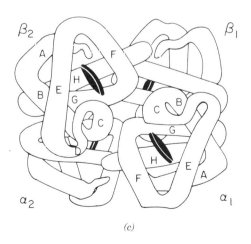

(c)

indicate that they arose by an ancient gene duplication. Moreover, studies of myoglobin in muscle have shown that it is also evolutionarily related to the various Hb subunits.

There is considerable information about the primary structure of globin subunits from a variety of species (52). These sequences have been used to determine ancestral relationships among taxa (72). For example, the amino acid sequences of Hb chains from a number of vertebrates have been used to construct evolutionary relationships that agree remarkably well with the fossil record (Fig. 2). This figure assumes a single myoglobin-like ancestral chain prior to 450 million years ago. Subsequently, there was a gene duplication, the α and β chains arose by divergence, and a tetramer molecule eventually evolved. About 350 million years ago an α chain gene duplication gave rise to α chain alternatives, including the ζ chain of human Hb. A number of gene duplications over the last 250 million years yielded a series of β chain derivatives including δ, τ, and ϵ chains.

Figure 1. (*a*) One of the resonance structures of the heme group. Note the four pyrrole rings, methylene bridges, and propionic acid side chains. (*b*) Diagrammatic illustration of myoglobin from sperm whale. A modification of the model in (52). The polypeptide backbone is enclosed by a tube to help illustrate the course of the basic structure. The straight stretches represent alpha-helices; labeled A, B, C, D, E, F, G, and H. The heme group is represented by a large disk in the center of the molecule with histidines that interact with the heme to the left and right. The oxygen molecule would be located at point W. (*c*) Schematic drawing of the arrangement of subunits in horse hemoglobin. The helical segments are lettered A through H for the α_1 and β_2 chains. The polypeptide course is indicated as a tube, but without the α-carbon positions. The flat darkened disks represent the heme group [(52)].

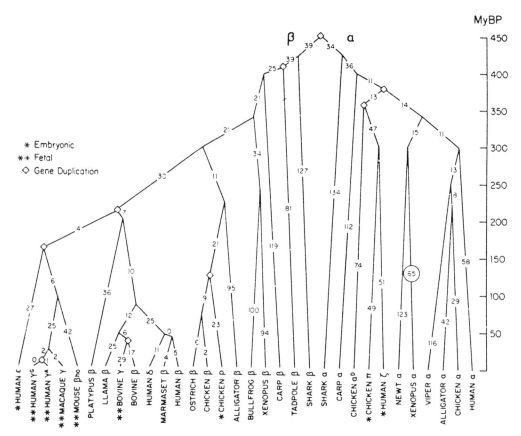

Figure 2. Phylogenetic tree of the hemoglobin family. The vertical axis is millions of years before present (MyBP), as deduced from the paleontological record. The diamonds represent postulated gene duplications. [From (73).]

Secondary Structure. The secondary structure of myoglobin as well as the subunits of Hb are highly helical. Myoglobin and the β chain of Hb have eight helical segments designated A through H (see Fig. 1c) but the D helix is missing from the α chains. These helices are connected by short nonhelical segments. Hemoglobin has clusters of hydrophobic residues on the outside of some helices. These residues are at interfaces with the other subunits of the Hb polymer and thus not exposed to solvent. Such subunit contacts are important in (a) stabilizing the tetrameric structure, (b) controlling oxygen-binding capability, and (c) cooperative binding of oxygen.

Tertiary and Quaternary Structure of Vertebrate Hemoglobins. The α and β chains of tetrameric vertebrate Hbs are arranged as a dimer of dimers—that is, a tetramer of two α chains and two β chains (58, 175). One dimer is designated $\alpha_1\beta_1$, and the other dimer $\alpha_2\beta_2$. Together these form the $\alpha_1\beta_1 : \alpha_2\beta_2$ tetramer (Fig. 1c). The contacts between these various polypeptides are thus referred to as the $\alpha_1\beta_1$ (or $\alpha_2\beta_2$) and $\alpha_1\beta_2$ (or $\alpha_2\beta_1$) contacts, respectively.

Comparisons of the oxy- and deoxy-Hb three-dimensional structures indicate that upon oxygenation the β chains move apart by approximately 7 Å. This is associated with a change in the position of the β chains relative to the α chains. These and other structural alterations upon oxygenation involve the movement of the subunits relative to each other such that there are changes in the amino acid contacts between complementary subunits. Although there are relatively few changes in the 35 subunit contacts at the $\alpha_1\beta_1$, (or $\alpha_2\beta_2$) interface, there are numerous changes at the $\alpha_1\beta_2$ (or $\alpha_2\beta_1$) interface. The energy changes that take place during the breaking and remaking of bonds at the $\alpha_1\beta_2$ interface are responsible for most of the subunit cooperativity of hemoglobin–oxygen binding (152).

Among the $\alpha_1\beta_2$ contacts, evolutionary substitutions have been limited in number and are generally conservative isopolar changes. Residues conserved in one chain tend to be different in the other chain but complimentary contacts tend to be conserved across taxa. The importance of the $\alpha_1\beta_2$ contacts, presumed by the evolutionary conservation of these residues (160), has been shown to be very important in subunit cooperativity (152).

While many vertebrate Hbs are tetrameric molecules of ~64,000 Da, there are polymeric forms. Estimates of molecular size have been made for many Hbs by determining the sedimentation coefficient using an ultracentrifuge. The sedimentation coefficient, $s_{20,w}$, is the sedimentation velocity in centimeters per second (cm/s) in a unit centrifugal field at 20°C in a medium reduced to the viscosity of water. Thus, the larger the sedimentation coefficient, the greater the molecular weight or size of the molecule. In general, sedimentation coefficients for vertebrate Hbs are 4.3–4.7 × 10^{-13} cm/s (referred to as 4.3–4.7 S*). Hemoglobins of amphibians and reptiles sometimes have higher $s_{20,w}$ values than those of mammals, birds, and fishes. In the toad *Bufo valliceps*, three components with $s_{20,w}$ of 4, 7, and 12 S suggest aggregation of subunits. Such ag-

gregation can often be minimized by the presence of reducing agents, which break disulfide bonds between subunits. However, sometimes aggregates are the result of other types of bonding. For example, sickle cell Hb of humans forms large polymers due to a valine residue at the sixth residue of the β chain, resulting in hydrophobic association between Hb tetramers. In a survey of 54 species of turtles, some had 4 S and 7 S components, and others had some greater than 8 S, probably the result of polymerization. The fact that frog and turtle Hbs tend to polymerize on hemolysis supports this hypothesis (168).

Human myoglobin contains one heme per molecule while Hb has four hemes per molecule. The circulating Hb of cyclostomes is ~17,800 D with a single heme per molecule (1.9 S). The Hb of *Petromyzon marinus* is a monomer when oxygenated and a tetramer when deoxygenated. The Hbs of three amphibians differ in capacity for polymerization. In the axolotl, deoxyhemoglobin is tetrameric over the entire pH range. In *Rana esculenta* oxyhemoglobin may be octomeric at pH 5–8 while in *Triturus* both oxy- and deoxyhemoglobin remain tetramers over the entire pH range.

In summary, the quaternary structure of vertebrate Hb is usually a 64,000-Da tetramer of four 16,000-Da subunits. While some species have Hbs that are polymers of these tetramers *in vitro*, such polymers may not exist *in vivo* except under rare circumstances.

Structure of Invertebrate Hemoglobins
Many of the polymeric Hbs of invertebrates form regular structures, and have been studied by electron microscopy. For example, Annelid Hbs form a hexagonal structure with a hollow center (65). The polymers are ~240 Å in diameter and ~160 Å high. They contain two stacks of six 70-Å spherical subunits. The molec-

*$s_{20,w}$ is abbreviated as S.

ular size of many invertebrate Hbs have also been studied by ultracentrifugation.

In those invertebrates in which the Hb is in solution in the hemolymph, the molecular weight (MW) is sometimes >1,000,000 Da. On the other hand, Hb inside cells tend to be much smaller molecules. In *Thyone* and *Gastrophilus* the Hb in cells is ~34,000 Da, probably in two units per molecule. However, the chironomid larva has a low molecular weight Hb circulating freely in the hemolymph. In the annelid *Glycera*, the Hb in the coelom has a MW of 18,200; the number of residues per helix is the same as in myoglobin. The Hb of *Arenicola cristata* has a MW of 2.85×10^6; the intact molecule consists of two six-membered rings, 12 subunits of MW 230,000 with 8 hemes each, for a total of 96 hemes per Hb molecule. In the clam shrimp *Cyzicus*, the Hb in the hemolymph has an $s_{20,w}$ of 11.4 S, corresponding to a MW of 2.2×10^5; the molecule consists of 12 or 13 subunits. In the perienteric fluid of *Ascaris* the sedimentation coefficient of the major component of the Hb is 11 S. Some molecules occur at 8 and at 2 S, while the body wall Hb has a value of 1.5 S. These correspond to 8 hemes per molecule in the perienteric fluid, or a MW for each heme of 40,600.

In summary, invertebrate Hbs can vary between one and several hundred subunits. Hemoglobins in the hemolymph tend to be large polymers while those confined to O_2-carrying cells (e.g., erythrocytes) tend to be of lower molecular weights. It is often argued that polymerization and confining Hbs to circulating cells may have a physiological advantage that relates to osmoregulation (see Chapter 2).

Hemerythrin

The structure of hemerythrin is unique, differing from that of the other respiratory pigments in that the iron atoms binding

O_2 are bound to amino acid side chains rather than associated with the porphyrin heme (103, 105). The O_2-binding site in hemerythrin consists of two nonheme Fe atoms linked to the protein by histidine and tyrosine side chains in an azidomet structure which is very similar in all hemerythrins studied (84, 107). Deoxygenated hemerythrins are colorless and have Mössbauer parameters characteristic of a high spin Fe(II) complex but take on a wine red color and change to an antiferromagnetically coupled high spin Fe(III) state on oxygenation (84).

Hemerythrin contains three times as much iron as hemoglobin but the stoichiometry between Fe and O_2 remains 2:1.

Hemerythrin is always located intracellularly, but different hemerythrins are found in vascular fluids, in coelomic fluids, and intramuscularly (myohemerythrin). All three forms may occur in a single species—for example, the sipunculid *Themiste zostericola* (130). Here, each molecule is distinctly different as evidenced by electrophoretic analysis, peptide mapping, amino acid composition, and amino acid sequencing (103). Myohemerythrins have the lowest molecular weight (~13,900 Da) and are considered monomers. Many vascular and coelomic hemerythrins are octomers with a MW of ~100,000 Da, but trimers, tetramers, and hexameric aggregates have been reported (84).

Amino acid sequencing has been completed for several coelomic hemerythrins and one myohemerythrin. The molecules are similar with the two coelomic hemerythrins showing ~80% homology. On the other hand, homology with the monomeric myohemerythrin is only ~40% (103). Under the electron microscope hemerythrin octomers are seen to be associated in two groups of four subunits arranged one above the other as a two tiered box. Up to 3 subunit types hybridizing to form 10 molecular species are known in the sipunculid *Phascolopsis agassizii*. It will be interesting to see if this preservation of structure will be maintained when sequences of non-sipunculid hemerythrins are available.

Chlorocruorin (Greenish Hb)

As mentioned previously, greenish Hb (i.e., chlorocruorins) appear similar in gross structure to annelid extracellular hemoglobins. Early authors treated them as distinct respiratory pigments, possibly because chlorocruorin becomes green, rather than red, upon oxygenation. The molecules, however, are similar in size (~3,000,000 Da), with a sedimentation coefficient of 54–61 S, and similar isoelectric point. The structure under the electron microscope is also similar with both molecules appearing as a two tiered hexagonal arrays of 12 subunits (194). In fact, the structures of the two heme molecules differ only in the substitution of one vinyl group by a formyl group. In chlorocruorin, O_2 is assumed to react with one iron atom bound to a protoporphyrin plus one histidine residue of the globin, as in hemoglobin.

Following the discovery of this close structural correspondence many authors have argued that chlorocruorins should be grouped along with other annelid extracellular hemoglobins, either under the term erythrocruorins (185, 199) or in a specific class (Class D) of invertebrate hemoglobins (194). Others have argued that the term erthrocuorin is unnecessary. Class D hemoglobins, which occur only in annelids and Vertimentiformes (and perhaps pogonophorans), are typified as single domain, multiple subunit hemoglobins, in which some of the subunits are disulfide bonded. A domain is taken to be a molecular mass of ~16,000 Da, which contains heme and binds O_2. Since the iron content of annelid extracellular

hemoglobins differ from all others, some have concluded that all the polypeptide chains do not contain heme (194); however this allegation has not been proven.

Within a genus, individual species may possess either chlorocruorin, or hemoglobin, or neither (Table 1). Other species may possess both hemoglobin in the tissues and chlorocruorin in the vascular fluid (*Potamilla* and *Sabellidae*) or may possess both in the same vascular fluid and perhaps even as part of the same molecule, as in *Serpula vermicularis* (Serpulidae). Chlorocruoheme has been reported in some tissues of *Sabella (Spirographis) spalanzini* (61), its reported presence there probably results from contamination from blood.

Hemocyanin

Hemocyanins (Hcs) are distinct from all other respiratory O_2 carriers in that O_2 binding occurs with copper rather than iron atoms (Fig. 3), and in that functional Hcs are always large extracellular molecules (14, 53, 193). Native Hcs are constructed from monomeric subunits of ~500,000 Da in molluscs and 68,000–82,000 Da in arthropods. These subunits are usually polymerized into much larger aggregates, with molecular mass exceeding 1–3 million daltons in the arthropods and 8 million in the molluscs. These giant molecules (exceeding the size of ribosomes!) can be visualized under the electron microscope where the structure of native arthropod Hc molecules appears as squared, or rectangular hexagonal arrays, while the larger subunits of molluscan Hc are assembled into 3–12 tiered cylinders (14, 53). Arthropodan and molluscan Hc's appear to have evolved from a molecule similar to the enzyme tyrosinase (67, 174), but they have diverged sufficiently such that the two proteins are structurally similar only at the O_2-binding site.

Under experimental conditions of high

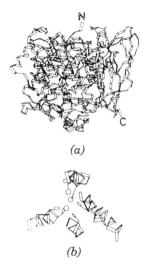

Figure 3. Three-dimensional structure of hemocyanin from the crustacean *Panulirus interruptus*. (*a*) Ribbon diagram of complete subunit. (*b*) Detailed stereo diagram of active (O$_2$-binding) site of Panulirus hemocyanin. The copper ions forming the O$_2$-binding site are represented by open circles (195).

pH and removal of divalent cations, Hc dissociates into its subunits. Reassociation usually results in nonnative aggregates. Arthropod subunits are the smallest aggregates that can bind O$_2$ reversibly and thus incorporate single O$_2$-binding domains. X-ray studies (67, 195) suggest that these monomeric subunits are kidney shaped structures consisting of a single polypeptide chain of ~670 amino acid residues organized into three domains of comparable size. Only the central (2nd) domain binds O$_2$. The smallest native associated Hc molecule of the arthropods is a hexamer apparently composed of two trimers joined together (Fig. 4*A*). In some crustaceans, for example, the euphausid *E. superba*, in some isopods, and in several decapod crusta-

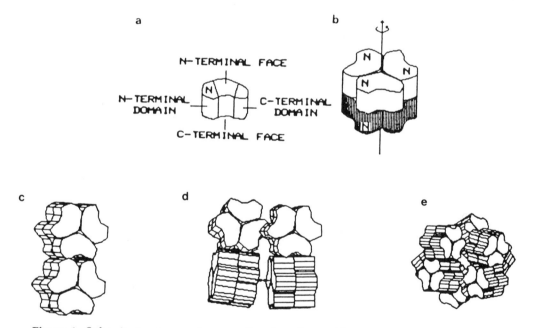

Figure 4. Subunit structure and aggregation in arthropod hemocyanin. (*a*) Aggregation of the three domains—N-terminal, central, and C-terminal—into the complete subunit (*Panulirus*). (*b*) Aggregation of subunits into the hexameric molecule. (*c–e*) Aggregation of hexamers into naturally occurring multiples of two (many crustaceans), four (many chelicerates), and eight (*Limulus*), respectively.

ceans, including the shrimp *Penaeus*, the lobster *Panulirus*, and the crab *Uca*, native Hc consists entirely or largely, of these simple hexameric units (133). However, in most decapod crustaceans the predominant natural unit incorporates two hexamers dimerized to form a dodecamer of 2 × 6 mer (Fig. 4c), while in *Callianassa* and *Upogebia* further dimerization into 4 × 6 multiples is seen (Fig. 4d). The association patterns of higher polymers of Hc are best known for chelicerates. Dodecamers, formed from two hexamers associating with their trigonal axes perpendicular to each other (Fig. 4c) are common in spiders. Aggregations of four hexamers (24 mers) common in many chelicerate groups and 48 mers, characteristic of the Limulidae, are formed from apposition of two or more dodecamers (108) (Fig. 4e and d).

The subunits of molluscan Hcs are large molecules (11 S, 400,000–450,000 Da) and contain 7–8 covalently linked O_2-binding domains (14, 53, 193). Native molluscan Hc are thus giant aggregates of multidomain subunits. Initially these may dimerize into 19 S units that in turn associate into pentamers, forming a 50- to 60-S molecule characteristic to the cephalopods. In the gastropods further dimerization occurs to form either 100-S molecules, which predominate in hemolymph of many species or, occasionally even larger aggregates (13). The 10 subunits of the cephalopod Hc molecule are helically wound to form the cylindrical structures (14). The cylinders are not perfectly symmetrical, and may have a specialized region (cap) at one end (53, 193). In some gastropod Hcs two such subunits are joined together with the cap structures outward.

Much diversity occurs within the subunits of Hc, especially in arthropods. For example, up to seven subunits are found in crustaceans and up to eight in chelicerates (14, 53, 112, 193). The reason for

the production of so many subunit types is unclear but reassembly of native Hc is prevented by the absence of single subunits suggesting that they may function in ensuring assembly of the correct multisubunit molecule (109, 112, 132).

Primary structures of some arthropod Hcs have been elucidated (4, 53, 112, 193). The primary structure of the A subunit from the scorpion *Eurypelma* shows a 42–70% similarity in amino acid composition when compared with equivalent subunits from *Limulus* and *Tachypleus* (112), despite the 400–500 million years of evolution that separate these two groups. The more distantly related crustacean Hc (*Panulirus*) shows similarity with Hc units in chelicerates (112, 180), suggesting that the basic three domain structure is common among arthropod Hc.

Molluscan subunits, separated from the native aggregate molecules as described above, are larger (400,000 Da) and although they can be cleaved proteolytically into smaller fragments capable of O_2 binding, these usually cannot reassemble into fully functional subunits. Fewer subunits may occur in Mollusca [e.g., two for the primitive gastropod *Megathura* and the neogastropod *Murex* (14) but up to seven are reported for *Octopus* (69)].

X-Ray crystallographic studies of the O_2-binding site of Hc shows that each of the copper atoms has three protein ligands all of which appear to be bound via histidine (4, 53). This structure is apparently similar for all arthropod Hcs. In addition, similarities can be seen between this binuclear copper site and that of molluscan Hcs on the one hand and that of other copper containing proteins such as tyrosinase from *Neurospora crassa* on the other, supporting the evolutionary relationship between these copper-containing proteins.

Oxygen binding by all Hcs occurs at a bis-copper active site. The pair of copper atoms that bind the oxygen atom probably form a bridged dioxygen complex that links the two copper atoms at a distance of ~3.4 Å. Release of O_2 is probably associated with a stereochemical change incorporating a movement apart of the two atoms. The blue color resulting from this change in state of the copper atoms gives the characteristic absorption peak of oxygenated Hc between 335 and 345 nm (14).

Oxygen Transport by Respiratory Pigments and Plasma

Oxygen Transport—Theory and Fundamentals

Although respiratory pigments vary in their molecular size and chemical constituents, all pigments share a common and crucial set of properties: (1) they form weak and reversible chemical bonds with O_2 and, (2) when all the possible binding sites are saturated, they considerably enhance the total O_2 capacity of the blood. These properties ensure that a large quantity of O_2 is bound to the pigment as it passes through the gas exchange surfaces, and that this O_2 in turn can be quickly and efficiently released upon the pigment's arrival at the respiring tissues.

The tremendous variation in respiratory pigment structure, discussed above, is reflected in significant differences in pigment function. In order to place this into a comparative context it is necessary to review the fundamental biochemistry and physiology of O_2 transport. Most of the discussion will revolve around vertebrate hemoglobins because they are the best studied, but the concepts that emerge are generally applicable to invertebrate hemoglobins (including chlorocruorins), hemerythrins, and hemocyanins.

Homotrophic Interactions: Oxygen Dissociation

The number of possible binding sites for O_2 on any particular respiratory pigment

depends on whether that pigment is a monomer, dimer, tetramer, and so on. Regardless of the form of the pigment, the number of binding sites that are actually occupied by molecular O_2 is fundamentally dictated by the O_2 concentration immediately surrounding the pigment and the intrinsic association constant. If the O_2 concentration decreases, the number of O_2 molecules that remain bound to the pigment also decreases. As the O_2 concentration increases, the number of binding sites that are occupied by O_2 increases until the pigment is saturated with O_2. Further increases in ambient O_2 concentration will have no additional effect on the amount of O_2 bound to the respiratory pigment, as all binding sites will be in the oxygenated state. The oxygenation state of a given population of pigment molecules can be represented as

$$Y = \frac{[\text{oxyprotein}]}{[\text{oxyprotein}] + [\text{deoxyprotein}]} \quad (1)$$

where Y is the fraction of molecules in the oxygenated state, ranging from 0 to 1. (Physiologists sometimes convert fractional saturation (Y) to percentage saturation by multiplying Y by 100.)

In polymeric proteins, oxygenation within the polymer often depends on the oxygenation state of the other subunits within the same macromolecule. In other words, the binding of O_2 at one site facilitates the binding of O_2 at other sites. This is a type of *homotrophic* interaction that is referred to as *cooperativity*. In a monomer like myoglobin (Mb), that binds one O_2 molecule for each pigment molecule, binding is noncooperative. In other words, only multisubunit molecules can be cooperative.

Since O_2 is a gas, it is convenient to express O_2 concentration as the partial pressure of O_2 (P_{O_2}). The P_{O_2} can be converted to O_2 concentration using Henry's law ($[O_2] = K\,P_{O_2}$, where K = Henry's

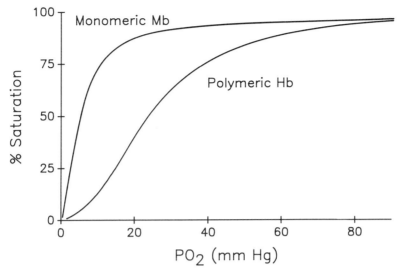

Figure 5. Oxygen equilibrium curves (OECs) depicting the relationship between O_2 saturation of monomeric and oxygen transport pigments like myoglobin and hemoglobin and partial pressure of oxygen (P_{O_2}). Modified from (28).

solubility constant). A convenient parameter used to quantify the tendency of O_2 to dissociate from a pigment is the partial pressure at which one-half of the molecules (in this case myoglobin) have bound oxygen (i.e., Y = 0.5). This parameter is called the P_{50}, and like P_{O_2} is expressed in conventional units of partial pressure (mmHg, torr, or kP).

The mathematical description of O_2 dissociation for a monomeric transport protein (e.g., Mb) is given by

$$Y = \frac{P_{O_2}}{P_{O_2} + P_{50}} \qquad (2)$$

Equation 2 can be used to generate an *oxygen equilibrium curve* (OEC) for that transport pigment. In practice, OECs are usually generated experimentally by measuring the fraction or percentage of O_2 saturation of a pigment at a series of controlled P_{O_2} values, and then plotting the results. Figure 5 shows the OEC for the monomeric pigment myoglobin, which typically is hyperbolic.

A simple model describing the OEC for

a polymorphic protein like Hb with n binding sites can be represented by

$$Y = \frac{(P_{O_2})^n}{(P_{O_2})^n + (P_{50})^n} \qquad (3)$$

This equation can be used to generate the OEC for a polymeric pigment like Hb, depicted in Figure 5. The OECs of polymeric pigments typically are sigmoidal to various degrees, with the exact shape of the curve being strongly influenced by the degree of cooperativity between subunits.

Equation 3 can be rearranged in the log form as

$$\log \frac{Y}{(1 - Y)} = n(\log P_{O_2} - \log P_{50}) \qquad (4)$$

Equation 4 is known as the *Hill equation*. A graphical plot of Eq. 4 yields essentially a straight line when 0.1 < Y < 0.9 (Fig. 6 inset). In actuality, the ends of the Hill plot curve at very low and very high P_{O_2} values. However, at the P_{50} the slope of the Hill plot is equal to n, which is re-

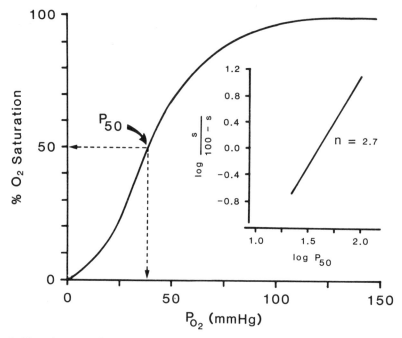

Figure 6. The shape and position of any oxygen equilibrium curve can be quantified. The P_{O_2} at which the pigment is 50% O_2 saturated is designated the P_{50}, and is an indicator of the pigment's O_2 affinity. The inset shows a Hill plot, which linearizes the O_2 equilibrium curve. The slope of the line defines Hill's coefficient, n. This value is an indicator of the shape of the curve, which is dictated by subunit interactions within the pigment. The P_{50} and Hill's n provide convenient values for comparing different O_2 equilibrium curves.

ferred to as the Hill coefficient and gives an index of cooperativity and the shape of the OEC. Monomeric pigments (e.g., myoglobin or myohemerythrin) have Hill values of $n = 1$, indicating no subunit cooperativity. The higher the value of n, the greater is the subunit cooperativity. Thus, the sigmoid curves of vertebrate Hbs or crustacean Hcs typically have n values of two to three, while the larger Hbs of invertebrates tend to have n values exceeding three. Although the value for the *Hill coefficient* (n) has a theoretical limit equivalent to the number of subunits making up the pigment (e.g., 4.0 for a tetrameric Hb, 6.0 for a hexameric pigment), it rarely approaches that limit. Values exceeding 4 have been recorded for

Hbs from amphibians (115), reptiles (119), birds (116), and exceeding 6 for molluscan Hcs (53, 193). In these cases n is found to vary with pigment concentration, which suggests that additional cooperativity occurs when high blood pigment concentrations promote interactions between subunits of adjacent pigment moieties (see the discussion of polymerization in the section on structure of invertebrate hemoglobins).

The molecular mechanisms responsible for the cooperativity among subunits of respiratory pigments is not only of biochemical interest but is also of great physiological significance. As O_2 is unloaded from a pigment molecule the ability to dissociate or unload O_2 increases. The converse is also true for O_2 binding. This cooperativity makes respiratory pigments very efficient at O_2 transport. The association and dissociation of O_2 is more responsive to changes in P_{O_2} because of subunit interaction than it would be if the subunits were independent. In fact, this cooperativity amplifies the ability of the pigment to deliver O_2 to metabolically active tissues.

Under defined conditions of pH, temperature, ionic strength, and so on, comparisons of purified Hbs have provided information about species-specific functional differences. That is, differences in the O_2 affinity of the pigment reflect differences in the amino acid sequences of the subunits that are based on genetic differences in the DNA coding for those polypeptides. It follows that subunit cooperativity differences, reflected by interaction between sites during oxygenation (i.e., homotrophic interactions) are also under genetic control. On the other hand, studies of whole blood, erythrocytes, and Hb solutions do not necessarily provide such information because there are a number of other molecules (ligands) present in blood that can bind to respiratory pigments and alter their O_2-binding ability. These molecules show some species differences but they also change in response to environmental changes. Thus, while intrinsic Hb—O_2 affinity is genetically determined, physiological plasticity is provided by regulating the intracellular concentrations and types of certain modifier molecules that are able to modify Hb—O_2 binding affinity.

Heterotrophic Interactions
A variety of molecules (H^+, CO_2, organic phosphates) bind to respiratory pigments and modify their function. The interactions between these modifiers and their effect on the oxygen affinity of the pigment are termed *heterotrophic interactions*.

Heterotrophic interactions provide animals with considerable physiological flexibility in gas transport. The binding of a ligand to the pigment influences not only O_2 binding but also the binding of other allosteric modifiers. In fact, modulators act as molecular amplifiers to promote O_2 uptake and delivery under changing internal and external environmental conditions. Since the binding of these ligands to the pigment is functionally interconnected, they are often referred to as *linked functions* (216).

Most allosteric modifiers act directly on the respiratory pigment. Binding of the ligand to the pigment causes adjustments in conformation that alter the pigment's O_2 affinity. Depending on the respiratory pigment and the particular modifier, increasing concentration of a ligand may shift the OEC either to the left or the right, raising or lowering O_2 affinity. A rightward shift of the OEC is generally assumed to facilitate O_2 unloading at the tissues.

Of the number of different heterotrophic interactions described, a few will now be discussed in greater detail.

The Bohr Effect. Almost all respiratory pigments show a decreased O_2 affinity when the partial pressure of CO_2 (P_{CO_2}) and/or proton (H^+) concentration increases in the pigment's immediate environment. The effect, termed the *Bohr effect* after its Swedish discoverer, is manifested as a rightward shift in the OEC (termed the *Bohr shift*) and consequently an increase in P_{50} of the pigment. Figure 7A illustrates the Bohr effect by showing a family of O_2-equilibrium curves generated by changing the pH surrounding the pigment.

The magnitude of the Bohr effect is given by the change in O_2 affinity per

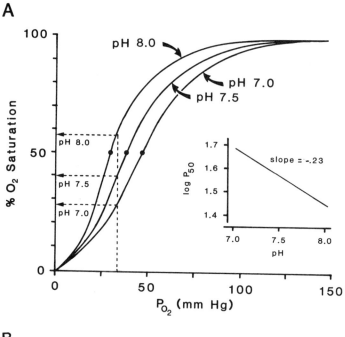

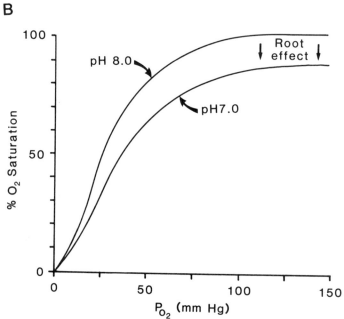

Figure 7. (*A*) Effect of changes in blood pH upon the O_2 equilibrium curve of a respiratory pigment at constant temperature. The points on the dissociation curves indicate the P_{50}. The dashed lines indicate the change in O_2 saturation at a given P_{O_2} that results from a change in blood pH. The inset, a plot of $\Delta\log P_{50}$ as a function of ΔpH, depicts the Bohr effect, given by the slope of the resulting line. (*B*) Effect of pH on blood showing a Root effect. While resembling an exaggerated Bohr shift, the Root effect differs in that the respiratory pigment never reaches full O_2 saturation.

change in pH unit:

$$\text{Bohr effect} = \frac{\Delta \log P_{50}}{\Delta \text{pH}} \qquad (5)$$

and typically ranges from near zero to values slightly below -1. At low pH (usually well below the physiological range), the Bohr effect may reverse and the O_2 affinity of the pigment actually increases; thus P_{50} is maximal at a particular pH.

The magnitude of the Bohr effect increases with increasing respiratory pigment concentration, and also with a rise in temperature or ionic strength (i.e., increased salt concentration). Hemocyanin from the xiphosuran arthropod *Limulus* (88) shows a reversed Bohr shift, while the Hcs of many gastropod molluscs [e.g., *Buccinum undatus* (25) and *Busycon canaliculatum* (127)] show a normal Bohr effect at high pH and a reversed Bohr effect at low pH. While the majority of fish hemoglobins have substantial Bohr efects, some do not (158).

The Root Effect. The extreme rightward shift of the OEC brought about by a pronounced Bohr effect can result in incomplete O_2 saturation of the pigment at the P_{O_2} values normally prevalent in the gas exchange organs. If the P_{O_2} is elevated sufficiently, however, the pigment ultimately becomes saturated. In some animals, increased [H^+] alters not only the O_2 affinity (Bohr effect), but also reduces the level of O_2 saturation that can be achieved at even very high P_{O_2} values. This reduction in the maximal O_2 saturation achieved by the pigment is referred to as the *Root effect*, and is illustrated in Fig. 7B. Recent studies suggest that the Root effect is an exaggerated Bohr effect (24). At the molecular level the Root effect depends on a substitution of a Ser residue at position 94 of the β chain. A Root effect is prevalent in the Hbs of teleost fishes

but not elasmobranchs (24). This phenomenon is important for respiration and the secretion of O_2 to the swim bladders of most teleost fish so that neutral buoyancy can be maintained. It is also important for secretion of O_2 to the vitelline fluid of the eyes of some fish. The Hcs of some molluscs show a reversed Root effect (in addition to a reversed Bohr effect) (25).

Haldane Effect. The *Haldane effect* describes the tendency for the affinity of *carbon dioxide* binding to respiratory pigments to decrease when the O_2 saturation of the pigment increases. An acid group present on each pigment subunit lowers its pK upon oxidation. That is, O_2 saturated pigments are stronger acids than nonsaturated pigments. Thus, as O_2 is given off in the tissues, the deoxygenated pigment becomes a stronger buffer. As O_2 is taken up in the gas exchange organs, there is an increase of negative charge and equivalent displacement of HCO_3^-, thus facilitating CO_2 loss (see the section on the chemistry of carbon dioxide). While the Haldane effect influences CO_2 transport in the same general way as the Bohr effect influences O_2 transport, they are independent phenomena.

Effects of Organophosphates and Other Modulators. A number of other naturally occurring cofactors can modulate pigment—O_2 affinity. Modulators act in a variety of ways. When they bind directly to the pigment, conformational adjustments occur that alter the pigment's O_2 affinity. Since all respiratory pigments are proteins; ions that favor rearrangement of protein tertiary or quatenary structure—for example, Ca^{2+}, Mg^{2+}, Cl^-—may act as cofactors altering O_2 affinity (12, 68, 121) as well as the binding of H^+ and other ligands. In addition, more specific cofactors have been described, including lactate for Hc (22, 191) and a variety of organophosphate compounds for vertebrate Hbs (5, 95). The major organic phosphate in fish erythrocytes is either adenosine triphosphate (ATP) (70, 71, 157) or guanosine triphosphate (GTP). Amphibian erythrocytes have ATP, GTP, and 2,3-diphosphoglycerate (2,3-DPG), while reptiles rely primarily on ATP but have small amounts of GTP. Birds have an array of organic phosphates that change during development, but the major allosteric modifier of Hb—O_2 affinity is a polyphosphate-like inositol pentaphosphate (IPP). The major organic phosphate in mammalian red blood cells is 2,3-DPG. In the case of respiratory pigments that are enclosed within blood cells, cofactor modulation of O_2 affinity may be particularly well regulated, since the pigment is enclosed within a microenvironment that can be chemically altered.

The binding of intraerythrocytic organic phosphates to Hb was first shown over 20 years ago (7, 40, 182) and is now known to be a potent effector of Hb—O_2 affinity (Fig. 8). Organic phosphates are in much higher concentration in the erythrocyte than in blood plasma. Consequently, a solution in which erythrocytes has been lysed, releasing the Hb into the plasma (and thereby greatly diluting the organophosphate cofactors), exhibits a much higher O_2 affinity than does whole blood. When all the organic phosphates and related modifiers are removed from the solution, the Hb—O_2 affinity of this "stripped" Hb is even higher. Unfortunately, much of the extensive research on Hb function prior to discovery of intraerythrocyte cofactors remains of biochemical interest, but is of lesser use in the study of *in vivo* O_2 transport because undefined amounts of cofactors are present.

The influence of organic phosphates on both the Bohr effect and on the pH dependence of the subunit cooperativity indicates that the reactions of many Hbs with O_2, organic phosphates, and protons are closely linked phenomena. Since the negatively charged organic phosphates form ionic interactions with positively charged amino acids in the hemoglobin, it should not be surprising that the affinity of Hb for organic phosphates is inversely related to pH. At low pH the presence of organic phosphates actually amplifies the Bohr effect. The interaction between Hb, O_2, H^+, and organic phosphates is diagrammatically illustrated in Fig. 9.

Intracorpuscular modulators can also exert an indirect effect on O_2 binding by the respiratory pigment. If the cofactor is produced in sufficient quantities, it can alter the Donnan equilibrium, that dynamic equilibrium that dictates the passive distribution of ions across the erythrocyte membrane (209, 211). For example, increased intraerythrocytic organic phosphate concentration produces a net influx of H^+ ions, resulting in a fall in intraerythrocytic pH and a right shift in the O_2-dissociation curve. Moreover, adaptive changes in fish Hb—O_2 affinity during acclimation to low O_2 or increased temperature are directly related to changes in red blood cell organic phosphate levels (158, 159, 209). That is to say, as environmental O_2 is reduced, fish often respond by lowering intraerythrocytic organic phosphates, thereby increasing oxygen affinity in an O_2-poor environment; this facilitates O_2 loading at the gills. However, mammals take the opposite approach, increasing 2,3-DPG (and thus decreasing P_{50}) to enhance O_2 unloading. This basic difference may arise from the fact that mammals are in a relatively O_2-rich environment while aquatic organisms live in an O_2-poor environment that varies greatly in concentration of O_2 in water.

Effect of Temperature on Oxygen—Pigment Binding

The binding of O_2, H^+, CO_2, and organic phosphates to a protein like Hb are linked

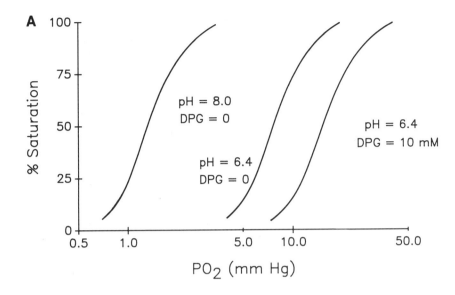

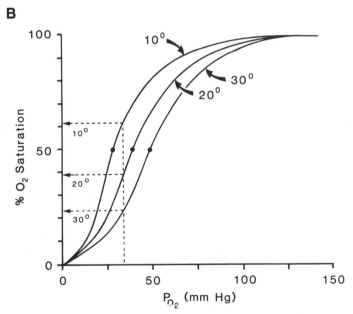

Figure 8. (*A*) Effect of pH and 2.3-diphosphoglyceride (DPG) on human hemoglobin–O$_2$ affinity. At high pH DPG has little or no effect, while at low pH, DPG amplifies the Bohr effect. Modified from (28). (*B*) Effect of changes in blood temperature upon the oxygen equilibrium curve of a respiratory pigment at constant pH. The dashed lines indicate the change in O$_2$ saturation at a given P_{O_2} that results from a change in blood temperature.

thermodynamic functions. These equilibria, like all such equilibria, are temperature dependent. As temperature increases, *dissociation* of the protein–ligand complexes are favored. Since ligand binding to respiratory pigments are exothermic (i.e., heat is released as the weak bonds are formed), an increase in envi-

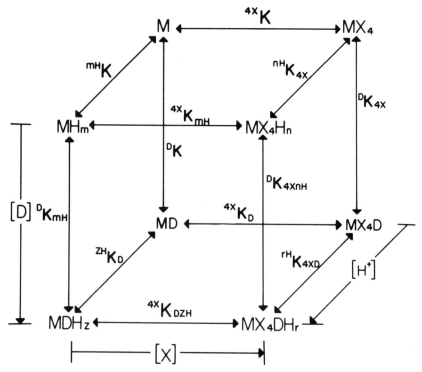

Figure 9. A generalized schematic of the linkage equilibria relating the binding of protons (H⁺), organic phosphate (D), and oxygen (X) to hemoglobin tetramers (M). The macro-binding constants (K) are described with presuperscripts indicating the number and kinds of ligands *in the process of being bound* and the postsubscripts representing the numbers and kind of ligands *already* bound to the hemoglobin macromolecules (M). [From (158).]

ronmental temperature will weaken these bonds and favor dissociation. When O_2 is the ligand, the result is a right shift in the OEC (increased P_{50}) with an increase in temperature. Figure 8*B* depicts a "family" of curves for a single pigment at three different temperatures.

The change in free energy (ΔG) of a ligand–protein interaction is

$$\Delta G = -RT \ln K \qquad (5)$$

where R is the gas constant, T is temperature in degrees Kelvin, and K is the dissociation constant for the ligand–protein complex. The enthalpy (ΔH) of the reac-

tion is defined by

$$\Delta H = \frac{RT^2 \, d \ln K}{dT} \qquad (6)$$

or for a gaseous ligand dissociation

$$\Delta H = \frac{-RT^2 \, d \ln P_{50}}{dT} \qquad (7)$$

which includes the heat of solution for the gas (about -3 kcal/mol) and ΔS is obtained from

$$\Delta G = \Delta H - T\Delta S \qquad (8)$$

The heat of combination of the pigment with O_2 varies not only between pigments but also between different species with the same pigment. However, differences in the thermal sensitivity of the oxygenation of blood often depends more on the ligands involved in heterotrophic interactions (e.g., protons, organic phosphates, CO_2, and Cl^-) than the species-specific hemoglobins and their homotrophic interactions (162).

Using the O_2 Equilibrium Curve— In vivo O_2 Transport

A quantitative analysis of *in vivo* O_2 transport requires knowledge of the shape of the OEC and values of arterial and venous blood P_{O_2} and/or O_2 saturation. Knowing only P_{O_2} without information on either the OEC or blood O_2 saturation conveys little information about blood O_2 transport.

Generally, the P_{O_2} of blood leaving the gas exchange organ is high enough to ensure full blood oxygenation (Fig. 10). At rest the P_{O_2} (and thus O_2 saturation) of blood leaving the metabolizing tissues (i.e., venous blood) generally falls towards the upper regions of the steep slope of the OEC (Fig. 10), although there can be considerable variation with time in any given individual. If the metabolic rate of perfused tissues is low, reflecting resting conditions, the relatively high P_{O_2} in these tissues results in removal of only a part of the total O_2 load on the respiratory pigment as it passes through the exchange vessels. If tissue metabolism increases during activity, the P_{O_2} surrounding the exchange vessels will fall accordingly. More O_2 will be released from the pigment to diffuse from the blood to the surrounding tissue, and the O_2 saturation of blood leaving the tissue and reentering the gas exchange organ will be lower for any given unit of blood.

The considerable quantity of O_2 that typically remains bound to the pigment even after passing through the tissues is commonly referred to as the venous O_2 reserve. If the tissue P_{O_2} falls further (e.g., because of increased metabolic rate) then the tissue P_{O_2} values slide down the OEC, and additional O_2 is released to the tissues.

In many species (e.g., vertebrates) the nonlinear shape of the OEC is important for efficient *in vivo* O_2 transport. Within limits, large changes in P_{O_2} in the gas exchanger will have little negative effect on O_2 loading of the respiratory pigment when the P_{O_2} values lie on the upper, shallow region of the OEC (Fig. 10). In the exchange vessels of the tissues, however, the range of prevailing P_{O_2} values falls on the steepest region of the OEC. Relatively small decreases in tissue P_{O_2} result in proportionately large increases in the amount of O_2 released to the tissues.

The Bohr effect plays an important role in O_2 transport of many species. As molecular CO_2 diffuses from metabolizing cells into the exchange vessels of the tissues, both plasma $[H^+]$ and P_{CO_2} increase. This shifts the OEC to the right, facilitating additional O_2 unloading without a change in the P_{O_2} of the surrounding tissues (Fig. 7A). If tissue O_2 consumption increases, then P_{CO_2} and $[H^+]$ in the tissue exchange vessels also increases, thus enhancing O_2 unloaded at the respiring tissues.

Within the gas exchange organ the process is reversed. As CO_2 is eliminated, blood pH increases, producing a leftward shift of the OEC and thus an increased blood O_2 affinity. The P_{O_2} in the gas exchanger is often higher than required to achieve full O_2 saturation of the pigment; however, the Bohr effect is generally more important in O_2 unloading.

A temperature-sensitive respiratory pigment is often considered of adaptive advantage in ectotherms. When body

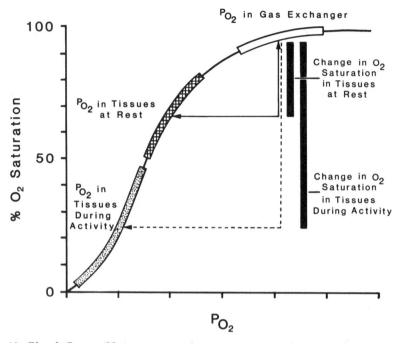

Figure 10. Blood O_2 equilibrium curve showing aspects of *in vivo* O_2 transport in a hypothetical animal with a blood-borne respiratory pigment. The empty boxed region at the upper end of the curve indicates typical combinations of P_{O_2} and O_2 saturation of blood leaving the gas exchange organ. The hatched region of the curve shows combinations of P_{O_2} and O_2 saturation of blood in the exchange vessels of metabolizing tissues at rest. The average change in O_2 saturation, which directly reflects the amount of O_2 transported to the tissues, is indicated by the smaller of the two vertical bars to the right of the curve. The dotted lower region of the curve shows combinations of P_{O_2} and O_2 saturation typical of blood in the exchange vessels in metabolizing tissues during periods of elevated O_2 consumption. The average change in O_2 saturation under these conditions is indicated by the larger of the two vertical bars. Note the large fall in O_2 saturation (and thus increase in O_2 transported to the tissues) produced by a relatively small decrease in P_{O_2} of blood in the exchange vessels of the tissues. Although the O_2 affinity, shape of the curve, and prevailing blood P_{O_2} values vary enormously in different animals, *in vivo* O_2 transport by blood-borne pigments follows these general principles.

temperature and thus O_2 consumption rise, the rightward shift of the OEC favors additional O_2 unloading at the tissues. This helps facilitate O_2 transport *provided* that the rightward shift is not so large as to adversely affect O_2 loading in the gas exchange organ. Although much attention has been paid to effects of temperature on the blood of ectotherms (137, 209), the effects of temperature on Hb in homeotherms is also significant in the context of large localized temperature differ-

ences in exercising muscle or in hibernating animals.

Increasing concentration of Hb modulators (cofactors) may shift the OEC either to the left or the right, depending on the respiratory pigment and the particular cofactor. A rightward shift is generally assumed to facilitate O_2 unloading at the tissues, while a leftward shift is assumed to facilitate O_2 loading at the gas exchange organ (see the section on blood adaptations resulting from environmen-

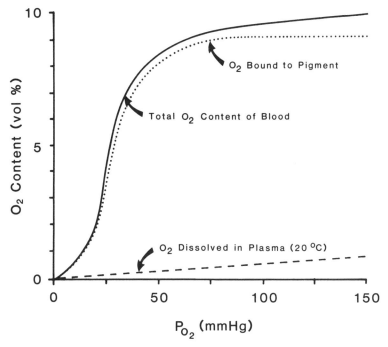

Figure 11. The O_2 content of whole blood of a turtle (20°C, pH 7.80) as a function of blood P_{O_2}. The dashed line indicates the quantity of O_2 physically dissolved in plasma. The dotted line indicates the quantity of O_2 chemically bound to hemoglobin. The solid line indicates the sum of dissolved and chemically bound O_2.

tal hypoxia). Again, it is important to emphasize that an apparent advantage of a shift in the OEC must be verified by examining the attendant changes in *both* O_2 loading and unloading!

In quantitative descriptions of blood O_2 transport, it becomes necessary to translate percent O_2 saturation of the pigment into actual O_2 content. Figure 11 replots an OEC such that O_2 content rather than saturation is presented as function of blood P_{O_2}. Commonly used units for O_2 capacity include ml of O_2/100 ml of blood (expressed as vol %) and mM O_2/L blood. The O_2 content increases as the pigment becomes progressively more loaded with O_2, until O_2 content reaches the total O_2 capacity (i.e., when all binding sites on the pigment are completely saturated). When O_2 content rather than O_2 saturation is plotted, the OEC contin-

ues to increase very slowly in a linear fashion as P_{O_2} continues to increase. Although at high P_{O_2} values the fully saturated pigment cannot bind additional O_2, molecular O_2 continues to become physically dissolved in plasma as long as plasma P_{O_2} increases. In animals with high blood O_2 capacity due to high pigment concentrations (e.g., most vertebrates), the dissolved O_2 component plays little or no role in the overall transport of O_2. In many invertebrates, however, the concentration of pigment is quite low, and dissolved molecular O_2 may be important under some circumstances.

As evident from the discussion above, the O_2 affinity of a respiratory pigment is the result of a complex interplay of P_{O_2}, $[H^+]$ (and sometimes P_{CO_2} specifically), temperature, and cofactor concentration.

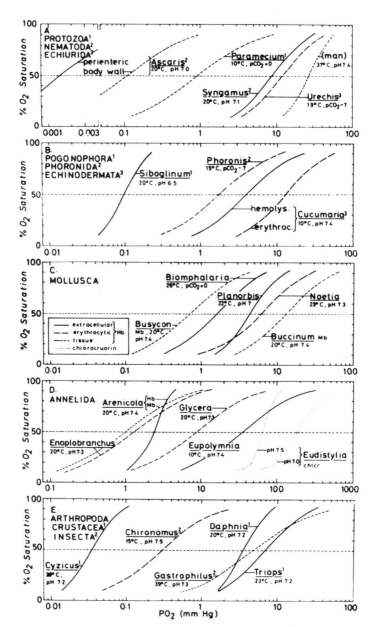

Figure 12. The O_2 equilibrium curves and magnitude of the Bohr shift in chlorocruorin and comparison with a range of invertebrate heme-containing pigments. Dotted curves, chlorocruorin from *Eudistylia infundibulum*. Solid curves, polymeric extracellular Hbs. Broken curves, oligomeric (generally erythrocytic) Hbs. Dashed curves, Muscle Hbs. [From (199).]

Thus, an OEC (or values for P_{50} and Hill's *n*) can be placed in a meaningful context *only* if the environment of the pigment is adequately specified. In practice, the variables most commonly reported are pH of the blood or pigment solution and its temperature—cofactor concentration is comparatively difficult to determine.

Respiratory Physiology of Hemerythrin

Hemerythrin-O_2 affinity varies, with P_{50} values ranging from 1 to 42 mmHg intracellularly and 1 to 15 mmHg for hemerythrin in solution (Fig. 12). Vascular, coelomic, and muscle variants of hemerythrin differ in their O_2 affinities, suggesting an affinity cascade that would facilitate oxygen diffusion from the external environment to the muscle. For example, in *Themiste zostericola* (84, 130) the O_2 affinity of the hemerythrins from the vascular system and coelomic fluid have P_{50} values of 42 and 4.5 mmHg, respectively, while the muscle myohemerythrin has a P_{50} between 0.9 and 1.0 mmHg (Fig. 13). Facilitation of O_2 delivery by myohemerythrin may compensate for the lack of capillary perfusion in the retractor muscles of this species.

Both vascular and coelomic hemerythrins are oligomeric. In other respiratory molecules the oligomeric state is usually associated with subunit cooperativity and ligand modulation of O_2 affinity. As yet, only one hemerythrin, that of the brachiopod *Lingula unguis* (P_{50} = 1.8), has shown cooperativity (130) and no evidence for modulation by H^+ or other factors has been observed. Although hemerythrins show little or no cooperativity *in vitro*, it has been suggested that greater cooperativity may occur *in vivo* because higher O_2 affinities of hemerythrins in solution are observed than for intact cells (126, 130) (Fig. 13). Finally, like most O_2 transport pigments, the O_2 affinity of

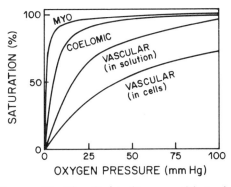

Figure 13. The O_2-binding equilibria for myohemorythrin (MYO) and hemerythrins from *Themiste zostericola* [modified from (130) and (84)]. Curves for vascular hemerythrin illustrate marked difference in O_2 affinity between stripped hemerythrin in phosphate buffer (in solution) and hemerythrin tested within the original blood cells.

hemerythrins changes with temperature. In fact, the specific heat of oxygenation is greater for hemerythrins than for hemoglobins.

Respiratory Physiology of Chlorocruorin

Compared with vertebrate hemoglobin little is known of the O_2 transport role of invertebrate hemoglobins (198, 199). This is particularly true in the case of the greenish hemoglobins (chlorocruorins) where physiological data are rare.

Chlorocruorins are present in the circulating body fluids of a considerable number of polychaete species. The few published oxygen affinity measurements reveal that O_2 capacity is usually moderately high (3.2 − 4.0 to 4.0 mmol/L) (198) and the degree of cooperativity and magnitude of the Bohr effect are also high. Oxygen equilibrium curves (198) (Fig. 12) indicate that the chlorocruorins have very low O_2 affinities, suggesting that the pigment may be rarely saturated under natural conditions. However, the *in vivo* pH and polymerization state of these mole-

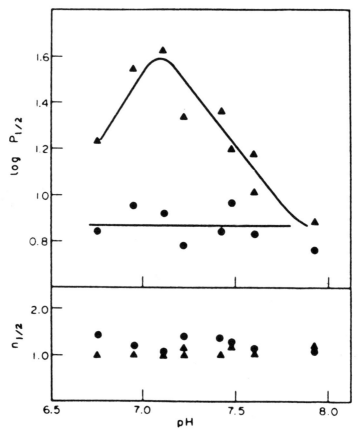

Figure 14. The O_2 equilibria of chlorocruorin (triangles) and hemoglobin (circles) from *Serpula vermicularis*. Upper panel—Variation in O_2 affinity with pH. $P_{1/2} = P_{50}$, the partial pressure for 50% O_2 saturation. Lower panel—Variation in cooperativity of O_2 binding with pH. Here $n_{1/2}$ = Hill's n at P_{50}. Sample contains 2.5×10^{-5} M heme and 3.5×10^{-5} M chlorocruoroheme. [From (184).]

cules may result in a higher O_2 affinity than that determined *in vitro*.

Evidence supporting a role in O_2 transport for chlorocruorin is provided by experiments where marked decrease in O_2 uptake results from poisoning the O_2 carrier with carbon monoxide (198). Chlorocruorins may also function in O_2 storage by providing O_2 for continued aerobic metabolism during periods of ventilatory shutdown (apnea). The magnitude of the chlorocruorin-bound O_2 stores of *Sabella pavonina* can be calculated to extend preapneic levels of O_2 consumption for 6–9 min (198). Although this is not sufficient

to extend aerobic metabolism greatly during the long periods of apnea characteristic of intertidal emmersion, it would suffice for short-term ventilatory disturbances [e.g., during withdrawal of the tentacular gas exchanger during defensive maneuvers or during the periodic ventilation patterns commonly seen in aquatic organisms (101, 143)].

The O_2 affinities of annelid extracellular (type D) hemoglobins are known to be modified by change in blood ions, particularly divalent cations (198). Preliminary evidence suggests that similar modification may also occur in chlorocruorins (3).

Vascular fluids of *Serpula vermicularis* apparently contain both chlorocruorin and hemoglobin (61). Chlorocruoroheme is reported to have low O_2-binding affinity (P_{50} = 40 mmHg, pH 7.0) which is pH sensitive, while the heme has higher oxygen affinity (P_{50} = 8 mmHg, pH 7.0) and is relatively pH insensitive (Fig. 14). Despite this marked functional difference it has not been possible to separate the two carriers electrophoretically (185). The possibility that two O_2-carrier systems with differing O_2-binding properties could be located on the same protein molecule implies a novel mechanism for adaptation to varying environmental O_2.

No study has yet addressed the role of chlorocruorins in CO_2 transport. In fact, the only data on CO_2 transport in annelids are from the hemoglobin-containing species *Arenicola marina* (198). A large Haldane effect has, however, been reported for the hemoglobin (erythrocruorin) from the giant Australian earthworm *Megascolex* (202) suggesting that facilitation between O_2 and CO_2 binding exists in at least one annelid's extracellular O_2 carrier.

Respiratory Physiology of Hemocyanin

Hemocyanin–Oxygen Binding

The O_2 binding of both molluscan and arthropod Hc indicates cooperativity greater than that observed in most O_2 carriers and it is often extremely sensitive to one or more modulator. A complex relationship occurs between O_2 binding and subunit aggregation of Hc. Native Hc from the tarantula, *Eurypelma californicum*, occurs predominantly as an aggregate of four hexamers, which exhibits relatively low affinity, very high cooperativity that changes with pH, and a strong negative Bohr shift (133). Each hexamer is composed of subunits of which seven distinct types are known (109, 134). Isolated subunits are not identical and their O_2-binding properties differ from those of the native molecule, exhibiting relatively high O_2 affinity, noncooperative O_2 binding, and no Bohr effect (47). Reaggregation studies confirm that these differences are correlated with the aggregation state of the molecule (172). A one-quarter molecule (hexamer, 1 × 6 mer) shows an O_2 affinity characteristic of the complete molecule and a full Bohr effect, but has only 25% of the maximum cooperativity. Maximum cooperativity is obtained with the complete molecule (4 × 6 mer) but 50% cooperativity can be reached with half molecules (2 × 6 mer). In molluscan Hcs the full native molecule has O_2-binding properties, which cannot be predicted from the responses of the constituent parts.

It is generally assumed that Hc molecules function in O_2 transport. Doubts concerning this assumption might be raised for some crustaceans, where Hc content may be insufficient to raise O_2 levels significantly. On the other hand, this may be counterbalanced by the large circulating blood volumes of Crustacea (138). Table 2 suggests that the total O_2 capacity of the open circulatory system of a crab and the closed circulatory system of a fish are similar; suggesting that dissolved O_2 may constitute a substantial fraction of the total O_2 carried in the circulatory system of crabs. Thus, dissolved O_2 may play an important role in the delivery of O_2 to their tissues.

Modulation of Hc—O_2 Binding

Adaptive Changes. To some extent the variability in O_2-binding properties seen in Hc from adult animals can be correlated with adaptation to the range of O_2 levels normally experienced in that animal's natural environment. Thus, animals with high O_2-affinity Hc often occupy environments that are typically hypoxic (137)

TABLE 2. A Comparison of Oxygen Storage Capacities of a Fish and a Crab

Species	Circulatory System	O_2 Carrier	Blood Volume (% Body Mass)	Blood O_2 Capacity (vol %)	mL of O_2 Stored
Platichthys (flounder)	Closed	Hb	6	6	1.1
Cancer (crab)	Open	Hc	30	1.3	1.3

while animals from eurythermal environments may show reduced temperature sensitivity (145, 207) (Fig. 20). However, the literature also contains exceptions (123).

Some animals may be able to synthesize several different types of Hcs. For example, developmental changes in both Hc structure and function have been reported in Crustacea (133, 183) (Fig. 26). Seasonal variation in O_2-binding properties has also been reported for several crustacean Hcs (135, 148). While the role of changes in subunit composition in such variation has been studied to some extent, it has not yet been widely demonstrated. Although functional variation in O_2-binding properties occurs between Hc subunits and between artificially constructed single subunit aggregates, very few studies have been able conclusively to link changes in subunit composition in Hc with adaptive change in O_2 properties *in vivo*.

Acclimatory Changes. In addition to adaptive selection which fits O_2 carrier function to a particular niche, O_2 binding by many Hc molecules can also be adjusted rapidly. This allows Hc function to be physiologically adjusted to match variation in oxygen demand whether brought about by changes in environmental conditions or activity regimes.

Rapid changes in O_2-binding characteristics can result from a variety of physicochemical and physiological factors involving both active and passive mechanisms. Passive mechanisms may predominate in acclimation to temperature change. Two factors, increase in specific heat of oxygenation and change in pH (Bohr shift) combine to reduce O_2 affinity as temperature rises. The result is an increased gradient between venous blood and tissues that helps satisfy the increased O_2 demand. The magnitude of the effect, however, varies markedly between species. The Hcs of several crab species may be relatively insensitive to temperature change, particularly over a range of temperatures approximating their normal habitat range (100, 145). Selection of such temperature-independent Hc may be characteristic of eurythermal animals. Seasonal changes, which usually act to reduce the magnitude of the passive effects outlined above have also been reported (135, 148, 190), but the mechanisms are not clear (124, 148).

Effects of Modulators on Hc Function. The O_2 affinity of Hc can also be regulated by changes in its chemical environment (22). Both inorganic and organic factors are known to be involved. In many Hc molecules, especially of cephalopods and many decapod crustaceans, these heterotrophic effects are larger than occurs in most other O_2 carriers.

The effects of variation in $[H^+]$ on Hc—O_2 affinity (Bohr effect) are complex, since increase in [H⁺] can either increase or decrease O_2 affinity (14). Usually, however, the physiological range of hemolymph pH experienced by a particular animal limits the expression of this biphasic Bohr shift.

The functional importance of decreased O_2 affinity has been demonstrated for crustacean Hc during activity (137), when additional acidosis amplifies the Bohr shift and allows displacement of additional O_2. An increase in O_2 affinity may be important in adjusting O_2 delivery as in animals exposed to environmental hypoxia (137). Two response patterns are seen. First, in many decapod crustaceans exposure to maintained moderate hypoxic conditions causes hyperventilation, CO_2 washout and respiratory alkalosis (139). The resulting reduction of the Bohr effect increases O_2 affinity, allowing better O_2 uptake from the reduced ambient P_{O_2}. Potentiation of this effect by lactate ion is discussed below.

The second response pattern is typified by Hc of *Limulus* and the mollusc *Buccinum*. In these animals hypoxia is associated with acidosis (25) but the Bohr effect is reversed. This allows acidosis to increase O_2 affinity, enhancing O_2 uptake from hypoxic water (123). A similar function for the reversed Bohr effect is postulated for Hc of the whelk *Buccinum undulatum* (25, 127), but this Hc additionally shows a reversed Root effect and marked changes in cooperativity with a change in [H⁺]. This combination of reversed effects allows an increase in both the O_2 capacity of hemolymph and its ability to load O_2 and thus can maintain O_2 delivery even in relatively severe hypoxia (25).

The O_2 affinity of Hc may also be adjusted by changes in the hemolymph contents of other ions such as Ca^{2+} and Mg^{2+} (53, 123) and, particularly in crustaceans, inorganic molecules. Organic modulators such as lactate and urate (22, 146) may be unique to Hc systems (122). The action of these modulators opposes the Bohr shift (Fig. 15*A*). Generally, their interactions with Hc are complex (147) and maximal effects (conformational change) may arise from a combination of factors. The functional importance of lactate has been demonstrated in crustaceans where it may serve to attenuate the normally beneficial Bohr effect under extreme conditions (e.g., strenuous activity and high temperature) (Fig. 15*B*). Here, a combination of a large Bohr effect and the cumulative effects of temperature and exercise-induced acidosis, reduce O_2 affinity sufficiently that O_2 uptake at the gills could be severely impaired in the absence of lactate (138). Secondly, where acidosis can be prevented, lactate and potentially other modulator substances, can increase O_2 affinity, that is, cause a left shift in the O_2 equilibrium curve, thereby increasing O_2 uptake. A specific effect of CO_2, the reverse of that reported for mammalian Hb, is reported for several crustacean and molluscan Hcs but its function is poorly understood (125, 148).

The hemocyanins, particularly of Crustacea, are thus richer in both the magnitude and variety of modulator effects compared with other O_2 carriers (128, 129). This may reflect both the complex multimeric structure of the pigments and their presence in the hemolymph as naked molecules unprotected by cellular membranes.

Adaptations of Oxygen Transport

While there have been numerous *in vitro* measurements of blood respiratory properties (164), comparatively few *in vivo* measurements of blood O_2 saturation and P_{O_2} have been made. However, to be effective in O_2 transport, a blood pigment must become substantially O_2 saturated at the P_{O_2} in the gas exchange organ(s) and then give up a substantial part of its

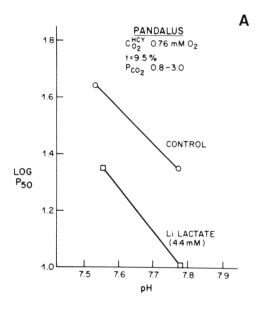

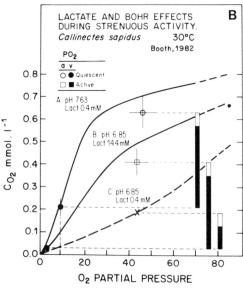

Figure 15. (*A*) Proton and lactate effects on oxygen binding by Hc of *Pandalus platyceros*. Li lactate = lithium lactate. (*B*) Modulation of hemocyanin O_2 binding *in vivo* in *Callinectes sapidus* during activity at high temperature. C_{O_2} = total O_2 content of hemolymph. Solid lines represent O_2-binding curves calculated at *in vivo* pH and lactate levels stated. Dotted line, O_2-binding curve under (hypothetical) conditions where no lactate effect occurs. Histograms illustrate O_2 delivered to tissues (post-prebranchial O_2 content). Clear areas repre-

bound O_2 at the P_{O_2} values prevalent in the systemic vessels. A common mistake in discussing the purported adaptive advantage of a particular pigment is to focus exclusively on one of these two functions, without recognizing that a change in pigment property that favors O_2 loading at the gas exchange organ may, under certain circumstances, actually be unfavorable for unloading of O_2 at the tissues. To be truly adaptive a given change in a pigment must enhance at least one aspect of blood transport without significantly diminishing the other.

Adaptive adjustments that alter blood O_2 transport—either the capacity to make short-term changes (acclimation) in an individual or long-term, evolutionary changes in a species—result primarily from limitations in available O_2. These limitations may arise from either reduced environmental O_2 or from internal hypoxia. Environmental O_2 concentrations are dependent on physical parameters such as altitude, temperature, salinity, as well as biological components (e.g., respiration of microorganisms in aquatic environments). Internal O_2 concentration varies with parameters such as metabolic rate, physical activity, body temperature, and breath holding (diving).

The most widely observed blood adjustment to hypoxia is an increase in concentration of blood respiratory pigment. Sometimes there is also a change in blood O_2 affinity and Bohr shift but the mechanisms responsible for these phenomena vary from species to species. Animals may maintain O_2 uptake in a hypoxic environment by increasing ventilation, which provides additional O_2 and helps eliminate CO_2 (see Chapter 5). The re-

sent O_2 delivered from solution, solid areas represent oxygen delivered from O_2 bound to Hc. Asterisk = maximum Hc bound O_2. [From (138), based on data from (16).]

sulting respiratory alkalosis increases both extracellular and intracellular pH. If there is a significant Bohr effect, then blood O_2 affinity will increase, facilitating O_2 loading at the gas exchange surface without compromising O_2 unloading at the tissues. This pH-dependent, ventilation-mediated adjustment in blood O_2 transport is rapid, and is useful in those animals exposed to very brief bouts of environmental hypoxia.

A more slowly developing response is a pH-independent increase in blood O_2 affinity, mediated by a variety of intra- or extracellular cofactors. Some of these mechanisms will now be examined in detail.

Blood Adaptations Resulting from Environmental Hypoxia

Aquatic Hypoxia

Animals inhabiting intertidal pools and stagnant freshwater habitats may be exposed to aquatic hypoxia on a tidal, diel, or seasonal basis. In some species, blood respiratory properties in individuals show short-term changes in response to a transient aquatic hypoxia. In other species, which appear to have a long evolutionary history of aquatic hypoxia, blood properties appear adapted compared to related species that have not evolved in hypoxic environments.

Short-term blood transport adaptations by individuals in response to aquatic hypoxia abound in the teleost fishes. For example, in the eel Anguilla anguilla 2 weeks of exposure to water with a P_{O_2} of 40–50 mmHg (air saturation = 150 mmHg) results in an increase in blood O_2 capacity from 6.6 to 9.8 vol% and a decrease in P_{50} from 17 down to 11 mmHg (pH 7.8, 20°C) (211) (Fig. 16). The increase in blood O_2 capacity results from an increase in hemoglobin concentration because of erythropoesis, while the increased Hb—O_2 affinity (fall in P_{50}) re-

sults indirectly from a decreased concentration of intraerythrocytic ATP. The decreased ATP concentration causes an efflux of H^+ ions and thus increases intracellular pH. This increase in pH is primarily (but not exclusively) responsible for the increased O_2 affinity via the Bohr effect. Although similar adaptive changes in blood respiratory properties in response to chronic aquatic hypoxia have been observed in other fish (74, 202), the balance between the indirect and direct effects of ATP is unknown.

The regulatory mechanisms responsible for these changes in the O_2 affinity of fish blood are not clearly understood, but appear to be controlled at both the cellular (75) and organ system level (19, 43). Under anoxic condition the erythrocytes of the fish Fundulus heteroclitus show a decrease in vitro in intracellular levels of ATP (75). The use of metabolic poisons showed that this ATP reduction resulted from a decreased supply of O_2 to the erythrocyte mitochondria (the erythrocytes of all vertebrates other than mammals have mitochondria). Thus, the O_2-consuming electron-transport system in fish red cell mitochondria is capable of regulating erythrocyte ATP levels in response to environmental O_2. This changes Hb—O_2 affinity via the direct allosteric effect of ATP on Hb and the indirect effect on intracellular pH.

Fish can rapidly increase or decrease blood O_2 affinity by employing hormonal and/or other factors in the plasma. Circulating catecholamine concentrations increase markedly in fish subjected to stresses such as hypoxia, physical disturbance, and confinement (43, 196). In several fish species, it has been shown that pharmacological doses of epinephrine increases blood O_2 affinity (Fig. 17) both in vivo and in vitro. Evidence from trout suggests that this phenomenon is the result of changes in erythrocyte volume and an intraerythrocytic alkalization arising from

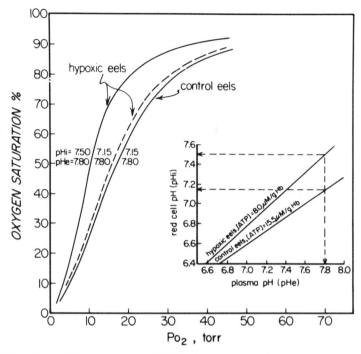

Figure 16. The O_2 equilibrium curves (20°C) of the blood of eels exposed to either normoxia or 2 weeks of hypoxia. Also indicated in the insert are *in vivo* values of extracellular and intracellular pH as affected by the concentration of ATP. [From (209).]

a proton efflux. However, these experiments (18, 19, 150) were based on *in vitro* studies that employed synthetic buffers, the use of which may be very misleading. While epinephrine-stimulated proton efflux and intraerythrocytic alkalization may be correct for trout blood (42), under physiological levels of CO_2 no alkalization of intracellular fluid occurred in either blood or erythrocytes from the killifish *F. heteroclitus* in CO_2/HCO_3^- buffered saline (43). Since increasing P_{CO_2} and thus decreasing pH amplified the epinephrine effect (Fig. 18), it would appear that changes in ATP and its linkage to pH and Hb—O_2 affinity were responsible for some of the observed epinephrine effect. Because epinephrine caused a 44% decrease in red cell ATP (43), the magnitude of the effect was amplified at high P_{CO_2}, that is, at low pH (Fig. 18). Moreover, the

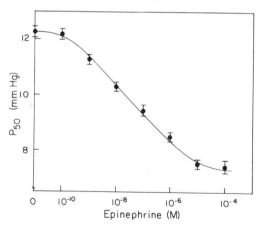

Figure 17. The effect of varying concentrations of epinephrine on the P_{50} values of washed erythrocytes from *Fundulus heteroclitus*. Bars indicate one standard error of the mean (SEM). [From (43).]

Na^+–K^+ and Cl^-–HCO_3^- exchanges appeared to be involved in the lowering of erythrocyte ATP concentrations (43).

These adrenergic responses would facilitate O_2 loading at the gills during transient acidosis by offsetting the Bohr effect. Such studies in other species also attribute the change in blood O_2 affinity to a decrease in intraerythrocyte ATP concentrations and its associated direct and indirect effects on Hb—O_2 affinity.

The blood of primarily water-breathing amphibians also has been shown to respond to chronic aquatic hypoxia. After 2 days of exposure to an aquatic P_{O_2} of 30 mmHg, the blood from larvae of the salamander *Ambystoma tigrinum* shows a sharp decline in both nucleotide triphosphate (NTP) and 2,3-DPG, and as a consequence the P_{50} decreases sharply from 41 to 25 mmHg (pH 7.74, 20°C) (210). Such a marked response is not ubiquitous among aquatic amphibians. Larval bullfrogs (*Rana catesbeiana*) exposed to 30 days of aquatic hypoxia P_{O_2} (70–75 mmHg) showed only a slight decrease in P_{50} and no change in blood-O_2 capacity (Fig. 19) (153). However, adult bullfrogs exposed to combined aquatic and aerial hypoxia showed a profound blood response, with P_{50} decreasing from 35 to 25 mmHg (P_{CO_2} = 7 mmHg, 23°C) and blood oxygen capacity rising from 4.5 to 7.5 vol%.

Not all adjustments in blood respiratory properties require days or weeks to develop. Many stagnant freshwater habitats and marine tide pools may have P_{O_2} values as high as 400 mmHg in the day when photosynthesis is highest, falling to below 10 mmHg at night when respiration prevails (23, 192). No single set of blood characteristics can provide for optimal O_2 transport under these widely varying conditions. Some species show short-term oscillations in blood respiratory properties. For example, the hematocrit and hemoglobin concentration of

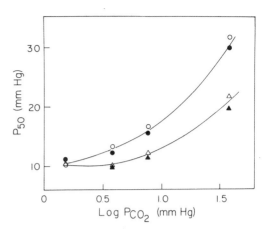

Figure 18. The effect of epinephrine and CO_2 concentration (i.e., log P_{CO_2}) on the oxygen affinity of *Fundulus heteroclitus* whole blood (open symbols) and erythrocytes in bicarbonate buffer (filled symbols). Circles represent control samples (i.e., without epinephrine) while the triangles represent whole blood and erythrocytes in the presence of epinephrine. The SEM of each point falls within the symbols in the figure. At the highest experimental P_{CO_2} values the pH was about 7 while at the lowest P_{CO_2} experimental values the pH was about 8.3. [From (43).]

carp (*Cyprinus carpio*) exposed to 12-h cycles of aquatic normoxia and extreme hypoxia (10 mmHg) show significant increases in phase with the period of hypoxia (116). Concomitantly, the intraerythrocytic concentration of ATP falls, presumably resulting in an increase in Hb—O_2 affinity and a facilitation of O_2 loading in the gills.

Amphibians, like fishes, may also be exposed to brief, regular bouts of aquatic hypoxia. The anuran amphibian *Rana berlandieri* (92) shows extremely large diurnal fluctuations in hematocrit (10–36%) and in the concentration of hemoglobin (2.6–7.9 g of Hb/100 mL of blood) and intraerythrocytic phosphates (3.7–15.6 mol of DPG/g of Hb). Although these changes were stimulated by changes in light intensity rather than by environmental O_2,

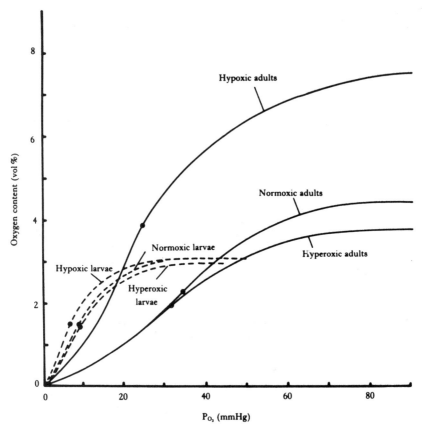

Figure 19. The O_2-equilibrium curves (P_{CO} = 7 mmHg, 23°C) from larval (dashed lines) and adult (solid lines) bullfrogs (*R. catesbeiana*) exposed to 4 weeks of normoxia (150 mmHg), hypoxia (70–75 mmHg), or hyperoxia (>275 mmHg). [From (153).]

the potential exists for rapid and large changes in blood respiratory properties.

Chronic hypoxic exposure changes the hemolymph O_2 transport properties in some aquatic invertebrates. In the crayfish *Orconectes virilis* (207), for example, exposure to $3\frac{1}{2}$ weeks of aquatic hypoxia (50–55 mmHg) resulted in a pH-independent decrease in P_{50} from ~8 down to 6 mmHg (pH 7.8, 15°C). Similar results have been observed in the lobster *Homarus vulgaris* (17, 35). Chronic aquatic hypoxia also stimulates an increase in extracellular hemoglobin concentration and a decrease in P_{50} in invertebrates with this pigment. Hemoglobin concentration is inversely correlated with environmental P_{O_2} in *Daphnia magna*, with a very mild hypoxia being sufficient to stimulate hemoglobin synthesis (106). Hemoglobin in *Daphnia* apparently serves a respiratory function only under relatively severe hypoxia or exercise, and under these conditions the increased concentration of Hb allows *Daphnia* to maintain O_2 uptake at a significantly lower P_{O_2}. Similarly, in larval *Chironomus plumosus* chronic hypoxia stimulates a small increase in Hb—O_2 affinity and a large increase in Hb concentration (197). These combined effects are manifested by a longer period of survival in severe environmental hypoxia.

Respiratory adaptations of the blood may also involve species-level changes

(137, 159, 209). Interspecific comparisons, purportedly showing changes related to aquatic hypoxia in the blood of one species are best made between closely related taxa, because factors such as environmental temperature, body size and activity levels can be as or more important than environmental O_2 levels as a selection pressure acting on blood respiratory properties. Thus, while the P_{50} under comparable conditions of blood from the goldfish (30) is ~2.5 compared to 20–30 mmHg in salmonid fishes, it should not be assumed that the lower P_{50} of goldfish blood is an adaptation to the lower ambient levels of O_2 potentially experienced by the goldfish, which is a much smaller, less active fish from a considerably different systematic lineage. The adaptive benefits of a certain set of blood properties become most evident either when comparisons under the same physiological conditions (e.g., temperature and pH) are made in phyletically similar lineages, or when relatively large numbers of species are included in the comparisons.

Fishes provide many examples of species-level adaptation of blood respiratory properties to aquatic hypoxia. Comparison of the blood from fishes occupying a variety of habitats with different degrees of environmental O_2 but similar environmental temperatures indicates a strong correlation between P_{50} and O_2 availability (Fig. 20). Thus, fishes inhabiting slowly moving, poorly oxygenated water have blood with much higher O_2 affinity than fishes either inhabiting relatively fast-moving, oxygenated fresh water or inhabiting consistently oxygenated seawater.

Although the comparison involves animals of widely differing structure and phylogeny, the magnitude of the differences between O_2 affinity seen in the Hcs of animals from hypoxic and normoxic environments (Fig. 21) suggests that adaptation to hypoxia also occurs in Hcs.

The respiratory properties of Hcs from aquatic decapods may also show similar adaptations to hypoxia (137). The O_2 affinity of freshwater crayfishes (137) or of the bathypelagic Mysids (63), many species of which frequently encounter hypoxia, are very high when compared with the O_2 affinity of marine decapods residing in a habitat that rarely, if ever, is hypoxic (Fig. 21). While there are important differences in body size, lifestyle, and systematics between the three decapods depicted in Fig. 21, the differences in the hemocyanin O_2 affinity are so large as to suggest adaptation to environmental O_2 levels similar to those observed in fishes.

The respiratory properties of invertebrate hemoglobins show a vast range (199)—for example, the P_{50} of invertebrate hemoglobins ranges from ~12 mmHg (P_{CO_2} = 19 mmHg, 19°C) in the echiurid *Urechis caupo* down to 0.0015 mmHg (pH 7.0, 20°C) in the perienteric fluid of *Ascaris lumbricoides!* Major complications in interpretation of apparently adaptive features of invertebrate hemoglobins result from (1) the enormous phyletic diversity of invertebrates containing Hb, (2) the currently unknown mechanisms by which hemoglobins are manufactured and their functions modulated, and (3) the fact that a major role of hemoglobin in many invertebrates may be to store O_2 as well as transport it. Thus, while invertebrate hemoglobins like their vertebrate counterparts have become adapted to facilitate O_2 transport especially in hypoxic environments (200), care must be taken to clearly delineate between adaptive changes involving *transport* and those involved in O_2 *storage*.

Air Breathing in Aquatic Animals
Aquatic hypoxia has been implicated by many investigators as a major selection pressure for the evolution of air breathing in fishes (97, 113, 166). With the evolution of the ability to avoid aquatic hypoxia by

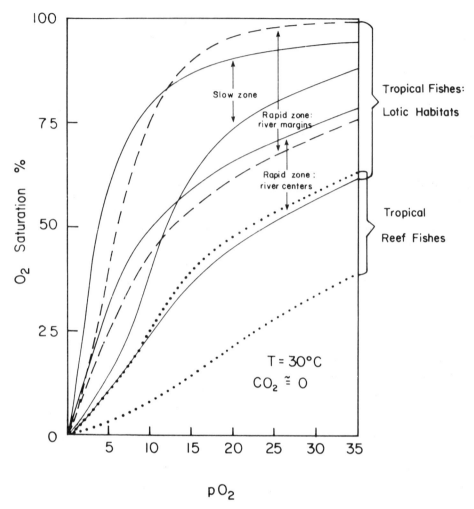

Figure 20. Blood O_2 equilibrium curves (P_{CO_2} = 0 mmHg, 30°C) from water breathers from a variety of lotic habitats. The range of curves from species inhabiting slowly moving water (slow zone) is indicated by thick solid lines, while species inhabiting the center of rapidly moving water (rapid zone) fall within the thin solid lines. The dashed lines represent the range of curves for rapid zone species that inhabits stream and river margins, while the dotted lines represent the range of curves from fishes inhabiting well-oxygenated reef habitats. [From (158).]

air breathing, it might be anticipated that the blood of air-breathing fishes might show a reduced O_2 affinity, given the larger concentration of O_2 in air. In some instances this expectation appears to be borne out. The blood of the air-breathing lungfishes, bowfin, and electric eel has a much lower O_2 affinity when compared with that of the strictly aquatic carp, cat-

fish, and mackerel (98). Similarly, a spectrum of amphibians showing progressively greater dependence on aerial respiration also shows a progressive decrease in blood O_2 affinity (98). Unfortunately, the diversity of lifestyle, habitat (e.g., temperature), and phylogeny between the groups chosen for comparison may complicate interpretation of respi-

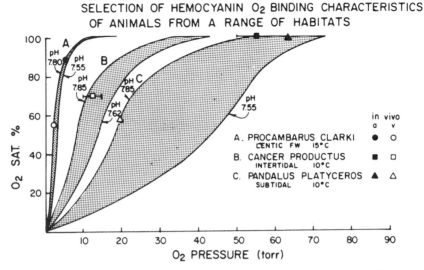

Figure 21. The O_2 equilibrium curves and magnitude of the Bohr shift of hemocyanin from three decapod crustaceans occupying habitats characterized by (*A*) potentially severe limitations in O_2 availability (*Procambarus*), (*B*) intermediate intertidal habitats (*Cancer*), and (*C*) full air saturation (*Pandalus*). [From (137).]

ratory adaptations specifically for air breathing. A very broad survey of 40 genera of aquatic and air-breathing fishes inhabiting the Amazon basin, where temperature is comparable, revealed no special adaptations of the blood related to breathing air (161).

Because many different factors (e.g., temperature and lifestyle) can potentially affect blood respiratory properties, it becomes particularly instructive to contrast species within narrow phyletic lineages inhabiting the same or similar habitats. *Osteoglossum bicirrhosum* is a strictly aquatic osteoglossid from the Amazonian basin, while *Arapaima gigas* is a closely related obligatory air-breathing species (99). The blood of the strictly water-breathing *Osteoglossum* has a much higher O_2 affinity than *Arapaima* (Fig. 22), and consequently the former can maintain its arterial blood at 80–100% air saturation even in severely hypoxic water. A high O_2 affinity can be disadvantageous in terms of O_2 unloading at the tissues, particularly if metabolic rate is high. How-

ever, the metabolic rate of *Osteoglossum* is considerably lower than in most freshwater fishes (99) and adequate O_2 unloading can be achieved even with a high affinity blood. *Arapaima*, which has blood with a much lower O_2 affinity, would not be able to survive if it had to depend strictly on branchial gas exchange with hypoxic water. In fact, by breathing air *Arapaima* is able to saturate fully the blood leaving its gas bladder. The lower affinity blood is not only adequate for gas transport when encountering the relative high P_{O_2} values in the gas bladder, but facilitates tissue unloading of O_2 to support a metabolic rate in *Arapaima* that is twice that of *Osteoglossum* (99).

Evolution of the air-breathing habit may have occurred more recently in decapod crustaceans than in vertebrates. Although O_2 capacity is consistently elevated (137) and sensitivity to modulators is low (145) in air-breathing decapods, a clear pattern of changes in Hc—O_2 affinity has not emerged.

In conclusion, while numerous studies

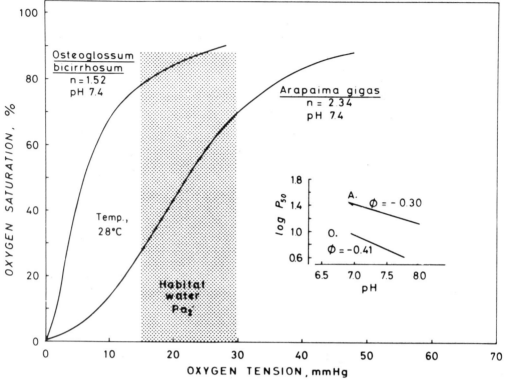

Figure 22. Blood O_2 equilibrium curves (pH 7.4, 28°C) and Bohr shifts (inset) from two osteoglossid fishes from the Amazon basin. *Osteoglossum bicirrhosum* is a strictly aquatic fish, while *Arapaima gigas* is an obligatory air-breathing species. The diurnal range of water P_{O_2} experienced by these fishes is shown by the rectangular hatched area. [From (99).]

on either closely coupled species or on a broad spectrum of species have indicated that the evolution of air breathing is accompanied by a decrease in blood O_2 affinity, there are also data inconsistent with that hypothesis. An increasing dependence on aerial respiration probably has influenced the evolution of blood characteristics, but air breathing must be regarded as only one of many different factors that influence O_2 affinity.

High Altitude
Atmospheric pressure decreases ~9 mmHg for every 100 m increase in elevation above sea level. Although the fractional concentration of O_2 remains

constant at ~0.21 regardless of altitude, the P_{O_2} of air decreases nearly 2 mmHg for every 100-m increase in elevation. Thus, at an altitude of 5400 m (near the upper limit for permanent human habitations) the P_{O_2} is only ~73 mmHg, or one-half that at sea level.

Temperature also decreases with increasing altitude. Thus, putative adaptations to high altitude in fact may represent combined responses to hypoxia and lower temperatures. The overall effect of these combined stressors may be quite complex. Guinea pigs exposed to either hypoxia or cold stress alone respond with a sharply increased rate of erythropoesis (red cell production) that

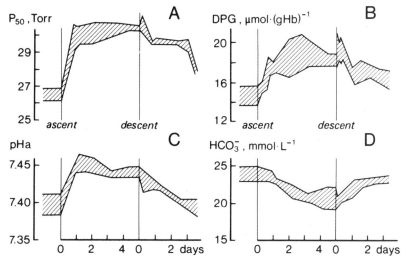

Figure 23. Changes in Hg–O$_2$ affinity (*A*), intraerythrocytic DPG concentration (*B*), arterial pH (*C*), and arterial HCO$_3^-$ in sea level native humans exposed for 5 days to an altitude of 4510 m and then returned to sea level. [From (21).]

considerably enhances blood hemoglobin concentration. Yet, the characteristic hematological responses to hypoxia in guinea pigs are considerably muted when cold stress occurs concomitantly (111). Elevated hematocrits in montane amphibian and reptiles are also thought to be primarily the result of cold stress and not hypoxia (85).

In spite of the complexity of cold stress, as well as taxonomic and genetic factors unrelated to physiological demands (179), high altitude imposes significant respiratory limitations on blood O$_2$ transport. Changes in blood hematology related specifically to O$_2$ transport at high altitude have been recognized and reported for both aquatic and terrestrial animals. Differentiation should be made between short-term responses (acclimitization), and long-term adaptations in species that have evolved at sea level or high altitude.

Short-Term Acclimitization. A multitude of studies have categorized the blood responses of vertebrates exposed to in-

creased altitude (21). Responses fall in two categories—increases in blood O$_2$ capacity and changes in blood O$_2$ affinity. The most extensive data are available for humans. Although increased erythropoesis begins within hours or days, the overall process is slow and takes weeks to months for full expression. Return to lower altitude reverses the process. Altitude-induced increases in red cell and hemoglobin concentration and thus in blood O$_2$ capacity have also been reported in rodents (110, 177), birds (11, 205), but are absent in some species of lizards (156). The hematology of few aquatic amphibians or fishes has been examined specifically in the context of high altitude (91). An increase in erythropoesis presumably occurs when aquatic hypoxia results from reduced P_{O_2} at high altitudes.

A second major acclimatory response of mammals and birds to short-term high altitude exposure involves a slight decrease in blood O$_2$ affinity (21). Unlike erythropoesis, which is relatively slow to develop, blood O$_2$ affinity changes much more rapidly (Fig. 23). As a result of hyperventilation, stimulated by hypoxia, blood P_{CO_2} falls and pH rises. The effect of the resulting Bohr shift decreases blood P_{50}. Within hours, however, there is a concomitant rapid increase in intraerythrocytic DPG, which overrides the Bohr effect and results in a net rise in blood P_{50}. A similar mechanism may account for the rise in P_{50} of bird blood at altitude (93).

Long-Term Adaptation. Great intra- and interspecific variation exists when comparing hematology and blood O_2 affinity of animals native to either high or low altitude. Comparisons should be made among closely related taxa with clearly defined evolutionary histories with respect to altitude (179).

Human populations that have lived for generations at high altitude (>3200 m) can have hematocrits and hemoglobin concentrations ~5–15% and 10–20% higher, respectively, than their sea level counterparts (21). Similar findings suggest enhanced erythropoesis for guinea pigs native to high altitude. Paradoxically, given that the blood O_2 affinity of humans changes during short-term exposure to high altitude, comparison of sea level and high altitude human populations reveal no consistent pattern of adaptation in blood P_{50} (22, 208). The llama, a high altitude camelid often used as a model to describe high altitude adaptations, has a blood P_{50} which is not significantly different from that of the closely related camel. This suggests that the high blood O_2 affinity of the llama is simply a genetic trait of the camelidae, rather than a high altitude adaptation (27); this emphasizes the need for caution in interpreting supposed cases of high altitude adaptation.

One of the more comprehensive studies of high altitude adaptations blood O_2 affinity investigated 10 subspecies of the deer mouse, *Peromyscus maniculatus*, na-

tive to a wide range of altitudes but acclimated in the lab to a common altitude to see if blood properties were innate (178, 179). Figure 24 indicates the interrelationships between P_{50}, altitude, and intraerythrocytic concentrations of DPG for these species. Within this single species, a decreased P_{50} is associated with high altitude, presumably resulting from a decrease in intracellular organic phosphate modulators. Moreover, these blood properties appear to be genetically dictated, since they persisted in second generation mice kept at sea level.

Few data are available on the O_2 affinity of the blood of terrestrial ectotherms at high altitude. The lizards *Uta stansburiana* (78) and *Liolaemus multiformis* (57) show higher Hb and hematocrit when collected at high altitude, but examination of 19 species of 8 genera from 3 families of lizards living at altitudes up to 2500 m showed no correlation between altitude and blood O_2 capacity (46, 156) or blood O_2 affinity (156). Such altitudes are relatively moderate compared to the much higher altitudes that endothermic vertebrates can inhabit, and may have environmental P_{O_2} values too high to stimulate adaptive changes.

Aquatic vertebrates and invertebrates living at high altitude encounter hypoxia even when the ambient water is air saturated. When larvae of tiger salamanders (*Ambystoma tigrinum*) from low altitudes were acclimated in the laboratory at levels of hypoxia simulating an altitude of 3353 m (the highest altitude at which they have been located), blood P_{50} (pH 7.74, 20°C) decreased from 41.3 to 24.8 mmHg (90). This change occurred within just 2 days, and was mediated by a sharp decrease in intraerythrocytic 2,3-DPG. The Lake Titicaca frog, *Telmatobius culeus*, lives at an altitude of 3812 m. The P_{50} is 15.6 mmHg (pH 7.65, 10°C) and the erythrocyte count is 729,000/mm^3 (91). The O_2 affinity and

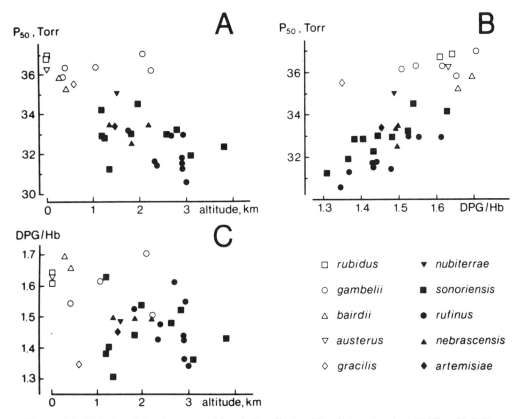

Figure 24. Relationships between blood O_2 affinity (P_{50} determined at 37°C, pH 7.4), intraerythrocytic DPG, and altitude in high and low altitude populations of the deer mouse, *Peromyscus maniculatus*. Ten subspecies of deer mice native to different altitudes were investigated after 3–30 months of acclimation to low altitude (340 m), to eliminate possible interference by short-term acclimation. [From (178, 179).]

erythrocyte count are among the highest recorded for anuran amphibians.

Effectiveness of Blood Adaptations to High Altitude. Increased red cell concentration is a widespread response to hypoxia, including that produced by high altitude exposure. An increase in blood O_2 capacity will help maintain O_2 transport to the tissues as arterial P_{O_2} decreases with increasing altitude. Indeed, experimentally reducing the hematocrit by removal of red cells in humans at high altitude decreases the maximum sustainable O_2 uptake in spite of an increase in maximum cardiac output (134).

Unfortunately, the interpretation of differences in blood P_{50} between animals native to high altitude is much more difficult, and many so-called adaptations have yet to be supported by physiological data. A major complication, which is frequently ignored, is that a change in position of the O_2-dissociation curve affects *both* O_2 loading in the gas exchanger and O_2 unloading in the tissues. Thus, while it might be tempting to view a strongly left-shifted curve (higher O_2 affinity) as an adaptation for ensuring full O_2 saturation in the face of lower P_{O_2} in the respiratory medium, a left-shifted curve will also release less O_2 in the tissues unless tissue

P_{O_2} falls. Similarly, a strongly right-shifted curve (lower P_{50}) would favor O_2 unloading in the tissues at a given tissue P_{O_2}, but might not allow full O_2 saturation in the gas exchanger.

Critical to interpretation of adaptability of blood O_2 affinity are data on O_2 partial pressures of arterial and venous blood. The most extensive comparative data are available for mammals. A detailed comparison of O_2 transport in llama, with a high affinity blood, and sheep, with low affinity blood, indicated that the low blood O_2 affinity of the sheep was clearly more beneficial at moderate altitudes (1600–2800 m), while at higher altitudes (4500–6400 m) the high blood oxygen affinity of the llama was more advantageous (89). Investigations on humans at sea level and various altitudes similarly indicate that there is no single adjustment in blood O_2 affinity that is universally adaptive (21). In fact, the small rightward shifts of the O_2-dissociation curve most often reported in humans moving to high altitude provide no net advantage and possibly even a slight disadvantage with respect to O_2 delivery to skeletal muscles during prolonged muscular work at 3100 m. The importance of measuring arterial and venous blood P_{O_2} values in assessing the utility of changes in Hc—O_2 affinity has also been emphasized for crustaceans (137). A decrease in P_{50} is a common response of invertebrates when exposed to aquatic hypoxia. Presumably such a response occurs in aquatic invertebrates living at elevated altitudes, though the effects of altitude on invertebrate O_2 transport has not been comprehensively investigated.

Embryonic–Fetal Development
The early stages of development of most animals are spent in a specialized microenvironment that is potentially hypoxic. The outer layers of eggs, both vertebrate and invertebrate, present a significant diffusion barrier to the free passage of respiratory gases, while the uterine environment of mammals and viviparous and ovoviviparous fishes and reptiles is generally less well oxygenated than maternal tissues. The blood of most vertebrate species undergoes marked ontogenetic changes, and these are often interpreted in terms of adaptations for providing adequate O_2 transport.

Many fishes show a decrease in blood O_2 affinity as development proceeds. The Hb from amnocoete larvae of the lamprey *Lampetra fluviatilis* has a P_{50} of 1.8 mmHg (pH 7.75, 10°C) compared with 11.8 in the adult (10). A lower P_{50} in the fetus (embryo) compared with the adult also occurs in the ovoviviparous spiny dogfish *Squalus* and in viviparous and oviparous teleosts (80, 94) (Fig. 25). Both O_2 affinity and temperature sensitivity of maternal blood are lower than that of the fetus.

Developmental changes in the blood of amphibians have been extensively investigated (26, 62). While the O_2 affinity of embryonic amphibian hemoglobins has not been determined, an increase in P_{50} upon metamorphosis from larva to adult occurs in urodeles (210), anurans (153), and apodans (188), though it does not occur universally (34, 210). A rise of nearly 25 mmHg in the P_{50} of whole blood (P_{CO_2} = 7 mmHg, 23°C) follows metamorphosis in the bullfrog *R. catesbeiana* (Fig. 19). The decrease in blood O_2 affinity of adult amphibians appears largely due to de novo synthesis of adult hemoglobins with lower O_2 affinity (82) rather than an increase in concentration of intraerythrocytic organic phosphates. In fact, intraerythrocytic organic phosphate levels in adult amphibians are generally lower than in larvae (98, 153).

A developmental shift to lower O_2 affinity of whole blood also occurs in a wide variety of reptiles (9, 155). Unlike in amphibians, the mechanism underlying this shift in reptiles apparently involves an in-

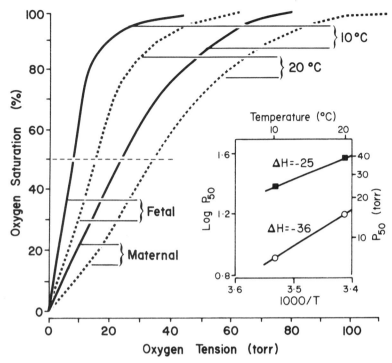

Figure 25. Whole blood O_2 equilibrium curves of maternal (broken lines) and fetal (continuous lines) blood from the viviparous teleost fish *Zoarces viviparous*, measured at pH 7.5 and 10 and 20°C. The inset shows the relationship between log P_{50} and temperature (T) in maternal (solid squares) and fetal (open circles) blood at pH 7.5. [From (80).]

crease in intraerythrocytic organic phosphates (155).

Development of bird embryos is accompanied by marked changes in blood O_2 affinity, but the pattern of change is complex (114) (Fig. 26). Like the pattern of affinity change, the mechanism is also complex. Whereas inositol pentaphosphate (IPP) is the major organic phosphate affecting O_2 affinity in the blood of adult birds, IPP remains in relatively low concentrations throughout embryonic development. Rather, ATP is the major erythrocytic organic phosphate of the embryo and, along with a brief pulse of 2,3-DPG around day 18, primarily modulates blood O_2 affinity in chicken and duck embryos (5). The concentration of these phosphates in chick blood is greatly increased in response to hyperoxic expo-

sure, resulting in a decrease in blood O_2 affinity (93).

An ontogenetic shift from high O_2 affinity fetal blood to low O_2 affinity adult blood occurs in almost all mammals, with the exception of the domestic cat. The increase in P_{50} from fetal to mature adult values is not necessarily progressive. For example, 65-day old dog pups actually have a slightly lower affinity than the adult (50).

Three different mechanisms account for these developmental changes in mammals. In ruminants both fetal and adult Hb are relatively unresponsive to 2,3-DPG and the erythrocytes of both contain relatively little of this organophosphate. The fetal–adult shift in ruminants is the result of fetal hemoglobin having an intrinsically higher O_2 affinity than adult

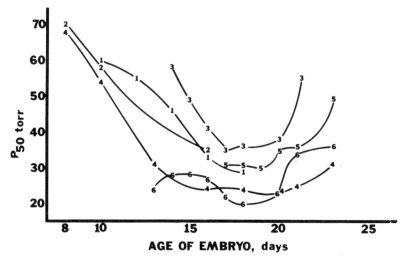

Figure 26. Changes in oxygen P_{50} values (pH 7.4, 37–39°C) during development of individual embryos of the domestic chicken. [From (114).]

hemoglobin. A second mechanism exists in primates where, as in ruminants, there are structural differences between fetal and adult hemoglobins. In human fetal hemoglobin, for example, a γ chain replaces each β chain. It should be emphasized that, in spite of structural differences, both fetal and adult hemoglobin from primates have a similar O_2 affinity—the P_{50}s of stripped solutions of adult and fetal hemoglobin are approximately the same at 14–15 mmHg. However, fetal hemoglobin is much less responsive to 2,3-DPG, which occurs in concentrations similar to those of adult primates. Thus, the P_{50} of adult whole blood is ~28 mmHg at pH 7.4 and 37°C, whereas that of fetal whole blood is ~20 mmHg. A third mechanism responsible for a fetal–adult shift is found in a diverse array of mammals including dog, horse, pig, guinea pig, rabbit, mouse, and opossum (28). In these animals there are no differences in structure, Hb—O_2 affinity or inorganic phosphate sensitivity. Rather, there is a lower 2,3-DPG concentration in the fetal erythrocytes that re-

sults in a higher O_2 affinity. Red cell 2,3-DPG starts to rise immediately after birth, resulting in blood of the newborn having an O_2 affinity equivalent to or even lower than the adult.

Structural changes in Hc accompany development in several crustaceans. In *Cancer magister, Carcinus maenas,* and *Hyas araneus* subunit composition of Hc from zooea and megalopa larval stages contains fewer subunits than that of the adult (133, 183). Additionally, larval Hc is predominantly hexameric while amounts of the dodecamer gradually increase in hemolymph after metamorphosis (133). At least in *C. magister* these structural differences are accompanied by changes in O_2 affinity (183) (Fig. 27).

Finally, it should be emphasized that immature animals are almost invariably smaller than the mature adult. Increased body size is often associated with increased blood O_2 affinity. Thus, in the same way that high altitude adaptations may be closely interwoven with adaptations to cold. Future studies should be directed towards the link between

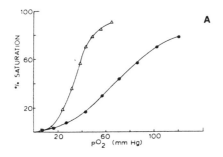

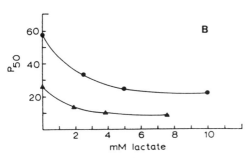

Figure 27. Functional differences in O_2 binding by hemocyanin during development. The O_2-binding curves (*A*) and magnitude of the lactate effect (*B*) in hemocyanins from larval (○) and adult (△) *Cancer magister*. [From (183).]

development and body size when examining the blood properties of larvae–embryos–fetuses.

Blood Adaptations Resulting from Internally Induced Hypoxia

Intermittent Breathing

Discontinuous ventilation of the gas exchange organs is characteristic of many fishes (169), of most amphibians and reptiles (176), of diving birds and mammals (55), and invertebrates (137). Most often, apnea (breath holding) is associated with a lifestyle that includes diving, but even terrestrial ectotherms may intermittently ventilate their gas exchange organs.

Prolonged apneic periods result in a reduced O_2 availability in the lungs, potentially disrupting O_2 loading. The blood of many intermittently breathing verte-

brates has been examined for evidence of increases in blood O_2 affinity on the assumption that such a change would favor O_2 loading in the lungs during apnea. Marine mammals show no difference in blood O_2 affinity compared with terrestrial, continuously breathing mammals (55). Whereas most birds have relatively low O_2 affinity blood (P_{50} ~40–50 mmHg) (114), the Adele penguin *Pygoscelis adeliae* has a P_{50} of <35 mmHg (98), which has been interpreted as of adaptive value in facilitating O_2 loading in the lungs of this diving bird. Reptiles present a great variety of diving species. Diving turtles have blood with the same O_2 affinity as continuously breathing tortoises (32). Among the snakes a much higher affinity blood is found in the file snake *Acrochordus javanicus* when compared with the continuously breathing terrestrial boa constrictor *Constrictor constrictor*, as indicated in Fig. 28. Yet the diving varanid lizard *Varanus niloticus* has blood with a

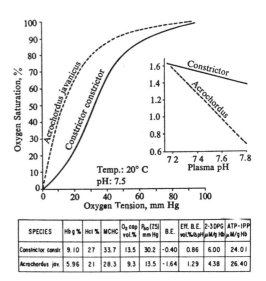

SPECIES	Hb g %	Hct %	MCHC	O₂ cap vol.%	P₅₀ (7.5) mm Hg	B.E.	Eff. B.E. vol.%/ΔpH	2-3DPG µM/g Hb	ATP-IPP µM/g Hb
Constrictor constr.	9.10	27	33.7	13.5	30.2	-0.40	0.86	6.00	24.01
Acrochordus jav.	5.96	21	28.3	9.3	13.5	-1.64	1.29	4.38	26.40

Figure 28. The O_2 equilibrium curves and Bohr effects in the diving snake *Acrochordus javanicus* and the terrestrial snake *Constrictor constrictor*. Also indicated below the figure are a variety of blood respiratory characteristics and hematological parameters. [From (98).]

P_{50} of ~40–45 mmHg (pH 7.45, 25°C) compared with only 30 mmHg in the regularly breathing terrestrial *V. exanthematicus* (213).

In diving vertebrates during periods of breathing, blood O_2 loading occurs at P_{O_2} values that are the same as continuously breathing terrestrial animals, and no particular adjustment in blood O_2 affinity would seem to be required. During apnea, however, lung P_{O_2} falls and O_2 loading may become more difficult. An increased O_2 affinity would be advantageous during prolonged apnea from the perspective of O_2 loading in the lungs. It is important to consider how this would affect O_2 unloading in the tissues. Importantly, apnea is accompanied by a buildup of blood CO_2 and fall in plasma pH. As pH falls during apnea, the OEC will shift to the right, facilitating O_2 unloading at the tissues. However, too large a rightward shift of the OEC may prevent full saturation of the blood as it leaves the lungs, especially since pulmonary P_{O_2} continues to fall. If the major O_2 stores are in the form of lung gas taken down during the dive, then a large Bohr shift could be maladaptive (214). On the other hand, if the primary O_2 stores are in blood rather than in lung gas, then during apnea, O_2 loading in the lungs is of little relevance compared to O_2 unloading in the tissues, and in this situation a large Bohr shift will be advantageous. Perhaps because of these complications, there is no clear relationship between the diving habit and the magnitude of the Bohr shift or P_{50} when looking across wide taxonomic groups. Selected species of diving reptiles, birds, and mammals tend to show larger Bohr effects than related nondiving species (55, 214), but there are numerous exceptions. Although diving animals can undergo long periods of apnea, most dives are highly aerobic and rarely result in severe depletion of O_2 stores and production of metabolic acids

associated with anaerobic metabolism (29, 55, 87).

While adaptations of the respiratory properties of the blood for diving remain equivocal, there clearly are increases in blood O_2 stores in many taxa. Diving mammals and birds, which show the greatest proportion of O_2 storage in blood and tissue rather than in lung gas (29), have a significantly elevated blood volume and blood O_2 capacity when compared with nondiving species (55). As one example, the total blood volume (per kilogram of body mass) and blood O_2 capacity (per kilogram of active tissue) of a Weddell seal are ~2 and 5.3 times greater than in humans (55). In diving reptiles and amphibians, which tend to rely more heavily on lung gas rather than blood and tissue O_2 stores, blood O_2 capacity is, nonetheless, frequently higher than in terrestrial, continually breathing forms, though once again many exceptions occur (32, 212).

Among the invertebrates with respiratory pigments there are many examples of noncontinuous ventilation of the gas exchange organs (140, 142). While the hemolymph of crustaceans (137) and annelids (198) has a significant O_2 storage role, specific blood respiratory properties have yet to be correlated with specific ventilatory patterns.

Metabolic Rate—Activity,
Body Mass, and Temperature
Metabolic rate, and thus the need for blood O_2 and CO_2 transport, increases with increased activity (e.g., running and flying). Metabolism approximately doubles with each 10°C increase in body temperature (Chapter 9), affecting O_2 transport not only in ectotherms, but in any animals where activity may generate locally large temperature gradients within active muscle. Additionally, larger animals have resting metabolic rates that tend to be lower than small animals. All

of these factors potentially have been important in the evolution of blood respiratory properties.

Activity. In vertebrates with a high aerobic scope of activity, ventilation usually rises in proportion with O_2 demand, maintaining arterial P_{O_2} values at or above resting levels (203). Nonetheless, localized hypoxia during exercise usually occurs in metabolically active tissues. Changes in blood O_2 affinity related to a regimen of chronic exercise resemble those induced by exposure to environmental hypoxia. For example, an intense training regimen in race horses results in a fall in P_{50} from 26.3 mmHg (pH 7.4, 37.9°C) to 24.3 mmHg, mediated by a precipitous fall in concentration of intraerythrocytic 2,3-DPG (118). Increased activity in other vertebrates, either as a result of training in an individual or an evolutionary difference between species, frequently is accompanied by increased blood O_2 capacity. Enhanced blood O_2 capacity, particularly when combined with a large Bohr shift, facilitates blood O_2 transport to metabolically active tissues *if* the P_{O_2} of blood draining the gas exchange organ is maintained during exercise (48).

In crustaceans an increase in ventilation during exercise (16, 136) allows for maintained or even increased arterial O_2 saturation and increased delivery of O_2 to tissues (137, 206). In exercising crabs increased release of O_2 to tissues results from the Bohr shift occurring in response to a substantial acidosis (137, 206). In some cases the magnitude of the Bohr shift may in turn be modified by an opposing lactate shift (16, 138) (Fig. 15*B*). Short-term modulation of Hc—O_2 binding during activity thus may play a major role in adjustment of O_2 delivery during activity in crustaceans. The long-term effects on respiratory pigments that result from activity are not known for crustaceans or any other invertebrate group.

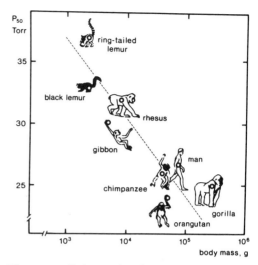

Figure 29. Relationship between adult body mass and blood O_2 affinity (pH 7.4, 37–38°C) in a variety of primates. [From (51).]

Body Size. The O_2 affinity of whole blood is negatively correlated with body mass when making interspecific comparisons of mammals and birds (36, 151, 173). Figure 29 illustrates this general relationship for primates. However, there is considerable variation when a wide range of mammals are examined (165). For example, some small burrowing mammals have very high blood O_2 affinities relative to their body mass (79), perhaps relating to the potentially hypoxic nature of burrows. Whereas reptiles generally (84, 155) show a similar relationship to that of homeotherms, a positive rather than negative correlation between P_{50} and body mass has been reported for some snakes (155). The effects of body mass on whole blood seem to be due to variations in intraerythrocytic phosphate concentrations, since stripped Hb solutions do not show a similar correlation.

A comparison of adults of different species yields a negative correlation between metabolic rate and body mass in a very wide range of vertebrates from highly diverse taxa (Chapter 9). The respiratory properties of blood play a pivotal role in supporting elevated rates of tissue metabolism, so it is hardly surprising to find that there is a correlation between blood P_{50} and body mass. The lower O_2 affinity of smaller animals is generally interpreted as an adaptive advantage in facilitating O_2 unloading at the tissues to support their elevated metabolic rate.

Special habits or habitats of a particular species may override the influence of body mass on blood O_2 affinity (48). While intraspecific comparisons may help to avoid the complications of interspecific comparisons, few such studies have been made. In the piranha *Serrasalmus rhombius*, blood O_2 affinity increases with increasing body mass (215), as for interspecific comparisons in mammals. However, in larvae of the tiger salamander *Ambystoma tigrinum* neither P_{50} nor the Bohr shift are significantly affected by body mass (31). Hematocrit, mean corpuscular volume, Hb concentration, mean cell Hb, O_2 capacity, and Hill's n in these larvae are all positively rather than negatively correlated with body mass, suggesting that larger rather than smaller salamander larvae may have greater O_2 transport needs.

The concept of scaling as it applies to metabolic rate and hemolymph O_2 transport applies equally well to invertebrates as to vertebrates, though few data are available. Complex changes occur in the hemoglobin concentration of *Daphnia* with changing body mass (106). Generally, larger *Daphnia* have lower hemoglobin concentrations, though factors such as egg laying and molting appear equally important.

Temperature—Influence of Ectothermy and Endothermy. The reaction between a respiratory pigment and O_2 is exothermic and, as a consequence, the O_2 affinity of almost all pigments decreases with increasing temperature. The blood of most vertebrates shows a large sensitivity to temperature change (209). The perceived adaptive significance of this temperature sensitivity is best illustrated in homeothermic mammals, in which the temperature of active muscle may be several degrees higher than core temperature. In heavily exercising humans, for example, femoral blood has a temperature of nearly 41°C and a pH of only 7.27 (186). The combined effects on arterial blood of increased temperature and the rightward Bohr shift result in a sharp increase in P_{50} from ~26 up to 38 mmHg. The resulting decrease in O_2 affinity facilitates tissue O_2 unloading to such an extent that the required blood flow to femoral muscle can fall by 40% compared to its requirement if no change in O_2 affinity occurred (209).

While it might be anticipated that temperature-related blood adaptations might occur in ectotherms where gas transport must function over a wide variety of temperatures, blood properties in ectotherms are not easily correlated with the extent of heterothermy. Perhaps the most convincing example is that of the tuna, *Thunnus,* with a nearly temperature insensitive blood (171) (Fig. 30). Because of a countercurrent arrangement of arterial and venous vessels in the tuna's swimming muscle, high rates of sustained swimming result in up to a 20°C temperature difference between arterial and capillary blood (38). If the tuna possessed blood with a temperature sensitivity similar to that of mammals or even of most other fishes, the abrupt increase in temperature might result in a rate of O_2 unloading into the warm plasma so high as to cause O_2 emboli (86)! Less dramatic examples of the advantages of temperature-insensitive blood when attempting to sustain O_2 transport during wide and rapid fluctuations in body temperature are evident in intertidal fishes (23) and the lizards *Varanus niloticus* and *Iguana iguana* (214).

The temperature sensitivity of whole

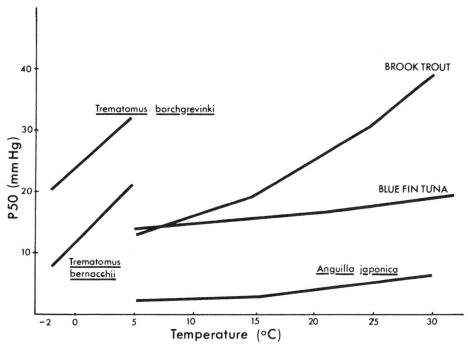

Figure 30. Temperature sensitivity, as indicated by the rise in hemoglobin P_{50} with increasing temperature, of the hemoglobins of several fishes. Note the temperature insensitivity (particularly when expressed as percentage change) of the blood from the tuna. [From (86).]

blood of fishes (76, 77, 98, 162), anuran amphibians (64), and reptiles (209) has been shown to vary between individuals acclimated to different environmental temperatures. In most instances the long-term change in O_2 affinity with temperature is considerably lower than the acutely measured effect in either acclimation group (Fig. 31*A*). In each instance, the O_2 affinity of warm-acclimated animals was lower when compared with cold-acclimated animals, suggesting a change facilitating O_2 unloading at the tissues to support the elevated metabolic rates at higher body temperatures (209). Similar findings exist for decapod Crustacea (Fig. 31*B*). While the exact mechanisms responsible for these adaptations in vertebrates have not been studied in detail, in one species (*Fundulus heteroclitus*) the Hb—O_2-binding effect of thermal acclimation could be explained by changes in the ratio of ATP to Hb (76).

Adaptive changes in blood O_2 affinity, including low-temperature sensitivities, have been reported for Hb and Hc from Mollusca (41), Annelida (198), and Crustacea (137). (Thermal insensitivity of O_2 binding of hemocyanins of eurythermal crabs has been discussed above.) Seasonal changes in O_2 affinity have also been reported for hemolymph containing Hc or Hb.

Temperature-induced shifts in O_2 affinity should be interpreted in the context of measured *in vivo* levels of oxygenation and modulator substances. For example, in the crab *Callinectes sapidus* blood O_2 affinity decreases with rising temperature *in vitro*, and *in vivo* measurements indicate a marked increase in circulating P_{O_2} (16, 137). As described above for activity,

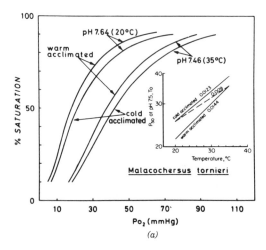

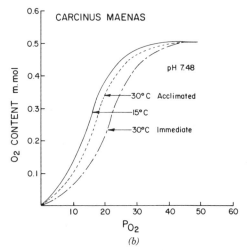

Figure 31. Temperature acclimation by the blood of ectotherms. (a) The O_2 equilibrium curves of warm and cold acclimated pancake tortoises (*Malacochersus tornieri*) measured at 20 and 35°C at *in vivo* values of pH. The effect of acclimation on reducing the effects of temperature on P_{50} (dashed line) is shown in the inset. [From (209).] (b) Effects of temperature acclimation on hemolymph from the intertidal crab *Carcinus maenas* measured at pH 7.48. The solid curve represents the OEC of 15°C acclimated animals, while the other two lines show the effect of acute and long-term exposure to 30°C. [From (190).]

these changes with temperature facilitate O_2 release because of a larger O_2 gradient between venous hemolymph and tissues (Fig. 32).

In summary, while a selective advantage for thermally insensitive blood O_2 affinity occurs in a few species, there appears not to have been a generalized evolutionary development of reduced thermal sensitivity of O_2 transport pigments for organisms in variable thermal environments. If evolution has favored decreased blood O_2 thermal sensitivity, then it must be primarily associated with the genetic and physiological regulation of intracellular pH and the levels and types of organic phosphates and other modifier molecules that modulate pigment O_2 affinity, rather than selection for thermally insensitive oxygen carriers per se (162).

Multiple Hemoglobins as Adaptations to Hypoxia

One way in which animals may adapt to changing internal or external hypoxia is to have several hemoglobins, some with unique functional properties. In contrast to the hemoglobins of most mammals and birds, the presence of multiple hemoglobins is common among fish (15, 59, 60, 168). Many of these hemoglobins are the result of several genetically unique globin chains that combine to yield a series of electrophoretically different hemoglobin tetramers. While structural and functional analyses of multiple fish hemoglobins are rare, they can generally be divided into three major classes (201). Class I contains species with one or more forms of Hb, all of which are sensitive to both temperature and pH (8, 81, 163, 217, 218). Class II has some hemoglobins that are functionally similar to Class I and others that are not strongly affected by either temperature or pH (8, 81, 163, 217, 218). Class III fishes have hemoglobins that are

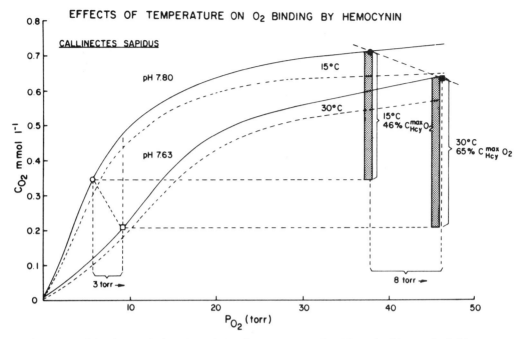

Figure 32. Circulating O_2 levels and O_2 affinity at 15 and 30°C in the blue crab *Callinectes sapidus*. Broken lines show O_2 bound to Hc alone, while solid lines show total hemolymph O_2. The bars to the right of the curves represent the amount of O_2 transported to the tissues at each temperature. [From (137).]

pH sensitive, but temperature insensitive (2, 171).

A typical Class I hemoglobin system is that of the teleost *Fundulus heroclitus* (144). It has four isohemoglobins (HbI, HbII, HbIII, and HbIV), each tetramer being ~64,000 D composed of two α and two β subunits. But there are four different globin chains: $α^a$, $α^b$, $β^a$, and $β^b$. The HbIV isohemoglobin is a homotetramer consisting of two $α^b$ and two $β^b$ subunits. Isohemoglobins II and III are heterotetramers consisting of all four chains. While the amino acid compositions of the chains are significantly different, the end groups of homologous chains (i.e., $α^a:α^b$ and $β^a:β^b$) are identical. Like most fish Hbs, the N-termini of the α chains are blocked with an acetyl group, while the β chains have free N-terminal Val. The C-termini are -Tyr-Arg and -Tyr-His for the α chains and β chains, respectively. The effect of

pH and ATP on the P_{50} of HbI, HbII, HbIII, and HbIV are similar. Moreover, O_2 equilibria studies done at high pH as a function of temperature yield identical thermal sensitivities of all four Hb components ($ΔH = -15.5 ± 0.5$ kcal/mol).

Suckers of the subgenus *Pantosteus* (genus *Catostomus*) are typical examples of fish with Class II hemoglobins (163). *Catostomus (Pantosteus) clarkii* has 10 isohemoglobins comprised of at least two α chains ($α^a$ and $α^b$) and four β chains ($β^a$, $β^b$, $β^c$, and $β^d$). The end groups of the α and β chains for the electrophoretically anodal hemoglobins are similar to the hemoglobins of *Fundulus heroclitus* and other fish with Class I Hbs. However, the β chains or the electrophoretic cathodal Hbs have C-terminal -Tyr-Phe instead of -Tyr-His.

This end group difference, which has also been found in trout HbI versus HbIV, is important in the fish's physiological ecology. An increase in lactic acid and drop in blood pH following violent exercise or transient hypoxia can result in death by asphyxiation. In the natural environment, many fish species hide after swimming stress until physiological parameters have reequilibrated. However, some species live in habitats that do not provide such an opportunity. Some, but by no means all, of these species, have some hemoglobins with normal alkaline Bohr effects and one or more Hb components that are relatively unaffected by changes in pH. The sympatric catostomid fishes are particularly good examples to illustrate this point. Catostomids are distributed so that only one species of a subgenus inhabits a given geographical region, although each species of one subgenus is usually found living with a member of the other subgenus (i.e., sympatric). The sympatric species, *Catostomus (Pantosteus) clarkii* and *C. (Catostomus) insignis*, can be distinguished from one another by the presence or absence, respectively, of cathodal components in the electrophorectic patterns of their hemoglobins (Fig. 33). The anodal Hbs from both species showed similar Hb–O$_2$ affinities and had similar Bohr and Root effects (158). However, the cathodal Hbs isolated from *C. clarkii* did not have a significant sensitivity to changes in pH (158, 163).

The NH protons in the imadazolium ring of C-terminal histidine in the β chains and N-termini of the α chains are known to be largely responsible for the Bohr effect in vertebrate Hbs. As indicated above, the anodal Hbs had the normal Tyr-His at the C-termini of the β chains, while the cathodal (non-Bohr effect) Hb had Tyr-Phe. The reduced pH sensitivity of these cathodal Hbs was primarily a result of this C-terminal substitution.

Cathodal Hbs, without the Bohr effect,

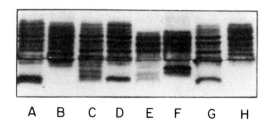

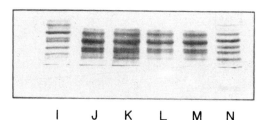

Figure 33. Starch gel electrophoresis patterns of hemoglobins from individual fish of various species of the genus *Catostomus*. The anode was at the top and the origin was in the middle of the gels. The samples, designated by letters, are listed with the subgenus in parentheses between the genus and species: (A) *Catostomus (Pantosteus) discobolus;* (B) *C. (Pantosteus) columbianus;* (C) *C. (Pantosteus) plebius;* (D) *C. (Pantosteus) clarkii;* (E) *C. (Pantosteus) platyrhynchus* (from the Humboldt River); (F) *C. (Pantosteus) platyrhynchus* (from the Sevier River, Utah); (G) *C. (Pantosteus) clarkii;* (H) *C. (Catostomus) ardens;* (I) *C. (Catostomus) insignis;* (J–M) *C. (Catostomus) catostomus;* (N) *C. (Pantosteus) clarkii.* [From (160).]

may provide an emergency back-up system to allow continued swimming in the fast water habitat during transient acidosis and internal hypoxia following emergency exertion (e.g., to escape a predator) (163). A similar molecular adaptation may occur in other hyperactive fishes that must maintain swimming during transient acidosis and hypoxia. The fact that fish from similar habitats have been found to have one or more Hbs without a Bohr effect tends to support this hypothesis.

Class III fish Hbs are best represented

by tuna (171) and related species whose mode of adaptation was discussed earlier.

Carbon Dioxide Transport by Blood

The effective elimination of CO_2 is as important to long-term respiratory homeostasis as is the efficient uptake of O_2 from the environment, particularly since the transport of CO_2 is intimately associated with both acid–base balance and ion–fluid balance regulation. Unfortunately, our knowledge of CO_2 transport lags behind that of O_2 transport for at least two important reasons. First, the accurate measurement of CO_2 in biological fluids is difficult compared with the measurement of O_2, and so the study of CO_2 transport is technically far more challenging. Second, CO_2 transport is qualitatively more complex than O_2 transport in that CO_2 can be transported bound to a respiratory pigment, as molecular CO_2 dissolved in solution, or in a variety of forms such as the bicarbonate ion HCO_3^-.

The Chemistry of Carbon Dioxide

CO$_2$ in Aqueous Solutions
Fundamental to understanding the transport of CO_2 is an appreciation for the chemistry of carbon dioxide—in particular how CO_2 behaves in aqueous solutions (for physiologically orientated reviews, see 48, 102, 181). When dissolved in water (or water-based solutions) molecular CO_2 reversibly combines with H_2O to form carbonic acid (H_2CO_3). The equilibrium between CO_2 and H_2O is strongly towards H_2CO_3. Carbonic acid is a weak acid (dissociation constant of 8.0×10^{-7} at 37°C, apparent pK 6.1), and readily dissociates into bicarbonate ions (HCO_3^-), carbonate ions (CO_3^{2-}) and hydrogen ions (H^+):

$$CO_2 + H_2O \rightleftharpoons H_2CO_3 \rightleftharpoons$$
$$H^+ + HCO_3^- \rightleftharpoons H^+ + CO_3^{2-} \quad (9)$$

The uncatalyzed reaction between CO_2 and H_2O requires several minutes near 0°C, and many seconds at the temperatures characteristic of the blood of birds and mammals. However, conversion between the various species of CO_2 represented in Eq. 9 is sped up thousands of times by the enzyme carbonic anhydrase, which has now been identified in a variety of gas exchange organs and blood cells.

The relation between the various species of CO_2 is mathematically described by the Henderson–Hasselbalch equation:

$$pH = pK + \log \frac{[HCO_3^-]}{[H_2CO_3]} \quad (10)$$

The proportions of CO_2, HCO_3^-, and CO_3^{2-} in solution vary with the temperature, pH, and ionic strength of the solution. At 37°C, pH 7.4, and an *apparent* pK 6.1 (i.e., under *in vivo* conditions for birds and mammals), the ratio in the blood of CO_2 to H_2CO_3 is ~1:1000, the ratio of CO_2 to HCO_3^- is ~1:20, and only insignificant amounts of the carbonate ion are present. At the lower body temperatures characteristic of many vertebrate and invertebrate ectotherms, however, plasma pH (83, 147) and pK (18) and CO_2 solubility (18) are all elevated, and the carbonate content may account for as much as 5% of the total CO_2 content of the blood. In spite of these effects of temperature, HCO_3^- is still the predominant form of CO_2 when considering the equilibrium of CO_2 in water, and the total CO_2 content is fairly accurately represented by the sum of the dissolved molecular CO_2 and HCO_3^-.

The total CO_2 content in a protein-free solution (i.e., no transport pigment) varies predictably with P_{CO_2}. The relation between P_{CO_2} and total CO_2 in water is nonlinear and complex, particularly if there are carbonates in the water and especially at low P_{CO_2} values (49).

Both pH and total CO_2 can be accurately measured. Although, there are no direct methods for distinguishing between dissolved CO_2, HCO_3^-, and CO_3^{2-} ions, they can be calculated. Dissolved CO_2 can be determined from the P_{CO_2} and the solubility coefficient (α) for CO_2 under the measurement conditions (this assumes that P_{CO_2} values are high enough to be operating in the linear region of the curve relating P_{CO_2} and total CO_2). Thus, the Henderson–Hasselbalch equation can be rewritten:

$$pH = pK + \log \frac{[HCO_3^-]}{P_{CO_2}} \quad (11)$$

Equation 11 contains three unknowns—if any two are measured the third can be calculated.

CO_2 and Respiratory Pigments
Carbon dioxide also reversibly binds with proteins, and with hemoglobins in particular. Whereas O_2 binds with the Fe^{2+} ion, CO_2 binds loosely with the —NH_2 groups as follows:

pigment—NH_2 + CO_2 $\rightleftharpoons$

pigment—$NHCO^-$ + H^+ (12)

For Hb this reaction proceeds very rapidly and does not require a catalyst *in vivo*.

Like all proteins, respiratory pigments can act as buffers, and contain groups that can dissociate as acids and bases—that is, respiratory pigments can both give off or take up H^+ ions. Under certain physiologically relevant circumstances, Hb can take up at one binding site a H^+ lost from a different site of the same molecule! Consequently, solutions with respiratory pigments tend to show far less change in pH with change in CO_2 content than would a water-based solution without a pigment.

In Vivo CO_2 Transport

Vertebrates
The CO_2 transport in vertebrates is crucially influenced by the fact that the respiratory pigment is enclosed within the microenvironment of the erythrocyte. This design feature results in regional concentration differences between the erythrocyte interior and the surrounding plasma. Moreover, the erythrocytes of vertebrates generally contain high levels of the enzyme carbonic anhydrase (33, 39, 66, 131), which effectively removes the rate of reaction of CO_2 and H_2O in the erythrocyte as a rate-limiting step in HCO_3^- formation. Both the general mechanism for CO_2 elimination and the role of carbonic anhydrase are understood for mammals, while basic information is still emerging for aquatic vertebrates (167).

CO_2 Carriage in Blood. The crucial pathways describing CO_2 transport in the blood of air-breathing vertebrates are presented in Fig. 34. Constant aerobic metabolism results in a perpetual P_{CO_2} gradient from the mitochondria in cells of systemic tissues to the blood within the systemic capillaries. As molecular CO_2 diffuses across the capillary wall and into capillary blood, some CO_2 dissolves directly in the plasma (Fig. 34A). Of this CO_2 dissolved in solution, the vast majority slowly combines with H_2O to form HCO_3^- according to Eq. 9. Most of the H^+ resulting from this reaction are buffered by plasma proteins such as albumin.

The vast majority of molecular CO_2 from the tissues diffuses through the plasma into the erythrocyte. Once in the erythrocyte, some small proportion of CO_2 remains in physical solution, but the formation of HCO_3^- is heavily favored (Eq. 9). Unlike conditions outside the erythrocyte, where CO_2 is slowly converted to HCO_3^- at the uncatalyzed rate, carbonic anhydrase within the erythro-

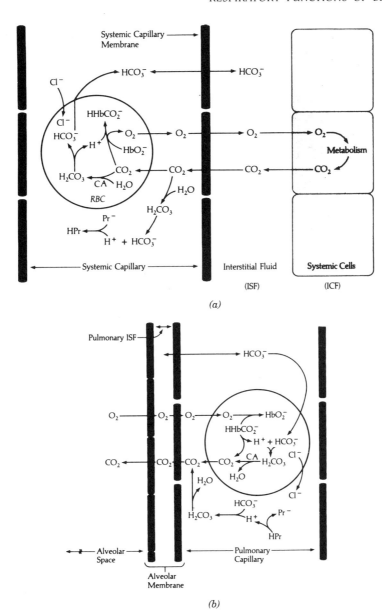

Figure 34. CO_2 transport in terrestrial vertebrates. (*a*) Gas exchange in systemic tissue capillaries. (*b*) Gas exchange in pulmonary capillaries. Hb, hemoglobin; CA, carbonic anhydrase; Pr^-, plasma protein; $HbCO_2^-$, carbaminohemoglobin. [From (102).]

cyte speeds this chemical conversion enormously. Consequently, the great majority of CO_2 initially in solution in the intracellular fluid is very rapidly converted to HCO_3^-. The buildup of HCO_3^- within the erythrocyte creates a large con-

centration gradient from the interior of the cell to the surrounding plasma, and as a result HCO_3^- diffuses out into the plasma. The deficit of negative ions within the erythrocyte is countered by the inward influx of Cl^- ions (96), thus main-

TABLE 3. Percentage Distribution of Total Blood CO_2 between Plasma and Erythrocytes in Human Whole Blood[a]

	Arterial Blood (% of Total CO_2)			Venous Blood (% of Total CO_2)			Arterial Venous Difference (% of Total CO_2 Change)		
	CO_2	HCO_3^-	CO_2—Hb	CO_2	HCO_3^-	CO_2—Hb	CO_2	HCO_3^-	CO_2—Hb
Plasma	3	71		3	70		5	57	
Red cells	2	21	3	2	22	4	3	11	24

[a] 8.9 g Hb/dL blood; Hct = 40%, respiratory quotient = 0.82; total CO_2 content = 21.53 mmol/L of arterial, 23.21 mmol/L of venous. Also shown is the percentage change when comparing arterial and venous blood. [From (44).]

taining electrical balance of the erythrocyte in the face of a large HCO_3^- efflux. This process, variously known as the "Hamburger" or "chloride" shift, has a half-time in human blood at 37°C of ~100 ms, allowing for almost all of the chloride shift to occur during the transit time within the capillary. Chloride shifts have also been described in the red cells of red snapper, rainbow trout, and the dogfish and a chloride shift is now generally perceived to exist in fish and other lower vertebrates (167). A notable exception is the hagfish *Eptatretus stouti*, which lacks a significant chloride shift (54).

Finally, some molecular CO_2 entering the erythrocyte binds directly to Hb. Some of the H^+ resulting from this process (Eq. 12) are immediately buffered by sites on the Hb molecule that accept protons. The more deoxygenated the respiratory pigment, the greater is its ability to accept protons. Thus, as O_2 dissociates from Hb in the capillaries of the tissues concomitant with the formation of carbaminohemoglobin, the net change in $[H^+]$ (and thus change in pH) within the erythrocyte remains very small. Complete deoxygenation of O_2-saturated Hb resulting in the release of 1 mol of O_2 will result in the binding of 0.7 mol of protons. Given respiratory quotients of from 0.7 to 1.0, almost all of the H^+ resulting from

both HCO_3^- and carbaminohemoglobin formation in the capillaries of the tissues are buffered by Hb and plasma proteins. Consequently, the difference between the pH of arterial and venous blood of a wide variety of vertebrates is generally only a few 100ths of a pH unit (44.83).

Thus, CO_2 can be carried in vertebrate blood in one of three forms: (1) dissolved in solution as molecular CO_2, in both plasma and intracellular fluid; (2) as HCO_3^- formed either at the uncatalyzed rate in plasma or at the catalyzed rate in the erythrocyte; and (3) as carbaminohemoglobin. The relative importance of these three modes of transport of CO_2 to actual CO_2 elimination in the gas exchange organ(s) depends greatly on plasma temperature, capillary, and venous transit times, presence and distribution of carbonic anhydrase, and even erythrocyte size.

The quantification of CO_2 distribution between intra- and extracellular compartments is a nontrivial matter, and the most extensive data exist for mammals. Table 3 provides approximate normal values for human blood at 37°C (44). Various estimates indicate that from 80 to 90% of the CO_2 carried in human systemic venous blood exists as the bicarbonate ion, primarily in the plasma following intra-erythrocytic conversion from molecular

CO_2. Note that Hb accounts for only 1–2% of CO_2 resident in systemic venous blood.

The movements of CO_2 from blood in the pulmonary, branchial, or cutaneous capillaries into the respiratory medium essentially represents a reversal of the pathways occurring in the systemic tissues (Fig. 34B). Given adequate levels of ventilation of the gas exchange organ, the P_{CO_2} in the respiratory medium is lower than in the capillary blood within the gas exchanger—thus, there is a gradient for the diffusion of molecular CO_2 from the interior of the erythrocyte to the environment. As CO_2 leaves the erythrocyte, Eq. 9 proceeds from right to left. The HCO_3^- in the erythrocyte is converted back to molecular CO_2 at the catalyzed rate, and then diffuses out of the cell into the plasma. As intraerythrocytic concentrations of HCO_3^- fall, HCO_3^- diffuses along its concentration gradient from the plasma back into the cell and the chloride shift reverses, once again maintaining electrical balance within the erythrocyte. As the P_{CO_2} inside the erythrocyte declines during passage through the capillaries of the gas exchange organ, CO_2 dissociates from Hb, also diffusing out into the plasma and eventually into the respiratory medium.

At least for human blood, the overall pattern of distribution of CO_2 with arterial blood following passage through the lungs still heavily favors CO_2 as HCO_3^-. However, the direct dissociation of molecular CO_2 from Hb accounts for nearly one-fourth of the CO_2 actually eliminated into alveolar gas, with most of the remainder being accounted for by HCO_3^- both in plasma and intraerythrocytic fluid. Little data on the partitioning of CO_2 in plasma and erythrocytes following passage through the gas exchange organs exist for lower vertebrates. In the anuran amphibian *Xenopus laevis*, plasma CO_2 (in its various forms) accounts for ~84% of the CO_2 in arterial whole blood during lung ventilation, whereas after a dive the plasma proportion falls to 65% (56). In the toad *Bufo marinus* only 11% of the total CO_2 content of arterial blood is resident in the erythrocytes (20). However, the relative contributions of carbaminohemoglobin, HCO_3^-, and dissolved CO_2 to CO_2 transport (i.e., how CO_2 distribution changes following passage through the gas exchange organs) have yet to be determined in anurans. Total blood CO_2 is reduced by 10–20% following passage through the gills of fishes (33, 167), compared with ~6–8% following passage of blood through the lungs of man (44). Most of the CO_2 elimination from fish blood is thought to result from the catalyzed conversion of HCO_3^- to molecular CO_2 (167).

Carbonic anhydrase occurs in endothelial cells of the gas exchange organs of many vertebrates; for example, lungs of mammals (42) and amphibians (187), skin of amphibians (170, 187), and the gills of aquatic fishes (33, 167), and some pedomorphic amphibians (187) (e.g., *Necturus*). The air-breathing organs of certain air-breathing fishes (e.g., *Trichogaster trichopterus*, *Calamoichthys calabaracus*, and *Clarias batrachus*) contain carbonic anhydrase, and this is associated with unusually high rates of aerial CO_2 elimination in these fishes (33). While carbonic anhydrase participates in interconversion between CO_2 species within the endothelial cells of mammals, its contribution to overall elimination of CO_2 remains uncertain.

In aquatic organisms CO_2 transport (and the intimately associated processes of acid–base balance and hydromineral regulation) is further complicated by the fact that ions directly involved in CO_2 transport (e.g., H^+, and HCO_3^-) also can be exchanged with the environment through active membrane transport (167) (Fig. 35). Whereas in the lungs of terres-

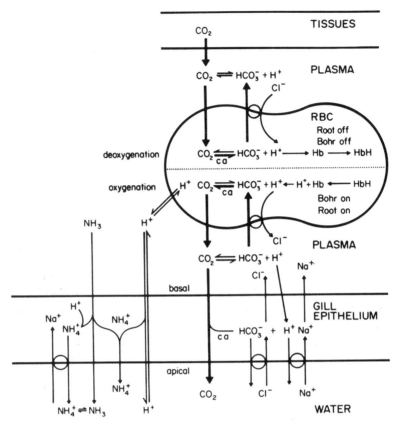

Figure 35. A schematic diagram indicating potential pathways for transport of ions and CO_2 across the gills of fishes. [From (167).]

trial vertebrates only molecular CO_2 is eliminated, in the gills of aquatic organisms CO_2 potentially can be eliminated as both molecular CO_2 and as HCO_3^-, while H^+ can be eliminated or taken up through stoichiometrically linked transport with both Na^+ and NH_4^+. While the role of carbonic anhydrase in the gill epithelium is thought to be relatively minor with respect to CO_2 elimination (167), the influence of active H^+ excretion linked to acid–base balance and hydromineral regulation may be more important (Chapter 2). In resting salmonid fish, for example, the uptake of NaCl in exchange for H^+ and HCO_3^- is thought to account for ~10% of the total CO_2 excretion (37). Quite obviously a system with interac-

tions between CO_2 elimination and ion exchange is exceptionally complex, particularly when aquatic organisms encounter changing external ion concentrations (e.g., anadromous fishes).

CO_2 Equilibrium Curves. CO_2 transport can be described quantitatively using the same methodological approach as for O_2 transport. Figure 36 presents the CO_2 equilibrium curve for human whole blood, and indicates the approximate contributions of CO_2 in its various forms. Figure 37 compares the CO_2 equilibrium curve with the O_2 equilibrium curve in the turtle *Pseudemys scripta.* Apparent from this figure is the much larger total CO_2 content of blood. The blood transports

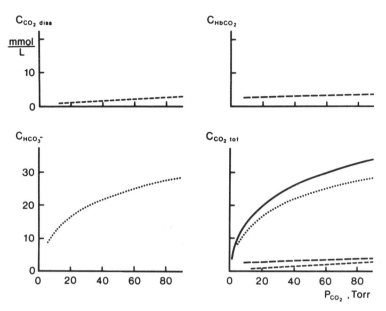

Figure 36. Concentrations of CO_2 in its various forms in deoxygenated human whole blood (38°C) as function of P_{CO_2}. Upper left; dissolved CO_2 (solubility coefficient = 0.029 mmol/L · mmHg). Upper right; carbaminohemoglobin. Lower left; HCO_3^- in whole blood. Lower right; total CO_2 concentration of whole blood, depicted by solid line. (The line depicting CO_2 dissolved in solution appears linear in apparent contrast to Fig. 37. This results simply from the fact that, because the great majority of CO_2 is carried in forms other than as molecular CO_2, the scale for CO_2 in Fig. 37 covers a considerably greater range.) [From (48).]

similar amounts of O_2 and CO_2—the difference being that even arterial blood still contains large quantities of CO_2, primarily in the form of HCO_3^-. The CO_2 equilibrium curve of most animals generally is curvilinear (as opposed to sigmoid for O_2), and is much steeper than the O_2 equilibrium curve. Thus, although the arteriovenous P_{CO_2} difference is only a few millimeters of Hg in those vertebrates in which it has been measured, the steep nature of the CO_2 equilibrium curves ensures adequate CO_2 transport. In mammals and birds CO_2 transport normally occurs over a narrow range of the CO_2 equilibrium curve at comparatively high P_{CO_2} values where the relation between P_{CO_2} and total CO_2 can be regarded as approximately linear. In ectothermic vertebrates for which CO_2 equilibrium curves

have been determined [e.g., the freshwater turtle *P. scripta* (1), and the trout *Salmo gairdneri* (82)], *in vivo* P_{CO_2} values are much lower and CO_2 transport occurs over the region of the curve that is distinctly curvilinear.

The release of O_2 from Hb (i.e., the formation of deoxyhemoglobin) results in an increase in the total CO_2 content of whole blood, a phenomenon termed the Haldane effect (Fig. 38). The Haldane effect has two components. First, the formation of deoxyhemoglobin directly increases CO_2 hemoglobin affinity, which in human blood accounts for ~70% of the total Haldane effect (102). The remainder of the Haldane effect results as additional H^+ are buffered by Hb as CO_2 is released. This drives Eq. 9 to the right, and allows the formation of additional HCO_3^- from

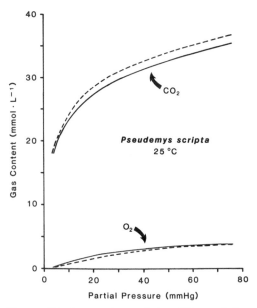

Figure 37. Comparison of the CO_2- and O_2-dissociation curves of whole blood from the turtle *Pseudemys scripta* plotted on the same coordinates for P_{O_2} and gas content. Curves determined at 25°C, P_{CO_2} = 37 mmHg.

molecular CO_2. The Haldane effect facilitates both CO_2 uptake in the capillaries of the systemic tissues, and CO_2 elimination into alveolar gas. A Haldane effect is not evident in the blood of elasmobranch fishes.

Temperature has a profound effect on the CO_2 equilibrium curves of most vertebrate ectotherms (Fig. 38). If large changes in temperature occur, changes in CO_2 solubility make a significant contribution to the overall change. However, the effect of temperature on CO_2-equilibrium curves results mainly from the effect of temperature on the pK of hemoglobin. Thus, increasing temperature decreases the pK of Hb, reducing the amount of H^+ that Hb can buffer. The net effect is a reduction in the amount of HCO_3^- that will form in the blood. This reflects an intrinsic chemical property of proteins in water-based solutions, rather than a property of a particular Hb that could be altered

through natural selection. Thus, a CO_2 transport system unaffected by temperature is unlikely to have evolved in heterotherms in the same way that temperature insensitive O_2 transport has.

While teleological arguments for adaptive changes in transport characteristics of vertebrate blood are as valid for CO_2 as for O_2, CO_2 transport has rarely been investigated from the same adaptive perspective that guides study of blood O_2 transport. Quantitative investigation of the comparative physiology of CO_2 transport remains as one of the more prominent unexploited areas of respiratory physiology.

In Vivo CO_2 Transport—Invertebrates

The mechanisms involved in CO_2 transport in invertebrates have been poorly studied, but appear basically similar to those described above for vertebrates. As in vertebrates, most of the molecular CO_2 formed in the tissues is converted to bicarbonate in the blood for transport to the external gas exchange surfaces. Smaller

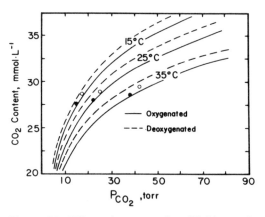

Figure 38. Effect of oxygenation (Haldane effect) and temperature on the CO_2 equilibrium curves of the turtle *Pseudemys scripta*. *In vivo* values of arterial (solid circles) and venous (open circles) blood are presented. [After (204).]

HALDANE EFFECTS ∶ CRUSTACEAN HEMOCYANINS

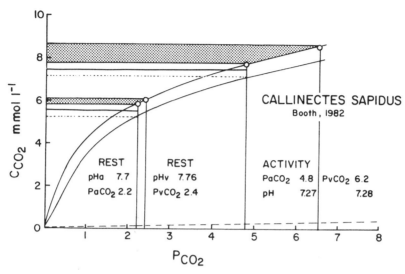

Figure 39. Carbon dioxide equilibrium curves and the role of the Haldane effect in CO_2 transport in *Callinectes sapidus*, a marine swimming crab. The hatched area represents CO_2 elimination indicated by the CO_2 partial pressure differential between arterial and venous hemolymph. Broken horizontal lines illustrate the maximum Haldane effect. Solid horizontal lines indicate the magnitude of the Haldane effect *in vivo*. [From (137).]

fractions are carried as molecular CO_2 and as carbonate. The existence of carbamino-bound CO_2 (i.e., CO_2 bound to pigment) in invertebrate blood is suspected but not proven and its role remains to be quantified. Levels of circulating CO_2 in body fluids of aquatic invertebrates at rest are generally similar to those of vertebrates of similar habitat. Circulating CO_2 levels in terrestrial invertebrates are higher but hemolymph pH values are similar and thus terrestrial invertebrates are perpetually in a state of fully compensated respiratory acidosis (48, 140).

The enzyme carbonic anhydrase, which facilitates mobilization of CO_2 as described above, is found in many invertebrate gills and has been shown to facilitate branchial CO_2 elimination in the crab *Cancer productus* (141). In invertebrates in which the respiratory carrier molecule is simply dissolved in hemolymph, the absence of carbonic anhydrase in circulating blood may prevent blood reaching equilibrium within the gas exchangers (1). Many invertebrates carry their O_2-carrying pigment packaged in erythrocytes, but a role in CO_2 transport similar to that of vertebrates is not confirmed. At least in freshwater and brackish water crabs a variable proportion of CO_2 excretion occurs by way of HCO_3^- exchanged for Cl^- during branchial ion regulation. Thus, the respiratory exchange mechanisms occurring at the crab gill are similar to the extraerythrocytic mechanisms illustrated in Fig. 35.

The amount of CO_2 eliminated from the gills of aquatic crabs has proven difficult to quantify. Invertebrate CO_2 transport is best studied by reference to CO_2 equilibrium curves. The P_{CO_2} difference across the gills (arterio-venous) may be as small as 0.2 mmHg at rest (Fig. 39). This corresponds, however, with the clearance of a substantial fraction (10%) of total

CO_2. With increasing activity the P_{CO_2} of circulating hemolymph also increases. The arterial-venous P_{CO_2} difference necessary for adequate CO_2 elimination increases as P_{CO_2} levels become located on the less steep regions of the CO_2 equilibrium curve (Fig. 39).

In invertebrates with Hc or Hb, elimination of CO_2 is probably aided by the Haldane effect as described above. This effect has been demonstrated in the blood of several invertebrates but again is often small and difficult to quantify (137, 148, 189, 202, 206). Figure 39 nonetheless allows an estimation of the magnitude of the Haldane effect in Hc from crustaceans from aquatic and aerial habitats at rest and following activity. Since O_2 and CO_2 transport are clearly linked by the Bohr and Haldane effects, it follows that the potentiating effects of modulator substances on O_2 transport discussed in the section on heterotrophic interactions will also be mirrored in the transport of CO_2.

Summary

Respiratory pigments for the transport and storage of O_2 (and, to a lesser extent, CO_2) are widely distributed throughout the animal kingdom. Three of these pigments—hemoglobin, hemerythrin, and cholorocruorin are iron based, while hemocyanin is copper based. Hemoglobin is the sole pigment in vertebrates, and is always enclosed within a blood cell. Respiratory pigments in invertebrates may be intra- or extracellular. All pigments can (and usually do) occur in complex monomeric forms.

All respiratory pigments reversibly bind O_2, this property making them effective at O_2 transport. The O_2-binding properties of a pigment can be described by the oxygen equilibrium curve (OEC). Pigment—O_2 affinity is influenced via a variety of heterotrophic interactions. Increased hydrogen ion concentration decreases pigment—O_2 affinity (displaces the OEC to the right) through a phenomenon known as the Bohr effect. The Root effect, also caused by increased [H^+] and common in fish blood, results additionally in a reduction in the total O_2 capacity of the blood. A variety of compounds (the organophosphates ATP, 2,3-DPG, and ITP for vertebrate hemoglobins and invertebrate iron-based pigments; various cations as well as lactate for hemocyanin) are also potent modulators of pigment—O_2 binding. The reaction between a pigment and O_2 is exothermic, and consequently increased temperature reduces the affinity of respiratory pigment for O_2.

Adaptive modifications in pigment—O_2 binding have been described for a wide variety of pigments from various taxa. Adaptations fall into two categories—those that result from environmental hypoxia as well as internally induced hypoxia produced by intermittent breathing or increased activity. In either category, adaptations can include increased pigment concentration as well as adjustments in the shape and position of the O_2 equilibrium curve. While there are numerous cases of clear adaptive advantage for a particular set of characteristics for a pigment, such adaptations must be viewed in the context of the dual function of a pigment—loading of O_2 in the gas exchange organ *and* unloading of O_2 at the tissues. To be truly adaptive, a change in pigment function must facilitate one of these functions without compromising the other.

Respiratory pigments in plasma play an important role in CO_2 transport between tissues and the gas exchange organ. However, the complex chemistry of CO_2 provides for alternative modes of transport in addition to being bound to a respiratory pigment. Thus, CO_2 can be transported bound to the pigment or in plasma or intraerythrocytic fluid as either the bicarbonate ion or as CO_2 physically dissolved in solution. Little is known of adaptive changes in CO_2 transport by blood, though effective CO_2 elimination is as important as effective O_2 uptake in the long term.

References

1. Aldridge, J. B. and J. N. Cameron. *J. Exp. Zool.* 207:321–328, 1979. CO_2 exchange in the blue crab *Callinectes sapidus* (Rathun).

2. Anderson, M. E., J. S. Olson, and Q. H. Gibson. *J. Biol. Chem.* 248:331–341, 1973. Studies on ligand binding to hemoglobins from teleosts and elasmobranchs.

3. Antonini, E., A. Rossi-Farelli, and A. Caputo. *Arch. Biochem. Biophys.* 97:336–342, 1962. Studies on chlorocruorin. I. The oxygen equilibrium of *Spirographis chlorocruorin*.

4. Bak, H. J., N. M. Soeter, J. M. Vereijken, P. A. Jekel, and P. Neuteboom. In *Invertebrate Oxygen Carriers* (B. Linzen, ed.), pp. 149–152. Springer-Verlag, Berlin, 1986. Primary structure of the a chain of *Panulirus interruptus* hemocyanin.

5. Bartlett, G. *Am. Zool.* 20:103–114, 1980. Phosphate compounds in vertebrate red blood cells.

6. Barton, P., S. Malcolm, C. Murphy, and M. A. Ferguson-Smith. *J. Mol. Biol.* 156:269–278, 1982. Localization of the human alpha-globin gene cluster to the short arm of chromosome 16 (16p12-16p-ter) by hybridization *in situ*.

7. Benesch, R. and R. E. Benesch. *Biochem Biophys. Res. Commun.* 26:162–167, 1967. The effect of organic phosphates from the human erythrocyte on the allosteric properties of hemoglobin.

8. Binotti, S., B. Giovenco, B. Giardina, E. Antonini, M. Brunori, and J. Syman. *Arch. Biochem. Biophys.* 142:274–280, 1971. Studies on functional properties of fish hemoglobins. II. The oxygen equilibrium of the isolated hemoglobin components of trout blood.

9. Birchard, G., C. Black, G. Schuett, and V. Black. *J. Exp. Biol.* 108:247–256, 1984. Foetal-maternal blood respiratory properties of an ovoviviparous snake the cottonmouth, *Agkistrodon piscivorus*.

10. Bird, D., P. Lutz, and I. Potter. *J. Exp. Biol.* 65:449–458, 1976. Oxygen dissociation curves of the blood of larval and adult lamprey (*Lampetera fluviatilis*).

11. Black, C. P. and S. M. Tenney. *Respir. Physiol.* 39:217–239, 1980. Oxygen transport during progressive hypoxia in high-altitude and sea-level waterfowl.

12. Bonaventura, C. and J. Bonaventura. *Am. Zool.* 20:131–138, 1980. Anionic control of function in vertebrate hemoglobins.

13. Bonaventura, C., J. Bonaventura, K. I. Miller, and K. E. van Holde. *Arch. Biochem. Biophys.* 211:589–598, 1981. Hemocyanin of the chambered *Nautilus:* Structure-function relationships.

14. Bonaventura, J. and C. Bonaventura. *Am. Zool.* 20:7–17, 1980. Hemocyanins: Relationships in their structure, function and assembly.

15. Bonaventura, J., C. Bonaventura, and B. Sullivan. *J. Exp. Zool.* 194:155–174, 1975. Hemoglobins and hemocyanins: Comparative aspects of structure and function.

16. Booth, C. E. Ph.D. Thesis, University of Calgary, Calgary, Alberta, Canada, 1982. Respiratory responses to activity in the blue crab, *Callinectes sapidus*.

17. Bouchet, J. and J.-P. Truchot. *J. Exp. Biol.* 118:461–465, 1985. Effects of hypoxia and L-lactate on the haemocyanin-oxygen affinity of the lobster, *Homarus vulgaris*.

18. Boutilier, R. G., T. A. Heming, and G. K. Iwama. In *Fish Physiology* (W. Hoar and D. Randall, eds.), Vol. 10A, pp. 403–430. Academic Press, Orlando, FL, 1984. Appendix: Physiochemical parameters for use in fish respiratory physiology.

19. Boutilier, R. G., G. K. Iwama, and D. J. Randall. *J. Exp. Biol.* 123:145–157, 1986. The promotion of catecholamine release in rainbow trout, *Salmo gairdneri*, by acute acidosis: Interactions between red cell pH and heamoglobin oxygen carrying capacity.

20. Boutilier, R. G., D. J. Randall, G. Shelton, and D. P. Toews. *J. Exp. Biol.* 82:331–344, 1979. Acid-base relationships in the blood of the toad, *Bufo marinus.*

21. Bouverot, P. *Adaptation to Altitude-Hypoxia in Vertebrates.* Springer-Verlag, Berlin, 1985.

22. Bridges, C. R. and S. Morris. In *Invertebrate Oxygen Carriers* (B. Linzen, ed.), pp. 341–352. Springer-Verlag, Berlin, 1986. Modulation of haemocyanin oxygen affinity by L-lactate—a role for other cofactors.

23. Bridges, C. R., A. C. Taylor, S. J. Morris, and M. K. Grieshaber. *J. Exp. Mar. Biol. Ecol.* 77:151–167, 1984. Ecophysiological adaptations in *Blennius pholis* (L.) blood to intertidal rockpool environments.

24. Brittain, T. *Comp. Biochem. Physiol. B* 86B:473–481, 1987. The Root effect.

25. Brix, O. *J. Exp. Zool.* 221:27–36, 1982. The adaptive significance of the reversed Bohr and Root shifts in blood from the marine gastropod, *Buccinum undatum.*

26. Broyles, R. In *Metamorphosis: A Problem in Developmental Biology,* (L. Gilbert and E. Frieden, eds.), pp. 461–490. Plenum, New York, 1981. Changes in the blood during amphibian metamorphosis.

27. Bullard, R. In *Physiological Adaptations; Desert and Mountain.* (M. K. Yousef, S. M. Horvath, and R. W. Bullard, eds.), pp. 209–225. Academic Press, New York, 1972. Vertebrates at altitudes.

28. Bunn, H. F. and B. G. Forget. *Hemoglobin: Molecular, Genetic and Clinical Aspects.* W. B. Saunders Co. Philadelphia, 1986.

29. Burggren, W. W. *Can. J. Zool.* 66:20–28, 1988. Cardiovascular response to diving and their relation to lung and blood stores in vertebrates.

30. Burggren, W. W. *Physiol. Zool.* 55(4):327–334, 1982. "Air-gulping" improves blood oxygen transport during aquatic hypoxia in the goldfish *Carassius auratus.*

31. Burggren, W. W., R. K. Dupré, and S. C. Wood. *Respir. Physiol.* 70:73–84, 1987. Allometry of red cell oxygen binding and hematology in larvae of the salamander, *Ambystoma tigrinum.*

32. Burggren, W. W., C. E. W. Hahn, and P.

Föex. *Respir. Physiol.* 31:39–50, 1977. Properties of blood oxygen transport in the turtle *Pseudemys scripta* and the tortoise *Testudo graeca*: Effects of temperature, CO_2 and pH.

33. Burggren, W. W. and M. S. Haswell. *J. Exp. Biol.* 82:215–226, 1979. Aerial CO_2 excretion in the obligate air breathing fish, *Trichogaster trichopterus*: A role for carbonic anhydrase.

34. Burggren, W. W. and S. C. Wood. *J. Comp. Physiol.* 144:241–246, 1981. Respiration and acid-base balance in the tiger salamander, *Ambystoma tigrinum;* influence of temperature acclimation and metamorphosis.

35. Butler, P. J., E. W. Taylor, and B. R. McMahon. *J. Exp. Biol.* 73:131–146, 1978. Respiratory and circulatory changes in the lobster (*Homarus vulgaris*) during long term exposure to moderate hypoxia.

36. Calder, W. A., III. *Size, Function and Life History.* Harvard Univ. Press, Cambridge, MA, 1984.

37. Cameron, J. N. *J. Exp. Biol.* 56:711–725, 1976. Branchial ion uptake in the Arctic grayling: Resting values and effects of acid-base disturbance.

38. Carey, F. G. *Sci. Am.* 228(2):36–44, 1973. Fishes with warm bodies.

39. Carter, M. J. *Biol. Rev. Cambridge Philos. Soc.* 47:465–513, 1972. Carbonic anhydrase: Isoenzymes, properties, distribution, and functional significance.

40. Chanutin, A. and R. R. Curnish. *Arch. Biochem. Biophys.* 121:96–102, 1967. Effect of organic and inorganic phosphates on the oxygen equilibrium of human erythrocytes.

41. Collett, L. C. and A. K. O'Gower. *Comp. Biochem. Physiol. A* 41A:843–850, 1972. Molluscan hemoglobins with unusual temperature-dependent characteristics.

42. Cossins, A. K. and P. A. Richardson. *J. Exp. Biol.* 118:229–246, 1985. Adrenaline-induced Na^+/H^+ exchange in trout erythrocytes and its effects upon oxygen-carrying capacity.

43. Dalessio, P. M. Ph.D. Thesis, George Washington University, Washington,

DC. Adrenergic control of blood oxygen affinity in the common Killifish.

44. Davenport, H. W. *The ABC of Acid-Base Chemistry.* Univ. of Chicago Press, Chicago, IL, 1974.

45. Davis, R. H. and E. Steers. *Comp. Biochem. Physiol. B* 54B:141–143, 1976. Myoglobin from the ciliate protozoan *Paramecium aurelia.*

46. Dawson, W. R. and T. L. Poulson. *Am. Midl. Nat.* 68:154–164, 1962. Oxygen capacity of lizard bloods.

47. Decker, H., J. Markle, R. Loewe, and B. Linzen. *Hoppe-Seyler's Z. Physiol. Chem.* 360:1505–1507, 1979. Hemocyanin in spiders. VIII. Oxygen affinity of the individual subunits isolates from *Eurypelma californicum* hemocyanin.

48. Dejours, P. *Principles of Comparative Respiratory Physiology.* Elsevier/North-Holland, Amsterdam, 1981.

49. Dejours, P., J. Armand, and G. Verriest. *Respir. Physiol.* 5:23–33, 1968. Carbon dioxide dissociation curves of water and gas exchange of water-breathers.

50. Dhindsa, D. S. and A. S. Hoversland. *Comp. Biochem. Physiol. A* 66A:173–180, 1980. Respiratory characteristics and 2,3-diphosphoglycerate concentrtion of dog blood during the postnatal period.

51. Dhindsa, D. S., J. Metcalfe, and A. S. Hoversland. *Respir. Physiol.* 15:331–342, 1972. Comparative studies of the respiratory functions of mammalian blood. IX. Ring-tailed lemur (*Lemur catta*) and black lemur (*Lemur macaco*).

52. Dickerson, R. E. and I. Geis. *Hemoglobin: Structure, Function, Evolution and Pathology.* Benjamin-Cummings, Menlo Park, CA, 1983.

53. Ellerton, H. D., N. F. Ellerton, and H. A. Robinson. *Prog. Biophys. Mol. Biol.* 41(3):143–248, 1983. Hemocyanin-a current perspective.

54. Ellory, J. C., M. W. Wolowyk, and J. D. Young. *J. Exp. Biol.* 129:377–383, 1987. Hagfish (*Eptatretus stouti*) erythrocytes show minimal chloride transport activity.

55. Elsner, R. and B. Gooden. *Diving and As-phyxia.* Cambridge Univ. Press, London, 1983.

56. Emilio, M. G. and G. Shelton. *J. Exp. Biol.* 85:253–262, 1980. Carbon dioxide exchange and its effects on pH and bicarbonate equilibria in the blood of the amphibian, *Xenopus laevis.*

57. Engbretson, G. and V. H. Hutchison. *Copeia* 1976:186, 1976. Erythrocyte count, hematocrit and hemoglobin content in the lizard *Liolaemus multiformis.*

58. Fermi, G., M. F. Perutz, B. Shaanan, and B. Fourme. *J. Mol. Biol.* 175:159–174, 1984. The crystal structure of human deoxyhemoglobin at 1.7 A resolution.

59. Fhyn, U., H. Fhyn, J. Davis, D. A. Powers, W. T. Fink, and R. Garlick. *Comp. Biochem. Physiol. A* 62A:39–66, 1979. Hemoglobin heterogeneity in Amazonian Fishes.

60. Fhyn, U. and B. Sullivan. *Biochem. Genet.* 11:373–385, 1974. Complex phenotypic patterns in the toadfish.

61. Fox, H. M. *Proc. R. Soc. London, Ser. B* 136:378–388, 1949. On chlorocruorin and haemoglobin.

62. Fox, H. M. *Amphibian Morphogenesis,* pp. 106–113. Humana, Clifton, NJ, 1983. The blood.

63. Freel, R. W. *J. Exp. Zool.* 204:267–274, 1978. Oxygen affinity of the hemolymph of the mesopelagic mysidacean *Gnathophausia ingens.*

64. Gahlenbeck, H. and H. Bartels. *Z. Vergl. Physiol.* 59:232–240, 1968. Temperature adaptation der Sauerstoff-Affinitatdes Blutes von *Rana esculenta* L.

65. Garlick, R. L. *Am. Zool.* 20:69–77, 1980. Structure of annelid high molecular weight Hb.

66. Gay, C. V., H. Schraer, D. J. Sharkey, and E. Rieder. *Comp. Biochem. Physiol. A* 70A:173–177, 1981. Carbonic anhydrase in developing avian heart, blood and chorioallantoic membrane.

67. Gaykema, W. P. J., W. G. J. Hol, J. M. Vereijken, N. M. Soeter, H. J. Bak, and J. J. Beintema. *Nature (London)* 309:23–29, 1984. A structure of the copper-contain-

ing, oxygen-carrying protein *Panulirus interruptus* haemocyanin.

68. Giardina, B., M. Coletta, L. Zolla, and M. Brunori. In *Respiratory Pigments in Animals* (J. Lamy, J.-P. Truchot, and R. Gilles, eds.). Springer-Verlag, Berlin, 1985. pp. 125–140. Oxygen transport proteins: A unitarian view based on thermodynamic kinetic and stereochemical consideration.

69. Gielens, C., C. Benoy, G. Preaux, and R. Lontie. In *Invertebrate Oxygen Carriers* (B. Linzen, ed.), pp. 223–226. Springer-Verlag, Berlin, 1986. Presence of only seven functional units in the polypeptide chain of the haemocyanin of the cephalopod *Octopus vulgaris*.

70. Gillen, R. C. and A. Riggs. *Comp. Biochem. Physiol. B* 38B:585–595, 1971. The hemoglobins of a freshwater teleost *Cichlasoma cyanoguttatum:* The effects of phosphorylated organic compounds upon oxygen equilibria.

71. Gillen, R. G. and A. Riggs. *J. Biol. Chem.* 247:6039–6046, 1972. Structure and function of the hemoglobins of the carp, *Cyprinus carpio*.

72. Goodman, M., A. E. Romero-Herrera, H. Dene, J. Czelusniak, and R. E. Tashian. In *Macromolecular Sequence in Systematic and Evolutionary Biology* (M. Goodman, ed.), pp. 115–191. Plenum, New York, 1982. Amino acid sequence evidence on the phylogeny of primates and other eutherians.

73. Goodman, M., M. L. Weiss, and J. Czelusniak. *Syst. Zool.* 31:376–399, 1982. Molecular evolution above the species level: Branching pattern, rates and mechanisms.

74. Greaney, G. S., A. Place, R. Cashon, G. Smith, and D. Powers. *Physiol. Zool.* 53(2):136–144, 1980. Time course of changes in enzyme activities and blood respiratory properties of killifish during long-term acclimation to hypoxia.

75. Greaney, G. S. and D. A. Powers. *Nature (London)* 270:73–74, 1977. Cellular regulation of an allosteric modifier of fish hemoglobin.

76. Greaney, G. S. and D. A. Powers. *J. Exp. Zool.* 203:339–350, 1978. Allosteric modifiers of fish hemoglobins: *In vitro* and *in vivo* studies of the effect of ambient oxygen and pH on erythrocyte ATP concentrations.

77. Grigg, G. C. *Comp. Biochem. Physiol.* 29:1203–1223, 1969. Temperature-induced changes in the oxygen 'equilibrium curve of the blood of the brown bullhead, *Ictalurus nebulosus*.

78. Hadley, N. F. and T. A. Burns. *Copeia* 1968:737–740, 1968. Intraspecific comparison of the blood properties of the side-blotched lizard.

79. Hall, F. G. *J. Appl. Physiol.* 21:375–378, 1966. Minimal utilizable oxygen and the oxygen dissociation curve of blood of rodents.

80. Hartvig, M. and R. E. Weber. In *Respiration and Metabolism of Embryonic Vertebrates* (R. Seymour, ed.), pp. 17–30, Junk, Dordrecht, The Netherlands, 1984. Blood adaptations for maternal-fetal oxygen transfer in the viviparous teleost, *Zoarces viviparus* L.

81. Hashimoto, K., Y. Yamaguchi, and F. Matsuura. *Bull. Jpn. Soc. Sci. Fish.* 26:827, 1960. Comparative studies on two hemoglobins of salmon. IV. Oxygen dissociation curve.

82. Hazard, E. S., III and V. H. Huthinson. *J. Exp. Zool.* 206:109–118, 1978. Ontogenetic changes in erythrocytic organic phosphates in the bullfrog, *Rana catesbeiana*.

83. Heisler, N. In *Fish Physiology* (W. Hoar and D. Randall, eds.), Vol. 10A. Academic Press, Orlando, FL, 1984. Acid-base regulation in fishes.

84. Hendrickson, W. A., J. L. Smith, and S. Sheriff. In *Respiratory Pigments in Animals: Relation Structure-Function* (J. Lamy, J.-P. Truchot, and R. Gilles, eds.), pp. 1–8. Springer-Verlag, Berlin, 1985. Structure and function of hemerythrins.

85. Hillyard, S. D. In *Environmental Physiology: Aging, Heat and Altitude* (S. M. Horvath and M. K. Yousef, eds.), pp. 362–377. Elsevier/North-Holland, New York,

1981. Respiratory and cardiovascular adaptations of amphibians and reptiles to altitude.

86. Hochachka, P. W. and G. N. Somero. *Strategies of Biochemical Adaptation*. Saunders, Philadelphia, PA, 1973.

87. Hochachka, P. W. *Living Without Oxygen*. Harvard Univ. Press, Cambridge, MA, 1980.

88. Hogben, L. T. and K. F. Pinhey. *J. Exp. Biol.* 5:55–65, 1927. Some observations on the hemocyanins of *Limulus*.

89. Horstman, D., R. Weiskopf, and R. E. Jackson. *J. Appl. Physiol.: Respir., Environ. Exercise Physiol.* 49:311–318, 1980. Work capacity during 3-wk sojourn at 4,3000 m: Effects of relative polysythemia.

90. Hoyt, R. W., S. C. Wood, and J. Garcia. *Am. Physiol. Soc., Abstr.* 1979.

91. Hutchison, V. H., H. B. Haines, and G. Engebretson. *Respir. Physiol.* 27:115–129, 1976. Aquatic life at high altitude: Respiratory adaptations in the lake Titicaca frog, *Telmatobius culeus*.

92. Hutchison, V. H. and E. S. Hazard, III. *Comp. Biochem. Physiol. A* 79A:533–538, 1984. Erythrocytic organic phosphates: Diel and seasonal cycles in the frog, *Rana berlandieri*.

93. Ingermann, R. L., M. K. Stock, J. Metcalfe, and T.-B. Shih. *Respir. Physiol.* 51:141–152, 1983. Effect of ambient oxygen on organic phosphate concentrations in erythrocytes of the chick embryo.

94. Ingermann, R. L. and R. C. Terwilliger. In *Respiration and Metabolism of Embryonic Vertebrates* (R. Seymour, ed.), pp. 1–16. Junk, Dordrecht, The Netherlands, 1983. Facilitation of maternal-fetal oxygen transfer in fishes: Anatomical and molecular specializations.

95. Isaacks, R. E. and D. R. Harkness. *Am. Zool.* 20:115–130, 1980. Erythrocyte organic phosphates and hemoglobin function in birds, reptiles, and fishes.

96. Jennings, M. L. *Annu. Rev. Physiol.* 47:519–533, 1985. Kinetics and mechanism of anion transport in red blood cells.

97. Johansen, K. In *Fish Physiology* (W. S. Hoar and D. J. Randall, eds.), Vol. 4. Academic Press, New York, 1970. Air-breathing in fishes.

98. Johansen, K. and C. Lenfant. In *Oxygen Affinity of Hemoglobin and Red Cell Acid-Base Status* (M. Rorth and P. Astrup, eds.), pp. 750–780. Munksgaard, Copenhagen, 1972. A comparative approach to the adaptibility of O_2-Hb affinity.

99. Johansen, K., C. P. Magnum, and R. E. Weber. *Can. J. Zool.* 56:891–897, 1978. Reduced blood O_2 affinity associated with air breathing in osteoglossid fishes.

100. Jokumsen, A. and R. E. Weber. *J. Exp. Zool.* 221:389–394, 1982. Haemocyanin oxygen affinity in hermit crab blood is temperature independent.

101. Jones, J. E. *Comparative Physiology of Respiration*. Arnold, London, 1972.

102. Keyes, J. L. *Fluid, Electrolyte and Acid-Base Regulation*. Wadsworth, Belmont, CA.

103. Klippenstein, G. L. *Am. Zool.* 20:39–51, 1980. Structural aspects of hemerythrin and myohemerythrin.

104. Klocke, R. A. *J. Appl. Physiol.* 44:882–888, 1978. Catalysis of CO_2 reactions by lung carbonic anhydrase.

105. Klotz, I. M. and D. M. Kurtz, Jr. *Acc. Chem. Res.* 17:16–22, 1984. Binuclear oxygen carriers: Hemerythrin.

106. Kobayashi, M. *Comp. Biochem. Physiol. A* 72A(3):599–602, 1982. Influence of body size on haemoglobin concentration and resistance to oxygen deficiency in *Daphnia magna*.

107. Kurtz, D. M., Jr. In *Invertebrate Oxygen Carriers* (B. Linzen, ed.), pp. 9–21. Springer-Verlag, Berlin, 1986. Structure, function and oxidation levels of hemerythrin.

108. Lamy, J. N., J. Lamy, P. Billiald, P.-Y. Sizaret, J. C. Taveau, and N. Boisset. In *Invertebrate Oxygen Carriers* (B. Linzen, ed.), pp. 185–201. Springer-Verlag, Berlin, 1986. Mapping of antigenic determinants in *Androctonus australis* hemocyanin: Preliminary results.

109. Lamy, J., J. Lamy, J. Weill, J. Bonaventura, C. Bonaventura, and M. Brenowitz. *Arch. Biochem. Biophys.* 196:324–339, 1979. Immunological correlates between

the multiple subunits of *Limulus polyphemus* and *Tachypleus tridentatus*.

110. Lechner, A. *J. Appl. Physiol.* 41(2):168–173, 1976. Respiratory adaptations in burrowing pocket gophers from sea level and high altitude.

111. Lechner, A., V. Salvato, and N. Banchero. *Comp. Biochem. Physiol. A* 70A:321–327, 1981. The hematological response to hypoxia in growing guinea pigs is blunted during concomitant coldstress.

112. Linzen, B., W. Schartau, and H.-J. Schneider. In *Respiratory Pigments in Animals* (J. Lamy, J.-P. Truchot, and R. Gilles, eds.), pp. 59–71. Springer-Verlag, Berlin, 1985. Primary structure of arthropod hemocyanins.

113. Little, C. *The Colonisation of Land*. Cambridge Univ. Press, London, 1983.

114. Lutz, P. L. *Am. Zool.* 20:187–198, 1980. On the oxygen affinity of bird blood.

115. Lutz, P. L., I. S. Longmuir, and K. Schmidt-Nielsen. *Respir. Physiol.* 20:325–330, 1974. Oxygen affinity of bird blood.

116. Lykkeboe, G. and R. E. Webber. *J. Comp. Physiol.* 128:117–125, 1978. Changes in the respiratory properties of the blood in the carp, *Cyprinus carpio*, induced by diurnal variation in the ambient oxygen tension.

117. Lykkeboe, G. and K. Johansen. *Respir. Physiol.* 35:119–122, 1978. An O_2-Hb 'paradox' in frog blood? (*n*-values exceeding 4.0).

118. Lykkeboe, G., H. Schougaard, and K. Johansen. *Respir. Physiol.* 29:315–325, 1977. Training and exercise change respiratory properties of blood in race horses.

119. MacMahon, J. A. and A. Hamer. *Comp. Biochem. Physiol. A* 51A:59–69, 1975. Effects of temperature and photoperiod onoxygention and other blood parameters of the sidewinder (*Crotalus cerastes*): Adaptive significance.

120. Malcolm, S., P. Barton, C. Murphy, and M. A. Ferguson-Smith. *Ann. Hum. Genet.* 45:135–141, 1981. Chrosomal localization of a single copy gene by in situ hybridization—human beta globin genes on the short arm of chromosome 11.

121. Mangum, C. P. *Am. Zool.* 20:19–38, 1980. Respiratory function of the hemocyanins.

122. Mangum, C. P. *Mar. Biol. Lett.* 4:139–149, 1983. On the distribution of lactate sensitivity among the hemocyanins.

123. Mangum, C. P. In *The Biology of Crustacea* (L. Mantel, ed.), Vol. 5, pp. 373–429. Academic Press, New York, 1983. Oxygen transport in the blood.

124. Mangum, C. P. *Life Chem Rep.* 4:335–352, 1983. Adaptability and inadaptability among HcO_2 transport systems: An apparent paradox.

125. Mangum, C. P. and L. E. Burnett, Jr. *Biol. Bull. (Woods Hole, Mass.)* 171:248–263, 1986. The CO_2 sensitivity of the hemocyanins and its relationship to Cl sensitivity.

126. Mangum, C. P. and M. Kondon. *Comp. Biochem. Physiol. A* 50A:777–785, 1975. The role of coelomic hemerythrin in the sipunculid worm *Phascolopsis gouldi*.

127. Mangum, C. P. and G. Lykkeboe. *J. Exp. Zool.* 207:417–430, 1979. The influence of inorganic ions and pH on the oxygenation properties of the blood in the gastropod mollusc *Busycon canaliculatum*.

128. Mangum, C. P., B. R. McMahon, P. L. deFur, and M. G. Wheatly. *J. Crustacean Biol.* 5(2):188–206, 1985. Gas exchange and acid-base balance and the oxygen supply to the tissues during a moult of the blue crab *Callinectes sapidus*.

129. Mangum, C. P. and J. S. Rainer. *Biol. Bull. (Woods Hole, Mass.)* 174:77–82, 1988. The relationship between subunit composition and O_2 binding of blue crab hemocoyanin.

130. Manwell, C. *Comp. Biochem. Physiol.* 1:277–285, 1960. Histological specificity of respiratory pigments. II. Oxygen transfer systems involving hemerythrins in sipunculid worms of different ecologies.

131. Maren, T. H. *Physiol. Rev.* 47:595–781, 1967. Carbonic anhydrase: chemistry, physiology and inhibition.

132. Markl, J., H. Decker, B. Linzen, W. G. Schutter, and E. F. S. Bruggen. *Hoppe-*

Seyler's Z. Physiol. Chem. 363:73–87, 1982. Hemocyanins in spiders. XV. The role of the individual subunits in the assembly of *Eurypelma hemocyanin.*

133. Markl, J., W. Stocker, R. Runzler, and E. Precht. In *Invertebrate Oxygen Carriers* (B. Linzen, ed.), pp. 281–292. Springer-Verlag, Berlin, 1986. Immunological correspondences between the hemocyanin subunits of 86 arthropods: Evolution of a multigene protein family.

134. Markl, J., W. Strych, W. Schartau, H.-J. Schneider, P. Schoberl, and B. Linzen. *Hoppe-Selyer's Z. Physiol. Chem.* 360:639–650, 1979. Hemocyanins in spiders. VI. Comparison of the polypeptide chains of *Eurypelma californicum* hemocyanin.

135. Mauro, N. A. and C. P. Mangum. *J. Exp. Zool.* 291:179–188, 1982. The role of the blood in the temperature dependence of oxidative metabolism in decapod crustaceans. I. Intraspecific responses to seasonal differences in temperature.

136. McMahon, B. R. and J. L. Wilkens. In *The Biology of Crustacea* (L. H. Mantel, ed.), Vol. 5, pp. 289–372. Academic Press, New York, 1983. Ventilation, perfusion and oxygen uptake.

137. McMahon, B. R. In *Respiratory Pigments in Animals* (J. Lamy, J.-P. Truchot, and R. Gilles, eds.), pp. 35–58. Springer-Verlag, Berlin, 1985. Functions and functioning of crustacean hemocyanin.

138. McMahon, B. R. In *Invertebrate Oxygen Carriers* (B. Linzen, ed.), pp. 299–319. Springer-Verlag, Berlin, 1986. Oxygen binding by hemocyanin: Compensation during activity and environmental change.

139. McMahon, B. R. *Am. Zool.* 28:39 53, 1987. Physiological responses to oxygen depletion in intertidal animals.

140. McMahon, B. R. and W. W. Burggren. In *Biology of the Land Crabs* (W. W. Burggren and B. R. McMahon, eds.), pp. 249–297. Cambridge Univ. Press, New York, 1988. Respiration.

141. McMahon, B. R., L. E. Burnett, and P. L. deFur. *J. Comp. Physiol. B* 154B:371–383, 1984. Carbon dioxide excretion and carbonic anhydrase function in the Red Rock Crab *Cancer productus.*

142. McMahon, B. R. and J. L. Wilkens. In *The Biology of Crustacea* (L. H. Mantel, ed.), Vol. 5, Chapter 6. Academic Press, New York, 1983. Ventilation, perfusion and oxygen consumption.

143. McMahon, B. R. and J. L. Wilkens. *Can. J. Zool.* 50:165–170, 1972. Simultaneous apnoea and bradycardia in the lobster *Homarus americanus.*

144. Mied, P. and D. A. Powers. *J. Biol. Chem.* 253:3521–3528, 1977. Hemoglobins of the killifish *Fundulus heteroclitus*: Separation, characterization and a model for the subunit composition.

145. Morris, S. and C. R. Bridges. *Physiol. Zool.* 59:606–615, 1986. Oxygen binding by the hemocyanin of the terrestrial hermit crab *Coenobita clypeatus* (Herbst)—the effect of physiological parameters *in vitro.*

146. Morris, S. and C. R. Bridges. In *Invertebrate Oxygen Carriers* (B. Linzen, ed.), pp. 353–356. Springer-Verlag, Berlin, 1986. Novel non-lactate cofactors of haemocyanin oxygen affinity in crustaceans.

147. Morris, S., C. R. Bridges, and M. K. Grieshaber. *J. Exp. Biol.* 133:339–352, 1987. The regulation of haemocyanin oxygen affinity during emersion of the crayfish *Austropotamobius pallipes*. III. The dependency of Ca^{2+}-haemocyanin binding on the concentration of L-lactate.

148. Morris, S., A. C. Taylor, C. R. Bridges, and M. K. Grieshaber. *J. Exp. Zool.* 233:175–186, 1985. Respiratory properties of the haemolymph of the intertidal prawn *Palaemon elegans* (Rathke).

149. Morse, P. M., E. Meyerhoffer, J. J. Otto, and A. Kuzirian. *Science* 231:1302–1302, 1986. Hemocyanin respiratory pigment in bivalve mollusks.

150. Nikinmaa, M. *J. Comp. Physiol.* 152:62–67, 1983. Adrenergic regulation of Hb-O_2 affinity in rainbow trout red cells.

151. Peters, R. H. *The Ecological Implication of Body Size.* Cambridge Univ. Press, New York, 1983.

152. Pettigrew, D. W., P. H. Romeo, A. Sapis, J. Thillet, M. L. Smith, B. W. Turner, and

G. K. Ackers. *Proc. Natl. Acad. Sci. U.S.A.* 79:1849–1853, 1982. Probing the energetics of proteins through structural perturbation: Sites of regulatory energy in human hemoglobin.

153. Pinder, A. and W. W. Burggren. *J. Exp. Biol.* 105:205–213, 1983. Respiration during chronic hypoxia and hyperoxia in larval and adult bullfrogs (*Rana catesbeiana*). II. Changes in respiratory properties of whole blood.

154. Portner, H. O., N. Heisler, and M. K. Grieshaber. *Respir. Physiol.* 59:361–377, 1985. Oxygen consumption and mode of energy production in the intertidal worm *Sipunculus nudus* L.: Definition and characterization of the critical PO$_2$ for an oxyconformer.

155. Pough, F. H. *Am. Zool.* 20:173–185, 1980. Blood oxygen transport and delivery in reptiles.

156. Pough, F. H. *Comp. Biochem. Physiol.* 31:885–901, 1969. Environmental adaptations in the blood of lizards.

157. Powers, D. A. *Ann. N.Y. Acad. Sci.* 241:472–490, 1974. Structure-function and molecular ecology of fish hemoglobins.

158. Powers, D. A. *Am. Zool.* 20:139–162, 1980. Molecular ecology of teleost fish hemoglobins: Strategies for adapting to changing environments.

159. Powers, D. A. In *Respiratory Pigments in Animals* (J. Lamy, J.-P. Truchot, and R. Gilles, eds.), pp. 97–124. Springer-Verlag, Berlin, 1985. Molecular and cellular adaptations of fish hemoglobin-oxygen affinity to environmental changes.

160. Powers, D. A. and A. B. Edmundson. *J. Biol. Chem.* 247:6686–6693, 1972. Multiple hemoglobins of catostomid fish: I. Isolation and characterization of isohemoglobins from *Catostomus clarkii* hemoglobins.

161. Powers, D. A., H. J. Fyhn, U. F. H. Fyhn, J. P. Martin, R. L. Garlick, and S. C. Wood. *Comp. Biochem. Physiol. A* 62A: 67–85, 1979. A comparative study of the oxygen equilibria of blood from 40 genera of Amazonian fishes.

162. Powers, D. A., J. P. Martin, R. L. Garlick,

H. J. Fyhn, and V. E. H. Fyhn. *Comp. Biochem. Physiol. A* 62A:87–94, 1979. The effect of temperature on the oxygen equilibrium of fish hemoglobins in relation to environmental thermal variability.

163. Powers, D. A. *Science* 177:360–362, 1972. Hemoglobin adaptation for fast and slow water habitats in sympatric catostomid fishes.

164. Prosser, C. L. *Comparative Animal Physiology*, 3rd ed. Saunders, Philadelphia, PA, 1976.

165. Prothero, J. *Comp. Biochem. Physiol. A* 67A:649–657, 1980. Scaling of blood parameters in mammals.

166. Randall, D. J., W. W. Burggren, A. P. Farrell, and M. S. Haswell. *The Evolution of Air Breathing in Vertebrates*. Cambridge Univ. Press, New York, 1981.

167. Randall, D. J. and C. Daxboeck. In *Fish Physiology* (W. Hoar and D. Randall, eds.), Vol. 10A. Academic Press, Orlando, FL, 1984. Oxygen and carbon dioxide transfer across fish gills.

168. Riggs, A. In *Fish Physiology* (W. S. Hoar and D. J. Randall, eds.), Vol. 4. Academic Press, New York, 1970. Properties of fish hemoglobins.

169. Roberts, J. L. and D. M. Rowell. *Can. J. Zool.* 66:182–190, 1988. Periodic respiration of gill-breathing fishes.

170. Rosen, S. and N. J. Friedley. *Histochemie* 36:1–4, 1973. Carbonic anhydrase activity in Rana pipiens skin: Biochemical and histochemical analysis.

171. Rossi-Fanelli, A. and E. Antonini. *Nature (London)* 188:895–896, 1960. Oxygen equilibrium of hemoglobin from *Thunnus thynnus*.

172. Savel, A., J. Markl, and B. Linzen. In *Invertebrate Oxygen Carriers* (B. Linzen, ed.), pp. 399–402. Springer-Verlag, Berlin, 1986. The spatial range of allosteric interaction in a 24-meric arthropod hemocyanin.

173. Schmidt-Nielsen, K. *Scaling: Why Animal Body Size is Important*. Cambridge Univ. Press, New York, 1984.

174. Schneider, H. J., R. Drexel, S. Sigmund, B. Linzen, C. Geilens, R. Luntser, G. Preaus, F. Lottspeich, and A. Henschen.

Union Biol. Sci. Ser. A 155, 1984. Partial amino acid sequence of *Helix pomatia* hemocyanin: Tyrosinase and hemocyanins have a common ancestor.

175. Shaanan, B. *J. Mol. Biol.* 171:31–59, 1983. The structure of human oxyhaemoglobin at 2.1 A resolution.

176. Shelton, G. and R. G. Boutilier. *J. Exp. Biol.* 100:245–273, 1982. Apnoea in amphibians and reptiles.

177. Smith, R., R. Kruszyna, and L. Ou. *Pfluegers Arch.* 380:65–70, 1979. Hemoglobinemia in mice exposed to high altitude.

178. Snyder, L. R. G. *Respir. Physiol.* 48:107–123, 1982. 2,3-diphosphoglycerate in high- and low-altitude populations of the deer mouse.

179. Snyder, L. R. G., L. S. Born, and A. Lechner. *Respir. Physiol.* 48:89–105, 1982. Blood oxygen affinity in high- and low-altitude populations of the deer mouse.

180. Soeter, N. M., J. J. Beintema, P. A. Jekel, H. J. Bak, J. J. Vereijken, and B. Neuteboom. In *Invertebrate Oxygen Carriers* (B. Linzen, ed.), pp. 153–163. Springer-Verlag, Berlin, 1986. Subunits a, b, and c of *Panulirus interruptus* hemocyanin and evolution of arthropod hemocyanins.

181. Stewart, P. A. *How to Understand Acid-Base.* Elsevier, New York, 1981.

182. Sugita, Y. and A. Chanutin. *Proc. Soc. Exp. Biol. Med.* 112:72–75, 1963. Electrophoretic studies of red cell hemolysates supplemented with phosphorylated carbohydrate intermediates.

183. Terwilliger, N. B., R. C. Terwilliger, and R. Graham. In *Invertebrate Oxygen Carriers* (B. Linzen, ed.), pp. 333–335. Springer-Verlag, Berlin, 1986. Crab hemocyanin function changes during development.

184. Terwilliger, R. C. *Comp. Biochem. Physiol.* B 61B:463–469, 1978. The respiratory pigment of the serpulid polychaete *Serpula vermicularis*. Structure of its hemoglobin and chlorocruorin.

185. Terwilliger, R. C. *Am. Zool.* 20:53–67, 1980. Structures of invertebrate hemoglobins.

186. Thomson, J. M., J. A. Dempsey, L. W. Chosy, N. T. Shahidiand, and W. G. Reddan. *J. Appl. Physiol.* 37:658–664, 1974.

Oxygen transport and oxyhemoglobin dissociation during prolonged muscular work.

187. Toews, D., R. Boutilier, L. Todd, and N. Fuller. *Comp. Biochem. Physiol. A* 59A:211–213, 1978. Carbonic anhydrase in the Amphibia.

188. Toews, D. and D. MacIntyre. *Nature (London)* 266:464–465, 1977. Blood respiratory properties of a viviparous amphibian.

189. Truchot, J.-P. *J. Comp. Physiol.* 112:283–293. 1976. Carbon dioxide combining properties of the blood of the shore crab *Carcinus maenas* (L.).

190. Truchot, J.-P. *Respir. Physiol.* 24:173–189, 1975. Factors controlling the *in vitro* and *in vivo* oxygen affinity of the hemocyanin in the crab *Carcinus maenas* (L.).

191. Truchot, J.-P. *J. Exp. Zool.* 214:205–208, 1980. Lactate increases the oxygen affinity of crab hemocyanin.

192. Truchot, J.-P. and A. Duhamel-Jouve. *Respir. Physiol.* 39:241–254, 1980. Oxygen and carbon dioxide in the marine intertidal environment: Diurnal and tidal changes in rockpools.

193. van Holde, K. E. and K. I. Miller. *Q. Rev. Biophys.* 15:1–129, 1982. Haemocyanins.

194. Vinogradov, S. M. *Comp. Biochem. Physiol. B* 82B:1–15, 1985. The structure of invertebrate extracellular hemoglobins (erythrocruorins and chlorocruorins).

195. Volbeda, A. and W. G. J. Hol. In *Invertebrate Oxygen Carriers* (B. Linzen, ed.), pp. 136–147. Springer-Verlag, Berlin, 1986. Three-dimensional structure of haemocyanin from the spiny lobster, *Panulirus interruptus*, at 3.2 A resolution.

196. Wahlqvist, I. and S. Nilsson. *J. Comp. Physiol.* 137:145–150, 1980. Adrenergic control of the cardiovascular system of the Atlantic cod, *Gadus morhua*, during stress.

197. Weber, R. E. Ph.D. Thesis, University of Leiden, The Hague, 1965. On the haemoglobin and respiration of Chironomus larvae, with special reference to *Chironomus plumosus* L.

198. Weber, R. E. In *Physiology of Annelids* (P. J. Mill, ed.), Chapter 10, pp. 393–446.

Academic Press, London, 1978. Respiratory pigments.

199. Weber, R. E. *Am. Zool.* 20:79–101, 1980. Functions of invertebrate hemoglobins with special reference to adaptations to environmental hypoxia.

200. Weber, R. E. In *Exogenous and Endogenous Influences on Metabolic and Neural Control,* (A. D. F. Addink and N. Spronk, eds.), pp. 87–102. Pergamon, Oxford, 1982. Intraspecific adaptation of hemoglobin function in fish to oxygen availability.

201. Weber, R. E., B. Sullivan, J. Bonaventura, and C. Bonaventura. *Biochim. Biophys. Acta* 434:18–31, 1976. The hemoglobin system of the primitive fish *Amia calva*: Isolation and functional characterization of the individual hemoglobin components.

202. Weber, R. E. and J. Baldwin. *Mol. Physiol.* 7:93–106, 1985. Blood and erythrocruorin of the giant earthworm, *Megascolides australis*: Respiratory characteristics and evidence for CO_2 facilitation of O_2 binding.

203. Weibel, E. R. *The Pathway for Oxygen.* Harvard Univ. Press, Cambridge, MA, 1984.

204. Weinstein, Y., R. A. Ackerman, and F. N. White. *Respir. Physiol.* 63:53–63, 1986. Influence of temperature on the CO_2 dissociation curve of the turtle *Pseudemys scripta*.

205. Weinstein, Y., M. Bernstein, P. Bickler, D. Gonzales, F. Samaniego, and M. Escobedo. *Am. J. Physiol.* 249:R765–R775, 1985. Blood respiratory properties in pigeons at high altitudes: Effects of acclimation.

206. Wheatly, M. G., B. R. McMahon, W. W. Burggren, and A. W. Pinder. *J. Exp. Biol.* 125:225–244, 1986. Haemolymph acid-base, electrolyte and blood gas status during sustained voluntary activity in the land hermit crab (*Coenobita compressus*—H. Milne-Edwards).

207. Wilkes, P. and B. R. McMahon. *J. Exp. Biol.* 98:139–150, 1982. Effect of maintained hypoxic exposure on the crayfish *Orconectes rusticus*. II. Modulation of haemocyanin oxygen affinity.

208. Winslow, R. M., C. C. Monge, N. J. Statham, C. G. Gibson, S. Charache, J. Whittembury, O. Moran and R. L. Berger. *J. Appl. Physiol.: Respir., Environ. Exercise Physiol.* 51:1411–1416, 1981. Variability of oxygen affinity of blood: human subjects native to high altitude.

209. Wood, S. C. *Am. Zool.* 20:163–172, 1980. Adaptation of red blood cell function to hypoxia and temperature in ectothermic vertebrates.

210. Wood, S. C., R. W. Hoyt, and W. W. Burggren. *Mol. Physiol.* 2:263–272, 1982. Control of hemoglobin function in the salamander, *Ambystoma tigrinum*.

211. Wood, S. C. and K. Johansen. *Neth. J. Sea Res.* 7:328–338. Organic phosphate metabolism in nucleated red cells: Influence of hypoxia on eel HbO_2 affinity.

212. Wood, S. C. and K. Johansen. *J. Comp. Physiol.* 89:145–158, 1974. Respiratory adaptations to diving in the nile monitor lizard, *Varanus niloticus*.

213. Wood, S. C., K. Johansen, and R. N. Gatz. *Am. J. Physiol.* 233(3):R89–R93, 1977. Pulmonary blood flow, ventilation/perfusion ratio, and oxygen transport in a varanid lizard.

214. Wood, S. C. and C. Lenfant. In *Evolution of Respiratory Processes* (S. C. Wood and C. Lenfant, eds.), pp. 193–224. Dekker, New York, 1979. Oxygen transport and oxygen delivery.

215. Wood, S. C., R. Weber, and D. Powers. *Comp. Biochem. Physiol. A* 62A:163–167, 1979. Respiratory properties of blood and hemoglobin solutions from the piranha.

216. Wyman, J., Jr. *Adv. Protein Chem.* 4:410–531, 1948. Hemeproteins.

217. Wyman, J., S. J. Gill, L. Noll, B. Giardina, A. Colosimo, and M. Brunori. *J. Mol. Biol.* 109:195–205, 1977. The balance sheet of a hemoglobin: Thermodynamics of CO binding by hemoglobin trout.

218. Yamaguchi, K., Y. Kochiyama, and F. Matsuura. *Bull. Jpn. Soc. Sci. Fish.* 28:192–198, 1962. Studies on multiple hemoglobins of eel. II. Oxygen dissociation curves and relative amounts of components F and S.

Chapter
11 | *Circulation of Body Fluids*

Anthony P. Farrell

In addition to diffusion, animals require mechanisms to transport essential gases, nutrients, and the products of metabolism. Six major kinds of circulatory mechanisms are recognized in animals (Fig. 1):

1. *Intracellular Movement.* Protoplasmic streaming, which may follow a definite course, supplements diffusion in protozoans and most, if not all, metazoan cells.

2. *Movement of External Medium.* Ciliary, flagellar, or muscular activity may propel external medium through definite body channels in, for example, sponges and coelentrates.

3. *Movement of Body Fluids by Somatic Muscles.* Body movements circulate pseudocoelemic fluid (in nematodes, bryozoans, and rotifers) and coelomic fluid (in echinoderms, annelids, sipunculids, and chordates), and assist blood flow in the tissue sinuses and veins of open and closed circulations.

4. *Movement of Hemolymph in an Open Circulatory System.* Most arthropods, many molluscs, and ascidians have an open circulatory system. Hemolymph is pumped by a heart at low pressures through vessels to tissue spaces (hemocoel) and sinuses. Hemolymph volume is high and circulation may be slow.

5. *Movement of Blood in a Closed Circulatory System.* Blood is pumped by one or more hearts within a closed vascular system in most annelids, phoronids, nemerteans, cephalopod molluscs, holothurian echinoderms, and vertebrates. Blood pressures and velocities are higher and blood volume is lower in closed compared with open circulatory systems. A network of thin-walled capillaries and sinuses ensure intimate association between the blood and the tissues.

6. *Movement of Lymph in Vertebrates.* Lymphatic flow is aided by interstitial fluid pressure, contraction of lymphatic vessels and, in some animals, lymph hearts.

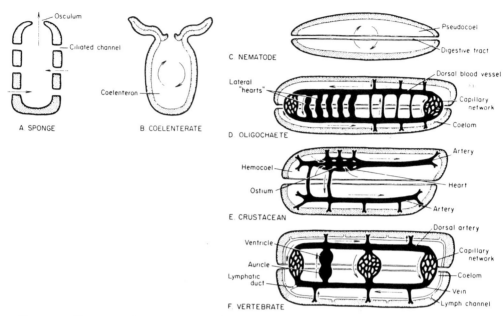

Figure 1. Diagram of systems of internal transport. (*A*) sponge type with flagellated channels; (*B*) coelenterate type; (*C*) nematode type with pseudocoel primary body cavity; (*D*) oligochaete type with coelom and closed blood system; (*E*) crustacean–molluscan type with blood system open to hemocoel, which is derived from primary cavity or blastocoel; (*F*) vertebrate type with closed vascular system, lymphatic channels and coelom.

Classification of Hearts

Morphologically, hearts can be divided into four types.

1. *Chambered hearts* can be single or double pumps with one to four atria and one or two ventricles. They are found in molluscs and all vertebrates. The heart beat is initiated in modified cardiac muscle cells that comprise *a myogenic pacemaker*.

2. *Tubular hearts* consist of contractile tubes. The systemic heart of most arthropods is tubular and valved, but in tunicates it is not valved and pumps blood in either direction. The heart beat may be either *neurogenic* (initiated in nerve cells, e.g., arthropods) or *myogenic* (e.g., tunicates).

3. *Ampullar hearts* usually serve as accessory hearts to boost fluid through peripheral channels. Lymph hearts in amphibians and reptiles, and caudal hearts of fish are valved and composed of striated, anastomosing muscle cells (214). The branchial hearts of cephalopods consist of spongy musculo–epithelial tissue. Accessory hearts are also common in insects and some decapod crustaceans.

4. *Pulsating vessels* propel blood with peristaltic waves in annelids and some holothurians. *Branchiostoma* has many pulsating vessels and the lack of a chambered heart makes it unique among chordates. While the distinction between tubular hearts and pulsating vessels is not always sharp in invertebrates, the pulsating veins in bat wings (51) and the con-

tractile conus arteriosus in elasmobranchs are conspicuous among vertebrates.

Fluid Compartments

Water content of an animal varies from 60 to 80% (Chapter 2). All animals, including unicellular ones, have multiple fluid compartments. In multicellular animals, fluid compartments are conveniently divided into extracellular and intracellular spaces.

In vertebrates the *extracellular space* (ECS) is 18–25% of body weight and is about one half the size of the *intracellular space* (ICS). In contrast, the ECS is larger than the ICS in some invertebrates (Table 1). Extracellular space is estimated from the dilution of marker substances that do not readily penetrate cells and are not rapidly excreted from the animal, for example, thiocyanate, mannitol, and sucrose. Different values for the ECS may be obtained with different markers. Thus, there is no single marker system that is ideal for all animals or for all organs. Intracellular space can be calculated from the difference between ECS and total body water (as determined by dessication or by using D_2O and H_3O as markers).

In animals with a closed circulation, the ECS is comprised of *plasma, interstitial fluid, lymph,* and *coelomic fluid.* The *blood volume* (plasma plus blood cell volume) is contained in the vascular network and interstitial fluid is formed from plasma filtered across the capillary wall. *Plasma volume* is estimated from the dilution of marker substances that do not pass into the interstitial space, for example, Evans blue dye and ^{131}I-labeled albumin. *Blood cell volume* is measured from packed cell volume (*hematocrit*) or with radiolabeled blood cells. Blood volume in mammals and birds is normally regulated at 7–10% of body weight and is generally lower in larger animals (Table 1). In reptiles and amphibia, blood volume is 3.5–8%, and in teleosts and elasmobranchs, estimates range from 1.2 to 9.8% (83, 244, 245). Blood volume may be larger in cyclostomes (8.5–16.9%) and those molluscs with a closed circulation (5.9–14.7%). An advantage of a small blood volume is a short circulation time. A large blood volume is advantageous for O_2 storage and for use as a hydrostatic skeleton. Increases in hematocrit occur during adaptation to altitude in most birds and mammals, and decreases in plasma volume occur during hibernation (62) (mammals), aestivation [amphibia (110) and lungfish (58)], and exercise (certain mammals and fish).

In animals with an open circulation, the distinction between blood, lymph, and interstitial fluid is lost, and the term *hemolymph* is adopted; hemolymph volume is essentially the same as total ECS. In animals with a heavy exoskeleton, ECS is better expressed in terms of the volume of the body rather than mass. The ECS for several crustaceans is 25–30% of body volume (Table 1), but on a mass basis the ECS is 30% just after moulting and 10% when the shell hardens. Many adult insects have a smaller hemolymph volume than their larval form. In the larvae of the hawk moth *Celerio* and the blowfly *Sarcophaga*, the hemolymph makes up 18.6% and 35 to 42% of body weight, respectively, compared to 7.2–7.8% and 24–33% in the adult. Hemolymph volume in gastropods and pelecypods has the greatest range and value, ranging from 40% in the land snail *Achatina* to 90% in *Chiton* (162).

Coelomic fluid is extensive in echinoderms, annelids, and insects, but it is restricted to the lumen of the gonads and kidneys in some molluscs, to the kidney in crustaceans, and to the peritoneal, pericardial, and pleural cavities in vertebrates.

TABLE 1. Cardiovascular Parameters from Selected Animals[a]

Animal	Reference	State	Plasma Volume (%body wt.)	Blood Volume (%body wt.)	Extracellular Space (%body wt.)	Pressure (mmHg) (sys/dias or mean)		Resting Heart Rate (beats/min)	Stroke Volume (mL)	Cardiac Output (mL/min)	Cardiac Output/kg (mL/kg · min)
						Systemic	Pulmonary or Branchial				
Mammals											
Man	(95)	Male(altitude)	4.8	7.8(10.0)	17.5	120/75 RA	25/10(28)	75	77	5500–6000	80–90
		exercise				120		180	130	25000	360
Horse		Rest(canter)	4.2–6.2	7.6–10.3		171/103		48(145)	1350(1450)	65000(210000)	150(505)
Cow			3.9	5.2–5.7		160/110		45–60	872	45800	113
Dog			5.0	8.6		112/56		90–100	9–18	2000–3000	209
Rat	(184)	37°C	4.0	8.5		130/91		360	0.18	60	60
		20°C						107	0.17	20	
Mouse						113/81		498			
Bat	(146)	37°C(5°C)	6.5	13.0		39–105		250–450(20)			
Marmota	(277)	37°C(5–8°C)	5.1	10.0		140(50)		140(15)			60(8)
Ground squirrel	(187)	37°C(7°C)				139/99(60/30)		276(3–10)		140(2)	313(4.6)
Elephant seal	(49)	Surface(diving)		20.7		120/90		60(4)	0.5(0.6–0.2)		
Phoca	(172)	Surface(diving)		15.9							5.2(0.6)
Tursiops	(235)		3.9	7.1		150/121 CA	55/35–76/50	84–140	109	12200	47–105
Birds											
Pigeon	(18,36)	Rest(exercise)	4.4	9.2		135/105		100(600)	(unchanged)		128–270
Chicken				9(7)		149/43		178–460			
Duck	(239)	Surface(diving)	6.5	10.2		100–175		175	1.1	400	200–400
Reptiles											
Turtle	(221,247)	Surface(apnoea)	7.4	9.1		25/10	24/19	23(11)	3.7	148	57.4(26.6)
	(106)	diving 3°C				10/2		0.4			
Crocodile	(115,247)		3.5	15.4		45/30	18/9	10–60			
Iguana	(254)		4.2	16.8		48/37	30/13	38.2	1.5	57	58
Varanus	(25,104)	30°C (apnoea)				83/58			1.0		35.3(26.9)
Amphibians											
Rana pipiens	(159,246)	(apnoea)	8.0	9.5	22.0	32/21(23/16)	31/16(26/15)	37.5–60.0	0.035		20–30
Xenopus laevis		20°C (apnoea)				32/21(37/22)	31/11(37/22)	43	0.056	2.4(2.6)	24.0(26.4)
Necturus	(246)		3.5	3.5	24.0						
Cryptobranchus	(246)				22.0						
Amphiuma	(127)					30/25			0.67		30

Species	Ref.	Temp.			VENT/DA	VA				
Fishes										
Ictalurus			1.2							
Gadus morhua	(6)	10°C	1.8		30/23 DA; 25 DA	40/30 VA; 38 VA	21.6; 43.0	0.75	28.8	11.3; 17.3
Salmo gairdneri	(136,193)	10°C(exercise)	3.4	6.4	32/27(41/30); 31/24 DA; 30/25 DA	45/33(72/48); 45/27 VA; 39/30 VA	37.8(51.4); 78.6; 29.4	0.46(1.03); 0.92; 1.60	17.6(52.6); 73.4; 47.0	17.6(52.6); 36.7; 11.2
Ophiodon elongatus	(273)	15°C			23(24) DA		30.0(46.0)	0.55(0.90)	21.1(59.7)	15.5(41.8)
Pseudopleuronectes	(65)	10°C			24.7 DA	28.6 VA	31.0	0.70	21.6	10.8
Hemitripterus	(44)	5°C(15°C)					40–50			25.0
Squalus	(68)	7°C	5.5	6.8; 21.2	17/16 DA	30/24 VA	16–50			21.0
Raja	(208)		5.9	7.2		16/14 VA	30–40			
Myxine	(46,126)				5	17/7 VA				
Lampetra	(21)		5.5	8.5						
Molluscs										
Eledone	(203)		14.7	21.8						
Sepia	(263)		8.5	14.5						
Octopus vulgaris	(263)	22°C	5.8	28	19–38/7.7–12	3–7/1–2	45.0	1.0	42.0	42.0
Logilo peali	(263)	22°C			56/21		102.0	0.13	13.0	130.0
Nautilus pompilus		17°C			27/12		12.0	0.21	2.5	5.0
Busycon		20°C					14.1	0.18	2.5	25.9
Haliotis cracherodii	(138)	15°C					20–28	1.0–2.1	25–40	98–153
Placopecten	(137)			57			7.0	0.15	1.1	
Helix	(137)			40	18–24/4.6 VENT		22.4	0.01–0.07	0.1–1.4	6.7
Halistus corrugata	(137)				6.0		16–20	0.16–0.18	2.1–3.6	
Littorina littorea	(137)			65	3.1		26	0.0025	0.065	
Tresus nuttalli	(101)				6.0/0		10–12	2.9–3.2	31	90
Arthropods										
Cancer productus	(170)	16°C	51.6		10.4/6 VENT		94–150	1.9–2.8	271–723	103–275
Carcinus	(203)	15°C	32.6		9.2/6.3		92	1.3	5.9	118
Homarus americanus	(170)	16–20°C	17.0		14/1.2 VENT		90	0.11–0.33	10–30	22–67
Panulirus	(199)	15°C	33.0		36/16 VENT		64	1.4	90	128
Locusta	(11,153)		18.1		6.3/2.3 DA		80	0.00019	0.015	
Annelids										
Sipunculus			50.0							
Dendrostomum			47.0							
Glossoscolex	(129)				18/10 DA	48/35 VA	20(6)			
Tunicates										
Ciona	(144)						20.0	0.045	0.9	66.0

aData from cited references, from previous editions of this book, or from the Handbook of Biological Data. RA = radial artery; CA = carotid artery; VA = ventral aorta; DA = dorsal aorta; VENT = ventricle. Values in parentheses refer to state in parentheses in column 3.

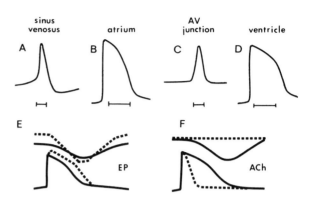

Figure 2. (A–D) Intracellular action potentials from different regions of heart of chick embryo. [Adapted from (154).] (E and F) Tension (upper trace) and action potentials (lower trace) in guinea pig atrium: (E) effect of epinephrine (EP ---) and (F), effect of acetylcholine (ACh ---). (Courtesy of W. Sleator.)

Vertebrate Cardiovascular Systems

The vertebrate cardiovascular system is characterized by a chambered heart and a closed vascular system. The basic principles governing excitation of the heart, how the heart functions as a pump, cardiac regulation, and blood flow, are similar for all vertebrates. There are, however, important differences with respect to the functional morphology of the heart, the pattern of the circulation, and the mechanisms that regulate the distribution of cardiac output among vertebrates.

Excitation of the Heart

Coordinated contraction of cardiac cells starts early in embryonic development (143). The heart beat is initiated by rhythmic action potentials in a myogenic pacemaker. This excitation spreads in a coordinated fashion to other cardiac cells, where it is coupled to the contractile events by an increase in intracellular $[Ca^{2+}]$.

MYOGENIC PACEMAKER. The pacemaker is located in the sinus venosus in fish and amphibians, and in the sinoatrial (SA) node in reptiles, birds, and mammals. Rhythmic action potentials in the sinus act as the pacemaker for the heart beat by

virtue of their faster intrinsic rate [about twice that of the ventricle (72, 99)] and the electrical coupling between the cardiac cells. Other regions of the heart can initiate a regular, but slower heart beat if the sinus rhythm is lost. Myocytes in culture can also contract rhythmically.

CARDIAC ACTION POTENTIALS. In the mammalian heart, the *pacemaker potential* (-55 to -60 mV) is more positive than the resting potential in other cardiac cells (-80 to -90 mV). Pacemaker cells typically show a slowly rising potential which, at a threshold of 10–20-mV depolarization, gives rise to a spike potential (Fig. 2). A steady, moderately high inward Na^+ conductance, combined with a low, time- and voltage-dependent K^+ conductance, permits the steady depolarization between spikes (100, 249). The rising potential activates an increase in Na^+ conductance to initiate the all or none spike. Atrial and ventricular cells also depolarize rapidly, but remain for 0.1–1.5 s partially depolarized in a plateau after a partial repolarization (Fig. 2). The Na^+ conductance is low during the plateau. The plateau is maintained by an increase in Ca^{2+} conductance and a delay in the increase of the K^+ conductance. The K^+ conductance increases (79, 182) during the repolarization following the plateau when Ca^{2+} conductance is switched off. A sarcolemmal Na$^+$-K$^+$ pump restores the intracellular Na$^+$ and K$^+$ balance. The action potential in cardiac muscle is 10–100 times slower than in skeletal muscle (Chapter 13).

EXCITATION–CONTRACTION COUPLING. The action potential increases Ca^{2+} permeability in the sarcoplasmic reticulum (SR) and inhibits the SR Ca^{2+} pump. These events, coupled with an inflow of extracellular Ca^{2+} via sarcolemmal Ca^{2+} channels, produce a 10-fold increase in intracellular Ca^{2+}, which initiates the final sequence of the contractile process by binding to the troponin–tropomyosin complex. Cardiac relaxation occurs when intracellular Ca^{2+} is reduced below 1 μM through removal by the Ca pump at SR, Ca^{2+}-H$^+$ exchange at mitochondria, and a sarcolemmal Na$^+$-Ca^{2+} carrier and Ca^{2+} pump. In addition to the pivotal role of Ca^{2+} in excitation–contraction coupling, the force of contraction (*inotropy*) is dependent on the intracellular [Ca^{2+}] in most, if not all vertebrate hearts.

TRANSMISSION OF EXCITATION. A wave of depolarization, initiated at the pacemaker, first spreads over the atrial myocardium (at 1 m/s in mammals) and then, after an atrioventricular (AV) delay, spreads more rapidly over ventricular muscle (1.5–4 m/s) via bundles of *Purkinje tissue*, which are muscle fibres modified for rapid conduction (202). The electrical connection between the atria and ventricles is only through the AV node. An AV delay (0.1 s) is produced in the mammalian heart by the slow conduction velocities in junctional fibers (0.05 m/s) and nodal fibers (0.1 m/s); this delay allows blood to move from the atria to the ventricles before the ventricles contract. Conduction velocities are generally slower in lower vertebrates. In the turtle, the spread of electrical activity over the ventricle changes direction

and decreases from 0.15–0.09 m/s with the onset of breathing (24).

Vertebrate hearts consist of branched, striated fibers that are joined by transverse intercalated disks. Intercalated disks include desmosomal regions, probably for mechanical coupling, and nexuses or gap membrane junctions, which are paths for intercellular electrical conduction since they appear to impose virtually no resistance (10, 116, 249). The space constant (length for an applied potential to fall to 1/e of its applied value) in rat atria is 130 μm parallel to the fiber axis; interfiber junctional resistance is ~1 Ω/cm^2 (261, 262, 275). In Purkinje fibers (in sheep) the space constant is two to four times longer than in striated muscle. The space constant of single fibers of right ventricle of sheep or calf is 880 μm. Acetylcholine decreases the space constant of cardiac muscle. Gap junctions are fewer and less convoluted in teleost fish; they have not yet been identified in hagfish, which may help explain why heart rate is two to three times slower in hagfish than in many teleosts (54, 213).

ELECTROCARDIOGRAM. The electrical wave front passing over the heart is of sufficient magnitude to be recorded extracellularly as an electrocardiogram (ECG) at some distance from the heart (Fig. 3). Typically, the ECG consists of negative upward deflections called P, R, and T, and positive downward deflections, Q and S. The P wave corresponds to depolarization of the atria, the PQ interval represents delay at the AV junction, and the QRS complex corresponds on depolarization of the ventricles. The T wave represents repolarization of the ventricular surface, an upward T indicating earlier repolarization of the left side and a downward T indicating earlier repolarization of the right side of the mammalian heart. In amphibians and teleosts, a V wave originating in the sinus venosus can precede the P

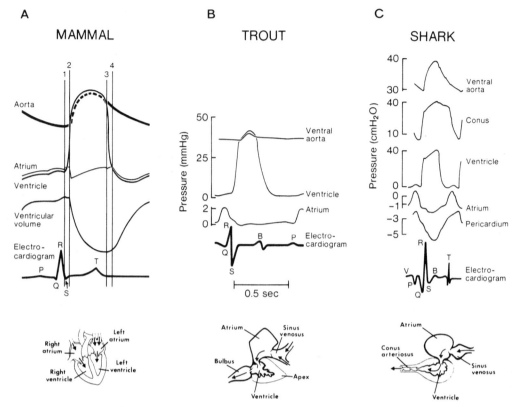

Figure 3. Correlated events of the cardiac cycle. (A) *Mammalian heart:* Aortic, atrial and ventricular pressures and ventricular volume adapted from the dog. Electrocardiogram from humans. 1 signifies closure of the AV valves and beginning of ventricular contraction; 2 signifies opening of the aortic valves; 2–3 signifies ejection phase; 3 signifies closure of aortic valves and atrial systole. (Adapted from H. E. Hoff, in Howell's Textbook of Physiology, 1949). (B) *Trout heart:* Ventricle filling occurs only during atrial systole. [Adapted from (195) and A. P. Farrell, unpublished.] (C) *Shark heart:* Subambient pressures pulses, found in the atrium and pericardium, indicate that atrial filling is *vis-a-fronte.* Contraction of the conus arteriosus prolongs systolic pressure in the conus. [Adapted from (212).]

wave; in elasmobranchs, a B wave resulting from depolarization of the conus arteriosus occurs between the S and T waves (Fig. 3). The time occupied by the P–R interval in the ECG is ~30–50% of the total duration, whether the heart rate is 36 beats per minute (bpm) (crocodile and horse) or over 600 bpm (mouse). However, a longer P–R interval is prominent in hibernating mammals.

The Heart as a Pump

This section focuses on cardiac pumping in mammals (a double pump) and fish (a single pump), in which the respiratory and systemic circulations lie in series. Functional differences in cardiac morphology are described later for air-breathing fishes, amphibians and reptiles in which the respiratory and systemic circulations are in parallel.

THE CARDIAC CYCLE IN MAMMALS. The mammalian heart is separated into right and left pumps, each with an atrium and a ventricle. The right ventricle pumps deoxygenated blood into the pulmonary circulation. The left ventricle pumps oxygenated blood to the systemic circulation. The two sides of the heart pump the same amount of blood, but at different pressures. The relatively thicker walled left ventricle and its higher systolic pressure reflect a higher systemic compared to pulmonary resistance.

The pattern of contractions in the mammalian heart (Fig. 3) is essentially similar for all chambered hearts. Contraction (*systole*) of the ventricle forces blood from the ventricle into the aorta, past the open semilunar valves at the base of the aorta. Flow stops when ventricular pressure falls below that in the aorta and the semilunar valves close. In humans, the *isometric* contraction with all valves closed lasts for 0.05 s, followed by *isotonic* contraction for 0.22 s during the ejection of blood. Ventricular relaxation (*diastole*) lasts 0.53 s. During ventricular contraction the atrial pressure is below that in the ventricle and the AV valves remain closed. As the atria fill, atrial pressure gradually rises while the ventricular pressure is falling. When atrial pressure exceeds ventricular pressure the AV valves open, blood flows into the ventricles such that 70% of ventricular filling occurs during atrial relaxation (diastole). Ventricular filling is completed by atrial contraction. Atrial systole lasts 0.11 s and diastole 0.69 s in humans.

THE CARDIAC CYCLE IN FISH. The fish heart is a single pump that generates sufficient pressure to pump deoxygenated blood first into the branchial circulation and then directly to the systemic circulation. Blood pressure (but not vascular resistance) in the respiratory circulation is therefore higher than in the systemic circulation; a situation that contrasts with most air breathers.

A single ventricle is separated from a thin-walled atrium by an AV valve. Ventricle filling occurs only during atrial systole (Fig. 3) and as a result ventricular volume is matched with atrial volume (48, 195). In birds and mammals atrial volume is smaller than ventricular volume. A thin-walled sinus venosus, which is separated from the atrium by the SA valve, contains a sparse arrangement of myocardial cells and acts as a venous reservoir. The sinus is contractile in the New Zealand hagfish (*Eptatretus cirrhatus*) and eel (*Anguilla anguilla*) (48, 54). Blood is ejected from the ventricle into a fourth chamber; an elastic bulbus arteriosus in teleosts or a conus arteriosus in elasmobranchs (Fig. 3). The elasmobranch conus contains its own cardiac muscle and is capable of producing systoles (212). Both the bulbus and conus apparently smooth flow to some degree, so that blood flows more evenly to the gills. At 10°C the trout atrium (beating at 46 bpm) contracts isometrically for 0.075 s and isotonically for 0.28 s; the ventricle contracts isometrically for 0.12 s and isotonically for 0.38 s. Thus, ventricular systole occupies about one third of the cardiac cycle in fish and mammals.

The cyclostome heart generates low pressures (Table 1) with a slow rate of contraction (54, 213). Peripheral circulation is assisted by muscular gills and caudal and portal accessory hearts. The portal heart develops pressures of 1–6 mmHg and its heart beat is not synchronous with the systemic heart (46, 126).

Characteristics of the Cardiac Pump

STROKE VOLUME. This is the volume ejected with each heart beat (SV_H) and is the difference between end-diastolic and end-systolic volumes. During diastole, ventricular volume increases to its *end-diastolic volume*. During systole, ventricular volume decreases to its *end-systolic volume*. In humans, end-diastolic volume is normally 120–130 ml, end-systolic volume is 50–60 ml, and thus SV_H is 70 ml, or ~1 ml/kg of body mass. In trout, SV_H is 0.3 ml/kg of body mass, but information on ventricular volumes during the cardiac cycle is lacking.

CARDIAC OUTPUT AND WORK. Like any pump, the heart generates flow and pressure. The volume of blood pumped per unit time from a ventricle is termed *cardiac output* (Q). The term Q is calculated from heart rate times SV_H. Direct measurements of Q are made with electromagnetic or Doppler flow probes. Indirect measurements of Q utilize the Fick method, dye dilution, thermal dilution, and foreign gas retention or elimination. The Fick method divides the rate of O_2 uptake of the animal by the difference in O_2 content of arterial and venous blood going to and from the respiratory surface. Fick calculations are inaccurate if there is a vascular shunt, if there is more than one respiratory organ, and if the respiratory organ consumes a significant amount of O_2.

Average Q for a 45-year-old man is 5.2 L/min or 3 L/min · m² of surface area at 68 bpm and SV_H of 74 ml/beat (95). Most comparative data are expressed as relative Q per unit mass (M), although surface area specific (cardiac index; $M^{2/3}$) is also used, since relative Q in mammals decreases with body size, $Q = 0.76 \, M^{0.776}$ (269). Comparisons between ectotherms should consider body temperature since Q varies directly with temperature (Table 1). Caution should also be used when comparing Q values between species since flow requirements are also dependent on O_2 carrying capacity and O_2 extraction. Cardiac output, arterial blood saturation, and body temperature are lower in the echidna than in comparable sized rabbits (185).

The *external work* performed by the heart is the product of flow times the pressure developed, plus any kinetic energy. Cardiac work per unit time, the myocardial *power output*, therefore varies with pressure *and* volume loading of the heart. Since cardiac tissue is primarily aerobic, myocardial O_2 consumption is linearly related to power output, even in fish hearts that rely on venous blood rather than a coronary circulation for an adequate O_2 supply (66, 174). However, isometric contraction is metabolically more costly than isotonic contraction. Stronger correlations are found between myocardial O_2 consumption and either the peak ventricular pressure, or the pressure–time integral, or the systolic pressure–volume area (240).

Myocardial power output can be expressed in terms of mL of O_2 and divided by O_2 uptake of the heart to determine the *mechanical efficiency* of the heart. Vertebrate hearts generally have a mechanical efficiency of 10–20% (66, 86, 174). Thus, only a small portion of the O_2 consumed is converted to mechanical work. Efficiency is related to work load; it is reduced at abnormally low work loads and is improved up to 25% at higher work loads.

The heart generates blood pressure by increasing its wall tension. *LaPlace's law* describes the relationship between pressure (P, dyn/cm²), wall tension (T, dyn/cm), and the principle radii of a simple sphere (R_1 and R_2, cm):

$$P = T(1/R_1 + 1/R_2)$$

A greater pressure development is possible in a ventricle with a thicker wall or smaller radii. Conversely, a larger heart must have a relatively thicker wall to develop the same pressure as a smaller heart. While the general principle of LaPlace's law may apply to hearts, hearts are not simple spheres. For example, the highly trabecular nature of the fish ventricle may facilitate pressure development by acting as many small pumps rather than a single, larger pump (131). Likewise, the high radius of curvature at the apex of the heart (in mammals and perhaps fast swimming fish) provides a mechanical advantage in terms of pressure development.

Cardiac Regulation

Cardiac output can be varied through stroke volume (SV_H) and heart rate (f_H), or a combination. Changes in SV_H are brought about by changes in either end-diastolic or end-systolic volume. End-diastolic volume is related to (a) venous filling pressure, (b) atrial distensibility and contractility, (c) ventricular wall distensibility, and (d) ventricular filling time. End-systolic volume is related to (a) ventricular contractility and (b) arterial diastolic pressure. The heart must also develop sufficient pressure to overcome vascular resistance. Pressure development and SV_H are determined by intrinsic mechanical properties of the heart and extrinsic factors that modify these properties. The intrinsic pacemaker frequency is modified by nerves, hormones, and temperature.

Mechanical Properties

In isolated vertebrate hearts an increase in filling pressure increases SV_H primarily by augmenting end-diastolic volume. An increase in arterial pressure produces little change in SV_H as both end-diastolic and end-systolic volumes increase. Thus, the heart has an intrinsic capability to in-

crease mechanical work (through either flow or pressure) in response to an increased stretch of the cardiac muscle during diastole and this is termed the *Frank–Starling mechanism*, for which the energy of contraction is a function of the length of the muscle fiber.

The functional ability of the ventricle to pump blood is shown by *ventricular function curves* (Fig. 4). *Stroke work* (the product of SV_H times pressure developed) increases up to a maximum as atrial filling pressure is increased. However, because extrinsic factors alter these curves by modifying contractility, stroke work can be varied independent of filling pressure. Increases in contractility (positive inotropy), as with β-adrenergic stimulation of the mammalian and fish hearts, can shift the curve upwards and to the left. Vagal inhibition has a negative inotropic effect and shifts the curve downwards. The observation in intact mammals that f_H increases Q to a greater degree than SV_H has led to the suggestion that the Frank–Starling mechanism may be more important in balancing the volume output from the left and right sides of the heart rather than increasing SV_H (31). This is not necessarily the case in some fish where changes in SV_H increase Q much more than changes in f_H (136).

Many extrinsic factors modify cardiac contractility by altering intracellular Ca^{2+}. The positive inotropic effect of epinephrine (EP) is associated with an increased probability that sarcolemmal Ca^{2+} channels are in an open state (201). Extracellular Ca^{2+} can also increase contractility (176, 200), but homeostasis of free plasma Ca^{2+} levels may preclude a normal role for this effect in many vertebrates. Negative inotropy is found in a variety of vertebrates with high extracellular Na^+, which affects Na^+-Ca^{2+} exchange (84, 110, 217), and with low extracellular pH, which affects intracel-

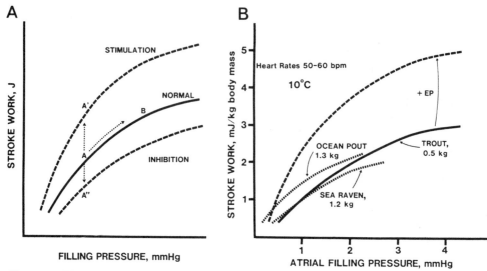

Figure 4. Ventricular function curves for vertebrate hearts. (*A*) A family of curves is generated by inhibition or stimulation of ventricular contractility. Stroke work can be varied by either increasing filling pressure (from A to B), or varying contractility at a constant filling pressure (from A to A′ or A″). [Adapted from (105).] (*B*) Ventricular function curves for perfused hearts from teleost fish. The effect of maximum stimulation by 1 μM EP (epinephrine) on the curve for the trout heart is indicated. [Adapted from (69–71).]

lular Ca^{2+} availability (85, 133). Plasma $[Na^+]$ varies significantly in euryhaline fish and dehydrated amphibians. Blood pH is reduced during anoxia and after exhaustive exercise, but cardiac function can be maintained in acidosis-resistant species, or restored in acidosis-sensitive species by release of intracellular calcium stores, intracellular pH regulation, or catecholamine stimulation (85, 90).

Lowered temperature reduces contractility and the positive inotropy produced by catecholamines (97). In human and chimpanzee atria, reduced temperatures lower the plateau voltage and increase its duration. Trout exposed to cold water compensate for reduced contractility by increasing their relative heart mass (69).

Pericardium and Venous Return
The vertebrate heart is contained in a pericardial cavity in which secreted pericardial fluid acts as a lubricant. The pericardial membrane in elasmobranchs,

lungfish, and tuna is relatively stiff (133, 148, 212). Therefore, ventricular systole creates intrapericardial pressures that are subambient (-1 to -5 mmHg) (Fig. 3). Subambient intrapericardial pressures increase the atrial transmural pressure gradient and assist atrial filling and venous return. This is termed *vis-a-fronte* filling since a portion of the energy of contraction of the ventricle is used for atrial filling. Venous pressures near to the heart may be subambient in sharks (43). If the pericardium is ruptured or pericardial pressure is raised, Q is reduced. When SV_H is increased in elasmobranchs, an inhibitory elevation of pericardial pressure is prevented by loss of pericardial fluid into the coelom through the pericardioperitoneal canal (219). In birds, mammals, and many teleost fish, the pericardium is thin and flexible and venous pressures are positive. Atrial filling is primarily *vis-a-tergo*. During exercise venous return is increased by venous massage by somatic muscles, venous valving, and splanchnic vasoconstiction to reduce venous capacitance. Even so, *vis-a-fonte* filling is found in trout (69), presumably because surrounding skeletal and muscular elements increase the overall rigidity of the thin pericardium. Also, elastic recoil provides a *vis-a-fronte* effect in mammalian hearts (204). *Vis-a-fronte* filling prevents cardiac dilation in failing, but not normal dog hearts (2) and may limit overdistension of the thin-walled atria found in fishes.

Heart Rate

Reflex regulation of f_H is essentially similar in mammals, birds, reptiles, amphibians, teleosts, and elasmobranchs (178). Cholinergic stimulation of muscarinic receptors shows f_H (*negative chronotopy*) and adrenergic stimulation of β-adrenoceptors increases f_H (*positive chronotropy*). Vagal inhibition is the more influential control in most ectotherms, whereas sympathetic excitation is more prominent in endotherms. In contrast, the lamprey heart has excitatory vagal innervation (nicotinic receptors for ACh that are blocked by tubocurarine). The hagfish heart has no extrinsic innervation and is unaffected by applied ACh and catecholamines (195, 213).

Vagal fibers generally terminate in the sinus venosus, the SA node, in the walls of the atrium, and near the AV junction, but not in the ventricles (178). Vagal stimulation or applied ACh hyperpolarizes and diminishes the rate of rise of the pacemaker potential (116), increases K^+ conductance, and reduces the spike overshoot. Acetylcholine shortens the duration of the plateau in atrial (248) and pacemaker cells (Fig. 2), and increases the AV delay. Tonic vagal activity normally keeps the heart depressed. Thus vagotomy or atropine injection increases f_H and blood pressure. Vagal inhibition can be particularly powerful in ectotherms during breath holding, hypoxia, and the fright response of fish (151, 191, 195), and during hibernation and diving in endotherms (see later sections on diving and hibernation).

Sympathetic nerve fibers from the stellate ganglion innervate the atrium and ventricle, as well as the SA node in reptiles, birds, and mammals. These posganglionic fibers liberate norepinephrine (NE) (198). Positive chronotropy is brought about by membrane actions on pacemaker cells (Fig. 2), improved conduction velocity, and increased rate of relaxation. In amphibians and fish, the sympathetic innervation to the heart is often in the same trunk as the vagus (151). In anurans, all sympathetic innervation is in the vagal trunk, while urodeles possess both a vagosympathetic and direct sympathetic cardiac innervation (28, 178). In trout the vagus contains adrenergic excitatory fibers as well as cholinergic inhibitory ones (82); the sinus and, to a lesser degree, the atrium receive sympathetic innervation via the vagus, while the ventricle is innervated by sympathetic nerves associated with the coronary vessels (151). β-Adrenoceptors generally mediate positive chronotropy in fish and amphibians, while α-adrenoceptors mediate negative chronotropy in the perch, *Perca fluviatilis* (45, 72, 178). Dipnoans and some teleosts (particularly pleuronectids) lack adrenergic innervation of the heart (151). Cardiac cells containing EP and NE are found in all vertebrates, including lampreys (16). Epinephrine stimulates the lamprey heart (123). Chromaffin cells or masses also occur along veins leading to the heart, and release catecholamines into the blood.

Stretch (increased cardiac filling) increases f_H by 10–15% in mammals because of changes in membrane permeability. Much larger changes in f_H (up to 75%) were found in isolated trout, shark, and hagfish hearts (123–125), but perfused hearts from several teleosts including trout showed only small changes in f_H (±10%) associated with physiological increases in filling pressure (69, 72). A marked temperature effect on f_H is particularly evident in ectotherms and hibernators (Table 1). Heart rate has a Q_{10} of 1.5–3 in most vertebrates. At lower temperatures vagal control of f_H in trout is more prominent than adrenergic control (188, 272), the reverse is true in the eel, *A. anguilla*.

In summary (1) cardiac innervation either is lacking in cyclostomes, or is excitatory, cholinergic, and stimulates nicotinic receptors; (2) all other vertebrates have inhibitory vagal cholinergic innervation that stimulates muscarinic receptors; and (3) excitatory adrenergic fibers are mixed with the vagus in some fish and amphibians, but are mainly in separate sympathetic trunks in reptiles, birds, and mammals. Cells containing catecholamines are found in all vertebrate hearts.

Influence of Size and Activity on Heart Mass, Q and f$_H$

Cardiac function and design are closely tied to the O_2 needs of the animal. Thus Q and basal metabolic rate scale similarly in mammals: $Q = 0.76 M^{0.776}$ and O_2 consumption $= 0.676 M^{0.75}$. However, heart mass increases with body mass (M) in a nearly proportional manner for mammals (heart mass $= 0.0059 M^{0.98}$) and birds (heart mass $= 0.0082 M^{0.91}$), and so SV_H increases in proportion with body size. In fact, Q increases with the end-diastolic volume (EDV) according to the relation $Q = 106 EDV^{0.77}$ in mammals (111). The corollary is that f_H scales negatively with body mass ($f_H = 241 M^{-0.25}$ in mammals) and therefore f_H is generally faster in small animals of a given kind (Table 1). The f_H term is 600/min in a 3-mg hummingbird heart and ~20/min in a 200-kg heart from a humpback whale. This generality also

extends to invertebrates, for example, *Daphnia* hearts beat at 250–450/min and crayfish at 30–60/min. The f_H value is usually slower in sluggish animals than in related active ones.

Variability about the linear relationships between heart mass and body mass reflect, in part, the larger heart found in more active species within a class. For example, the relative heart weight in tuna is about four times greater than that in less active fish. Heart mass is also greater in more active classes of vertebrates. For an equivalent size animal, a bird's heart is ~0.82% of body mass, a mammal's 0.59%, a reptile's 0.51%, and an amphibian's 0.46%. Fish hearts are much smaller (0.06–0.2%) and show more variability between species.

In summary, large or growing animals increase their Q by increasing SV_H rather than f_H. More active animals have a relatively larger heart with a faster heart rate.

Distribution of Blood Flow

Hemodynamics

The principles of blood flow and pressure (*hemodynamics*) are extensively covered in mammalian physiology texts (31, 168) and only an overview of some basic principles is presented here. Flow occurs from a region of high *total fluid energy* to one of lower total fluid energy. Total fluid energy (E) is given by the sum of pressure energy, gravitational potential energy, and kinetic energy:

$$E = P + \rho g h + \tfrac{1}{2}\rho v^2$$

Where P is the pressure (dyn/cm^3), ρ is the density (g/cm^2), g is the gravitational acceleration (980 cm/s^2), h is the height (cm), and v is the velocity (cm/s).

In a closed tubular system, a pump develops a pressure head that declines with frictional loss in the tubes. Mean blood pressure falls from arteries to veins (Fig. 5) and the pulse pressure (systolic minus diastolic pressure) varies according to the vessel stiffness and wave reflections. In mammals the descending aorta becomes stiffer along its length and with age; in teleost fish the dorsal aorta is stiffer than the ventral aorta.

Gravitational potential energy (ρgh) is important because of its effect on the transmural pressure gradient across vessel walls, especially in tall animals (or long arboreal snakes). When a giraffe stands, the veins in the head are in danger of collapse because the transmural pressure is below zero, whereas the vessels of the lower limbs are in danger of overexpansion because of high transmural pressures. These problems are offset by valving, reducing vascular distensibility, tight skin, and high arterial blood pressures. Hydrostatic pressure in aquatic environments offsets the gravitational effect on vascular transmural pressure and may have been important for large aquatic dinosaurs. Antigravity suits, worn by pilots who perform high-g maneuvers, employ a similar principle. Regional distribution of pulmonary blood flow is also influenced by gravitational effects.

Kinetic energy ($\frac{1}{2}mv^2$) is negligible in most of the circulation because blood velocity is relatively slow. Mean blood velocity is determined by Q divided by vascular cross-sectional area. Velocity is greatest in the aorta, and kinetic energy may represent 1% of the total fluid energy in mammals, increasing to 17% with exercise. Velocity is reduced considerably in the capillaries because of a greater total cross-sectional area (Fig. 5); ~800 times that of the aorta in mammals. Slower blood flow and lower pressure in capillaries is important for effective gas and fluid exchange between the blood and tissues. Kinetic energy in the vena cava and pulmonary artery of mammals can be an

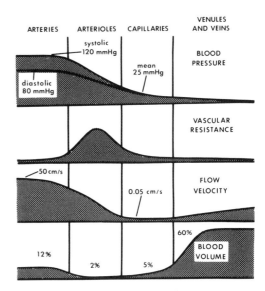

Figure 5. Distribution of blood pressure, vascular resistance, flow velocity, and blood volume in the mammalian cardiovascular system. [Adapted from (105).]

appreciable component of the total fluid energy (17–50%) because of their relatively low blood pressures (31).

Poiseuille's law describes the relationship between pressure and steady laminar flow in straight rigid tubes. These conditions are not strictly satisfied for pulsatile, non-Newtonian blood flow in compliant vessels, but the main factors that affect blood flow are described by Poiseuille's law, which states that flow (ml/s) is directly proportional to the pressure difference (ΔP, mmHg) along the length of the vessel and the fourth power of the vessel radius (R, cm), and inversely proportional to the tube length (L, cm) and fluid viscosity (η, cP):

$$\text{flow} = \Delta P \left(\frac{R^4}{L}\right)\left(\frac{1}{\eta}\right)\left(\frac{\pi}{8}\right)$$

While flow is directly proportional to the blood pressure generated by the heart, vessel caliber is clearly a critical determinant of blood flow. This explains why the small arterioles are the major resistance site in circulation (Fig. 5), and why arteriolar constriction or dilatation (*vasomotion*) is important in controlling the distribution of blood.

Blood viscosity is dependent on temperature, vessel diameter, hematocrit, and blood velocity and is therefore an important variable. Water has a viscosity of 1 cP at 20°C and the viscosity of blood fluids is usually expressed relative to this value. The relative viscosity of plasma is 1.5–2 cP at 37°C. Blood viscosity in humans is normally 3–4 cP and increases exponentially with hematocrit. High altitude adaptation, spleen constriction during exercise, and plasma skimming all increase hematocrit. High altitude sickness is associated with a blood viscosity of 10 cP. Deep diving mammals have high hematocrits (55–65%) and blood viscosities (8.9 cP), which may compromise cardiac performance (102). Blood viscosity is inversely related to blood velocity, but in small vessels, blood displays an anomalous viscosity, which approaches that of the plasma. An inverse and exponential relationship between blood viscosity and temperature is of great importance in those endotherms that allow their extremities to cool appreciably and in ectotherms in general. For example, fish exposed to colder water must either reduce blood viscosity (91), or compensate for increased blood viscosity in other ways (e.g., increased blood pressure, reduced blood flow, reduced vascular resistance, or some combination).

Vascular resistance is determined by vessel geometry, but detailed knowledge of vascular dimensions is limited to certain mammalian microcirculations and the fish gill. Consequently, Poiseuille's law is often simplified to an ohmic relationship where blood flow is determined by pressure difference divided by vascular resistance [e.g., a resistance of 1 peripheral resistance unit (PRU) exists when a pressure of 1 mmHg results in a flow of 1 mL/s]. Vascular resistance obeys the same laws as electrical resistances. In series, vascular beds have a higher combined resistance, while in parallel, vascular beds have a lower combined resistance. Most vascular arrangements in vertebrates are in parallel. Peripheral vascular resistance is conveniently divided into the systemic and respiratory resistances. Normal systemic resistance in humans is 1.1 PRU; pulmonary resistance is ~1/7th of this value. Branchial resistance in fish is ~ one fourth of the systemic resistance (136). Interspecific comparisons of vascular resistance have limited value since blood flow, but not necessarily blood pressure, is size dependent.

In summary, Poiseuille's law indicates that animals can modulate either blood pressure or vessel diameter to vary flow. Normally, regulation of arterial blood pressure provides a constant driving pressure for blood flow while selective distribution of blood flow is brought about by vasomotion in the many parallel vascular beds. Blood viscosity is also an important determinant of blood flow.

Blood Pressure and Its Control

Arterial blood pressure is determined by (1) the total peripheral resistance and (2) the pressure developed by the heart. Blood volume, through its effect on vessel dimensions and hence peripheral resistance, also influences blood pressure. The mean arterial pressure is between 25–150 mmHg for most vertebrates (Table 1). In water breathers, blood pressure in the branchial arteries is greater than in the systemic arteries. In contrast, the pulmonary circulation of air breathers is either perfused by a lower pressure pump (birds or mammals) or guarded by a high arterial input resistance (lungfish, amphibians, and reptiles. Pulmonary blood

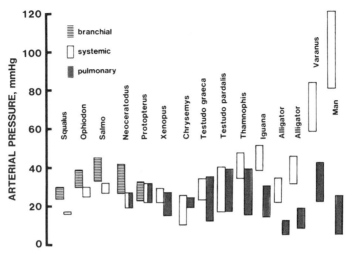

Figure 6. Blood pressures in the branchial (lined bar) versus systemic (open bar) and pulmonary (stipled bar) versus systemic (open bar) arteries of selected vertebrates. Height of vertical bar is equivalent to pulse pressure. Note the evolutionary trend to increase systemic blood pressure and reduce respiratory blood pressure. [Adapted from (131,132).]

pressure is therefore lower than systemic blood pressure. There is a clear evolutionary trend towards (a) higher pressures in the systemic circulation and (b) lower pressures in the respiratory circulation (Fig. 6). Lower pressures in the pulmonary circuit allow a thinner diffusion barrier between blood and air and less risk of excess plasma filtration. Higher systemic pressures permit the higher flow and renal ultrafiltration that are needed with increased aerobic scope.

Evidence exists for most vertebrate groups that blood pressure is regulated by a negative-feedback loop, the *barostatic reflex,* which modifies Q and vascular resistance. In mammals, *baroreceptors,* which increase their discharge frequency in proportion to the arterial pulse and mean pressure (198), are located in the carotid sinus, the atrium (112, 127), and the adrenal glands (177). A *vasomotor center,* located in the ventrolateral aspect of the medulla oblongata, integrates this afferent information and sends efferent signals in autonomic nerves to the heart and vascular smooth muscle. Elevated blood pressure results in a reflex reduction of Q (reduced f_H and contractility) and peripheral vasodilatation. A reduction in blood pressure has the opposite effect. Vagally mediated bradycardia in response to elevated blood pressure is found in amphibians, reptiles, and birds (7, 135, 155), although the associated baroreceptors have not always been identified and characterized. Baroreceptors are found near the origin of the pulmocutaneous vessels in amphibians. Evidence of a vagally mediated barostatic reflex in teleosts has been provided by studies with perfused gills, adrenergic agonists, and antagonists, hemorrhage, hypoxia, tilting, and vascular compression (68, 273, 274). Baroreceptors are probably located in gill or pseudobranch tissue of fish (151). Early studies with elasmobranchs that demonstrated reflex bradycardia in response to elevated intravascular pressure (121) have not been supported by more recent studies using hemorrhage, adrenergic blocking agents, and tilting (7). Since a barostatic reflex is present in teleosts, it cannot represent a general mechanism which evolved solely to compensate for the gravitational stresses on the circulatory system (135). Indeed, the normalized gain for the barostatic reflex obtained in fish (1.5–4.5%/mmHg) compares favorably with values obtained in terrestrial and semiaquatic vertebrates (1–2%/mmHg for the toad; 7.2%/mmHg for the lizard; 2–5%/mmHg for the dog, rabbit, and human) (68, 135). Nevertheless, the relative importance of gravitational stress is indicated by the fact that the barostatic reflex is more prominent in arboreal snakes compared with terrestrial or aquatic snakes (155).

The mammalian vasomotor center integrates other afferent information. For example, an increase in blood CO_2, and a decrease in pH or O_2 (detected by chemoreceptors) can produce peripheral vasoconstriction and bradycardia directly or indirectly through stimulation of baroreceptors. An interplay between baroreceptor and chemoreceptor reflexes is also found in fish (273) since branchial O_2 receptors that detect hypoxic water elicit bradycardia and changes in peripheral resistance (56, 65, 227).

Baroreceptors and chemoreceptors therefore provide rapid, beat-to-beat feedback control of blood pressure. Long-term regulation of blood pressure is achieved in mammals by the renin–angiotensin system and renal regulation of extracellular and blood volume (Chapter 2). Long-term pressure regulation has not been studied extensively in lower vertebrates, but renin is produced in the kidneys of teleosts (181), and angiotensin-II injections cause hypertension through catecholamine release in *Squalus* (183).

Control of Blood Distribution
Blood flow is controlled at the local level and by the vasomotor center. Routine fluctuations in capillary blood flow can be regulated locally, whereas abrupt changes in flow to an organ are largely controlled by the autonomic nervous system through either direct innervation or the release of vasoactive hormones.

NEURAL AND HUMORAL REGULATION. Sympathetic fibers innervate arteriolar smooth muscle and usually release NE, which constricts vessels via stimulation of α-adrenoceptors. Arterial pressure is reduced in many vertebrates by agents that block α-adrenoceptors or deplete adrenergic nerve endings of neurotransmitter (28, 178). This clearly indicates that all vascular beds are not fully perfused at all times. Consequently, tonic sympathetic vasoconstriction of systemic vessels controls blood distribution as well as blood pressure. Adrenergic innervation of veins is quite limited in fish and amphibians, compared with reptiles, birds, and mammals.

Vascular smooth muscle also responds to circulating catecholamines. Resting catecholamine concentrations in the blood are normally 10 pM to 1 nM and may provide a cardiovascular tonus. Following stress, (fright, exhaustive exercise, or hypoxia), catecholamines (NE and EP) are released from chromaffin tissue and the adrenal medulla under sympathetic control. The dominant hormone in teleost fish is EP and blood levels may reach 50–200 nM. In elasmobranchs, both NE and EP reach 100 nM, while in lungfish NE is the dominant hormone and may reach 1.7 μM (32, 178). Circulating catecholamines are thought to be more important than sympathetic nerves in circulatory control in elasmobranchs, but not telosts (178).

The response of a given vascular bed to adrenergic stimulation varies according to the relative density and distribution of adrenoceptor subtypes, the type of catecholamine, the temperature, and the involvement of local controls (17). For example, the mammalian coronary circulation is dilated by low doses of EP, but constricted by high doses of EP. However, metabolic vasodilation can override adrenergic coronary vasoconstriction (74). β-Adrenergic vasodilatation is generally limited or absent in fish, amphibians, and reptiles (81). Increased temperature relaxes vascular smooth muscle and reduces its sensitivity to EP and NE in endotherms and ectotherms (5, 69, 89).

Sympathetic nerves may also carry cholinergic fibers that release ACh and produce vasodilatation in some muscle beds and in the facial skin of humans. With the notable exceptions of vasodilatation in the genitalia during sexual arousal, pulmonary vasoconstriction in amphibians and reptiles and branchial vasoconstriction in fish, cholinergic nerves are more important in cardiac regulation than vasomotion.

LOCAL REGULATION. In mammals, local regulation (*autoregulation*) of arteriolar diameter varies from tissue to tissue. Arteriolar vasomotion is initiated by either metabolic (*metabolic autoregulation*) and pressure (*myogenic autoregulation*) signals, or the release of *autocoids* (222).

Inadequate capillary blood flow produces a localized decrease in O_2 tension, an increase in CO_2 tension, and an accumulation of other metabolites and ions. Changes in the local environment of vascular smooth muscle produce vasodilatation (134). Coronary, cerebral, skin, and splanchnic circulations dilate with increased CO_2. Metabolic-related increases in $[K^+]$ and $[H^+]$ produce a marked increase in blood flow to skeletal muscle after a period of occlusion (*reactive hyperemia*) or after exercise (*active hyperemia*). Hypoxia dilates coronary vessels, but constricts pulmonary vessels such that pulmonary blood pressure increases considerably at altitude (9) (Table 1). Vascular smooth muscle contracts in response to

stretch. Thus, if blood pressure increases without a change in metabolism, myogenic vasoconstriction will tend to keep blood flow constant. Myogenic responses are found in mammalian coronary, kidney, and skeletal vascular beds (134). Local controls are probably present in lower vertebrates but rarely tested (42).

Various autocoids are present in cells or can be rapidly formed by cell constituents (222). Histamine, released from injured tissue, relaxes arterioles and constricts venules when sympathetic vasoconstrictor tone is withdrawn in skeletal muscle. Serotonin is important in vasoconstriction after vascular injury and during blood clotting, but dilates arterioles, interrupts adrenergic neurotransmission, and constricts venules in the gastrointestinal wall. The vascular endothelium also can influence vasomotion. For example, ACh stimulates the release of endothelium derived relaxation factor(s) which produce vasodilatation; ACh constricts if the endothelium is damaged or experimentally removed.

In summary, the importance of pressure regulation and vasomotion is highlighted by the control of blood flow to the two priority circulations. The cerebral circulation has little vasomotor activity and so blood flow is relatively constant provided arterial blood pressure is maintained. In contrast, coronary flow increases in proportion with myocardial work; increases of up to fivefold are brought about by vasomotion and increased perfusion pressure.

Functional Cardiac Morphology and the Control of Respiratory Blood Flow
Air-breathing fish, lungfish, amphibians, and reptiles are intermittent air breathers and, unlike birds, mammals, and water-breathing fish, they have their systemic and pulmonary circulations in parallel. Reflex cardiovascular events, which are associated with each intermittent air

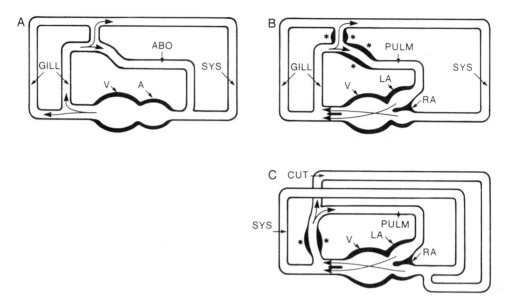

Figure 7. Comparative aspects of circulatory patterns and major vasoactive sites (*) for control of blood flow to air-breathing organs. (*A*) *Hoplerythrinus*, an air-breathing fish; (*B*) *Protopterus*, a lungfish; (*C*) *Xenopus*, an amphibian. Symbols: V = ventricle; A = atrium; LA = left atrium; RA = right atrium; GILL, CUT, ABO, PULM, and SYS indicate the branchial, cutaneous, accessory air-breathing organ, pulmonary, and systemic circulations, respectively.

breath produce large increases in lung perfusion to match intermittent lung ventilation (194). Typically, the cardiovascular events associated with air breathing are more pronounced following longer periods of apnoea. While these animals (except crocodiles) have a single ventricle, partial to almost complete separation of the pulmonary and systemic venous returns is still possible by means of laminar flow within the heart (Reynolds numbers of 200–300) and modifications to cardiac anatomy. Anatomical changes included the development of (1) two atria, (2) partial septa and chambers within the ventricle, (3) a septal ridge dividing the single aorta in lungfish and amphibians, (4) separate pulmonary and systemic openings from the single ventricle in reptiles, and ultimately (5) two ventricles in crocodiles. The evolutionary transition in vertebrates from a single to a double circulation was accompanied by the transition from

aquatic to aerial respiration. The following considers the control of respiratory blood flow in water-breathing fish, air-breathing fish, lungfish, amphibians and reptiles.

Water-breathing fish direct the entire Q to the gill (195). There may be nonrespiratory shunts through the gills and the pattern of flow through the respiratory exchange area (secondary lamella) may change passively as a result of changes in blood pressure and flow, or as a result of vasomotion (73, 192, 196). The branchial vascular resistance is increased by ACh and α-adrenergic stimulation, and it is reduced by β-adrenergic stimulation (178–180).

Air-breathing fish use their gills for breathing water and intermittently ventilate an accessory air-breathing organ. Usually the circulation to the accessory air-breathing organ is derived postbranchially (Fig. 7). Thus the entire Q is directed to the gills but only a portion of Q perfuses the accessory air-breathing organ. Modifications to the gill circulation also improve the regulation of flow to the accessory air-breathing organ. Nonrespiratory branchial shunts, selective perfusion of gill arches, and a reduction in the number of secondary lamellae are common in air-breathing fish, for example, *Amia, Channa, Hoplerythrinus, Electrophorus,* and *Synbranchus* (194, 211). *Channa argus* has two ventral aortae that selectively perfuse the anterior and posterior gill arches (122). Branchial shunts, which bypass the secondary lamellae, probably prevent an excessively high gill resistance from overtaxing the heart when lamellae collapse in air. Lamellae are better supported in fish that use the gills as an air-breathing organ, for example, *Anguilla* and *Amia* (194). Associated with each air breath are reflex increases in perfusion of the accessory air-breathing organ and heart rate, with or without a concomitant increase in Q (131, 194). Cholinergic and adrenergic vasoactivity is not significant in the accessory air-breathing organ and selective perfusion may depend on α-adrenergic vasoconstriction of the systemic circulation.

Lungfishes and *Lepisosteus* have a single ventricle and a divided atrium (Fig. 7). Nevertheless, laminar flow through the ventricle, a partial spongy septum extending from the apex of the ventricle, and a spiral ridge that extends along most of the length of the bulbus cordis allow *Protopterus* to direct 65–95% of pulmonary venous blood to the systemic circulation (130, 131). Lungfish intermittently breathe air in water and during estivation in mud cocoons. Up to a fourfold increase in lung perfusion, reflex tachycardia, and increased Q are associated with each air breath (130). Control of pulmonary blood flow involves branchial shunts in the posterior arches, hypoxic vasodilatation of the ductus Botalli,

and cholinergic vasoconstriction of the pulmonary artery (75).

Amphibians. In the frog, blood from the right and left atria enters and leaves the ventricle as simultaneous streams that are spatially separated by the trabecular and laminar flow (220). A spiral ridge maintains separation in the single outflow vessel, the conus arteriosus (Fig. 7), such that 91% of left atrial blood can go to the systemic circulation, while 84% of the right atrial blood goes to the pulmocutaneous artery (242). Control of pulmonary flow in the frog and toad is achieved by a strong vagal cholinergic constriction of the pulmocutaneous artery that is extrinsic to the lung (41, 139, 194, 265). Adrenergic constriction of the cutaneous circulation contributes secondarily to increase pulmonary flow (265). During apnoea progressive vasoconstriction reduces flow and increases diastolic pressure in the pulmonary artery. After a breath, a reflex increase in Q and pulmonary vasodilatation increase pulmonary flow. Thus, intermittent pulmonary flow matches intermittent lung ventilation without compromising systemic flow (151, 198, 220).

Amphibian groups that are less committed to lung breathing have a different cardiac morphology. Some aquatic urodeles and the Apoda, have a reduced spiral ridge. Lungless salamanders, which rely on cutaneous gas exchange, have a single atrium.

Reptiles have two atria and separate openings for the pulmonary and systemic arteries. The ventricle is partially divided in noncrocodilian reptiles (lizards, snakes, turtles, and tortoises) (Fig. 8) and fully divided in crocodiles (Fig. 9). However, the pulmonary artery *and* the left systemic aorta arise from the right ventricle in crocodiles. Therefore, reptiles use different strategies to separate pulmonary and systemic blood flows.

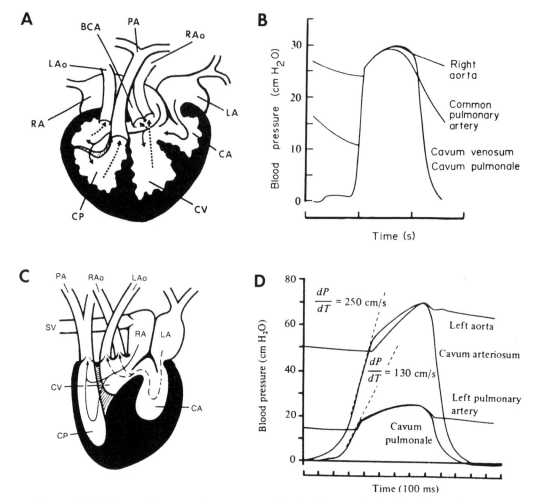

Figure 8. Cardiac anatomy and pressure profiles in the ventricular chambers of noncrocodilian reptiles. (A) Flow patterns through the three chambers of the turtle ventricle that function as a single pump because of their similar pressure profiles (B) during systole. (C) Flow patterns through the varanid ventricle where the cavum venosum (CV) is reduced in size and is separated from the cavum pulmonale (CP) during systole by the horizontal septum. Hence, the heart acts as a double pump, generating a lower pressure (D) in the cavum pulmonale and pulmonary artery (PA). Symbols: CA = cavum arteriosum; LAo and RAo = left and right aortae; BCA = branchiocephalic artery; LA and RA = left and right atria; SV = sinus venosus. [Adapted from (23,104,131).]

THE NONCROCODILIAN STRATEGY. The ventricle has three interconnected chambers, the *cavum pulmonale, cavum venosum,* and *cavum arteriosum.* The *cava arteriosum* and *venosum* have direct atrial input, while the *cava venosum* and *pulmonale* have direct aortic outlets. This leaves the *cavum pulmonale* without a direct atrial input and

the *cavum arteriosum* without a direct aortic outlet. During ventricular diastole, the right atrium pumps deoxygenated blood primarily *through* the *cavum venosum,* across the horizontal septum and into the *cavum pulmonale.* The left atrium pumps oxygenated pulmonary return primarily into the *cavum arteriosum.* During ventric-

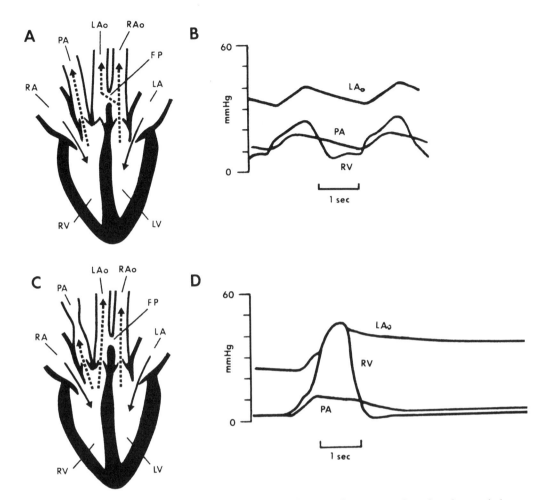

Figure 9. Cardiac anatomy and pressure profiles in the ventricular chambers of the crocodile. (*A*) Normally, left ventricle (LV) output enters the right systemic aorta (RAo) and left systemic aorta (LAo), via the foreamen Panizzae (FP). Right ventricle (RV) output enters only the pulmonary artery (PA). The heart acts as a double pump with lower pulmonary pressures (*B*). (*C*) during prolonged apnoea, vasoconstriction of the PA produces an increase in systolic pressure in the RV (*D*) and RV output is divided between the LAo and PA. [Adapted from (61 and 267).]

ular systole, the *cavum pulmonale* empties first because of a lower pulmonary compared to systemic resistance. Thus, even though the ventricle acts as a single pump, with similar pressure profiles in the three chambers (Fig. 8), early flow in the pulmonary artery and laminar flow during systole provide effective separation of pulmonary and systemic venous blood; 60–90% of right atrial blood is delivered to the pulmonary circulation and 70–90% of left atrial blood is delivered to the systemic vessels in the turtle, *Chrysemys scripta*. Intracardiac shunting is influenced by increased pulmonary outflow resistance, a change in the spread of depolarization across the ventricle, and mechanical events related to the relative SV_H of the heart (23, 104, 131).

Control of pulmonary blood flow is

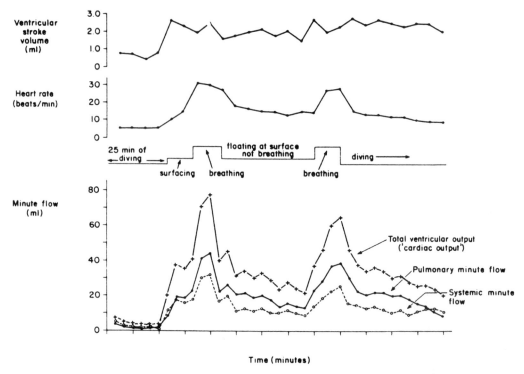

Figure 10. Cardiovascular events during intermittent breathing in *Chrysemys*. Note the pronounced tachycardia and increase in pulmonary flow associated with the bout of breathing following a prolonged dive. [From (221).]

achieved by vagal cholinergic constriction of the pulmonary artery, either proximal to the lung parenchyma or immediately proximal to the *cavum pulmonale* (23). During prolonged apnoea in *C. scripta*, vagally mediated bradycardia and pulmonary vasoconstriction can reduce Q by 50% and lung perfusion by 3.5-fold with only a small reduction (26%) in systemic blood flow (see Fig. 10).

In varanid lizards, the ventricle acts as two pumps during systole. The *cavum venosum* is greatly reduced and systolic pressure in the *cavum arteriosum* is greater than in the *cavum pulmonale* (Fig. 8). Separation of the ventricle by the horizontal septum is not maintained during diastole and right-to-left intracardiac shunting is still possible (104, 131), as found in *Iguana*, which has a 20–25%

right to left shunt to favor blood flow to the skin when heat stressed (8).

THE CROCODILIAN STRATEGY. Complete separation of pulmonary and systemic blood flow is possible during normal ventilation because of the low resistance of the pulmonary circulation and the low systolic pressure in the right ventricle. Blood is pumped at high pressure from the left ventricle into the right aorta and, via the foreamen of Panizzae, into the left aorta (Fig. 9). Thus, the valves at the base of the left aorta remain closed and the right ventricle only empties into the pulmonary artery (93, 131). However, during prolonged apnoea, vagal cholinergic constriction of the pulmonary artery increases pulmonary resistance and systolic pressure in the right ventricle rises to match that in the left ventricle (Fig. 9). Therefore right ventricular output is distributed between the lungs and left systemic aorta according to their relative resistances. This right-to-left shunt favors blood flow to the brain and is probably advantageous in extending apnoea (266, 268). In an analogous situation, the right ventricle of the mammalian fetus pumps blood into the systemic circulation (via the ductus arteriosus that connects the pulmonary artery and aorta) because of the high vascular resistance of the collapsed lung.

In summary, lungfish, amphibians, and noncrocodilian reptiles have a single ventricle and intracardiac blood flow separated by partial separation of the ventricle, laminar blood flow and separate atria. In crocodiles and varanid lizards the heart acts as a double pump. Ectothermic vertebrates that breathe air intermittently vary lung perfusion by vagally mediated changes in pulmonary resistance, thereby matching ventilation with perfusion without compromising systemic flow.

Cardiovascular Responses

Exercise

In exercising mammals, the vasomotor center integrates afferent information from active muscles (proprioceptors), arterial and cardiac mechanoreceptors, chemoreceptors, the cortex, and the hypothalamus. Autonomic nervous output ensures increased blood flow to working muscle, adequate blood flow to the skin for elimination of heat, and adequate coronary and cerebral blood flow. Sympathetic stimulation of the heart and improved venous return enables up to a threefold increase in f_H and a small increase in SV_H (Table 1). Training improves cardiac performance; hypertrophy of ventricular muscle allows a larger SV_H and lower f_H at rest, such that during exercise

f_H can increase almost fourfold, with an almost twofold increase in SV_H. Oxygen transport is also aided by increased tissue O_2 extraction (from 25 to 80–90%) and hematocrit (by reduced plasma volume and red blood cell recruitment). Sympathetic vasodilatation of the skin to increase heat loss slightly decreases (10%) blood pressure and SV_H, with a compensatory increase in f_H.

Blood flow in active skeletal muscle increases severalfold through local vasodilatation coupled with a loss of tonic sympathetic vasoconstriction; sympathetic vasoconstriction reduces flow to the kidney, gut, and nonworking muscle (156). Peripheral resistance is reduced, but less than the increase in Q and so mean arterial blood pressure increases to 120 mmHg. Blood flow in working skeletal (and cardiac) muscle is greatest during relaxation, when vascular compression is least. Flow can be reduced when skeletal muscle develops 15% of its maximum tension, and completely prevented at 70% of maximum tension (157). Consequently, mean arterial blood pressures exceed 160 mmHg during heavy isometric exercise.

Cardiovascular studies with exercising birds and reptiles are limited. Birds have larger hearts, a lower resting f_H and a higher Q for a given oxygen consumption than similar sized mammals (94). Pigeons flying at 10 m/s increase f_H sixfold and O_2 extraction 1.8-fold, but there is no change in SV_H. Running in birds and swimming in ducks involves increased f_H, mediated in part by β-adrenergic stimulation (37) Treadmill exercise in reptiles produces a two- to threefold increase in Q, primarily through increased f_H; SV_H increases 20% in *Iguana* and 30% in *Varanus* (87, 88, 254).

Cardiovascular changes during sustained exercise in fish are better documented (136, 193, 271). In trout, O_2 transport is augmented by a two- to threefold increase in Q, increased hematocrit (release of stored red blood cells from the spleen), reduced blood volume (from 6 to 4%), and increased O_2 extraction by the tissues (from 30 to 90%) (6, 271). Blood flow is increased 14-fold to red muscle and twofold to white muscle, and is reduced to the intestine, spleen, and liver (193). Unlike higher vertebrates, increased SV_H [aided by tail movements to promote venous return (212), a small increase in venous pressure and sympathetic stimulation of contractility (72)] is the primary means of increasing Q. In actively swimming trout, SV_H may increase threefold, whereas f_H increases 53% at 5°C and 30% at 12–20°C. The increase in f_H is brought about primarily through reduced vagal inhibition and secondarily through sympathetic stimulation (188, 272). Moderate swimming increases f_H (19%) and SV_H (26%) in Atlantic cod (*Gadus morhua*), but has no effect on f_H or Q in the eel (*Anguilla australis*) (6, 53). Arterial blood pressure increases with swimming in the trout and Atlantic cod. Systemic vascular resistance decreases in the trout, but not the Atlantic cod or eel (6, 53, 136). Sympathetic vasoconstrictor tone is apparently responsible for maintenance of blood pressure during exercise.

Stressful or exhaustive exercise in fish is initially characterized by vagal mediated bradycardia, and reduced Q and blood pressure (65, 238). Subsequently f_H, Q, blood pressure (65), circulating levels of catecholamines (32) and blood flow to red and white muscle all increase (175), while systemic resistance decreases.

In summary, exercise involves adrenergic stimulation of the cardiovascular system, perhaps to a lesser degree in fish. Increased f_H is the primary means of increasing Q in mammals, birds, and reptiles, but not in fish.

Diving

Breathhold diving underwater is normally a self-imposed state of hypoxia. The cardiovascular response to diving reflects (1) the duration of the dive, (2) the level of stress associated with the dive, and (3) the level of exercise underwater. For example, long or forced dives are associated with a marked bradycardia, whereas short voluntary dives do not generally initiate a bradycardia.

In birds and mammals, the classic *diving reflex* occurs when animals (a) naturally perform long (+20 min) dives (e.g., Weddel and elephant seals), (b) are forced to dive, and (c) are prevented from surfacing (33, 34). The diving reflex conserves O_2 stores by maintaining blood flow only to essential tissues (lung, brain, heart, and adrenal glands) (15). Even so, anaerobic metabolism is essential for longer dives (Chapter 8). The diving reflex involves vagally mediated bradycardia to reduce Q and peripheral sympathetic vasoconstriction to maintain blood pressure and selectively distribute blood flow (15, 19, 33, 34, 76, 128). In mammals, facial receptor input to the respiratory and vasomotor centers overrides the proprio- and chemoreceptor inputs, which normally drive the cardiovascular responses to exercise and hypoxia. In birds, the bradycardia and about one half of the increase in peripheral resistance is due to chemoreceptor input (33). Enlarged sinuses and arterial rete structures are found in whales and seals; these increase blood volume and presumably reduce the pulsatility of blood flow (15, 173).

Many birds and mammals routinely perform short, shallow aerobic dives without reflex bradycardia (33, 140, 237). For example, voluntary diving ducks show a transient bradycardia, but f_H increases to a more or less steady rate similar to the resting rate for the remainder of the dive (Fig. 11). Since ducks are exercising underwater, there is a compromise between the cardiovascular reflexes to diving and exercise. This compromise

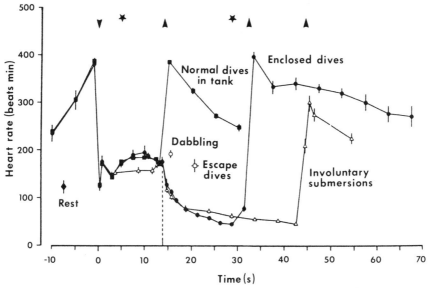

Figure 11. Heart during various types of breath hold diving in the duck which illustrate (a) pre- and post-dive tachycardia, (b) heart rate during a normal dive is similar to rest, but (c) a pronounced diving reflex bradycardia is evoked under stress. V = dive, A = surface, ** = period when access to surface was prevented. [From (237).]

is evident in harbor seals that increase f_H when they swim faster underwater (270). Blood flow is not necessarily reduced in the skeletal muscle groups responsible for underwater swimming. An important psychogenic component of the diving response is evidenced by the anticipatory pre-dive tachycardia in ducks (Fig. 11) and by the diving bradycardia evoked with forced diving.

Crocodiles, turtles, amphibians, lung-fish, and air-breathing fish also perform short dives without bradycardia (23, 67, 130). With prolonged diving, they develop a vagally mediated bradycardia and generally reduce Q, as in birds and mammals (5, 12, 13, 131, 150, 159, 194, 268). Furthermore, pulmonary flow can be reduced independent of the systemic circulation (see earlier section on functional cardiac morphology), unlike birds and mammals. This strategy is clearly advantageous in prolonging apnoea. During a remarkable submersion of 129 days in an-

oxic water at 3°C by *Chrysemys*, f_H decreased to 0.1–1.0 bpm and blood pressure to 10 mmHg (106). Exercise also influences cardiovascular responses to diving since sea turtles can maintain pulmonary flow during sustained underwater swimming (35).

In summary, the range of cardiovascular events associated with diving represents compromises between reflexes associated with exercise and the classic diving reflex. The diving reflex is characterized by bradycardia and preserves O_2 delivery to priority tissues during dives that are naturally long or forced. Shorter voluntary dives do not usually elicit a bradycardia.

Hibernation, Torpor, and Aestivation
Certain birds and mammals reduce their O_2 consumption overnight (*torpor*) or overwinter (*hibernation*) by regulating their body at a low temperature (5–15°C). Hibernators reduce Q, f_H, and arterial

pressure (Table 1) (57, 160, 161, 216), and maintain blood flow to vital organs (brain, heart, lungs, brown fat, and adrenals). Direct temperature effects on the heart and blood viscosity are primarily responsible for the reduction in f_H and up to fivefold increase in peripheral resistance (161). Torpid animals have higher body temperatures (12–15°C). Cardiovascular changes during torpor are less profound. The heart of the hibernator, independent of season and unlike nonhibernators does not fibrillate at temperatures below 17–20°C. This is probably because of reduced sensitivity to adrenergic stimulation (60) and because coronary vasomotion (26) and membrane ion pumps remain functional.

Arousal from torpor or hibernation involves an adrenergic-mediated increase in f_H and blood flow to brown fat. Brown fat thermogenesis, which is also under adrenergic control, warms the heart directly or through warmed venous return and allows further increases in Q. Increased blood flow to shivering skeletal muscle enhances the positive-feedback warming cycle until normal body temperature is attained.

Certain semiaquatic vertebrates burrow, for example, *Protopterus* and *Bufo*, and reduce their O_2 consumption over summer to survive drought conditions (*aestivation*). Typically Q, f_H, and blood pressure are reduced (58). Elevated plasma Na^+ because of dehydration may compromise cardiac contractility in burrowing amphibians (110). Dehydrated desert mammals decrease f_H when their body temperature falls overnight.

Lymph and Lymph Hearts

Lymph is generated by the filtration of plasma through intercapillary spaces in much the same way as the glomerular ultrafiltrate is formed (see Chapter 2). Its formation is dependent on capillary blood pressure and interstitial fluid colloid os-

motic pressure and is opposed by the colloid osmotic pressure of the blood. There is a net driving force in arterial capillaries, which becomes a net uptake force as blood pressure decreases in venous capillaries. Endothelial permeability varies considerably between tissues, for example, mammalian lung < liver < kidney, and it is increased by some autocoids. Endothelial permeability is relatively high in fish capillaries and in turtle lungs, and colloid osmotic pressure is generally lower in ectotherms compared to endotherms (22). In the turtle lung, lymph is formed 10–30 times faster than in the mammalian lung and reduced lung perfusion during apnoea may be important in restricting lymph formation (23).

In mammals and birds, blind-ending lymphatic vessels drain into the jugular vein via a single thoracic duct. Lymphatic pressures are low and pulsatile, and flow is aided by somatic muscles, intrinsic contractions of lymph vessels and arterial pulsation. In amphibians and reptiles lymph hearts return fluid to veins at many points (214). These lymph hearts, and the caudal heart of the hagfish, eel, and shark are chambered organs normally controlled by spinal nerves (52, 209, 210). The toad lymph heart shows an oscillatory action potential, apparently preceded by several motor impulses (59). In intact frogs, ACh may increase contraction amplitude and tonus; high concentrations of ACh may stop the heart (57).

True lymphatics may not be present in elasmobranchs and teleosts (257). A *secondary vascular system*, previously termed the venolymphatic or lymphatic system, exists which has narrow input connections from the arterial side of the primary branchial and systemic circulation. It is apparently a low pressure (64), low flow, and high volume system.

Invertebrate Cardiovascular Systems

It is difficult to make generalizations about circulatory systems in invertebrates because of their diversity. A chambered or tubular heart usually pumps oxygenated blood and its pacemaker is either myogenic or neurogenic, although the distinction is not always clear. The circulatory system typically involves in-series circuits, but it is not necessarily valved for one-way flow. Even the distinction between open and closed circulations is not always clear. For example, *Helix* have an open circulation but some vessels are lined with a discontinuous endothelium. The reason for this diversity undoubtedly reflects the fact that, unlike vertebrates, the primary role of the invertebrate circulatory system is not necessarily respiratory transport. Nutrient transport and hydraulic function are often more important. In soft-bodied animals, body movement can significantly alter vascular resistance and hence blood distribution. Nonetheless, whenever the primary role is respiratory transport, the circulatory system is generally more sophisticated and is similar to vertebrate systems. For example, squid and octopus have closed circulations and powerful hearts; less active molluscs such as *Mytilus*, *Geukensia*, and *Placopecten* have open circulations. The following describes the cardiovascular systems within the major invertebrate phyla. This phylogenetic approach is used more for convenience and does not imply an evolutionary hierarchy between phyla.

Molluscs

The anatomy of the chambered heart in molluscs is highly variable (228, 263). Cephalopods have three hearts; a systemic heart, with two atria and one ventricle, and two accessory branchial hearts. *Nautilus* has a systemic heart with four atria and one ventricle. Bivalves, gastropods, and polyplacophora have a systemic heart with one or two atria and one ventricle. Monoplacophora have two ventricles, each with paired atria.

Open and closed circulatory systems are found among molluscs (228). The closed circulation in cephalopods is characterized by low blood volumes (5–6%), high systolic blood pressures (27–56 mmHg), and a 1–2 min circulation time (232). The systemic heart is thick walled, with the relative heart weight in *Octopus vulgaris* being similar to that in some teleost fish. A coronary circulation is derived from the lumen of the ventricle and drains into the pericardium. Coronary flow is propelled by ventricular systole (114). The bilateral branchial hearts are thin walled, generate lower pressures (Table 1) and their contraction is influenced by visceral nerves, respiratory movements, and possibly peristaltic contraction of the lateral vena cava (263). Contraction of the branchial heart aids blood flow through the gills and provides ultrafiltration across its wall for excretory function. *Nautilus* lack a branchial heart and so vein peristalsis and branchial irrigation must be important for gill blood flow.

The open circulation found in gastropods and bivalves is characterized by larger blood volumes (36–83%), lower systolic pressures (1.5–18 mmHg), and longer circulation times (4–55 min) (Table 1) (137). However, circulation time is 4–6 min and blood pressure and ventricular thickness are several times greater in the terrestrial pulmonate *Helix* than in bivalves and marine pulmonates (Fig. 12). Pericardial pressure in *Helix* is lower than in the heart chambers at all stages of contraction and aids heart filling (50). Ultrafiltration also occurs across the ventricle or auricle into the pericardial cavity. The renopericardial duct connects the pericardial cavity to the nephridia.

Molluscan hearts are myogenic and function as a syncitium; cardiac fibers are connected by intercalated disks, but may lack T-tubules, for example, osyter (117, 119, 207). The pacemaker lacks morphological specialization, and cardiac fibers

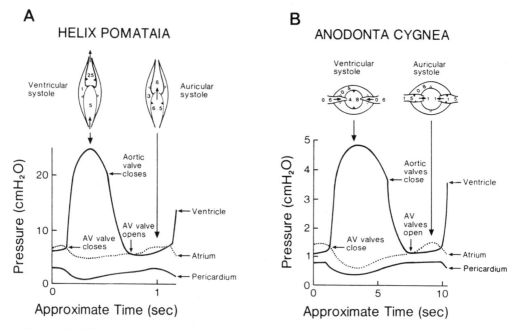

Figure 12. The cardiac cycles in (A) *Helix pomataia,* a terrestrial pulmonate and (B) *Anodonta cygnea,* a freshwater bivalve. The higher cardiac frequency and pressure development in *Helix* reflects its terrestrial life style and a more "closed" circulation. [Adapted from (137).]

have dimensions similar to the mammalian pacemaker. The AV junction may be more important as a pacemaker because most nerve endings are located there. The prepotential (-40 to -60 mV), mainly resulting from K^+ distribution, slowly depolarizes to generate an action potential that rarely goes positive. Three general forms of action potential are recognized; *spike potentials* with rapid depolarization and hyperpolarization (e.g., *Aplysia*), *slow potentials* with slow depolarization and hyperpolarization (e.g., *Mya*), and *spike-plateau potentials* (e.g., *Dolabella*) (263). The spike potential is Ca^{2+} dependent, and the slow potential and plateau appear to be Na^+ dependent (117, 119, 120). Force of contraction is increased by plateau height which is increased by stretch and 5-hydroxytryptamine (5-HT) and decreased by ACh.

Cardiac Regulation
Mass specific Q is comparable to vertebrates (Table 1). *Octopus* increase Q, SV_H, and arterial pressure approximately threefold during activity; f_H and O_2 extraction change little (113). Hypoxia produces a progressive decrease in Q, f_H, and arterial pressure (264). The swimming scallop, *Placopecten*, doubles both f_H and SV_H immediately after swimming (243). In *Aplysia*, food arousal elicits a neurally mediated increase in f_H and blood pressure, and rhythmic biting produces a cyclic alternation of blood flow between the head and gut; flow to the gut is restricted by vasoconstriction of the abdominal artery during the protraction phase of biting (141). Penis eversion, proboscis eversion, siphon extension, and burrowing also require higher blood pressures and blood redistribution (47, 252, 253). Regulation of

Q (SV_H and f_H) and pressure development in molluscs involve intrinsic, neural, and humoral mechanisms.

INTRINSIC CONTROL. The intrinsic properties of the molluscan heart show similarities with those of the vertebrate heart (77, 232, 263). Isolated hearts from bivalves, gastropods, and cephalopods increase SV_H and f_H with increased filling pressure (Fig. 13). With an increase in output pressure, isolated hearts from *Busycon* and *O. vulgaris*, but not *Eledone* maintain Q and f_H. However, the functional significance of these intrinsic mechanisms is unclear since the filling pressure of isolated hearts is often very high and the isolated heart may not beat unless stretched or treated with 5-HT (80, 109). Such observations strongly support an important role for the pericardium (50) and tonic humoral or neural stimulation (234). As with vertebrate hearts, f_H is dependent on temperature, for example, *Noetia* 14.5 bpm at 25°C and 5 bpm at 5°C, and myocardial mechanical efficiency is directly related to performance; efficiency is 6–26% in *Helix* (107) and 2.5–6.5% in *O. vulgaris* (114).

EXTRINSIC CONTROL. Combined inhibitory and excitatory cardiac innervation may be universal in molluscs since it has always been revealed by selective blockade of neurotransmitters. In *Aplysia*, inhibitory nerve fibers terminate on the atrium and excitatory fibers on the ventricle. In pelecypods, gastropods, and *Octopus* both types of fiber originate in the visceral ganglion and either inhibition (e.g., *Mya*, *Anodonta*, *Mercenaria*, and *Octopus*), or excitation (e.g., chitons, *Aplysia*, *Haliotis*, and *Ariolimax*) may dominate with nerve stimulation (147). In *Helix*, *Limax*, and a nudibranch, *Triopha*, inhibitory and excitatory fibers arise in the pleural ganglion. In the cockle *Clinocardium*, the cerebral ganglion excites and the visceroparietal ganglion inhibits the heart (226).

Acetylcholine is usually the inhibitory neurotransmitter, sometimes with a very low threshold (1–100 pM in *Mercenaria*). The ACh receptors are blocked by tetraethylammonium and benzoquinonium, but not atropine (205, 206) and therefore differ from those in vertebrate hearts. Acetylcholine usually produces negative inotropy, but a weak negative chronotropy is possible (225). It decreases f_H and SV_H in isolated perfused *Busycon* hearts and the action potential shows a decreased rate of rise of the prepotential, a reduced depolarization, and a longer plateau phase (231). In contrast, ACh produces a strong positive chronotropy in *Mytilus* (163).

The excitatory neurotransmitter is usually 5-HT, which produces positive inotropy and chronotropy in many bivalves and gastropods (Fig. 13). 5-Hydroxytryptamine is blocked by methysergide and depleted by reserpine (92, 158, 163). It is favored as the excitatory neurotransmitter in cephalopods (93, 205, 233). Dopamine, EP, and NE are found in cardiac nerves, for example, *Helix*, *Aplysia*, and *Lymnaea*, and are usually excitatory with dopamine being the more potent (163, 206).

Several cardioactive peptides have been identified in molluscs, but studies of these peptides are in their early days. Currently, the most prominent of these is FMRFamide, a tetrapeptide (Phe-Met-Arg-Phe-NH$_2$) (93). It is released into the heart by neurosecretion and levels of 1 nM are found in molluscan blood. Generally, FMRFamide has similar excitatory effects as 5-HT, but they are not blocked by methysergide. FMRFamide accelerates perfused *Busycon* hearts at low concentrations, but reduces SV_H at high concentrations by reducing end-diastolic volume (234). FMRFamide increases the rate of rise of the prepotential and the amplitude of the plateau. The duration of the plateau is shortened at low concentrations and

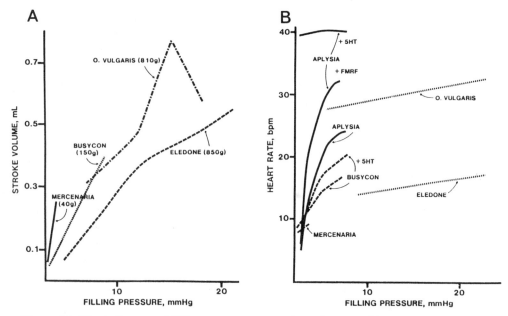

Figure 13. The influence of filling pressure on stroke volume and heart rate in isolated, perfused hearts from various molluscs. Unusually high filling pressures are generally required to generate normal stroke volume. Also the response of heart rate to stretch is greatly influenced by pretreatment with cardioactive drugs (5-HT, FMRFamide). [Adapted from (230,231,233).]

extended at high concentrations. FMRFamide may be inhibitory in oysters in the summer and inactive in the winter. Thus, FMRFamide is not universally excitatory or even active among molluscs.

In summary, cardiovascular control in the intact mollusc is far from completely understood and probably shows considerable species variability. Nonetheless, the control is highly integrated, probably with neural and humoral stimulation providing a variable cardiac tonus against which intrinsic and other extrinsic factors are expressed (234). For example, stretching the heart modulates the effect of ACh in *Busycon* (233). Also, in *Aplysia*, FRMFamide increases f_H if the initial rate is low, whereas, if f_H is first increased by raising perfusion pressure, FRMFamide decreases f_H.

Arthropods

An open circulation of hemolymph is found in arthropods. Small crustaceans, for example, copepods and ostracods, may lack a heart and vessels, but larger malacostracans (particularly the decapods) have a muscular heart and an arterial system that clearly function for respiratory transport. Insects, spiders, and scorpions have a heart and an open circulation. In insects a tracheal system provides gas transport.

Crustaceans and Limulus

Typically the dorsal heart is suspended in a pericardial sinus by *alary ligaments*, which restore diastolic volume of the heart after systole by elastic recoil. The tubular heart may extend over many segments and pumps hemolymph to tissue sinuses via many paired arteries (169). Malacostraca may have an enlarged dorsal anterior artery, which acts as an accessory heart (*cor-frontale*) (236). There are no veins and the hemolymph is oxygenated at the gills before returning to the pericardial sinus through valved *ostia*. The number of paired ostia is variable, for example, one in *Daphnia*, three in most decapods, and 16–18 in Anostraca. Ventilatory pumping, which generates oscillations in water pressure (-4 to $+7$ mmHg) may assist hemolymph flow through the gills particularly since heart and scaphognathite frequencies are often synchronized (170). Barnacles lack a heart, but *Lithotrya* has a contractile rostral sinus and *Pollicipes* uses thoracic muscles.

Arterial pressure and Q are comparable to other aquatic invertebrates and some vertebrates (Table 1). Systolic pressures range from 6 to 36 mmHg and are greatly influenced by locomotion. Leg movement in the lobster, *Homarus*, increases ventricular pressures from 13/0.7 to 37/17 mmHg. Circulatory pressures in *Carcinus* are presented in Fig. 14. Fick estimates of Q, typically ranging from 30 to 130 ml/min · kg, have been confirmed by thermal dilution methods in some crustaceans (27). The high Q probably relates to the low O_2 carrying capacity and large volume of hemolymph (~30%) which must be moved. These animals lack fine control of flow distribution. Circulation time in lobster is 1–8 min. In *Homarus* and *Libina*, Q increases and f_H decreases with hypoxia, and Q increases much more than f_H with activity. The mechanism regulating SV_H is unclear but may involve stretch (170, 241).

The heart beat in crustaceans and the horseshoe crab, *Limulus*, is *neurogenic* and f_H is determined by the rate of bursting of a *cardiac ganglionic pacemaker* (Fig. 15). The

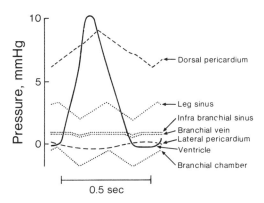

Figure 14. Pressures in various regions of the circulatory system of the crab *Carcinus*. [Adapted from (14).]

cardiac ganglion consists of *small pacemaker* and *large follower cells*, which are electrically coupled. The cardiac ganglion of a lobster has five large anterior follower and four smaller posterior pacemaker neurons (96–98). The follower or motor cells integrate activity from the small pacemaker cells and axons of the follower cells distribute it to the heart muscle. Follower cells show small graded waves, which are postsynaptic potentials (psps) initiated by pacemaker spikes (6–40 per heart beat in *Squilla*). These waves may give rise to spikes which originate at the base of the axon and spread back to the soma. The nerve cell somas apparently do not spike. One or two pacemaker spikes may trigger trains of spikes in follower cells (149). In *Limulus*, recordings from the somas of large neurons show a train of spikes, which may vary in amplitude and correspond to the total ganglionic discharge. Occasionally spikes occur between bursts, and sometimes a slow potential, probably an excitatory psp, precedes the first spike (Fig. 15).

Contractile force is dependent on spike frequency and the number of spikes within a burst. Each beat is evidently tetanic, unlike the nontetanizing contrac-

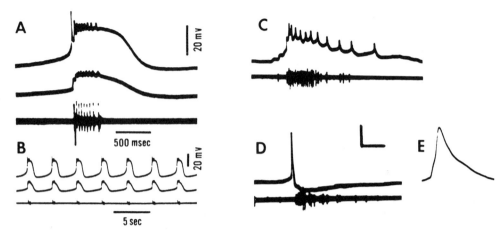

Figure 15. (*A–B*) Electrocardiograms from nerve cells of cardiac ganglion of crustacean *Squilla*. (*A*) top record from neuron 5, middle record from neuron 6, lower record from extracellular recording. (*B*) Records from same source at lower sweep speed. [From (258).] (*C–E*) Records from neurons in the cardiac ganglion of *Limulus*. In (*C* and *D*) upper records are intracellular and lower records are extracellular. (*C*) Bursts from large follower neurons. (*D*) Single spikes from a small pacemaker neuron. Each follower neuron gives multiple spikes on a sustained depolarization in response to one spike in a driver (pacemaker) cell. (*E*) Expanded record of pacemaker cell, to show pacemaker potential preceding spike. Calibration: (*C*) Top 10 mV, bottom 20 μV, 500 ms (*D*) top 20 mV, bottom 20 μV, 500 ms (*E*) 20 mV, 40 ms. (Courtesy of F. Lang.)

tion of the myogenic vertebrate heart. Each muscle fiber may have multiple innervation and shows a series of fused junction potentials, each corresponding to one nerve impulse. Cardiac muscle potentials are caused by increasing Ca^{2+} conductance (256).

Cardioregulatory nerves, humoral factors, stretch, and metabolic factors all modulate f_H at the level of the ganglion (170). Inhibitory (SNII) and excitatory (SNIII) nerves, originating in the CNS, innervate the cardiac ganglion and alter the pacemaker frequency. The heart of *Callinectes* receives two pairs of accelerator and one pair of inhibitory nerve fibers. In *Limulus*, inhibitory nerves arise from the posterior part of the brain; in crustaceans, both inhibitory and accelerator nerves arise from the subesophageal ganglion, the inhibitors lying anterior to the accelerators. The threshold firing rate for accelerators is 5–10 Hz and they increase

the rate of rise of pacemaker potentials (258). The threshold firing rate for inhibitors is 15–20 Hz and they hyperpolarize cardiac neurons, changing their threshold for a spike from 7 to 15 mV depolarization in *Squilla* (29,186). Inhibitory nerves usually override accelerators, but adapt faster than accelerators with continuous stimulation.

The physiological cardiac neurotransmitters in crustaceans are not well defined. Acetylcholine and γ-aminobutyric acid (GABA) are inhibitory and glutamate is excitatory (1). But the glutamate high concentrations (up to 0.1 mM) required to produce these responses raises some uncertainty about their physiological nature. Octopamine, 5-HT, and dopamine increase the rate and force of contraction in *Astacus, Eriphia*, and *Homarus* (170). In *Limulus*, octopamine, dopamine, EP, and NE increase the force and rate of contraction (259, 260). The pentapeptide proctolin increases amplitude, not frequency, of *Limulus* heart; it may have transmitter action on the muscle (129).

Insects, Spiders, and Scorpions

Scorpions lack a tracheal system and hemocyanin is found in the blood of some species. Spiders have a fewer number of heart segments, which is associated with a reduction in book lungs and increased complexity of the tracheal system. The arterial system to the legs is more developed in spiders and exceedingly high fluid pressures (400 mmHg) extend the legs (63). The tarantula and scorpion hearts are probably neurogenic. *Eurypelma* has inhibitory and excitatory cardiac ganglion cells and an oscillatory ECG (30). The cardiac ganglion is excited by ACh, EP, and NE at 1 μM and glutamate and 5-HT at 0.1 mM and is inhibited by GABA at 10 μM (223, 224, 279).

Insects have a tracheal system for gas transport. Their spherical abdominal heart has alary muscles and valved ostia, and develops low hemolymph pressures. Flow along the anterior part of the aorta may be aided by peristaltic waves. In *Locusta* the aortic pressure is normally 6.6/2.5 mmHg at a heart rate of 79 bpm, and increases with body or leg movement. Accessory pumping is important for circulation in appendages where pressures may be negative (-1.9 to -2.7 mmHg) (11). Pulsating organs are found at the base of the wing and antennae and an accessory heart aids circulation in the legs. The f_H value commonly ranges between 20–100 bpm (max 350 bpm) and decreases with successive developmental stages. Light and temperature can set up a circadian rhythm for f_H, which may be related to a circulatory role in temperature regulation (103).

Some insect hearts are neurogenic and others seem to be myogenic. Intracellular recordings from heart muscle of the

blowfly *Sarcophaga* show summed oscillations like those of *Limulus* (20). Oscillatory ECGs occur in *Melanoplus, Galleria,* and *Dytiscus*. Ganglionic pacemakers are located in segments 2–7 of the cicada heart (118). The cockroach, *Periplaneta,* has a cardiac ganglion of six neurons that discharges with each heartbeat. However, rhythmic heartbeats persist after the ganglion is removed or blocked by tetrodotoxin. The cardiac ganglion is stimulated by distension of the heart (229). Acetylcholine is excitatory, evidently stimulating the ganglionic control (171). The hearts of the cockroach, grasshopper, cricket, and honeybee are accelerated by ACh and are stretch sensitive.

The heart of the moth *Hyalophora cecropia* is myogenic. It lacks nerve cells, has no localized pacemaker region, can beat when stretched, and its beat is not affected by ACh or EP (166). The resting potential (-47 mV) is lower than the equilibrium potential for K^+ (-11.6 mV) (164). The action potential is an overshooting spike with delayed repolarization and is dependent on Ca^{2+}, but not Na^+ (165, 167).

In summary, the arthropod heart is spherical or tubular with valved ostia. The origin of the heart beat is neurogenic in crustaceans and *Limulus*, but in spiders, scorpions and insects it may be myogenic or neurogenic.

Annelids

Annelida have a closed circulatory system of coelomic origin which serves a respiratory function since hemoglobin or chlorocruorin are dissolved in the blood. A *ventral longitudinal vessel* distributes blood into segmental arteries and capillaries. A *dorsal longitudinal vessel* acts as the main collecting vessel. Both vessels are contractile and valving ensures one-way flow (78). Paired lateral hearts or heart tubes, which have valves or sphincters, are

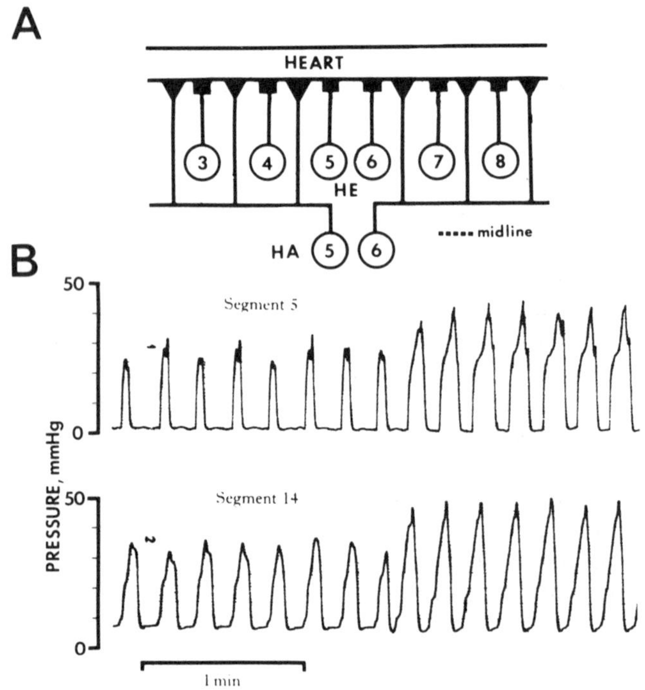

A

Figure 16. (A) Schematic diagram showing the innervation of one of the leeches bilaterally paired hearts in the third to eighth body segments. Circles indicate cell bodies, lines are processes, and the numbers are body segments. Filled squares and triangles indicate neuromuscular contacts of the HE and HA neurones. (B) Simultaneous recordings of pressures in the right lateral vessel of the leech showing the transition in pressure pulse. [Adapted from (40 and 142).]

found in Oligochaeta, Hirudinea, and some Polychaeta. The hearts are considered myogenic (163).

Oligochaetes and Polychaetes
Among polychaetes, *Nereis* and *Nephthys* lack heart tubes and contraction frequency of the dorsal and ventral vessels is independent. *Arenicola* has two tubular hearts and *Flabellidema* has one. Oligochaetes have paired tubular hearts linking the dorsal and lateral vessels in the anterior segments. In *Magoscolex giganticus*, a synchronous heartbeat of 20 bpm at 20°C is accompanied by dorsal vessel pulsations of 8/min that apparently aid heart function. The tubular hearts and dorsal vessel are innervated. Acetylcholine and EP increase the rate of contraction in *Arenicola* and *Lumbricus* (189). Blood pressure can increase with activity from 54 to 76 mmHg in the ventral vessel (129). Coelomic pressures also increase considerably with burrowing to provide rigidity, for example, from 1.5 to 85 mmHg in *Ar-*

enicola (250, 251), from 1.5 to 13 mmHg in *Nereis* and *Glycera*, from 0.2 to 9 mmHg in *Lumbricus* (218), and up to 460 mmHg in *Sipunculus* (278).

Leeches
Unlike other segmented worms, leeches have paired *lateral vessels* that act as tubular hearts. High- and low-pressure pulsations alternate in a coordinated and antiphasic fashion between the lateral vessels (Fig. 16). The high-pressure pattern (48/5 mmHg) results from a peristaltic wave mainly in the anterior segments. Systolic pressures may reach 100 mmHg. The low-pressure pattern (26/4 mmHg) is a more widespread, nonperistaltic synchronous constriction. Transition between the high- and low-pressure patterns occurs after 20–60 pulses and systolic pressure is independent of contraction frequency. These myogenic pulsations are normally modulated by the CNS in *Hirudo* (38, 39).

Spiral vascular muscle cells receive direct innervation (Fig. 16). Pairs of heart motor (HE) neurons are found in segments 3–18. Pairs of heart accessory (HC) neurons are found in segments 5 and 6, but innervate several other segments. The heart motor neurons regulate heart rhythm with rhythmic bursts of action potentials that produce unitary excitatory potentials in heart muscle and are mediated by nicotinic cholinoceptors (40, 255). Heart interneurons (HN) provide bursts of inhibitory psps to the HE neurons to set the pattern of excitation (4). Heart accessory neurons modulate cardiac force and myogenic properties without generating junction potentials in muscle cells through a noncholinergic, perhaps FMRFamide-like transmitter (40). Temperature, water O_2 tension, swimming, and mechanical stimulation all influence the rhythm of the heart tubes through CNS effects (55).

Echinoderms

Echinoderms have three separate tubular complexes, the *water vascular, perihemal,* and *hemal systems,* which lie in parallel to each other within the perivisceral coelom. They consist basically of a circumoral vessel, five radial vessels, and an axial component that runs from the circumoral vessel to the dorsal surface. The axial component of the hemal system forms the spongy *axial organ,* which may connect to a small pulsating "heart." The three tubular components all probably serve a circulatory function, but the perivisceral coelom, which supports respiratory exchange across the tube feet, is probably the principle circulatory and hydraulic system (152).

Sea cucumbers have a highly developed hemal system in which endothelial-lined dorsal and ventral vessels parallel the gut and left respiratory tree. Peristalsis in the dorsal vessel propels hemal fluid in an anterior direction to numerous single-chambered, peristaltic hearts (120–150 in *Isostichopus badionotus*) (108). These hearts pump hemal fluid to the ventral vessel via specialized lamellar channels located inside the anterior gut. Peristaltic activity in the dorsal vessel is increased by EP and reduced by ACh (190, 276). The respiratory tree has a remarkable arrangement of vascular follicles and shunt lamellae extending between the dorsal and ventral vessel. This intricate closed vascular system of the sea cucumber clearly serves a respiratory function.

Urochordates and Cephalochordates

The sea squirt, *Ciona,* has a tubular dorsal heart that is myogenic, consists of spiral striated muscle, and lacks valves (144, 145, 197). Independent pacemakers, located at either end, alternately dominate to alter the direction of a peristaltic wave, which travels at ~3–6 mm/s (3). Neither flow direction brings blood directly from the gills to the heart. (Bidirectional flow is also found in the closed circulation of Nemerteans, where lateral vessels contract irregularly.) The stiff pericardium promotes expansion of one region of the heart when another is contracted. The cardiac resting potential is about -71 mV and a spike potential hyperpolarizes the cell. Muscle fibers are electrically coupled and the action potential is conducted at ~77 mm/s. Excitation–contraction coupling probably involves Ca^{2+} influx, since extracellular Ca^{2+} applied to the lumen of the heart increases the force of contraction. Heart rate in *Ciona* is ~50 bpm and is increased by internal pressure; the effects of EP and ACh are conflicting (144, 215).

The cephalochordate *Branchiostoma* has a well-defined circulation often connected by discrete vessels with a discontinuous endothelium. *Branchiostoma* lacks a heart, but three contractile blood vessels, the branchial artery, subintestinal vein and hepatic vein, contract myogenically in succession about once per minute. They deliver blood to the systemic circulation via the gills, as in fish.

Concluding Remarks

To be of adaptive value, a circulatory system must supply blood where it is needed, in adequate amounts, and when required. Open systems (coelom, pseudocoel, or hemocoel) without hearts are adequate where respiratory and nutritional demands are not great and where diffusion distances are short. Tubular or chambered hearts came into use in some open systems, as in crustaceans and molluscs, with increased body size and activity. These hearts generally develop low pressures; circulatory flow is assisted by contraction of somatic muscles and circulation time is long since blood volume is high. Chambered hearts are associated with closed systems, as in vertebrates and cephalopods, and develop high systemic pressures and flow rates; circulation time is short because blood volume is reduced. Higher systemic pressures were accompanied by the evolution of a low-pressure pulmonary circulation and a lymphatic circulation. Accessory hearts assist flow in the lymphatic and secondary circulations found in lower vertebrates. Vascular smooth muscle controls regional blood flow distribution and arterial pressure through a variety of mechanisms that may be reflex, humoral, and local. These mechanisms show organ and species specificity. The nature of pressure regulating systems in invertebrates is unknown and control of peripheral circulation is less precise with many organs lying in series rather than in parallel as in the vertebrates.

Heart muscle in all animal groups has some of the properties of somatic muscle and some of visceral muscle. The vertebrate and molluscan hearts contract in an all-or-none fashion and the heart beat is myogenic. Many adult arthropods have neurogenic hearts; a small number of neurons constitute the crustacean pacemaker ganglian. Cardiac muscles generally have prolonged action potentials and may be modified as pacemaker cells, fast conducting cells and muscles with varying durations of depolarization. Membrane channels are varied; most have sodium conductance and others calcium.

In most animals, the intrinsic properties of the cardiac muscle are modulated by extrinsic factors, (neural or humoral). Nervous regulation of heart beat and contractility is possible through inhibitory cholinergic and excitatory adrenergic fibers in vertebrates. Excitatory control is more prominent in higher than in lower vertebrates.

The evolution of circulatory mechanisms shows many parallel developments. Yet, varied patterns of peripheral circulation and cardiac physiology and pharmacology exist. Each circulatory strategy is well suited to the needs of a given animal.

Acknowledgments

Constructive comments provided by Dr. P. Belton, Dr. M. Graham, Dr. S. Holmgren, Dr. D. Jorgensen, Dr. L. Milligan, Dr. P. Mladenov, Dr. S. Nilsson, and Dr. M. Smith were greatly appreciated.

References

1. Abbott, B. C., F. Lang, and I. Parnas. *Comp. Biochem. Physiol.* 28:149–158, 1969; in *Comparative Physiology of the Heart* (F. McCann, ed.), pp. 232–243. Birkhaeuser, Basel, 1969. Muscle and ganglion of the heart of *Limulus*.

2. Allard, J. R., E. W. Gertz, E. D. Verrier, J. D. Bristow, and J. I. E. Hoffman. *Cardiovasc. Res.* 17:595–603, 1983. Role of the pericardium in the regulation of myocardial blood flow and its distribution in the

normal and acutely failing left ventricle of the dog.

3. Anderson, M. *J. Exp. Biol.* 49:363–385, 1968. Initiation and reversal of heart beat in *Ciona*.

4. Arbas, E. A. and R. L. Calabrese. *J. Comp. Physiol. A* 155A:783–794, 1984. Rate modification in the heartbeat central pattern generator of the medicinal leech.

5. Ask, J. A., G. Stene-Larsen, and K. B. Helle. *J. Comp. Physiol.* 143:161–168, 1981. Temperature effects on the β_2-adrenoceptors of the trout atrium.

6. Axelsson, M. and S. Nilsson, *J. Exp. Biol.* 126:225–236, 1986. Blood pressure control during exercise in the Atlantic cod, *Gadus morhua*.

7. Bagshaw, R. J. *Biol. Rev. Cambridge Philos. Soc.* 60:121–162, 1985. Evolution of cardiovascular baroreceptor control.

8. Baker, L. A. and F. N. White. *Comp. Biochem. Physiol.* 35:253–262, 1970. Cardiac output in *Iguana*.

9. Banchero, N., R. F. Grover, and J. A. Will. *Am. J. Physiol.* 220:422–427, 1971. Pulmonary hypertension in llama at high altitude.

10. Barr, L. and W. Berger. *Pfluegers Arch.* 279:192–194, 1964; Barr, L. et al. *J. Gen. Physiol.* 48:797–823, 1965. Conduction and structure in cardiac muscle.

11. Bayer, R. *Z. Vergl. Physiol.* 58:76–135, 1968. Circulation in *Locusta*.

12. Belkin, D. A. *Copeia* 1964:321–330, 1964. Changes in heart rate during diving in *Psuedemys*.

12a. Benson, J., R. Sullivan, W. Watson, and G. Augustine. *Brain Res.* 113:449–454, 1981. Proctolin as potential transmitter to Limulus heart.

13. Berkson, H. *Comp. Biochem. Physiol.* 21:507–524, 1967. Physiology of diving in *Chelonia*.

14. Blatchford, J. G. *Comp. Biochem. Physiol. A* 39A:193–202, 1971. Hemodynamics of *Carcinus*.

15. Blix, A. S. and B. Folkow. In *Handbook of Physiology* (J. T. Shephard and F. M. Abboud, eds.), Sect. 3, pp. 917–945. Am.

Physiol. Soc., Bethesda, MD, 1983. Cardiovascular adjustments to diving in mammals and birds.

16. Bloom, G., E. Östlund, U. S. v Euler, F. Lishajko, M. Ritzen, and J. Adams-Ray. *Acta Physiol. Scand.* 53(Suppl. 185):1–38, 1961. Catecholamines in hearts of cyclostomes.

17. Bohr, D. F. In *Electrolytes and Cardiovascular Diseases* (E. Basjusz, ed.), pp. 342–355. Karger, Basel, 1965. Differences among vascular smooth muscles.

18. Bond, C. F. and P. W. Gilbert. *Am J. Physiol.* 194:519–521, 1958. Blood volume of pigeon.

19. Bron, K. M., H. V. Murdaugh, Jr., J. E. Millen, R. Lenthall, P. Raskin, and E. D. Robin, *Science* 152:540–542, 1966. Vasomotor responses in diving mammals.

20. Bruen, J. P. and R. C. Ballard. *Comp. Biochem. Physiol.* 32:227–236, 1970. Action potentials in heart of fly *Sarcophaga*.

21. Bull, J. M. and R. Morris. *J. Exp. Biol.* 471:485–494, 1967. Blood volume in lamprey.

22. Burggren, W. W. *Science* 215:77–78, 1982. Pulmonary plasma filtration in the turtle: A wet vertebrate lung?

23. Burggren, W. W. In *Cardiovascular Shunts* (K. Johansen and W. Burggren, eds.), pp. 121–136. Munksgaard, Copenhagen, 1985. Hemodynamics and regulation of central cardiovascular shunts in reptiles.

24. Burggren, W. W. *J. Physiol. (London)* 278:349–364, 1978. Influence of intermittent breathing on ventricular depolarisation patterns in chelonian reptiles.

25. Burggren, W. W. and K. Johansen. *J. Exp. Biol.* 96:343–354, 1982. Ventricular hemodynamics in the monitor lizard, *Varanus exanthematicus:* Pulmonary and systemic pressure separation.

26. Burlington, R. F., M. S. Dean, and A. Darvish. In *Living in the Cold* (H. C. Heller et al., eds.), pp. 549–556. Elsevier, New York, 1986. Cardiovascular responses to low temperature in isolated hearts from rats and 13-lined ground squirrels (*Spermophilus tridecemlineatus*).

27. Burnett, L. E., P. L. de Fur, and D. D. Jorgenson. *J. Exp. Zool.* 218:165–173, 1981. Application of the thermal dilution technique for measuring cardiac output and assessing stroke volume in crabs.

28. Burnstock, G. *Pharmacol. Rev.* 21:247–313, 1969. Evolution of autonomic innervation of visceral and cardiovascular systems in vertebrates.

29. Bursey, C. R. and R. A. Pax. *Comp. Biochem. Physiol.* 35:41–48, 1970. Cardioregulation in *Limulus.*

30. Bursey, C. R. and R. G. Sherman. *Comp. Gen. Pharmacol.* 1:160–170, 1970. Spider cardiac physiology.

31. Burton, A. C. *Physiology and Biophysics of Circulation.* Yearbook Medical Publishers, Chicago, IL, 1972.

32. Butler, P. J. In *Fish Physiology: Recent Advances* (S. Nilsson and S. Holmgren, eds.), pp. 102–118. Croom Helm, London, 1986. Exercise.

33. Butler, P. J. *J. Exp. Biol.* 100:195–221, 1982. Respiratory cardiovascular control during diving in birds and mammals.

34. Butler, P. J. and D. R. Jones. *Adv. Comp. Physiol. Biochem.* 8:179–364, 1982. The comparative physiology of diving in vertebrates.

35. Butler, P. J., W. K. Milson, and A. J. Woakes. *J. Comp. Physiol. B* 154B:167–174, 1984. Respiratory, cardiovascular and metabolic adjustments during steady state swimming in the green turtle, *Chelonia mydas.*

36. Butler, P. J., N. H. West, and D. R. Jones. *J. Exp. Biol.* 71:7–26, 1977. Respiratory and cardiovascular responses of the pigeon to sustained, level flight in a wind tunnel.

37. Butler, P. J. and A. J. Woakes. In *Circulation, Respiration and Metabolism* (R. Gilles, ed.), pp. 39–55. Springer-Verlag, Berlin, 1986. Exercise in normally ventilating and apnoeic birds.

38. Calabrese, R. L. *J. Comp. Physiol.* 122:111–143, 1977. The neural control of alternate heartbeat coordination states in the leech, *Hirudo medicinalis.*

39. Calabrese, R. L. *Am. Zool.* 19:87–102, 1979. Neural regulation of the peristaltic and non-peristaltic heartbeat coordination modes in the leech *Hirudo medicinalis.*

40. Calabrese, R. L. and A. R. Maranto, *J. Exp. Biol.* 125:205–224, 1986. Cholinergic action on the heart of the leech, *Hirudo medicinalis.*

41. Campbell, G. *Comp. Gen. Pharmacol.* 2:287–294, 1971. Pulmonary vascular bed in toad *Bufo marinus.*

42. Canty, A. A. and A. P. Farrell. *Can. J. Zool.* 63:2013–2020, 1985. Intrinsic regulation of flow in an isolated tail preparation of the ocean pout (*Macrozoarces americanus*).

43. Capra, M. F. and G. H. Satchell. *Comp. Biochem. Physiol. C* 58C:41–47, 1977. The differential hemodynamic response of the elasmobranch, *Squalus acanthias,* to the naturally occurring catecholamines adrenaline and noradrenaline.

44. Cech, J. J., D. W. Bridges, D. M. Rowell, and P. J. Balzer. *Can. J. Zool.* 54:1383–1388, 1976. Cardiovascular responses of winter flounder, *Pseudopleuronectes americanus,* to acute temperate increase.

45. Chan, D. K. O. and P. H. Chow. *J. Exp. Zool.* 196:13–26, 1976. The effects of acetylcholine, biogenic amines and other vasoactive agents on the cardiovascular functions of the eel, *Anguilla anguilla.*

46. Chapman, C. B., D. Jensen, and K. Wildenthal. *Circ. Res.* 12:427–440, 1963. Circulatory control of hagfish.

47. Chapman, G. *J. Exp. Biol.* 49:657–667, 1968. Hydraulic system of *Urechis.*

48. Chow, P. H. and D. K. O. Chan. *Comp. Biochem. Physiol. C* 52C:41–45, 1975. The cardiac cycle and the effects of neurohumors on myocardial contractility in the Asiatic eel, *Anguilla japonica,* Timm. and Schle.

49. Citters, R. L. van, D. Franklin, O. Smith, Jr., N. Watson, and R. Elsner. *Comp. Biochem. Physiol.* 16:267–276, 1965. Cardiovascular adaptations to diving in elephant seal.

50. Civil, G. W. and T. E. Thompson. *J. Exp. Biol.* 56:239–247, 1972. Experiments with

the isolated heart of the gastropod *Helix pomatia* in an artificial pericardium.

51. D'Agrossa, L. S. *Am. J. Physiol.* 218:530–535, 1970. Venous vasomotion in bat wing.

52. Davie, P. S. *J. Exp. Biol.* 96:195–208, 1982. Changes in vascular and extravascular volumes of eel muscle in response to catecholamines: The function of the caudal lymphatic heart.

53. Davie, P. S. and M. E. Forster. *Comp. Biochem. Physiol. A* 67A:367–373, 1980. Cardiovascular responses to swimming in eels.

54. Davie, P. S., M. E. Forster, B. Davison, and G. H. Satchell. *Physiol. Zool.* 60:233–240, 1987. Cardiac function in the New Zealand hagfish, *Eptatretus cirrhatus.*

55. Davis, R. L. *J. Exp. Biol.* 123:401–408, 1986. Influence of oxygen on the heartbeat rhythm of the leech.

56. Daxboeck, C. and G. F. Holeton. *Can. J. Zool.* 56:1254–1259, 1978. Oxygen receptors in the rainbow trout, *Salmo gairdneri.*

57. Day, J. B., R. H. Rech, and J. S. Robb. *J. Cell. Comp. Physiol.* 62:33–42, 1963. Electrophysiology of frog lymph heart.

58. Delaney, R. G., S. Lahiri, and A. P. Fishman. *J. Exp. Biol.* 61:111–128, 1974. Aestivation of the African lungfish *Protopterus aethiopicus:* Cardiovascular and respiratory functions.

59. Del Castillo, J. and V. Sanchez. *J. Cell. Comp. Physiol.* 57:29–46, 1961. Bioelectrics of amphibian lymph heart.

60. Duker, G., P.-O. Sjoquist, O. Svensson, B. Wohlfart, and B. W. Johansson. In *Living in the Cold* (H. C. Heller et al., eds.), pp. 565–572. Elsevier, New York, 1986. Hypothermic effects on cardiac action potentials: Difference between a hibernator, hedgehog, and a non-hibernator, guinea pig.

61. Eckert, R. and D. J. Randall. *Animal Physiology.* Freeman, San Francisco, CA, 1977.

62. Eliassen, E. *Nature* (*London*) 192:1047–1049, 1961. Blood volume in hibernating hedgehog.

63. Ellis, C. H. *Biol. Bull.* (*Woods Hole, Mass.*) 86:41–50, 1944. Hydraulics of leg of spider.

64. Farrell, A. P. and D. J. Smith. *J. Exp. Zool.* 216:341–344, 1981. Microvascular pressures in gill filaments of lingcod (*Ophiodon elongatus*).

65. Farrell, A. P. *Can J. Zool.* 60:933–941, 1982. Cardiovascular changes in the unanaesthetised lingcod (*Ophiodon elongatus*) during short-term progressive hypoxia and spontaneous activity; *J. Exp. Biol.* 91:293–305. Cardiovascular changes in the lingcod (*Ophiodon elongatus*) following adrenergic and cholinergic drug infusions.

66. Farrell, A. P., S. Wood, T. Hart, and W. R. Driedzic. *J. Exp. Biol.* 117:237–250, 1985. Myocardial oxygen consumption in the sea raven, *Hemitripterus americanus:* The effects of volume loading, pressure loading, and progressive hypoxia.

67. Farrell, A. P. *Can. J. Zool.* 56:953–958, 1978. Cardiovascular events associated with air breathing in two teleosts, *Hoplerythrinus unitaeniatus* and *Arapaima gigas.*

68. Farrell, A. P. *J. Exp. Biol.* 122:65–80, 1986. Cardiovascular responses in the sea raven, *Hemitripterus americanus*, elicited by vascular compression.

69. Farrell, A. P. *J. Exp. Biol.* 129:107–123, 1987. Coronary flow in a perfused rainbow trout heart; *Physiol. Zool.* 61:213–221, 1988. The role of the pericardium in cardiac performance of the trout (*Salmo gairdneri*).

70. Farrell, A. P., K. R. MacLeod, and B. Chancey. *J. Exp. Biol.* 125:319–345, 1986. Intrinsic mechanical properties of the perfused rainbow trout heart and the effects of catecholamines and extracellular calcium under control and acidotic conditions.

71. Farrell, A. P., K. R. MacLeod, and W. R. Driedzic. *Can. J. Zool.* 60:3165–3171, 1982. The effects of preload, after load and epinephrine on cardiac performance in the sea raven, *Hemitripterus americanus.*

72. Farrell, A. P. *Can. J. Zool.* 62:523–536,

1984. A review of cardiac performance in the teleost heart: intrinsic and humoral regulation; in *Circulation, Respiration and Metabolism* (R. Gilles, ed.), pp. 377–385. Springer-Verlag, Berlin, 1985. Cardiovascular and hemodynamic energetics of fishes.

73. Farrell, A. P. and S. S. Sobin. In *Cardiovascular Shunts* (K. Johansen and W. Burggren, eds.), pp. 270–281. Munksgaard, Copenhagen, 1985. Sheet blood flow: Its application in predicting shunts in respiratory systems; Farrell, A. P. et al. *Am. J. Physiol.* 239:R428–R436, 1980. Sheet blood flow in the secondary lamellae of teleost gills.

74. Feigl, E. O. *Physiol. Rev.* 63:1–205, 1983. Coronary physiology.

75. Fishman, A. P., R. G. DeLaney, and P. Laurent. *J. Appl. Physiol.* 59:285–294, 1985. Circulation adaptation to bimodal respiration in the dipnoan lungfish; Farrell, A. P. pp. 1–22. In *The Pulmonary Circulation, Normal and Abnormal* (A. P. Fishman, ed.). Univ. of Pennsylvania Press, Philadelphia, 1989. Comparative biology of the pulmonary circulation.

76. Folkow, B., N. J. Nilsson, and N. R. Yonce. *Acta Physiol. Scand.* 70:347–361, 1967. Cardiac output of duck.

77. Foti, L., I. T. Genoino, and G. Agnisola. *Comp. Biochem. Physiol.* 82(2):483–488, 1985. In vitro cardiac performance in *Octopus vulgaris*.

78. Fourtner, C. R. and R. A. Pax. *Comp. Biochem. Physiol. A* 42A:627–638, 1972. The contractile blood vessels of the earthworn, *Lumbricus terrestris*.

79. Fozzard, H. O. and W. Sleator. *Am. J. Physiol.* 212:945–952, 1967. Ionic basis of conductance in guinea pig atrial muscle.

80. Fredericq, H. and Z. Bacq. *Arch. Int. Physiol. Biochim.* 49:490–496, 1939; 50:169–184, 1940. Effects of drugs on cephalopod hearts.

81. Furness, J. B. and J. Moore. *Z. Zellforsch. Mikrosk. Anat.* 108:150–176, 1970. Adrenergic innervation of cardiovascular system in lizard *Trachysaurus*.

82. Gannon, B. J. and G. Burnstock. *Comp. Biochem. Physiol.* 29:765–773, 1969. Excitatory innervation of fish heart.

83. Garey, W. *Comp. Biochem. Physiol.* 33:181–189, 1970. Blood volume and cardiac output of fishes.

84. Gesser, H. and E. Jorgensen. *Comp. Biochem. Physiol. C* 76C:199–202, 1983. Effect of vanadate and of removal of extracellular Ca^{2+} and Na^+ on tension development and ^{45}Ca efflux in rat and frog myocardium.

85. Gesser, H. and O. Poupa. *Comp. Biochem. Physiol. A* 76A:559–566, 1983. Acidosis and cardiac muscle contractility: Comparative aspects.

86. Gibbs, C. L. and J. B. Chapman. In *Handbook of Physiology* (R. M. Berne, ed.), Sect. 2, Vol. I, pp. 775–804. Am. Physiol. Soc., Bethesda, MD, 1979. Cardiac energetics.

87. Gleeson, T. T. and A. F. Bennett. In *Circulation, Respiration and Metabolism* (R. Gilles, ed.), pp. 23–38. Springer-Verlag, Berlin, 1985. Respiratory and cardiovascular adjustments to exercise in reptiles.

88. Gleeson, T. T., G. S. Mitchell, and A. F. Bennett. *Am. J. Physiol.* 239:R174–R179, 1980. Cardiovascular responses to graded activity in the lizards *Varanus* and *Iguana*.

89. Glover, W. E., D. H. Strangeways, and W. F. M. Wallace. *J. Physiol. (London)* 194:78P–79P, 1967. Cooling of rabbit arteries.

90. Gonzalez, N. C. and R. L. Clancy. *Respir. Physiol.* 56:289–300. 1984. Catecholamine effect on HCO_3^-/Cl^- exchange and rabbit myocardial cell pH regulation.

91. Graham, M. S., G. L. Fletcher, and R. L. Haedrich. *J. Exp. Zool.* 234:157–160, 1985. Blood viscosity in Arctic fishes.

92. Greenberg, M. J. *Br. J. Pharmacol. Chemother.* 15:365–374, 375–388, 1960. Pharmacology of heart of *Venus*.

93. Greenberg, M. J. and D. A. Price. In *Peptides: Integrators of Cell and Tissue Function* (F. E. Bloom, ed.), pp. 107–125. Raven Press, New York, 1980. Cardioregulatory peptides in Molluscs.

94. Grubb, B. R., D. D. Jorgensen, and M. Conner. *J. Exp. Biol.* 104:193–201, 1983.

Cardiovascular changes in the exercising emu.

95. Guyton, A. C. *Cardiac Physiology.* Saunders, Philadelphia, PA, 1980.

96. Hagiwara, S. *Ergeb. Biol.* 24:287–311, 1961. Electrophysiology of heart in crustaceans; Hagiwara, S., A. Watanabe, and N. Saito. *J. Neurophysiol.* 22:554–572, 1959. Cardiac ganglion of lobster.

97. Hartline, D. K. *J. Exp. Biol.* 47:327–340, 1967. Mapping of cardiac ganglion of lobster.

98. Hartline, D. K. *Am. Zool.* 19:53–66, 1979. Integrative neurophysiology of the lobster cardiac ganglion.

99. Hartline, D. K. and I. M. Cooke. *Science* 164:1080–1082, 1969. Control of cardiac ganglion in lobster.

100. Hauswirth, O., D. Noble, and R. W. Tsien. *Science* 162:916–917, 1968. Action of epinephrine on cardiac muscle.

101. Heath, C. H. and D. D. Jorgensen. *Am. Zool.* 26:50A, 1986. Circulatory function in the horse clam, *Tressus nuttali.*

102. Hedrick, M S., D. A. Duffield, and L. H. Cornell. *Can. J. Zool.* 64:2081–2085, 1986. Blood viscosity and optimal hematocrit in a deep diving mammal, the northern elephant seal (*Mirounga angustirostris*).

103. Heinrich, B. *Science* 169:606–607, 1970. Nervous control of heart of sphinx moth.

104. Heisler, N. and M. L. Glass. In *Cardiovascular Shunts* (K. Johansen and W. Burggren, eds.), pp. 334–347. Munksgaard, Copenhagen, 1985. Mechanisms and regulation of central vascular shunts in reptiles.

105. Heller, L. J., and D. E. Mohrman. *Cardiovascular Physiology.* McGraw-Hill, New York, 1981.

106. Herbert, C. V. and D. C. Jackson. *Physiol. Zool.* 58:670–781, 1985. Temperature effects on the response to prolonged submergence in the turtle *Chrysemys Picta Bellii.* II. Metabolic rate, blood acid-base and ionic changes, and cardiovascular function in aerated and anoxic water.

107. Herold, J. P. *Comp. Biochem. Physiol. A* 53A:435–440, 1975. Myocardial efficiency

108. Herreid, C. F., V. F. Larussa, and C. R. Defesi. *J. Morphol.* 150:423–452, 1976. Blood vascular system of the sea cucumber, *Stichopus moebii.*

109. Hill, R. B. and H. Irisawa. *Life Sci.* 6:1691–1696, 1967. Pressure relations in heart of marine gastropod.

110. Hillman, S. S. *J. Comp. Physiol. B* 154B:325–328, 1984. Inotropic influence of dehydration and hyperosmolal solutions on amphibian cardiac muscle.

111. Holt, J. P., E. A. Rhode, S. A. Peoples, and H. Kines. *Circ. Res.* 10:798–806, 1962. Cardiac output in animals of different sizes.

112. Homma, S. and S. Suzuki. *Jpn. J. Physiol.* 16:31–41, 1966. Sensory discharge from aortic and atrial receptors.

113. Houlihan, D. F., G. Duthie, P. J. Smith, M. J. Wells, and B. J. Wells. *J. Comp. Physiol. B* 156B:683–689, 1986. Ventilation and circulation during exercise in *Octopus vulgaris.*

114. Houlihan, D. F., C. Agnisola, N. M. Hamilton, and I. T. Genoino. *J. Exp. Biol.* 131:175–187. Oxygen consumption of the isolated heart of *Octopus:* Effects of power output and hypoxia.

115. Huggins, S. E. and R. A. Percoco. *Proc. Soc. Exp. Biol. Med.* 119:678–682, 1965. Blood volume in alligators.

116. Hutter, O. F. and W. Trautwein. *J. Gen. Physiol.* 39:715–733, 1956. Bioelectrics of the sinus venosus.

117. Irisawa, H. and M. Kobayashi. *Jpn. J. Physiol.* 13:421–430, 1963. Molluscan heart.

118. Irasawa, H., A. F. Irasawa, and T. Kadotani. *Jpn. J. Physiol.* 6:150–161, 1956. Electrocardiogram of cicada.

119. Irisawa, H. In *Comparative Physiology of the Heart* (F. McCann, ed.), pp. 176–191. Birkhaeuser, Basel, 1969. Ion effects on bivalve heart.

120. Irisawa, H., N. Shigeto, and M. Otani. *Comp. Biochem. Physiol.* 23:199–212, 1969; *Jpn. J. Physiol.* 18:157–168, 1968. Effects

of calcium and sodium on oyster and *Mytilus heart*.

121. Irving, L. et al. *J. Physiol. (London)* 84:187–190, 1935. Branchial pressure receptors in dogfish.

122. Ishimatsu, A., K. Johansen, and S. Nilsson. *Comp. Biochem. Physiol.* C 84C:55–60, 1986. Autonomic nervous control of the circulatory system in the air-breathing fish, *Channa argus*.

123. Jensen, D. *Comp. Biochem. Physiol.* 2:181–201, 1961. Cardiac regulation in aneural heart of hagfish.

124. Jensen, D. *Comp. Biochem. Physiol.* 30:685–690, 1969. Cardiac regulation in lamprey and trout.

125. Jensen, D. *Comp. Biochem. Physiol.* 34:289–296, 1970. Intrinsic cardiac rate regulation in elasmobranch: The horned shark, *Heterodontus pranciseci*, and thornback ray, *Platyrhinoidis triseriata*.

126. Johansen, K. *Biol. Bull. (Woods Hole, Mass)* 118:289–295, 1960. Circulation in hagfish *Myxine*.

127. Johansen, K. *Acta Physiol. Scand.* 60(Suppl.217):1–82, 1963. Cardiovascular dynamics in *Amphiuma*.

128. Johansen, K. *Acta Physiol. Scand.* 62:1–17, 1964; Johansen, K. and T. Aarhus. *Am. J. Physiol.* 205:1167–1171, 1963. Blood flow during submersion in duck.

129. Johansen, K. and A. W. Martin. *J. Exp. Biol.* 43:337–347, 1965. Circulation of giant earthworm *Glossoscolex*.

130. Johansen, K., C. Lenfant, and D. Hanson. *Z. Vergl. Physiol.* 59:157–186, 1968. Cardiovascular dynamics in lungfishes.

131. Johansen, K., and W. W. Burggren. In *Hearts and Heart-like Organs* (G. H. Bourne, ed.), Vol. 1, pp. 61–117. Academic Press, New York, 1980. Cardiovascular function in the lower vertebrates.

132. Johansen, K., C. Lenfant, and D. Hanson. *Fed. Proc. Fed. Am. Soc. Exp. Biol.* 29:1135–1140, 1970. Phylogenetic development of pulmonary circulation.

133. Johansen, K. and H. Gesser. In *Fish Physiology: Recent Advances* (S. Nilsson and S. Holmgren, eds.), pp. 71–85. Croom Helm, London, 1986. Fish cardiology: Structural, hemodynamic, electromechanical and metabolic aspects.

134. Johnson, P. C. *Peripheral Circulation.* Wiley, Toronto, 1978.

135. Jones, D. R. and W. K. Milsom. *J. Exp. Biol.* 100:59–91, 1982. Peripheral receptors affecting breathing and cardiovascular function in non-mammalian vertebrates.

136. Jones, D. R. and D. J. Randall. In *Fish Physiology* (W. S. Hoar and D. J. Randall, eds.), Vol. 7, pp. 425–501. Academic Press, New York, 1978. The respiratory and circulatory systems during exercise.

137. Jones, H. D. In *The Mollusca* (A. S. M. Saleuddin and K. M. Wilbur, eds.), Vol. 5, pp. 189–238. Academic Press, New York, 1983. The circulatory systems of gastropods and bivalves.

138. Jorgensen, D. D., S. K. Ware and J. R. Redmond. *J. Exp. Zool.* 231:309–324, 1984. Cardiac output and tissue blood flow in the Abalone, *Haliotis cracherodii* (Mollusca, Gastropoda).

139. Kadatz, R. *Pfluegers Arch.* 252:1–16, 1949. Effects of acetylcholine and noradrenaline on lung vessel in amphibians.

140. Kawakami, Y., B. H. Natelson, and A. B. DuBois. *J. Appl. Physiol.* 23:964–970, 1967. Cardiovascular reflexes of diving in man.

141. Koch, U. T. and J. Koester. *J. Comp. Physiol.* 149:31–42, 1982. Time sharing of heart power: Cardiovascular adaptations to food-arousal in *Aplysia*.

142. Krahl, B. and I. Zerbst-Boroffka. *J. Exp. Biol.* 107:163–168, 1983. Blood pressure in the leech *Hirudo medicinalis*.

143. Kreski, V. and W. W. Sleator. *Life Sci.* 5:1441–1446, 1966. Development of electrical activity in chick embryo hearts.

144. Kriebel, M. E. *J. Gen. Physiol.* 50:2097–2107, 1967; 52:46–59, 1968; *Biol. Bull. (Woods Hole, Mass)* 134:434–455, 1968; *Life Sci.* 7:181–186, 1968; *Am. J. Physiol.* 218:1194–1200, 1970. Action potentials of pacemakers and conduction in tunicate heart.

145. Kriebel, M E. *Biol. Bull.* (*Woods Hole, Mass*) 135:166–173, 1968. Pacemaker properties of tunicate heart.

146. Kulzer, E. *Z. Vergl. Physiol.* 56:63–94, 1967. Heart beat in hibernating bats.

147. Kuwasawa, K. *Sci. Rep. Tokyo Kyoiku Daigaku, Sect. B* 13:111–128, 1967. Nervous regulation in molluscan heart.

148. Lai, N C., J. B. Graham, W. R. Lowell, and R. Shabetai. *Am. Zool.* 26:124A, 1986. Comparative cardiac function in teleosts and elasmobranchs.

149. Lang, F. *J. Exp. Biol.* 54:815–826, 1971. Intracellular studies on pacemaker and follower neurons in *Limulus* cardiac ganglion.

150. Larimer, J. L. and J. T. Tindel. *Anim Behav.* 14:239–245, 1966. Reflex regulation of heart in crayfish.

151. Laurent, P., S. Holmgren, and S. Nilsson. *Comp. Biochem. Physiol. A* 76A:525–542, 1983. Nervous and humoral control of the fish heart: Structure and function.

152. Lavoie, M. E. *Biol. Bull.* (*Woods Hole, Mass.*) 111:114–122, 1956. Hydrostatic movement in starfish.

153. Lee, R. M. *J. Insect Physiol.* 6:36–51, 1961. Blood volume in locust.

154. Lieberman, M. and A. P. de Carvalho, *J. Gen. Physiol.* 49:351–363, 1965. Electrocardiogram of chick embryo.

155. Lillywhite, H. B. *Am. Zool.* 27:81–95, 1987. Circulatory adaptations of snakes to gravity.

156. Lind, A. R. *Circulation* 41:173–176, 1970. Cardiovascular response to exercise in man.

157. Lind, A. R. and G. W. McNicol. *Can. Med. Assoc. J.* 96:706–713, 1967. Muscular factors which determine the cardiovascular responses to sustained and rhythmic exercise.

158. Loveland, R. E. *Comp. Biochem. Physiol.* 9:95–104, 1963. Serotonin as excitatory transmitter to heart of *Mercenaria*.

159. Lund, G. F. and H. Dingle. *J. Exp. Biol.* 48:265–277, 1968. Vagal control of diving bradycardia in frog.

160. Lyman, C. P. and R. C. O'Brien. *Bull.*

Mus. Comp. Zool. 124:353–370, 1960. In Mammalian hibernation. Circulatory changes in the thirteen-lined ground squirrel during the hibernation cycle.

161. Lyman, C. P., J. S. Willis, A. Malan, and L. C. H. Wang. *Hibernation and Torpor in Mammals and Birds,* pp. 54–76. Academic Press, New York, 1982.

162. Martin, A. W. *J. Exp. Biol.* 35:260–279, 1958. Blood volume in molluscs.

163. Martin, A. W. In *Hearts and Heart-Like Organs* (G. H. Bourne, ed.), Vol. 1, pp. 1–39. Academic Press, New York, 1980. Some Invertebrate myogenic hearts: The hearts of worms and molluscs.

164. McCann, F. V. *J. Gen. Physiol.* 46:803–821, 1963. Electrophysiology of insect heart.

165. McCann, F. V. and C. R. Wira. *Comp. Biochem. Physiol.* 22:611–615, 1967. Ionic gradients in lepidopteran hearts.

166. McCann, F. V. *Comparative Physiology of the Heart.* Birkaeuser, Basel, 1969.

167. McCann, F. V. *Comp. Biochem. Physiol. A* 40A:353–357, 1971. Ionic basis of moth heart potentials.

168. McDonald, D. A. *Blood Flow in Arteries.* W. and J. Mackay & Co. Ltd., London, 1960.

169. McLaughlin, P. A. In *The Biology of Crustacea* (L. H. Mantel, ed.), Vol. 5, pp. 1–52. Academic Press, New York, 1983. Internal anatomy.

170. McMahon, B. R. and J. L. Wilkens. In *The Biology of Crustacea* (L. H. Mantel, ed.), Vol. 5, pp. 289–372. Academic Press, New York, 1983. Ventilation, perfusion and oxygen uptake.

171. Miller, T. *J. Insect Physiol.* 14:1265–1275, 1968. Neuronal control of cockroach heart.

172. Murdaugh, H. V., E. D. Robin, J. E. Millen, W. F. Drewry, and E. Weiss. *Am. J. Physiol.* 210:176–180, 1966. Cardiac output in seal.

173. Nagel, E. L., P. J. Morgane, W. L. McFarland, and R. E. Galliano. *Science* 161:898–900, 1968. Cerebral circulation in dolphin.

174. Neely, J. R., H. Liebermeister, E. J. Bat-

tersby, and H. E. Morgan. *Am. J. Physiol.* 212:804–814, 1967. Effect of pressure development on oxygen consumption by isolated rat heart.

175. Neumann, P., G. F. Holeton, and N. Heisler. *J. Exp. Biol.* 105:1–14, 1983. Cardiac output and regional blood flow in gills and muscles after strenuous exercise in rainbow trout (*Salmo gairdneri*).

176. Niedergerke, R. *J. Physiol.* (*London*) 143:486–503, 1958; 167:515–550, 551–586, 1963. Movements of calcium ions in frog heart.

177. Nijima, A. and D. L. Winter. *Science* 159:434–435, 1968. Baroreceptors in adrenal gland.

178. Nilsson, S. *Autonomic Nerve Function in the Vertebrates.* Springer-Verlag. Berlin, 1983.

179. Nilsson, S. In *Fish Physiology* (W. S. Hoar and D. J. Randall, eds.), Vol. 10A, pp. 185–227. Academic Press, Orlando, FL, 1984. Innervation and pharmacology of the gills.

180. Nilsson, S. In *Fish Physiology: Recent Advances* (S. Nilsson and S. Holmgren, eds.), pp. 86–101. Croom Helm, London, 1986. Control of gill blood flow.

181. Nishimura, H. and J. Bailey. *Kidney Int.* 22:5185–5192, 1982. Intrarenal renin-angiotensin system in primitive vertebrates.

182. Noble, P. and R. W. Tsien. *J. Physiol.* (*London*) 195:185–214, 1968. Membrane properties of cardiac muscle.

183. Opdyke, D. F., J. Bullock, N. E. Keller, and K. Holmes. *Am. J. Physiol.* 244:R641–R645, 1983. Dual mechanism for catecholamine secretion in the dogfish shark *Squalus acanthias*.

184. Ormond, A. P. and J. M. Rivera-Velez. *Proc. Soc. Exp. Biol. Med.* 118:600–602, 1965. Blood volume in rat.

185. Parer, J. T. and J. Metcalfe. *Respir. Physiol.* 3:151–159, 1967. Cardiac output, monotremes.

186. Pax, R. A. *Comp. Biochem. Physiol.* 28:293–305, 1969. Cardioregulation in *Limulus*.

187. Popovic, V. *Am. J. Physiol.* 207:1345–1348, 1964. Cardiac output of hibernating ground squirrels.

188. Priede, I. G. *J. Exp. Biol.* 60:305–319, 1974. The effects of swimming activity and section of the vagus nerves on heart rate in rainbow trout.

189. Prosser, C. L. *J. Cell. Comp. Physiol.* 21:295–305, 1943; *Biol. Bull.* (*Woods Hole, Mass.*) 98:254–257, 1950. Pacemaker activity in *Limulus* and *Arenicola*.

190. Prosser, C. L. and C. L. Judson. *Biol. Bull.* (*Woods Hole, Mass.*) 102:249–251, 1952. Pharmacology of haemal vessels of *Stichopus californius*.

191. Randall, D. J. and G. Shelton. *Comp. Biochem. Physiol.* 9:229–239, 1963. Respiratory and cardiovascular reflexes in fish.

192. Randall, D. J. *J. Exp. Biol.* 100:275–288, 1982. The control of respiration and circulation in fish during exercise and hypoxia.

193. Randall, D. J. and C. Daxboeck. *Can. J. Zool.* 60:1135–1140, 1982. Cardiovascular changes in rainbow trout (*Salmo gairdneri*) during exercise.

194. Randall, D. J., W. Burggren, A. P. Farrell, and M. S. Haswell. *The Evolution of Air Breathing in Vertebrates.* Cambridge Univ. Press, London, 1981.

195. Randall, D. J. In *Fish Physiology* (W. S. Hoar and D. J. Randall, eds.), Vol. 4, pp. 132–172. Academic Press, New York, 1970. The circulatory system.

196. Randall, D. J. In *Cardiovascular Shunts* (K. Johansen and W. Burggren, eds.), pp. 71–82. Munksgaard, Copenhagen, 1985. Shunts in fish gills.

197. Randall, D. J. and P. S. Davie. In *Hearts and Heart-Like Organs* (G. H. Bourne, ed.), Vol. 1, pp. 41–59. Academic Press, New York, 1980. The hearts of Urochordates and Cephalochordates.

198. Randall, W. C. *Nervous Control of Cardiovascular Function.* Oxford Univ. Press, New York, 1984.

199. Redmon, J. R. *J. Cell. Comp. Physiol.* 46:209–247, 1955. Cardiac output in crustaceans.

200. Reuter, H. *J. Physiol. (London)* 192:479–492, 1967; 197:233–253, 1968. Membrane properties of Purkinje fibers.

201. Reuter, H. *Nature (London)* 301:569–574, 1983. Calcium channel modulation by neurotransmitters, enzymes and drugs.

202. Rhodin, J., A. G. P. Missier, and L. C. Reid. *Circulation* 24:349–367, 1961. Conduction in beef heart.

203. Robertson, J. D. *Biol. Bull. (Woods Hole, Mass.)* 138:157–183, 1970. Inulin space in arthropods.

204. Robinson, T. F., S. M. Factor, and E. H. Sonnenblick. *Sci. Am.* 6:84–91, 1986. The heart as a suction pump.

205. Rozsa, K. and I. Z. Nagy, *Comp. Biochem. Physiol.* 23:373–382, 1967. Neuroendocrines of heart of snail.

206. Rozsa, K. and L. Perenyi. *Comp. Biochem. Physiol.* 19:105–113, 1966. Serotonin as excitor in *Helix* heart.

207. Sanger, J. W. *Am. Zool.* 19:9–28, 1979. Cardiac fine structure in selected arthropods and molluscs.

208. Satchell, G. H. *Fed. Proc., Fed. Am. Soc. Exp. Biol.* 29:1120–1123, 1970; Satchell, G. H. et al. *J. Exp. Biol.* 52:721–726, 1970. Regulation of peripheral circulation in fish.

209. Satchell, G. H. *Acta Zool.* 65:125–134, 1984. On the caudal heart of *Myxine* (Myxinoidea: Cyclostomata).

210. Satchell, G. H. and L. J. Weber. *Physiol. Zool.* 60:692–698, 1987. The caudal heart of the carpet shark, *Cephaloscyllium isabella*.

211. Satchell, G. H. In *Respiration of Amphibious Vertebrates* (G. M. Hughes, ed.), pp. 105–124. Academic Press, London, 1976. The circulatory system of air breathing fish.

212. Satchell, G. H. *Circulation in Fishes.* Cambridge Univ. Press, London and New York, 1971.

213. Satchell, G. H. *Acta. Zool.* 67:115–122, 1986. Cardiac function in the hagfish, *Myxine glutinosa* (Myxinoidea: Cyclostomata).

214. Satoh, Y. and T. Nitatori. In *Hearts and Heart-Like Organs* (G. H. Bourne, ed.), Vol. 1, pp. 149–169. Academic Press, New York, 1980. On the fine structure of lymph hearts in Amphibia and Reptiles.

215. Scudder, C. L., T. K. Akers, and A. G. Karczmar. *Comp. Biochem. Physiol.* 9:307–312, 1963. Electrocardiogram of tunicate.

216. Senturia, J. B., T. A. Campbell, and D. R. Caprette. In *Living in the Cold* (H. C. Heller et al., eds.), pp. 557–564. Elsevier, New York, 1986. Cardiovascular changes in hibernation and hypothermia.

217. Seyama, I. *Am. J. Physiol.* 216:687–692, 1969; Seyama, I. and H. Irisawa. *J. Gen. Physiol.* 50:505–518, 1967. Effect of sodium and calcium on heart of skate.

218. Seymour, M. K. *J. Exp. Biol.* 51:47–58, 1969. Hydrodynamics of *Lumbricus.*

219. Shabetai, R., D. C. Abel, J. B. Graham, V. Bhargava, R. S. Keyes, and K. Witztrium. *Am. J. Physiol.* 248:H198–H207, 1985. Function of the pericardium and pericardioperitoneal canal in elasmobranch fishes.

220. Shelton, G. In *Cardiovascular Shunts* (K. Johansen and W. Burggren, eds.), pp. 100–116. Munksgaard, Copenhagen, 1985. Functional and evolutionary significance of cardiovascular shunts in the Amphibia.

221. Shelton, G. and W. W. Burggren. *J. Exp. Biol.* 64:323–343, 1976. Cardiovascular dynamics of the chelonia during apnoea and lung ventilation.

222. Shepherd, J. T. and P. M. Vanhoutte. *The Human Cardiovascular System: Fact and Concepts.* Raven Press, New York, 1979.

223. Sherman, R. D. and R. A. Pax. *Comp. Biochem. Physiol.* 26:529–536, 1968; *Comp. Gen. Pharmacol.* 1:171–184, 185–195, 1970. Spider cardiac physiology.

224. Sherman, R. O. and R. A. Pax. *Comp. Biochem. Physiol.* 28:487–489, 1969. Electrical activity in spider hearts.

225. Shigeto, N. *Am. J. Physiol.* 218:1773–1779, 1970. Acetylcholine excitation and inhibition in molluscan heart.

226. Silvey, G. E. *Comp. Biochem. Physiol.* 25:257–269, 1968. Nervous regulation of heart in cockle.

227. Smith, F. M. and D. R. Jones. *Can. J. Zool.* 56:1260-1265, 1978. Localization of receptors causing hypoxic bradycardia in trout (*Salmo gairdneri*).

228. Smith, L. S. and J. C. Davis. *J. Exp. Biol.* 43:171–180, 1965. Hemodynamics in molluscs.

229. Smith, N. A. *Experientia* 15:200–205, 1969. Rhythmicity in cardiac ganglion of cockroach.

230. Smith, P. J. S. *J. Exp. Biol.* 93:243–255, 1981. The role of venous pressure in regulation of output from the heart of the octopus, *Eledone cirrhosa* (Lam.).

231. Smith, P. J. S. *J. Exp. Biol.* 119:301–320, 1985. Cardiac performance in response to loading pressures in *Busycon canaliculatum* (Gastropoda) and *Mercenaria mercenaria* (Bivalvia).

232. Smith, P. J. S. In *Circulation, Respiration and Metabolism* (R. Gilles, ed.), pp. 344–355. Springer-Verlag, Berlin, 1985. Molluscan circulation: Hemodynamics and the heart.

233. Smith, P. J. S. and R. B. Hill. *J. Exp. Biol.* 123:243–253, 1986. Cardiac performance in response to loading pressures and perfusion with 5-hydroxytryptamine in the isolated heart of *Busycon canaliculatum* (Gastropoda).

234. Smith, P. J. S. and R. B. Hill. *J. Exp. Biol* 127:105–120, 1987. Modulation of output from an isolated gastropod heart: Effects of acetylcholine and FMRFamide.

235. Sommer, L. S., W. L. McFarland, R. E. Galliano, E. L. Nagel, and P. J. Morgane. *Am. J. Physiol.* 215:1498–1505, 1968. Hemodynamics and coronary flow in dolphin *Tursiops*.

236. Steinacker, A. *Am. Zool.* 19:67–76, 1979. Neural and neurosecretory control of the decopod crustacean auxiliary heart.

237. Stephenson, R., P. J. Butler, and A. J. Woakes. *J. Exp. Biol.* 126:341–361, 1986. Diving behaviour and heart rate in tufted ducks (*Aythya fuligula*).

238. Stevens, E. D., G. R. Bennion, D. J. Randall, and G. Shelton. *Comp. Biochem.*

Physiol. 43:681–695, 1972. Factors affecting arterial blood pressures and blood flow from the heart in intact, unrestrained ling cod, *Ophiodon elongatus*.

239. Sturkie, P. D. *Proc. Soc. Exp. Biol. Med.* 123:487–488, 1966. Cardiac output in ducks.

240. Suga, H., Y. Igarashi, O. Yamada, and Y. Groto. *Basic Res. Cardiol.* 81(Suppl. 1):39–50, 1986. Cardiac oxygen consumption and systolic pressure volume area.

241. Taylor, E. W. *J. Exp. Biol.* 100:289–319, 1982. Control and coordination of ventilation and circulation in crustaceans: responses to hypoxia and exercise.

242. Tazawa, H., M. Mochizuki, and J. Piiper. *Respir. Physiol.* 36:77–95, 1979. Respiratory gas transport by the incompletely separated double circulation in the bullfrog, *Rana catesbeina*.

243. Thompson, R. J., D. R. Livingston, and A. de Zwaan. *J. Comp. Physiol.* 137:97–104, 1980. Physiological and biochemical aspects of valve closure in the giant scallop *Placeopecten magellanicus*.

244. Thorson, T. B. *Science* 138:99–100, 688–690, 1959. Partitioning of body fluids in sea lamprey, marine sharks, and freshwater shark.

245. Thorson, T. B. *Biol. Bull.* (*Woods Hole, Mass.*) 120:234–254, 1961. Partitioning of body water in fishes.

246. Thorson, T. B. *Physiol. Zool.* 37:395–399, 1964. Partitioning of body water in amphibians.

247. Thorson, T. B. *Copeia* 1968:592–601, 1968. Body fluid partitioning in reptiles.

248. Toda, N. and T. C. West. *Nature* (*London*) 205:808–809, 1965. Effect of vagal stimulation on rabbit atrium.

249. Trautwein, W. and D. G. Kassebaum. *J. Gen. Physiol.* 45:317–330, 1961. Pacemaker activity of heart.

250. Trueman, E. R. *Biol. Bull.* (*Woods Hole, Mass.*) 131:369–377, 1966. Hydrodynamics of burrowing in *Arenicola*.

251. Trueman, E. R. *J. Exp. Biol.* 44:93–118, 1966. Burrowing reactions in *Arenicola*.

252. Trueman, E. R. *Science* 152:523–525, 1966. Fluid dynamics of burrowing.

253. Trueman, E. R., A. R. Brand, and P. Davis. *J. Exp. Biol.* 44:469–492, 1966. Burrowing in bivalves.

254. Tucker, V. A. *J. Exp. Biol.* 44:77–92, 1966. Circulation in *Iguana*.

255. Van der Kloot, W. G. *Fed. Proc., Fed. Am. Soc. Exp. Biol.* 26:975–980, 1967. Regulation of heart of leech.

256. Van der Kloot, W. G. *J. Exp. Zool.* 174:367–380, 1970. Electrophysiology of cardiac muscle of crustaceans.

257. Vogel, W. O. P. In *Cardiovascular Shunts* (K. Johansen and W. Burggren, eds.), pp. 143–151. Munksgaard, Copenhagen, 1985. Systemic vascular anastomoses, primary and secondary vessels in fish, and the phylogeny of lymphatics.

258. Watanabe, A., S. Obara, and T. Akiyama. *J. Gen. Physiol.* 50:813–838, 839–862, 1967; 52:908, 1968; 54:212–231, 1969. Cardiac ganglion of *Squilla*.

259. Watson, W. H., G. J. Augustine, J. A. Benson, and R. E. Sullivan. *J. Exp. Biol.* 103:55–73, 1983. Proctolin and an endogenous proctolin-like peptide enhance contractility of the *Limulus* heart.

260. Watson, W. H., T. Hoshi, J. Colburne, and G. J. Augustine. *J. Exp. Biol.* 118:71–84, 1985. Neurohormonal modulation of the *Limulus* heart: Amine actions on neuromuscular transmission and cardiac muscle.

261. Weidmann, S. *J. Physiol. (London)* 118:348–360, 1952. Electrical constants of Purkinje fibers.

262. Weidmann, S. *J. Physiol. (London)* 210:1041–1054, 1970. Electrical constants of mammalian heart.

263. Wells, M. J. In *The Mollusca* (A. S. M. Saleuddin and K. M. Wilbur, eds.), Vol. 5, pp. 239–290. Academic Press, London, 1983. Circulation in cephalopods.

264. Wells, M. J. and J. Wells. *J. Exp. Biol.* 122:345–353, 1986. Blood flow in acute hypoxia in a cephalopod.

265. West, N. H. and W. W. Burggren. *Am. J. Physiol.* 247:R884–R894, 1984. Factors influencing pulmonary and cutaneous arterial blood flow in the toad, *Bufo marinus*.

266. White, F. N. *Fed. Proc., Fed. Am. Soc. Exp. Biol.* 29:1149–1153, 1970. Vascular shunt in reptiles.

267. White, F. N. In *Biology of the Reptilia* (C. Gans, ed.), Vol. 5, pp. 275–334. Academic Press, London, 1976. Circulation.

268. White, F. N. and G. Ross. *Am. J. Physiol.* 211:15–18, 1966. Circulatory changes in diving in turtle.

269. White, L., H. Haines, and T. Adams. *Comp. Biochem. Physiol.* 27:559–566, 1968. Cardiac output of various mammals.

270. Williams, T. M. *Am. Zool.* 26:99A, 1986. Heart rate and voluntary breathhold duration in swimming harbor seals.

271. Wood, C. M. and S. F. Perry. In *Circulation, Respiration and Metabolism* (R. Gilles, ed.), pp. 2–22. Springer-Verlag, Berlin, 1985. Respiratory, circulatory and metabolic adjustments to exercise in fish.

272. Wood, C. M., P. Pieprzak, and J. N. Trott. *Can. J. Zool.* 57:2440–2447, 1979. The influence of temperature and anaemia on the adrenergic and cholinergic mechanisms controlling heart rate in the rainbow trout.

273. Wood, C. M. and G. Shelton. *J. Exp. Biol.* 87:247–270, 1980a. Cardiovascular dynamics and adrenergic responses of the rainbow trout *in vivo*.

274. Wood, C. M. and G. Shelton. *J. Exp. Biol.* 87:271–284, 1980b. The reflex control of heart rate and cardiac output in the rainbow trout: Interactive influences of hypoxia, hemorrhage and systemic vasomotor tone.

275. Woodbury, J. W. and W. E. Crill. In *Nervous Inhibition, Proceedings of the International Symposium* (E. Florey, ed.), pp. 124–135. Pergamon, New York, 1961. Conduction in the atrium.

276. Wyman, L. C. and B. R. Lutz. *J. Exp. Zool.* 57:441–453, 1930. The action of adrenaline and certain drugs on the isolated holothurian cloaca.

277. Zatzman, M. L. and G. V. Thornhill. In *Living in the Cold* (H. C. Heller et al., eds.), pp. 453–459. Elsevier, New York, 1986. Seasonal changes in blood pressure and cardiac output in marmots.

278. Zuckerkandl, E. *Biol. Bull. (Woods Hole, Mass.)* 98:161–173, 1950. Pressure in holothurians and sipunculids.

279. Zwicky, K. T. *Nature (London)* 207:778–779, 1965; *Comp. Biochem. Physiol.* 24:799–808, 1968. Pharmacology of scorpion heart.

INDEX